Lecture Notes in Computer Science   6978

Commenced Publication in 1973
Founding and Former Series Editors:
Gerhard Goos, Juris Hartmanis, and Jan van Leeuwen

Giuseppe Maino   Gian Luca Foresti (Eds.)

Image Analysis and Processing – ICIAP 2011

16th International Conference
Ravenna, Italy, September 14-16, 2011
Proceedings, Part I

 Springer

Volume Editors

Giuseppe Maino
Università di Bologna
Facoltà di Conservazione dei Beni Culturali
Via Mariani 5, 48100 Ravenna, Italy
E-mail: giuseppe.maino@unibo.it

Gian Luca Foresti
Università di Udine
Dipartimento di Matematica e Informatica
via delle Scienze 206, 33100 Udine, Italy
E-mail: gianluca.foresti@uniud.it

ISSN 0302-9743   e-ISSN 1611-3349
ISBN 978-3-642-24084-3   ISBN 978-3-642-24085-0 (eBook)
DOI 10.1007/978-3-642-24085-0
Springer Heidelberg Dordrecht London New York

Library of Congress Control Number: 2011936230

CR Subject Classification (1998): I.4, I.5, I.3.5, I.2.10, I.2.6, H.3, F.2.2

LNCS Sublibrary: SL 6 – Image Processing, Computer Vision, Pattern Recognition, and Graphics

Typesetting: Camera-ready by author, data conversion by Scientific Publishing Services, Chennai, India

Printed on acid-free paper

Springer is part of Springer Science+Business Media (www.springer.com)

Preface

This volume collects the papers accepted for presentation at the International Conference on Image Analysis and Processing (ICIAP 2011), held in Ravenna, Italy, September 14–16, 2011. ICIAP 2011 was the 16th event in a series of conferences organized biennially by the Italian Member Society of the International Association for Pattern Recognition (IAPR). The aim of these conferences is to bring together international researchers for the presentation and discussion of the most recent advances in the fields of pattern recognition, image analysis, and image processing. Following the successful 2009 conference in Vietri sul Mare, ICIAP 2011 was held in the magnificent city of Ravenna, an historical city famous for its artistic and cultural heritage. The 16th ICIAP conference was organized jointly by the Faculty of Preservation of Cultural Heritage of the University of Bologna and the Department of Mathematics and Computer Science (DIMI) of the University of Udine.

Topics for ICIAP 2011 included Image Analysis and Processing, Pattern Recognition and Vision, Multimodal Interaction and Multimedia Processing, Cultural Heritage, and Applications.

There were 175 submissions. Each submission was reviewed by two Program Committee members. The committee decided to accept 121 papers, divided into 10 oral sessions (44 papers) and three poster sessions (77 papers).

The program included a special session on "Low Level Color Image Processing" (organized by M. Emre Celebi, Bogdan Smolka, Gerald Schaefer, and Raimondo Schettini), a demo session, and four invited talks by Jake K. Aggarwal (University of Texas, Department of Electrical and Computer Engineering, USA) on *Recognition of Human Activities*, Horst Bunke (University of Bern, Institute of Computer Science and Applied Mathematics, Switzerland) on *Bridging the Gap between Structural and Statistical Pattern Recognition*, Roberto Cipolla (University of Cambridge, Department of Engineering, UK), on *Novel Applications of 3D Shape from Uncalibrated Images*, and Kevin Karplus (University of California, Santa Cruz, Department of Biomolecular Engineering, USA) on *Protein Structure and Genome Assembly Tools*. These lectures survey established approaches, recent results and directions of future works of different topics of recognition of human activities, structural and statistical pattern recognition, computational vision, bioinformatics, and biomolecular engineering.

Three tutorials were offered, on "Image and Video Descriptors" (by Abdenour Hadid), on "Beyond Features: Similarity-Based Pattern Analysis and Recognition" (by Edwin R. Hancock, Vittorio Murino, and Marcello Pelillo), and on "Video Analytics on Reactive Camera Networks" (by Christian Micheloni).

ICIAP 2011 will also host the First International Workshop on Pattern Recognition in Proteomics, Structural Biology and Bioinformatics, PR PS BB 2011, organized by Virginio Cantoni and Giuseppe Maino.

During the conference, the Caianiello Prize, in memory of Prof. E. Caianiello, was awarded to the best paper by a young author, as at previous events. Also, a prize was awarded to the best paper presented to the conference.

We wish to thank the Italian group of researchers affiliated to the International Association for Pattern Recognition (GIRPR) for giving us the opportunity to organize this conference. We also thank the International Association for Pattern Recognition for the endorsement of ICIAP 2011. A special word of thanks goes to the Program Chairs, to the members of the Program Committee and to the reviewers, who contributed with their work to ensuring the high-quality standard of the papers accepted to ICIAP 2011.

Special thanks go to Claudio Piciarelli, who made a fundamental contribution to this conference, helping in managing, working on, and resolving those many problems that a large event like this presents.

Local organization for events and accommodation was managed by Carla Rossi of the Fondazione Flaminia and Daniela Raule of the NEREA-AIDA spin-off. We are indebted to the Fondazione Flaminia for financial and organization support. A special thanks goes to the members of the Local Organizing Committee, Roberta Menghi and Mariapaola Monti, who also took care of the graphic aspects of the event, Elena Nencini, Lorenza Roversi, and Lisa Volpe for their indispensable contribution to the organization and their help and availability to solve the many practical problems arising during the preparation of ICIAP 2011. Finally, Sara Armaroli, Donatella Lombardo, Mariapaola Monti, and Liu Wan are the young artists that have lent themselves to realize the Vision&Art exhibition accompanying ICIAP 2011.

September 2011   Giuseppe Maino
                                                       Gian Luca Foresti

Organization

Organizing Institutions

Alma Mater Studiorum, Università di Bologna
Università degli Studi di Udine

General Chairs

Giuseppe Maino	ENEA and University of Bologna, Italy
Gian Luca Foresti	University of Udine, Italy

Program Chairs

Sebastiano Battiato	University of Catania, Italy (Image Analysis and Processing)
Donatella Biagi Maino	University of Bologna, Italy (Cultural Heritage and Applications)
Christian Micheloni	University of Udine, Italy (Pattern Recognition and Vision)
Lauro Snidaro	University of Udine, Italy (Machine Learning and Multimedia)

Publicity Chair

Claudio Piciarelli	University of Udine, Italy

Steering Committee

Virginio Cantoni, Italy
Luigi Cordella, Italy
Alberto Del Bimbo, Italy
Marco Ferretti, Italy
Fabio Roli, Italy
Gabriella Sanniti di Baja, Italy

Program Committee

Jake K. Aggarwal, USA
Maria Grazia Albanesi, Italy
Hlder J. Araújo, Portugal
Edoardo Ardizzone, Italy
Prabir Bhattacharya, USA
Alessandro Bevilacqua, Italy
Giuseppe Boccignone, Italy
Gunilla Borgefors, Sweden
Alfred Bruckstein, Israel
Paola Campadelli, Italy
Elisabetta Canetta, UK
Andrea Cavallaro, UK
Rémy Chapoulie, France
M. Emre Celebi, USA
Rita Cucchiara, Italy
Leila De Floriani, Italy
Claudio De Stefano, Italy
Pierre Drap, France
Jean Luc Dugelay, France
Ana Fred, Portugal
Maria Frucci, Italy
André Gagalowicz, France
Giorgio Giacinto, Italy
Edwin Hancock, UK
Francisco H. Imai, USA
Rangachar Kasturi, USA
Walter Kropatsch, Austria
Josep Lladòs, Spain
Brian C. Lovell, Australia
Rastislav Lukac, Canada
Angelo Marcelli, Italy
Simone Marinai, Italy

Stefano Messelodi, Italy
Vittorio Murino, Italy
Mike Nachtegael, Belgium
Michele Nappi, Italy
Hirobumi Nishida, Japan
Jean-Marc Ogier, France
Marcello Pelillo, Italy
Alfredo Petrosino, Italy
Maria Petrou, Greece
Matti Pietikäinen, Finland
Giuseppe Pirlo, Italy
Fabio Remondino, Switzerland
Hanan Samet, USA
Carlo Sansone, Italy
Silvio Savarese, USA
Gerard Schaefer, UK
Raimondo Schettini, Italy
Linda Shapiro, USA
Filippo Stanco, Italy
Massimo Tistarelli, Italy
Alain Trémeau, France
Roberto Tronci, Italy
Adrian Ulges, Germany
Cesare Valenti, Italy
Mario Vento, Italy
Daniele Visparelli, Italy
Domenico Vitulano, Italy
Yehezkel Yeshurun, Israel
Marcel Worring, The Netherlands
Lei Zhang, Hong Kong, China
Primo Zingaretti, Italy
Galina I. Zmievskaya, Russia

Additional Reviewers

Lamberto Ballan
Silvia Bussi
Elena Casiraghi
Paul Ian Chippendale
Luca Didaci
Giovanni Maria Farinella
Francesco Fontanella
Alessandro Gherardi

Cris Luengo Hendriks
Michela Lecca
Paola Magillo
Iacopo Masi
Carla Maria Modena
Daniele Muntoni
Gabriele Murgia
Paolo Napoletano

Francesca Odone
Federico Pernici
Maurizio Pili
Giovanni Puglisi
Ajita Rattani
Elisa Ricci
Reza Sabzevari
Riccardo Satta

Giuseppe Serra
Nicola Sirena
Lennart Svensson
Francesco Tortorella
Ingrid Visentini
Erik Wernersson
Matteo Zanotto

Local Organizing Committee

Roberta Menghi
Mariapaola Monti
Carla Rossi
Lorenza Roversi
Lisa Volpe
Basilio Limuti

Endorsing Institutions

Italian Member Society of the International Association for Pattern
    Recognition – GIRPR
International Association for Pattern Recognition – IAPR

Sponsoring Institutions

Fondazione Flaminia, Ravenna
Ordine della Casa Matha, Ravenna

Table of Contents – Part I

Image Analysis and Representation

High Order Structural Matching Using Dominant Cluster Analysis 1
  Peng Ren, Richard C. Wilson, and Edwin R. Hancock

A Probabilistic Framework for Complex Wavelet Based Image
Registration 9
  Florina-Cristina Calnegru

Image De-noising by Bayesian Regression 19
  Shimon Cohen and Rami Ben-Ari

Image Segmentation

A Rough-Fuzzy HSV Color Histogram for Image Segmentation 29
  Alessio Ferone, Sankar Kumar Pal, and Alfredo Petrosino

Multiple Region Categorization for Scenery Images 38
  Tamar Avraham, Ilya Gurvich, and Michael Lindenbaum

Selection of Suspicious ROIs in Breast DCE-MRI 48
  *Roberta Fusco, Mario Sansone, Carlo Sansone, and
  Antonella Petrillo*

Regions Segmentation from SAR Images 58
  Luigi Cinque and Rossella Cossu

Adaptive Model for Object Detection in Noisy and Fast-Varying
Environment 68
  *Dung Nghi Truong Cong, Louahdi Khoudour,
  Catherine Achard, and Amaury Flancquart*

Shadow Segmentation Using Time-of-Flight Cameras 78
  Faisal Mufti and Robert Mahony

Pattern Analysis and Classification

Uni-orthogonal Nonnegative Tucker Decomposition for Supervised
Image Classification 88
  Rafal Zdunek

A Classification Approach with a Reject Option for Multi-label
Problems 98
  Ignazio Pillai, Giorgio Fumera, and Fabio Roli

Improving Image Categorization by Using Multiple Instance Learning
with Spatial Relation   108
    Thanh Duc Ngo, Duy-Dinh Le, and Shin'ichi Satoh

Shaping the Error-Reject Curve of Error Correcting Output Coding
Systems   118
    Paolo Simeone, Claudio Marrocco, and Francesco Tortorella

Sum-of-Superellipses – A Low Parameter Model for Amplitude Spectra
of Natural Images   128
    Marcel Spehr, Stefan Gumhold, and Roland W. Fleming

Dissimilarity Representation in Multi-feature Spaces for Image
Retrieval   139
    Luca Piras and Giorgio Giacinto

Forensics, Security and Document Analysis

Discrete Point Based Signatures and Applications to Document
Matching   149
    Nemanja Spasojevic, Guillaume Poncin, and Dan Bloomberg

Robustness Evaluation of Biometric Systems under Spoof Attacks   159
    Zahid Akhtar, Giorgio Fumera, Gian Luca Marcialis, and Fabio Roli

A Graph-Based Framework for Thermal Faceprint Characterization   169
    Daniel Osaku, Aparecido Nilceu Marana, and João Paulo Papa

Video Analysis and Processing

Reflection Removal for People Detection in Video Surveillance
Applications   178
    Dajana Conte, Pasquale Foggia, Gennaro Percannella,
    Francesco Tufano, and Mario Vento

The Active Sampling of Gaze-Shifts   187
    Giuseppe Boccignone and Mario Ferraro

SARC3D: A New 3D Body Model for People Tracking and
Re-identification   197
    Davide Baltieri, Roberto Vezzani, and Rita Cucchiara

Sorting Atomic Activities for Discovering Spatio-temporal Patterns in
Dynamic Scenes   207
    Gloria Zen, Elisa Ricci, Stefano Messelodi, and Nicu Sebe

Intelligent Overhead Sensor for Sliding Doors: A Stereo Based Method
for Augmented Efficiency   217
    Luca Bombini, Alberto Broggi, Michele Buzzoni, and Paolo Medici

Robust Stereoscopic Head Pose Estimation in Human-Computer
Interaction and a Unified Evaluation Framework   227
  *Georg Layher, Hendrik Liebau, Robert Niese, Ayoub Al-Hamadi,
  Bernd Michaelis, and Heiko Neumann*

Biometry

Automatic Generation of Subject-Based Image Transitions   237
  *Edoardo Ardizzone, Roberto Gallea, Marco La Cascia, and
  Marco Morana*

Learning Neighborhood Discriminative Manifolds for Video-Based Face
Recognition   247
  John See and Mohammad Faizal Ahmad Fauzi

A Novel Probabilistic Linear Subspace Approach for Face
Applications.   257
  Ying Ying and Han Wang

Shape Analysis

Refractive Index Estimation of Naturally Occurring Surfaces Using
Photometric Stereo.   267
  Gule Saman and Edwin R. Hancock

Synchronous Detection for Robust 3-D Shape Measurement against
Interreflection and Subsurface Scattering   276
  Tatsuhiko Furuse, Shinsaku Hiura, and Kosuke Sato

Unambiguous Photometric Stereo Using Two Images   286
  Roberto Mecca and Jean-Denis Durou

Low-Level Color Image Processing

Von Kries Model under Planckian Illuminants   296
  Michela Lecca and Stefano Messelodi

Colour Image Coding with Matching Pursuit in the Spatio-frequency
Domain.   306
  Ryszard Maciol, Yuan Yuan, and Ian T. Nabney

Color Line Detection   318
  Vinciane Lacroix

A New Perception-Based Segmentation Approach Using Combinatorial
Pyramids   327
  Esther Antúnez, Rebeca Marfil, and Antonio Bandera

Automatic Color Detection of Archaeological Pottery with Munsell
System   337
    *Filippo Stanco, Davide Tanasi, Arcangelo Bruna, and
    Valentina Maugeri*

Image Retrieval Based on Gaussian Mixture Approach to Color
Localization   347
    Maria Luszczkiewicz-Piatek and Bogdan Smolka

A Method for Data Extraction from Video Sequences for Automatic
Identification of Football Players Based on Their Numbers   356
    Dariusz Frejlichowski

Real-Time Hand Gesture Recognition Using a Color Glove   365
    Luigi Lamberti and Francesco Camastra

Applications

Improving 3D Reconstruction for Digital Art Preservation   374
    *Jurandir Santos Junior, Olga Bellon, Luciano Silva, and
    Alexandre Vrubel*

Exploring Cascade Classifiers for Detecting Clusters of
Microcalcifications   384
    Claudio Marrocco, Mario Molinara, and Francesco Tortorella

A Method for Scribe Distinction in Medieval Manuscripts Using Page
Layout Features   393
    *Claudio De Stefano, Francesco Fontanella, Marilena Maniaci, and
    Alessandra Scotto di Freca*

Medical Imaging

Registration Parameter Spaces for Molecular Electron Tomography
Images   403
    *Lennart Svensson, Anders Brun, Ingela Nyström, and
    Ida-Maria Sintorn*

A Multiple Kernel Learning Algorithm for Cell Nucleus Classification
of Renal Cell Carcinoma   413
    *Peter Schüffler, Aydın Ulaş, Umberto Castellani, and
    Vittorio Murino*

Nano-imaging and Its Applications to Biomedicine   423
    Elisabetta Canetta and Ashok K. Adya

Image Analysis and Pattern Recognition

IDEA: Intrinsic Dimension Estimation Algorithm 433
  Alessandro Rozza, Gabriele Lombardi, Marco Rosa,
  Elena Casiraghi, and Paola Campadelli

Optimal Decision Trees Generation from *OR*-Decision Tables 443
  Costantino Grana, Manuela Montangero, Daniele Borghesani, and
  Rita Cucchiara

Efficient Computation of Convolution of Huge Images 453
  David Svoboda

Half Ellipse Detection 463
  Nikolai Sergeev and Stephan Tschechne

A Robust Forensic Hash Component for Image Alignment.............. 473
  Sebastiano Battiato, Giovanni Maria Farinella,
  Enrico Messina, and Giovanni Puglisi

Focus of Expansion Localization through Inverse C-Velocity 484
  Adrien Bak, Samia Bouchafa, and Didier Aubert

Automated Identification of Photoreceptor Cones Using Multi-scale
Modelling and Normalized Cross-Correlation 494
  Alan Turpin, Philip Morrow, Bryan Scotney, Roger Anderson, and
  Clive Wolsley

A Finite Element Blob Detector for Robust Features 504
  Dermot Kerr, Sonya Coleman, and Bryan Scotney

Reducing Number of Classifiers in DAGSVM Based on Class
Similarity 514
  Marcin Luckner

New Error Measures to Evaluate Features on Three-Dimensional
Scenes 524
  Fabio Bellavia and Domenico Tegolo

Optimal Choice of Regularization Parameter in Image Denoising 534
  Mirko Lucchese, Iuri Frosio, and N. Alberto Borghese

Neighborhood Dependent Approximation by Nonlinear Embedding for
Face Recognition... 544
  Ann Theja Alex, Vijayan K. Asari, and Alex Mathew

Ellipse Detection through Decomposition of Circular Arcs and Line
Segments   554
   Thanh Phuong Nguyen and Bertrand Kerautret

Computing Morse Decompositions for Triangulated Terrains:
An Analysis and an Experimental Evaluation   565
   Maria Vitali, Leila De Floriani, and Paola Magillo

Spot Detection in Images with Noisy Background   575
   Denis Ferraretti, Luca Casarotti, Giacomo Gamberoni, and
   Evelina Lamma

Automatic Facial Expression Recognition Using Statistical-Like
Moments   585
   Roberto D'Ambrosio, Giulio Iannello, and Paolo Soda

Temporal Analysis of Biometric Template Update Procedures in
Uncontrolled Environment   595
   Ajita Rattani, Gian Luca Marcialis, and Fabio Roli

Biologically Motivated Feature Extraction   605
   Sonya Coleman, Bryan Scotney, and Bryan Gardiner

Entropy-Based Localization of Textured Regions   616
   Liliana Lo Presti and Marco La Cascia

Evaluation of Global Descriptors for Large Scale Image Retrieval   626
   Hai Wang and Shuwu Zhang

Improved Content-Based Watermarking Using Scale-Invariant Feature
Points   636
   Na Li, Edwin Hancock, Xiaoshi Zheng, and Lin Han

Crop Detection through Blocking Artefacts Analysis   650
   A.R. Bruna, G. Messina, and S. Battiato

Structure from Motion and Photometric Stereo for Dense 3D Shape
Recovery   660
   Reza Sabzevari, Alessio Del Bue, and Vittorio Murino

Genetic Normalized Convolution   670
   Giulia Albanese, Marco Cipolla, and Cesare Valenti

Combining Probabilistic Shape-from-Shading and Statistical Facial
Shape Models   680
   Touqeer Ahmad, Richard C. Wilson, William A.P. Smith, and
   Tom S.F. Haines

Visual Saliency by Keypoints Distribution Analysis................... 691
    Edoardo Ardizzone, Alessandro Bruno, and Giuseppe Mazzola

From the Physical Restoration for Preserving to the Virtual Restoration
for Enhancing 700
    Elena Nencini and Giuseppe Maino

Author Index.. 711

Table of Contents – Part II

Image and Video Analysis and Processing

A Visual Blindspot Monitoring System for Safe Lane Changes   1
  Jamal Saboune, Mehdi Arezoomand, Luc Martel, and
  Robert Laganiere

Extracting Noise Elements while Preserving Edges in Spatial Domain . . .   11
  Jalil Bushra, Fauvet Eric, and Laligant Olivier

Automatic Human Action Recognition in Videos by Graph
Embedding   19
  Ehsan Zare Borzeshi, Richard Xu, and Massimo Piccardi

Human Action Recognition by Extracting Features from Negative
Space   29
  Shah Atiqur Rahman, M.K.H. Leung, and Siu-Yeung Cho

Edge-Directed Image Interpolation Using Color Gradient Information . . .   40
  Andrey Krylov and Andrey Nasonov

Path Analysis in Multiple-Target Video Sequences   50
  Brais Cancela, Marcos Ortega, Alba Fernández, and
  Manuel G. Penedo

Statistical Multisensor Image Segmentation in Complex Wavelet
Domains   60
  Tao Wan and Zengchang Qin

Activity Discovery Using Compressed Suffix Trees   69
  Prithwijit Guha, Amitabha Mukerjee, and K.S. Venkatesh

A Continuous Learning in a Changing Environment   79
  Aldo Franco Dragoni, Germano Vallesi, and Paola Baldassarri

Human-Computer Interaction through Time-of-Flight and RGB
Cameras   89
  Piercarlo Dondi, Luca Lombardi, and Marco Porta

Handling Complex Events in Surveillance Tasks   99
  Daniele Bartocci and Marco Ferretti

Face Analysis Using Curve Edge Maps   109
  Francis Deboeverie, Peter Veelaert, and Wilfried Philips

Statistical Patch-Based Observation for Single Object Tracking   119
  Mohd Asyraf Zulkifley and Bill Moran

Exploiting Depth Information for Indoor-Outdoor Scene
Classification   130
  Ignazio Pillai, Riccardo Satta, Giorgio Fumera, and Fabio Roli

A Multiple Component Matching Framework for Person
Re-identification   140
  Riccardo Satta, Giorgio Fumera, Fabio Roli, Marco Cristani, and
  Vittorio Murino

Improving Retake Detection by Adding Motion Feature   150
  Hiep Van Hoang, Duy-Dinh Le, Shin'ichi Satoh, and
  Quang Hong Nguyen

RDVideo: A New Lossless Video Codec on GPU   158
  Piercarlo Dondi, Luca Lombardi, and Luigi Cinque

A New Algorithm for Image Segmentation via Watershed
Transformation   168
  Maria Frucci and Gabriella Sanniti di Baja

Supervised Learning Based Stereo Matching Using Neural Tree   178
  Sanjeev Kumar, Asha Rani, Christian Micheloni, and
  Gian Luca Foresti

Pre-emptive Camera Activation for Video-Surveillance HCI   189
  Niki Martinel, Christian Micheloni, and Claudio Piciarelli

Space-Time Zernike Moments and Pyramid Kernel Descriptors for
Action Classification   199
  Luca Costantini, Lorenzo Seidenari, Giuseppe Serra,
  Licia Capodiferro, and Alberto Del Bimbo

A Low Complexity Motion Segmentation Based on Semantic
Representation of Encoded Video Streams.........................   209
  Maurizio Abbate, Ciro D'Elia, and Paola Mariano

Audio-Video Analysis of Musical Expressive Intentions...............   219
  Ingrid Visentini, Antonio Rodà, Sergio Canazza, and Lauro Snidaro

Image Segmentation Using Normalized Cuts and Efficient Graph-Based
Segmentation   229
  Narjes Doggaz and Imene Ferjani

Applications

Stability Analysis of Static Signatures for Automatic Signature
Verification 241
Donato Impedovo and Giuseppe Pirlo

Segmentation Strategy of Handwritten Connected Digits (SSHCD) 248
Abdeldjalil Gattal and Youcef Chibani

An Experimental Comparison of Different Methods for Combining
Biometric Identification Systems 255
Emanuela Marasco and Carlo Sansone

Using Geometric Constraints to Solve the Point Correspondence
Problem in Fringe Projection Based 3D Measuring Systems 265
*Christian Bräuer-Burchardt, Christoph Munkelt, Matthias Heinze,
Peter Kühmstedt, and Gunther Notni*

Retrospective Illumination Correction of Greyscale Historical Aerial
Photos 275
Anders Hast and Andrea Marchetti

Multibeam Echosounder Simulator Applying Noise Generator for the
Purpose of Sea Bottom Visualisation 285
Wojciech Maleika, Michał Pałczyński, and Dariusz Frejlichowski

Automatic Segmentation of Digital Orthopantomograms for Forensic
Human Identification 294
Dariusz Frejlichowski and Robert Wanat

Common Scab Detection on Potatoes Using an Infrared Hyperspectral
Imaging System 303
*Angel Dacal-Nieto, Arno Formella, Pilar Carrión,
Esteban Vazquez-Fernandez, and Manuel Fernández-Delgado*

Automatic Template Labeling in Extensible Multiagent Biometric
Systems 313
Maria De Marsico, Michele Nappi, Daniel Riccio, and Genny Tortora

Automatic Bus Line Number Localization and Recognition on Mobile
Phones—A Computer Vision Aid for the Visually Impaired 323
Claudio Guida, Dario Comanducci, and Carlo Colombo

The Use of High-Pass Filters and the Inpainting Method to Clouds
Removal and Their Impact on Satellite Images Classification 333
*Ana Carolina Siravenha, Danilo Sousa, Aline Bispo, and
Evaldo Pelaes*

Hybrid Filter Based Simultaneous Localization and Mapping for a
Mobile Robot 343
    Amir Panah and Karim Faez

Mitotic HEp-2 Cells Recognition under Class Skew 353
    Gennaro Percannella, Paolo Soda, and Mario Vento

Error Compensation by Sensor Re-calibration in Fringe Projection
Based Optical 3D Stereo Scanners................................... 363
    Christian Bräuer-Burchardt, Peter Kühmstedt, and Gunther Notni

Advanced Safety Sensor for Gate Automation 374
    Luca Bombini, Alberto Broggi, and Stefano Debattisti

Using Blood Vessels Location Information in Optic Disk
Segmentation 384
    Alexander S. Semashko, Andrey S. Krylov, and A.S. Rodin

Orthophotoplan Segmentation and Colorimetric Invariants for Roof
Detection 394
    *Youssef El Merabet, Cyril Meurie, Yassine Ruichek,
    Abderrahmane Sbihi, and Rajaa Touahni*

A Simulation Framework to Assess Pattern Matching Algorithms in a
Space Mission 404
    Alessandro Gherardi and Alessandro Bevilacqua

A Novel T-CAD Framework to Support Medical Image Analysis and
Reconstruction 414
    Danilo Avola, Luigi Cinque, and Marco Di Girolamo

Fast Vision-Based Road Tunnel Detection 424
    *Massimo Bertozzi, Alberto Broggi, Gionata Boccalini, and
    Luca Mazzei*

A New Dissimilarity Measure for Clustering Seismic Signals 434
    *Francesco Benvegna, Antonino D'Alessando, Giosuè Lo Bosco,
    Dario Luzio, Luca Pinello, and Domenico Tegolo*

Character Segmentation for License Plate Recognition by K-Means
Algorithm.. 444
    Lihong Zheng and Xiangjian He

A Video Grammar-Based Approach for TV News Localization and
Intra-structure Identification in TV Streams 454
    Tarek Zlitni, Walid Mahdi, and Hanène Ben-Abdallah

Multispectral Imaging and Digital Restoration for Paintings
Documentation 464
    Marco Landi and Giuseppe Maino

Virtual Reality Models for the Preservation of the Unesco Historical
and Artistical Heritage 475
   Roberta Menghi, Giuseppe Maino, and Marianna Panebarco

Image Processing and a Virtual Restoration Hypothesis for Mosaics
and Their Cartoons 486
   Mariapaola Monti and Giuseppe Maino

Author Index... 497

High Order Structural Matching Using Dominant Cluster Analysis*

Peng Ren, Richard C. Wilson, and Edwin R. Hancock

Department of Computer Science, The University of York, York, YO10 5GH, UK
{pengren,wilson,erh}@cs.york.ac.uk

Abstract. We formulate the problem of high order structural matching by applying *dominant cluster analysis* (DCA) to a direct product hypergraph (DPH). For brevity we refer to the resulting algorithm as DPH-DCA. The DPH-DCA can be considered as an extension of the game theoretic algorithms presented in [8] from clustering to matching, and also as a reduced version of reduced version of the method of ensembles of affinity relations presented in [6]. The starting point for our method is to construct a K-uniform direct product hypergraph for the two sets of higher-order features to be matched. Each vertex in the direct product hypergraph represents a potential correspondence and the weight on each hyperedge represents the agreement between two K-tuples drawn from the two feature sets. Vertices representing correct assignment tend to form a strongly intra-connected cluster, i.e. a dominant cluster. We evaluate the association of each vertex belonging to the dominant cluster by maximizing an objective function which maintains the K-tuple agreements. The potential correspondences with nonzero association weights are more likely to belong to the dominant cluster than the remaining zero-weighted ones. They are thus selected as correct matchings subject to the one-to-one correspondence constraint. Furthermore, we present a route to improving the matching accuracy by invoking prior knowledge. An experimental evaluation shows that our method outperforms the state-of-the-art high order structural matching methods[10][3].

1 Introduction

Many problems in computer vision and machine learning can be posed as that of establishing the consistent correspondences between two sets of features. Traditional matching approaches are usually confined to structures with pairwise relations. Recently, a number of researchers have attempted to extend the matching process to incorporate higher order relations. Zass *et al.* [10] are among the first to investigate this problem by introducing a probabilistic hypergraph matching framework, in which higher order relationships are marginalized to unary order. It has already been pointed out in [1] that this graph approximation is just a low pass representation of the original hypergraph and causes information loss and inaccuracy. On other hand, Duchenne *et al.* [3] have developed the spectral technique for graph matching [4] into a higher order matching

---

* We acknowledge the financial support from the FET programme within the EU FP7, under the SIMBAD project (contract 213250). Edwin R. Hancock is supported by a Royal Society Wolfson Research Merit Award.

G. Maino and G.L. Foresti (Eds.): ICIAP 2011, Part I, LNCS 6978, pp. 1–8, 2011.

framework using the so called *tensor power iteration*. Although they adopt an L_1 norm constraint in computation, the original objective function is subject to an L_2 norm and does not satisfy the basic probabilistic properties.

We present a framework based on applying *dominant cluster analysis* (DCA) to a direct product hypergraph (DPH). The idea is to extend the main cluster method of Leordeanu and Hebert [4] for graphs and its generalization for higher order matching [3], using dominant cluster analysis. Furthermore, we present a method for initializing our algorithm that can be used to suppress outliers. This improves the matching performance of our method, and comparable results can not be achieved by using alternative high order matching algorithms [3][10]. Similar ideas have recently been presented in [6]. Our method however, generalises the methods descrbibed in [3][10] from graphs to hypergraphs, and is more pricipled in its formulation.

2   Problem Formulation

We represent the set of Kth order feature relationships by a K-uniform hypergraph $HG(V, E)$, whose hyperedges have identical cardinality K. Each vertex $v_i \in V$ in the K-uniform hypergraph $HG(V, E)$ represents one element in the feature set. Each hyperedge $e_i \in E$ represents one K-tuple $\{v_{i_1}, \cdots, v_{i_K}\} \in V$ and the weight attached to each hyperedge represents the similarity measure on the K-tuple encompassed by the hyperedge. For simplicity, we denote a vertex v_i by its index i in the remainder of our work. The K-uniform hypergraph $HG(V, E)$ can be represented as a Kth order tensor $\mathcal{H}$, whose element $H_{i_1, \ldots, i_K}$ is the hyperedge weight if there is a hyperedge encompassing the vertex subset $\{i_1, \cdots, i_K\} \in V$, and zero otherwise. The problem of matching two feature sets both constituted by Kth order relationships can then be transformed to that of matching the two associated K-uniform hypergraphs $HG(V, E)$ and $HG'(V', E')$. To this end, we establish the high order compatibility matrix $\mathcal{C}$, i.e. compatibility tensor, for $HG(V, E)$ and $HG'(V', E')$. The elements of the Kth order compatibility tensor $\mathcal{C}$ are defined as follows

$$
C_{i_1 i'_1, \ldots, i_K i'_K} = \begin{array}{l} 0 \text{ if } H_{i_1, \ldots, i_K} = 0 \text{ or } H'_{i'_1, \ldots, i'_K} = 0; \\ s(H_{i_1, \ldots, i_K}, H'_{i'_1, \ldots, i'_K}) \text{ otherwise;} \end{array} \tag{1}
$$

where $s(\cdot, \cdot)$ is a function that measures hyperedge similarity. We define the hyperedge similarity using a Gaussian kernel $s(H_{i_1, \ldots, i_K}, H'_{i'_1, \ldots, i'_K}) = \exp(-\|H_{i_1, \ldots, i_K} - H'_{i'_1, \ldots, i'_K}\|^2_2/\sigma_1)$ where σ_1 is a scaling parameter. Many alternative similarity measures can be used instead. Each element of the compatibility tensor $\mathcal{C}$ represents a similarity measure between the two corresponding hyperedges. The hyperedge pair $\{i_1, \cdots, i_K\}$ and $\{i'_1, \cdots, i'_K\}$ with a large similarity measure has a large probability $\Pr(\{i_1, \cdots, i_K\} \leftrightarrow \{i'_1, \cdots, i'_K\}|H, H')$ for matching. Here the notation $\leftrightarrow$ denotes a possible matching between a pair of hyperedges or a pair of vertices. Under the conditional independence assumption of the matching process [10], the hyperedge matching probability can be factorized over the associated vertices of the hypergraphs as $\Pr(\{i_1, \cdots, i_K\} \leftrightarrow \{i'_1, \cdots, i'_K\}|HG, HG') = \prod_{n=1}^{K} \Pr(i_n \leftrightarrow i'_n|HG, HG')$ where $\Pr(i_n \leftrightarrow i'_n|HG, HG')$ denotes the probability for the possible matching $i_n \leftrightarrow i'_n$

to be correct. For two hypergraphs $HG(V, E)$ and $HG(V', E')$ with $|V| = N$ and $|V'| = N'$ respectively, we denote their $N \times N'$ matching matrix by $\mathbf{P}$ with entries $P_{ii'} = \Pr(i \leftrightarrow i'|HG, HG')$. High order matching problems can be formulated as locating the matching probability that most closely accords with the elements of the compatibility tensor, i.e. seeking the optimal $\mathbf{P}$ by maximizing the objective function

$$
f(\mathbf{P}) = \sum_{i_1=1}^{N} \sum_{i'_1=1}^{N'} \cdots \sum_{i_K=1}^{N} \sum_{i'_K=1}^{N'} C_{i_1 i'_1, \; , i_K i'_K} \Pr(\{i_1, \cdots, i_K\} \leftrightarrow \{i'_1, \cdots, i'_K\} | HG, HG')
$$

$$
= \sum_{i_1=1}^{N} \sum_{i'_1=1}^{N'} \cdots \sum_{i_K=1}^{N} \sum_{i'_K=1}^{N'} C_{i_1 i'_1, \; , i_K i'_K} \prod_{n=1}^{K} P_{i_n i'_n} \tag{2}
$$

subject to $\forall i, j,\ P_{ii} \geq 0$ and $\sum_{i=1}^{N} \sum_{i'=1}^{N'} P_{ii'} = 1$. Let $\Pr(i \leftrightarrow i'|HG, HG') = P_{ii'}$ where $P_{ii'}$ is the (i, i')th entry of $\mathbf{P}$. We refer to $\Pr(i \leftrightarrow i'|HG, HG')$ as the matching probability for vertex i and i', and the set of matching probabilities $\{\Pr(i \leftrightarrow i'|HG, HG')|i \in V; i' \in V'\}$ obtained by maximizing (2) reveal how likely it is that each correspondence is correct according to structural similarity between the two hypergraphs HG and HG'. This formulation has also been adopted in tensor power iteration for higher order matching [3]. However, the difference between our method and the existing algorithms is that we restrict the solution of (2) to obey the the fundamental axioms of probability, i.e. positiveness and unit total probability mass. This constraint not only provides an alternative probabilistic perspective for hypergraph matching, but also proves convenient for optimization.

Once the set of matching probabilities satisfying (2) are computed, correspondences between vertices drawn from HG and HG' can be established. Matchings with a zero probability are the least likely correspondences, and matchings with nonzero probabilities tend to be those with significant similarity between their structural contexts. Our aim is to seek the subset of possible matchings with nonzero probabilities which satisfy (2) and that are subject to the one-to-one matching constraint.

3 High Order Matching as Dominant Cluster Analysis on a Direct Product Hypergraph

In this section we pose the high order relational matching problem formulated in (2) as one of *dominant cluster analysis* on a *direct product hypergraph*. We commence by establishing a direct product hypergraph for the two hypergraphs to be matched. Optimal matching can be achieved by extracting the dominant cluster of vertices from the direct product hypergraph.

3.1 Direct Product Hypergraph

The construction of a direct product hypergraph for two K-uniform hypergraphs is a generalization of that of the direct product graph [9], which can be used to construct kernels for graph classification. We extend the concept of a direct product graph to encapsulate high order relations residing in a hypergraph and apply this generalization to hypergraph matching problems. For two K-uniform hypergraphs $HG(V, E)$ and $HG'(V', E')$, the direct product $HG_\times$ is a hypergraph with vertex set

$$V_\times = \{(i, i')|i \in V, i' \in V'\}; \tag{3}$$

and edge set

$$E_\times = \{\{(i_1, i'_1) \cdots (i_K, i'_K)\}|\{i_1, \cdots, i_K\} \in E, \{i'_1, \cdots, i'_K\} \in E'\}. \tag{4}$$

The vertex set of the direct product hypergraph $HG_\times$ consists of Cartesian pairs of vertices drawn from HG and HG' separately. Thus the cardinality of the vertex set of $HG_\times$ is $|V_\times| = |V||V'| = NN'$. The direct product hypergraph $HG_\times$ is K-uniform, and each K-tuple of vertices in $HG_\times$ is encompassed in a hyperedge if and only if the corresponding vertices in HG and HG' are both encompassed by a hyperedge in the relevant hypergraph. Each hyperedge in a direct product hypergraph is weighted by the similarity between the two associated hyperedges from HG and HG'.

Furthermore, from our definition of direct product hypergraph, it is clear that the compatibility tensor $\mathcal{C}$ defined in (1) is in fact the tensor $\mathcal{C}_\times$ associated with the direct product hypergraph $HG_\times$ for HG and HG'. Every possible matching $i \leftrightarrow i'$ is associated with the vertex (i, i') in $HG_\times$. For simplicity we let α denote a vertex in $HG_\times$ instead of (i, i'), and let $\mathbb{D}$ denote the subset of vertices in $HG_\times$ which represent the correct vertex matching for HG and HG'. We denote the probability for the vertex α belonging to $\mathbb{D}$ by $\Pr(\alpha \in \mathbb{D}|HG_\times)$. For a direct product hypergraph with $N_\times$ vertices, we establish a $N_\times \times 1$ vector $\mathbf{p}$ with its αth element $p_\alpha = \Pr(\alpha \in \mathbb{D}|HG_\times)$. With these ingredients the optimal model satisfying the condition (2) reduces to

$$\mathbf{p} = \underset{\mathbf{p}}{\mathrm{argmax}} \sum_{\alpha_1=1}^{N_\times} \cdots \sum_{\alpha_K=1}^{N_\times} \mathcal{C}_{\alpha_1, \ldots, \alpha_K} \prod_{n=1}^{K} p_{\alpha_n} \tag{5}$$

subject to the constraints $\forall \alpha, \ p_\alpha \geq 0$ and $\sum_{\alpha=1}^{N_\times} p_\alpha = 1$. Following the construction of a direct product hypergraph, the objective function (5) is a natural extension of that in [8] from clustering to matching. It is also a reduced version of the objective function of ensembles of affinity relations [6], with no manual threshold on the optimization.

According to (5), zero probability will be assigned to the vertices that do not belong to $\mathbb{D}$. We refer to the probability $\Pr(\alpha \in \mathbb{D}|HG_\times) = p_\alpha$ where p_α is the αth element of the vector $\mathbf{p}$ satisfying the optimality condition in (5) as the association probability for the vertex α. Therefore, the matching problem can be solved by extracting the cluster of vertices with nonzero association probabilities in the direct product hypergraph.

3.2 Dominant Cluster Analysis

In this subsection, we formulate the problem of high order structural matching by applying *dominant cluster analysis* (DCA) to a direct product hypergraph (DPH). A dominant cluster of a hypergraph is the subset of vertices with the greatest average similarity, i.e. average similarity will decrease subject to any vertex deletion from or vertex addition to

the subset. Drawing on the concept of the dominant set in a graph [7] and its game theoretic generalization [8], we can easily perform DPH-DCA by applying the following
update until convergence is reached [2]

$$p_\alpha^{new} = \frac{p_\alpha \sum_{\alpha_2=1}^{N_\times} \cdots \sum_{\alpha_K=1}^{N_\times} C_{\alpha,\alpha_2,\,\ldots,\alpha_K} \prod_{n=2}^{K} p_{\alpha_n}}{\sum_{\beta=1}^{N_\times} p_\beta \sum_{\beta_2=1}^{N_\times} \cdots \sum_{\beta_K=1}^{N_\times} C_{\beta,\beta_2,\,\ldots,\beta_K} \prod_{n=2}^{K} p_{\beta_n}} \tag{6}$$

At convergence the weight p_α is equal to the association probability $\Pr(\alpha \in \mathbb{D}|HG_\times)$,
i.e. the probability for the corresponding potential matching $i \leftrightarrow i'$ to be correct.

4   Matching with Prior Rejections

The high order structural matching algorithm described in Section 3 is a unsupervised
process. The weight of each vertex in the direct product hypergraph can be initialized
by using a uniform distribution of probability. However, if two vertices in a hypergraph
have the same structural context, i.e. their interchange does not change the hypergraph
structure, they can cause ambiguity when matching is attempted. Two alternative state-
of-the-art methods, namely probabilistic hypergraph matching [10] and tensor power
iteration [3], also suffer from this shortcoming.

However, if prior knowledge about outliers (i.e. hypergraph vertices for which no
match exists) is available, we can to a certain extent avoid the ambiguity and improve
matching accuracy by using a different weight initialization strategy. We refer to the
vertex subset $V_\times^o \subseteq V_\times$ (i.e. possible correspondences) associated with available outliers as prior rejections, and the adopted initialization in the light of prior rejections is
as follows

$$w(\alpha) = \begin{array}{ll} 0 & \text{if } \alpha \in V_\times^o; \\ 1/(N_\times - N_\times^o) & \text{otherwise;} \end{array} \tag{7}$$

where $N_\times^o$ is the cardinality of $V_\times^o$.

The initialization scheme (7) improves the matching accuracy within the DPH-DCA
framework because the vertex weight $w(\alpha)$ in the numerator of the update formula (6)
plays an important role in maintaining the initial rejection. It enables the prior rejections to maintain a zero weight and does not affect the matching scores for other possible correspondences at each update until converged. The extent to which the matching
accuracy can be improved depends on the amount of prior rejections available. The
more prior knowledge concerning the outliers that is available, the more accurate the
matching that can be obtained. This will be verified in our experimental section.

In [6], the authors have described the same initialization step as a disadvantage. On
the other hand, we argue that the initialization scheme (7) does not apply to the alternative methods[10][3] even when identified outliers are available. The probabilistic
hypergraph matching method [10] initializes a matching score by a fixed value obtained
from the marginalization of the compatibility tensor, and thus can not accommodate the
prior rejections by using (7). The tensor power iteration method [3], though manually
initialized, converges to a fixed matching score for different initializations.

5   Experiments

We test our algorithm for high order structural matching on two types of data. Firstly, we test our method on synthetic data to evaluate its robustness to noise and outliers. Secondly, we conduct experiments to match features extracted from images. Prior rejections are considered for both types of data to improve the matching accuracy. We compare our method with two state-of-the-art methods, i.e. probabilistic hypergraph matching (PHM) [10] and tensor power iteration (TPI) [3].

5.1   Matching Synthetic Data

We commence with the random generation of a structural prototype with 15 vertices. The distance d_{ij} between each pair of vertices i and j of the prototype is randomly distributed subject to the Gaussian distribution $N(1, 0.5)$. We test our method by establishing correspondences between the prototype structure and a modified structure. The alternative modifications include a) noise addition, b) vertex deletion, c) rescaling and d) rotation. Since neither the probabilistic hypergraph matching method nor the tensor power iteration method relies upon a specific initialization, we test our DPH-DCA matching method without prior rejections to make a fair comparison with these two alternative methods. To test the performance of different methods for hypergraph matching we re-scaled the distance between each of vertex pairs by a random factor and rotate the structure by a random angle. In this case, the pairwise relationships no longer holds for the matching task. We use the sum of polar sines presented in [5] as a high order similarity measure for point tuples. We measure the similarity of every 3-tuple within the vertex set and thus establish a weighted 3-uniform hypergraph for the structure. The compatibility tensor $\mathcal{C}$ for two structures is computed according to (1) with $\sigma_1 = 0.1$. Figure 1(a) illustrates the results of the matching accuracy as a function of noise level. It is clear that our DPH-DCA framework outperforms the two alternative methods at each noise level. To take the investigation one step further, we study the performance of our method for matching structures of different vertex cardinality. To this end, we extract a substructure from a prototype and slightly perturb the distance between each vertex pair by adding random noise normally distributed according to $N(0, 0.04)$. The cardinality of the vertex set of the substructure varies from 14 down to 5. Vertices not in the substructure are outliers for the matching process. For each vertex cardinality of a substructure, 100 trials are performed. Figure 1(b) illustrates the matching accuracy as a function of outlier number for the three methods. It is clear that our DPH-DCA framework outperforms the two alternative methods at each number of outliers. We have also evaluated the matching accuracy of our DPH-DCA framework at different levels of available prior rejection. To this end, we have extracted a 5-vertex substructure from a prototype and slightly perturb the distance between each vertex pair by adding random noise normally distributed according to $N(0, 0.04)$. We involve prior rejections by rejecting the matchings associated with a varying number of outliers. Figure 1(c) illustrates the matching accuracy as a function of the number of rejected outliers. It is clear that the matching accuracy grows monotonically as the number of rejected outliers increases.

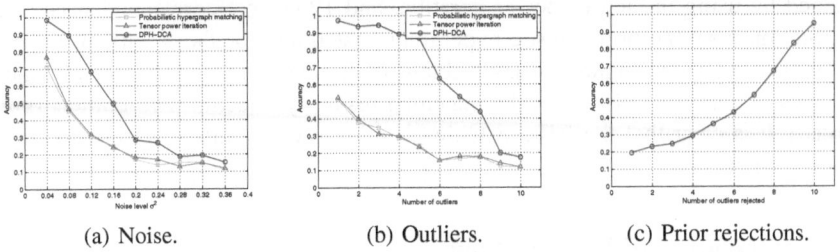

(a) Noise.   (b) Outliers.   (c) Prior rejections.

Fig. 1. Matching performance

5.2   Image Correspondences

To visualize the matching for real world images we test the alternative methods on frames of video[1]. We use the Harris detector to extract corner points from the first and 30th frames. We use the sum of polar sines presented in [5] to measure the similarity of every 3-tuple within the corner points and thus establish a weighted 3-uniform hypergraph for each image. Figure 2 illustrates the matching performances for alternative methods. The matching results for the two comparison methods are visualized in Figures 2(a) and 2(b), where 11 correct correspondences and 4 incorrect ones are

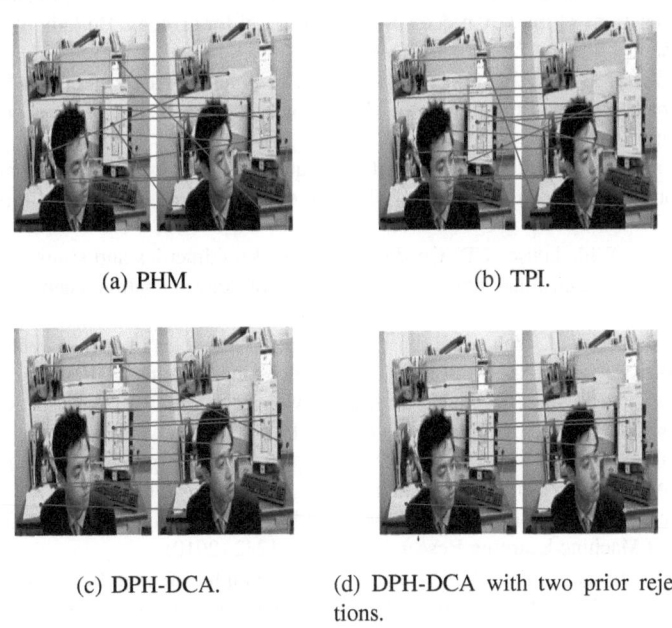

(a) PHM.   (b) TPI.

(c) DPH-DCA.   (d) DPH-DCA with two prior rejections.

Fig. 2. Image correspondences

---

[1] http://www.suri.it.okayama-u.ac.jp/e-program-separate.html

obtained by using the tensor power iteration, and 12 correct correspondences and 3 incorrect ones by the probabilistic hypergraph matching. For DCA without prior rejections (visualized in Figure 2(c)), we obtain 14 correct correspondences and 1 incorrect ones. Figure 2(d) visualizes the matching result by rejecting two outliers (green marked). It is clear that the false matching is eliminated by incorporating the proper prior rejections.

6   Conclusion and Future Work

We have presented a novel approach to high order structural matching. We have transformed the matching problem to that of extracting the dominant cluster from the direct product hypergraph for two feature sets with high order relationships. Prior knowledge about outliers can be easily involved in our framework by initializing the matchings associated with the outliers by a zero weight. Experiments have shown that our method outperforms the state-of-the-art methods.

References

1. Agarwal, S., Lim, J., Zelnik-Manor, L., Perona, P., Kriegman, D., Belongie, S.: Beyond pairwise clustering. In: Proceedings of IEEE Conference on Computer Vision and Pattern Recognition (2005)
2. Baum, L.E., Eagon, J.A.: An inequality with applications to statistical estimation for probabilistic functions of markov processes and to a model for ecology. Bulletin of the American Mathematical Society 73, 360–363 (1967)
3. Duchenne, O., Bach, F.R., Kweon, I.S., Ponce, J.: A tensor-based algorithm for high-order graph matching. In: Proceedings of IEEE Conference on Computer Vision and Pattern Recognition (2009)
4. Leordeanu, M., Hebert, M.: A spectral technique for correspondence problems using pairwise constraints. In: Proceedings of IEEE International Conference on Computer Vision (2005)
5. Lerman, G., Whitehouse, J.T.: On d-dimensional d-semimetrics and simplex-type inequalities for high-dimensional sine functions. Journal of Approximation Theory 156(1), 52–81 (2009)
6. Liu, H., Latecki, L.J., Yan, S.: Robust clustering as ensembles of affinity relations. In: Proceedings of Advances in Neural Information Processing Systems (2010)
7. Pavan, M., Pelillo, M.: Dominant sets and pairwise clustering. IEEE Transactions on Pattern Analysis and Machine Intelligence 29(1), 167–172 (2007)
8. Rota-Bulo, S., Pelillo, M.: A game-theoretic approach to hypergraph clustering. In: Proceedings of Advances in Neural Information Processing Systems (2009)
9. Vishwanathan, S.V.N., Borgwardt, K.M., Kondor, I.R., Schraudolph, N.N.: Graph kernels. Journal of Machine Learning Research 11, 1201–1242 (2010)
10. Zass, R., Shashua, A.: Probabilistic graph and hypergraph matching. In: Proceedings of IEEE Conference on Computer Vision and Pattern Recognition, pp. 234–778 (2008)

A Probabilistic Framework for Complex Wavelet Based Image Registration

Florina-Cristina Calnegru

University of Pitesti, Department of Computer Science
calnegru_florina@yahoo.com

Abstract. The aim of this article is to introduce a computationally tractable mathematical model of the relation between the complex wavelet coefficients of two different images of the same scene. Because the two images are acquisitioned at distinct times, from distinct viewpoints, or by distinct sensors, the relation between the wavelet coefficients is far too complex to handle it in a deterministic fashion. This is why we consider adequate and present a probabilistic model for this relation. We further integrate this probabilistic framework in the construction of a new image registration algorithm. This algorithm has subpixel accuracy, and is robust to noise and to a large class of local variations like changes in illumination and even occlusions. We empirically prove the properties of this algorithm using synthetic and real data.

Keywords: Image registration, probabilistic similarity measure, complex wavelet transform.

1 Introduction

In Visual Computing, next to the problem of analyzing a single image, one often encounters the problem of combining information contained in more images [1]. The first step in integrating the information comprised in a set of images is the registration of each pair of images from that set.

Image registration is the process of geometrically overlapping two images of the same scene, obtained at different moments in time, or from different view angles, or with different sensors [2]. The two images involved in the process of registration are the reference image and the target image or the sensed image. With this terminology, the registration can be defined as the process of finding a transformation such that the target image becomes similar with the reference image [3].

Largely speaking there are two kinds of registration methods: parametric and non-parametric. In the case of the parametric registration, the transformation is parametric, i.e. can be expanded in terms of some basis functions. In the case of non-parametric registration, the transformation is no longer restricted to a parametrizable set. As the algorithm that we propose is from the category of parametric image registration, we will no further insist on the non-parametric image registration. Parametric image registration can be divided into: landmark based parametric image registration, principal axes-based registration, and optimal parametric registration [1].

Landmark based parametric registration is a type of registration based on the features extracted at an initial stage in the process of registration. Those features can be

G. Maino and G.L. Foresti (Eds.): ICIAP 2011, Part I, LNCS 6978, pp. 9–18, 2011.
© Springer-Verlag Berlin Heidelberg 2011

lines intersections, road crossings, inflection points of curves, corners [2], local extremes of wavelet transform [4], etc. The quality of landmark-based registration is highly dependent on the performances of the feature detector that is used. If the feature detector is not reliable enough, the registration will be low quality.

This is why sometimes is better to use a registration method that relays on features that can be automatically deduced from the image. Such intrinsic features are for example the principal axes [5]. Although principal axes registration is fast and necessitates very few parameters, it needs the moment matrix and the eigenvalue decomposition of two large matrixes, it is not suitable for multimodal registration, and its results can be ambiguous [1].

The disadvantages of the landmark-based registration and principal axes registration led to the emergence of a more general and flexible class of parametric image registration. This category consists of optimal parametric registration algorithms.

The basic idea of optimal parametric registration is to define a distance (similarity) measure between the reference image and the target image, and then to find the parameters of the transformation that optimize this similarity measure. The most known similarity measures are the sum of squared differences, the correlation, and the mutual information [6]. The domain of these similarity measures is either the intensity space like in [7], or another feature space like the wavelet coefficients with a magnitude above a certain threshold [4], the energy map [8], and the wavelet coefficients from the first decomposition level [9].

The algorithm that we introduce is an optimal registration algorithm. The similarity measure that we use is defined on the complex wavelet coefficients space. This makes our similarity measure more robust to noise than the similarity measures defined on the intensity space. The later mentioned similarity functions are affected by the noise, that usually corrupts the image intensities.

This article is structured as following. In section 2 we present an introduction to the dual-tree complex wavelet transform. In section 3 we expose the probabilistic model underlying the registration algorithm. In section 4 we present the registration algorithm and introduce different modalities to integrate it into real world registration systems. In section 5 we justify the necessity for our mathematical model in the present context, which is that of the existence of a related well-known probabilistic model in the intensity domain. In section 6 we present experimental results on artificial data as well as on real data. In section 7 we present our conclusions.

2   Complex Wavelet Transform

Discrete wavelet transform (DWT) is a modality to project a signal onto an orthogonal wavelet basis. By using the DWT one can obtain local information about a signal both in the spatial domain and in the frequency domain. For a 2-D signal the DWT coefficients are obtained by passing the signal through a cascade of orthogonal high pass and low pass filters. The original image is decomposed at any scale j, into 4 components: HH_j (contains the diagonal details), HL_j (contains the horizontal details), LH_j (comprised of vertical details), and LL_j (contains the approximation coefficients). For more information on DWT see [10].

Any signal f(x, y) can be reconstructed via the inverse discrete wavelet transform from its detail and approximation coefficients as in (1)

$$f(x, y) = \frac{1}{\sqrt{MN}} \sum_{m} \sum_{n} W_{\varphi}(j_0, m, n) \varphi_{j_0, m, n}(x, y)$$

$$+ \frac{1}{\sqrt{MN}} \sum_{k=1}^{3} \sum_{j=j_0}^{\infty} \sum_{m} \sum_{n} W_{\psi}^{k}(j, m, n) \psi_{j, m, n}^{k}(x, y)$$

(1)

In (1) $\varphi_{j,m,n}$ represents the scaling function scaled with a factor of j and translated with m and n, $\psi^{k}_{j,m,n}$ represents k-th mother wavelet function, scaled with a factor of j and translated with m on Ox, and n on Oy, $W_{\varphi}(., .)$ represents the aproximation coefficients, and $W_{\psi}^{k}(., .)$ represents the detail coefficients.

For the 2-D DWT there are 3 mother wavelet functions: one that permits the extraction of horizontal details, one for the vertical details, and one for the diagonal details. So we can say that $W_{\varphi}(j, ., .)$ corresponds to $LL_j(., .)$ and that, for example, $W_{\psi}^{1}(j, ., .)$ corresponds to $HL_j(., .)$, $W_{\psi}^{2}(j, ., .)$ corresponds to $LH_j(., .)$, and that $W_{\psi}^{3}(j, ., .)$ corresponds to $HH_j(., .)$.

Unfortunately, the DWT has some major drawbacks that make it less appropriate for registration. Among those drawbacks, we mention poor directional selectivity, as the HH coefficients cannot differentiate between edges at 45 degrees and edges at 135 degrees and rotation and translation variance. Complex wavelet transform constitutes a remedy for these problems.

One can observe that by taking in (1), instead of a real scaling function and real wavelet functions, a complex scaling function, and complex wavelet functions, for which the real and the imaginary part form a Hilbert pair, the drawbacks of the DWT are eliminated [11].

In our article, we employed dual tree complex wavelet transform to obtain the complex wavelet decomposition for our images. The dual tree complex wavelet transform uses 6 complex mother wavelets that distinguish spectral features oriented at {75°, 45°, 15°, -75°, -45°, -15°}. By projecting the image onto the 6 complex wavelet functions, we obtain 6 complex wavelet coefficients for each scale and translation.

To facilitate the presentation, from now on, every time we mention wavelet transform, we refer to the dual tree complex wavelet transform.

3   Construction of the Probabilistic Framework

The research underlying this article is driven by the desire to understand the relation between corresponding complex wavelet coefficients of two images of the same scene. What happens with the complex wavelet coefficients when the two images are captured at different times, from different viewpoints, or with different sensors? In the following, we propose a mathematical model for this relation.

The intuition behind this model is that for each level of wavelet decomposition, the layers of magnitudes of the coefficients from one image, should have the same configuration as the layers of magnitudes of the coefficients from the other image. (For every image, each level of decomposition has 6 layers of coefficient magnitudes, one for each of the 6 mother wavelets.) This means that if a layer, from one image, has a certain area with large (or small) coefficient magnitudes, the corresponding layer from the other image has the same area (i.e. the area located at identical coordinates) populated with large (or small) coefficient magnitudes. More than that, we expect this to happen even at the finest granularity level, i.e. we expect that large (or small) coefficient magnitudes from one image correspond to large (or small) coefficient magnitudes from the other image. We base our expectations on the fact that wavelet coefficients are "the wavelet's response" to the structures from the image. If two images depict the same scene, then the wavelet response should be similar. This observation is sustained by a known property of this type of wavelet transform. That property says that the coefficient magnitudes should be large for wavelets that overlap singularities and should be small on smooth regions [11].

Let us denote by $M_{j,k,\psi}(m, n)$ the magnitude of the coefficient $W_\psi^k(j, m, n)$. Then, by denoting $M^1_{j,k,\psi}$, and $M^2_{j,k,\psi}$. the coefficient magnitudes for the two images, we propose the following mathematical model for their relation:

$$M^2_{j,k,\psi}(m, n) = M^1_{j,k,\psi}(m, n) + N(0, \sigma_{j,k}), \forall k, j, m, n \qquad (2)$$

$N(0, \sigma_{j,k})$ is a Gaussian random variable of mean 0 and variance $\sigma_{j,k}$, and can be viewed as an admissible difference between 2 magnitudes of coefficients, that contain information about the same scene structures. Those differences can be thought to be organized in layers and levels, in the same way in which the associated magnitudes are organized. We use this model to define a probabilistic similarity measure between two images of the same scene, that differ one from another by a parametric coordinate transformation, described by the set of parameters θ. From now on, we mention that transformation as $T\theta$. We define the probability that a coefficient from image 1 contains information about the same scene structures as a coefficient from image 2, when image 2 is transformed with $T\theta$:

$$p^\theta_{j,k,m,n} = \frac{1}{\sqrt{2\pi\sigma_{j,k}}} e^{-\frac{1}{2}\left(\frac{M^{2,\theta}_{j,k,\psi}(m,n) - M^1_{j,k,\psi}(m,n)}{\sigma_{j,k}}\right)^2}, \forall k, j, m, n \qquad (3)$$

We assume that the differences, between the magnitudes of 2 correspondent layers, (i.e. of any layer of differences) are independent. We define the probability that one layer of coefficients, from one image, has the same configuration as the correspondent layer of coefficients from the other image, when this image is transformed by $T\theta$:

$$p\left(M^{2,\theta}_{j,k,\psi} \approx M^1_{j,k,\psi}\right) = \prod_{m,n} p^\theta_{j,k,m,n}, \forall j, k \qquad (4)$$

We also assume that every 2 layers of differences from a level and every 2 levels of differences are independent. We obtain the probability that the coefficient magnitude of image 1 represent the same structures of the scene as the coefficients of the image 2, when image 2 is transformed by $T\theta$:

$$p\left(M_{\psi}^{2,\theta} \approx M_{\psi}^{1}\right) = \prod_{j,k} p\left(M_{j,k,\psi}^{2,\theta} \approx M_{j,k,\psi}^{1}\right) \tag{5}$$

Since any image can be reconstructed from its wavelet coefficients, (5) can be thought as the probability that image 1 is similar with image 2, when image 2 is transformed with Tθ.

It can be shown as in [12] that the maximization of (4), $\forall\, j,k$ is equivalent to the maximization of the cross-correlation (6) between the coefficients of the layer k from level j, $\forall\, j,k$.

$$CC_{j,k}^{\theta} = \frac{\displaystyle\sum_{m=1}^{M}\sum_{n=1}^{N}\left(M_{j,k,\psi}^{2,\theta}(m,n)-\overline{M_{j,k,\psi}^{2,\theta}}\right)\left(M_{j,k,\psi}^{1}(m,n)-\overline{M_{j,k,\psi}^{1}}\right)}{\sqrt{\displaystyle\sum_{m=1}^{M}\sum_{n=1}^{N}\left(M_{j,k,\psi}^{2,\theta}(m,n)-\overline{M_{j,k,\psi}^{2,\theta}}\right)^{2}}\displaystyle\sum_{m=1}^{M}\sum_{n=1}^{N}\left(M_{j,k,\psi}^{1}(m,n)-\overline{M_{j,k,\psi}^{1}}\right)}, \tag{6}$$

where $\overline{M_{j,k,\psi}^{2,\theta}}$ is the mean of the layer $M_{j,k,\psi}^{2,\theta}$ and $\overline{M_{j,k,\psi}^{1}}$ is the mean of the layer $M_{j,k,\psi}^{1}$.

More, it can be shown that the probability, that the 2 layers are equivalent, increases with the increase of $CC_{j,k}^{\theta}$. This means that $CC_{j,k}^{\theta}$ can be considered an indicator for the probability of equivalence between the 2 layers. Because of that and because $CC_{j,k}^{\theta}$ can be negative, we make the following approximation, and define the similarity between image 1 and image 2, when image 2 is transformed by Tθ:

$$p^{\theta} = \prod_{j,k} d_{j,k} \text{ , where } d_{j,k} = \begin{cases} \dfrac{1}{1-CC_{j,k}^{\theta}} & , \text{ if } CC_{j,k}^{\theta} \neq 1 \\[2mm] c & , \text{ otherwise} \end{cases} \tag{7}$$

In the definition for $d_{j,k}$ from (7), c is a large real number.
We name the similarity measure from (7) *wavelet layers correlation*.

4   Registration Algorithm

The registration algorithm consists in finding $\theta = \left(\theta_{1},...,\theta_{n}\right)_{n>0}$ such that p^{θ} is maximized. To avoid that the algorithm outputs a local maximum instead of a global maximum, we used simulated annealing [7].

We have used this algorithm to register images that differ by an affine transform or by a similarity transform (n is 6 respectively 4). For these images, the search space of the parameters is quite large. Fortunately, p^θ allows us to find the optimizing parameters, not by searching on the Cartesian product of the spaces of all parameters, but by searching, in turn, on the space of each parameter. This process necessitates several iterations, depending how far the solution is from the initial guess.

We have empirically found that it is sometimes better not to use, in (7), all the levels of decomposition. Actually the number of levels it is dependent on the type (for example retinal or outdoor) of images to register.

For high temperature the generation function, for our variant of the simulated annealing algorithm, is the product between a Gaussian centered in the current value, x, and the function h from (8). For smaller temperatures, the generation function is simply a Gaussian centered in x.

$$h(x) = \begin{cases} 1, & r < 0.5 \\ -1, & r \geq 0.5 \end{cases} \text{, where } r \in (0,1) \text{ is a uniform random number} \qquad (8)$$

The speed of our algorithm depends on three factors. The first is the initial value for the set of parameters θ. The second is the stopping criterion. We have used as the stopping criterion the value of p^θ as long as a number of iterations is not reached. The third speed factor is the number of parameters from the set θ. If this number is large, then the algorithm is more time costly.

Our algorithm can be used in many situations:

1. One can use it to estimate the vector θ, starting from a random value for this vector. When the vector θ has a single parameter, for example the rotation angle, the speed is reasonable and the algorithm can be considered even for time dependent applications. If the registration task consists in finding a transformation with a large number of parameters, we recommend our algorithm when the time is not crucial, but instead the registration accuracy is. This recommendation is supported by the fact that our algorithm permits the finding of a solution with high accuracy.
2. One can use the algorithm for tuning the solution vector θ. This assumes of course that a less accurate solution was already found by a different method. For example as in [13], one can find a first alignment from low frequency components. In this situation, the speed of convergence to a highly accurate solution is reasonable for time dependent applications even when the number of parameters is large.

5   Wavelet Layers Correlation versus Classical Cross-Correlation

We have arrived at wavelet layers correlation by imposing the condition (2) on corresponding wavelet coefficients. If one imposes a similar condition on the intensity values of corresponding pixels, then one obtains as a similarity measure the normalized cross-correlation. The question is why should anyone impose a condition of type (2) on wavelets coefficients rather than on intensities? Therefore, why should anyone use wavelet layers correlation rather than use the classical cross-correlation? We will endeavor to answer this question in the current section.

There are two major reasons for which wavelet layers correlation is superior to classical correlation:

1. A relation of type (2) is, in general, non-realistic for image intensities, as most of the times the noise is non-Gaussian. In exchange, because we imposed a different relation of type (2) for every layer of wavelet coefficients, our model can tackle a more general and realistic type of noise than the Gaussian noise. This phenomenon is more obvious in the case of multimodal images. Fig. 2 shows cross-correlation and wavelet layer correlation as functions of the shift on Ox-axis, for a multimodal retina pair of images. One can see in Fig. 2 that wavelet layers correlation attains the maximum around the correct value, which is 1.03, while cross correlation loses the right solution completely.

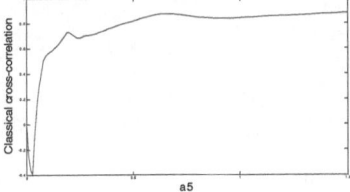 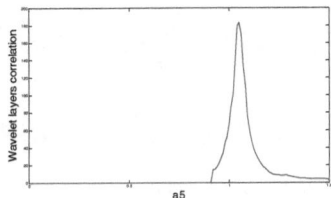

Fig. 1. The cross correlation (left) and the wavelet layers correlation (right) as functions of the 5-th parameter of the set θ = (a1, a2, a3, a4, a5, a6). θ describes the affine transform that registers a pair of multimodal retina images. The values of the other five parameters are fixed to their correct values. These correct values are computed from the ground truth.

2. Even when a model of type (2) can be applied to intensities, cross-correlation is a flatter similarity function than the wavelet layers correlation. This means that the classical correlation has more local maxima than the wavelet layers correlation. It also means that in the case of classical cross-correlation, the global maximum is less conspicuous than in the case of wavelet layers correlation (see Fig. 3). This increases, for an optimization method, the risk of being caught in a local maximum.

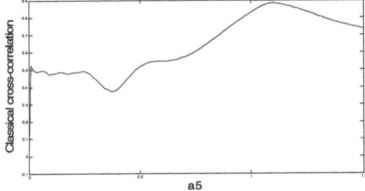 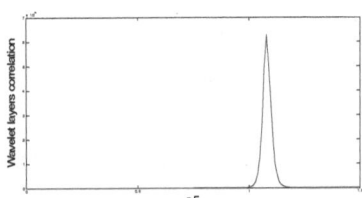

Fig. 2. The cross correlation (left) and the wavelet layers correlation (right) as functions of the 5-th parameter of the set θ = (a1, a2, a3, a4, a5, a6). θ describes the affine transform that registers a pair of single modality retina images. The values of the other five parameters are fixed to their correct values. These correct values are computed from the ground truth.

6 Experimental Results

6.1 Synthetic Data

In order to test the algorithm on synthetic data we have performed two kinds of tests:

1. We took some images and transformed them randomly by a single parameter transform (i.e. either by a rotation, with a random angle between -90° and 90°, either by a random translation between 0 pixels and half the image size, and either by a scaling with a random scaling factor between 0.1 and 3). For every pair of images we chose randomly the type of the transform. We have produced 100 pairs of images. Fig. 3 (left) shows an example of such a pair. The inaccuracy for this pair, as defined in (9), is 0.21 pixels.

Fig. 3. Example of artificial image pairs. The second image from the left pair is the first image from the left, rotated with -9.84°. The first image from the right side is the second image from the right, transformed by a similarity transform.

2. We took the same images and transformed them randomly by a similarity transform. This consists in scaling, together with translation and rotation. The scaling factor is between 0.2 and 2, the rotation angle is between -75° and 75 °, and the translation is between 0 and half the image size. We obtained 50 pairs. You can see in Fig. 3 (right) an example of such a pair. The inaccuracy for this pair, computed accordingly to (9), is 0.32 pixels.

If we denote, for a pair of images, (I_1, I_2), by $T\theta_t$ the real transform (the transform that is used to obtain I_2 from I_1) and by $T\theta_a$ the transform outputted by the algorithm, then the inaccuracy for that pair is given in (9):

$$\sum_{i=1}^{P} \frac{\left\| T\theta_t(x_i, y_i) - T\theta_a(x_i, y_i) \right\|_2}{P}, \ (x_i, y_i)_{i=\overline{1,P}} \text{ are random points in } I_1 \quad (9)$$

Table 1. The results of the tests on synthetic data

Type of the artificial transform	Percent of image pairs registered by wavelet layers correlation with an inaccuracy < 1 pixel
Single parameter	100%
Multiple parameter	98%

6.2 Real Data

We have used three categories of real world images: single modality retina images, multimodal retina images, and outdoor images. The retinal images are considered to differ by an affine transform and the outdoor images by a similarity transform.

The left side of Fig. 4 represents an example of outdoor pair on which we tested the algorithm. The right side of Fig. 4 shows the overlap of the two images from the left side, before registration, and after registration, by means of our registration algorithm. From Fig. 4 we can see that the algorithm is robust to small occlusions, as the man in front of the exit door from the second image, from the left side, is absent in the first image. This visual satisfactory behavior of the algorithm is supported by the accuracy test, which outputs for this pair an inaccuracy of 3.96 pixels.

Fig. 4. The first two images from the left side represent an example of outdoor pair of images that we employed to test the algorithm. The second image from the right side represents the two images, from the left side, overlapped without registration. The first image from the right side represents the two images, from the left side, overlapped after registration with our algorithm.

In order to test the accuracy on a pair of real images, we employed the formula (9) in which $T\theta_t$ (the true transform) is computed from the ground truth (i.e. some correspondent manual chosen points in every pair of images). We used in our tests 2000 random points (i.e. $P = 2000$ in formula (9)) in order to probe the accuracy.

We tested the algorithm on 50 outdoor image pairs, on 50 single modality image pairs, and on 30 multimodal image pairs. The outdoor images were captured with an Olympus Camedia C-500 Zoom camera. The retina images were provided by OD-OS GmbH.

We implemented the algorithm in MATLAB on a Intel(R) Core (TM) 2 CPU T5300, 1.73 GHz processor. We obtained the results summarized in Table 2.

Table 2. Results of the tests on real data

Image cathegory	Percent of image pairs registered by *wavelet layers correlation* with an inaccuracy < 5 pixels
Oudoor	98%
Single modality retinal	94%
Multimodal retinal	93.3%

7 Conclusions

In this paper we have proposed a probabilistic framework for modeling the relation between correspondent wavelet coefficients. We have used this framework for creating a new registration algorithm. The main qualities of this algorithm are: large applicability (can be employed for a variety of images (as section 6 shows), and in many situations (described in section 4), robustness to small occlusions (as the tests on outdoor images show), robustness to a large category of noise (as can be seen from the tests on single modality and multiple modality image), the ellimination of any neccesity for preprocessing like for example noise reduction (this can be seen from all our tests since we have used no preprocessing), and sub-pixel accuracy (this was proved only on synthetic images since the accuracy on real images (as is outputed by the tests) depends on the limited precision of the human individuals that created the ground truth).

References

1. Modersitzki, J.: Numerical Methods for Image Registration. Oxford University Press, New York (2004)
2. Zitova, B., Flusser, J.: Image registration methods: a survey. Image and Vision Computing 21(11), 977–1000 (2003)
3. Goshtasby, A.: 2-D and 3-D Image Registration for medical, remote sensing, and industrial applications. John Wiley and Sons, Inc., Hoboken (2005)
4. Moigne, J.L., Campbell, W.J., Cromp, R.F.: An automated parallel image registration technique based on the correlation of wavelet features. IEEE Trans. On Geoscience and Remote Sensing 40(8), 1849–1864 (2002)
5. Alpert, N.M., Bradshaw, J.F., Kennedy, D., Correia, J.A.: The principal axes transformation - A method for image registration. Journal of Nuclear Medicine 31(10), 1717–1722 (1990)
6. Viola, P., Wells III, W.: M.: Alignment by maximization of mutual information. In: International Conference on Computer Vision, pp. 16–23 (1995)
7. Ritter, N., Owens, R., Cooper, J., Eikelboom, R.H., van Saarloos, P.: Registration of Stereo and Temporal Images of the Retina. IEEE Transactions on Medical Imaging 18(5) (1999)
8. Pauly, O., Padoy, N., Poppert, H., Esposito, L., Navab, N.: Wavelet energy map: A robust support for multi-modal registration of medical images. In: IEEE Conference on Computer Vision and Pattern Recognition, pp. 2184–2191 (2009)
9. Li, S., Peng, J., Kwok, J.T., Zhang, J.: Multimodal registration using the discrete wavelet frame transform. In: Proc. of ICPR Conf., pp. 877–880 (2006)
10. Gonzalez, R., Richard, R.: Digital Image Processing. Prentice Hall, Upper Saddle River (2002)
11. Selesnick, I.W., Barniuk, R.G., Kingsbury, N.G.: The Dual-Tree Complex Wavelet Transform. IEEE Signal Processing Magazine (2005)
12. Nixon, M.S., Aguado, A.S.: Feature Extraction and Image Processing. Butterworth-Heinemann/Newnes, Oxford (2002)
13. Indian Institute of Information Technology, Allahabad,
   http://mtech.iiita.ac.in/Agrade/
   Sukriti-MedicalImageRegistrationusingNextGenerationWavelets.pdf

Image De-noising by Bayesian Regression

Shimon Cohen and Rami Ben-Ari

Orbotech Ltd., Yavneh, Israel
{Shimon.Cohen,Rami.Ben-Ari}@orbotech.com
http://www.orbotech.co.il/

Abstract. We present a kernel based approach for image de-noising in the spatial domain. The crux of evaluation for the kernel weights is addressed by a Bayesian regression. This approach introduces an adaptive filter, well preserving edges and thin structures in the image. The hyper-parameters in the model as well as the predictive distribution functions are estimated through an efficient iterative scheme. We evaluate our method on common test images, contaminated by white Gaussian noise. Qualitative results show the capability of our method to smooth out the noise while preserving the edges and fine texture. Quantitative comparison with the celebrated total variation (TV) and several wavelet methods ranks our approach among state-of-the-art denoising algorithms. Further advantages of our method include the capability of direct and simple integration of the noise PDF into the de-noising framework. The suggested method is fully automatic and can equally be applied to other regression problems.

Keywords: Image de-noising, Bayesian regression, Adaptive filtering.

1   Introduction

The need for efficient image restoration methods has grown with the massive production of digital images and movies of all kinds, often taken in poor conditions. No matter how good cameras are, an image improvement is always desirable to extend their range of action. The valid challenge of denoising methods is removing the noise without creating artifacts, while preserving the image edges and fine structures. Such denoising attempts are referred as *adaptive* filtering.

Many methods have been suggested in the past for image adaptive denoising. One class of methods filters the image via the frequency domain using wavelets [17,4,6,12]. Although yielding excellent results in terms of PSNR, yet these methods often produce particular visual artifacts such as *ringing*.

Another approach performs the filtering in the spatial domain, known as *steerable* filters. In the domain of PDE and diffusion methods the *inhomogeneous* (i.e. shift variant yet isotropic) and *anisotropic* diffusion approaches gained popularity for their desired properties on edge preserving [9,11,13]. In this framework the so called Total Variation regularization (TV) attracted special attention [9,5,13]. These PDE based approaches among other non-linear methods can be approximated by a *kernel* based filtering where the kernel is *shift variant* [15]. However,

G. Maino and G.L. Foresti (Eds.): ICIAP 2011, Part I, LNCS 6978, pp. 19–28, 2011.

in these methods the adjustment of the model to the noise distribution is highly non-trivial and practically hidden under the distance measure (norm) used in the model. Furthermore, these methods often need a tuning parameter, adjusted by the user, in order to obtain adequate performance.

Non-deterministic modeling approaches [2,7,16] rely on the data for adaptively varying the kernel. Our method belongs to this latter approach, where the noise PDF is an explicit part of our statistical model. The key ingredient of our approach is the use of simple but efficient Bayesian estimation model. Bayesian estimation seeks a predictive probability distribution function (PPDF), i.e prediction is done through model averaging. We view the image denoising as a *regression* problem in the spirit shown by Tipping [10]. While in [10] a Taylor expansion (to the second degree) is used to obtain an exact solution, we avoid this approximation by utilizing an efficient iterative procedure for solution of the regression problem.

Often measurement models are considered to be linear in the unknown image. Yet, more successful methods have taken a nonlinear estimation approach to this inverse problem [2,16]. In this work, we perform our regression on a transformed domain by Radial Basis Functions (RBF) [3] and impose a *prior* on the RBF's weights in order to avoid over fitting.

Experiments show that the proposed approach can reduce noise from corrupted images while preserving edge components efficiently. Despite the simplicity of our method both in its concept and implementation, the denoising results are among the best reported in the literature. It is further executed without any parameter setting or user intervention. The suggested regression scheme is not limited to image de-noising and can be employed for other regression problems.

2   Problem Formulation

Usually in a data-model matching procedure, one tries to optimize the likelihood of the data, i.e, the probability of the data given the model $p(D|M)$. We assume the model has parameters (a.k.a *weights*) arranged in the vector $\mathbf{w}$ and consider the method of Maximum *a-posterior* probability (MAP) with automatic inference of the regularization parameters. We then deduce the probability of the model given the data, i.e. $p(M|D)$ and use it to construct the *predictive distribution*.

2.1   Bayesian Regression

In this work, we assume that image pixels are corrupted by additive white Gaussian noise with an *unknown* variance σ^2:

$$y_i - t_i \sim N(0, \sigma^2) \tag{1}$$

where, t_i denotes the targets namely, the observed intensities and y_i the output of a regression method. The likelihood of the data under identically independent distribution (i.i.d) is given by:

$$p(D|M) = \prod_{i=1}^{K} \frac{1}{\sqrt{2\pi}\sigma} \exp\left(-\frac{(y_i - t_i)^2}{2\sigma^2}\right) \tag{2}$$

where K denotes the kernel size. The first stage of the restoration is based on regression of a pixel value based on the data in a kernel. To this end often linear regression model is employed. However, a linear model imposes a severe limitation on the allowed relation between the pixels in the kernel. We therefore consider a set of non-linear functions $\phi_i(x)$ mapping the input vector (observations) to a new space, allowing a more flexible model:

$$y(\mathbf{x}; \mathbf{w}) = \sum_{i=1}^{K} w_i \phi_i(\mathbf{x}) + w_0 = \mathbf{w}^T \phi(\mathbf{x}) + w_0, \tag{3}$$

where $\mathbf{w} := [w_1, w_2, ..., w_K]^T$ is the parameter vector of the model to estimate in a certain kernel and K denotes the kernel size (number of training samples). We hereby describe a new Bayesian probabilistic approach for learning $p(M|D)$, where D in this case presents the corrupted image. As the basis functions $\phi(\mathbf{x}) := [\phi_1(\mathbf{x}), \phi_2(\mathbf{x}), ..., \phi_K(\mathbf{x})]$ we choose the following RBF:

$$\phi_i(x) := \exp\left(-\frac{||x - t_i||^2}{2r^2}\right). \tag{4}$$

In the context of image de-noising, we define a regressor imposed on a *training sample set* at the size of the kernel. For a kernel size $k \times k$ the regressor will therefore have a training sample size of $K = k^2$. The width of the Gaussian functions, r, depends on the size k. We set this width to be $r = k/2.5$. Note that the regression (3) produces a model for prediction of the intensities for *all* the pixels in the kernel domain.

Estimation of the weights (model) is the crux of the proposed denoising algorithm. It is well known that that maximum likelihood estimation of $\mathbf{w}$ and σ^2 from (3) will lead to severe over-fitting [1]. To avoid this, we impose an additional constraint on the parameters, through the addition of "complexity" penalty term to the likelihood or error function. Here, though, we adopt a Bayesian perspective, and "constrain" the parameters to obey a *prior* probability distribution. The preference for smoother (less complex) functions is made by the popular choice of zero mean Gaussian prior distribution over the weights $\mathbf{w}$. Note that when $\mathbf{w} = 0$, then w_0 will obtain the mean intensity value in the kernel. We also apply a prior on w_0 as normal distribution around the mean value m, i.e. $w_0 \sim N(m, \sigma_0^2)$ and define the corresponding precision parameter as $\alpha_0 = 1/\sigma_0^2$. Assuming conditional independence in $\mathbf{w}$ components and characterization by individual *precisions* $\alpha_k = 1/\sigma_k^2$, namely *hyper-parameters* (with corresponding std σ_k), yields the following *prior* probability function:

$$p(M) = \frac{1}{(2\pi)^K} \prod_{i=1}^{K} \alpha_i^{1/2} \exp\left(-\frac{\alpha_i w_i^2}{2}\right) \cdot \alpha_0^{1/2} \exp\left(-\frac{\alpha_0(w_0 - m)^2}{2}\right) \tag{5}$$

Importantly, there is an individual hyper-parameter associated with every weight. This mechanism moderates the strength of the prior in space, adaptively.

Having defined the prior, Bayesian inference proceeds by computing, from Bayes' rule, the posterior over all unknowns given the data. To this end, we wish to optimize the posterior probability for model parameters M given the data D:

$$M^* = \arg\max_M \; p(M|D) = \arg\max_M \; p(D|M)p(M) \tag{6}$$

This condition is equivalent to maximization of the corresponding log term substituting Eq. (2) and Eq. (5) in Eq. (6), ignoring the constants:

$$L(\mathbf{w}, \boldsymbol{\alpha}, w_0, \alpha_0, \sigma) = -K\ln(\sigma) - \sum_{i=1}^{K} \frac{(y_i - t_i)^2}{2\sigma^2} - \sum_{i=1}^{K} w_i^2 \alpha_i/2 + \sum_{i=1}^{K} 1/2\ln(\alpha_i)$$

$$- \frac{(w_0 - m)^2}{2}\alpha_0 + \frac{1}{2}\ln(\alpha_0). \tag{7}$$

The optimal model is then a result of maximization of (7). Note that when a certain point in the kernel obtains a high mismatch value (e.g. $\frac{(y_i-t_i)^2}{2\sigma^2} \gg 1$), then the associated weight will reduce significantly due to the maximization process. This effectively "switches off" the influence of the corresponding input in regression, as desired. Nevertheless, in areas where the image is nearly constant (in sense of local *mean*), a limited data discrepancy is expected and maximization of (7) yields weights with homogeneous distribution (i.e. $\mathbf{w} \to 0$, and $w_0 \neq 0$). This is a highly effective mechanism for preserving the image structure in the restoration process while filtering out the noise, and it is a direct by product of our Bayesian formulation. Note that the suggested regression scheme doesn't make any use of spatial relations between points. Therefore this model is not limited to images and can be used for other regression problems.

Maximization of L is pendent to the following necessary conditions:

$$\frac{\partial L}{\partial w_k} = -\frac{\sum_{i=1}^{K}(y_i - t_i)\phi_k(t_i)}{\sigma^2} - w_k\alpha_k \tag{8}$$

$$\frac{\partial L}{\partial w_0} = -\frac{\sum_{i=1}^{K}(y_i - t_i)}{\sigma^2} - (w_0 - m)\alpha_0 \tag{9}$$

$$\frac{\partial L}{\partial \alpha_i} = \frac{-w_i^2}{2} + \frac{1}{2\alpha_i} = 0 \tag{10}$$

$$\frac{\partial L}{\partial \alpha_0} = \frac{1}{2\alpha_0} - \frac{(w_0 - m)^2}{2} = 0 \tag{11}$$

As for estimation of the noise std σ we use the none biased estimate as follows:

$$\sigma^2 = \frac{\sum_{i=1}^{K}(y_i - t_i)^2}{K - 1} \tag{12}$$

The unknown values of $w_i, \alpha_i, w_0, \alpha_0, \sigma$ which satisfy Eq. (8-12) cannot be obtained in a closed form, and hereby we suggest an iterative scheme for their estimation. Each iteration step is composed of three stages. The *conjugate gradient* [8] scheme is used for estimation of w_k in Eq.(8). This step is followed by evaluation of w_0 in (9) and the precision parameters α_i and α_0 in Eq. (10,11). The next iteration step is then applied after updating the noise std in (12). At each phase only one set of variables are updated to satisfy the equation, while the others are kept constant. The algorithm flow is then:

Algorithm Flow:

1) Initialization
2) Estimate **w** from Eq.(8).
3) Estimate w_0 from Eq.(9).
4) Update precision values α_k and α_0 using Eqs.(10) and (11).
5) Update noise std evaluation σ by Eq.(12).
6) Repeat steps 1-5 until convergence.

We experienced convergence of the scheme in just two iterations. The above model now can be used to produce a set of hypotheses for Bayesian inference as described in the following section.

2.2   Bayesian Inference

In this work we consider a set of $\mathcal{K}$ hypotheses for each data point, governed from the kernel centralized on the considered point and another $\mathcal{K} - 1$ overlapping kernels. We seek to compute the expected value $E(\hat{t}|\hat{x})$ associated with a new test point $\hat{x}$ mapped to the target $\hat{t}$. In Bayesian framework this is conducted by calculation of the *predictive distribution* [1]:

$$E(\hat{t}|\hat{x}) = \quad \hat{t} \cdot p(\hat{t}, \mathfrak{m}|\hat{x}, D)d\mathfrak{m} = \quad \hat{t} \cdot p(\hat{t}|\mathfrak{m}, \hat{x}, D)p(\mathfrak{m}|D)d\mathfrak{m}, \qquad (13)$$

where $\mathfrak{m}$ presents the model, in this case the regressor outcome. Evaluation $p(\mathfrak{m}|D)$ is based on computation of the marginal probability (see (6)). Since the the posterior $p(\hat{t}|\mathfrak{m}, \hat{x}, D)$ can not be directly computed we use the following decomposition [1,10]:

$$p(\hat{t}|\mathfrak{m}, \hat{x}, D) = \quad p(\hat{t}, w|\hat{x}, \mathfrak{m}, D)dw = \quad p(\hat{t}|w, \hat{x}, \mathfrak{m}) \cdot p(w|\mathfrak{m}, D)dw, \qquad (14)$$

The term $p(\hat{t}|w, \hat{x}, \mathfrak{m})$ has a normal distribution where the distance is measured by absolute difference in *intensity* values. As for $p(w|\mathfrak{m}, D)$, the distribution is also normal having the posterior mean obtained from Eq.(8-11) as its mode. Thus the expression under the integral sign in (14) presents convolution of two Gaussian distributions resulting in a new Gaussian characterized by the parameters of the distributions it was composed of [1]. Note however that our prediction is based on several hypotheses, obtained here from neighbouring kernels. This paradigm is inferred from Bias-Variance decomposition concept. Since the error is composed from bias and variance, Bayesian model averaging significantly improves the prediction by coping in fact, with the notorious variance part of the error.

3   Results

In this section, we will demonstrate the performance of the proposed approach on popular test images, contaminated with white Gaussian noise. Our data set incorporates images with different characteristics, from piecewise smooth to highly textured images with fine structures. Figure 1 shows our test bed comprised of the celebrated "Barbara", "Lena", "Boat" and "Pepper" images all 512×512 in size. For sake of illustration, the insets were corrupted with noise having standard deviation of 10, 20 and 30 grey levels. Figure 2 shows the restoration results for the Barbara and Pepper test cases with two different kernel sizes and noise level of 30 std. The high quality of restorations in both of these disparate cases, the piecewise smooth Pepper image and the fine textured Barbara, demonstrate the high capability of the proposed approach. Despite the excessive noise level, the denoised images are visually appealing. There are slight artifacts in the Barbara result when restored with a small kernel size of 3×3. This effect is vanished when the kernel is enlarged to 5×5, but yields a slightly lower PSNR. The larger kernel size produces visually improved denoising in both cases and higher PSNR for the Pepper image having a subtle texture. One can observe carefully the restoration of the delicate texture pattern in Barbara image, shown in the

Fig. 1. The test data set prior to noise contamination. From left to right: Barbara, Lena, Boat and Peppers.

---

[1] The interested reader is referred to [1] for details in the derivation.

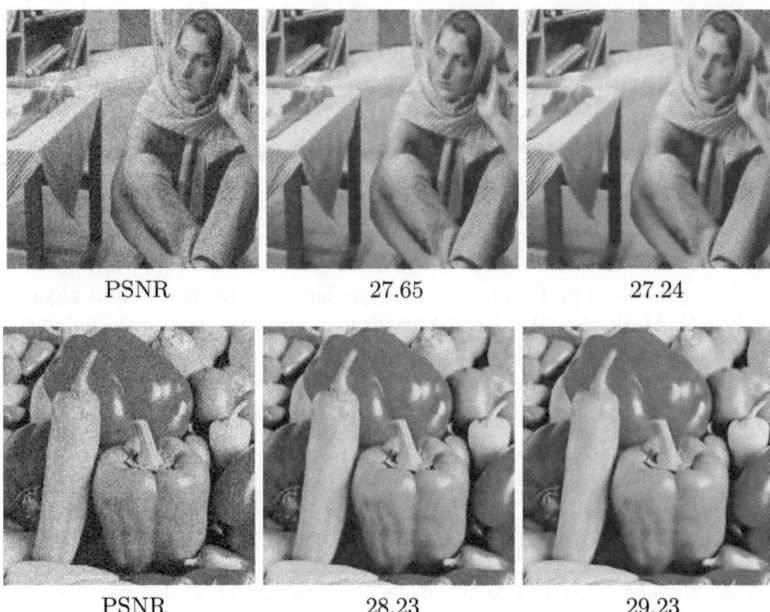

Fig. 2. Image denoising results for Barbara (top) and Pepper (down) images. Left: Image corrupted with 30 std Gaussian noise. Middle: Restored image with kernel size 3×3. Right: Restored image with kernel size 5×5. subfigures indicate PSNR values in dB.

Fig. 3. Barbara zoomed in. Left: Original patch. Middle: Input image with 30 std noise. Right: Restoration result with the proposed method, kernel size 3.

cropped patches in Figure 3. We further evaluate our method quantitatively by the common PSNR measure:

$$PSNR = 20 \log_{10} \frac{255}{\sqrt{\text{Mean Square Error}}} . \tag{15}$$

For comparison we present results from 5 methods in the literature, comprised of two popular and three recently published methods based on wavelet transforms. The PSNR values of the restored images are listed in Table 1 [2]. From this table it is clear that published algorithms [12,14,17] all substantially outperform the TV and the hard-threshold (HT) methods, with the MVM presenting the best results. The suggested Bayesian method shows comparable results and performs superiorly under high noise levels.

Table 1. Comparison of performance for image denoising using different algorithms in terms of PSNR. Methods: TV: Total Variation minimization in variational approach [13], HT: Hard Threshold, BS: Bivariate Shrinkage [14], GSM: Gaussian Scale Mixtures [12], MVM: Multivariate Statistical Model [17], Bayes.: Our Bayesian approach with 3×3 kernel size. Best results are in bold. Note: HT, GSM and MVM act in wavelet transform domain.

Method	Barbara			Lena			Boat			Pepper		
	10	20	30	10	20	30	10	20	30	10	20	30
TV	30.28	26.54	24.60	33.00	30.26	28.35	32.30	29.23	27.36	32.24	29.84	28.05
HT	31.99	27.68	25.43	34.48	31.29	29.31	32.63	29.31	27.42	33.59	29.73	27.55
BS	32.73	28.73	26.51	34.51	31.38	29.54	32.62	29.30	27.48	32.95	29.21	27.01
GSM	33.11	29.18	27.02	34.85	31.65	29.82	33.02	29.62	27.78	33.16	29.50	27.40
MVM	**33.25**	29.40	27.23	**34.95**	**31.83**	**29.97**	**33.01**	**29.75**	27.84	33.20	29.47	27.32
Bayes.	32.70	**29.67**	**27.65**	34.35	31.52	29.71	32.90	29.71	**27.98**	**33.37**	**31.11**	**28.23**

The complexity of our method is dominated by the kernel size and the scheme used for approaching the solution of the linear system (8-11). Considering the Conjugate Gradient method used the complexity of our scheme is $O(K \cdot n)$ when K indicates the size of the kernel and n the image size.

4   Summary

We hereby present a Bayesian approach for image de-noising in the spatial domain. The proposed method is based on kernel estimation, while the kernel weights are evaluated by regression using radial basis functions allowing a more flexible data modeling. The kernel weights are imposed with a prior probability distribution with individual hyper-parameters, moderating adaptively the strength of the prior (regularization), in space. The model unknowns are then evaluated efficiently through an iterative procedure. Finally, the intensity values

---

[2] Results for HT, GSM and MVM correspond to orthonormal wavelet transform as reported in [17].

are restored by Bayesian inference over hypotheses obtained from neighboring kernels.

The suggested method was illustrated on 4 popular test images contaminated with white additive Gaussian noise. Comparison to several recently published methods show comparative results, and introduces superior performance in the presence of large noise levels. Another advantage of the proposed approach is in the capability to explicitly incorporate the noise PDF in the model. It is therefore able to cope with non-Gaussian noise distributions *e.g*, a Poissonian PDF model obtained in SPECT medical images.

Finally, the proposed Bayesian approach is simple, generic, free of adjusting parameters and can be used in other regression based applications.

References

1. Bishop, C.M.: Pattern Recognition and Machine Learning. Springer Science+Business Media, LLC (2006)
2. Buades, A., Coll, B., Morel, J.M.: A review of image denoising methods, with a new one. Multiscale Model. Simul. 4(2), 490–530 (2005)
3. Buhman, M.D.: Radial Basis Functions, Theory and Implementations. Cambridge University Press, Cambridge (2003)
4. Chang, S.G., Vetterli, M.: Adaptive wavelet thresholding for image denoising and compression. IEEE Transaction on image processing 9(9), 1532–1546 (2000)
5. Combettes, P.L., Pesquet, J.C.: Image restoration subject to a total variation constraint. IEEE Trans. on Image Processing 13(9), 1213–1222 (2010)
6. Donoho, D.L.: De-noising by soft thresholding. IEEE Transactions on Information Theory 41(3), 613–627 (1995)
7. Halder, A., Shekhar, S., Kant, S., Mubarki, M., Pandey, A.: A new efficient adaptive spatial filter for image enhancement. In: Proc. Int. Conference on Computer Engineering and Applications, vol. 1
8. Hestenes, M., Stiefel, E.: Methods of conjugate gradients for solving linear systems. Journal of Research of the National Bureau of Standards 49(6), 409–436 (1952)
9. Loeza, C.B., Chen, K.: On high-order denoising models and fast algorithms for vector valued images. IEEE Transactions on Image Processing 19(6), 1518–1527 (2010)
10. Michael, T., Alex, S.: Sparse bayesian learning and relevance vector machine. IEEE Journal of Machine Learning research 1, 211–244 (2001)
11. Perona, P., Malik, J.: Scale space and edge detection using anisotropic diffusion. International Journal of Computer Vision 12(7), 629–639 (1990)
12. Portilla, J., Strela, V., Wainwright, M.J., Simoncelli, E.P.: Image denoising using scale mixtures of gaussians in the wavelet domain. IEEE Trans. on Image Processing 12(11), 1338–1351 (2003)
13. Rudin, L.I., Osher, S., Fetami, E.: None-linear total varaition based noise removal algorithms. Physica. D: Non-linear Phenomena 60, 259–268 (1992)

14. Sendur, L., Selesnick, I.W.: Bivariate shrinkage with local variance estimation. IEEE Signal Processing Letters 9, 428–441 (2002)
15. Spira, A., Kimmel, R., Sochen, N.A.: A short- time beltrami kernel for smoothing images and manifolds. IEEE Transactions on Image Processing 16(6), 1628–1636 (2007)
16. Takeda, H., Farsiu, S., Milanfar, P.: Kernel regression for image processing and reconstruction. IEEE Transaction on Image Processing 16(2), 349–366 (2007)
17. Tan, S., Jiao, L.: Multivariate statistical models for image denoising in the wavelet domain. IEEE Transaction on Pattern Analysis and Machine Intelligence 75(2), 209–230 (2007)

A Rough-Fuzzy HSV Color Histogram
for Image Segmentation

Alessio Ferone[1], Sankar Kumar Pal[2], and Alfredo Petrosino[1]

[1] DSA - Universitá di Napoli Parthenope, 80143 Napoli, Italy
[2] Machine Intelligence Unit, Indian Statistical Institute, Kolkata 700 108, India

Abstract. A color image segmentation technique which exploits a novel definition of rough fuzzy sets and the rough–fuzzy product operation is presented. The segmentation is performed by partitioning each block in multiple rough fuzzy sets that are used to build a lower and a upper histogram in the HSV color space. For each bin of the lower and upper histograms a measure, called index, is computed to find the best segmentation of the image. Experimental results show that the proposed method retains the structure of the color images leading to an effective segmentation.

Keywords: Image segmentation, Color Image Histogram, Rough Sets, Fuzzy Sets.

1 Introduction

Color image segmentation is one of the most challenging tasks in image processing, being the basic pre-processing step of many computer vision and pattern recognition problems.

Among the others, the most used approaches are represented by histogram based techniques due to the fact they need no a–priori information about the image. The task consists in finding clusters corresponding to regions of uniform colors, identified by peaks in the histogram. The task is complicated in color images, being characterized by three dimensional scatterograms, that make more difficult the search for peaks, either in the whole histogram or in each color channel independently. Also, typically they do not take into account the spatial correlation between adjacent pixels, while images usually show this property.

The approach reported here bases its rationale on Granular Computing, based on the concept of information granule, that is a set of similar objects that can be considered as indistinguishable. Partition of an universe in granules gives a coarse view of the universe where concepts, represented as subsets, can be approximated by means of granules. In this framework, rough set theory can be regarded to as a family of methodologies and techniques that make use of granules [8,9]. The focus of rough set theory is on the ambiguity caused by limited discernibility of

G. Maino and G.L. Foresti (Eds.): ICIAP 2011, Part I, LNCS 6978, pp. 29–37, 2011.

objects in the domain of discourse. Granules are formed as objects and are drawn together by the limited discernibility among them. Granulation is of particular interest when a problem involves incomplete, uncertain or vague information. In such cases, precise solutions can be difficult to obtain and hence the use of techniques based on granules can lead to a simplification of the problem at hand.

At the same time, multivalued logic can be applied to handle uncertainty and vagueness in information system, the most famous of which is fuzzy sets theory [16]. In this framework, uncertainty is modelled by means of functions that define the degree of belonginess of an object to a given concept. Hence membership functions of fuzzy sets enable efficient handling of overlapping classes.

Some researches already follow this approach. Cheng et al. [2] employed a fuzzy homogeneity approach to extract homogeneous regions in a color image. The proposed method introduces the concept of homogram built considering intensity variation in pixel neighborhood. In [5] the concept of encrustation of the histogram (histon), which is a contour plotted on the top of each primary color histogram, is presented. In a rough-set theoretic sense, the histon represents the upper approximation of the color regions, that is a collection of pixels possibly belonging to the same region, while the histogram represents the lower approximation. An histogram-based technique is employed on the histon to obtain the final segmentation. Mushrif and Ray [6] presented a segmentation scheme, based on the concept of histon [5], which employs the roughness index. Roughness is large when the boundary contains a large number of elements, hence it will be smaller in the boundary between two objects and larger in region with uniform color.

The novelty of our approach resides on the hybrid notion of rough fuzzy sets that comes from the combination of these two models of uncertainty (fuzzy and rough) to exploit, at the same time, properties like coarseness, by handling rough sets [8], and vagueness, by handling fuzzy sets [16]. In this framework, rough sets embody the idea of indiscernibility between objects in a set, while fuzzy sets model the ill-definition of the boundary of a sub class of this set. Marrying both notions leads to consider, as instance, approximation of sets by means of similarity relations or fuzzy partitions. The rough fuzzy synergy is hence adopted to better represent the uncertainty in granular computation. Specifically, we present a histogram based technique tat exploits a generalized definition of rough–fuzzy sets, i.e. an hybridization of rough sets and fuzzy sets, and a particular operation called rough–fuzzy product in the HSV color space.

2   Rough Fuzzy Color Histogram

Let us consider an image I defined over a set $U = [0, ..., H-1] \times [0, ...W-1]$ of picture elements, i.e. $I : u = (u_x, u_y) \in U \rightarrow [h(u), s(u), v(u)]$. We shall introduce the *Image Partition* as

Definition 1. *Let us consider a grid, superimposed on the image, whose cells Y_i are of dimension $w \times w$. Given a pixel u, whose coordinates are u_x and u_y, and a cell Y_i of the grid, whose coordinates of its upper left point are $x(Y_i)$ and $y(Y_i)$, u belongs to Y_i if $x(Y_i) \leq u_x \leq x(Y_i) + w - 1$ and $y(Y_i) \leq u_y \leq y(Y_i) + w - 1$. The set of all Y_i constitutes an* **Image Partition,** $\mathcal{Y}$, *over I.*

Different values of w yield different partitions $\mathcal{Y}$ of the same image. For instance, given a partition $\mathcal{Y}^i$, other partitions can be obtained by a rigid translation in the directions of $0°$, $45°$ and $90°$ degrees of $w-1$ pixels, so that for each partition a pixel belongs to a shifted version of the same cell Y_j^i.

If we consider four cells, Y_j^1, Y_j^2, Y_j^3 and Y_j^4 belonging to four partitions $\mathcal{Y}^1$ $\mathcal{Y}^2$ $\mathcal{Y}^3$ $\mathcal{Y}^4$, then there exists a pixel u with coordinates (u_x, u_y) such that u belongs to the intersection of Y_j^1, Y_j^2, Y_j^3 and Y_j^4 [11].

$$Y_j^{1,2,3,4} = Y_j^1 \cap Y_j^2 \cap Y_j^3 \cap Y_j^4 \tag{1}$$

The image is firstly partitioned in non–overlapping k blocks X_h of dimension $m \times m$, such that $m \geq w$, that is $X = \{X_1, \ldots, X_k\}$ and $k = H/m + K/m$. Considering each image block X_h, a pixel in the block can be characterized by two values $h_{\text{inf}}(u)$ and $h_{\text{sup}}(u)$ computed, for each pixel u belonging to a block X_h, as

$$h_{\text{sup}}(u) = \sup\{h_m^1(u), h_m^2(u), h_m^3(u), h_m^4(u)\}$$
$$h_{\text{inf}}(u) = \inf\{h_M^1(u), h_M^2(u), h_M^3(u), h_M^4(u)\}$$

where $h_m^i(u)$ $i = 2, 3, 4$ are obtained by translating $h_m^1(u)$ in the direction of 0, 45 and 90 degrees. For instance, for $w = 2$ and a generic $j - th$ cell of the $i - th$ partition, we have:

$$h_m^i(u) = \inf\{(u_x + a, u_y + b)|a, b = 0, 1\}$$
$$h_M^i(u) = \sup\{(u_x + a, u_y + b)|a, b = 0, 1\}$$

Let us now consider the HSV color space represented by a cone and a segment $[\theta, \theta + \Delta\theta - 1]$ on the maximum circumference, where $0 \leq \theta \leq 359$ and $[\Delta\theta_{\min} \leq \Delta\theta \leq \Delta\theta_{\max}]$ is the segment dimension. This interval contains a certain amount of colors. In particular, if we imagine to cut the HSV cone in wedges, each one contains all the possible combination of saturation and value given a portion of hue. Our goal is to describe each wedge using the blocks of the image, under the assumption that blocks with similar colors will fall in the same wedge.

Definition 2. *Each block X_h, of dimension $m \times m$, is characterized by a minimum and a maximum hue value*

$$h_m = \min\{h_{\sup}(u)|u \in X_h\}$$
$$h_M = \max\{h_{\inf}(u)|u \in X_h\}$$

defining a **hue Interval** $[h_m, h_M]$ *that can be:*

1. *totally contained into a wedge of dimension* $\Delta\theta$ *(i.e.* $\theta \le h_m \le h_M < \Delta\theta + \theta$*),*
2. *partially contained into a wedge (i.e.* $\theta \le h_m$ *or* $h_M < \Delta\theta + \theta$*),*
3. *not contained at all.*

Hence, we can describe the wedge by means of two sets of blocks.

Definition 3. *The* **L–set** *is the set of blocks whose interval* $[h_m, h_M]$ *are totally contained into the wedge. The* **U–set** *is the set of blocks whose* $[h_m, h_M]$ *are partially contained into the wedge.*

Now consider a wedge of dimension $[\theta_i, \theta_i + \Delta\theta - 1]$, $i = 0, \ldots, 359$ moving on the hue circle towards increasing hue values, starting from $\theta_1 = 0$. At each step the wedge is shifted by an offset x, i.e. $\theta_{i+1} = \theta_i + $ x, and the *L–set* and *L–set* of the wedge are computed. This procedure, shown in Algorithm 1, yields two histograms, the *L–Histogram* and the *U–Histogram* of the image.

Algorithm 1. Procedure to build *L–Histogram* and *U–Histogram*

1: **for all** θ by step x **do**
2:     **for all** blocks X_h **do**
3:         compute h_m and h_M
4:         **if** θ h_m h_M $\theta + $ θ **then**
5:             $L\text{-}Histogram[\theta] = L\text{-}Histogram[\theta] + 1$
6:             $U\text{-}Histogram[\theta] = U\text{-}Histogram[\theta] + 1$
7:         **else if** θ h_m $\theta + $ θ OR θ h_M $\theta + $ θ **then**
8:             $U\text{-}Histogram[\theta] = U\text{-}Histogram[\theta] + 1$
9:         **end if**
10:     **end for**
11: **end for**

Repeating the same procedure for each wedge dimension $\theta_{\min} \le \theta \le \theta_{\max}$, many histograms are produced according to the possible values of θ. Figure 1 and 2 depict respectively the *L–Histogram* and the *U–Histogram* of Figure 1.

It should be reminded that, if for a given pixel the saturation equals 0, the hue component is undefined and the pixel is characterized only by the value component, i.e. only by its gray level intensity. To overcome this problem, it is possible to exclude all the pixels with a saturation value lower than a given threshold ϵ and segment them separately (for instance employing a segmentation algorithm for gray scale images).

Fig. 1. Example image

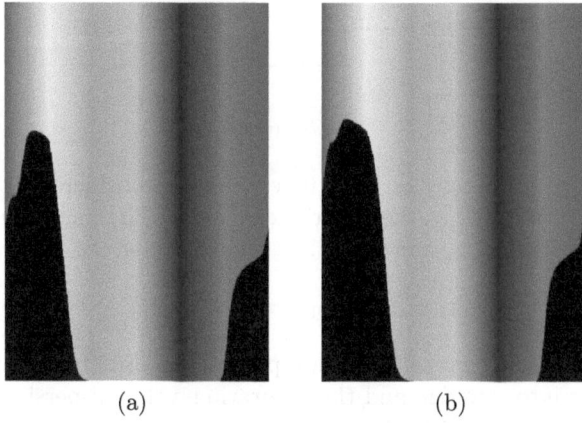

(a)   (b)

Fig. 2. a) *L–Histogram* and b) *U–Histogram*

3   Image Segmentation by Rough Fuzzy Color Histogram

The segmentation of a color image is performed in the HSV color space by choosing the wedges that are better represented employing the blocks of the image. The choice is guided by the accuracy of the wedge, i.e. the i-th wedge gets an accuracy computed by means of the corresponding bin in the *L–Histogram* and *U–Histogram*

$$\alpha_i = \frac{L\text{--}Histogram(i)}{U\text{--}Histogram(i)} \tag{2}$$

Clearly, this can not be the only discriminant index to obtain a good segmentation. First of all due to the accuracy, as computed in eq. 2, that does not take into account the number of blocks, and hence the number of pixels contained into the wedge, but only their ratio. Moreover, using only the accuracy does not take into account saturation and value of each pixel. The first problem is tackled by weighting the accuracy of each wedge by the fraction of pixels whose hue value belongs to the wedge, i.e.

$$\gamma_i = 1 - \frac{N_{wedge}(i)}{N_{tot}(I)} \tag{3}$$

where $N_{wedge}(i)$ represents the number of pixels whose hue value belongs to the wedge and $N_{tot}(I)$ represents the number of pixels of the image I.

Provided that regions of uniform colors are searched into the image, we need an index to measure the color uniformity of the pixels belonging to the the the wedge and then use this index to weight the accuracy. To this aim we propose to employ a measure of the dispersion of the pixels falling into a wedge with respect to saturation and value. A region characterized by uniform color will present a narrow scatter, while a region characterized by non uniform colors will have a sparse scatter. To compute the compactness of saturation and value into the i-th wedge, we propose the following index

$$\delta_i = \frac{1}{N_{wedge}(i)} \times \overline{\sum_{x \in i-thwedge} (x - \mu_i)^T (x - \mu_i)} \tag{4}$$

where $x = [x_{saturation}, x_{value}]$. This index can be considered as the weighted squared root of the track of the covariance matrix. The final index, τ_i, is computed by composing α_i, γ_i and δ_I indices (eqs. 2, 3 and 4)

$$\tau_i = \alpha_i \times (w_1 \times \gamma_i + w_2 \times \delta_i) \tag{5}$$

where w_1 and w_2, with $w_1 + w_2 = 1$, are parameters used to weight the fraction of pixels falling into a wedge and the saturation–value dispersion, respectively. A higher value for w_1 will lead to wedges comprising few pixels characterized by a low saturation–value dispersion, whilst a higher value for w_2 will produce wider wedges, with a larger number of pixels presenting a lower saturation–value dispersion. The index τ, computed for all the wedges, is used to segment the image. Firstly, the wedge with the highest τ value is selected as the region better represented into the image. Next, all the wedges that intersect the first one are removed to avoid overlapping regions. For instance, consider s_i the wedge with the highest τ value corresponding to the hue segment $q_{s_i}, q_{s_i} + qt - 1$, then all the wedges s_j such that $q_{s_i} \le q_{s_j} + \tilde{q}t - 1 < q_{s_i} + qt - 1$, with $\tilde{q}t$ varying in $[qt_{min}, qt_{max}]$, are removed. Next, the wedge with the highest τ value, among those not removed in the previous step, is selected, and so on until no more wedges are left.

4 Experimental Results

To assess the performance of the proposed method, we employed the Probabilistic Rand Index (PRI) [15] that counts the fraction of pairs of pixels whose labellings are consistent between the computed segmentation and the ground truth, averaging across multiple ground truth segmentations to account for scale variation in human perception. For each image, the quality of the segmentation is evaluated by comparing it with all the available segmentations of the same image.

The performance of the proposed algorithm were tested on the 100 color test images of "The Berkeley Segmentation Dataset" [4]. Threshold has been fixed to $\epsilon = 0.2$; all the pixels presenting a saturation value lower than ϵ have been segmented by employing another threshold $\eta = 0.5$, i.e., pixels are labelled as "white" if their value component is greater than η, as "black" otherwise. A larger granule dimension allows to produce wedges able to enclose more similar hues so to suppress small hue variations, while smaller granule dimension tends to better differentiate between similar hues. A larger granule size can be useful to segment images that show larger hue variance and hence obtain better PRI. Parameters w_1 and w_2 can be used to obtain distinct segmentations by weighting the importance of the number of pixels into the wedge with respect to the saturation–value dispersion. Higher values of w_1 mean that wedges enclosing few pixels are privileged, while higher values of w_2 privilege wedges characterized by higher saturation–value dispersion.

Figure 3 shows an example of segmentation of two test images of the BSD. Segmentation in Figure 3(a), obtained with parameters $w_1 = 0.6$ $w_2 = 0.4$

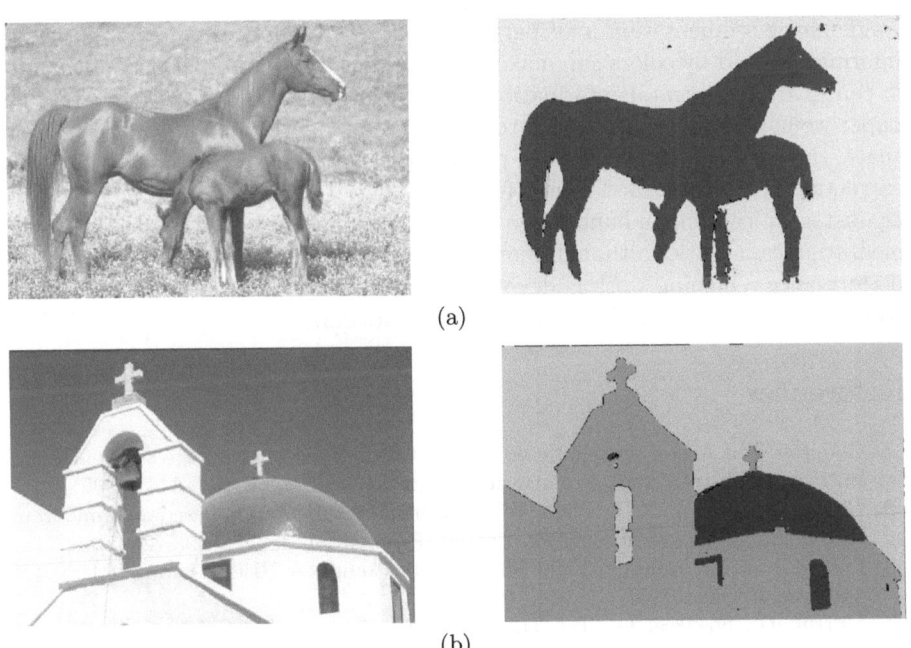

(a)

(b)

Fig. 3. Segmentation for images 113044 (a) and 118035 (b)

and granule dimension $w = 2$, produces $PRI = 0.774117$. Segmentation in Figure 3(b), obtained with parameters $w_1 = 0.7$ $w_2 = 0.3$ and granule dimension $w = 2$, produces $PRI = 0.870635$.

Table 1 summarizes results obtained with different parameter configurations in terms of mean PRI computed over the 100 color images adopted for testing the algorithm. It turns out that best results are obtained using small granule dimensions and giving importance to the number of pixels over the saturation–value dispersion. Here we want to point out that, although this configuration gives the best results on average, this does not imply that good results could not be obtained for single images employing different values.

Table 1. Mean PRI values for the 100 test images of the BSD

Granule dimension w	$w_1 = 0.8$, $w_2 = 0.2$	$w_1 = 0.6$, $w_2 = 0.4$	$w_1 = 0.5$, $w_2 = 0.5$
2	0.678028	0.663410	0.654179
4	0.661959	0.636016	0.624948
8	0.640885	0.621997	0.619233
16	0.623986	0.613345	0.609314
32	0.618413	0.601521	0.590546

5   Conclusions

Color image segmentation is of particular interest because the huge amount of information held by colors can make the task very difficult to perform, although it can give fundamental information about the image to be analyzed. In this paper we have presented a segmentation technique, performed in the HSV color space, that exploits peculiarities of rough–fuzzy sets and, in particular, a feature extraction operation called rough–fuzzy product. The proposed method, tested against a typical human hand made segmentation dataset, have shown good segmentation capabilities although more research is needed to obtain good average performance. Ongoing work is devoted to consider spatial relationship between blocks to increase the performance of the algorithm.

References

1. Chapron, M.: A new chromatic edge detector used for color image segmentation. In: Proc. 11th Int. Conf. on Pattern Recognition, vol. 3, pp. 311–314 (1992)
2. Cheng, H.D., Jiang, X.H., Wang, J.: Color image segmentation based on homogram thresholding and region merging. Pattern Recognition 35, 373–393 (2002)
3. Li, S.Z.: Markov Random Field Modeling in Computer Vision, Kunii, T.L. (ed.). Springer, Berlin (1995)
4. Martin, D., Fowlkes, C., Tal, D., Malik, J.: A Database of Human Segmented Natural Images and its Application to Evaluating Segmentation Algorithms and Measuring Ecological Statistics. In: Proc. 8th Int Conf. Computer Vision, vol. 2, pp. 416–423 (2001)

5. Mohabey, A., Ray, A.K.: Rough set theory based segmentation of color images. In: Proc. 19th Internat. Conf. NAFIPS, pp. 338–342 (2000)
6. Mushrif, M.M., Ray, A.K.: Color image segmentation: Rough-set theoretic approach. Pattern Recognition Letters 29, 483–493 (2008)
7. Panjwani, D.K., Healey, G.: Markov random field models for unsupervised segmentation of textured color images. IEEE Trans. Pattern Anal. Mach. Intell. 17(10), 939–954 (1995)
8. Pawlak, Z.: Rough sets. Int. J. of Inf. and Comp. Sci. 5, 341–356 (1982)
9. Pawlak, z.: Granularity of knowledge, indiscernibility and rough sets. In: Proceedings of IEEE International Conference on Fuzzy Systems, pp. 106–110 (1998)
10. Sen, D., Pal, S.K.: Generalized Rough Sets, Entropy, and Image Ambiguity Measures. IEEE Trans. Sys. Man and Cyb. 39(1), 117–128 (2009)
11. Petrosino, A., Ferone, A.: Feature Discovery through Hierarchies of Rough Fuzzy Sets. In: Chen, S.M., Pedrycz, W. (eds.) Granular Computing and Intelligent Systems: Design with Information Granules of Higher Order and Higher Type (to appear, 2011)
12. Shafarenko, L., Petrou, M., Kittler, J.V.: Histogram based segmentation in a perceptually uniform color space. IEEE Trans. Image Process. 7(9), 1354–1358 (1998)
13. Trémeau, A., Colantoni, P.: Regions adjacency graph applied to color image segmentation. IEEE Trans. Image Process. 9(4), 735–744 (2000)
14. Uchiyama, T., Arbib, M.A.: Color image segmentation using competitive learning. IEEE Trans. Pattern Anal. Mach. Intell. 16(12), 1197–1206 (1994)
15. Unnikrishnan, R., Pantofaru, C., Hebert, M.: A Measure for Objective Evaluation of Image Segmentation Algorithms. In: Proc. CVPR WEEMCV (2005)
16. Zadeh, L.A.: Fuzzy Sets. Information and Control 8, 338–353 (1965)

Multiple Region Categorization
for Scenery Images

Tamar Avraham, Ilya Gurvich, and Michael Lindenbaum

Computer Science Department, Technion - I.I.T., Haifa 32000, Israel
tammya@cs.technion.ac.il, ilya.gurvich@gmail.com, mic@cs.technion.ac.il

Abstract. We present two novel contributions to the problem of region classification in scenery/landscape images. The first is a model that incorporates local cues with global layout cues, following the statistical characteristics recently suggested in [1]. The observation that background regions in scenery images tend to horizontally span the image allows us to represent the contextual dependencies between background region labels with a simple graphical model, on which exact inference is possible. While background is traditionally classified using only local color and textural features, we show that using new layout cues significantly improves background region classification. Our second contribution addresses the problem of correct results being considered as errors in cases where the ground truth provides the structural class of a land region (e.g., mountain), while the classifier provides its coverage class (e.g., grass), or vice versa. We suggest an alternative labeling method that, while trained using ground truth that describes each region with one label, assigns both a structural and a coverage label for each land region in the validation set. By suggesting multiple labels, each describing a different aspect of the region, the method provides more information than that available in the ground truth.

Keywords: region annotation, multiple categorization, exact inference, scenery/landcape, boundary shape, contextual scene understanding.

1   Introduction

The incorporation of context into object detection and region labeling has recently come into the mainstream of computer vision (e.g., [2,3,4,5,6,7]). In these methods the identity of an image region depends both on its local properties and on the labels and appearance of the neighboring regions. To solve the region labeling problem generally, approximation methods (e.g., loopy belief propagation) are required. In our work we focus on context based region annotation for scenery images. It turns out that this more simple problem can be modeled by a rather simple graphical model on which exact inference is possible.

We follow [1], where statistical properties of scenery images were analyzed. It was observed that background regions in scenery images tend to horizontally span the image, making it possible to define a one-dimensional top-bottom order of the background regions. Moreover, it was observed that the label of a

G. Maino and G.L. Foresti (Eds.): ICIAP 2011, Part I, LNCS 6978, pp. 38–47, 2011.

background region correlates with the shape of the upper part of its boundary. It was shown that by using only those two *layout* properties, it is possible to capture the general appearance variability of scenery images: those cues enabled the generation of semantic sketches of scenes. However, [1] left an open question about whether those cues can assist in region annotation. In the first part of this work we answer this question; see Fig. 1(a). We suggest an exact inference model for annotating background regions that combines layout cues with texture and color cues. We show that this combination significantly improves classification over methods that rely only on local color and texture, as each type of cue contributes to a different dichotomy.

The second part of this paper considers a related but different problem. We observe that results counted as errors are not always wrong. Often, regions are associated with different labels which correspond to different aspects of them, while annotators usually provide only one label corresponding to one of these aspects. For instance, a mountain with trees may be classified by the ground truth as mountain and classified by our algorithm as trees, or vice versa. That is, a land region's annotation can either describe its structure (mountain, plain, valley) or the overlying land-cover (trees, grass, sand, rocks, etc.)

This relates to recent work on problems with large numbers of categories [8,9]. In order to get a more informative accuracy score, it was suggested that the cost of misclassification be associated with the relative location of the true classification and the estimated classification in the wordnet tree [10]. This solution is good for foreground objects for which the categorization can be for different semantic details (e.g., crow, bird or animal). However, we found this method unsuitable for background categories, as the structural descriptors and the coverage descriptors do not appear in close wordnet sub-trees. Another recent related work [11] suggests that data can be organized by several non-redundant clustering solutions, each providing a different facet.

To support multiple categorization of land regions, we suggest an alternative labeling method that, while trained using ground truth that describes each region with one label, assigns two types of annotation for test data; see examples in Fig. 3. The annotations in the training data allow us to generalize both the appearance of land coverage categories (using the regions in the training set that are labeled by their coverage) and to generalize the appearance of land structure categories (learning from regions labeled by their structure). Given a new test image, each of its land regions can now be classified by both characteristics.

Sec. 2 overviews background region classification cues. Sec. 3&4 discuss the region classification algorithm and its results. Sec. 5&6 discuss the multiple labeling algorithm and its results. Sec. 7 concludes.

2   Cues for Background Region Classification

2.1   Color and Texture

The most natural choice for classifying a background region is by its color and textural attributes. In [12], such attributes were used for classifying patches

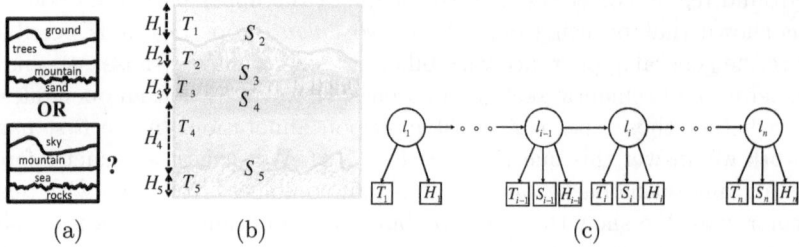

(a)   (b)   (c)

Fig. 1. (a) Demonstrating the contribution of 'layout': given this general image layout (without the texture), we can easily decide between the two suggested annotations. (b) The image is divided into n horizontal background regions, $R = (R_1, ..., R_n)$. Each region $R_i = (H_i, T_i, S_i, O_i)$ is described by its average height H_i, its texture&color descriptor T_i, by the curve separating it from the region above it, S_i, and by its relative location (order), $O_i = i$. (c) the HMM (Hidden Markov Model) representing the ORC model. The regions' labels, $l_1, ..., l_n$, are the hidden variables/states. The observed variables are $H_1, ..., H_n$, $T_1, ..., T_n$, and $S_2, ..., S_n$.

of images, as a first step in a method for scene categorization. We adapt the description suggested there and describe each region by: 1. Color histograms: a histogram for each of the components of the HSV color representation (36, 32, 16 bins for H, S, and V, respectively). 2. Edge direction histograms: we apply a Canny edge detector and collect the gradient directions in the edge locations (72 bins). 3. GLCM (Gray Level Co-occurrence Matrix [13]): the region is quantized to 32 gray levels. A GLCM is computed for 4 offsets (-1,1), (0,1), (1,1) and (1,0). For each, we compute the contrast, energy, entropy, homogeneity, inverse difference moment, and the correlation, for a total of 24 additional components.

2.2   Relative Location Statistics

In [1] it was shown that types of background regions tend to have typical relative locations. For example, a sand region will usually appear below a sea region, mountains are usually higher in the image than fields, and of course the sky is usually above all. These top-bottom relations were modeled by a Markov network. Let $\{L_1, ..., L_m\}$ be the possible background labels. The network has $m + 2$ nodes. The first m are associated with the m labels. In addition, there is a starting status denoted 'top' and a sink status denoted 'bottom'. $M(L_i, L_j)$ is the probability that a region labeled L_i appears above a region labeled L_j in an image. $M('top', L_i)$ and $M(L_i, 'bottom')$ are the probabilities that a region with label L_i is at the top/bottom of an image, respectively. The transition probabilities are estimated from the training image set [1].

---

[1] We specify the top-bottom order by the height of the highest pixel in each region. Therefore, the background regions do not always have to horizontally span the image.

2.3   Boundary Shape Characteristics

In [1], the characteristics of a contour separating two background regions were shown to correlate with the lower region's identity. The boundary on top of a sea, grass or field region is usually smooth and horizontal, resembling a DC signal. The boundary on top of a region of trees or plants can be considered as a high frequency 1D signal. The boundary on top of a mountain region usually resembles 1D signals of rather low frequency and high amplitude.

Following this observation, it was suggested that a signal representation be adopted. For each background labeled region, the upper part of its contour is extracted and cut to chunks of 64-pixel length. Each such chunk is actually a descriptor vector of length 64. In the model described in Sec. 3 we use these descriptor vectors as cues for region annotation, and use an SVM that provides probability estimates [14]. Let S_i describe the boundary above a region indexed i with identity $l \in \{L_1, ..., L_m\}$. S_i is cut to K_i chunks, $S_{i_1}, ..., S_{i_{K_i}}$. $K_i = \lfloor \frac{S_i}{64} \rfloor$. When the region belongs to a training image, the K_i chunks, each labeled l, are members of the training set. When the region is being classified, the SVM classifier returns a probability estimate for each of the K_i chunks, $p(l = L_j | S_{i_k})$, $k = 1, ..., K_i$, $j = 1, ..., m$. The class probability for the whole signal (boundary) is then $p(l = L_j | S_i) = \frac{1}{z_1} \sum_{k=1}^{K_i} p(l = L_j | S_{i_k})$, where z_1 is a normalizing factor.

3   Background Region Classification: The ORC Algorithm

In [1], it was shown that it is possible to capture the general appearance variability of scenery images using only the cues described in sections 2.2&2.3. Those cues enabled the generation of semantic sketches of scenes. However, [1] left an open question about whether those cues can assist in region annotation. In this section we answer this question by proposing a mechanism for combining all the cues described in Sec. 2: the ORC (Ordered Region Classification) algorithm.

Let $R = (R_1, ..., R_n)$ be n background regions in an image I, ordered by their top-bottom location. Each $R_i = (H_i, S_i, T_i, O_i)$ is characterized by its size, H_i^2, its color&texture, T_i, the '1D signal' S_i describing the boundary separating it from R_{i-1} ($S_1 = \emptyset$), and by its order in the image, $O_i = i$; see Fig. 1(b).

Taking a contextual approach, the identity l_i of region R_i depends on its appearance, its location, and on the appearance and relative location of the other image regions. Therefore, the probability for l_i to be L_j is a marginalization over all joint assignments $(l_1, ..., l_n)$ in which $l_i = L_j$:

$$p_{ORC}(i, L_j) = p(l_i = L_j | R) = \sum_{\substack{(k_1 ... k_{i-1} \, k_{i+1} ... k_n) \\ \in \{1 ... m\}^{n-1}}} p(l_1 = L_{k_1}, ..., l_i = L_j, ..., l_n = L_{k_n} | R) .$$

We use the Markovian property described in Sec. 2.2, i.e., the identity of region R_i directly depends only on the identity of region R_{i-1}. Also, we ignore the

---

[2] Since we are discussing horizontal patches that usually span the image from side to side, their size is described only by their average height.

direct dependency between the color, texture, and height of the different regions inside an image, and between the appearance of separating boundaries inside an image. The probability for a joint assignment is then

$$
p(l_1 = L_{k_1}, ..., l_n = L_{k_n} | R) = p(l_1 = L_{k_1} | R_1) \prod_{i=2}^{n} p(l_i = L_{k_i} | l_{i-1} = L_{k_{i-1}}, R_i) .
$$
(1)

Assuming T_i, S_i, H_i, and O_i are independent, and that T_i, S_i, H_i are independent in R_{i-1}, every term in the product can be expressed as

$$
p(l_i = L_{k_i} | l_{i-1} = L_{k_{i-1}}, R_i) = \frac{p_{T_i} p_{S_i} p_{H_i} p(l_i = L_{k_i} | l_{i-1} = L_{k_{i-1}}, O_i)}{p(T_i, S_i, H_i)} =
$$
(2)
$$
\frac{p(l_i = L_{k_i} | T_i) p(l_i = L_{k_i} | S_i) p_{H_i} p(l_i = L_{k_i} | l_{i-1} = L_{k_{i-1}}, O_i)}{z_2 [p(l_i = L_{k_i})]^2} ,
$$

where $p_{T_i} = p(T_i | l_i = L_{k_i})$, $p_{S_i} = p(S_i | l_i = L_{k_i})$, $p_{H_i} = p(H_i | l_i = L_{k_i})$ and $z_2 = \frac{p(T_i \ S_i \ H_i)}{p(T_i) p(S_i)}$. Since z_2 is not a function of the labels, we can infer it by normalization.

Given a label, the distribution $p(H_i | l_i = L_{k_i})$ is modeled by a simple Gaussian distribution (as suggested in [1]). For modeling the dependency of the label in the color and texture, $p(l_i = L_{k_i} | T_i)$, and in the boundary shape, $p(l_i = L_{k_i} | S_i)$, we use an extension of SVM that provides probability estimates [14] (the SVMs used for color&texture and for boundary shapes are separate). The prior probability $p(l_i = L_{k_i})$ is computed from the occurrences of labels in the training set. Finally,

$$
p(l_i = L_{k_i} | l_{i-1} = L_{k_{i-1}}, O_i) = \begin{cases} M(\text{'top'}, l_i) & i = 1 \\ M(l_{i-1}, l_i) & 1 < i < n \\ M(l_{i-1}, l_i) M(l_i, \text{'bottom'}) & i = n , \end{cases}
$$
(3)

where M is the transition matrix described in Sec. 2.2.

Eqs. (1)-(3) are equivalent to describing the problem of estimating the class probabilities for the region labels as the problem of calculating the (multi-class) marginals in the HMM (Hidden Markov Model) in Fig. 1(c). We estimate the class probabilities by the sum product algorithm [15]. ORC classifies each land region by $\text{ORC}(i) = \text{argmax}_{L_j} p_{\text{ORC}}(i, L_j)$.

4   Experiments: ORC

We experiment on the *coast, mountain,* and *open country* datasets from La-belme [16] presented in [17]. This provides a set of 1144 256X256 images of natural scenery. With the Labelme toolbox, a Web user marks polygons in the image and freely provides a textual annotation for each. This freedom encourages the use of synonyms and spelling mistakes. Following [16], synonyms were grouped together and spelling mistakes were corrected.

In this work we do not deal with automatic segmentations and rely on manual ones provided with the dataset. We select all regions whose annotation describes background. This gives us 4979 regions annotated by 19 background labels: sky (1120), mountain (1489), sea (401), trees (622), field (366), river (150), sand

Table 1. Left: Accuracy of ORC using different combinations of cues. A significant improvement from 61.5% to 68.2% is achieved by utilizing the new cues. Right: Accuracy of M-ORC demonstrating its ability to recognize separately and simultaneously structural categories and coverage categories of land regions.

Cue	Accuracy	cue	Acc. Sky-Water-Land	Acc. Land Structure	Acc. Land Cover
Color&Texture	0.615	Color&Texture	0.865	0.849	0.570
Relative Location	0.503	Relative Location	0.862	0.750	0.468
Boundary Shape	0.452	Boundary Shape	0.866	0.836	0.517
Relative Loc. + Boundary Shape	0.573	Relative Loc. + Boundary Shape	0.881	0.835	0.530
Color&Texture + Relative Loc.	0.676	Color&Texture + Relative Loc.	0.905	0.860	0.612
Color&Texture + Boundary Shape	0.641	Color&Texture + Boundary Shape	0.887	0.862	0.567
All (ORC)	0.682	All (M-ORC)	0.909	0.876	0.605

(182), ground (94), grass (36), land (41), rocks (201), plants (143), snow (50), plateau (28), valley (20), bank (20), lake (9), beach (3), and cliff (4).

For extracting the color&textural features describing a background region, we first compute a mask that includes all its parts that do not intersect with other annotated regions (e.g., foreground objects that occlude part of it). We compute the descriptors described in Sec. 2.1 over the pixels inside this mask.

To compute the probability estimates from color&textural features and boundary appearance, we use SVM with an RBF kernel (using LIBSVM [18]). To test *ORC* we perform a 5-fold cross-validation at the image level. Each iteration starts with a parameter selection stage (c and γ; see [18]) in which the training set is split into a training and validation test (also at image level).

We found that the height cue is noninformative and report results without using it. Fig. 2 demonstrates ORC results using each cue alone and all cues together. We can clearly see that the incorporation of new cues gives better results than using only color&texture. For instance, colored sky that is misclassified using only color&textural features is classified correctly using the relative location cue. Sea and mountain regions that are sometimes misclassified using color&texture, are correctly classified using the boundary shape cue. In the bottommost example, a narrow sea region that is missed in the ground truth is recognized. A region is sometimes misclassified when each cue is used separately, but classified correctly when all cues act as a committee. The total accuracies are reported in Table 1(a). A significant improvement from 61.5% to 68.2% is achieved by utilizing the new cues. The color&texture cues are better for classifying trees, field, rocks, plants, and snow. The new cues alone give more accurate results for sky, mountain, sea, and sand. The performance for the other classes is low for all cues due to their low occurrences in the dataset.

5   Multiple Categorization: The M-ORC Algorithm

In the second part of this work we examine the causes for errors. It turns out that results counted as errors are not always wrong. Some ambiguities are a matter of synonyms. A mountain is sometimes annotated as a cliff and vice versa. Other

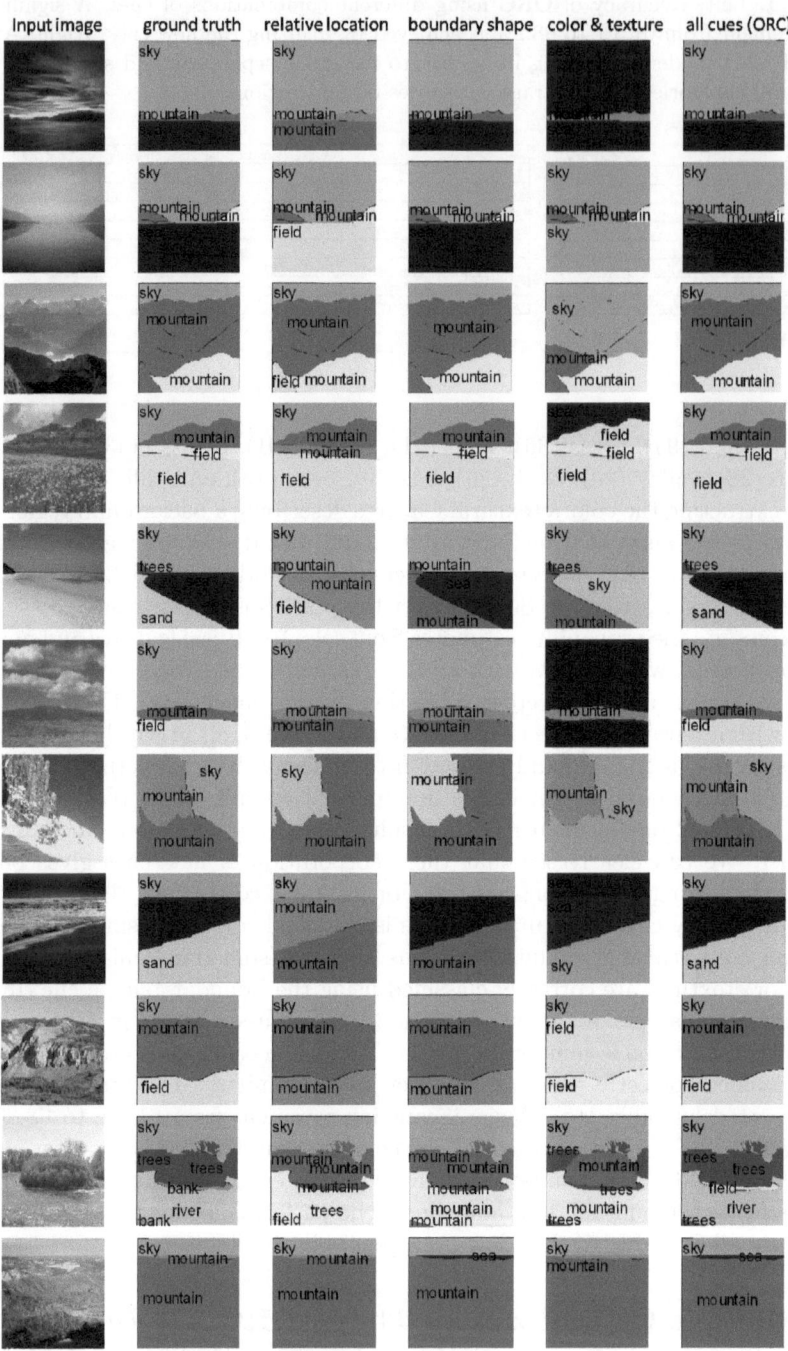

Fig. 2. Demonstrating region classification results using each cue alone and all cues together, using the ORC algorithm

ambiguities, however, are created by annotations that refer to different aspects of the labeled object. In particular, it seems that land regions are sometimes annotated with a description of the land structure (mountain, plain, valley) and sometimes annotated with a description of the overlying coverage (trees, grass, sand, rocks, etc.). For instance, a mountain with trees may be classified by the ground truth as mountain and classified by ORC as trees, or vice versa. A snowy plain can be correctly labeled both as 'snow' and as 'field' or 'plateau'.

These observations lead us to suggest an alternative labeling method, denoted *M-ORC* (Multiple Categorization Ordered Region Classifier). With this method, the correct labeling for land regions is multi-valued. Each region is first classified into one of the three categories SKY, WATER or LAND. Each region classified as LAND is further classified by a structural category and a coverage category.

More formally, $\text{M-ORC}(i) = \{c_{i_1}, c_{i_2}, c_{i_3}\}$, where $c_{i_1} \in \{\text{`SKY'}, \text{`WATER'}, \text{`LAND'}\}$, $c_{i_2} \in C_{\text{struct}} = \{\text{`MOUNTAIN'}, \text{`PLAIN'}, \text{`VALLEY'}, \text{`BANK'}, \emptyset\}$, and $c_{i_3} \in C_{\text{cover}}$, where $C_{\text{cover}} = \{\text{`SAND'}, \text{`GROUND'}, \text{`ROCKS'}, \text{`PLANTS'}, \text{`TREES'}, \text{`GRASS'}, \text{`SNOW'}, \emptyset\}$. When $c_{i_1} \in \{\text{`SKY'}, \text{`WATER'}\}$, $c_{i_2}, c_{i_3} = \emptyset$. To assign c_{i_1} a score r_{main} is defined: $r_{\text{main}}(i, \text{`SKY'}) = p_{\text{ORC}}(i, \text{`sky'})$, $r_{\text{main}}(i, \text{`WATER'}) = p_{\text{ORC}}(i, \text{`sea'}) + p_{\text{ORC}}(i, \text{`lake'}) + p_{\text{ORC}}(i, \text{`river'})$, and $r_{\text{main}}(i, \text{`LAND'}) = 1 - r_{\text{main}}(i, \text{`SKY'}) - r_{\text{main}}(i, \text{`WATER'})$. The selection of $c_{i_1}$ is according to the maximal $r_{\text{main}}$. For regions for which $c_{i_1} = \text{`LAND'}$, c_{i_2} and c_{i_3} are set. A score r_{struct} is defined for $c \in C_{\text{struct}}$: $r_{\text{struct}}(i, c) = \sum_{l \in G_c} p_{\text{ORC}}(i, l)$, where $G_c = \{\text{`mountain'}, \text{`cliff'}\}$ for $c = \text{`MOUNTAIN'}$, $G_c = \{\text{`land'}, \text{`field'}, \text{`plateau'}\}$ for $c = \text{`PLAIN'}$, $G_c = \{\text{`valley'}\}$ for $c = \text{`VALLEY'}$, and $G_c = \{\text{`bank'}, \text{`beach'}\}$ for $c = \text{`BANK'}$. The selection of c_{i_2} is according to the maximal r_{struct}. For $c \in C_{\text{cover}}$, $r_{\text{cover}}(i, c) = p_{\text{ORC}}(i, l_c)$, where $l_c = \text{lowercase}(c)$. The selection of c_{i_3} is according to the maximal r_{cover}.

6   Experiments: M-ORC

Each land region training example is provided with ground-truth (human labeling) of one type of label, structural or coverage. Nevertheless, by the categorization scheme suggested here, each test region, if recognized as a land region, is assigned with both a coverage and a structural label, as can be seen in Fig. 3.

When evaluating the categorization accuracy, we can only check if the available label is correct. Out of the total of 4979 regions, 1120 are annotated sky, 560 are annotated by a word describing water, and 3299 by words describing land. Out of the 3299 land regions, 1328 are described by their cover and 1971 by their structure. For land regions that are hand labeled with a structural label, we check whether it matches the structural label assigned by M-ORC, and for land regions hand labeled with a coverage label, we check whether it matches the coverage label assigned by M-ORC; see Table 1(b). Again we see the advantage of incorporating the new cues in comparison to using color&texture alone. The accuracy for the SKY-WATER-LAND categorization grows from 86% to 91%. The accuracy for classifying land structure grows from 85% to 88%, and the accuracy for classifying land coverage grows from 57% to 60%. The lower accuracy of the latter is probably due to the larger number of cover categories.

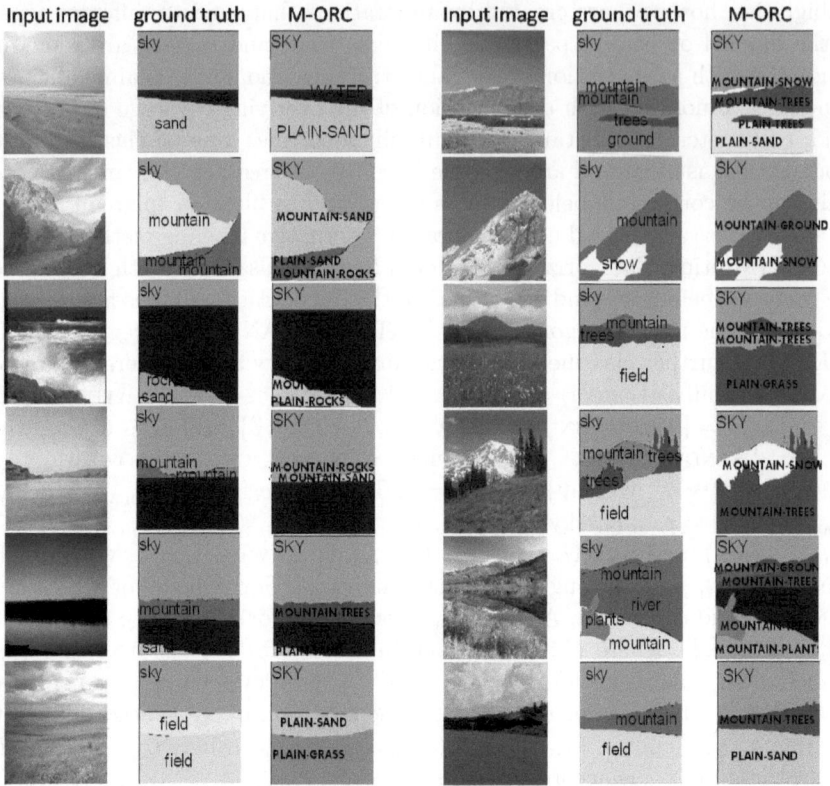

Fig. 3. Demonstrating M-ORC results: land regions are assigned with both a coverage and a structural label, while the ground-truth provides only one label for each region

7   Discussion

This paper presents: 1. A new model for contextual background region classification that uses statistical characteristics of scenery images and suggests an exact inference solution. 2. A novel method for categorizing land regions separately by their structural and their coverage categories, while learning from a training set in which each region is annotated only by one label type.

Note that an alternative solution for this scheme would be to use mixed class labels covering all combinations (e.g., "tree covered mountain") . However, this will lead to a quadratic number of classes, which implies lower performance and much greater efforts to label the training set.

One may argue that the images dealt with here are rather simple, and that modeling more complex images (street scenes, indoor scenes) is a more appropriate challenge. We agree. However, while computer vision has taken major steps forward in object detection and recognition also in complex scenes (e.g., vehicle and pedestrian detection), computer vision still has a long way to go before it can provide full interpretation and understanding of scenes. Our focus on scenery/ landscape images can be considered a step backwards to deal with the tasks that occupied the human vision system in the early stages of its evolution. Only after the visual system was able to cope with such scenery did it gradually evolve to cope with more complex scenes. Nevertheless, it is of course desirable to extend the models suggested here to more complex image classes and objects.

In this work we used pre-segmented images. In future work we intend to check our model's stability given automatic segmentation results, and to investigate the usability of the layout cues for the task of semantic segmentation.

References

1. Avraham, T., Lindenbaum, M.: Non-local characterization of scenery images: Statistics, 3D reasoning, and a generative model. In: Daniilidis, K., Maragos, P., Paragios, N. (eds.) ECCV 2010. LNCS, vol. 6315, pp. 99–112. Springer, Heidelberg (2010)
2. Torralba, A.B.: Contextual priming for object detection. IJCV 53(2), 169–191 (2003)
3. Kumar, S., Hebert, M.: A hierarchical field framework for unified context-based classification. In: ICCV (2005)
4. He, X., Zemel, R.S., Ray, D.: Learning and incorporating top-down cues in image segmentation. In: Leonardis, A., Bischof, H., Pinz, A. (eds.) ECCV 2006. LNCS, vol. 3951, pp. 338–351. Springer, Heidelberg (2006)
5. Rabinovich, A., Vedaldi, A., Galleguillos, C., Wiewiora, E., Belongie, S.: Objects in context. In: ICCV (2007)
6. Desai, C., Ramanan, D., Fowlkes, C.: Discriminative models for multi-class object layout. In: ICCV (2009)
7. Galleguillos, C., Belongie, S.: Context based object categorization: A critical survey. Comput. Vis. Image Understand (2010)
8. Deng, J., Berg, A.C., Li, K., Fei-Fei, L.: What does classifying more than 10,000 image categories tell us? In: Daniilidis, K., Maragos, P., Paragios, N. (eds.) ECCV 2010. LNCS, vol. 6315, pp. 71–84. Springer, Heidelberg (2010)
9. Fergus, R., Bernal, H., Weiss, Y., Torralba, A.: Semantic label sharing for learning with many categories. In: Daniilidis, K., Maragos, P., Paragios, N. (eds.) ECCV 2010. LNCS, vol. 6311, pp. 762–775. Springer, Heidelberg (2010)
10. Feebaum, C.: Wordnet: An Electronic Lexical Database. Bradford Books (1998)
11. Niu, D., Dy, J.G., Jordan, M.I.: Multiple non-redundant spectral clustering views. In: ICML (2010)
12. Vogel, J., Schiele, B.: Semantic modeling of natural scenes for content-based image retrieval. IJCV 72(2), 133–157 (2007)
13. Haralick, R.M., Shanmugam, K., Dinstein, I.: Textural features for image classification. IEEE Transactions on Systems, Man, and Cybernetics, 610–621 (1973)
14. Wu, T.F., Lin, C.J., Weng, R.C.: Probability estimates for multi-class classification by pairwise coupling. Journal of Machine Learning Research 5, 975–1005 (2004)
15. Kschischang, F.R., Frey, B.J., Loeliger, H.A.: Factor graphs and the sum-product algorithm. IEEE Tran. on Information Theory 47(2), 498–519 (2001)
16. Russell, B.C., Torralba, A.: Labelme: a database and web-based tool for image annotation. IJCV 77, 157–173 (2008)
17. Oliva, A., Torralba, A.: Modeling the shape of the scene: a holistic representation of the spatial envelope. IJCV 42(3), 145–175 (2001)
18. Chang, C., Lin, C.: LIBSVM: a library for support vector machines. (2001) Software available at http://www.csie.ntu.edu.tw/~cjlin/libsvm

Selection of Suspicious ROIs in Breast DCE-MRI

Roberta Fusco[1,3], Mario Sansone[1], Carlo Sansone[2], and Antonella Petrillo[3]

[1] Dipartimento di Ingegneria Biomedica, Elettronica e delle Telecomunicazioni
Universitá 'Federico II' di Napoli, Italia
[2] Dipartimento di Informatica e Sistemistica
Universitá 'Federico II' di Napoli, Italia
[3] Dipartimento di Diagnostica per Immagini
Istituto Nazionale dei Tumori 'Fondazione Pascale' Napoli, Italia

Abstract. Dynamic Contrast Enhanced Magnetic Resonance Imaging (DCE-MRI) could be helpful in screening high-risk women and in staging newly diagnosed breast cancer patients. Selection of suspicious regions of interest (ROIs) is a critical pre-processing step in DCE-MRI data evaluation. The aim of this work is to develop and evaluate a method for automatic selection of suspicious ROIs for breast DCE-MRI. The proposed algorithm includes three steps: (i) breast mask segmentation via intensity threshold estimation; (ii) morphological operations for hole-filling and leakage removal; (iii) suspicious ROIs extraction. The proposed approach has been evaluated, using adequate metrics, with respect to manual ROI selection performed, on ten patients, by an expert radiologist.

Keywords: DCE-MRI, Breast, ROI selection, segmentation.

1 Introduction

Breast cancer is the most common cancer type among women in the Western world. It is the second leading cause of cancer death in women today (after lung cancer) and is estimated to cause 15 % of cancer deaths [1].

The currently widespread screening method is RX mammography [2]; however, Dynamic Contrast Enhanced Magnetic Resonance Imaging (DCE-MRI) has demonstrated a potential in screening of high-risk women, in staging newly diagnosed breast cancer patients and in assessing therapy effects [1] thanks to its minimal invasiveness and to the possibility to visualize functional information not available with conventional imaging.

Breast DCE-MRI examination involves the imaging of the breast before and after the injection of a contrast agent (a commonly used tracer is Gd-DTPA, administered intravenously). DCE-MRI data consist in one pre-contrast series of T1-weighted images (images with greater signal intensity from fat-containing tissues and where most of the contrast between tissues is due to differences in the Spin-lattice relaxation time known as T1 value [3]) spanning both breasts, followed by a fixed number of post-contrast series.

G. Maino and G.L. Foresti (Eds.): ICIAP 2011, Part I, LNCS 6978, pp. 48–57, 2011.
© Springer-Verlag Berlin Heidelberg 2011

Highly vascularised regions, such as tumors, exhibit typical patterns of signal enhancement vs. time as described in [4]: typically, a quick tracer uptake (III,IV,V), is followed by a plateau (IV) or washout (V) (Fig.1); normal or benign tissues are characterised by either no enhancement (I) (especially in predominantly adipose regions) or slower enhancement with delayed washout (II).

By analyzing signal intensity-time curves, it is possible to characterize each voxel and detect abnormalities within the breast [5].

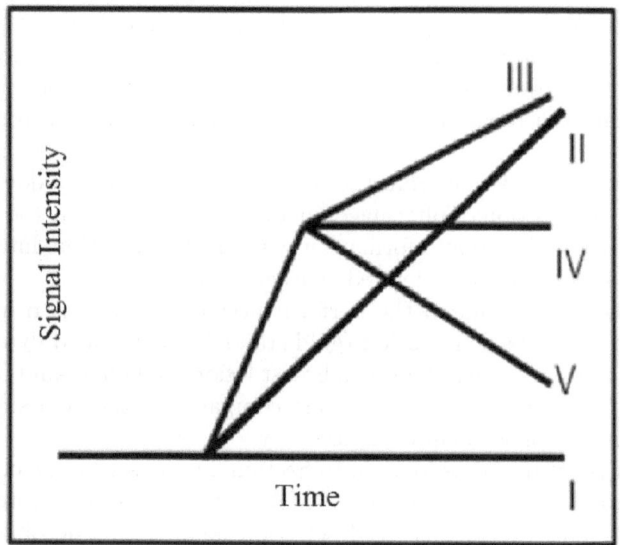

Fig. 1. Types of dynamic signal enhancement curves

Manual selection of suspicious regions of interest (ROIs), which is the critical first step in lesion detection and evaluation, is operator-dependent and time consuming. Moreover, given the vast quantity and multidimensionality of data to be analysed in a DCE MRI data set, the possibility exists that diagnostically significant regions of enhancement may be overlooked [6,7].

Therefore, lesion detection for breast DCE-MRI is a difficult task that can be supported by automatic procedures for identification of suspicious ROIs.

For example, while the main objective of Tzacheva et al. [8] was classification of malignant breast lesions, they used a very simple ROI selection procedure based only on high signal intensity but they did not specify how the corresponding threshold has been chosen; moreover in their study only static features were used without taking advantage of the whole dynamic information.

As another example, Lucht et al. [9] used pharmacokinetic modelling to produce parametric maps which were used for manual ROI selection; however, the computation of tracer kinetic parameters is still a lengthy task when performed on entire images.

Gal et al. [7] proposed a region growing approach based both on original image intensity values and fitted pharmacokinetic parameters. This approach involves the automatic identification of a seed voxel. Their results indicate a high sensitivity. However, ROI selection using kinetic parameters only (e.g. via thresholding) is an improper approach because tumor heterogeneity, which has been shown to be an important factor to be accounted for, could be missed.

From previous considerations, it emerges that an optimal automatic algorithm for selection of suspicious ROIs should have the following characteristics: it should take advantage of the whole dynamic information (not only static images), it should not miss tumor heterogenity and it should be fast enough to be applied as a first step to more sophisticated classification procedures.

The aim of this work is to propose an automatic method for suspicious ROI selection within the breast using dynamic-derived information from DCE-MRI data.

Our approach is different from the previous ones because it does not involve the lengthy computation of pharmacokinetic parameters but at the same time it exploits the whole dynamic information contained in the time-intensity curves by means of simple dynamic-derived characteristics.

In this study we evaluated the performances of the proposed method using the results of manual segmentation (gold standard) performed by an expert radiologists on ten histologically proven breast lesions (5 benign and 5 malignant). The results of the proposed method were compared to the gold standard using opportune metrics of segmentation accuracy.

The paper is organized as follows. In Section 2 we describe the characteristics of recruited patients, the breast DCE-MRI data acquisition protocol and the proposed automatic suspicious ROI selection algorithm. The obtained results are presented in Section 3 and discussed in Section 4, where we also draw some conclusions.

2   Materials and Methods

2.1   Patient Selection

Ten women (average age 40 years) with benign or malignant lesions histopathologically proven were enrolled (Table 1). Five cases were malignant (2 ductal carcinoma in situ, DCIS; 2 invasive ductal carcinoma, IDC; 1 invasive lobular carcinoma, ILC) and five cases were benign (4 fibroadenomata, 1 atypical ductal hyperplasia).

2.2   Data Acquisition

The patients underwent imaging with a 1.5 T scanner (Magnetom Symphony, Siemens Medical System, Erlangen, Germany) equipped with a phased-array body coil. DCE T1-weighted FLASH 3-D coronal images were acquired (TR/TE: 9.8/4.76 ms; flip angle: 25 degrees; field of view 330x247 mmxmm; matrix:

Table 1. Patients characteristics

Patient ID	Age	Pathology	Cancer Type
1	37	Malignant	IDC
2	39	Benign	Fibroadenomata
3	47	Malignant	ILC
4	67	Malignant	DCIS
5	27	Malignant	DCIS
6	37	Benign	Fibroadenomata
7	41	Benign	Fibroadenomata
8	36	Benign	ADH
9	33	Benign	Fibroadenomata
10	40	Malignant	IDC

256x128; thickness: 2 mm; gap: 0; acquisition time: 56 s; 80 slices spanning entire breast volume). One series was acquired before and 9 series after intra-venous injection of 2 ml/kg body weight of a positive paramagnetic contrast medium (Gd-DOTA, Dotarem, Guerbet, Roissy CdG Cedex, France). Automatic injection system was used (Spectris Solaris EP MR, MEDRAD, Inc.,Indianola, PA) and injection flow rate was set to 2 ml/s followed by a flush of 10 ml saline solution at the same rate.

2.3   Manual Segmentation

The manual segmentation was performed by an expert radiologist on the fat-suppressed image obtained subtracting the basal pre-contrast image from the 5th post-contrast image. Per each patient all the slices including the lesion have been used. The segmentation was performed by means of the OsiriX v.3.8.1 3 software 3.

2.4   ROI Selection

The proposed algorithm includes three steps.

The first step involves Breast Mask (BM) extraction by means of automatic intensity threshold estimation (Otsu Thresholding) [10] on the parametric map obtained considering the sum of intensity differences (SOD) calculated pixel by pixel. In fact, this parameter describes the dynamic information of the whole curve and reflects the history of contrast agent enhancement with time [11]:

$$SOD_p = Pre_p + \sum_{i=1}^{T} |Post_p(i) - Post_p(i-1)| \tag{1}$$

where SOD_p is the SOD for the p pixel; Pre_p is the pre-contrast intensity; $Post_p(i)$ is i-th post-contrast scan and T is the total number of scans.

The second step includes hole-filling and leakage removal by means of morphological operators: 'closing' is required to fill the holes on the boundaries of breast mask; 'filling' is required to fill the holes within the breasts; 'erosion' is required to reduce the dilation obtained by the closing operation [12].

The third step includes suspicious ROIs extraction. The dynamical features of each pixel are analysed. A pixel is assigned to suspicious ROI if it satisfies two conditions: the maximum of its normalized time-intensity curve (as Figure 2 shows) should be grater than 0.3 and the maximum signal intensity should be reached before the end of the scan time. The first condition assures that the pixels within the ROI have a significant contrast agent uptake (thus excluding type I and type II curves) and the second condition is required for the time-intensity pattern to be of type IV or V (thus excluding type III curves) [13,14,15].

The choice of the threshold 0.3 was based on the findings by [16]: in fact, in their study lesions with TIC enhancement less than 50% above the baseline were considered non-tumoral; they also noticed that lowering the threshold to 40% improved the accuracy of diagnosis. We proposed a threshold of 30% in order to reduce the number of false negatives.

All procedures were implemented in Matlab R2008a using Image toolbox.

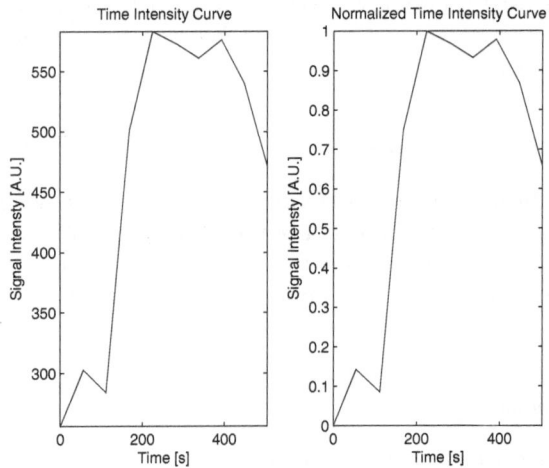

Fig. 2. Time Intensity Curve vs Normalized Time Intensity Curve: an example of a time intensity curve was reported on the left, while an example of a normalized time intensity curve with respect to the maximum signal intensity was reported on the right

2.5 Evaluation

According to the approach proposed by [17] the accuracy of a segmentation method can be evaluated calculating the following quantities (BM is the Breast Mask): the Overlap (O) and the Union (U) between the ground truth S_m and the automatic segmentation S_c; the true positive fraction (TPF); the false negative fraction (FNF); the true negative fraction (TNF); the false positive fraction (FPF); the accuracy (ACC):

$$O = S_m \quad S_c$$

$$U = S_m \quad S_c$$

$$TPF = \frac{O}{S_m}$$

$$TNF = \frac{BM - U}{BM - S_m}$$

$$FNF = \frac{S_m - S_c}{S_m} = 1 - TPF$$

$$FPF = \frac{S_m - O}{BM - S_m} = 1 - TNF$$

$$ACC = \frac{TPF + TNF}{TPF + TNF + FPF + FNF}$$

It is clear that only two measures are independent and are required to quantify the accuracy of the method.

3 Experimental Results

Figure 3 shows the result of a manual ROI lesion selection onto a fat-suppressed image.

Figure 4 (a) shows the results of the first step, Breast mask selection after Otsu thresholding on SOD feature. Figure 4 (b) shows the breast mask after morphological operators (closing, filling and erosion). Figure 4 (c) shows the result of automatic suspicious ROI selection.

Table 2 reports the results of the evaluation study. Per each patient the TPF, TNF and ACC have been reported.

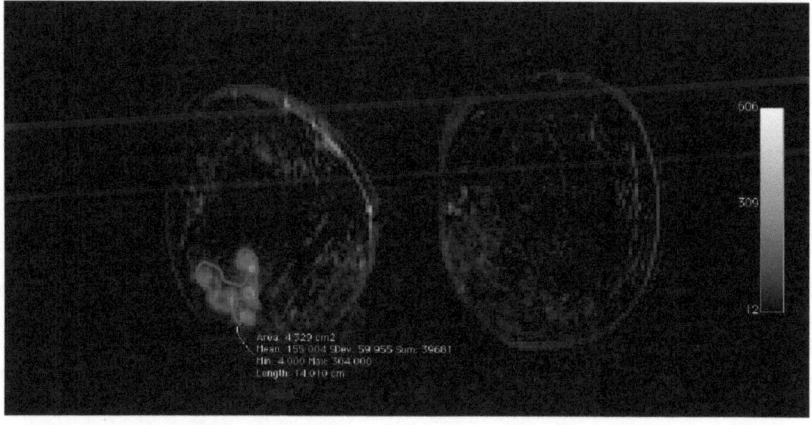

Fig. 3. Manual selection of suspicious ROI

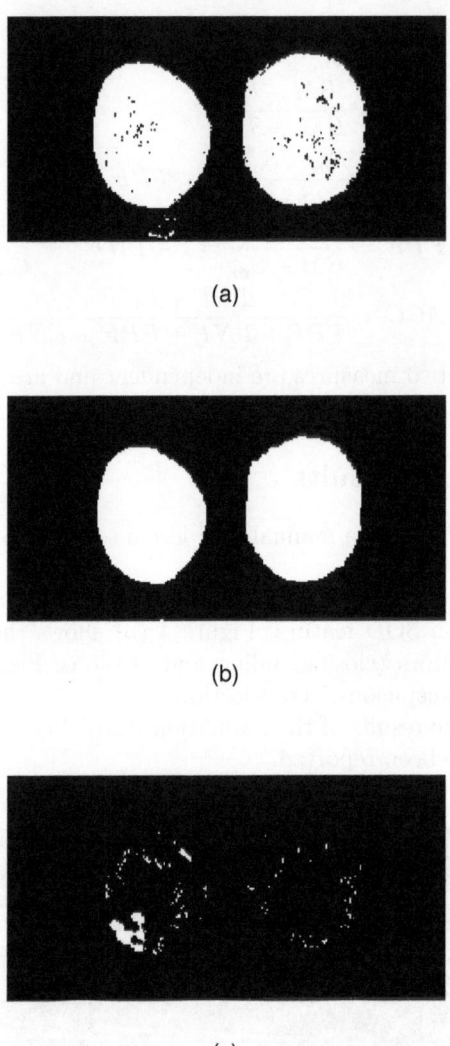

(a)

(b)

(c)

Fig. 4. Results of automatic ROI selection: (a) breast mask selection via Otsu thresholding on SOD; (b) breast mask after morphological operators (closing, filling and erosion); (c) automatic suspicious ROI selection

Table 2. ROI selection results

Patient ID	TPF	TNF	ACC [%]
1	0.887	0.999	93.8
2	0.799	0.999	90.0
3	0.750	0.999	87.7
4	0.670	0.999	83.0
5	0.750	0.999	87.6
6	0.580	0.999	78.8
7	0.750	0.999	87.0
8	0.827	0.999	90.9
9	0.730	0.999	86.9
10	0.879	0.990	93.0
Average	0.762	0.998	87.9

4 Discussion and Conclusions

In this study we have proposed an algorithm for selection of suspicious ROIs on breast DCE-MRI data. The proposed algorithm is based on dynamic-derived features. The performances of the proposed method have been evaluated with respect to manual selection of an expert radiologist on a dataset of ten breast lesions.

It is worth noting that previous studies [18,19,20] have also performed the breast DCE-MRI segmentation on restricted population (from thirteen to four patients) because of the difficulty to enroll patients with the same data acquisition protocol and the absence of a public database.

Our results indicate that the TPF of our automatic selection varied in the range 0.580 to 0.887 (average 0.762) and then the accuracy of authomatic suspicious ROI selection increases when invasive lobular o ductal cancer are considered. In fact, in those cases the lesion is larger with a greater contrast agent uptake: this determines a lower number of misclassified pixels (suspicious or not suspicious). On the contrary, the accuracy decreases when ductal carcinoma in situ (because of their size) and benign lesion are considered (because of their lower contrast agent uptake).

Selection of suspicious ROIs is to be considered as a preliminary step in tumour evaluation before more sophisticated algorithms for tissue malignancy classification. Our preliminary results show that selection of suspicious ROIs in the breast is feasible using a simple and fast algorithm based on the whole dynamic information contained in DCE-MRI. While the fraction of false positives is almost zero, further investigation is required in order to reduce the number of lost pixels potentially suspicious: this could be accomplished, for example, introducing the analysis of non-dynamic images. In the future, our preliminary study will be extended on a larger number of patients and manual segmentation will be done by multiple readers.

References

1. Lehman, C.D., Gatsonis, C., Kuhl, C.K., Hendrick, R.E., Pisano, E.D., Hanna, L., Peacock, S., Smazal, S.F., Maki, D.D., Julian, T.B., DePeri, E.R., Bluemke, D.A., Schnall, M.D.: MRI evaluation of the contralateral breast in women with recently diagnosed breast cancer. The New England Journal of Medicine 356, 1295–1303 (2007)
2. Olsen, O., Gøtzsche, P.C.: Screening for breast cancer with mammography, Cochrane Database of Systematic Reviews, p. CD001877 (2001) (Online)
3. McRobbie, D., et al.: MRI, From picture to proton (2003)
4. Daniel, L., et al.: Breast Disease: Dynamic Spiral MR Imaging. Radiology 209, 499–509 (1998)
5. Eyal, E., Degani, H.: Model based and Model free parametric analysis of breast dynamic contrast enhanced MRI. NMR Biomed. (2007)
6. Liney, G.P., Sreenivas, M., Gibbs, P., Garcia-Alvarez, R., Turnbull, L.W.: Breast lesion analysis of shape technique: semiautomated vs. manual morphological description. Journal of Magnetic Resonance Imaging: JMRI 23, 493–498 (2006)
7. Gal, Y., Mehnert, A., Bradley, A., McMahon, K., Crozier, S.: Automatic Segmentation of Enhancing Breast Tissue in Dynamic Contrast-Enhanced MR Images. In: 9th Biennial Conference of the Australian Pattern Recognition Society on Digital Image Computing Techniques and Applications (DICTA 2007), Glenelg, Australia, pp. 124–129 (2007)
8. Tzacheva, A.A., Najarian, K., Brockway, J.P.: Breast cancer detection in gadolinium-enhanced MR images by static region descriptors and neural networks. J. Magn. Reson. Imaging 17(3), 337–342 (2003)
9. Lucht, R.E.A., Knopp, M.V., Brix, G.: Classification of signal time curves from dynamic MR mammography by neural networks. Magn. Reson. Imaging, 51–57 (2001)
10. Lee, S.H., Kim, J.H., Park, J.S., Chang, J.M., Park, S.J., Jung, Y.S., Moon, W.K.: Computerized Segmentation and Classification of Breast Lesions Using Perfusion Volume Fractions in Dynamic Contrast-enhanced MRI. In: International Conference on BioMedical Engineering and Informatics, pp. 58–62. IEEE Computer Society, Los Alamitos (2008)
11. Twellmann, T., Lichte, O., Nattkemper, T.W.: An adaptive tissue characterization network for model-free visualization of dynamic contrast-enhanced magnetic resonance image data. IEEE Trans. Med. Imaging 3, 21 (2005)
12. Chen, W., Giger, M.L., Li, H., Bick, U., Newstead, G.M.: Volumetric texture analysis of breast lesions on contrast-enhanced magnetic resonance images. Magn. Reson. Med. (2007)
13. Issa, Improved discrimination of breast lesions using selective sampling of segmented MR images, MAGMA (2006)
14. Meinel, L.A., Buelow, T., Huo, D., Shimauchi, A., Kose, U., Buurman, J., Newstead, G.: Robust segmentation of mass-lesions in contrast-enhanced dynamic breast MR images. Journal of Magnetic Resonance Imaging: JMRI 32, 110–119 (2010)
15. Sinha, U., Sinha, S., Lucas-Quesada, F.A.: Segmentation strategies for breast tumors from dynamic MR images. Journal of Magnetic Resonance Imaging 6, 753–763
16. Torricelli, P., Pecchi, A., Luppi, G., Romagnoli, R.: Gadolinium-enhanced MRI with dynamic evaluation in diagnosing the local recurrence of rectal cancer. Abdominal Imaging 28, 19–27 (2003)

17. Udupa, J.K., Leblanc, V.R., Zhuge, Y., Imielinska, C., Schmidt, H., Currie, L.M., Hirsch, B.E., Woodburn, J.: A framework for evaluating image segmentation algorithms. Computerized Medical Imaging and Graphics: The Official Journal of the Computerized Medical Imaging Society 30, 75–87 (2006)
18. Woods, B.J., Clymer, B.D., Kurc, T., Heverhagen, J.T., Stevens, R., Orsdemir, A., Bulan, O., Knopp, M.V.: Malignant-lesion segmentation using 4D co-occurrence texture analysis applied to dynamic contrast-enhanced magnetic resonance breast image data. J. Magn. Reson. Imaging 25(3), 495–501 (2007)
19. Hill, A., Mehnert, A., Crozier, S.: Edge intensity normalization as a bias field correction during balloon snake segmentation of breast MRI. In: Conf. Proc. IEEE Eng Med Biol Soc., pp. 3040–3043 (2008)
20. Cui, Y., Tan, Y., Liberman, L., Parbhu, R., Kaplan, J., Schwartz Malignant, L.H.: lesion segmentation in contrast-enhanced breast MR images based on the marker-controlled watershed. Med. Phys. 36, 4359 (2009)

Regions Segmentation from SAR Images

Luigi Cinque[1] and Rossella Cossu[2]

[1] Università degli Studi *Sapienza*, Roma
[2] Istituto per le Applicazioni del Calcolo-CNR, Roma
r.cossu@iac.cnr.it

Abstract. In this paper, we propose an approach based on the level set method for segmenting SAR (Synthetic Aperture Radar) images. In particular, the segmentation process presented consists in the evolution of an initial curve, including the interested region, until it reaches the boundary of the area to be extracted. The procedure proposed allows to obtain the same result of segmentation independently of the initial position of the curve. The results are shown on both synthetic and real images. The analyzed images are SAR PRI (Precise Images), acquired during the mission ERS2.

Keywords: Level set, Image segmentation, SAR, Speckle noise.

1 Introduction

Image segmentation plays a fundamental role in the SAR image interpretation. The segmentation of SAR images is usually recognized as a complex problem, because of multiplicative noise, called speckle, which produces grainy images. Recently, a number of methods based on curve evolution have been proposed for segmentation of SAR images. In this paper we present a procedure based on the evolution of curves, described by the level set method, to extract distinct regions from SAR images. This method, proposed by Osher and Sethian [1,2,3], is based on the identification of an area of interest as the zero level set of an implicit function that evolves according to a PDE (partial differential equation) model with an appropriate speed function.

SAR images present advantages and disadvantages; the acquired images provide information under varied weather conditions, during night as well as day. On the other hand SAR images present a granular effect due to the presence of the speckle noise.

The segmentation methods based on edge detection filters, often show edges which may not form a set of closed curves surrounding connected regions. The traditional techniques of histogram thresholding and region-based need a pre-processing based on speckle reduction. Moreover, the region-growing techniques have the limit of depending on the selection of the starting points.

The segmentation process proposed starts from an initial curve (zero level set) defined on the image that evolves until it stops at the contour of the interest object. The evolution of the initial curve is determined by a speed function, which is a fundamental choice to achieve a good segmentation.

G. Maino and G.L. Foresti (Eds.): ICIAP 2011, Part I, LNCS 6978, pp. 58–67, 2011.

In this paper two different speed functions are introduced and their results compared with a series of tests on synthetic SAR images. In particular a first approach developed is based on the assumption that each region to segment through level set is modeled by a Gamma distribution. In this case an expression of propagation speed of the front is obtained by computing intensity averages of the regions; the method does not need to reduce speckle noise [5].

In the second approach developed the speed function is based on the computation of image gradient and takes into account the problem of filtering speckle noise in the image, which was faced with the application of the SRAD (Speckle Reducing Anisotropic Diffusion) technique to SAR image [6] [7] [8].

Finally we propose a combined speed, based on the contributes of the previous speeds functions. In this case the proposed procedure has the important peculiarity to extract the same contour from an image also starting from different initial contours. Our approach is validated and compared using a series of tests on synthetic SAR images. These tests demonstrate our method allows to obtain the same result of segmentation independently of the initial position of the curve.

The SAR PRI image here segmented has been acquired during ERS2 mission. ERS2 SAR system is capable of 25 m. resolution from an altitude of 800 km, at radar wavelength of 5.7 cm.

The paper is organized as follows. In Section 2, the level set method is briefly described. In Section 3, speed computation related to level set method and noise reducing are presented. In Section 4 experimental results and applications to SAR image are shown. Some conclusions are drawn in Section 5.

2   Level Set Approach

It should be mentioned that the segmentation through the level set method has become very popular over the last decades. In particular, this methodology describes the evolution of an initial curve, including the interested region, until it reaches the boundary of the area to be extracted.

Let $I\colon \Omega \to \Re^n$ be the intensity image function where $\Omega \subset \Re^2$.

The goal of image segmentation is to partition Ω, moving from image I, in order to extract disjoint regions covering Ω.

The boundary of the region of interest may be considered as a curve belonging to a family in which the time evolution is described by the following level set equation

$$\frac{\partial \Phi(\mathbf{x}(t))}{\partial t} + F(\mathbf{x}(t))|\nabla \Phi(\mathbf{x}(t))| = 0 \qquad (1)$$

where $F(\mathbf{x}, t)$, representing the curve speed in the normal direction, is related to the image features. The main advantages of using the level set is that complex shaped regions can be detected and handled implicitly. The initial curve (zero level set) evolves until it stops at the contour of the interest object.

For the numerical approximation of the level set equation in a domain $\Omega \subset \Re^2$ we introduce the computational domain Ω^* obtained by considering a uniform partition of Ω in $(N-1) \times (M-1)$ disjoint rectangles Ω_{ij} with edges $\Delta x = \Delta y$, usually in an image $\Delta x = \Delta y = 1$. Let $P_{i,j} \equiv P(x_i, y_j)$ $i = 1, ..., N; j = 1, ..., M$ a point in Ω^* and $\phi_{i,j}^n$ the value of the function $\phi(\mathbf{x}(t))$ at $P_{i,j}$ at time t^n. Let $v(\mathbf{x}(t))$ be the speed function: the algorithm starts by initializing $\phi(\mathbf{x}(t))$ as a signed distance function

$$\phi(\mathbf{x}(0)) = \pm d$$

where

$$d(\bar{x}) = \min_{\bar{x}\, \in \gamma} |\bar{x} - \bar{x}_\gamma|.$$

Now, known the value of $\phi_{i,j}^n$, the value $\phi_{i,j}^{n+1}$ is computed by a 2-order ENO scheme with the TVD (Total Variation Diminishing) Runge Kutta scheme for the time integration.

We underline that the definition of $\phi(\mathbf{x}(t))$ as a signed distance function is crucial. In fact, during the evolution the level set function does not remain a signed distance function; so that it is necessary to re-initialize the algorithm at regular intervals in order to limit numerical dissipation. Moreover the choice of speed function is a fundamental task for this segmentation approach.

3   Combined Speed

As mentioned above, the level set method starts from the definition of an initial curve in the domain of the image. In our case, the initial curve on the SAR images is placed in the background zone, so that it surrounds the object of interest. The evolution of the initial curve is determined by a speed function, of fundamental importance to achieve a good segmentation.

In this section we present an combined speed obtained by the contributes of the two speed functions, here introduced and compared. The first, called average-based speed, is a function based on modeling the intensity of image by a Gamma distribution. The second, called gradient-based speed, is a function based on the computation of image gradient.

In this work the goal of the segmentation process is to extract two types of regions R_i $i \in \{1, 2\}$) representing objects and background.

Let be $I(\mathbf{x}(t))$ the SAR image intensity which we model by a Gamma distribution. After some probabilistic considerations and algebraic manipulations we obtain the average-based speed given

$$v(\mathbf{x}(t)) = \frac{d\gamma}{dt} = -\ \log \mu_{R_1} + \frac{I(\mathbf{x}(t))}{\mu_{R_1}} - \log \mu_{R_2} - \frac{I(\mathbf{x}(t))}{\mu_{R_2}} + \lambda k \qquad (2)$$

where a λk is a regularization term, with λ a positive real constant and k the mean curvature function, μ_{R_i} is the mean intensity given

$$\mu_{R_i} = \frac{\int_{R_i} I(\mathbf{x}(t))d\mathbf{x}}{a_{R_i}} \tag{3}$$

and where the area a_{R_i} is given

$$a_{R_i} = \int_{R_i} d\mathbf{x}$$

The implementation of the level set method with the speed based on regions means detects the object with more precision in terms of pixels than the next one since the image is not dealt with filters for noise reduction. We observed that the best result is obtained by locating the initial curve as nearly as possible to the region of interest, so that the final result depends on the position of the starting curve.

It is well known that in images corrupted by strong noise, the computation of gradient could detect false edges. Because the SAR images are affected by speckle noise, they are pre-processed by means of the SRAD algorithm which is an extension of Perona-Malik algorithm [6,7]

$$\begin{cases} \frac{\partial I(\mathbf{x}(t))}{\partial t} = \nabla \cdot [c(q)\nabla I(\mathbf{x}(t))] \\ I(\mathbf{x}(0)) = I_0 \end{cases} \tag{4}$$

where the diffusion coefficient is

$$c(q) = \frac{1}{1 + [q^2(\mathbf{x}(t)) - q_0^2(t)]/[q_0^2(t)(1 + q_0^2(t))]}$$

or

$$c(q) = exp\ \left\{-[q^2(\mathbf{x}(t)) - q_0^2(t)]/[q_0^2(t)(1 + q_0^2(t))]\right\}$$

and $q(\mathbf{x}(t))$ is named instantaneous coefficient of variation and $q_0(t)$ is the speckle scale function. The speckle scale function $q_0(t)$ effectively controls the amount of smoothing applied to the image by SRAD.

We observe that in the case of N-looks SAR image, we can assume $q_0 = \frac{1}{\sqrt{N}}$.

The gradient-based speed is computed on the filtered image

$$v(\mathbf{x}(t)) = -\frac{1}{1 + |\nabla I'(\mathbf{x}(t))|^2} - \lambda k$$

where $I'(\mathbf{x}(t))$ is the image $I(\mathbf{x}(t))$ filtered by SRAD, k is the curvature and $\lambda \in (0,1)$ is a constant. So, the speed term is defined in such a way that the curve proceeds rather fast in low gradient zones, while it wades through to high gradient ones. This strategy allows the contour to propagate until it achieves the limits of the object of interest in the image and then goes slowly close to those limits. The implementation obtained by the speed based on image gradient is less accurate in terms of pixels than the previous one, because it works on the filtered image and not on the original one. However, this last approach is independent from the position of the initial curve.

The new velocity is constituted by the mean of the two terms corresponding to the average-based speed and gradient based speed and it is given by a following expression:

$$v(\mathbf{x}(t)) = -\frac{1}{2} \, \log \frac{\mu_{R_2}}{\mu_{R_1}} + I(\mathbf{x}(t)) \frac{\mu_{R_1} - \mu_{R_2}}{\mu_{R_1}\mu_{R_2}} - \frac{1}{2} \, \frac{1}{1 + |\nabla I'(\mathbf{x}(t))|^2} - \lambda k. \quad (5)$$

As it is possible to see in the (5) the second term is computed on the image filtered by SRAD $I'(\mathbf{x}(t))$, while the first term is computed on the original image $I(\mathbf{x}(t))$.

The procedure developed by using the combined speed improves the results obtained by the two speeds separately and it saves the property to be independent of the position of the initial curve.

4   Experimental Results

To validate the efficiency of the proposed approach, the results obtained applying the combined procedure are compared with ones obtained employing the segmentation process based on the two single speed functions. We tested the procedure on synthetic SAR images to have an exact reference of the contours to detect. Since the location of the edges is not known in the real images and moreover there are not benchmark ones, tests have been synthesized from an original image without noise copying SAR patterns. These tests (150×150 pixels) are shown in Figures 1 (a), (b) and (c).

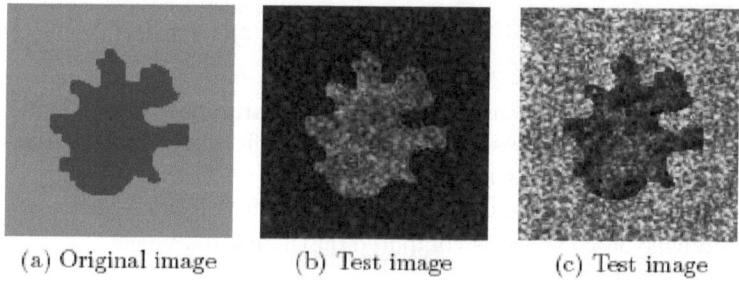

(a) Original image   (b) Test image   (c) Test image

Fig. 1. Test images

Figures 2 (a), (b), (c), (d) show the initial curves and the corresponding results obtained applying the procedure, using the average-based speed, to the test of Figure 1 (b). We obtain different results starting from different initial contours. In Figures 2 (e), (g) we show the initial contours and the corresponding results in (f) and (h), based on the gradient-based speed. In this case the segmentation results do not depend on the initial contours. The final results have been obtained applying the procedure to the smoothed images, by SRAD filter. However, results obtained from both these processes are wrong.

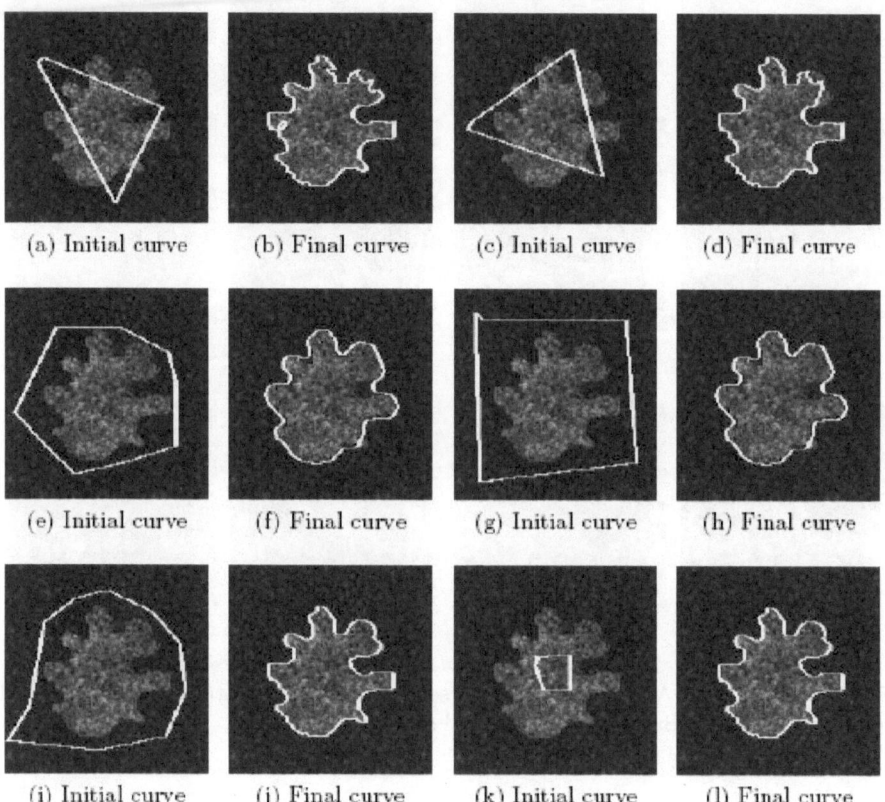

Fig. 2. Subplot(a-d) Average-based speed segmentation. Subplot(e-h) Gradient-based speed segmentation. Subplot(i-l) Combined speed segmentation.

The last four Figures show the results, based on the combined speed. In these images the final contours of (j), (l) have been computed starting by both internal and external initial contours of (j), (k). An important peculiarity of this procedure is that different initial contours lead to the same results.

Analogously, Figure 3 underlines the results obtained applying the segmentation to the image in Figure 1 (c). In particular, Figures 3 (a), (b), (c), (d) show the initial curves and the corresponding results, using the average-based speed. In this case different initial contours lead to different results. Initial contours and the corresponding results, obtained by the gradient-based speed, are shown in Figures 3 (e), (g) and (f), (h) respectively. Here the segmentation result is the

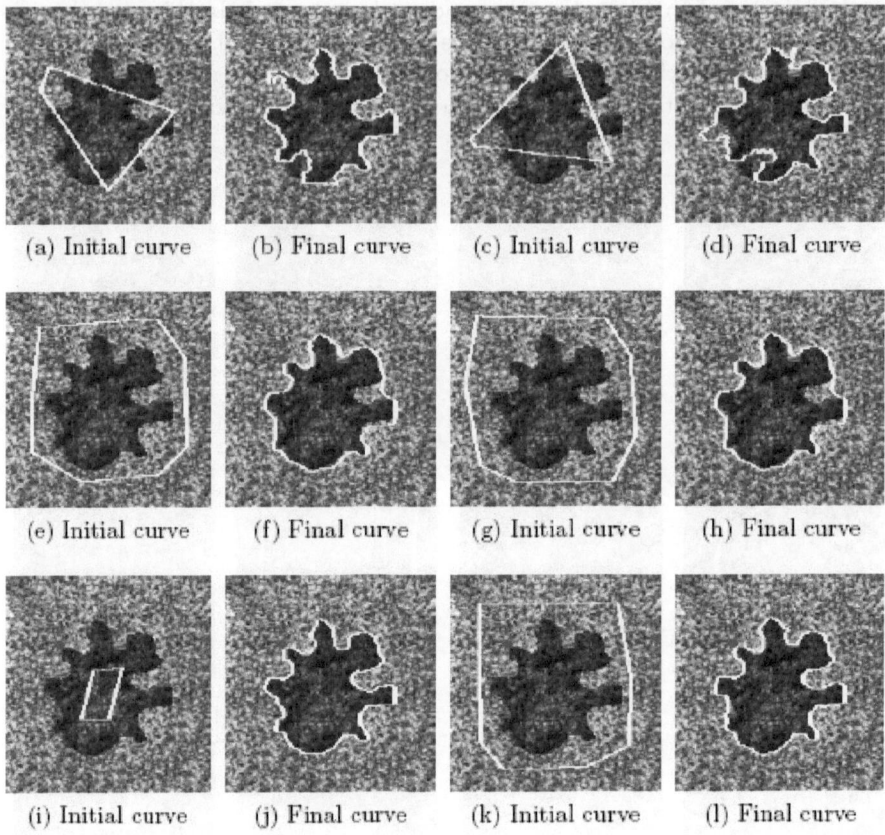

(a) Initial curve   (b) Final curve   (c) Initial curve   (d) Final curve

(e) Initial curve   (f) Final curve   (g) Initial curve   (h) Final curve

(i) Initial curve   (j) Final curve   (k) Initial curve   (l) Final curve

Fig. 3. Subplot(a-d) Average-based speed segmentation. Subplot(e-h) Gradient-based speed segmentation. Subplot(i-l) Combined speed segmentation.

same even starting from different initial contours. The final results have been obtained applying the procedure to the filtered images. However, results obtained from both these processes are wrong. The last line of Figure 3 shows the results, based on the combined speed. In these images the final contours of (j), (l) have been computed starting from both internal and external initial contours of (j), (k). In this case the same results are obtained starting from different initial contours.

In Figure 4 we present the final contours in (b) and (d) obtained applying segmentation, based on the combined speed, to synthetic images, constituted by disjoint regions, characterized by the same gray levels in (a) and different gray levels in (c).

The segmentation process has also been applied to real images, acquired during ERS2 mission, to extract the contours of the coastlines. In particular, Figure 5 (a) shows the coastline obtained by a 500 × 700 pixels SAR image representing the Capraia Island. The convergence is achieved after 3340 iterations.

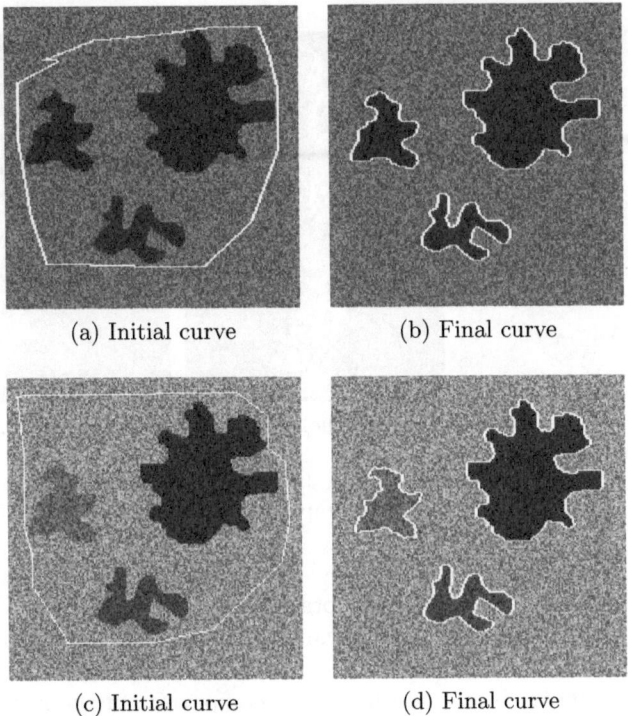

(a) Initial curve   (b) Final curve

(c) Initial curve   (d) Final curve

Fig. 4. Synthetic image constituted by regions of same gray level (a), regions of different gray level (c), results (b) (d)

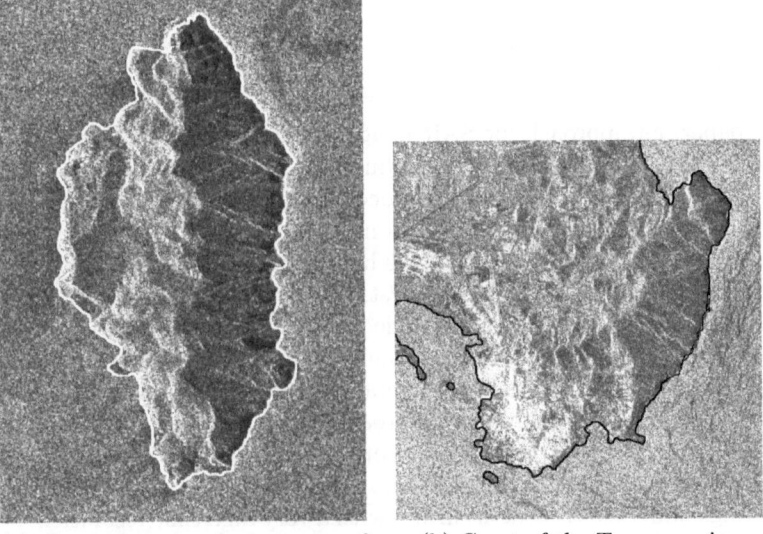

(a) Coast detection from image of   (b) Coast of the Tuscan region
Capraia island

Fig. 5. Original images

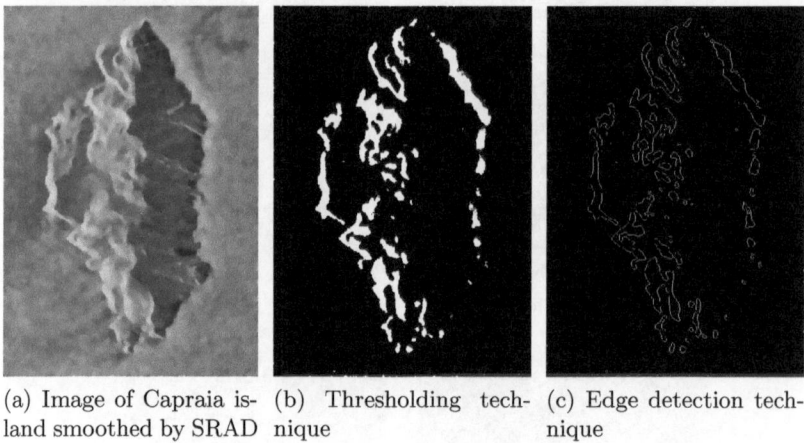

(a) Image of Capraia is- (b) Thresholding tech- (c) Edge detection tech-
land smoothed by SRAD nique   nique

Fig. 6. Visual and qualitative comparison with result of Figure 5 (a)

Figure 5 (b) shows the the coastline obtained by a 750×750 pixels SAR image representing part of the coast of Tuscany. The result shown is obtained after 6500 iterations.

Now we present a visual and qualitative comparison of the results obtained by using segmentation traditional methods. In Figure 6 we show the images produced by applying the technique of segmentation based on the Otzu algorithm and one of edge detection based on the algorithm of Canny. The used images have been smoothed by the SRAD technique.

5   Conclusions

In this paper, an approach for SAR image segmentation based on the level set method has been proposed. The procedure has been applied on both synthetic and real SAR images. Segmentation process proposed by us allows the obtained result is independent of the initial location of the curve and moreover, it automatically stops when the curve achieves the boundary. Two distinct speed evolution functions have been examined. The first, based on the mean intensities of the regions, does not need to reduce speckle noise; the second, based on the image gradient, takes into account the problem to filter speckle noise by the SRAD technique. Finally our proposal combining the previous functions improves the results obtained by two individual approaches.

In the future we plan to apply the proposed methodology to images obtained from the constellation of satellites Cosmo Sky-Med. These satellites are useful in monitoring changes in the Earth surface with a very high time resolution, because they can observe the same area several times a day in all weather conditions. Moreover we are interested to extend the procedure to detect the components of the cryosphere such as frozen soil, snow, sea ice, ice sheets from SAR images.

Acknowledgments. The authors would like to thank Maria Mercede Cerimele for her contribution to this article. Then the authors wish to thank the Consortium for Informatics and Telematics "Innova" of Matera, which has provided the PRI images of ERS Mission.

References

1. Sethian, J.A.: Level Set Methods and Fast Marching Methods. Cambridge University Press, Cambridge (1999)
2. Sethian, J.A.: Evolution, implementation and application of level set and fast marching methods for advancing front. Journal of Computational Physics 169, 503–555 (2001)
3. Osher, S., Fedkiw, R.: Level Set Methods and Dynamic Implicit Surfaces. Springer, New York (2002)
4. Dellepiane, S., De Laurentiis, R., Giordano, F.: Coastline extraction from SAR images and a method for the evaluation of the coastline precision. Pattern Recognition Letters 25, 1461–1472 (2004)
5. Ben Ayed, I., Mitiche, A., Belhadj, Z.: Multiregion level-set partitioning of synthetic aperture radar images. IEEE Trans. Pattern Analysis and Machine Intelligence 27, 793–800 (2005)
6. Yu, Y., Acton, S.T.: Speckle reducing anisotropic diffusion. IEEE Trans. on Image Processing 11, 1260–1270 (2002)
7. Perona, P., Malik, J.: Scale space and edge detection using anisotropic diffusion. IEEE Trans. Pattern Analysis and Machine Intelligence 12, 629–639 (1990)
8. Cerimele, M., Cinque, L., Cossu, R., Galiffa, R.: Coastline detection from SAR images by level set model. In: Foggia, P., Sansone, C., Vento, M. (eds.) ICIAP 2009. LNCS, vol. 5716, pp. 364–373. Springer, Heidelberg (2009)

Adaptive Model for Object Detection in Noisy and Fast-Varying Environment

Dung Nghi Truong Cong[1], Louahdi Khoudour[1],
Catherine Achard[2], and Amaury Flancquart[1]

[1] IFSTTAR, LEOST, F-59650 Villeneuve d'Ascq, France
[2] UMPC Univ Paris 06, ISIR, UMR 7222, France
truong@ifsttar.fr, louahdi.khoudour@ifsttar.fr,
catherine.achard@upmc.fr, amaury.flancquart@ifsttar.fr

Abstract. This paper presents a specific algorithm for foreground object extraction in complex scenes where the background varies unpredictably over time. The background and foreground models are first constructed by using an adaptive mixture of Gaussians in a joint spatio-color feature space. A dynamic decision framework, which is able to take advantages of the spatial coherency of object, is then introduced for classifying background/foreground pixels. The proposed method was tested on a dataset coming from a real surveillance system including different sensors installed on board a moving train. The experimental results show that the proposed algorithm is robust in the real complex scenarios.

Keywords: Background subtraction, foreground segmentation, mixture of Gaussians, spatio-color feature space.

1 Introduction

Detecting foreground objects from a video sequence is a critical task in many computer-vision applications. It can be considered as the basic level of processing to achieve higher level vision tasks. Even though there exist numerous algorithms in the literature, foreground object detection in complex environments, including non-stationary background motion, illumination variations, and camera vibration, is still far from being completely solved.

As surveyed in [1], there exists a vast literature on background subtraction. Most proposed methods are based on the pixel-level background model, which construct a background representation for each pixel location. One of the simplest approaches consists in modeling each pixel intensity with a single Gaussian distribution [2]. However, such a model is unsuitable for noisy sequences and multi-modal scenes. More complex models are based on a mixture of Gaussians [3], or a probability density function estimated by kernel function [4]. The background can also be modeled by a group of clusters which represent a compressed form of background model [5].

In contrast to pixel-wise approach, interest has grown recently in region-level methods which employ regional models representing spatial relationships between pixels. Sheikh et al. [6] used Kernel Density Estimation to build full background model as a single distribution, in conjunction with a MAP-MRF decision framework. In [7], Heikkila et al. used a group of weighed adaptive local binary pattern histograms to capture the

G. Maino and G.L. Foresti (Eds.): ICIAP 2011, Part I, LNCS 6978, pp. 68–77, 2011.

background statistics of each image block, and produced a coarse detection of foreground object. Chen et al. [8] extended this idea to obtain more detailed foreground by using a contrast histogram to describe each block. More recently, Dickinson et al. [9] modeled the background as an adaptive mixture of Gaussians in color and space, and used this model to probabilistically classify new pixels observations.

In this article, we consider the problem of foreground object detection in a complex environment where the background varies unpredictably over time. This work is carried out in the framework of the BOSS European project (on BOard wireless Secured video Surveillance), whose objective is to set up an onboard surveillance system. Indeed, the complex environments inside a moving train make the detection task extremely difficult. Therefore, in order to deal with such particular problems, we propose an approach based on an adaptive spatio-colorimetric background and foreground model coupled with a dynamic decision framework. The proposed method has three novel contributions. Firstly, in order to handle multi-modal uncertainties of the background, a joint spatio-colorimetric region based representation is employed to model the observed scene. The statistical regions of the background, which share common homogeneity properties, are modeled by using an adaptive mixture of Gaussians in a five-dimensional spatio-colorimetric feature space. Secondly, both background and foreground are modelized in order to better distinguish foreground and background pixels. Thirdly, instead of directly applying a threshold to classify background/foreground pixels, we propose a dynamic decision framework based on cellular automata which enforces the spatio-colorimetric context in the detection process.

The outline of the paper is as follows: after this introduction, we present in Section 2 the proposed approach to extract foreground objects. Section 3 presents global performances of the proposed system on different real datasets. Finally, in Section 4, conclusions and important short-term perspectives are given.

2   The Proposed Approach

2.1   Modeling the Background

The initial representation of the background is constructed from the first frame of the sequence by using a region merging technique [10] coupled with an adaptive mixture of Gaussians. The homogeneous regions of the observed scene are first extracted by iteratively combining smaller pixels or regions sharing homogeneous color properties.

Let I be the observed image containing N pixels; (p, p') be a couple of adjacent pixels in 4-connexity and A_I be the set of these couples. We first compute the local gradient between each couple of pixels defined as:

$$g(p, p') = \max_{c \in \{R,G,B\}} \overline{R_p(p')}_c - \overline{R_{p'}(p)}_c \tag{1}$$

where $R_p(p')$ is the set of neighborhood pixels of p' which satisfies the condition: $R_p(p') = \{q \in I : \|q - p'\|_1 < \delta \ \& \ \|q - p'\|_1 < \|q - p\|_1\}$ (δ is a predefined radius depending on the noise corruption of images; it is set to $\delta = 5$ in our experimentations), $\overline{R_p(p')}_c$ is the mean color value of channel c of all pixels belonging to $R_p(p')$. The principle of region merging process is that the couples of A_I are first sorted in increasing order of $g(p, p')$. For each couple of pixels $(p, p') \in A_I$, let $r(p)$ and $r(p')$ be the current regions to which pixels p and p' belong respectively. These two regions are merged if the following condition is verified:

$$\overline{r(p')}_c - \overline{r(p)}_c \leq \kappa \quad \frac{1}{N_{r(p)}} + \frac{1}{N_{r(p')}} , \quad \forall c \in \{R, G, B\} \tag{2}$$

where $\overline{r(p)}_c$ is the mean color value of the channel c of region $r(p)$, $N_{r(p)}$ is the number of pixels of region $r(p)$ and κ is a parameter defined as:

$$\kappa = C \quad \frac{2 \log(N)}{\Phi} \tag{3}$$

where C is the maximum value of color space, N is the number of pixels of the observed image and Φ is a parameter modifying the coarseness of the segmentation.

The region merging process is resumed in Algorithm 1.

Algorithm 1. Region merging algorithm

---

1 **Initialization**: the set A_I and the local gradient value of each couple $g(p, p')$.

2 Sorting the couples of A_I in increasing order of $g(p, p')$

3 **for** *each couple* $(p, p') \in A_I, r(p) \neq r(p')$ **do**

4   **if** $\overline{r(p')}_c - \overline{r(p)}_c \leq \kappa \quad \dfrac{1}{N_{r(p)}} + \dfrac{1}{N_{r(p')}} \quad \forall c \in \{R, G, B\}$ **then**

5     $\lfloor$ merging $r(p)$ and $r(p')$

After this region merging procedure, the observed scene is segmented into K_B homogeneous regions. Each region is now modeled by a Gaussian distribution in the joint spatio-colorimetric feature space $\mathbf{x} = [x, y, R, G, B]^T$ where x and y are spatial coordinates in the two-dimensional image and R, G, B are color coordinates:

$$\eta(\mathbf{x}|\mu, \Sigma) = \frac{\exp -\frac{1}{2}(\mathbf{x} - \mu)^T \Sigma^{-1}(\mathbf{x} - \mu)}{(2\pi)^5 |\Sigma|} \tag{4}$$

Here μ and Σ are the mean vector and covariance matrix estimated on the pixels belonging to the current region. The background model is finally defined by:

$$f(\mathbf{x}|BG) = \sum_{i=1}^{K_B} w_i \eta(\mathbf{x}|\mu_i, \Sigma_i) \tag{5}$$

where $w_i = N_i/N$ is the weight of the ith component, N_i is the number of pixels of the region i.

2.2   Modeling the Foreground

Since foreground objects tend to have smooth motion from frame to frame, the temporal persistence property is a powerful tool to increase the accuracy of object detection. Here, we propose to model the foreground objects in order to employ simultaneously background and foreground models to improve the detection.

The foreground model is initialized as a uniform function. Once a foreground region is detected, the foreground model is constructed in the same manner as the background one and is expressed by:

$$f\left(\mathbf{x}|FG\right) = \alpha + (1 - \alpha) \sum_{i=1}^{K_F} w_i^F \eta\left(\mathbf{x}|\mu_i^F, \Sigma_i^F\right) \tag{6}$$

where α is a constant which yields robustness when foreground is not observed ($\alpha < 0.5$), K_F is the number of components of the foreground model.

2.3   Foreground Object Segmentation

In this section, we propose a particular algorithm for foreground object segmentation based on the principle of cellular automata introduced by von Neumann [11]. The idea of this algorithm is that each pixel p is considered as a cellular automaton characterized by a triplet (l_p, N_p, Δ), where l_p and N_p are respectively the label and the set of neighborhood pixels of the current pixel p, Δ is the local transition function. The pixel label at instant $k + 1$ is estimated based on the states of the neighborhood pixels at instant k and the local transition rule.

Each pixel p of the new captured frame is first classified into one of three classes (foreground, background or undefined) based on the likelihood ratio $\Gamma = -\log \dfrac{f\left(\mathbf{x}_p|BG\right)}{f\left(\mathbf{x}_p|FG\right)}$. The label of pixel p is defined as:

$$l_p = \begin{cases} -1 \ (BG) & \text{if } \Gamma < T_{BG} \\ 1 \ (FG) & \text{if } \Gamma > T_{FG} \\ 0 & \text{otherwise} \end{cases} \tag{7}$$

where T_{BG} and T_{FG} are two parameters a priori defined for classifying foreground and background pixels.

The confidence score of each pixel is also estimated by using the probability of observing a background/foreground pixel:

$$\begin{cases} C_p = f\left(\mathbf{x}_p|BG\right) & \text{if } l_p = -1 \\ C_p = f\left(\mathbf{x}_p|FG\right) & \text{if } l_p = 1 \\ C_p = 0 & \text{if } l_p = 0 \end{cases} \tag{8}$$

Algorithm 2 describes the foreground object segmentation procedure.

Algorithm 2. Foreground object segmentation

1 $k = 0$: $l_p^0 = l_p$ and $C_p^0 = C_p$ for all $p \in I$
2 **while** *not converged* **do**
3   **for** *each pixel* $p \in I$ **do**
4          $l_p^{k+1} = l_p^k$, $C_p^{k+1} = C_p^k$
5   **for** *each pixel* $q \in N(p)$ **do**
6   **if** $\Delta(p, q).C_q^k > C_p^k$ **then**
7                  $l_p^{k+1} = l_q^k$
8                  $C_p^{k+1} = \Delta(p, q).C_q^k$

9      $k = k + 1$

Here, the local transition function Δ is defined as:

$$\Delta(p, q) = 1 - \exp\left(\frac{-\beta}{\varepsilon + \|\mathbf{I}_p - \mathbf{I}_q\|_2}\right) \tag{9}$$

where β and ε are the predefined parameters, $\mathbf{I}_p$ and $\mathbf{I}_q$ are the color vectors of pixels p and q.

Thus, instead of applying a single threshold to classify background/foreground pixels, we try to exploit the spatial coherency of object in order to obtain an optimal segmentation. Each pixel is represented by its confidence score. The higher the confidence score C_p, the stronger the influence on the neighborhood pixels. If two neighbor pixels have similar color, $\Delta(p, q)$ is big and the label of pixel with lower score will be replaced.

2.4   Updating the Background Model

The background model is updated by first assigning the new background pixels to their corresponding components, and then re-estimating the component parameters. The pixel $\mathbf{x}$ is assigned to component C if:

$$C = \arg\max_i \{\eta(\mathbf{x}|\mu_i, \Sigma_i)\} \tag{10}$$

Let μ_i^* and Σ_i^* be the mean vector and covariance matrix estimated from the N_i^* new pixels assigned to component i. The parameters of component i are re-estimated with:

$$w_i^t = \frac{w_i^{(t-1)}N + N_i^*}{N + N^*}$$

$$\mu_i^t = \frac{w_i^{(t-1)}N\mu_i^{(t-1)} + N_i^*\mu_i^*}{Nw_i^{(t-1)} + N_i^*}$$

$$\Sigma_i^t = \frac{w_i^{(t-1)}N\Sigma_i^{(t-1)} + N_i^*\Sigma_i^*}{Nw_i^{(t-1)} + N_i^*} - \mu_i^t[\mu_i^t]^T + \frac{w_i^{(t-1)}N\mu_i^{(t-1)}\left[\mu_i^{(t-1)}\right]^T + N_i^*\mu_i^*[\mu_i^*]^T}{Nw_i^{(t-1)} + N_i^*}$$

$$\tag{11}$$

Unlike the background, the foreground model is reconstructed by using the new extracted foreground objects in the same manner as initializing the background model. Thus, the foreground model adapts rapidly from frame to frame, which makes the detection task in the next frame more robust.

Figure 1 presents the detection results for an image of the sequence. Images 1(b) and 1(c) represent the likelihood maps for both background and foreground while image 1(f) is the final result of detection.

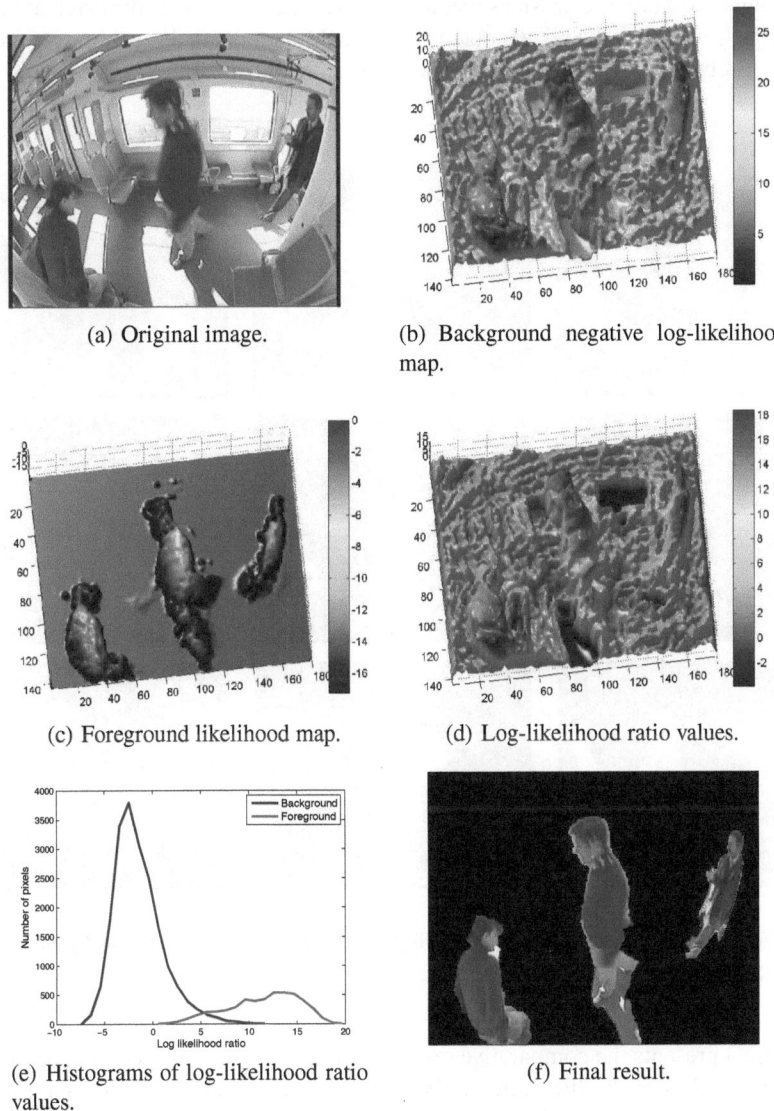

(a) Original image.

(b) Background negative log-likelihood map.

(c) Foreground likelihood map.

(d) Log-likelihood ratio values.

(e) Histograms of log-likelihood ratio values.

(f) Final result.

Fig. 1. Different steps of the proposed algorithm for foreground object extraction

3   Results and Discussion

The performance of the proposed method is evaluated using the real dataset collected by the cameras installed on board a moving train in the framework of the BOSS European project [12]. This dataset is really difficult, since the captured video is influenced by many factors including fast illumination variations and non-static background due to the movement of the train, reflections, vibrations of the cameras...

Figure 2 illustrates the effectiveness of the proposed method in comparison with the well-known Mixture of Gaussians method (GMM). The first row is the original images, the second row shows the results obtained by GMM method, and the third row presents the results obtained by the proposed method. Note that no post-processing is used in the results.

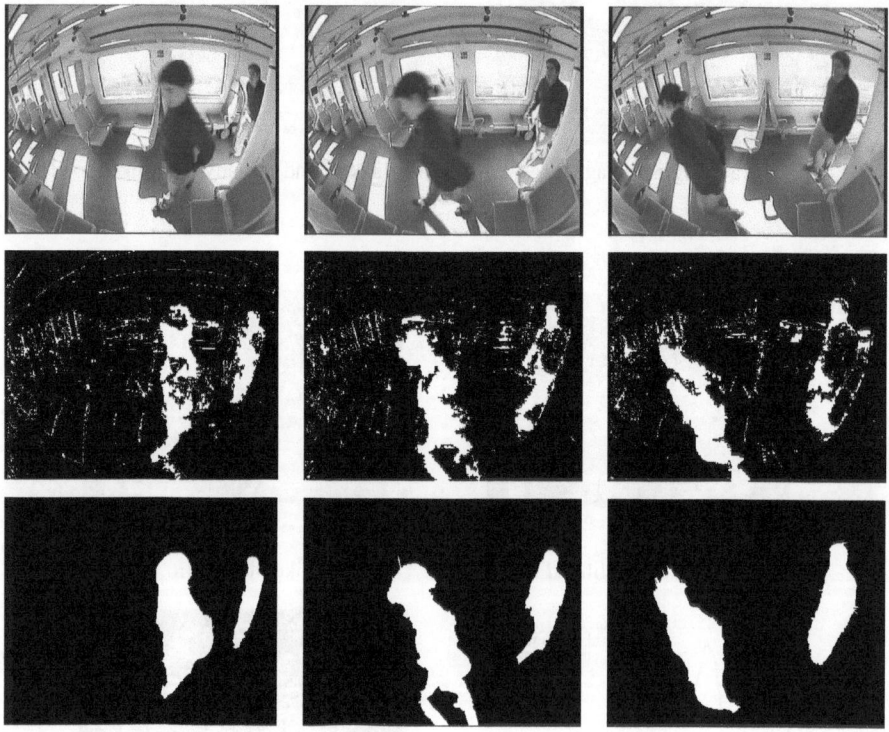

Fig. 2. Foreground object extraction results. Top to bottom: original images, GMM method, proposed algorithm.

Figure 3 presents the comparative results obtained by the proposed method and two other approaches of the literature: a pixel-based approach using GMM and a region-based method proposed by Sheikh and Shah [6]. We can notice that the results obtained by GMM are very noisy due to sudden illumination changes of the scene, while the detection accuracy of two region-based methods is still high.

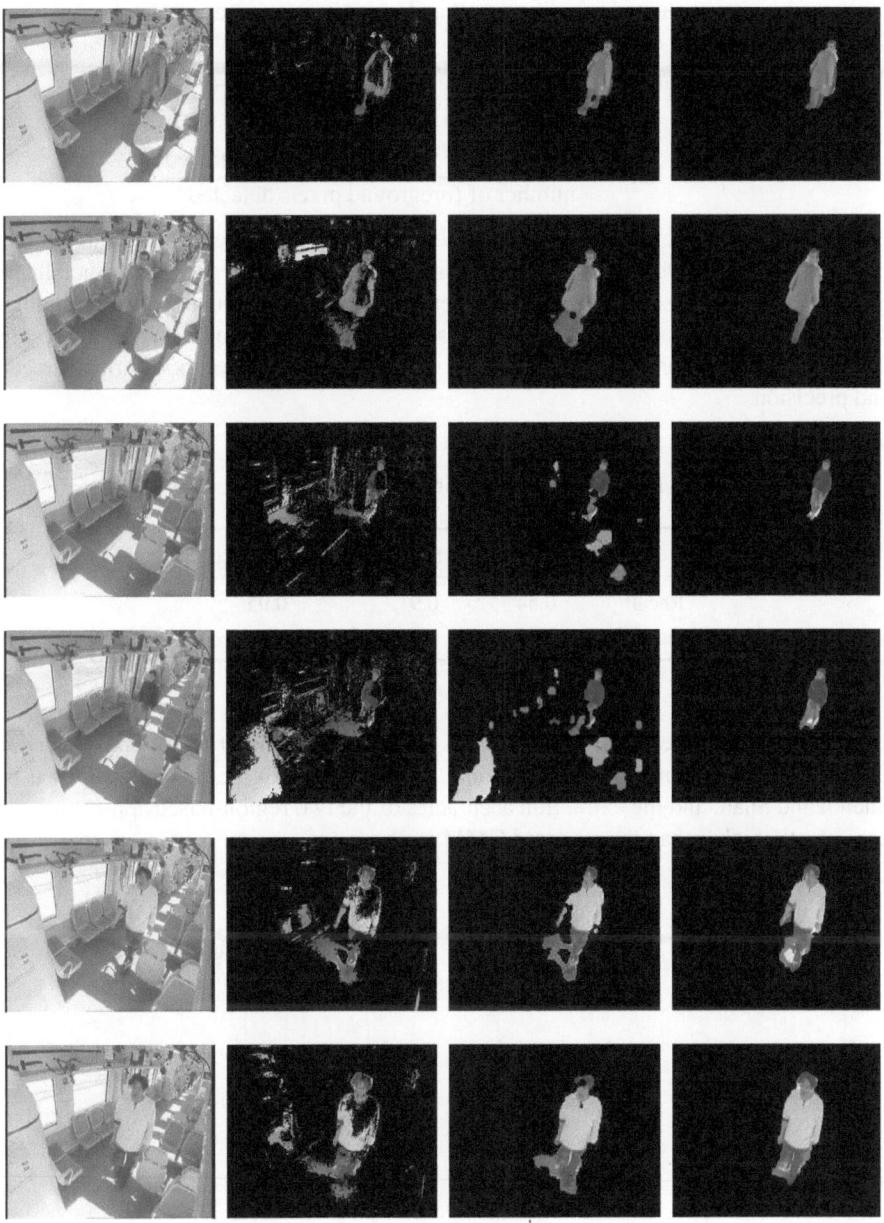

Fig. 3. Foreground object extraction results. Left to right: original images, GMM method, method proposed by Sheikh and Shah, our proposed algorithm.

In order to perform a quantitative analysis of the proposed approach, we have manually segmented 700 frames of a long sequence illustrated in Figure 3. The performances of the system are evaluated by using recall and precision measurements, where

$$\text{recall} = \frac{\text{number of true foreground pixels detected}}{\text{number of true foreground pixels}}$$

$$\text{precision} = \frac{\text{number of true foreground pixels detected}}{\text{number of foreground pixels detected}}$$

Table 1 presents the evaluation results in terms of recall and precision of three methods: the GMM algorithm with optimal parameters, the method proposed by Sheikh and Shah, and our proposed approach. In order to make a fair comparison, morphological operations are also used in the tests of standard GMM method. Clearly, the results demonstrate that the proposed approach obtains best performance in both terms of recall and precision.

Table 1. Comparative results in terms of recall and precision

	GMM	SS05	Proposed approach
Recall	0.54	0.91	0.95
Precision	0.73	0.91	0.94

Figure 4 shows the per-frame detection accuracy in terms of recall and precision. One can notice that our method is slightly more robust than the method proposed by Sheikh and Shah, and the extraction accuracies of the two region-based approaches are consistently higher than the standard GMM method.

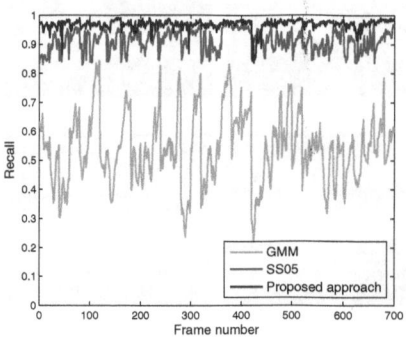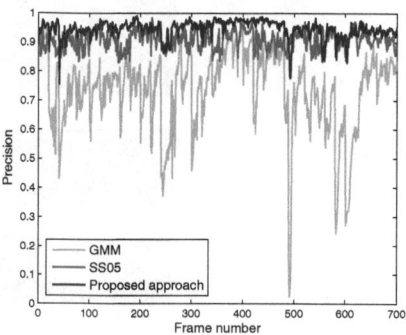

Fig. 4. Recall and precision curves obtained from the tested sequence

4   Conclusion

In this paper, we have presented a specific algorithm for foreground object extraction in complex scenes with non-stationary background. Several originalities are introduced to manage this difficult problem. A region-wise model of background and foreground is first proposed by using an adaptive mixture of Gaussians in a joint spatio-colorimetric feature space. A great robustness is introduced thanks to the simultaneous exploitation of background and foreground models. A dynamic decision framework, which is able to take advantages of both spatial and temporal coherency of object, is introduced for classifying background and foreground pixels. The proposed method was tested on a dataset coming from a real surveillance system including different sensors installed on board a moving train. The experimental results show that the proposed algorithm is robust in these real difficult scenarios.

In order to further improve the performance of the system and to reduce false detections caused by shadows, a normalized color space could be used instead of the RGB space. Moreover, several features (texture, edge,...) should be considered and integrated to the system to manage the cases where foreground and background colors are very similar.

References

1. Elhabian, S.Y., El-Sayed, K.M., Ahmed, S.H.: Moving object detection in spatial domain using background removal techniques - state-of-art. Recent Patents on Computer Science 1(1), 32–54 (2008)
2. Wren, C.R., Azarbayejani, A., Darrell, T., Pentland, A.P.: Pfinder: Real-time tracking of the human body. IEEE Transactions on Pattern Analysis and Machine Intelligence 19(7), 780–785 (1997)
3. Stauffer, C., Grimson, W.E.L.: Adaptive background mixture models for real-time tracking. In: Proceedings of the IEEE Computer Society Conference on Computer Vision and Pattern Recognition, vol. 2, pp. 246–252 (1999)
4. Elgammal, A., Harwood, D., Davis, L.: Non-parametric model for background subtraction. In: Vernon, D. (ed.) ECCV 2000. LNCS, vol. 1843, pp. 751–767. Springer, Heidelberg (2000)
5. Kim, K., Chalidabhongse, T.H., Harwood, D., Davis, L.: Background modeling and subtraction by codebook construction. In: International Conference on Image Processing, vol. 5, pp. 3061–3064 (2004)
6. Sheikh, Y., Shah, M.: Bayesian modeling of dynamic scenes for object detection. IEEE Transactions on Pattern Analysis and Machine Intelligence 27(11), 1778–1792 (2005)
7. Heikkila, M., Pietikainen, M.: A texture-based method for modeling the background and detecting moving objects. IEEE Transactions on Pattern Analysis and Machine Intelligence 28(4), 657–662 (2006)
8. Chen, Y.T., Chen, C.S., Huang, C.R., Hung, Y.P.: Efficient hierarchical method for background subtraction. Pattern Recognition 40(10), 2706–2715 (2007)
9. Dickinson, P., Hunter, A., Appiah, K.: A spatially distributed model for foreground segmentation. Image and Vision Computing 27(9), 1326–1335 (2009)
10. Nock, R., Nielsen, F.: Statistical region merging. IEEE Transactions on Pattern Analysis and Machine Intelligence 26(11), 1452–1458 (2004)
11. Von Neumann, J., Burks, A.W.: Theory of Self-Reproducing Automata. University of Illinois Press Champaign, IL (1966)
12. http://www.multitel.be/boss

Shadow Segmentation Using Time-of-Flight Cameras

Faisal Mufti[1] and Robert Mahony[2]

[1] Center for Advanced Studies in Engineering
faisal.mufti@ieee.org
[2] Australian National University
robert.mahony@anu.edu.au

Abstract. Time-of-flight (TOF) cameras are primarily used for range estimation by illuminating the scene through a TOF infrared source. However, additional background sources of illumination of the scene are also captured in the measurement process. This paper uses radiometric modelling of the signals emitted from the camera and a Lambertian reflectance model to develop a shadow segmentation algorithm. The proposed model is robust and is experimentally verified using real data.

Keywords: Time-of-flight, Radiometric Modelling, Shadow Segmentation, Reflectance Modelling.

1 Introduction

The presence of shadows due to lighting conditions complicates the process of shape and behaviour estimation of objects, especially where there is a significant association between the non-background points and shadow. Shadow segmentation algorithms are an important component in solutions to many computer vision scenarios such as video surveillance, people tracking and traffic monitoring [3, 19]. Background subtraction methods and colour space techniques are widely exploited for shadow segmentation [4, 8]. The drawback with such techniques is that they tend to under perform when there is insufficient colour information or when the scene is dynamic. Similarly, model based adaptive and Bayesian methods [1] suffer from same problems.

Conventional CCD camera are susceptible to dynamic range and illumination conditions. On the other hand, research in imaging devices in recent years has lead to development of range sensing cameras, especially 3D time-of-flight (TOF) [12] cameras, and is used in a number of fields; such as; for example, detection and recognition [5], 3D environment reconstruction [11] and tracking [15], etc. In general, 3D TOF cameras work on the principle of measuring time of flight of a modulated infrared light signal as phase offset after reflection from the environment and provide range and intensity data over a full image array at video frame rate [9].

This paper presents a novel algorithm for shadow detection using TOF camera technology by exploiting the additional measurement capability of a TOF camera compared to a standard CCD camera. The proposed algorithm is based on a

G. Maino and G.L. Foresti (Eds.): ICIAP 2011, Part I, LNCS 6978, pp. 78–87, 2011.

radiometric range model derived from TOF measurements and the background light source. We assume a statistical noise model of TOF measurements [16] and a Lambertian reflectance model to derive a radiometric range model that is independent of the coefficient of diffuse reflectivity of the environment. The radiometric range model is used to formulate a criterion to identify shadows and highlights using TOF camera data. For the purpose of this conference paper, we restrict our attention to the case of planar surfaces. More complex environment models are straight forward to develop based on the proposed approach. However, experiments show that the planar environment model is sufficient to obtain robust results for typical non-planar environments.

This paper is organised as follows: Section 2 describes TOF signal measurement, Section 3 describes Lambertian reflectance modelling from TOF perspective by exploiting TOF measurement and background light sources. In Section 4, the dependencies between measurement parameters of amplitude, range and intensity are used along with the reflectance model for sensor response to derive radiometric range model. Section 5 provides details of shadow segmentation based on radiometric range model using real data and is followed by its experimental verification.

2   Time-of-Flight Signal Measurement

Time-of-flight (TOF) sensors estimate distance to a target using the time of flight of a modulated infrared (IR) wave between the target and the camera. The sensor illuminates/irradiates the scene with a modulated signal of amplitude A (*exitance*) and receives back a signal (*radiosity*) after reflection from the scene with background signal offset I_o that includes non-modulated DC offset generated by TOF camera as well as ambient light reflected from the scene. The amplitude, intensity offset I and phase of a modulated signal can be extracted by demodulating the incoming signal $A_i = A\cos(\omega t_i + \varphi) + I$; ($t_i = i \cdot \frac{\pi}{2\omega}, i = 0, \ldots 3$) [9],

$$A := \frac{\overline{(A_3 - A_1)^2 + (A_0 - A_2)^2}}{2},$$

$$I := \frac{A_0 + A_1 + A_2 + A_3}{4},$$

$$\varphi := \tan^{-1} \frac{A_3 - A_1}{A_0 - A_2}.$$

With known phase φ, modulation frequency f_{mod} and knowledge of speed of light c, it is possible to measure the un-ambiguous distance r from the camera [17].

3   Reflectance Model

The measurement parameters of amplitude A, intensity I, and range r are not independent but depend on the reflectance characteristics of the scene [17]. In the following discussion we consider a near-field IR point source for the camera's active LED array, a far-field source for background illumination and ambient illumination. The primary source of illumination in TOF cameras is an IR source that produces a modulated IR signal offset and a non-modulated DC signal.

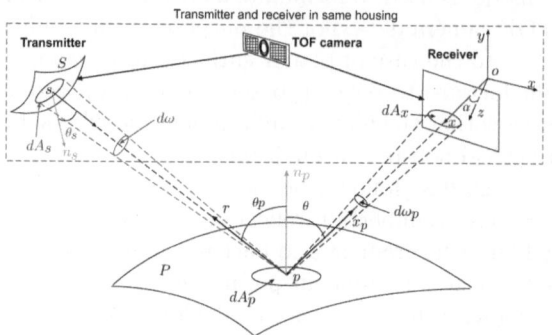

Fig. 1. Geometry of reflectance model for time-of-camera. Note that although the LED source and receiver of a physical TOF camera are co-located, it is difficult to provide a visualisation of this geometry. Here the source is shown separately to make is easier to see notation. However, in practice the directional vectors r and x_p are equal. Note that time variation (discussed in Section 2) of $A(s)$ does not need to be modelled as only the relative magnitude of $A(s)$ is of interest.

Let P be a Lambertian surface in space with n_p denoting the normal to each point $p \in P$ on the surface as shown in Figure. 1. Following the laws of radiometry [18] the amplitude of total radiance $= A(p)$ (called *radiosity*) leaving point p due to illumination by the modulated signal $A(s)$ is proportional to the *diffuse reflectance* or *albedo* $\rho_d(p)$ scaled by the cosine of arrival angle θ_p. In the present analysis, the LED point sources of the camera are part of the compact IR array of the TOF camera, and can be approximated by a single virtual modulated point source [7, p. 78] with the centre of illumination aligned with the optical axis of the camera [11]. In this case, the integration for illuminating sources can be written as a function of the exitance of a single point source at S as [7, p. 77] [17]

$$A(p) := \frac{1}{\pi}\rho_d(p)\frac{A(s)\cos\theta_p\cos\theta_s}{r^2}, \tag{1}$$

where θ_s is the angle between the normal to the source point $s \in S$ and the ray of the modulated IR signal reaching point p and r is the distance between source and the point p.

The irradiance of an image point x is obtained as

$$A(x) = \Upsilon A(p), \tag{2}$$

where $\Upsilon := \Upsilon(x)$ is the lens function [18] representing the vignetting due to aperture size and irradiance fall-off with cosine-fourth law.

The TOF camera IR source produces a DC signal from the same IR source LEDs. This signal will have the same reflectance model as has been derived for the modulated IR source (see (1)). The received signal $I_c(x)$ is given by [17]

$$I_c(x) = \Upsilon I_c(p). \tag{3}$$

The effect of this signal is an added offset to the modulated signal that provides better illumination of the scene.

For a point source $q \in Q$ that is far away compared to the area of the target surface, the exitance $I_b(q)$, does not depend on the distance from the source or the direction in which the light is emitted. Such a point source can be treated as constant [7, p. 76]. The radiosity perceived by a TOF image plane as a result of this IR source is given by [17]

$$I_b(x) = \frac{\Upsilon}{\pi} \rho_d(p) I_b(q) \cos \theta_q$$
$$= \Upsilon I_b(p). \tag{4}$$

where θ_q is the angle between normal to the surface point p.

Now consider an ambient background illumination of the scene i.e an illumination that is constant for the environment [7, p. 79] and produces a diffuse uniform lighting over the object [6, p. 273]. Let I_a be the intensity (called *exitance*) of the ambient illumination, then the received intensity $I_a(p)$ from a point p is expressed in an image plane as [17]

$$I_a(x) = \frac{\Upsilon}{\pi} \rho_a(p) I_a$$
$$= \Upsilon I_a(p), \tag{5}$$

where ρ_a is the *ambient reflection coefficient* which is often estimated empirically instead of relating it to the properties of a real material [6, p. 723]. Since it is an empirical convenience, for all practical purposes $\rho_a \approx \rho_d$.

4 Radiometric Range Model

From the principles of TOF camera (see Section 2) signals one knows that intensity component of TOF carries information for both, amplitude of the modulated signal and the background offset I_o [12]. The radiometric intensity of TOF camera is then

$$I := A + I_o. \tag{6}$$

The background offset I_o is composed of a DC offset I_c, due to the DC component of the illumination by the TOF camera LED array and background illumination that are modelled by an ambient illumination I_a and a background illumination I_b due to an infrared far field source present in the environment such as the Sun or other light source. Indexing the point p in the scene by the TOF receiving pixel x, one has

$$I_o(x) = I_c(x) + I_a(x) + I_b(x). \tag{7}$$

Dividing (6) by $A(x)$ and using the local shading model for IR signal and the illumination of point sources (2), (3), (4), (5), after substituting (7), one obtains

$$\frac{I(x)}{A(x)} = 1 + \frac{I_c(s)}{A(s)} + \frac{I_a r^2(x)}{A(s) \cos \theta_p \cos \theta_s} + \frac{I_b(q) \cos \theta_q r^2(x)}{A(s) \cos \theta_p \cos \theta_s}, \tag{8}$$

where $\theta_s := \theta_s(x)$ is a known function of pixel.

Define κ_a as the ratio of background ambient light I_a to modulated TOF IR source $A(s)$. Observe that κ_a does not depend upon scene or camera geometry and hence is a constant parameter over the full image array. Similarly, define κ_b as a constant ratio of far-field illumination I_b to the TOF IR source $A(s)$. Finally define κ_c as the ratio of TOF non-modulated IR source $I_c(s)$ and TOF modulated IR source $A(s)$. Since the two sources of illumination originating from the TOF camera IR LED source have the same ray geometry, then $\kappa_c(x)$ is a pixel dependent [17].

Thus, using the parameters $(\kappa_a, \kappa_b, \kappa_c, \theta_p, \theta_q)$ of sources, one obtains the radiometric relationship as

$$\boxed{\frac{I(x)}{A(x)} = 1 + \kappa_c(x) + \kappa_a \frac{r^2(x)}{\cos \theta_p \cos \theta_s} + \kappa_b \frac{\cos \theta_q r^2(x)}{\cos \theta_p \cos \theta_s}.} \tag{9}$$

Note that θ_s, is the angle measured from the camera, is stored in a look up table (projected angle of the IR beam of TOF camera with respect to each camera pixel) and angle θ_p is measurable from camera data. For any surface patch[1], it is possible to numerically compute an estimate of the angle $\theta_p := \theta_p(x)$ from the set of range measurements $r(x_i)$ associated with that patch based on an estimate of the normal vector to the surface [10].

The radiometric range model for a far-field point source of illumination depends upon the angle θ_q that is the angle between the normal to the surface and the direction of the far-field source. Hence, an estimate of the direction of background IR source (in terms of azimuth and elevation) is required and is a problem of source estimation in computer vision [2].

A specialized case of considerable practical interest is that of a planar surface. For a single planar surface it follows that θ_q is constant, while θ_p is nearly constant for a small field of TOF optical sensor over the surface, and the parameters (κ_a, κ_b) are constant over the image plane. As a result several parameters can be combined into a single constant κ_o. An approximate model is presented in [17] as

$$\kappa_o := \frac{\kappa_a}{\cos \theta_p} + \frac{\kappa_b \cos \theta_q}{\cos \theta_p} \approx \text{constant}. \tag{10}$$

As a consequence of (10), a pixel measurement of κ_o is given by

$$\kappa_o(x) = \frac{I(x)}{A(x)} - \kappa_c(x) - 1 \cdot \frac{\cos \theta_s}{r^2(x)}. \tag{11}$$

---

[1] The surface patch must be sufficiently large to be imaged by a small window of pixels.

A detailed analysis of κ_o provides a statistical distribution of $\kappa_o(x)$ for a planar surface [17] shown in Figure

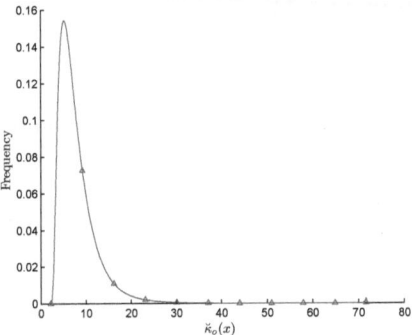

Fig. 2. Normalized histogram of $\check{\kappa}_o(x)$ of a flat surface. Pixels with $\check{\kappa}_o(x) \rightarrow \quad$ (due to amplitude and phase) have been scaled down to finite values. The heavy tail is associated with noisy data.

5 Shadow Segmentation

Shadow segmentation based on radiometric range data is a natural outcome of the radiometric framework proposed in Section 4. An important contribution of performing shadow segmentation using TOF cameras is the ability to provide an algorithm that can be used in dynamic scenes as well as for surfaces with insufficient colour information. A close examination of (2) and (6) reveals that an object in shadow does not effect the amplitude value received by the TOF camera. However the intensity varies with the background illumination or shadowing of the object. Based on this fact, it is possible to segment the scene into shadow and highlights. The proposed shadow segmentation model is derived from the radiometric model (9) using far-field background illumination parameter κ_b. Thus, one obtains a new $\kappa_b(x)$ for this application as

$$\kappa_b(x) := \frac{I(x)}{A(x)} - \kappa_c(x) - 1 \quad \frac{\cos\theta_s \cos\theta_p}{r^2(x)\cos\theta_q} - \frac{\kappa_a(x)}{\cos\theta_q}. \tag{12}$$

In terms of the measured value, one defines a measured value of $\check{\kappa}_b(x)$ as

$$\check{\kappa}_b(x) := \frac{I(x)}{A(x)} - \hat{\kappa}_c(x) - 1 \quad \frac{\cos\theta_s \cos\theta_p}{\check{r}^2(x)\cos\theta_q} - \frac{\check{\kappa}_a(x)}{\cos\theta_q}, \tag{13}$$

where $\hat{\kappa}_c(x) \in \mathbb{R}^2$ is an estimate of camera based pixel parameter for an entire image, since $\kappa_c(x)$ is scene independent and can be measured offline in a set of calibration experiments. The proposed algorithm uses a k-means clustering algorithm [14] on the full set of measured $\check{\kappa}_b(x)$ obtained for a full frame. Here the term involving $\check{\kappa}_a(x)$ is a scalar value for ambient illumination and is computed empirically assuming constant angle θ_q for a planar surface. The objective function $\Psi := \Psi(c)$ for $c = \{c_1, \cdots, c_k\}$ is defined as

$$\Psi(c) := \sum_{i=1}^{k} \sum_{x_n \in S_i} |\check{\kappa}_b(x_n) - c_i|. \tag{14}$$

The c_i values are the geometric centroids of the data points S_i, where x_n is a vector representing nth data point from radiometric range model $\check{\kappa}_b$ and $|\check{\kappa}_b(x_n) - c_i|$ is the L_1 norm. The minimisation of this problem for segmentation (partitions) of n data points into k disjoint subset S_i containing n_i data points is given by $\hat{c}$

$$\hat{c} = \arg\min_c(\Psi). \tag{15}$$

The proposed shadow segmentation based on $\kappa_b(x)$ value represents the ratio of background source of illumination to the camera source. The angle θ_q is assumed constant for planar cases and θ_p is relatively constant for the small field of view of a TOF camera. Under these assumption the probability distribution of $\check{\kappa}_b$ behaves like $\check{\kappa}_o$ (11).

Remark: *When shadow segmentation due to the Sun is considered then geographical knowledge of the camera pose and the Sun position can be used to estimate θ_p and θ_q given an estimate of the scene geometry [13].*

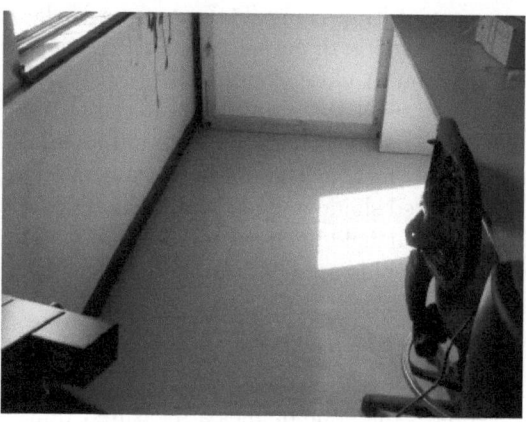

Fig. 3. CCD image of the setup taken from a 2D camera with TOF camera positioned in the lower left corner of the image

Experiments: A TOF camera was placed to capture an environment with object (flat floor) in sunlight and the remaining background appeared shadowed compared to bright light coming from the window (see Figure 3). The intensity image captured by the TOF camera is shown in Figure. 5(a). A histogram of

$\check{\kappa}_b(x)$ with three marked regions is shown in Figure 4. The central blue region corresponds to the highlighted area in image space. The region on the right hand side (in a red rectangle) corresponds to noisy values. Both the highlighted and the noisy regions produced significantly higher $\check{\kappa}_b(x)$ values compared to shadowed regions (higher numerator of the right hand side of (13) either due to $I \rightarrow \infty$ or noisy signal where $A \rightarrow 0$). The values on the left side (in a red elliptical) have two peaks. These two regions are associated with the angle approximation θ_q when applied over an entire frame including non-planar regions and θ_p (sharp corners and bends in scene with respect to camera viewing direction).

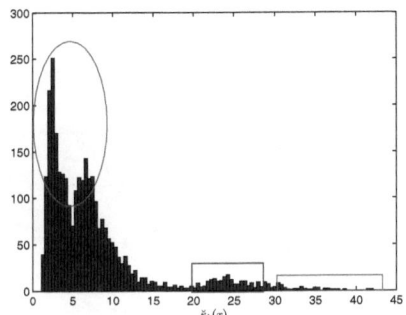

Fig. 4. Histogram of a $\check{\kappa}_b(x)$ for a single frame

In the initial experiment the algorithm was applied to a single planar surface. The shadow segmentation was performed using the k-means clustering. The $\check{\kappa}_b(x)$ data is segmented into two clusters of highlights and shadows and shows quite precise segmentation (see Figure 5(c)) with a precision of 99%. The proposed algorithm was also applied to an entire single frame and observed that the segmentation works reasonably well even when applied to a non-planar scene (see Figure 5(b)). However, there were certain regions of false positive due to failure of planar assumption (precision factor of 75%). These regions can either be segmented using more that two clusters or eliminated using further noise filtering [17] based on signal-to-noise ratio (SNR) estimation. Since in the case of highlight areas the estimated $\widehat{\text{SNR}}$ value is high while the noisy regions have a lower $\widehat{\text{SNR}}$ value, the image was further refined, using this two step approach as illustrated in Figure 5(d) and achieved a precision of 96%. The algorithm, when implemented in MATLAB on an Intel Core2 Duo 2.2GHz machine with 4GB RAM, performs region segmentation of a complete frame in less than a second.

A precise modelling of scene geometry (discussed in Section 4) for angle θ_q in the model (13) would result into a single region of shadow in a single step. Despite this simplification, the algorithm is effective in region segmentation based on illumination condition and unlike background subtraction and color cues, this method is independent of scene movement and works equally well on gray scale and range image data.

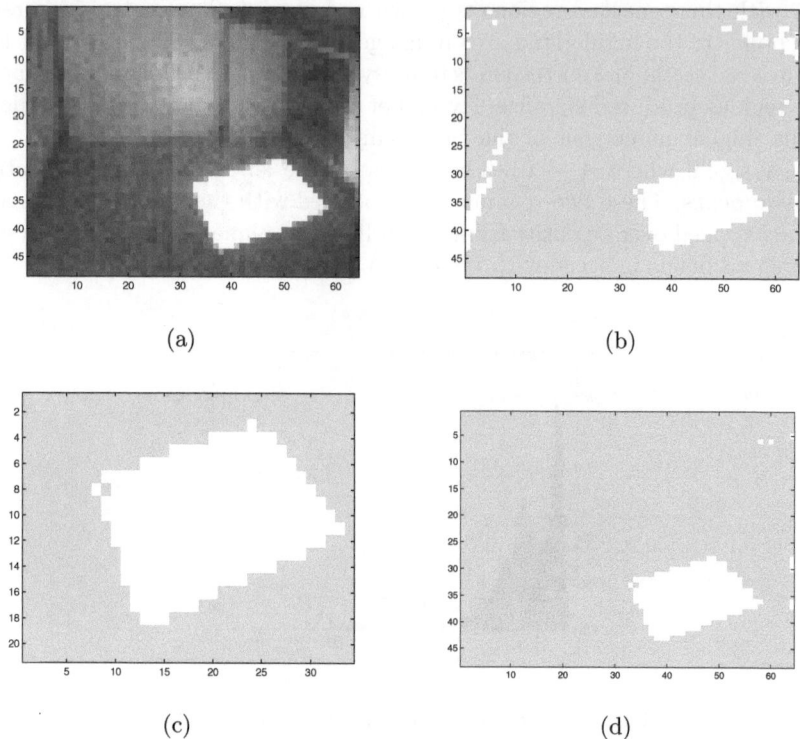

Fig. 5. (a) Intensity image observed from a TOF camera (b) segmentation of shadow and highlighted regions (non shadow) of a complete frame (c) shadow segmentation of a planar surface only. (d) Refined shadow segmentation after two step approach.

6   Summary

Time-of-flight (TOF) cameras are primarily used for range estimation by illuminating the scene through a TOF IR source. Unlike conventional cameras where only a single parameter is measured as intensity, TOF camera's additional measurement facilitate in deriving a radiometric range model for the TOF camera. The model comprising light sources and scene geometry is independent of reflectivity for all practical purposes. Despite the fact that the radiometric model is based on much simpler assumption, the framework proved robust and effective. Experimental results prove the effectiveness of this model for TOF cameras in vision based algorithms.

References

1. Benedek, C., Sziranyi, T.: Bayesian foreground and shadow detection in uncertain frame rate surveillance videos. IEEE Trans. Image Process. 17(4), 608–621 (2008)
2. Cao, X., Shah, M.: Camera calibration and light source estimation from images with shadows. In: Proc. IEEE Computer Society Conference on Computer Vision and Pattern Recognition CVPR 2005, vol. 2, pp. 918–923 (2005)

3. Cucchiara, R., Grana, C., Piccardi, M., Prati, A.: Detecting moving objects, ghosts, and shadows in video streams. IEEE Trans. Pattern Anal. Mach. Intell. 25(10), 1337–1342 (2003)

4. Elgammal, A., Duraiswami, R., Harwood, D., Davis, L.: Background and foreground modeling using nonparametric kernel density estimation for visual surveillance. Proc. IEEE 90(7), 1151–1163 (2002)

5. Fardi, B., Dousa, J., Wanielik, G., Elias, B., Barke, A.: Obstacle detection and pedestrian recognition using a 3D PMD camera. In: Proc. IEEE Intell. Vehicles Symp., pp. 225–230 (2006)

6. Foley, J.D., Da, A.V., Feiner, S.K., Hughes, J.F.: Computer Graphics: Principles and Practices. Addison-Wesley Publishing Company, Inc., Reading (1997)

7. Forsyth, D.A., Ponce, J.: Computer Vision: A Modern Approach. Prentice-Hall, Englewood Cliffs (2003)

8. Jacques, J., Jung, C., Musse, S.: Background subtraction and shadow detection in grayscale video sequences. In: Proc. 18th Brazilian Symposium on Computer Graphics and Image Processing SIBGRAPI 2005, pp. 189–196 (2005)

9. Kahlmann, T., Remondino, F., Guillaume, S.: Range imaging technology: new developments and applications for people identification and tracking. In: Proc. SPIE-IS&T Electronic Imaging, San Jose, CA, USA, vol. 6491 (January 2007)

10. Klasing, K., Althoff, D., Wollherr, D., Buss, M.: Comparison of surface normal estimation methods for range sensing applications. In: Proc. IEEE International Conference on Robotics and Automation, ICRA 2009, May 12-17, pp. 3206–3211 (2009)

11. Kuhnert, K.D., Stommel, M.: Fusion of stereo-camera and PMD-camera data for real-time suited precise 3D environment reconstruction. In: Proc. IEEE/RSJ Int. Conf. Intell. Robot. Systs. (2006)

12. Lange, R., Seitz, P.: Solid-state time-of-flight range camera. IEEE J. Quantum Electron. 37, 390–397 (2001)

13. Leroy, M., Roujean, J.L.: Sun and view angle corrections on reflectances derived from NOAA/AVHRR data. IEEE Trans. Geosci. Remote Sens. 32(3), 684–697 (1994)

14. MacQueen, J.: Some methods for classification and analysis of multivariate observations. In: Proc. Fifth Berkeley Symp. on Math. Statist. and Prob., vol. 1, pp. 281–297. Univ. of Calif. Press (1967)

15. Meier, E., Ade, F.: Tracking cars in range image sequnces. In: Proc. IEEE Int. Conf. Intell. Trans. Systs., pp. 105–110 (1997)

16. Mufti, F., Mahony, R.: Statistical analysis of measurement processes for time-of-flight cameras. In: Proc. SPIE Videometrics, Range Imaging, and Applications X, vol. 7447-21 (2009)

17. Mufti, F., Mahony, R.: Radiometric range image filtering for time-of-flight cameras. In: Proc. Int. Conf. on Computer Vision Theory and Applications (VISAPP 2010), Angers, France, vol. 1, pp. 143–152 (May 2010)

18. Sillion, F.X., Puech, C.: Radiosity and Global Illumination. Morgan Kaufmann, San Francisco (1994)

19. Tsai, V.: A comparative study on shadow compensation of color aerial images in invariant color models. IEEE Trans. Geosci. Remote Sens. 44(6), 1661–1671 (2006)

Uni-orthogonal Nonnegative Tucker Decomposition for Supervised Image Classification

Rafal Zdunek

Institute of Telecommunications, Teleinformatics and Acoustics,
Wroclaw University of Technology, Wybrzeze Wyspianskiego 27,
50-370 Wroclaw, Poland
rafal.zdunek@pwr.wroc.pl

Abstract. The Tucker model with orthogonality constraints (often referred to as the HOSVD) assumes decomposition of a multi-way array into a core tensor and orthogonal factor matrices corresponding to each mode. Nonnegative Tucker Decomposition (NTD) model imposes non-negativity constraints onto both core tensor and factor matrices. In this paper, we discuss a mixed version of the models, i.e. where one factor matrix is orthogonal and the remaining factor matrices are nonnegative. Moreover, the nonnegative factor matrices are updated with the modified Barzilai-Borwein gradient projection method that belongs to a class of quasi-Newton methods. The discussed model is efficiently applied to supervised classification of facial images, hand-written digits, and spectrograms of musical instrument sounds.

1 Introduction

The Tucker model [1] decomposes a multi-way array into a core tensor multiplied by a factor matrix along each mode. When orthogonality constraints are imposed onto all the factor matrices, the model is referred to as Higher-Order Singular Value Decomposition (HOSVD), and can be regarded as a multi-linear extension to SVD [2]. When factor matrices and a core tensor are nonnegatively constrained, the model is referred to as Nonnegative Tucker Decomposition (NTD), and it can be considered as a generalization to Nonnegative Tensor Factorization (NTF) or nonnegativity constrained PARAFAC model [3, 4]. In Semi-NTD (SNTD) models, nonnegativity constraints are relaxed for a core tensor or selected factor matrices [5].

In this paper, we assume a special case of SNTD, where one factor matrix is orthogonal, the others are nonnegative, and a core tensor is unsigned. This approach combines NTD with HOSVD, which is particularly useful when the model is applied to image classification. Assuming the images to be classified are arranged to form a three-way array, the orthogonality constraint should be imposed onto the factor matrix that corresponds to the mode along which the images are stacked. The orthogonal column vectors in that factor can be regarded as discriminant vectors, especially as there are as many vectors as classes. The

G. Maino and G.L. Foresti (Eds.): ICIAP 2011, Part I, LNCS 6978, pp. 88–97, 2011.

core tensor multiplied along all but that mode contains lateral slices that can be considered as feature images.

There are many applications of the Tucker-based models. A survey of the applications can be found, e.g. in [5, 6, 7]. Vasilescu and Terzopoulos [8] applied the Tucker model to extract TensorFaces in computer vision. Feature representations in TensorFaces are considerably more accurate than in EigenFaces that can be obtained from the standard PCA technique. The Tucker model has been also used for analyzing facial images by Wang and Ahuja [9], and Vlasic *et al* [10]. Savas and Elden [11] applied the HOSVD for identifying handwritten digits. NTD has been applied to image feature extraction [12], image clustering [5], and supervised image segmentation [13, 14].

The Tucker decomposition can be obtained with many numerical algorithms. The factor matrices in HOSVD are typically estimated by finding leading left singular vectors of a given data tensor unfolded along each mode. However, a number of algorithms for estimating NTD is considerably greater. Similarly as for Nonnegative Matrix Factorization (NMF) [15], NTD can be estimated with multiplicative updates, projected gradient descent, projected least squares, and active set methods [5, 7, 12, 13, 14, 16].

In this paper, we attempt to estimate the nonnegatively constrained factor matrices with the modified GPSR-BB method that was originally proposed by Figueiredo, Nowak, and Wright [17] for reconstruction of sparse signals. The GPSR-BB is based on a similar approximation to the inverse Hessian as in the Barzilai-Borwein gradient projection method [18, 19]. This method has been extended in [5, 20] to efficiently solve nonnegatively constrained systems of linear equations with multiple right-hand sides, and then applied for NMF problems.

The paper is organized as follows: the next section reviews the selected Tucker models and the related basic algorithms. The uni-orthogonal NTD is discussed in Section 3. Section 4 is concerned with the modified GPSR-BB method for estimating nonnegative factor matrices. The classification results are presented in Section 5. Finally, the conclusions are given in the last section.

2   Tucker Models

Given a N-way tensor $\mathcal{Y} \in \mathbb{R}^{I_1 \times I_2 \times \dots \times I_N}$, the Tucker model has the following form:

$$
\mathcal{Y} = \mathcal{G} \times_1 \boldsymbol{U}^{(1)} \times_2 \boldsymbol{U}^{(2)} \times_3 \dots \times_N \boldsymbol{U}^{(N)}
$$
$$
= \sum_{j_1=1}^{J_1} \sum_{j_2=1}^{J_2} \dots \sum_{j_N=1}^{J_N} g_{j_1,j_2,\dots,j_N} \boldsymbol{u}_{j_1}^{(1)} \circ \boldsymbol{u}_{j_2}^{(2)} \circ \dots \circ \boldsymbol{u}_{j_N}^{(N)}, \tag{1}
$$

where $\mathcal{G} = [g_{j_1,j_2,\dots,j_N}] \in \mathbb{R}^{J_1 \times J_2 \times \dots \times J_N}$ is the core tensor of rank-$(J_1, J_2, \dots, J_N)$ with $J_n \leq I_n$ for all $n = 1, \dots, N$ and $1 \leq j_n \leq J_n$. The matrices $\boldsymbol{U}^{(1)} = [\boldsymbol{u}_1^{(1)}, \dots, \boldsymbol{u}_{J_1}^{(1)}] = [u_{i_1,j_1}] \in \mathbb{R}^{I_1 \times J_1}$, $\boldsymbol{U}^{(2)} = [\boldsymbol{u}_1^{(2)}, \dots, \boldsymbol{u}_{J_2}^{(2)}] = [u_{i_2,j_2}] \in \mathbb{R}^{I_2 \times J_2}$, $\boldsymbol{U}^{(N)} = [\boldsymbol{u}_1^{(N)}, \dots, \boldsymbol{u}_{J_N}^{(N)}] = [u_{i_N,j_N}] \in \mathbb{R}^{I_N \times J_N}$ are factor matrices, where

$i_n = 1, \ldots, I_n$, $j_n = 1, \ldots, J_n$, and $n = 1, \ldots, N$. The symbol $\times_n$ denotes the n-mode tensor product, and $\boldsymbol{u}^{(k)} \circ \boldsymbol{u}^{(l)} = \boldsymbol{u}^{(k)}(\boldsymbol{u}^{(l)})^T \in \mathbb{R}^{M \times N}$ is the outer product of the vectors $\boldsymbol{u}^{(k)} \in \mathbb{R}^M$ and $\boldsymbol{u}^{(l)} \in \mathbb{R}^N$.

In the original Tucker model [1] and in HOSVD [2], the factor matrices are column-wise orthogonal, i.e. $(\boldsymbol{U}^{(1)})^T \boldsymbol{U}^{(1)} = \boldsymbol{I}_{J_1}$, $(\boldsymbol{U}^{(2)})^T \boldsymbol{U}^{(2)} = \boldsymbol{I}_{J_2}$, $\ldots$, $(\boldsymbol{U}^{(N)})^T \boldsymbol{U}^{(N)} = \boldsymbol{I}_{J_N}$, where $\boldsymbol{I}_{J_1} \in \mathbb{R}^{J_1 \times J_1}$, $\boldsymbol{I}_{J_2} \in \mathbb{R}^{J_2 \times J_2}$, $\boldsymbol{I}_{J_N} \in \mathbb{R}^{J_N \times J_N}$ are identity matrices. In contrary to SVD of a matrix, the core tensor $\mathcal{G}$ in HOSVD is not a super diagonal tensor but it is rather a dense tensor. For all $n = 1, \ldots, N$, the column-wise orthogonal factor $\boldsymbol{U}^{(n)}$ can be computed from the SVD of the n-mode unfolded tensor $\mathcal{Y}$. Let $\boldsymbol{Y}_{(n)} \in \mathbb{R}^{I_n \times \prod_{p \neq n} I_p}$ be a matrix that is obtained from the tensor $\mathcal{Y}$ by unfolding it along n-mode. Thus $\boldsymbol{U}^{(n)} = [\boldsymbol{u}_1^{(n)}, \ldots, \boldsymbol{u}_{J_n}^{(n)}] \in \mathbb{R}^{I_n \times J_n}$, where $\boldsymbol{u}_j^{(n)}$ is the j-th left singular vector of $\boldsymbol{Y}_{(n)}$ or the j-th leading eigenvector of the symmetric semi-positive defined matrix $\boldsymbol{Y}_{(n)}(\boldsymbol{Y}_{(n)})^T \in \mathbb{R}^{I_n \times I_n}$. Having the factor matrices $\{\boldsymbol{U}^{(n)}\}$, the core tensor can be readily updated with the formula $\mathcal{G} \leftarrow \mathcal{Y} \times_1 (\boldsymbol{U}^{(1)})^T \times_2 (\boldsymbol{U}^{(2)})^T \times_3 \ldots \times_N (\boldsymbol{U}^{(N)})^T$.

In NTD [3, 4], the core tensor and factor matrices are all nonnegative, i.e. $g_{j_1, j_2, \ldots, j_N} \geq 0$ and $u_{i_n, j_n} \geq 0$ for $i_n = 1, \ldots, I_n$, $j_n = 1, \ldots, J_n$, and $n = 1, \ldots, N$. The nonnegative factor matrices $\boldsymbol{U}^{(n)}$ are updated alternatingly – similarly as the factors in NMF [15]. To apply the alternating optimization procedure, note that the mode-n unfolding of the model (1) is as follows:

$$\boldsymbol{Y}_{(n)} = \boldsymbol{U}^{(n)} \boldsymbol{G}_{(n)} \left[\boldsymbol{U}^{(N)} \otimes \ldots \otimes \boldsymbol{U}^{(n+1)} \otimes \boldsymbol{U}^{(n-1)} \otimes \ldots \otimes \boldsymbol{U}^{(1)} \right]^T \quad (2)$$

$$= \boldsymbol{U}^{(n)} \boldsymbol{Z}^{(n)},$$

where $\boldsymbol{G}_{(n)} \in \mathbb{R}^{J_n \times \prod_{p \neq n} J_p}$ is the unfolded tensor $\mathcal{G}$ along the n-mode, and the symbol $\otimes$ denotes the Kronecker product. Applying the projected ALS algorithm to (2), we have:

$$\boldsymbol{U}^{(n)} = \left[\boldsymbol{Y}_{(n)}(\boldsymbol{Z}^{(n)})^T (\boldsymbol{Z}^{(n)}(\boldsymbol{Z}^{(n)})^T)^{-1} \right]_+, \quad n = 1, \ldots, N, \quad (3)$$

where $[\xi]_+ = \max\{0, \xi\}$ is the projection of ξ onto the nonnegative orthant of $\mathbb{R}$. The core tensor $\mathcal{G}$ can be updated with the formula:

$$\mathcal{G} \leftarrow \left[\mathcal{Y} \times_1 (\boldsymbol{U}^{(1)})^\dagger \times_2 (\boldsymbol{U}^{(2)})^\dagger \times_3 \ldots \times_N (\boldsymbol{U}^{(N)})^\dagger \right]_+, \quad (4)$$

where $(\boldsymbol{U}^{(n)})^\dagger = \left((\boldsymbol{U}^{(n)})^T (\boldsymbol{U}^{(n)}) \right)^{-1} (\boldsymbol{U}^{(n)})^T \in \mathbb{R}^{J_n \times I_n}$ is the Moore-Penrose pseudoinverse of $\boldsymbol{U}^{(n)}$ for $n = 1, \ldots, N$. The columns of the nonnegative factor matrices $\boldsymbol{U}^{(n)}$ are often normalized to the unit l_p norm, i.e. $\boldsymbol{u}_l^{(n)} \leftarrow \frac{\boldsymbol{u}_l^{(n)}}{\|\boldsymbol{u}_l^{(n)}\|_p}$, where $p = 1$ or $p = 2$, and $l = 1, \ldots, J_n$, $n = 1, \ldots, N$.

3   Uni-orthogonal NTD

We assume that the training and testing images have the same resolution ($I_1 \times I_2$), and the training images arranged along the mode-3 form the 3-way tensor

$\mathcal{Y} \in \mathbb{R}^{I_1 \times I_2 \times I_3}$, where I_3 is the number of training images. Thus $N = 3$, and the N-way Tucker model (1) simplifies to the Tucker3 model [6].

In our approach, we have the following model:

$$\mathcal{Y} = \mathcal{G} \times_1 U^{(1)} \times_2 U^{(2)} \times_3 U^{(3)}, \tag{5}$$

where $\mathcal{G} \in \mathbb{R}^{J_1 \times J_2 \times J_3}$ is the core tensor, the factor matrices $U^{(1)} = [u_{i_1,j_1}^{(1)}] \in \mathbb{R}^{I_1 \times J_1}$ and $U^{(2)} = [u_{i_2,j_2}^{(2)}] \in \mathbb{R}^{I_2 \times J_2}$ are nonnegative ($u_{i_1,j_1}^{(1)}, u_{i_2,j_2}^{(2)} \geq 0$), and the factor matrix $U^{(3)} \in \mathbb{R}^{I_3 \times J_3}$ is column-wise orthogonal, i.e. $(U^{(3)})^T U^{(3)} = I_{J_3}$. The number J_3 should be equal to the number of classes, and the numbers J_1 and J_2 should satisfy the conditions: $1 \leq J_1 << I_1$ and $1 \leq J_2 << I_2$. Thus, our Uni-Orthogonal NTD (UO-NTD) is given by Algorithm 1.

Algorithm 1. UO-NTD

Input : $\in \mathbb{R}^{I_1 \times I_2 \times I_3}$, $_1$, $_2$, $_3$ - lower ranks, k_{max} - number of inner iterations

Output: Factor matrices $U^{(1)} \in \mathbb{R}^{I_1 \times}$ $_1$, $U^{(2)} \in \mathbb{R}^{I_2 \times}$ $_2$ and $U^{(3)} \in \mathbb{R}^{I_3 \times}$ $_3$, $\mathcal{G} \in \mathbb{R}$ $_1 \times$ $_2 \times$ $_3$ - core tensor

1   Initialize (randomly) $U^{(1)}$ and $U^{(2)}$ with positive numbers, and $U^{(3)}$ and $\mathcal{G}$ with real numbers ;

2   **repeat**

3     $Z^{(1)} = G_{(1)}(U^{(3)} \quad U^{(2)})^T$;

4     $U^{(1)} \leftarrow \text{gpsrbb}(Y_{(1)}, Z^{(1)}, U^{(1)}, k_{max})$;   // Update for $U^{(1)}$

5     $u_l^{(1)} \leftarrow \dfrac{u_l^{(1)}}{||u_l^{(1)}||_2}$, where $l = 1, \dots,$ $_1$;   // Normalization of $U^{(1)}$

6     $Z^{(2)} = G_{(2)}(U^{(3)} \quad U^{(1)})^T$;

7     $U^{(2)} \leftarrow \text{gpsrbb}(Y_{(2)}, Z^{(2)}, U^{(2)}, k_{max})$;   // Update for $U^{(2)}$

8     $u_l^{(2)} \leftarrow \dfrac{u_l^{(2)}}{||u_l^{(2)}||_2}$, where $l = 1, \dots,$ $_2$;   // Normalization of $U^{(2)}$

9     $Z^{(3)} = G_{(3)}(U^{(2)} \quad U^{(1)})^T$;

10    $U^{(3)} = Y_{(3)}(Z^{(3)})^T(Z^{(3)}(Z^{(3)})^T)^{-1}$;   // Update for $U^{(3)}$

11    $u_l^{(3)} \leftarrow \dfrac{u_l^{(3)}}{||u_l^{(3)}||_2}$, where $l = 1, \dots,$ $_3$;   // Normalization of $U^{(3)}$

12    $U^{(3)} \leftarrow U^{(3)} \left((U^{(3)})^T U^{(3)} \right)^{-1/2}$;   // Column-wise orthogonalization

13    $\mathcal{G} \leftarrow$ $\times_1 (U^{(1)})^\dagger \times_2 (U^{(2)})^\dagger \times_3 (U^{(3)})^T$;   // Update for $\mathcal{G}$

14 **until** Stop criterion is satisfied ;

The GPSR-BB algorithm given in Steps 4 and 7 in Algorithm 1 is described in Section 4. The stop criterion in Algorithm 1 can be determined with many rules. It might be a fixed number of iterations (usually less than 50) or the truncation of iterations when the normalized residual error drops below a certain threshold.

Each lateral slice (along mode-3) of the tensor $\mathcal{G} \times_1 U^{(1)} \times_2 U^{(2)} \in \mathbb{R}^{I_1 \times I_2 \times J_3}$ can be considered as a basis image that has rather holistic nature (similarly as in

PCA) than a part-based representation. For classification of handwritten digits, each base image is expected to represent each digit. Each lateral image in the tensor $\mathcal{Y}$ is therefore a linear combination of the basis images. The coefficients of that linear combination are given by row vectors of the factor matrix $\boldsymbol{U}^{(3)}$ that can be regarded as encoding vectors. Classification can be performed in the low-dimensional space of encoding vectors.

Unsupervised classification can be obtained by clustering the encoding vectors. For supervised classification the encoding vectors for testing images should be computed using the basis images that have been already estimated for training images. Then, each testing image can be classified according to the highest similarity in the space of encoding vectors.

The algorithm of supervised classification is given by Algorithm 2. The testing images are collected in the tensor $\mathcal{T} \in \mathbb{R}^{I_1 \times I_2 \times R}$ in the similar way as training images in $\mathcal{Y}$, where R is the number of testing images. The unfolded tensor $\mathcal{T}$ with respect to the mode-3 is given by the matrix $\boldsymbol{T}_{(3)}$. Indices of the classes to which training images belong are given in the vector $\boldsymbol{c}^{(train)} \in \mathbb{R}^{I_3}$. Algorithm 2 returns the vector $\boldsymbol{c}^{(test)} \in \mathbb{R}^R$ that contains the classes of testing images.

Algorithm 2. Supervised classification

Input : $\mathcal{G} \in \mathbb{R}^{J_1 \times J_2 \times J_3}$, $\boldsymbol{U}^{(1)} \in \mathbb{R}^{I_1 \times J_1}$, $\boldsymbol{U}^{(2)} \in \mathbb{R}^{I_2 \times J_2}$ and $\boldsymbol{U}^{(3)} \in \mathbb{R}^{I_3 \times J_3}$, $\boldsymbol{c}^{(train)} \in \mathbb{R}^{I_3}$ - classes of training images, $\mathcal{T} \in \mathbb{R}^{I_1 \times I_2 \times R}$ - testing images

Output: $\boldsymbol{c}^{(test)} \in \mathbb{R}^R$ - classes of testing images

1 Initialize (randomly) $\boldsymbol{U}^{(1)}$ and $\boldsymbol{U}^{(2)}$ with positive numbers, and $\boldsymbol{U}^{(3)}$ and $\mathcal{G}$ with real numbers ;

2 $\boldsymbol{Z}^{(3)} = \boldsymbol{G}_{(3)}(\boldsymbol{U}^{(2)} \otimes \boldsymbol{U}^{(1)})^T$;

3 $\boldsymbol{U}^{(test)} = \boldsymbol{T}_{(3)}(\boldsymbol{Z}^{(3)})^T(\boldsymbol{Z}^{(3)}(\boldsymbol{Z}^{(3)})^T)^{-1}$;   // Encoding vectors

4 $\boldsymbol{u}_l^{(test)} \leftarrow \dfrac{\boldsymbol{u}_l^{(test)}}{||\boldsymbol{u}_l^{(test)}||_2}$, where $l = 1, \ldots, J_3$;   // Normalization of $\boldsymbol{U}^{(test)}$

5 $\boldsymbol{c}^{(test)} \leftarrow \text{knnclassify}(\boldsymbol{U}^{(test)}, \boldsymbol{U}^{(3)}, \boldsymbol{c}^{(train)}, 1, 'cosine')$;   // Matlab function

---

The knnclassify function in Step 5 of Algorithm 2 comes from the *Bioinformatics Toolbox* in Matlab 2008. It uses the nearest-neighbor method for classification of the rows in the matrix $\boldsymbol{U}^{(test)}$ into one of the classes of the the matrix $\boldsymbol{U}^{(3)}$. We used only one nearest neighbor, and the cosine measure to determine the similarity between samples (row vectors).

4   Modified GPSR-BB Algorithm

To solve the system (2) with respect to the nonnegativity constrained $\boldsymbol{U}^{(n)}$, we formulate the Nonnegative Least Squares (NNLS) problem:

$$\min_{\boldsymbol{U}^{(n)} \geq 0} \Psi(\boldsymbol{U}^{(n)}), \quad \text{where} \quad \Psi(\boldsymbol{U}^{(n)}) = \frac{1}{2}||\boldsymbol{Y}_{(n)} - \boldsymbol{U}^{(n)}\boldsymbol{Z}^{(n)}||_F^2 \qquad (6)$$

which can be solved with the modified GPSR-BB method [17] that is based on the Spectral Projected Gradient (SPG) method [21]. For (6), the SPG takes the form of the following updates:

$$U^{(n)} \leftarrow U^{(n)} - \mathrm{diag}\{\lambda^{(n)}\} D^{(n)}, \tag{7}$$

with the search direction defined by

$$D^{(n)} = \left[U^{(n)} - \mathrm{diag}\{\alpha^{(n)}\} \nabla_{U^{(n)}} \Psi(U^{(n)}) \right]_+ - U^{(n)}, \tag{8}$$

where the step length $\lambda^{(n)} \in [0,1]^{I_n}$ minimizes $\Psi(U^{(n)} - \mathrm{diag}\{\lambda^{(n)}\} D^{(n)})$ and $\alpha^{(n)} \in \mathbb{R}^{I_n}$ should be selected such that the matrix $H^{(n)} = \mathrm{diag}\{\alpha^{(n)}\}$ approximates the inverse to the Hessian of $\Psi(U^{(n)})$. This approach comes from the Barzilai-Borwein gradient projection method [18, 19], thus the updates for the search direction (8) have the quasi-Newton nature.

The factors $\alpha^{(n)}$ can be computed from the secant equation which for the quasi-Newton update in (8) takes the form: $H_{k+1}^{(n)} S_k^{(n)} = W_k^{(n)}$, where $S_k^{(n)} = U_{k+1}^{(n)} - U_k^{(n)}$ and $W_k^{(n)} = \nabla_{U^{(n)}} \Psi(U_{k+1}^{(n)}) - \nabla_{U^{(n)}} \Psi(U_k^{(n)}) = -\mathrm{diag}\{\lambda^{(n)}\} D^{(n)}$, and k is the number of an iterative step. From the secant equation, we have:

$$
\begin{aligned}
\bar{\alpha}_{k+1}^{(n)} &= \frac{\mathrm{diag}\ W_k^{(n)} (S_k^{(n)})^T \langle}{\mathrm{diag}\ S_k^{(n)} (S_k^{(n)})^T \langle} = \frac{\mathrm{diag}\ S_k^{(n)} Z^{(n)} (Z^{(n)})^T (S_k^{(n)})^T \langle}{\mathrm{diag}\ S_k^{(n)} (S_k^{(n)})^T \langle} \\[2mm]
&= \frac{\mathrm{diag}\ D^{(n)} Z^{(n)} (Z^{(n)})^T (D^{(n)})^T \langle}{\mathrm{diag}\ D^{(n)} (D^{(n)})^T \langle} \\[2mm]
&= \frac{D^{(n)} \circledast (D^{(n)} Z^{(n)} (Z^{(n)})^T)\ 1_{J_n}}{D^{(n)} \circledast D^{(n)}\ 1_{J_n}},
\end{aligned}
\tag{9}
$$

where the symbol $\circledast$ denotes the Hadamard multiplication, $\mathrm{diag}\{X\}$ is a vector created from main diagonal entries of the matrix X, and $1_{J_n} = [1, \ldots, 1]^T \in \mathbb{R}^{J_n}$.

Inserting (7) to (6), and from $\frac{\partial}{\partial \lambda^{(n)}} \Psi(U^{(n)}) \triangleq 0$, we get the update for $\lambda^{(n)}$ in the closed-form:

$$\bar{\lambda}^{(n)} \leftarrow \frac{D^{(n)} \circledast \nabla_{U^{(n)}} \Psi(U^{(n)})\ 1_{J_n}}{D^{(n)} \circledast (D^{(n)} Z^{(n)} (Z^{(n)})^T)\ 1_{J_n}}. \tag{10}$$

The final form of the modified GPSR-BB algorithm is given by Algorithm 3.

5   Classification Results

The proposed UO-NTD algorithm has been tested for various supervised image classification problems. For the reference, the ALS-NTD algorithm that updates

---

Algorithm 3. GPSR-BB Algorithm

Input : $Y_{(n)}$, $Z^{(n)}$, $U^{(n)}$ - initial guess, k_{max}, α_{min}, α_{max}

Output: $U^{(n)}$ - mode-n factor matrix,

1 **for** $k = 1, 2, \ldots, k_{max}$ **do**

2     $F^{(n)} = \nabla_{U^{(n)}} (U^{(n)}) = (U^{(n)} Z^{(n)} - Y_{(n)})(Z^{(n)})^T$;   // Gradient

3     $D^{(n)} = \left[U^{(n)} - \text{diag}\{\alpha^{(n)}\}F^{(n)} \right]_{+} - U^{(n)}$;   // Search direction

4     $\lambda^{(n)} \leftarrow \max\{0, \min\{1, \bar{\lambda}^{(n)}\}\}$;   // where $\bar{\lambda}^{(n)}$ is given by (10)

5     $U^{(n)} \leftarrow U^{(n)} - \text{diag}\{\lambda^{(n)}\}D^{(n)}$;

6     $\alpha^{(n)} \leftarrow \max\{\alpha_{min}, \min\{\alpha_{max}, \bar{\alpha}^{(n)}\}\}$;   // where $\bar{\alpha}^{(n)}$ is given by (9)

---

the factor matrices with the rule (3) and the core tensor with the update (4) is chosen. We analyze three classification problems of: (A) musical instruments, (B) hand-written digits, (C) facial images.

For analyzing the problem A, the audio recordings of 6 musical instruments (cello, soprano saxophone, violin, bassoon, flute, and piano) are selected from the MIS database[1] of the University of Iowa. Each audio recording at the sampling rate of 44.1kHz is restricted to contain meaningful information of about 4 sec long. The training and testing sets contain totally 56 and 12 samples, respectively. All the samples are transformed to log-magnitude spectrograms into the frequency range from 86Hz to 10.9kHz, and the time window from 0 do 4 seconds. Then, the spectrogram are downsampled to 64 frequencies × 128 time intervals. Thus $\mathcal{Y} \in \mathbb{R}^{64 \times 128 \times 56}$ and $\mathcal{T} \in \mathbb{R}^{64 \times 128 \times 12}$. For this case, we set $J_1 = J_2 = 20$, and $J_3 = 6$. The spectrograms of the testing samples are depicted in Fig. 1(a).

The samples for the problem B are images of hand-written digits. Each class in the training set is represented by 8 images of the resolution downsampled to 64 × 64 pixels. This gives $\mathcal{Y} \in \mathbb{R}^{64 \times 64 \times 80}$ for 10 digits from 0 to 9. The testing set consists of 20 images (2 by each class) - thus $\mathcal{T} \in \mathbb{R}^{64 \times 64 \times 20}$. We set $J_1 = J_2 = 20$, and $J_3 = 10$. The testing samples are illustrated in Fig. 1(b).

The problem C is concerned with classification of facial images from the ORL database [2] that contains 400 frontal face images of 40 people (10 pictures per person). The images were taken at different times (between April 1992 and April 1994 at the AT&T Laboratories Cambridge), varying the lighting, facial expressions (open / closed eyes, smiling / not smiling) and facial details (glasses / no glasses). All the images have a dark homogeneous background with the subjects in an upright, frontal position. The whole set is randomly divided into 320 training images containing all the classes and 80 testing images. The resolution of the images is 112 × 92 pixels. Despite the number of classes is 40, we noticed that setting $J_1 = J_2 = J_3 = 20$ gives nearly the same recognition rate as for $J_3 = 40$ but in a considerably shorter time.

---

[1] http://theremin.music.uiowa.edu

[2] http://people.cs.uchicago.edu/ dinoj/vis/orl/

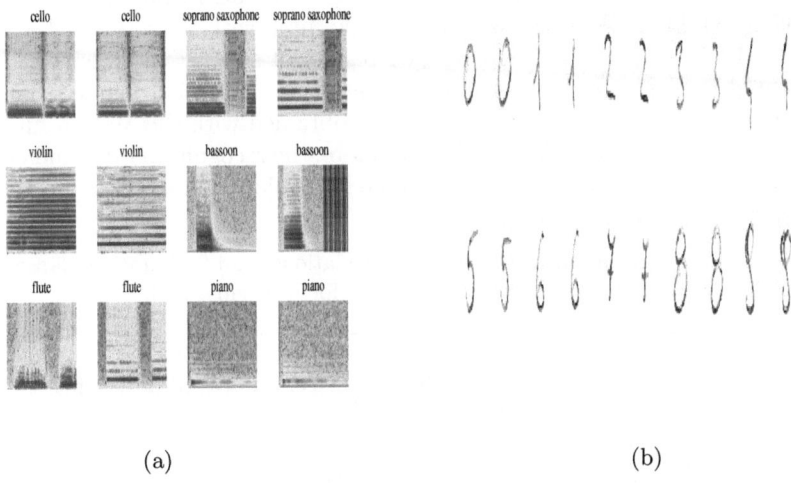

(a)   (b)

Fig. 1. Testing samples: (a) spectrograms for the problem C; (b) hand-written digits

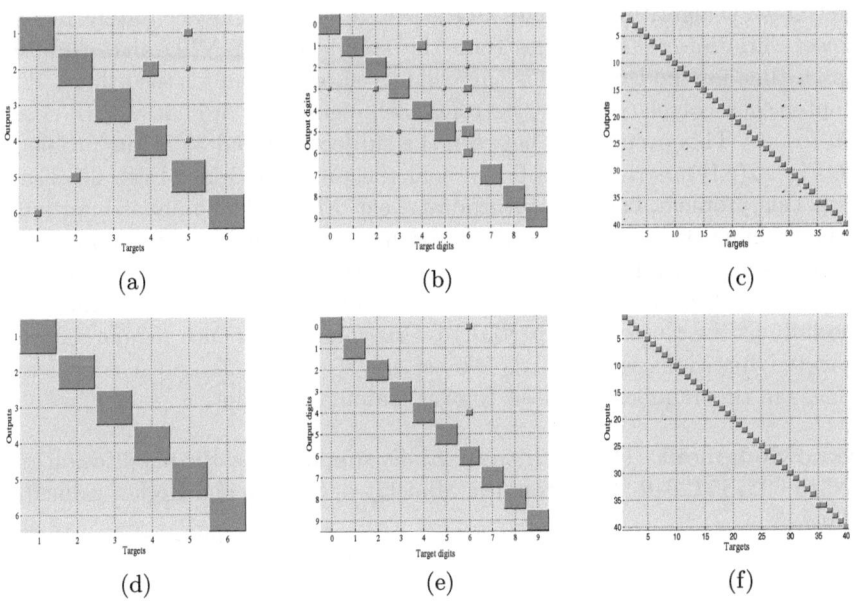

Fig. 2. Confusion matrices: (a) ALS-NTD: 6 musical instruments; (b) ALS-NTD: 10 hand-written digits; (c) ALS-NTD: 40 facial images; (d) UO-NTD: 6 musical instruments; (e) UO-NTD: 10 hand-written digits; (f) UO-NTD: 40 facial images

Each tested algorithm is initiated 100 times, starting from a random initial guess for factor matrices and a core tensor. Both algorithms are run for 20 iterations, and the GPSR-BB algorithm is executed with the settings: $k_{max} = 2$, $\alpha_{min} = 10^{-8}$, $\alpha_{max} = 1$. The averaged results of supervised classification are illustrated in Fig. 2 with the Hinton graph of the confusion matrix. Both the Hinton graph and the confusion matrix are obtained with the Matlab functions from the *Neural Network Toolbox*. Furthermore, the recognition rates and elapsed time averaged over 100 trials are presented in Table 1.

Table 1. Mean recognition rates, standard deviations, and 20 iteration elapsed time averaged over 100 runs of the tested algorithms for problems A, B, and C

Problem	ALS-NTD			UO-NTD		
	Rec. rate [%]	Std. [%]	Time [sec]	Rec. rate [%]	Std. [%]	Time [sec]
Problem A	91.92	8.41	3.67	99.67	1.64	3.27
Problem B	86.35	6.85	2.57	98.25	2.86	2.34
Problem C	92.5	2	20.8	97.36	0.39	19.8

6   Conclusions

The UO-NTD algorithm classifies all the tested images more accurately and with a lower variation of the results than the standard ALS-NTD algorithm. The elapsed time for the UO-NTD is only slightly shorter than for the others. When an outer-class correlation is very strong both algorithms are not able to classify such results. This occurs, e.g. between the subjects 34 and 35 in the problem C (see Figs. 2(c,f)) or the digits 6 and 4 in the problem B (see Figs. 2(b,e)). To tackle this problem, one may incorporate, e.g. Fisher discriminant information to the training process, however, this results in difficulty in determining inner- and outer-class correlations. The underlying iterative algorithms (GPSR-BB or ALS) update the factors with inconsistent and usually ill-posed training data, especially as the clusters are partially overlapping, and hence regularization by truncated iterations is essential. We set up 20 iterations and a considerable increase in this number may lead to over-training behavior.

Acknowledgment. This work was partially supported by the habilitation grant N N515 603139 (2010-2012) from the Ministry of Science and Higher Education, Poland.

References

[1] Tucker, L.R.: Some mathematical notes on three-mode factor analysis. Psychometrika 31, 279–311 (1966)
[2] De Lathauwer, L., de Moor, B., Vandewalle, J.: A multilinear singular value decomposition. SIAM Journal of Matrix Analysis and Applications 21, 1253–1278 (2001)

[3] Kiers, H.A.L.: A three-step algorithm for CANDECOMP/PARAFAC analysis of large data sets with multicollinearity. Journal of Chemometrics 12(3), 155–171 (1998)

[4] Smilde, A., Bro, R., Geladi, P.: Multi-way Analysis: Applications in the Chemical Sciences. John Wiley and Sons, New York (2004)

[5] Cichocki, A., Zdunek, R., Phan, A.H., Amari, S.I.: Nonnegative Matrix and Tensor Factorizations: Applications to Exploratory Multi-way Data Analysis and Blind Source Separation. Wiley and Sons, Chichester (2009)

[6] Kolda, T.G., Bader, W.: Tensor decompositions and applications. SIAM Review 51(3), 455–500 (2009)

[7] Mørup, M., Hansen, L.K., Arnfred, S.M.: Algorithms for sparse nonnegative Tucker decompositions. Neural Computation 20(8), 2112–2131 (2008)

[8] Vasilescu, M.A.O., Terzopoulos, D.: Multilinear analysis of image ensembles: TensorFaces. In: Heyden, A., Sparr, G., Nielsen, M., Johansen, P. (eds.) ECCV 2002. LNCS, vol. 2350, pp. 447–460. Springer, Heidelberg (2002)

[9] Wang, H., Ahuja, N.: Facial expression decomposition. In: Proc. of the Ninth IEEE International Conference on Computer Vision, vol. 2, pp. 958–965 (2003)

[10] Vlasic, D., Brand, M., Phister, H., Popovic, J.: Face transfer with multilinear models. ACM Transactions on Graphics 24, 426–433 (2005)

[11] Savas, B., Eldén, L.: Handwritten digit classification using higher order singular value decomposition. Pattern Recognition 40(3), 993–1003 (2007)

[12] Kim, Y.D., Choi, S.: Nonnegative tensor decomposition. In: Proc. IEEE Conference on Computer Vision and Pattern Recognition (CVPR 2007), Minneapolis, MN, pp. 1–8 (2007)

[13] Phan, A.H., Cichocki, A.: Tensor decompositions for feature extraction and classification of high dimensional datasets. IEICE Nonlinear Theory and Its Applications 1(1), 37–68 (2010)

[14] Phan, A.H., Cichocki, A., Vu-Dinh, T.: Classification of scenes based on multiway feature extraction. In: Proc. 2010 International Conference on Advanced Technologies for Communications, Ho Chi Minh City, Vietnam, pp. 142–145 (2010)

[15] Lee, D.D., Seung, H.S.: Learning of the parts of objects by non-negative matrix factorization. Nature 401, 788–791 (1999)

[16] Kim, H., Park, H., Elden, L.: Non-negative tensor factorization based on alternating large-scale non-negativity-constrained least squares. In: Proc. BIBE 2007, pp. 1147–1151 (2007)

[17] Figueiredo, M.A.T., Nowak, R.D., Wright, S.J.: Gradient projection for sparse reconstruction: Application to compressed sensing and other inverse problems. IEEE Journal of Selected Topics in Signal Processing 1(4), 586–597 (2007)

[18] Barzilai, J., Borwein, J.M.: Two-point step size gradient methods. IMA Journal of Numerical Analysis 8(1), 141–148 (1988)

[19] Dai, Y.H., Fletcher, R.: Projected Barzilai-Borwein methods for large-scale box-constrained quadratic programming. Numer. Math. 100(1), 21–47 (2005)

[20] Zdunek, R., Cichocki, A.: Fast nonnegative matrix factorization algorithms using projected gradient approaches for large-scale problems. In: Computational Intelligence and Neuroscience 2008(939567) (2008)

[21] Birgin, E.G., Martínez, J.M., Raydan, M.: Nonmonotone spectral projected gradient methods on convex sets. SIAM Journal on Control and Optimization 10, 1196–1211 (2000)

A Classification Approach with a Reject Option for Multi-label Problems

Ignazio Pillai, Giorgio Fumera, and Fabio Roli

Deparment of Electrical and Electronic Engineering, Univ. of Cagliari
Piazza d'Armi, 09123 Cagliari, Italy
{pillai,fumera,roli}@diee.unica.it

Abstract. We investigate the implementation of multi-label classification algorithms with a reject option, as a mean to reduce the time required to human annotators and to attain a higher classification accuracy on automatically classified samples than the one which can be obtained without a reject option. Based on a recently proposed model of manual annotation time, we identify two approaches to implement a reject option, related to the two main manual annotation methods: browsing and tagging. In this paper we focus on the approach suitable to tagging, which consists in withholding either all or none of the category assignments of a given sample. We develop classification reliability measures to decide whether rejecting or not a sample, aimed at maximising classification accuracy on non-rejected ones. We finally evaluate the trade-off between classification accuracy and rejection rate that can be attained by our method, on three benchmark data sets related to text categorisation and image annotation tasks.

Keywords: Multi-label classification, Reject option.

1 Introduction

In a multi-label classification problem each sample can belong to more than one class, contrary to traditional, single-label problems. Multi-label problems occur in several applications related to retrieval tasks [13], notably text categorisation [12] and scene categorization [2], and are receiving an increasing interest in the pattern recognition and machine learning literature. Nevertheless, in many tasks automatic classification techniques do not achieve a satisfactory performance yet [14]. As an example in a text categorisation task, the best results obtained through the automatic "Medical Text Indexer" tool at the U.S. National Library of Medicine database (MEDLINE), is a recall of about 0.53 and a precision of about 0.30 [1]. In the recent ImageCLEF 2010 image annotation contest, the best automatic system attained a mean average precision of 0.45 [9]. Therefore, manual categorisation remains the only reliable solutions for many practical applications, although it is a tedious and labour-intensive procedure. This is also confirmed by the proliferation of manual image annotation tools [14], and by the use of "Medical Text Indexer" only as a *recommendation* tool by MEDLINE's human indexers [11].

G. Maino and G.L. Foresti (Eds.): ICIAP 2011, Part I, LNCS 6978, pp. 98–107, 2011.

Based on the above premises, in this paper we investigate a hybrid manual–automatic annotation approach inspired by the *reject option* used in single-label classifiers. The reject option consists in withholding the automatic classification of a sample, if the decision is not considered reliable enough. It is a mean to limit excessive misclassifications, at the expense either of a manual post-processing of rejections, or of their automatic handling by a more accurate but also computationally more costly classifier, and requires therefore a trade-off between the accuracy attainable on non-rejected samples and the amount (cost) of rejections [4,10]. Analogously, in multi-label problems a classifier with a reject option could automatically take decisions on category assignments deemed reliable for a given sample, and could withheld and leave to a manual annotator only the ones deemed unreliable. This could allow a classifier to attain a high classification performance on non-withheld decisions, which should be traded for the cost of manual annotation of withheld decisions.

However, the theory and implementations of the reject option proposed so far in the pattern recognition literature have been developed only for single-label classifiers and only under the framework of the minimum risk theory. They can not be applied to multi-label classifiers, whose performance measures are based on precision and recall, and do not take into account the cost of correct/incorrect decisions. Therefore, in this paper we will first discuss how a reject option can be implemented in multi-label classifiers, based on the analysis of the cost (time) of manual labelling given in [14]. In Sect. 2 we show that this analysis suggests two possible implementations: rejecting all the category assignments of a sample, or only a subset of them. The latter option has already been proposed in previous works by the authors, although it was not not rigorously motivated [6,7]. Therefore, in this paper we focus on the former option. In Sect. 3 we discuss how classification accuracy on non-rejected samples can be measured in terms of precisions and recall, and in Sect. 4 we derive two methods to maximise such accuracy for a given fraction of rejected samples, namely a given cost of manual annotation. The trade-off between classification accuracy and the fraction of rejected samples is experimentally evaluated in Sect. 5 on three benchmark data sets related to a text categorisation and to an image annotation task.

2   Rejection Criteria for Multi-label Problems

Single-label classification problems with a reject option were formalized under the framework of the minimum risk theory in [4]. Denoting the costs of correct classifications, rejections, and misclassifications respectively as λ_C, λ_R and λ_E (with $\lambda_C < \lambda_R < \lambda_E$), the expected classification cost is minimized by assigning a sample to the class with the maximum a posteriori probability, if such probability is higher than $(\lambda_E - \lambda_R)/(\lambda_E - \lambda_C)$, and otherwise in rejecting it. This framework does not fit multi-label problems, whose performance measures are given in terms of precision and recall, which are not related to classification costs (see Sect. 3). To devise an implementation of a reject option in multi-label problems, one issue to address is how to evaluate the cost of manually handling withheld category assignments.

The cost of manual annotation clearly depends on the annotation time. A model of the annotation time has been proposed in [14], for two possible annotation procedures: *tagging*, which consists in labelling a sample according to a given set of categories (keywords or "tags"), and *browsing*, in which the relevance has to be decided for a whole set of samples to one category at a time. According to [14], the annotation time of tagging and browsing is:

$$t_{\text{tagging}} = M \cdot (\overline{K} \cdot t_f + t_s), \qquad t_{\text{browsing}} = \sum_{k=1...N} (M_p^k \cdot t_p + M_n^k \cdot t_n) ,$$

where N is the number of categories, M is the number of samples, t_s is the so called "initial setup" time to analyse a sample, t_f is the time to assign one label, $\overline{K}$ is the average number of labels per sample, M_p^k and M_n^k are respectively the number of samples which belong and do not belong to the k-th category, while t_p and t_n are the time for deciding whether or not a sample belongs to a category. The most efficient procedure among tagging and browsing can be made on the basis of the values of the above parameters, according to the task at hand [14].

The above model of manual annotation time suggests two main approaches to implement a reject option in multi-label classifiers, aimed at trading the classification accuracy on automatically assigned labels for the cost of manually processing category assignments withheld by a classifier:

1. In tasks where tagging is used, the manual annotation time of withheld category assignments can be directly controlled by setting a constraint to the number of samples which contain withheld assignments, which corresponds to the term M. Accordingly, in this case it make sense to reject either *all* or none of the assignments of a sample, rather than only a subset of them.
2. In tasks where browsing is used, the manual annotation time of withheld category assignments can be controlled by setting a constraint on the number of samples which contain withheld assignments, *independently* for each category, which correspond to M_p^k and M_n^k. In this case it makes sense to withheld for each sample only a subset of its category assignments (not necessarily all of them), obviously the most unreliable ones.

In the former approach, the objective is clearly to maximise classification accuracy on non-rejected samples, with a constraint on the maximum fraction of rejected samples. In the latter approach, the objective is instead to maximise classification accuracy on non-withheld decisions, with a constraint on the maximum fraction of withheld decisions for each individual category. In both cases, the effectiveness of a reject option has to be evaluated in terms of the attainable trade-off between the accuracy of the classifier on non-withheld category assignments, and the cost (annotation time) of withheld ones, taking into account the application requirements of the task at hand.

An implementation of the latter approach has already been investigated by the authors in [6,7], although it was not motivated by the above arguments. Therefore, in this paper we focus on the former implementation.

3 Accuracy of Multi-label Classifiers with a Reject Option

In this section we discuss how to evaluate the accuracy of a multi-label classifier in presence of withheld category assignments. To this aim we first introduce accuracy measures based on precision and recall.

In the field of information retrieval, *precision* is the probability that a retrieved document is relevant to a given query or topic, while *recall* is the probability that a relevant document is retrieved. In a multi-label classification problem, each class corresponds to a distinct topic. Denoting the set of categories as $\Omega = \{\omega_1, \ldots, \omega_N\}$, and the feature vector of a sample as $\mathbf{x} \in X \subseteq \mathbb{R}^n$, where n is the size of the feature space X, a multi-label classifier implements a decision function $f : X \to \{+1, -1\}^N$, where the value $+1$ (-1) in the k-th element of $f(\mathbf{x})$ means that the sample $\mathbf{x}$ is labelled as (not) belonging to ω_k. Accordingly, precision for the k-th class, denoted as p_k, is the probability that a sample belongs to ω_k, given that it is labelled as such: $p_k = \mathrm{P}(\mathbf{x} \in \omega_k \mid f_k(\mathbf{x}) = 1)$. Recall (r_k) is the probability that a sample is correctly labelled as belonging to ω_k: $r_k = \mathrm{P}(f_k(\mathbf{x}) = 1 \mid \mathbf{x} \in \omega_k)$. Ideally, both precision and recall should equal 1. However, in practice a higher precision can be attained only at the expense of a lower recall, and vice versa. As limit cases, labelling all samples as belonging to ω_k leads to $p_k = 0$ and $r_k = 1$, why labelling all samples as not belonging to ω_k leads to $p_k = 1$ and $r_k = 0$.

To obtain a scalar performance measure, the Van Rijsbergen's F measure is often used. For a class ω_k it is defined as:

$$F_{\beta,k} = \frac{1 + \beta^2}{\beta^2/p_k + 1/r_k},\tag{1}$$

where the parameter $\beta \in [0, +\infty]$ weigh the relative importance of precision and recall: $\beta < 1$ gives a higher weight to recall, while the opposite happens for $\beta > 1$.

Precision and recall can be estimated from a multi-label data set as:

$$\hat{p}_k = \frac{TP_k}{TP_k + FP_k}, \quad \hat{r}_k = \frac{TP_k}{TP_k + FN_k},\tag{2}$$

where TP_k (true positive) and FP_k (false positive) are respectively the number of samples correctly and erroneously labelled as belonging to ω_k, while FN_k (false negative) is the number of samples erroneously labelled as not belonging to ω_k. The F measure can be estimated by replacing the estimates of precision and recall of eq. (2) into eq. (1).

For a multi-label classifier, the global precision and recall over all categories can be computed either by macro- or micro-averaging the class-related values, depending on application requirements [12]. We will denote macro- and micro-averaged values respectively with the superscripts 'M' and 'm'. Macro- and micro-averaged precision and recall are defined as:

$$\hat{p}^{\mathrm{M}} = \frac{1}{N} \sum_{k=1\ldots N} \hat{p}_k, \qquad \hat{r}^{\mathrm{M}} = \frac{1}{N} \sum_{k=1\ldots N} \hat{r}_k, \tag{3}$$

$$\hat{p}^{\mathrm{m}} = \frac{\sum_{k=1\ldots N} TP_k}{\sum_{k=1\ldots N} (TP_k + FP_k)}, \qquad \hat{r}^{\mathrm{m}} = \frac{\sum_{k=1\ldots N} TP_k}{\sum_{k=1\ldots N} (TP_k + FN_k)}. \tag{4}$$

The corresponding F measure is defined as [15]:

$$\hat{F}_\beta^{\mathrm{M}} = \frac{1}{N} \sum_{k=1\ldots N} \hat{F}_{\beta,k} = \frac{1}{N} \sum_{k=1\ldots N} (1+\beta^2)/ \left[(1+\beta^2) + \frac{FP_k + \beta^2 FN_k}{TP_k} \right], \tag{5}$$

$$\hat{F}_\beta^{\mathrm{m}} = \frac{1+\beta^2}{\beta^2/\hat{p}^{\mathrm{m}} + 1/\hat{r}^{\mathrm{m}}} = (1+\beta^2)/ \left\{ (1+\beta^2) + \frac{\sum_{k=1}^{N}(FP_k + \beta^2 FN_k)}{\sum_{k=1}^{N} TP_k} \right\}. \tag{6}$$

Let us now consider how to extend the above performance measures to a multi-label classifier with a reject option. A withheld decision for a sample $\mathbf{x}$ and a category ω_k can be denoted with the value 0 as the output of $f_k(\mathbf{k})$. In single-label problems the accuracy attained by a classifier with a reject option is evaluated as the conditional probability that a pattern is correctly classified, given that it has not been rejected. Analogously, precision and recall for a given category, when a reject option is used, can be defined only with respect to non-withheld decisions. It is easy to see that their corresponding probabilistic definition remains the standard one given at the beginning of this section, which only considers the case $f_k(\mathbf{k}) = 1$, thus excluding withheld assignments (namely, the case when $f_k(\mathbf{k}) = 0$). Consequently, also the F measure can still be defined as in Eq. (1). The estimate of these measures on a given data set can be obtained using again Eq. (2), but taking into account only non-withheld category assignments in the computation of TP_k, FN_k and FP_k. The micro- and macro-averaged values can be computed in the same way using Eqs. (3)–(6).

In the rest of this paper we will consider only the F measure (both macro- and micro-averaged), as it is widely used in multi-label tasks, is easier to handle being a scalar measure, and can be used to find a trade-off between precision and recall [15].

4   Maximising the F Measure for a Given Cost of Rejections

In this section we address the issue of how to define a decision function $f(\mathbf{x}) = \{f_1(\mathbf{x}), \ldots, f_N(\mathbf{x})\} \in \{-1, 0, +1\}^N$ for a N-category multi-label classifier with a reject option, with the constraint that either each or none of the $f_k(\mathbf{x})$ equals 0, according to the approach discussed in Sect. 2. As explained in Sect. 2, the goal is to maximise the classification accuracy on non-rejected samples, with the constraint that up to a given fraction of samples can be rejected. We will denote such fraction as r_{max}.

To decide whether a given sample $\mathbf{x}$ has to be rejected or not, by analogy with approaches widely used in single-label problems we would like to define a measure of "classification reliability" $R(\mathbf{x})$ and a rejection threshold T, such that a sample $\mathbf{x}$ is rejected if $R(\mathbf{x}) < T$, and is automatically classified otherwise. The value of T has to be set according to the desired rejection rate $r_{\max}$, usually from validation data. To define a classification reliability measure, one could estimate the effect of rejecting a sample on the F measure: intuitively, the higher is the F measure obtained after rejecting a given sample $\mathbf{x}$ belonging to any set of samples S, the less reliable is its automatic classification. Formally, the sample $\mathbf{x}^* \in S$ which is classified with the lowest reliability is given by:

$$\mathbf{x}^* = \arg\max_{\mathbf{x} \in S} \hat{F}_\beta(S - \{\mathbf{x}\}) \,, \tag{7}$$

where $\hat{F}_\beta(A)$ denotes the value of the F measure (either macro- or micro-averaged) evaluated on the set of samples A. Accordingly, $R(\mathbf{x})$ could be defined as a monotonic decreasing function of (an estimate of) $\hat{F}_\beta(S - \{\mathbf{x}\})$: the higher $\hat{F}_\beta(S - \{\mathbf{x}\})$, the less reliable the classification of $\mathbf{x}$.

Consider first the micro-averaged F measure of Eq. (6). Maximising $\hat{F}_\beta^m(S - \{\mathbf{x}\})$ amounts to maximise the term

$$\frac{TP(S) - TP(\mathbf{x})}{(FP(S) + \beta^2 FN(S)) - (FP(\mathbf{x}) + \beta^2 FN(\mathbf{x}))} \,, \tag{8}$$

where $FP(Z)$ denotes the number of false positive errors made by the classifier on the set of samples Z, while the meaning of $TP(Z)$ and $FN(Z)$ is similar. Unfortunately, while in single-label problems the contribution of the classification outcome (either correct or wrong) of a given sample to the expected risk does not depend on the outcome of the other samples, it turns out that this does not hold for multi-label problems when classification performance is evaluated using the F measure. Indeed, it is easy to see that the value of Eq. (8) depends not only on the rejected sample $\mathbf{x}$, but also on all the other samples.

Nevertheless, the analysis of Eq. (8) reveals that, under some conditions on the values of its terms, the contribution of a sample $\mathbf{x}$ *does not* depend on the other samples. In particular, under such conditions it can be shown that the individual sample $\mathbf{x}^*$ whose rejection maximises $\hat{F}^m$ can be found as follows:

$$\mathbf{x}^* = \min_{\mathbf{x} \in S} \frac{TP(\mathbf{x}) + A}{FP(\mathbf{x}) + \beta^2 FN(\mathbf{x}) + B} \,, \tag{9}$$

where A and B are two arbitrary positive constants.[1] Whether or not the conditions mentioned above hold is however unknown in practice. Therefore, (9) can be used to define only a suboptimal classification reliability measure to be used for any $\mathbf{x}$. In this paper we define $R(\mathbf{x})$ exactly as the right-hand side of (9):

$$R(\mathbf{x}) = \frac{TP(\mathbf{x}) + A}{FP(\mathbf{x}) + \beta^2 FN(\mathbf{x}) + B} \,. \tag{10}$$

---

[1] Due to lack of space, the proof of these properties is reported here, and can be found at http://prag.diee.unica.it/pra/bib/pillai_iciap2011_rj.

As explained above, any pair of values $A > 0$ and $B > 0$ can be used, if the conditions mentioned above hold. To take into account the cases when they do not hold, we can set A and B such that $R(\mathbf{x})$ approximates the value of expression (8). Namely, we can set $A = \hat{TP}(S)$, $B = \hat{FP}(S) + \beta^2 \hat{FN}(S)$, where $\hat{TP}(S)$, $\hat{FP}(S)$ and $\hat{FN}(S)$ are estimated from validation data. The values $TP(\mathbf{x})$, $FP(\mathbf{x})$ and $FN(\mathbf{x})$ can be estimated from validation data as well. For instance, $TP(\mathbf{x})$ (true positives) can be estimated as the number of correct category assignments on the subset of the K validation samples nearest to $\mathbf{x}$, for some K. An alternative method can be used for classifiers which provide a score $s_k(\mathbf{x}) \in \mathbb{R}$ for each class ω_k, as many multi-label classifiers do: one can consider for each class ω_k the K validation samples whose scores are closest to $s_k(\mathbf{x})$.

Consider now the macro-averaged F measure of Eq. (5). It is not difficult to see that using the criterion (7) to define a classification reliability measure, the contribution of a sample $\mathbf{x}$ is not independent on the other samples, similarly to the case of the micro-averaged F measure. However, note that F_β^{M} is defined as the mean of $F_{\beta,k}$ of the N classes. It turns out that under some conditions on $TP_k(\mathbf{x})$, $FP_k(\mathbf{x})$ and $FN_k(\mathbf{x})$, the analogous of Eq. (9) holds for the individual $F_{\beta,k}$, for any $A_k, B_k > 0$. As a suboptimal classification reliability measure we chose therefore to use:

$$R(\mathbf{x}) = \frac{1}{N} \sum_{k=1...N} \frac{TP_k(\mathbf{x}) + A_k}{FP_k(\mathbf{x}) + \beta^2 FN_k(\mathbf{x}) + B_k} , \tag{11}$$

where the values A_k, B_k, $FP_k(\mathbf{x})$, $FN_k(\mathbf{x})$ and $TP_k(\mathbf{x})$ can be estimated form validation data as explained above, independently for each category ω_k.

5   Experimental Evaluation

The aim of our experiments was to evaluate the trade-off between automatic classification accuracy and the fraction of rejected samples that can be attained by a multi-label classifier using the approach proposed in this paper.

The experiments were carried out on three widely used benchmark data sets, related to two text categorisation and one image annotation task: the "ModApte" version of "Reuters 21578",[2] the Heart Disease subset of the Ohsumed data set [8], and the Scene data set[3]. Their main characteristics are reported in Table 1.

The *bag-of-words* feature model with the term frequency–inverse document frequency (tf-idf) features [12] was used for Reuters and Ohsumed. A feature selection pre-processing step was carried out for Reuters and Ohsumed, through a four-fold cross-validation on training samples, by applying stemming, stopword removal and the information gain criterion. This lead to the selection of $15,000$ features for both data sets.

To implement a N-class multi-label classifier we used the well known *binary relevance* approach. It consists in independently constructing N two-class classifiers using the one-vs-all strategy [12,13]. We used as the base two-class classifier

---

[2] http://www.daviddlewis.com/resources/testcollections/reuters21578/
[3] http://www.csie.ntu.edu.tw/~cjlin/libsvmtools/datasets/multilabel.html

Table 1. Characteristics of the three data sets used in the experiments

Data set	Reuters	Ohsumed	Scene
N. of training samples	7769	12775	1211
N. of testing samples	3019	3750	1196
Feature set size	18157	17341	295
N. of classes	90	99	6
Distinct sets of classes	365	1392	14
N. of labels per sample (avg./max.)	1.23 / 15	1.492 / 11	1.06 / 3
N. of samples per class (min./max.)	1.3E-4 / 0.37	2.4E-4 / 0.25	0.136 / 0.229

a support vector machine (SVM) implemented by the libsvm software [3]. A SVM linear kernel was used for Reuters and Ohsumed, as it is considered the state of the art classifier for text categorisation tasks. A radial-basis function (RBF) kernel was used for Scene instead. The C parameter of the SVM learning algorithm was set to the libsvm default value of 1. The σ parameter of the RBF kernel, defined as $K(\mathbf{x}, \mathbf{y}) = \exp\ -||\mathbf{x} - \mathbf{y}||^2/2\sigma$, was estimated by a four-fold cross-validation on training samples.

Since the output of a SVM is a real number, a threshold has to be set to decide whether labelling or not an input sample as belonging to the corresponding class. The N threshold values can be chosen as the ones which optimise the considered performance measure. In these experiments we used the F measure with $\beta = 1$. To maximise the macro-averaged F measure of Eq. (5), it is known that the threshold can be set by independently maximising the individual F measure of each class, Eq. (1) [15]. No optimal algorithm exists for maximising the micro-averaged F measure instead. We used the suboptimal iterative maximisation algorithm recently proposed in [5]. In both cases the thresholds were estimated through a five-fold cross-validation on training data.

In the experiments several values of the rejection rate r_{max} were considered, ranging in $[0, 0.3]$ with a step of 0.05. For each r_{max} value, we implemented a decision rule with the reject option by using the reliability measures $R(\mathbf{x})$ of Eq. (10) and Eq. (11), respectively when the micro- and macro-averaged F measure was used. For any input sample $\mathbf{x}$, the values of TP, FP and FN in $R(\mathbf{x})$ were estimated using the scores $s_k(\mathbf{x})$, $k = 1 \ldots N$ of the SVMs, as described in Sect. 4. To this aim, we estimated the score distribution for each class on training samples, using 20 bins histograms, where the bins correspond to disjoint intervals of the score range. The rejection threshold T was set to the value that lead to the desired rejection rate r_{max} on training samples. Note that for $r_{max} = 0$ we obtain a standard multi-label classifier without a reject option.

For each data set ten runs of the experiments were carried out, using 80% of the patterns of the original training set. To this aim, ten different training sets were obtained by randomly partitioning the original one into ten disjoint subsets of identical size, and using at each run only eight partitions as the training set. The original test set was used at each run.

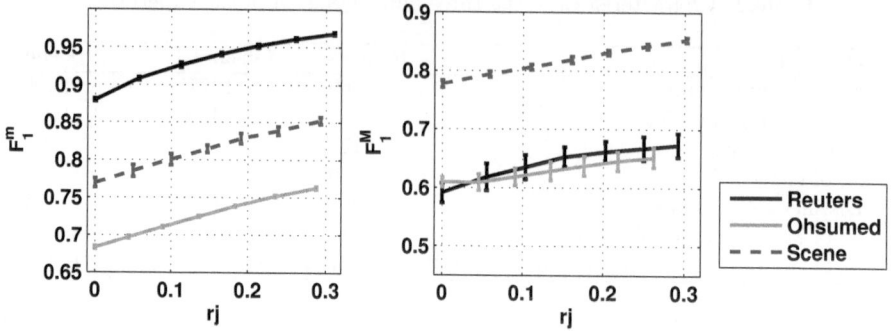

Fig. 1. Test set averaged F_1^m (left) and F_1^M (right) versus the rejection rate on the three data sets. The standard deviation is denoted by vertical bars.

In Fig. 1 we report the average micro- and macro-averaged F measure over the ten runs, as a function of r_{max}. The standard deviation is also reported as vertical bars. Note that the decision thresholds of the N two-class SVM classifiers were computed by optimising the same performance measure (either the micro- or macro-averaged F measure) used to evaluate the classifier performance.

The results in Fig. 1 show that the classification accuracy attained on non-rejected samples always increases as the rejection rate increases. In particular, rejecting up to 30% of the samples, the accuracy improvements are quite remarkable for the micro-averaged F measure, and also for the macro-averaged one in the Scene data set, taking also into account the small standard deviation. Another relevant result is that the rejection rate observed in the test set was always very close to the desired rejection rate ($r_{max} = 0.5, 1.0, 1.5, 2.0, 2.5, 3.0$), which was set on training samples through the choice of the threshold T. This can be seen in Fig. 1, where the rejection rates correspond to the position of the standard deviation bars.

6   Conclusions

We proposed two approaches to implement a reject option in multi-label classifiers, aimed at reducing the manual annotation time in tasks like text categorisation and image annotation (either by using the tagging or browsing approach), attaining at the same time a higher classification accuracy on automatically classified samples than the one which can be obtained without a reject option. We also derived a classification reliability measure to decide whether a sample has to be rejected or not, for the case when the tagging approach is used, with the aim of maximising both the macro- and micro-averaged F measure on non-rejected samples. Reported experimental results related to text categorisation and image annotation tasks provided evidence that the proposed approach can allow to significantly improve the accuracy of an automatic classifier, even when only 30% of samples are rejected and must be manually labelled.

We mention two issues to be further investigated. One is the definition of more accurate reliability measures, especially for the macro-averaged F measure. The other one stems from the novel manual annotation approach proposed in [14], which combines tagging and browsing, and is more efficient than both of them for some applications. Accordingly, a hybrid rejection approach obtained as the combination of the two ones identified in Sect. 2 can be devised for this hybrid tagging-browsing approach.

Acknowledgements. This work was partly supported by a grant from Regione Autonoma della Sardegna awarded to Ignazio Pillai, PO Sardegna FSE 2007-2013, L.R.7/2007 "Promotion of the scientific research and technological innovation in Sardinia".

References

1. Aronson, A., Rogers, W., Lang, F., Névéol, A.: 2008 report to the board of scientific counselors (2008), http://ii.nlm.nih.gov/IIPublications.shtml
2. Boutell, M.R., Luo, J., Shen, X., Brown, C.M.: Learning multi-label scene classification. Pattern Recognition 37(9), 1757–1771 (2004)
3. Chang, C.C., Lin, C.J.: LIBSVM: a library for support vector machines (2001), software available at http://www.csie.ntu.edu.tw/~cjlin/libsvm
4. Chow, C.K.: On optimum recognition error and reject tradeoff. IEEE Transactions in Information Theory 16(1), 41–16 (1970)
5. Fan, R.E., Lin, C.J.: A study on threshold selection for multi-label. Tech. rep., National Taiwan University (2007)
6. Fumera, G., Pillai, I., Roli, F.: Classification with reject option in text categorisation systems. In: Int. Conf. Image Analysis and Proc. (2003)
7. Fumera, G., Pillai, I., Roli, F.: A Two-Stage Classifier with Reject Option for Text Categorisation. In: Structural, Syntactic, and Statistical Patt. Rec. (2004)
8. Lewis, D.D., Schapire, R.E., Callan, J.P., Papka, R.: Training algorithms for linear text classifiers. In: SIGIR, pp. 298–306 (1996)
9. Nowak, S., Huiskes, M.: New strategies for image annotation: Overview of the photo annotation task at imageclef 2010. Working Notes of CLEF 2010 (2010)
10. Pudil, P., Novovicova, J., Blaha, S., Kittler, J.V.: Multistage pattern recognition with reject option. In: ICPR, pp. II:92–II:95 (1992)
11. Ruiz, M., Aronson, A.: User-centered evaluation of the medical text indexing (mti) system (2007), http://ii.nlm.nih.gov/IIPublications.shtml
12. Sebastiani, F.: Machine learning in automated text categorization. ACM Computing Surveys 34(1), 1–47 (2002)
13. Tsoumakas, G., Katakis, I., Vlahavas, I.: Mining multi-label data. In: Data Mining and Knowledge Discovery Handbook, pp. 667–685 (2010)
14. Yan, R., Natsev, A., Campbell, M.: An efficient manual image annotation approach based on tagging and browsing. In: Workshop on Multimedia Inf. Retr. on The Many Faces of Multimedia Semantics, pp. 13–20 (2007)
15. Yang, Y.: A study of thresholding strategies for text categorization. In: Int. Conf. on Research and Development in Information Retrieval, New York, USA (2001)

Improving Image Categorization by Using Multiple Instance Learning with Spatial Relation

Thanh Duc Ngo[1,2], Duy-Dinh Le[2,1], and Shin'ichi Satoh[2,1]

[1] The Graduate University for Advanced Studies (Sokendai), Japan
[2] National Institute of Informatics, Tokyo, Japan
{ndthanh,ledduy,satoh}@nii.ac.jp

Abstract. Image categorization is a challenging problem when a label is provided for the entire training image only instead of the object region. To eliminate labeling ambiguity, image categorization and object localization should be performed simultaneously. Discriminative Multiple Instance Learning (MIL) can be used for this task by regarding each image as a bag and sub-windows in the image as instances. Learning a discriminative MI classifier requires an iterative solution. In each round, positive sub-windows for the next round should be selected. With standard approaches, selecting only one positive sub-window per positive bag may limit the search space for global optimum; meanwhile, selecting all temporal positive sub-windows may add noise into learning. We select a subset of sub-windows per positive bag to avoid those limitations. Spatial relations between sub-windows are used as clues for selection. Experimental results demonstrate that our approach outperforms previous discriminative MIL approaches and standard categorization approaches.

Keywords: Image Categorization, Multiple Instance Learning, Spatial Relation.

1 Introduction

We investigated image categorization using Multiple Instance Learning (MIL). Image categorization is a challenging problem especially when a label is provided for a training image only instead of the object region. Low categorization accuracy may result because the object region and background region within one training image share the same object label. To eliminate labeling ambiguity, image categorization and object localization should be simultaneously performed. In order to do that, one can use MIL, which is a generalization of standard supervised learning. Unlike standard supervised learning in which the training instances are definitely labeled, in the MIL setting, labels are only available for groups of instances called bags. A bag is positive if it contains at least one positive instance. Meanwhile, all instances in negative bags must be negative. Given training bags and instances that satisfy MIL labeling constraints, MIL approaches can learn to classify unlabeled bags as well as unlabeled instances

G. Maino and G.L. Foresti (Eds.): ICIAP 2011, Part I, LNCS 6978, pp. 108–117, 2011.

in the bags. Thus, if we regard each image as a bag and sub-windows in images as instances, we can perform image categorization and object localization simultaneously using MIL.

Several MIL approaches have been proposed [1,2,3,4,5,6,7]. Empirical studies [2,4,7] demonstrate that generative MIL approaches perform worse than discriminative MIL approaches on benchmark datasets, because of their strict assumption on compact clusters of positive instances in the feature space. Thus, it is more appealing to tackle image categorization by using discriminative MIL approaches. In a brief overview, discriminative MIL approaches can be found in [5,4,6,7]. Andrews et al. [5] introduce a framework in which MIL is considered in different maximum margin formulations. A similar formulation of [5] can be found in [9]. DD-SVM presented in [4] trains an SVM for bags in a new feature space constructed from a mapping model defined by the local extremums of the Diverse Density function on instances of positive bags. In contrast, MILES [6] uses all instances in all training bags to construct the mapping model without applying any instance selection method explicitly. IS-MIL [7] then propose an instance selection method to tackle large-scale MIL problems. Because [4,6,7] heavily rely on bag-instance mapping process which is out of scope, we address our work to the framework proposed in [5].

In this paper, we extend the framework in [5] using spatial relations between sub-windows. Although spatial relation information have shown their important role in computer vision tasks [10,11,13,14], there is a few of MIL works utilizing such information. Zha et al. [12] introduced a MIL approach which captures the spatial configuration of the region labels. However, their work target to multi-label MIL problem and spatial relations between segmented regions. Instead of that, we investigate single-label MIL problem and overlapping relations between sub-windows. In the framework [5], learning a discriminative MI classifier is formulated as a non-convex problem and requires an iterative solution. In each round, positive training sub-windows (i.e. instances) for the next round should be selected with certain criteria. With original criteria, selecting only one positive sub-window per positive bag may limit the search space for the global optimum; meanwhile, selecting all temporal positive sub-windows may add noise into learning. We propose to select a subset of sub-windows per positive bag to avoid those limitations. Spatial relations between sub-windows are used as clues for selection. We directly enforce sub-windows spatial relations into learning by selecting sub-windows of the subset based on their overlapping degree with the most discriminative sub-window. Experimental results demonstrate the effectiveness of our approach.

2   Support Vector Machine for Multiple Instance Learning

In statistical pattern recognition, given a set of labeled training instances coupled with manual labels $(x_i, y_i) \in \mathcal{R}^d \times \mathcal{Y}$, the problem is how to obtain a classification function going from instances to labels $f : \mathcal{R}^d \to \mathcal{Y}$. In the binary case,

$\mathcal{Y} = \{-1, 1\}$ indicates positive or negative labels associated with instances. MIL generalizes this problem by relaxing the assumption on instance labeling. Labels are given for bags, which are groups of instances. A bag is assigned a positive label if and only if at least one instance of the bag is positive. Meanwhile, a bag is negative if all instances of the bag are negative. Formally, given a set of input instances $x_1, \ldots, x_n$ grouped into non-overlapping bags $B_1, \ldots, B_m$, with $B_I = \{x_i : i \in I\}$ and index sets $I \subseteq \{1, \ldots, n\}$. Each bag B_I is then given a label Y_I. Labels of bags are constrained to express the relation between bag and instances in the bag as follows: if $Y_I = 1$ then at least one instance $x_i \in B_I$ has label $y_i = 1$, otherwise, if $Y_I = -1$ then all instances $x_i \in B_I$ are negative: $y_i = -1$. A set of linear constraints can be used to formulate the relation between bag labels Y_I and instance labels y_i:

$$\sum_{i \in I} \frac{y_i + 1}{2} \geq 1, \forall I : Y_I = 1 \quad \text{and} \quad y_i = -1, \forall I : Y_I = -1, \tag{1}$$

or compactly represented as: $Y_I = \max_{i \in I} y_i$.

Learning the discriminative classifiers entails finding a function $f : \mathcal{X} \to \mathcal{R}$ for a multiple-instance dataset with the constraint $Y_I = \operatorname{sgn} \max_{i \in I} f(x_i)$.

3   The Former Approaches of SVM-Based Multiple Instance Learning

Andrews et al. [5] proposed two learning approaches based on SVM with different margin notions. The first approach, called mi-SVM, aims at maximizing the instance margin. Meanwhile, the second approach, called MI-SVM, tries to maximize the bag margin. Both mi-SVM and MI-SVM can be formed as mixed integer quadratic programs and need heuristic algorithms to be solved. The algorithms have an outer loop and an inner loop. The outer loop sets the values for the integer variables. Meanwhile, the inner loop trains a standard SVM. The outer loop stops if none of the integer variables changes in consecutive rounds.

The mixed integer formulation of mi-SVM based on the generalized soft-margin SVM can be presented as:

$$\begin{aligned}
\min_{\{y_i\}} \min_{\{w, b, \xi\}} \quad & \frac{1}{2} \|w\|^2 + C \sum_i \xi_i \\
\text{subject to} \quad & \forall i : y_i (\langle w, x_i \rangle + b) \geq 1 - \xi_i, \, \xi_i \geq 0, \\
& y_i \in \{-1, 1\}, \text{ and (1) hold.}
\end{aligned} \tag{2}$$

In (2), labels y_i of instances x_i not belonging to any negative bag are treated as unknown integer variables. The target here is to find a linear discriminative *MI-separating* that satisfies the constraint wherein at least one positive instance from each positive bag lies in the positive half-space, while all instances belonging to all negative bags are in the negative half-space.

In MI-SVM, Andrews et al. introduce an alternative approach to the MIL problem. The notion of a margin is extended from individual instances to bags.

The margin of a positive bag is defined as the margin of *"the most positive"* instance of the bag. Meanwhile, the margin of a negative bag is defined by the margin of *"the least negative"* instance of the bag. Let $x_{mm(I)}$ be the instance of bag B_I and has maximum margin to the hyper-plane. Then, MI-SVM can be formulated as follows:

$$\min_{\{y_i\}} \min_{\{w,b,\xi\}} \quad \frac{1}{2}\|w\|^2 + C \sum_I \xi_I$$

$$\text{subject to} \quad \forall I : Y_I = -1 \wedge -\langle w, x_i \rangle - b \geq 1 - \xi_I, \, \forall i \in I,$$

$$\text{or} \quad Y_I = 1 \wedge \langle w, x_{mm(I)} \rangle + b \geq 1 - \xi_I, \text{ and } \xi_I \geq 0 \tag{3}$$

4 Support Vector Machine with Spatial Relation for Multiple Instance Learning

MI-SVM and mi-SVM can be applied to image categorization by regarding each image as a bag and sub-windows in images as instances. However, their formulations and heuristic solutions do not involve spatial relations of sub-windows despite such information being extremely meaningful. Surrounding sub-windows always contain highly related information with respect to visual perception. If a sub-window in image is classified as a positive instance, it is supposed to be associated with the object label given to the class. In that sense, its neighboring sub-windows should be positive also. For example, if a sub-window tightly covers an object, its slightly surrounding sub-windows also contain that object.

Moreover, in terms of learning, the original approaches require a heuristic iterative solution to obtain the final discriminative classifier. In each learning round, candidate positive instances must be selected for the next round. Thus, positive instance selection criterion is the key step in the learning process. With mi-SVM, selecting all positive instances in the current round may add noisy instances to learning. Meanwhile, selecting only the most positive instance which has largest margin in the current round, as in MI-SVM, may limit the search space for the global optimum. To avoid such limitations, we propose to select a subset of instances as candidate positive instances for the next learning round. Spatial relations between instances (i.e. sub-windows) can be used as clues for selection. Therefore, we extend the framework proposed by Andrews et al. to take the spatial relation between sub-windows into account. Positive candidate selection criteria of the approaches are illustrated in Figure 1 .

In our extension, the notion of a bag margin is used as in the MI-SVM formulation. This means the margin of a positive bag is defined as the margin of *"the most positive"* instance of the bag. However, we directly enforce the spatial relations between *"the most positive"* instance with its spatially surrounding instances by adding constraints to the optimization formulation. Here, let $x_{mm(I)}$ be the instance of bag B_I has maximum margin with respect to the hyper-plane, and $\mathcal{SR}(x_{mm(I)}, T)$ denotes the set of $x_{mm(I)}$ and instances that surround $x_{mm(I)}$ with respect to the overlap parameter T. An instance belongs to $\mathcal{SR}(x_{mm(I)}, T)$ if its overlap degree with $x_{mm(I)}$ is greater or equal to T, where

$0 < T \leq 1$. The overlap degree between two instances (i.e. sub-windows) is the fraction of their overlap area over their union area. To this end, our formulation can be expressed as follows:

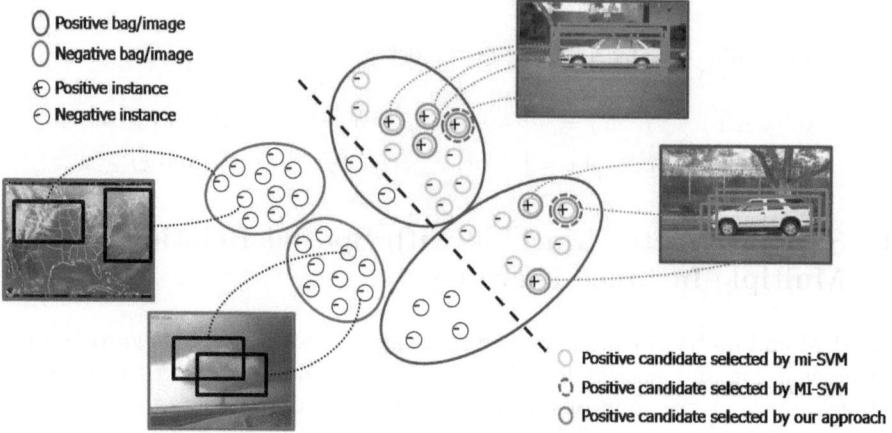

Fig. 1. Illustration of positive candidate selection for the next learning round by different approaches. mi-SVM selects all temporal positive instances (*orange*). MI-SVM selects only the most positive instance per positive bag (*dash-purple*). Meanwhile, our approach selects a subset of spatially related instances (*green*) per positive bag based on their overlap degree with the most positive instance of the bag.

$$\min_{\{y_i\}} \min_{\{w,b,\xi\}} \quad \frac{1}{2}\|w\|^2 + C \sum_I \xi_I$$

$$\text{subject to} \quad \forall I: Y_I = -1 \wedge -\langle w, x_i \rangle - b \geq 1 - \xi_I, \forall i \in I, \tag{4}$$
$$\text{or} \quad Y_I = 1 \wedge \langle w, x^* \rangle + b \geq 1 - \xi_I,$$
$$\forall x^* \in \mathcal{SR}(x_{mm(I)}, T), 0 < T \leq 1, \text{ and } \xi_I \geq 0$$

This formulation can be cast as a mixed integer program in which integer variables are the selectors of $x_{mm(I)}$ and instances in $\mathcal{SR}(x_{mm(I)}, T)$. This problem is hard to solve for the global optimum. However, we exploit the fact that if integer variables are given, the problem reduces to a quadratic programming (QP) that can be solved. Based on that insight, our solution is as follows.

Pseudo code for heuristic algorithm

```
Initialize: for every positive bag B_I
    Compute x_I =   ∑_{i ∈ I} x_i/|I|.
    SR_I  =  x_I.
REPEAT
```

```
    - Compute QP solution w, b for dataset with positive
          samples {SR_I : Y_I = 1} and negative samples {x_i : Y_I = -1}.
    - Compute outputs f_i = ⟨w, x_i⟩ + b for all x_i in positive bags.
    - FOR (every positive bag B_I)
          Set x_I = x_mm(I),  mm(I) = arg max_{i∈I} f_i
          SR_I  =  FindSurround(x_I, T)
    - END
WHILE ({mm(I)} have changed)
OUTPUT (w, b)
```

In our pseudo code, $FindSurround(x_I, T)$ is the function to find instances (i.e. sub-windows) surrounding x_I and have an overlap degree with x_I greater than or equal to T. The greater T is chosen, the fewer instances (i.e. sub-windows) surrounding x_I are selected. Thus, T can be considered as a trade-off parameter for expanding the search space as well. T is a predefined number and is fixed throughout learning iteration. The optimal T is obtained automatically by cross validating on the training set. Additionally, negative candidates of all learning rounds are instances of the negative bags.

5   Experiments

5.1   Dataset

We perform experiments on Caltech benchmark datasets.

- **Caltech 4** contains images of 4 object categories: airplanes (1,075 images), cars_brad (1,155 images), faces (451 images), motorbikes (827 images), and a set of 900 clutter background images.
- **Caltech 101** consists of images in 101 object categories and a set of clutter background images [8]. Each object category contains about 40 to 800 images.

Ground-truth annotations indicating object's locations in images are available for all object categories (but cars_brad category in Caltech 4). These are challenging datasets because of their large variations in object appearance and background. Some example images are shown in Figure 2.

We evaluate the performance of the approaches on binary categorization tasks which are distinguishing images of each object category from background images. On the Caltech 101 dataset, with each binary classification task, a set of 15 positive images taken from one object category and 15 negative images from the background category are given for training; 30 other images from both categories are used for testing. The correlative numbers of positive images, negative images and testing images on Caltech 4 dataset are 100, 100 and 200 respectively. All images are randomly selected.

Fig. 2. Example images taken from Caltech 101. From top to bottom are images of airplanes, cellphones, faces and motorbikes respectively.

5.2   Bag and Instance Representation

In order to apply the MIL approaches, we treat bags as images and instances of a bag as sub-windows in the image. We employ the standard Bag-of-Word (BoW) approach for feature representation. First, on each image, we sample a set of points using a grid. The sampling grid has an 8-pixel distance between adjacent points. Then, we use the SIFT descriptor to extract SIFT feature at each point. The SIFT descriptor frame has a 16-pixel width. All descriptors are then quantized using a visual codebook with 100 visual words obtained by applying K-Means to 100,000 training descriptors. Finally, the sub-windows of the image are represented by using a histogram of visual words appearing inside the sub-window region.

5.3   Evaluated Approaches

We compare our approach with the original SVM-based MIL approaches - mi-SVM and MI-SVM - and two other standard approaches called GH and MA. GH denotes a traditional approach in which SVM is used to classify images represented by a histogram of visual words on the whole image region (GH stands for Global Histogram). Meanwhile, MA is an approach that uses tight object rectangles given manually as positive examples and a set of randomly selected windows from negative images - ten windows per negative image - as negative examples for training (MA stands for Manual Annotation). The measure for comparison is the accuracy ratio with respect to image classification performance. To obtain the best performance of the approaches for fairness, all parameters are optimized. Kernel parameters for SVM and overlap threshold T of our approach are automatically obtained by using the grid-search approach together with 5-fold cross validation.

5.4   Experimental Results

Table 1 and Table 2 list the classification performances of the approaches on Caltech 4 and Caltech 101. Our proposed approach is superior to the others in most object classes. This means the most discriminative instances found by our approach are more meaningful than the one selected by MI-SVM and is also more discriminative than the object regions classified by MA. Moreover, these results prove that our arguments on the effectiveness of using the spatial relation and the limitations of the instance selection criteria of mi/MI-SVM are valid. Because of adding all possible positive instances, mi-SVM also adds more noise to learning and its performance consequently suffers. MI-SVM has a better accuracy than mi-SVM, but it is still worse than ours because of its limited search space.

Table 1. Average classification accuracy of the evaluated approaches on Caltech 4. MA: trains SVM using manual annotation of object region in images. GH: trains SVM using global histogram of images. mi/MI-SVM: MIL approaches proposed by Andrews et al [5]. Note that the performance of MA is computed on 3 categories (airplanes, faces and motorbikes) due to the lack of ground-truth object box of the category cars_brad.

Approaches	Average Classification Rate(%)
MA	90.73
GH	94.46
mi-SVM	72.54
MI-SVM	95.74
Ours	**96.28**

Table 2. Average classification accuracy of the evaluated approaches on Caltech 101

Approaches	Average Classification Rate(%)
MA	78.32
GH	83.37
mi-SVM	60.49
MI-SVM	84.25
Ours	**86.89**

Table 3. Average classification accuracy of the evaluated approaches on 10 categories of Caltech 101

	MA	GH	mi-SVM	MI-SVM	Ours
Butterfly	76.7	76.7	53.3	86.7	**93.3**
Camera	70.0	80.0	53.3	73.3	**86.7**
Ceiling_fan	70.0	**80.0**	53.3	66.7	**80.0**
Cellphone	80.0	**90.0**	63.3	83.3	**90.0**
Laptop	80.0	76.7	66.7	76.7	**86.7**
Motorbikes	73.3	**93.3**	63.3	80.0	90.0
Platypus	83.3	90.0	53.3	86.7	**100.0**
Pyramid	**90.0**	90.0	63.3	76.7	**90.0**
Tick	76.7	83.3	56.7	80.0	**90.0**
Watch	**80.0**	**80.0**	53.3	73.3	**80.0**

6   Conclusion and Future Work

We proposed an extension of the SVM-based Multiple Instance Learning framework for image categorization by integrating spatial relations between instances into the learning process. Experimental results on the benchmark dataset show that our approach outperforms state-of-the-art SVM-based MIL approaches as well as standard categorization approaches. To the best of our knowledge, this is the first MIL approach that considers sub-window overlapping relations on image space rather than feature space only. For future work, we want to extend our MIL framework so it can be applied to weakly supervised object localization and recognition.

References

1. Dietterich, T., Lathrop, R., Lozano-Perez, T.: Solving the Multiple-Instance Problem with Axis-Parallel Rectangles. Artificial Intelligence, 31–71 (1997)
2. Maronand, O., Lozano-Perez, T.: A Framework for Multiple Instance Learning. In: Advances in Neural Information Processing Systems, pp. 570–576 (1998)
3. Zhang, Q., Goldman, S.: EM-DD: An Improved Multiple Instance Learning Technique. In: Advances in Neural Information Processing Systems, pp. 1073–1080 (2002)
4. Chen, Y., Wang, J.Z.: Image Categorization by Learning and Reasoning with Regions. Journal of Machine Learning Research, 913–939 (2004)
5. Andrews, S., Tsochantaridi, I., Hofmann, T.: Support Vector Machines for Multiple-Instance Learning. In: Advances in Neural Information Processing Systems, pp. 561–568 (2003)
6. Chen, Y., Bi, J., Wang, J.: MILES: Multiple-Instance Learning via Embedded Instance Selection. IEEE Transactions on Pattern Analysis and Machine Intelligence, 1931–1947 (2006)
7. Fu, Z., Robles-Kelly, A.: An Instance Selection Approach to Multiple Instance Learning. In: IEEE Conference on Computer Vision and Pattern Recognition, pp. 911–918 (2009)

8. Fei-Fei, L., Fergus, R., Perona, P.: Learning generative visual models from few training examples: an incremental Bayesian approach tested on 101 object categories. In: Workshop on Generative-Model Based Vision, IEEE Conference on Computer Vision and Pattern Recognition (2004)

9. Nguyen, M.H., Torresani, L., Torre, F., Rother, C.: Weakly Supervised Discriminative Localization and Classification: A Joint Learning Process. In: IEEE Conference on Computer Vision and Pattern Recognition (2009)

10. Galleguillos, C., Belongie, S.: Context Based Object Categorization: A Critical Survey. In: Computer Vision and Image Understanding (2010)

11. Marques, O., Barenholtz, E., Charvillat, V.: Context Modeling in Computer Vision: Techniques, Implications, and Applications. Journal of Multimedia Tools and Applications (2010)

12. Zha, Z.J., Hua, X.S., Mei, T., Wang, J., Qi, G.J., Wang, Z.: Joint Multi-Label Multi-Instance Learning for Image Classification. In: IEEE Conference on Computer Vision and Pattern Recognition, pp. 1–8 (2008)

13. Divvala, S.K., Hoiem, D., Hays, J.H., Efros, A., Hebert, M.: An Empirical Study of Context in Object Detection. In: IEEE Conference on Computer Vision and Pattern Recognition, pp. 1271–1278 (2009)

14. Wolf, L., Bileschi, S.: A Critical View of Context. International Journal of Computer Vision (2006)

Shaping the Error-Reject Curve of Error Correcting Output Coding Systems

Paolo Simeone, Claudio Marrocco, and Francesco Tortorella

DAEIMI, Università degli Studi di Cassino,
Via G. Di Biasio 43, 03043 Cassino, Italy
{paolo.simeone,c.marrocco,tortorella}@unicas.it

Abstract. A common approach in many classification tasks consists in reducing the costs by turning as many errors as possible into rejects. This can be accomplished by introducing a reject rule which, working on the reliability of the decision, aims at increasing the performance of the classification system. When facing multiclass classification, Error Correcting Output Coding is a diffused and successful technique to implement a system by decomposing the original problem into a set of two class problems. The novelty in this paper is to consider different levels where the reject can be applied in the ECOC systems. A study for the behavior of such rules in terms of Error-Reject curves is also proposed and tested on several benchmark datasets.

Keywords: Error-Reject Curve, reject option, multiclass problem, Error Correcting Output Coding.

1 Introduction

The reduction of misclassification errors is a key point in Pattern Recognition. Such errors, in fact, can have a heavy impact on the applications accomplished by a classification system and can lead to serious consequences. Typically, error costs are defined and are helpful in defining which kind of error is convenient to avoid. However, those costs can be so high that the best choice could be an abstention from the decision so as to demand the last decision to a further and more efficient test. Since even the decision to abstain brings along some costs (e.g. the intervention of a human expert), the best approach is to find the optimal trade off between the numbers of errors and rejects.

The reject option in a classification system was introduced by Chow in [4] which demonstrated how the optimality could be reached when the prior probabilities and the conditional densities for each class were known. Since then, many approaches have been tried to introduce a reject rule for tuning the performance of a classification scheme. For neural networks a criteria to evaluate the reliability of the decision can be fixed as shown in [5,11]. Similar approaches were followed for Support Vector Machines as proposed in [12,3]. Thus, depending on the implementation of the system a criterion has to be found to evaluate the reliability of the decision and to fix a threshold for the application of the

G. Maino and G.L. Foresti (Eds.): ICIAP 2011, Part I, LNCS 6978, pp. 118–127, 2011.

reject rule. A profitable instrument to assess the performance of the reject rule independently from the classification costs is the Error-Reject curve that plots the percentage of errors versus rejects for each decision threshold.

Our work focuses on the application of a rejection rule in a system which face a multiple class problem through a pool of dichotomizers arranged according to an Error Correcting Output Coding (ECOC). Such technique was introduced by [6] to split a multiclass problem in many binary subproblems and has proved to be an efficient way to increase the performance attainable by a single monolithic classifier able to produce multiple outputs. The rationale lies on the capability of the code to correct errors and on the stronger theoretical roots and the better comprehension which characterize popular dichotomizers like Decision Trees or Support Vector Machines. Moreover, ECOC systems have been currently used as a starting point to extend boosting techniques to multiclass problems [7].

An analysis of reject for ECOC systems has been analyzed in our previous paper [10] where a reject rule evaluating the reliability of the final decision was proposed. In this paper to improve the performance of the classification system two consecutive thresholds are applied to focus more on the reliability of each level of the system. Each classifier, in fact, has an *internal* decision level where a first reject can be applied; then, in the decoding stage, an *external* reject threshold can be fixed uniquely based on the observation of the output of the ensemble of classifiers. Meanwhile, since each internal threshold induces an Error-Reject curve plotted according to an external threshold, we also show how to obtain a proper description of the system from this range of curves by using the Error-Reject curve given by their convex hull.

The paper is organized as follows: in section 2 we briefly analyze ECOC framework while in section 3 we introduce the two proposed reject rules. An extended analysis on how such rule modifies the Error-Reject curve is done in 4. Experimental results on many benchmarks data are reported in section 5 while the final section 6 presents some conclusions and some possible future developments.

2   The ECOC Classification System

Several multiclass classification systems use a decomposition of the original problem in many binary subproblems. Among them ECOC has been proved to be one of the more efficient and flexible to the application needs. Each original class ω_i with $i = 1, \ldots, n$ is associated to a *codeword* of length L. The collection of these codewords in a matrix, as shown in table 1, represents a coding matrix $\mathbf{C} = \{c_{hk}\}$ where $c_{hk} \in \{-1, +1\}$. Such matrix maps the original multiple class classification task in n different binary tasks defined by the matrix columns. Binary classifiers can be trained on each of these new binary data sets.

The classification is then performed by feeding each sample $\mathbf{x}$ to all the dichotomizers and collecting their outputs in a vector $\mathbf{o}$ (*output vector*) that is compared with the original coding matrix words. Several decoding rules have been proposed in literature and it has been largely proved that a *loss decoding rule* is the most sensitive and outperforming one [1]. Such rule takes into account

Table 1. Example of a coding matrix of length 15 for a 4 classes problem

classes	codewords						
1	+1	+1	+1	+1	+1	+1	+1
2	−1	−1	−1	−1	+1	+1	+1
3	−1	−1	+1	+1	−1	−1	+1
4	−1	+1	−1	+1	−1	+1	−1

the reliability of the decision evaluating the loss function on the margin of the classifier. For ECOC the margins related to a particular codeword $\mathbf{c_i}$ are given by $c_{ih} f_h(\mathbf{x})$ with $h = 1, \ldots, L$. If we know the original loss function $\mathcal{L}(\cdot)$ of the employed dichotomizers, a global *loss-based distance* can be evaluated as:

$$D_{\mathcal{L}}(\mathbf{c}_i, \mathbf{f}) = \sum_{h=1}^{L} \mathcal{L}(c_{ih} f_h(\mathbf{x})). \tag{1}$$

and thus, the following rule can be defined to predict the k-th class:

$$\omega_k = \arg\min_i D_{\mathcal{L}}(\mathbf{c}_i, \mathbf{f}). \tag{2}$$

3   Two Levels of Rejection for ECOC System

In the description of ECOC technique, it is possible to observe that there are two different levels where a reliability parameter can be evaluated and thus a reject applied. The first simple option is at the output of the classification system where a threshold can be externally set on the output value without any assumption neither on the dichotomizers nor on the coding matrix. We already analyzed this approach [10] by applying a reject option for a loss decoding technique which proved to work sensibly better than other traditional decoding techniques, like ones based on Hamming distance. If we assume a loss value normalized in the range $[0, 1]$, such a criterion (indicated as *Loss Decoding*) can be formalized as:

$$r(\mathbf{f}, t_l) = \begin{cases} \omega_k & \text{if } D_{\mathcal{L}}(\mathbf{c}_k, \mathbf{f}) < t_l, \\ reject & \text{if } D_{\mathcal{L}}(\mathbf{c}_k, \mathbf{f}) \geq t_l. \end{cases} \tag{3}$$

where ω_k is the class chosen according to eq. (2) and $t_l \in [0, 1]$.

A second level of decision can be, instead, localized for each single dichotomizer before grouping the outcomes in the output vector. In fact, each dichotomizer outcome $f_h(\mathbf{x})$ is tipically compared with a threshold τ_h to decide to which of the two classes the sample belongs. This means that $\mathbf{x}$ is assigned to class $+1$ if $f_h(\mathbf{x}) \geq \tau_h$ and to class -1 otherwise.

Independently from the choice of each threshold the most unreliable outcomes will be on its proximity and thus, it could be convenient to reject those samples on each dichotomizer. This can be accomplished by choosing two different thresholds

τ_{h1} and τ_{h2} (with $\tau_{h1} \leq \tau_{h2}$) and defining a reject rule on each binary classifier as:

$$r(f_h, \tau_{h1}, \tau_{h2}) = \begin{cases} +1 & \text{if } f_h(\mathbf{x}) > \tau_{h2}, \\ -1 & \text{if } f_h(\mathbf{x}) < \tau_{h1}, \\ reject & \text{if } f_h(\mathbf{x}) \in [\tau_{h1}, \tau_{h2}]. \end{cases} \tag{4}$$

It is worth remarking that the choice of the thresholds should be made so as to encapsulate the class overlap region into the *reject interval* $[\tau_{h1}, \tau_{h2}]$ and turn most of the errors into rejects. However, to avoid the bias that can occur because of the high differences between each classifier outcomes, instead of choosing the same pair of thresholds for all the dichotomizers, we chose to let all dichotomizers work at the same level of reliability by fixing a common rejection rate ρ. Accordingly, the ROC curve of each dichotomizer has been used to evaluate the pair of thresholds (τ_{h1}, τ_{h2}) such that f_h abstains for no more than ρ samples at the lowest possible error rate [9].

Therefore, considering that the value produced by the classifier in the case of a reject is assumed to be 0, we can apply a reject rule defined as:

$$f_h^{(\rho)}(\mathbf{x}, \tau_{h1}, \tau_{h2}) = \begin{matrix} 0 & \text{if } f_h(\mathbf{x}) \in [\tau_{h1}, \tau_{h2}] \\ f_h(\mathbf{x}) & \text{otherwise} \end{matrix}. \tag{5}$$

It is worth noting that the null value is a possible outcome also for the dichotomizer without the reject option. It corresponds to the particular case when the sample falls on the decision boundary and thus it is assigned neither to the positive nor to the negative label. In this case, the loss calculated on the margin is $\mathcal{L}(c_{ih} f(\mathbf{x})) = \mathcal{L}(0)$, whichever the value of c_{ih}, while it is higher or lower than the "don't care" loss value $\mathcal{L}(0)$ if $f_h(\mathbf{x}) \neq 0$ (depending on whether $c_{ih} f(\mathbf{x})$ is positive or negative). The reject option actually extends such behavior to all the samples whose outcome $f_h(\mathbf{x})$ falls within the reject interval $[\tau_{h1}, \tau_{h2}]$. In this way, a part of the values assumed by the loss is not considered in the final decision procedure, which we indicate as *Trimmed Loss Decoding*.

Loss distance is now modified by the presence of the zero values in the output word $\mathbf{f}^{(\rho)}$ and it is given by:

$$D_{\mathcal{L}}(\mathbf{c}_i, \mathbf{f}^{(\rho)}) = \sum_{h \in I_n} \mathcal{L}(c_{ih} f_h(\mathbf{x})) + |I_z| \cdot \mathcal{L}(0) \tag{6}$$

where I_{nz} and I_z are the sets of indexes of the nonzero values and zero values in the output word, respectively. In practice the loss is given by two contributions, where the second one is independent from the codeword that is compared to the output word. Through this new loss distance, a rejection rule can be immediately applied at the output of the ECOC system by choosing a threshold value t_l:

$$r(\mathbf{f}^{(\rho)}, t_l) = \begin{matrix} \omega_k & \text{if } D_{\mathcal{L}}(\mathbf{c}_k, \mathbf{f}^{(\rho)}) < t_l, \\ reject & \text{if } D_{\mathcal{L}}(\mathbf{c}_k, \mathbf{f}^{(\rho)}) \geq t_l. \end{matrix} \tag{7}$$

where ω_k is the class chosen according to eq. (2).

4   How to Evaluate the Reject Rule

Generally speaking, a reject option is accomplished by evaluating the reliability of the decision taken by the classifier and rejecting such decision if it is lower than some given threshold. A complete description of the classification system with the reject option is given by the *Error-Reject (ER) curve* which plots the error rate $E(t)$ against the reject rate $R(t)$ when varying the threshold t on the reliability estimate. In the ECOC approach when the threshold is applied at the output of the classification system as in eq. (3), it is very simple to build the error-reject curve by varying the threshold t_l and observing the errors and rejects obtained (see fig. 1).

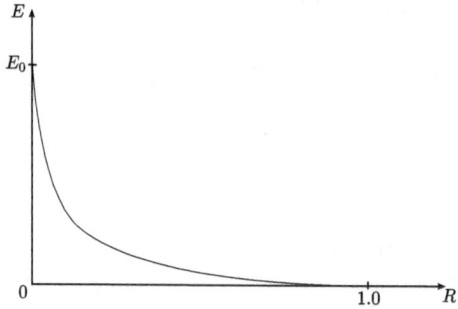

Fig. 1. A typical Error-Reject curve for an external reject rule. E_0 is the error at 0-reject while the error rate becomes null for a reject rate equal to 1.0.

Some remarks have, instead, to be done on the resulting decision rule which depends on two different thresholds (ρ and t_l) while the previously described rules depend only on one parameter. In the previous cases the reject option generates a unique curve where each point is function of only one threshold, while now we have a family of ER curves, each produced by a particular value of ρ. There is thus some ambiguity in defining the ER-curve representative of the performance of the whole system. To solve this problem, let us consider two ER-curves corresponding to two different values ρ_1 and ρ_2 of the internal reject threshold. They can be arranged into two different ways: one of the curve can be completely below the other one (see fig. 2.a) or they can intersect (see fig. 2.b). In the first case, the lower curve (and the corresponding ρ) must be preferred because it achieves a better error rate at the same reject rate (and vice versa). The second case shows different regions in which one of the curves is better than the other one and thus, there is not a curve (and an internal reject threshold value) definitely optimal. Therefore, to obtain an optimal ECOC system under all circumstances, the ER-curve should include the locally optimal parts of the two curves. This is obtained if we assume as ER-curve of the ECOC system the convex hull of the two curves (see fig. 2.c).

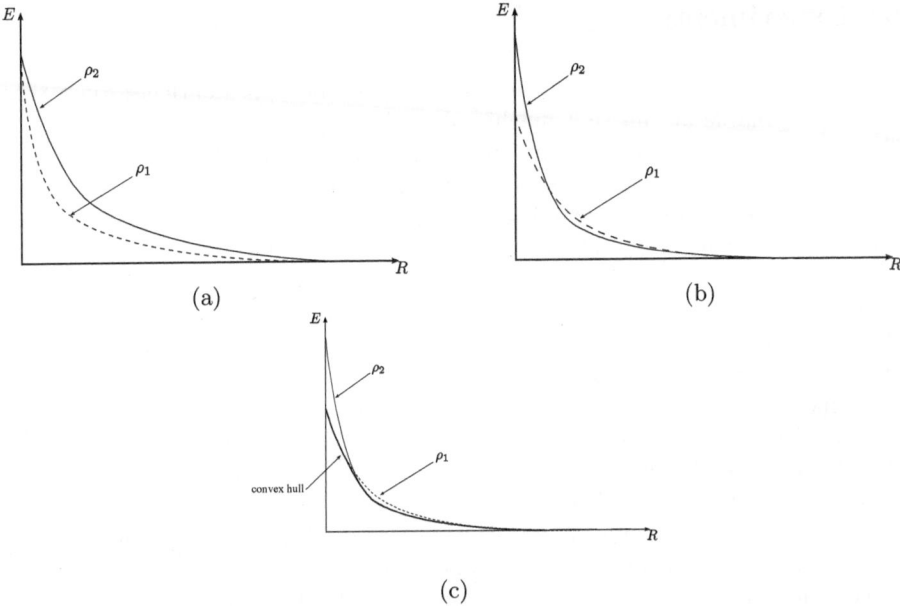

(a)   (b)

(c)

Fig. 2. Different cases for two ER-curves produced by two rejection rates $_1$ and $_2$ when varying the external threshold t_l. (a) The ER-curve produced by $_1$ dominates the curve produced by $_2$. (b) There is no dominating ER-curve. (c) The convex hull of the ER-curves shown in (b) including the locally optimal parts of the two curves.

This can be easily extended to the curves related to all the values considered for ρ so as to assume as the ER-curve of the ECOC system the convex hull of all the curves (see fig. 3).

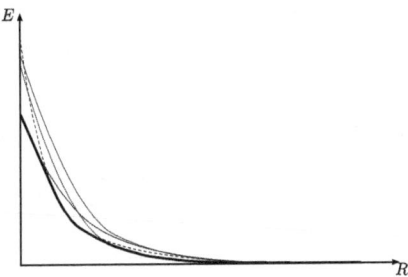

Fig. 3. The ER-curves generated for different   and their convex hull assumed as the ER-curve of the ECOC system

5   Experiments

To test the performance of the proposed reject rules, six multiclass data sets publicly available at the UCI machine learning repository [2] have been used. To avoid any bias in the comparison, 12 runs of a multiple hold out procedure have been performed on all the data sets. In each run, the data set has been split into three subsets: a training set (containing the 70% of the samples of each class), a validation set and a test set (each containing the 15% of the samples of each class). The training set is used to train the base classifiers, the validation set to normalize the outputs into the range $[-1, 1]$ and to calculate the thresholds (τ_{h1}, τ_{h2}) and the test set to evaluate the performance of the classification system. A short description of the data sets is given in table 2. In the same table we also report the number of columns of the coding matrix chosen for each data set according to [6]. As base dichotomizer Support Vector Machines (SVM) have been implemented through the SVM^{Light} [8] software library using a linear kernel and an RBF kernel with $\sigma = 1$. In both cases the C parameter has been set to the default value calculated by the learning procedure and the "hinge" loss $\mathcal{L}(z) = \max\{1 - z, 0\}$ has been adopted.

We show in fig. 4 and fig. 5 the results obtained for the two classifiers in terms of the Error-Reject curves calculated by averaging the Error-Reject curves obtained in the 12 runs of the multiple hold out procedure. The range for the reject rate on the x-axis has been limited to $[0, 0.30]$ since higher reject rates are typically not of interest in real applications. Both figures reports a comparison of the two considered reject rules: Loss Decoding (LD) and TLD (Trimmed Loss Decoding). For the LD the loss output was normalized in the range $[0, 1]$ and consequently the thresholds were varied in this interval with the step 0.01. In the case of TLD we have varied the parameter ρ from 0 to 1 with step 0.05 and, as in the LD case, the loss output has been normalized in the range $[0, 1]$ and the external threshold varied with step 0.01 into the same range.

In the majority of the analyzed cases, the two figures show that the ER-curves generated by the TLD method dominate those of LD. Only in two cases, i.e., Abalone and Vowel with an SVM with RBF kernel, there is a complete equivalence of the two curves. Consequently, we can say that TLD approach

Table 2. Data sets used in the experiments

Data Set	Classes	Features	Length (L)	Samples
Abalone	29	8	30	4177
Ecoli	8	7	62	341
Glass	6	9	31	214
Letter	26	16	63	5003
Vowel	11	10	14	435
Yeast	10	8	31	1484

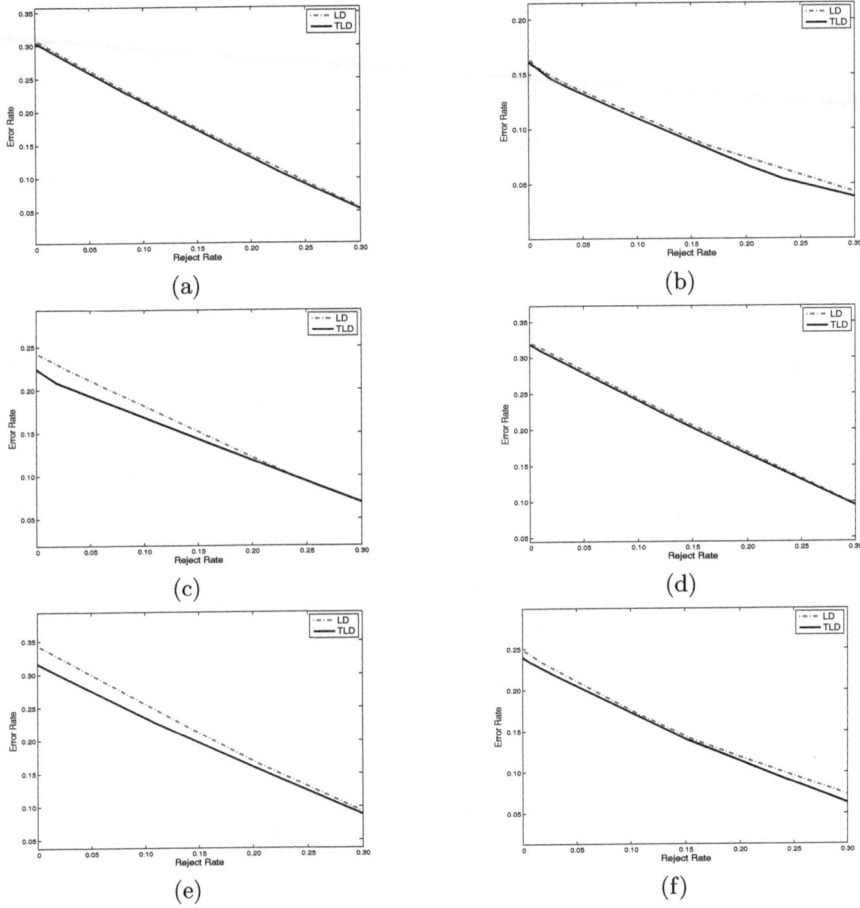

Fig. 4. Error-Reject curves obtained on Abalone (a), Ecoli (b), Glass (c), Letter, (d), Vowel (e) and Yeast (f) with linear SVM

is superior than LD since it allows us to control the individual errors of the base classifiers. Thus, the use of an internal reject rule can be profitably used to improve the performance of the ECOC systems. In conclusion, knowing the architectural details of the base classifiers (i.e. the nature of their outputs and their loss functions), the system has the possibility to face the uncertainty of wrong predictions in a more precise and effective way than a simple external technique.

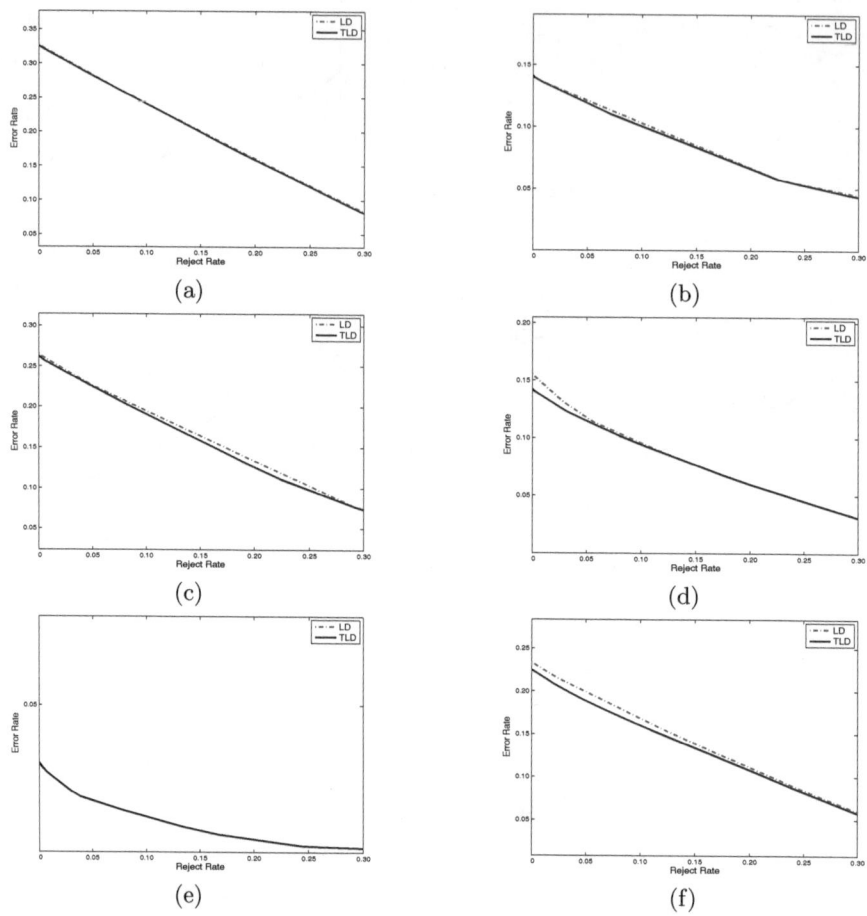

Fig. 5. Error-Reject curves obtained on Abalone (a), Ecoli (b), Glass (c), Letter, (d), Vowel (e) and Yeast (f) with RBF SVM

6   Conclusions

In this paper we have analyzed two different techniques to enrich an ECOC classification system with a reject option. The main difference is in the simplicity of the external approach that only requires an intervention on the decoding stage of the ECOC system while the Trimmed Loss Decoding requires to manage more parameters. However, the geometrical method described for the Error-Reject curve simplifies the use of the internal technique and this can be particularly useful in improving the error/reject trade-off, as shown by the experiments. A possible development of this work can be focused on the investigation of the relation between the rejection rule with the characteristics of the coding matrix.

References

1. Allwein, E.L., Schapire, R.E., Singer, Y.: Reducing multiclass to binary: A unifying approach for margin classifiers. Journal of Machine Learning Research 1, 113–141 (2000)
2. Asuncion, A., Newman, D.J.: UCI machine learning repository (2007)
3. Bartlett, P.L., Wegkamp, M.H.: Classification with a reject option using a hinge loss. J. Mach. Learn. Res. 9, 1823–1840 (2008)
4. Chow, C.: On optimum recognition error and reject tradeoff. IEEE Transactions on Information Theory 16(1), 41–46 (1970)
5. Cordella, L.P., De Stefano, C., Tortorella, F., Vento, M.: A method for improving classification reliability of multilayer perceptrons. IEEE Transactions on Neural Networks 6(5), 1140–1147 (1995)
6. Dietterich, T.G., Bakiri, G.: Solving multiclass learning problems via error-correcting output codes. Journal of Artificial Intelligence Research 2, 263–286 (1995)
7. Guruswami, V., Sahai, A.: Multiclass learning, boosting, and error-correcting codes. In: Proceedings of the Twelfth Annual Conference on Computational Learning Theory, COLT 1999, pp. 145–155. ACM, New York (1999)
8. Joachims, T.: Making large-scale SVM learning practical. In: Schölkopf, B., Burges, C., Smola, A. (eds.) Advances in Kernel Methods - Support Vector Learning, ch. 11. MIT Press, Cambridge (1999)
9. Pietraszek, T.: On the use of ROC analysis for the optimization of abstaining classifiers. Machine Learning 68(2), 137–169 (2007)
10. Simeone, P., Marrocco, C., Tortorella, F.: Exploiting system knowledge to improve ecoc reject rules. In: International Conference on Pattern Recognition, pp. 4340–4343. IEEE Computer Society, Los Alamitos (2010)
11. De Stefano, C., Sansone, C., Vento, M.: To reject or not to reject: that is the question-an answer in case of neural classifiers. IEEE Transactions on Systems, Man, and Cybernetics, Part C 30(1), 84–94 (2000)
12. Tortorella, F.: Reducing the classification cost of support vector classifiers through an roc-based reject rule. Pattern Anal. Appl. 7(2), 128–143 (2004)

Sum-of-Superellipses – A Low Parameter Model for Amplitude Spectra of Natural Images

Marcel Spehr[1], Stefan Gumhold[1], and Roland W. Fleming[2]

[1] Technische Universität Dresden
Inst. of Software- and Multimedia-Technology,
Computer Graphics and Visualization (CGV) lab,
Nöthnitzer Straße 46,
01187 Dresden, Germany
{marcel.spehr,stefan.gumhold}@tu-dresden.de
www.inf.tu-dresden.de/cgv
[2] University of Giessen
Dept. of Psychology,
Otto-Behaghel-Str 10/F,
35394 Giessen, Germany
roland.w.fleming@psychol.uni-giessen.de
www.allpsych.uni-giessen.de

Abstract. Amplitude spectra of natural images look surprisingly alike. Their shape is governed by the famous $1/f$ power law. In this work we propose a novel low parameter model for describing these spectra. The *Sum-of-Superellipses* conserves their common falloff behavior while simultaneously capturing the dimensions of variation—concavity, isotropy, slope, main orientation—in a small set of meaningful illustrative parameters. We demonstrate its general usefulness in standard computer vision tasks like scene recognition and image compression.

Keywords: Natural images, image statistics, amplitude spectrum, Lamé curve, superellipse, image retrieval.

1 Introduction

Because the world is highly structured, natural images constitute only a small subset of all possible images. Their pixel statistics exhibit several properties which distinguish them from artificially generated images, and which lead to a number of statistical redundancies ([BM87], [RB94], [HBS92], [TO03]). Our visual system is shaped to make use of these regularities ([Fie87], [OF96], [HS98]). Therefore, it is important to study these characteristics to enhance understanding of its functioning. Additionally, new insights about the manifold inhabited by natural images within the set of all images can lead to more efficient image compression schemes ([BS02]), improve computer graphics algorithms ([RSAT04]) or serve as priors for image enhancing tasks ([LSK+07], [KK10]). More recently, links were proposed that connect natural image statistics with aesthetic principles in artworks ([Red07]).

G. Maino and G.L. Foresti (Eds.): ICIAP 2011, Part I, LNCS 6978, pp. 128–138, 2011.

Probably the best known statistical regularity of natural images is the similarity between their Fourier amplitude spectra. As an example, consider the images in figure 1, which shows four images in a), with their corresponding amplitude spectra in b) and c). Although clear differences between the spectra are visible, these differences appear to be highly systematic, and there are also some notable similarities. All four spectra are strongly dominated by low spatial frequencies, with spectral energy falling off following the well-known $1/f$ power law. The spectra also tend to have only a few dominant orientations, with a characteristic shape between the dominant orientations, which varies from concave to convex. In this work we propose a novel empirical model designed to capture these variations across natural amplitude spectra in a small set of meaningful parameters, while identifying key common shape features.

The main contributions of our work are twofold. First, we describe our simple low parameter model for natural image spectra that preserves sufficient detail to reconstruct images and recognize scenes. Second, we demonstrate the feasibility of an image compression scheme based on the model. We also show how our model can be used to successfully recognize different scene categories with just a few parameters, and that it can describe the scale invariant property of natural images by also being an appropriate model for image patches.

The rest of the paper is structured as follows. First we review earlier work on modeling approaches for amplitude spectra of natural images. Then we develop and discuss our *Sum-of-Superellipses* model S^* guided by empirical data, show its general usefulness in standard computer vision tasks and summarize our work in the conclusion.

2   Previous Work

For many decades the $1/f$ falloff behavior of amplitude spectra of natural images has been well known ([MCCN77], [BM87], [TTC92], [RB94]). [Rud97] investigated the reason for this and concluded that it is caused by the distribution of statistically independent objects that typically occur in real world scenes. However, individual images often vary substantially from the average [TTC92]. Additional, image specific, parameters α and A were introduced (A/f^α [Rud97]) to capture this. The spectral energy is also often anisotropically distributed. [SH96] proposed an orientation specific variation from the postulated $1/f^\alpha$ slope by taking samples at orientations $\theta \in [0...\pi]$ of both slope and spectral energy. [TO03] summarize the final model. Let the spectral decomposition of image i be

$$I(f_x, f_y) = \sum_{x=0}^{M-1} \sum_{y=0}^{N-1} i\,(x,y)\,e^{-j2\pi((f_x x/M + f\ y/N))}\,.$$ I can be indexed either by the spatial frequency (f_x, f_y) or polar coordinates (f, θ) ($f = \|(f_x\ f_y)\|_2$ and $\theta = \arctan(f_y/f_x)$). Then

$$I\,(f, \theta) \simeq \frac{A(\theta)}{f^{\alpha(\theta)}} \tag{1}$$

where $A(\theta)$ is an orientation dependent amplitude scaling factor that represents the energy of the spectral components in a certain direction and $\alpha(\theta)$ describes the descent in each direction. The jagged lines in figure 1g) and h) show

$\log{(A(\theta))}$ and $\alpha(\theta)$ for an orientation sampling of 180 samples between $-\pi/2$ and $\pi/2$. Increasing the sampling resolution of the orientations improves the accuracy of the orientation information, but at the same time, it also increases the number of parameters required.

Amplitude spectra can also be described model free. A popular method is the usage of cumulative statistics to summarize the amplitude spectrum, which can be done in either frequency or space domain. The *homogeneous texture* descriptor—as defined in the *MPEG 7* standard [MSS02]—constitutes a frequency domain representation. Mean energy and deviation from a set of 30 frequency band channels are summarized into a feature vector with 60 elements.

By contrast, [OT01] use a set of oriented Gabor-like bandpass filters $G_{\theta,f}$ as a wavelet basis. To capture the spectral content in space domain for orientation θ and frequency f they filter an image with $G_{\theta,f}$ and take the mean over the filter responses. Both of the latter approaches can describe arbitrary images but lack intuitive descriptive power for the similarities and differences between the amplitude spectra of natural images.

3   The *Sum-of-Superellipses* Model

The aim of our work was to create an empirically based model that (i) incorporates the $A(\theta)/f^{\alpha(\theta)}$ behavior in a closed formula, (ii) is parameterized with a small set of intuitive parameters and (iii) captures orientation related image structure. Consider the similarities between the spectra in figure 1b) and c). As, mentioned earlier, the spectra are dominated by low frequencies and a small number of dominant directions. Inspection of large numbers of natural spectra such as these reveals that the shape of the function spanning the regions between the dominant orientations tends to vary systematically. Specifically, the shape ranges from highly concave, like a sharply pointed star, through isotropic to convex. We are seeking a function that behaves in this way, so that we can capture intuitively the variations between spectra. One such function is the superellipse, or as it is sometimes known, a Lamé curve.

$$S(f_x, f_y) = \left|\frac{f_x}{v_x}\right|^n + \left|\frac{f_y}{v_y}\right|^n \quad S: \mathbb{R}^2 \to \mathbb{R} \tag{2}$$

v_x and v_y reflect scaling along the cardinal axis and $n < 0$ denotes the isotropic falloff behavior. When we fix S to particular values, we can find parameter values for v_x, v_y and n that produce isolines like shown in figure 2a) to d).

However, we are not yet able to describe rotated spectra. Therefore we introduce a rotation matrix R_ϕ with parameter ϕ.

$$\begin{pmatrix} f'_x \\ f'_y \end{pmatrix} = R_\phi \begin{pmatrix} f_x \\ f_y \end{pmatrix} \quad R = \begin{pmatrix} \cos(\phi) & -\sin(\phi) \\ \sin(\phi) & \cos(\phi) \end{pmatrix} \tag{3}$$

Furthermore we want to be able to describe anisotropic falloff behavior and therefore make a distinction between n_x and n_y. Due to the spectrum's exponential nature—having extremely large values close to the origin and small

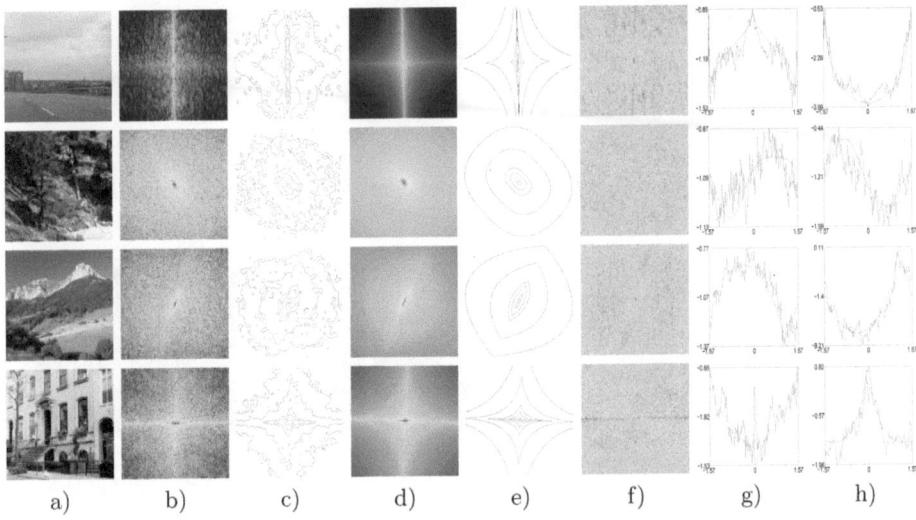

a)   b)   c)   d)   e)   f)   g)   h)

Fig. 1. a) Example images from our image database ([Tor11]. Resolution: 256×256). b) Associated $\log\left(I\left(f_x, f_y\right)\right)$ as 2D color map and c) contour plot with isolines displayed at 20, 60, 90, 95, 99 and 99.9 percent of spectral image energy (I slightly smoothed to reduce clutter). d) and e) Superellipse S fit with 6 parameters as color map and contour plot. f) Quadratic pixel-wise difference between b) and d) (Color table scaled by factor 6 to enhance error visibility with respect to b) and d)). g) Jagged line corresponds to $\alpha(\theta)$ for b) and c) in eq. 1. Smooth line corresponds to $\alpha(\theta)$ for fitted S in d) and e). h) Jagged line corresponds to $\log(A(\theta))$ in b) and c) in eq. 1. Smooth line corresponds to $\log\left(A\left(\theta\right)\right)$ for fitted S model in d) and e).

values at its periphery—fitting the original amplitude spectrum is a numerically ill-conditioned problem. When visualizing spectra it is usual to take the logarithm. To stabilize the subsequent fitting procedure we integrate the logarithm function into our model and achieve a model that describes the logarithm of the amplitude spectrum of natural images. Additionally, we restrict the scope of the $|.|$ function to the nominator in order to decouple the scaling parameters $v_{x\ y}$ and slope parameters $n_{x\ y}$.

Finally we want to couple the dominant directions with more than a simple sum. We introduce an exponent m that leads to mixed terms of f_x and f_y along oblique directions. This is expressed as a multiplicative factor preceding the logarithm.

$$S(f_x, f_y) = m \log \left\{ \frac{|f'_x|^{n_x}}{v_x} + \frac{f'_y}{v_y}^n \right. \tag{4}$$

Visualizations of the influences of different parameter combinations are shown in figure 2. This simple model with 6 parameters offers sufficient modes of variation to describe the images in figure 1 and the patches in figure 6. The interpretation of the parameters is straightforward. The ratio between v_x/v_y indicates anisotropy and the dominant direction. The values n_1 and n_2 describe convexity of the spectrum and m reflects the spectral energy.

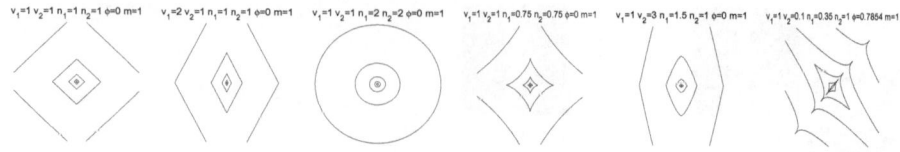

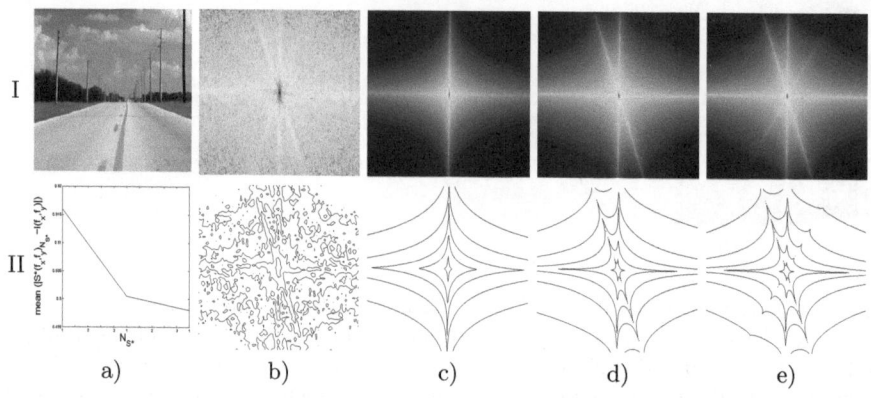

Fig. 2. Degrees of freedom that the superellipse offers for describing amplitude spectra.

Fig. 3. *Sum-of-Superellipses* results for N_S 1. Ia) example image. Ib) associated amplitude spectrum as color coded plot and IIb) isoline plot. Columns c), d) and e) depict the associated *Sum-of-Superellipses* with $N = 1, 2$ and 3 as color map and isoline plot. IIa) shows the mean model error per spatial frequency component (f_x, f_y).

The smooth lines in figure 1g) and h) show the corresponding $\alpha(\theta)$ and $A(\theta)$ (see eq. 1) to our superellipse. Note how well they interpolate the orientation dependent scaling and slope of the "real" images' amplitude spectra even though the fitting procedure of S did not take equation 1 into account but was solely based on I.

However, eq. 4 only serves well for spectra with two dominant directions. Figure 3.Ia) shows a more complex example. Observe how its amplitude spectrum can be described by adding different superellipses in 3c), d) and e) together. A comparison of the average model error (shown in figure 3.IIa)) for increasing N_S supports the intuitive impression one has when looking at the isoline plots. As expected, the mean error per estimated spatial frequency component (f_x, f_y) decreases. We conclude this section with the final formulation of the *Sum-of-Superellipses* model function.

$$S^\star(f_x, f_y)_{N_S} = \sum_{i=1}^{N_S} m_i \log \left\{ \frac{|f_x'|^{n_{x,i}}}{v_{x,i}} + \frac{f_y'^{\,n\,,i}}{v_{y,i}} \right. \tag{5}$$

N_S is chosen on an image per image basis at the elbow of the mean error curve shown in figure 3.IIa). Hence, the final model complexity equals $6N_S$.

3.1   Fitting of Our Model

We fit a superellipse S to the amplitude spectrum $I(f_x, f_y)$ of image $i(x, y)$ $x, y \in \Omega \subset \mathbb{R}^2$ using the Levenberg-Marquardt optimization algorithm included in the Matlab Optimization Toolbox^{TM} and used finite differences as gradient approximizations. We ignore $I(0,0)$ during the fitting process because it only reflects an image's global brightness offset and carries no meaning for the frequency distribution. The spectrum is point symmetric if $i\forall x, y : i(x, y) \in \mathbb{R}$ so we need to include only half of the spectrum's values for the fitting procedure. We also experienced that the fitting stays stable if as few as 10% of all availabe values of I are used.

For $N_S = 1$ we initialize ϕ to the direction that includes the maximum spectral energy. $v_{x\ y}$ and $n_{x\ y}$ were set to 1 and m to -1. If N_S is chosen larger than 1 then first a superellipse S_1 with 6 parameters is fitted. Subsequently S_2 is fitted to $I' = I - S_1$. After an additional optimization run over $v_{x\ y,1\ 2}, n_{x\ y,1\ 2}, m_{1,2}$ and $\phi_{1\ 2}$ we get $S_2^\star$. This procedure runs iteratively until the user selected $S_{N_S}^\star$ is defined.

4   Applications

The parameters of the *Sum-of-Superellipses* model describe properties of images that become apparent in their amplitude spectra. Hence, it should become useful in applications that make use of these characteristics. Lossy image compression for example tries to identify properties that are shared by the images that are to be compressed. Only discriminative differences are encoded. This procedure mimics the approach we followed when defining our model—that is we analyzed the spectra of natural images, identified similarities and designed model parameters to encode the differences.

Scene recognition is largely guided by the image signal's spectral amplitude values ([GCP$^+$04]). We trained a *Support Vector Machine* on a standard scene database using our parameters as feature values to show that our model successfully captures scene specific information. Additionally, we thereby present evidence that our model reflects the scale invariance property of natural images since it also works for image patches.

4.1   Image Compression

We have found that the *Sum-of-Superellipses* model can often reconstruct an image to a recognizable level with only 6 fitted parameters. We support this statement by showing reconstruction results using the original phase spectrum combined with our fitted model. Additionally, we compare our model with a data-driven approach. We applied *PCA* to all amplitude spectra in our image database ([Tor11]) and reconstruct the amplitude spectrum from 6 coefficients. We follow the naming conventions from [TO03]. Let

$$I(f_x, f_y) = \mu + \sum_{n=1}^{P} u_n SPC_n(f_x, f_y) \tag{6}$$

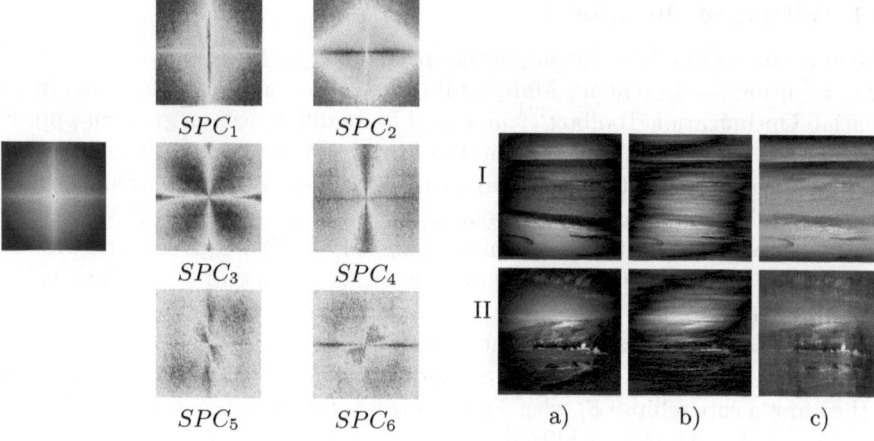

SPC_1 SPC_2

SPC_3 SPC_4

SPC_5 SPC_6

I

II

a)   b)   c)

Fig. 4. Mean   and first 6 SPCs of all amplitude spectra in the given image dataset.

Fig. 5. Two reconstruction examples (I+II) using original phase and a) original amplitude spectrum, b) amplitude spectrum reconstructed using $SPC_1...SPC_6$ and c) our fitted model superellipse with 6 parameters. Dark borders result from applying the Hann window function before performing the Fourier transformation.

be the reconstruction of I from P decorrelated principle components (*Spectral Principle Components*) and the sample mean μ. For a visualization of $SPC_1...$ SPC_6 and μ see figure 4. Figure 5 shows the reconstruction results using the image's original phase information. 5a) combines original phase with the original amplitude spectrum. The dark borders result from applying the Hann window function before performing the Fourier transformation. 5b) and c) compare reconstruction results using equation 6 with $P = 6$ and equation 4.

We only describe the amplitude spectra with 6 parameters thus state-of-the-art results like [TMR02] cannot be expected. Also the phase spectrum reflects much of the image's detail content and meaning. However, our approach produces identifiable reconstructions and the compression rate is enormous compared to the alternative where the SPCs and μ must be saved. Importantly, we believe that an analysis of the distribution of the *Sum-of-Superellipses* model parameters can contribute useful insights into the causes of the regularities in natural images, something that PCA and other data-driven approaches cannot.

4.2   Scene Recognition

[TO99] demonstrated that certain shape characteristics of amplitude spectra correspond to different scene categories. In [OTDH99] the same authors identified 5 distinct characteristics that distinguish amplitude spectra of natural images. 1.

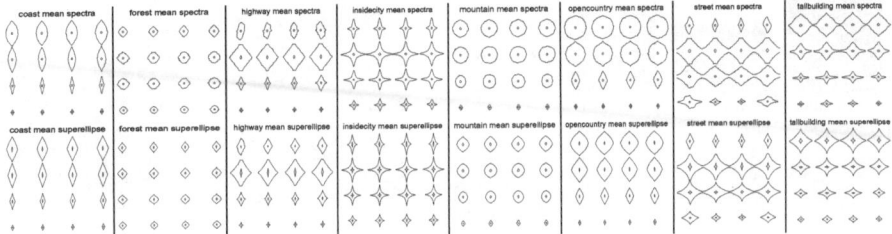

Fig. 6. First row shows mean local amplitude spectra for each image patch for each class. Second row shows mean superellipse estimations.

Horizontal shape, 2. Cross shape, 3. Vertical shape, 4. Oblique shape, 5. Circular shape. Note, that the *Sum-of-Superellipses* model can capture these characteristics in only 6 easily comprehensible parameter values. In addition [GCP+04] provides experimental evidence that although phase is vital for image details the amplitude properties alone provide ample cues for rapid scene categorization.

[OT01] extend this finding to the spatial layout of natural images. Their *Gist* descriptor applies Gabor-like filters to image patches that divide the image into equally sized areas. The filter responses describe each patch's frequency and orientation content. The union of all patch descriptors form an image's *spatial envelope*. The first row of figure 6 shows the mean amplitude spectra of 16 image patches sorted by scene category. It indicates that the spatial layout of the frequency distribution can serve as an effective scene descriptor. We found that our model also serves as an appropriate descriptor for patch amplitude spectra. Row two of figure 6 shows the mean of all superellipses that were fitted to the image patches.

We tested the model's ability to capture scene specific information in a standard classification task, using a publicly available standard database. [Tor11] provides 8 scene categorization labels (see figure 7) and classification benchmark results. We used the quadratic difference between the parameter values of the superellipses that were fitted to the image patches as a distance measure. Figure 8 presents our classification results as a confusion table 8a) and recall values 8b). Due to the fact that we tried to distinguish 8 classes the chance recall level would be at 12.5%. The results show that we are far above chance level. Even though the results are somewhat below those of state-of-the-art scene descriptors like the *Gist* it proves that our model parameters capture scene specific information with few feature values (No patches × 6 = 96 compared to 512).

Another often reported phenomenon of natural images is their apparent invariance regarding the similarity of visual information on different scales ([RB94], [Rud97]). Our classification results indicate that our model—by also suitably describing image patches—reflects this prominent quality on global and local image scale.

Fig. 7. Examples for each scene classes. n   330 images per class.

Fig. 8. Classification results as a) confusion matrix and b) recall rate for each class

5   Conclusion

In this work we proposed a novel empirically derived low parameter model for amplitude spectra of natural images. We believe it to be superior to older models because it represents characteristic properties in a closed formula with a set of intuitive parameters. It also proved adequate for describing local patches of natural images.

We demonstrated its usefulness in two application scenarios. We classified scene categories with recall results far above chance and we can reconstruct images using only 6 fitted parameter values to a recognizable level. The automatic deduction of an image's tilt angle for horizon estimation could be another useful use case. The *Sum-of-Superellipses* was already successfully employed for clustering paintings according to their visual appearance ([SWF09]).

Further research must be conducted how measures like the *Akaike-* or *Bayesian-information-criterion* could be used to automatically restrict the necessary model complexity governed by $N^\star$ for an image. Additionally, we plan to investigate if the model parameters can serve as low level image descriptors for more sophisticated image retrieval tasks.

The research was financed by the DFG SPP 1335 project "The Zoomable Cell". Please send inquires per email to `marcel.spehr@tu-dresden.de`.

References

[BM87]   Burton, G.J., Moorhead, I.R.: Color and spatial structure in natural scenes. Applied Optics 26(1), S. 157–S. 170 (1987) ISSN 1539–4522

[BS02]   Buccigrossi, R.W., Simoncelli, E.P.: Image compression via joint statistical characterization in the wavelet domain. IEEE Transactions on Image Processing 8(12), S. 1688–S. 1701 (2002) ISSN 1057–7149

[Fie87]   Field, D.J.: Relations between the statistics of natural images and the response properties of cortical cells. J. Opt. Soc. Am. A 4(12), S. 2379–S. 2394 (1987)

[GCP+04]   Guyader, N., Chauvin, A., Peyrin, C., Hérault, J., Marendaz, C.: Image phase or amplitude? Rapid scene categorization is an amplitude-based process. Comptes Rendus-Biologies 327(4), S. 313–S. 318 (2004)

[HBS92]   Hancock, P.J.B., Baddeley, R.J., Smith, L.S.: The principal components of natural images. Network: Computation in Neural Systems 3(1), S. 61–S. 70 (1992)

[HS98]   van Hateren, J.H., van der Schaaf, A.: Independent component filters of natural images compared with simple cells in primary visual cortex.. Proceedings of the Royal Society B: Biological Sciences 265(1394), S. 359 (1998)

[KK10]   Kim, K.I., Kwon, Y.: Single-Image Super-Resolution Using Sparse Regression and Natural Image Prior. IEEE Transactions on Pattern Analysis and Machine Intelligence 32(6), S. 1127–S. 1133 (2010) ISSN 0162–8828

[LSK+07]   Liu, C., Szeliski, R., Kang, S.B., Zitnick, C.L., Freeman, W.T.: Automatic estimation and removal of noise from a single image. IEEE Transactions on Pattern Analysis and Machine Intelligence 30(2), S. 299–S. 314 (2007) ISSN 0162–8828

[MCCN77]   Mezrich, J.J., Carlson, C.R., Cohen, R.W.: NJ, RCA LABS P.: Image descriptors for displays (1977)

[MSS02]   Manjunath, B.S., Salembier, P., Sikora, T.: Introduction to MPEG-7: multimedia content description interface. John Wiley & Sons Inc., Chichester (2002) ISBN 0471486787

[OF96]   Olshausen, B.A., Field, D.J.: Natural image statistics and efficient coding*. Network: Computation in Neural Systems 7(2), S. 333–S. 339 (1996) ISSN 0954–898X

[OT01]   Oliva, A., Torralba, A.B.: Modeling the Shape of the Scene: A Holistic Representation of the Spatial Envelope. International Journal of Computer Vision 42(3), S. 145–S. 175 (2001)

[OTDH99]   Oliva, A., Torralba, A., Dugue, A.G., Herault, J.: Global semantic classification of scenes using power spectrum templates (1999)

[RB94]   Ruderman, D.L., Bialek, W.: Statistics of natural images: Scaling in the woods. Physical Review Letters 73(6), S. 814–S. 817 (1994)

[Red07]   Redies, C.: A universal model of esthetic perception based on the sensory coding of natural stimuli. Spatial Vision, 21 1(2), S. 97–S. 117 (2007) ISSN 0169–1015

[RSAT04]   Reinhard, E., Shirley, P., Ashikhmin, M., Troscianko, T.: Second order image statistics in computer graphics. In: Proceedings of the 1st Symposium on Applied Perception in Graphics and Visualization, pp. S. 99–S. 106 (2004)

[Rud97]   Ruderman, D.L.: Origins of scaling in natural images. Vision Research 37(23), S. 3385–S. 3398 (1997)

[SH96]   van der Schaaf, A., van Hateren, J.: Modelling the power spectra of natural images: Statistics and information. Vision Research 36(17), S. 2759–S. 2770 (1996)

[SWF09]   Spehr, M., Wallraven, C., Fleming, R.W.: Image statistics for clustering paintings according to their visual appearance. In: Deussen, O., et al. (eds.) Computational Aesthetics 2009, 2009, pp. 1–8. Eurographics, Aire-La-Ville (2009)

[TMR02]   Taubman, D.S., Marcellin, M.W., Rabbani, M.: JPEG2000: Image compression fundamentals, standards and practice. Journal of Electronic Imaging 11, S. 286 (2002)

[TO99]   Torralba, A.B., Oliva, A.: Semantic Organization of Scenes Using Discriminant Structural Templates. In: IEEE International Conference on Computer Vision, vol. 2, p. S. 1253 (1999)

[TO03]   Torralba, A., Oliva, A.: Statistics of natural image categories. Network: Computation in Neural Systems 14(3) (2003)

[Tor11]   Torralba, Antonio: Spatial Envelope, Version: März 2011, http://people.csail.mit.edu/torralba/code/spatialenvelope/

[TTC92]   Tolhurst, D.J., Tadmor, Y., Chao, T.: Amplitude spectra of natural images. Ophthalmic Physiol Opt. 12(2), S. 229–S. 232 (1992)

Dissimilarity Representation in Multi-feature Spaces for Image Retrieval

Luca Piras and Giorgio Giacinto

Department of Electrical and Electronic Engineering University of Cagliari,
Piazza D'armi 09123 Cagliari, Italy
luca.piras@diee.unica.it, giacinto@diee.unica.it

Abstract. In this paper we propose a novel approach to combine information form multiple high-dimensional feature spaces, which allows reducing the computational time required for image retrieval tasks. Each image is represented in a "(dis)similarity space", where each component is computed in one of the low-level feature spaces as the (dis)similarity of the image from one reference image. This new representation allows the distances between images belonging to the same class being smaller than in the original feature spaces. In addition, it allows computing similarities between images by taking into account multiple characteristics of the images, and thus obtaining more accurate retrieval results. Reported results show that the proposed technique allows attaining good performances not only in terms of precision and recall, but also in terms of the execution time, if compared to techniques that combine retrieval results from different feature spaces.

1 Introduction

One of the peculiarities of content-based image retrieval is its suitability to a huge number of applications. Image retrieval and categorization is used to organize professional and home photos, in the field of fashion, for retrieving paintings from a booklet of a picture gallery, for retrieving images from the Internet, and the number of applications is growing every day. The increasing use of images in digital format causes the size of visual archives of becoming bigger and bigger, thus increasing the difficulty of image retrieval tasks. In particular, the main difficulties are related to the increase of the computational load, and to the reduction of the separation between different image categories. In fact, a larger number of images belonging to different categories may be represented as close points in each one of the low-level feature spaces that can be used to represent the visual content.

The combination of multiple image representations (colors, shapes, textures, etc.) has been proposed to effectively cope with the reduced inter-class variation. As a drawback, the use of multiple image representations with a high number of components increases the computational cost of retrieval techniques. As a consequence, the response time of the system might become an issue for interactive applications (e.g., web searching). Over the years, the pattern recognition

G. Maino and G.L. Foresti (Eds.): ICIAP 2011, Part I, LNCS 6978, pp. 139–148, 2011.

community proposed a number of solutions for combining the output of different sources of information [10]. The most popular and effective techniques for output combination are based on fusion techniques, such as the mean rule, the maximum rule, the minimum rule, and weighted means. In the field of content-based image retrieval, similar approaches can be employed by considering the value of similarity between images as the output of a classifier. In particular, combination approaches have been proposed for fusing different feature representations, where the appropriate similarity metric is computed in each feature space, and then all the similarities are fused through a weighted sum [15]. Another approach to combine different image representations is to stack all the available feature vectors into a single feature vector, and then computing the similarity between images by using this high-dimensional representation. As the computational cost increases with the size of the database and the size of the feature space, it is easy to see that the use of a unique feature vector made up by stacking different feature representations might not be a feasible solution.

The issue of combining different feature representations is also relevant when relevance feedback mechanisms are used. In this case, at each iteration, similarities have to be computed by exploiting relevance feedback information, for example by resorting to Nearest-Neighbor or Support Vector Machine [17] techniques. In particular, when the combination is attained by computing a weighted sum of distances, the cost of the estimation of the weights related to relevance feedback information have to be also taken into account. It is easy to see that the effectiveness of a given representation of the images is strictly related to the retrieval method employed. In this viewpoint, an approach that has been recently proposed in the pattern recognition field is the so called "dissimilarity space". This approach is based on the creation of a new space where patterns are represented in terms of their (dis)similarities to some reference prototypes. Thus the dimension of this space does not depend on the dimensions of the low-level features employed, but it is equal to the number of reference prototypes used to compute the dissimilarities This technique has been used recently to exploit Relevance feedback in content-based image retrieval field [5,12], where relevant images play the role of reference prototypes. In addition, dissimilarity spaces have been also proposed for image retrieval to exploit information from different multi-modal characteristic [2].

In this paper we propose a novel use of the dissimilarity representation for improving relevance feedback based on the Nearest-Neighbor approach [4]. Instead of computing (dis)similarities by using different prototypes (e.g., the relevant images) and a single feature space, we propose to compute similarities by using just one prototype, and multiple feature representations. Each image is thus represented by a very compact vector that summarizes different low-level characteristics, and allows images that are relevant to the user's goals to be represented as near points. The resulting retrieval system is both accurate and fast, because, at each relevance feedback iteration, retrieval performances can be significantly improved with a low computational time compared to the number of low-level features considered.

The rest of the paper is organized as follows. Section 2 briefly reviews the dissimilarity space approach, and introduces the technique proposed in this paper. Section 3 shows the integration of the proposed approach in the learning process of a relevance feedback mechanism based on the Nearest-Neighbor paradigm. Section 4 illustrates some approaches proposed in the literature to combine different feature spaces. Experimental results are reported in Section 5. Reported results show that the proposed approach allows outperforming other methods of feature combination both in terms of performances and execution time. Conclusions are drawn in Section 6.

2 From Multi-spaces to Dissimilarity Spaces

Dissimilarity spaces are defined as follows [13]. For a given classification task, let us consider a set $P = \{\mathbf{p}_1, \ldots, \mathbf{p}_L\}$ made up of L patterns selected as *prototypes*, and let us compute the distances $d(\cdot)$ between each pattern and the set of prototypes. These distances can be computed in a low-level feature space. Each pattern is then represented in terms of a L-dimensional vector, where each component is the distance between the pattern itself and one of the L prototypes. If we denote with $d(\mathbf{I}_i, \mathbf{p}_j)$ the distance between pattern $\mathbf{I}_i$ and the prototype $\mathbf{p}_j$, the representation of pattern $\mathbf{I}_i$ in the dissimilarity space will be:

$$\mathbf{I}_i^P = [d(\mathbf{I}_i, \mathbf{p}_1), \ldots, d(\mathbf{I}_i, \mathbf{p}_P)]. \tag{1}$$

It should be quite clear that the performances depend on the choice of the prototypes, especially when this technique is used to transform a high-dimensional feature space into a lower dimensional feature space. The literature clearly shows that the choice of the most suitable prototypes is not a trivial task [13]. In this paper we use the basic idea of dissimilarity spaces to produce a new vector from different feature spaces.

Before entering into the details of the proposed technique, let us recall that the goal is to produce an effective way of combining different feature representations of images in the context of a Nearest-Neighbor relevance feedback approach for content-based image retrieval [4]. Relevance feedback provides the systems a number of images that are relevant to the user's needs at each iteration. It is quite easy to see that if we consider different image representations, usually different sets of images are found in the nearest neighborhood of relevant images. Which strategy can be employed to assess which of the images can be considered as relevant? One solution can be the use of combination mechanisms based on the weighted fusion of similarity measures computed in different feature spaces. As an alternative, strategies based on the computation of the *max*, or the *min* similarity measure can be employed. Finally, the computation of similarity can be carried out using a vector where the components from different representations are stacked. The fusion of similarities requires some heuristics to compute the weights of the combination, while the *max* and *min* rules can be more sensitive to "semantic" errors in the evaluation of similarity due to the so-called semantic gap [9]. Finally, the use of stacked vectors can be computationally expensive, and can suffer from the so-called "curse of dimensionality", as the dimension of the resulting space may be too large compared to the number of available samples of relevant images.

In order to provide a solution to the computation of the relevance of an image with respect to the user's goal by exploiting information from different image representations, we propose to construct a dissimilarity space by computing the dissimilarities from a single prototype using multiple feature representations. This approach results in a very compact feature space, as the dimension is equal to the number of feature representations. In order to formalize the proposed technique, let $F = \{f_1, \ldots, f_M\}$ be the set of low-level feature spaces extracted from the images, and let $d_{f_m}\left(\mathbf{I}_i^{f_m}, \mathbf{I}_j^{f_m}\right)$ be the distance between the images $\mathbf{I}_i$ and $\mathbf{I}_j$ evaluated in the feature space f_m. Given a reference image $\mathbf{q}$, the new representation of a generic image $\mathbf{I}_i$ in the *dissimilarity multi-space* is

$$\mathbf{I}_i' = \left[d_{f_1}\left(\mathbf{q}^{f_1}, \mathbf{I}_i^{f_1}\right), \ldots, d_{f_M}\left(\mathbf{q}^{f_M}, \mathbf{I}_i^{f_M}\right) \right]. \tag{2}$$

Summing up, while dissimilarity space are usually constructed by stacking dissimilarities from multiple prototypes, we propose to stack multiple dissimilarities originated by considering a single reference point, and measuring the distances from this point in different feature representations.

Let us have a close look on the choice of the reference image to be used in equation (2). When the first round of retrieval is performed, i.e., no feedback is available, we use the query image as the reference point. At each round of relevance feedback, the reference point is computed according to a "query shifting mechanisms", i.e., a mechanism designed to exploit relevance feedback by computing a new vector in the feature space such that its neighborhood contains relevant images with high probability [15]. In particular, we used a modified Rocchio formula, that has been proposed in the framework of the Bayes decision theory, namely Bayes Query Shifting (BQS) [6].

$$\mathbf{q}_{BQS} = \mathbf{m}_r + \frac{\sigma}{\|\mathbf{m}_r - \mathbf{m}_n\|}\left(1 - \frac{r-n}{\max(r,n)}\right)(\mathbf{m}_r - \mathbf{m}_n) \tag{3}$$

where $\mathbf{m}_r$ and $\mathbf{m}_n$ are the mean vectors, in each feature space, of relevant and non-relevant images respectively, σ is the standard deviation of the images belonging to the neighborhood of the original query, and r and n are the number of relevant and non relevant images retrieved after the latter iteration, respectively.

The choice of the query, and the BQS as the reference prototypes is twofold. First of all, as we are taking into account retrieval tasks in which the user performs a "query by example" search, and the BQS technique is aimed to represent the concept that the user is searching for by definition. On the other hand, the use of multiple images as prototypes can introduce some kind of "noise" because not all the images may exhibit the same "degree" of relevance to the user's needs. The second reason is that the use of a single prototype makes the search independent from the number of images in the database that are relevant to the user's query.

In the literature of content-based image retrieval, few works addressed the use of dissimilarity spaces to provide for a more effective representation. Some of the approaches proposed so far employed the original definition of dissimilarity space, where dissimilarities are computed by taking into account multiple prototypes of relevant images [12,5]. Other authors have proposed to use the "dissimilarity space" technique for combining different feature space representations [2]. However, their approach is based on the computation of dissimilarity relationships between all the patterns in the dataset. Then, a number of prototypes are selected in each feature space, and the resulting dissimilarity spaces are then combined to attain a new multi-modal dissimilarity space. Thus the components of the resulting space are not related to the number of the original feature spaces, but they are related to the number of patterns used as prototypes to create the different dissimilarity spaces.

3 Nearest-Neighbor Relevance Feedback in the Dissimilarity Multi-space

The generation of the dissimilarity space is strictly related to the use of a Nearest-Neighbor approach to exploit relevance feedback in multiple feature spaces. In fact, the new space provides for a compact representation of patterns that ease the computation of nearest-neighbor relationships in multiple low-level feature representations. The dissimilarity representation computed with respect to a set of prototypes basically assumes that patterns belonging to the same category are represented as close points. Analogously, we expect that relevant images are represented as close points in the space made up of dissimilarities computed with respect to one reference point in multiple low-level feature spaces. The Nearest-Neighbor technique employed to exploit relevance feedback is based on the computation of a relevance score for each image according to its distance from the nearest relevant image, and the distance from the nearest non relevant image [4]. This score is further combined to a score related to the distance of the image from the point computed according to the BQS (Eq. 3), that is the likelihood that the image is relevant according to the users' feedback. The combined relevance score is computed as follows:

$$rel(\mathbf{I}_i')_{stab} = \frac{n/k}{1 + n/k} \quad rel_B \; s(\mathbf{I}_i') + \frac{1}{1 + n/k} \quad rel_{NN}(\mathbf{I}_i') \tag{4}$$

where n and k are the number of non-relevant images, and the whole number of images retrieved after the latter iteration, respectively. The two terms rel_{NN} and rel_{BQS} are computed as follows:

$$rel_{NN}(\mathbf{I}_i') = \frac{\mathbf{I}_i' - NN^{nr}(\mathbf{I}_i')}{\mathbf{I}_i' - NN^r(\mathbf{I}_i') + \mathbf{I}_i' - NN^{nr}(\mathbf{I}_i')} \tag{5}$$

where $NN(\mathbf{I}_i')$ denotes the Nearest-Neighbor of $\mathbf{I}_i'$, and $\| \cdot \|$ is the Euclidean distance,

$$rel_B \; s(\mathbf{I}_i') = \frac{1 - e^{1 - \; d'(\mathbf{q}_B' \; s \; \mathbf{I}_i') / \max_i d'(\mathbf{q}_B' \; s \; \mathbf{I}_i')}}{1 - e} \tag{6}$$

where i is the index of all images in the database and $d' \; \mathbf{q}_{BQS}', \mathbf{I}_i'$ is the distance of image $\mathbf{I}_i'$ from the point computed according to Eq. 3.

The dissimilarity multi-space is included in a content-based retrieval system with Nearest-Neighbor relevance-feedback according to the following algorithm:

i) the user submits a query image $\mathbf{q}$. The distances d_{f_m} $\mathbf{q}^{f_m}, \mathbf{I}_i^{f_m}$, $m = 1, \ldots, M$, and $i = 1, \ldots N$ are computed, where M is the number of low-level features used to represent the images, and N is the number of the images in the database;

ii) in each feature space these distances are normalized between 0 and 1 and they are used to create, for each image $\mathbf{I}_i$, the new dissimilarity representation $\mathbf{I}'_i$, $i = 1, \ldots N$, according to Eq. 2;

iii) the Euclidean distances $d'\left(\mathbf{q}', \mathbf{I}'_i\right)$ between the dissimilarity representation of the query, and the dissimilarity representation of all the images are computed, and then sorted from the smallest to the largest;

iv) the first k images are labelled by the user as being relevant or not;

v) after the relevance feedback, the new reference point $\mathbf{q}_{BQS}$ is computed according to Eq. 3 in each feature space;

vi) the distances d_{f_m} $\mathbf{q}_{BQS}^{f_m}, \mathbf{I}_i^{f_m}$ are computed and normalized in each low-level feature space analogously to steps **i)** and **ii)**, where the query $\mathbf{q}$ is substituted with the new point $\mathbf{q}_{BQS}$. These distances are then used to create a new dissimilarity representation according to Eq. 2 where, again, the query $\mathbf{q}$ is substituted with the new point $\mathbf{q}_{BQS}$;

vii) in this new space, a score for all the images in the dataset is evaluated according to Eq. 4, where all the distances are computed according to the dissimilarity representation;

viii) all the images are sorted according to the value of the relevance score, and the first k images are labelled by the user as in step **iv)**;

ix) the algorithm starts again from step **iv)** until the user is satisfied.

4   Techniques for Combining Different Feature Spaces

In the previous sections we have mentioned a number of techniques that can be used to combine different image representations. In the following we will briefly review the six combination techniques that have been used in the experimental section for comparison purposes. Four combination methods aims to combine the relevance scores computed after relevance feedback, while the other two methods aim at combining distances.

The four techniques used to combine the relevance scores computed separately in each of the available feature spaces are the following:

$$score_{MA}\ \left(\mathbf{I}_i\right) = \max_{f \in F}\left(score_f\left(\mathbf{I}_i\right)\right) \tag{7}$$

$$score_{MIN}\left(\mathbf{I}_i\right) = \min_{f \in F}\left(score_f\left(\mathbf{I}_i\right)\right) \tag{8}$$

$$score_{MEAN}\left(\mathbf{I}_i\right) = \frac{\sum_{f \in F} score_f(\mathbf{I}_i)}{|F|} \tag{9}$$

where F is the set of the feature spaces and $score_f\left(\mathbf{I}_i\right)$ is the relevance score evaluated in the feature space f. RR weight is the weighted sum of the relevance scores, where the Relevance Rank Weights are obtained as in the following equation [7]

$$
wRR_f = \frac{\sum\limits_{j \in R} \frac{1}{rank_f(\mathbf{I}_j)}}{\sum\limits_{f' \in F} \sum\limits_{j \in R} \frac{1}{rank_{f'}(\mathbf{I}_j)}} \tag{10}
$$

and

$$
score_{RRW}(\mathbf{I}_i) = \sum\limits_{f \in F} wRR_f \; score_f(\mathbf{I}_i) \tag{11}
$$

where $f \in F$, $score_f(\mathbf{I}_i)$ is the relevance score evaluated in the feature space f, $rank_f(\mathbf{I}_j)$ is the rank of the image $\mathbf{I}_j$ according to $score_f$, and R is the set of the relevant images.

The other two combination methods are used to combine the distances computed in different feature spaces. One method computes the sum of the normalized distances (SUM), while the other method computes a "Nearest-Based" weighted sum (NBW) where the weights are computed in a similar way as in [14]:

$$
w_f = \frac{\sum\limits_{i \in R}\sum\limits_{j \in R} d_f(I_i, I_j)}{\sum\limits_{i \in R}\sum\limits_{j \in R} d_f(I_i, I_j) + \sum\limits_{i \in R}\sum\limits_{h \in N} d_f(I_i, I_h)} \tag{12}
$$

where $f \in F$, $d_f(\cdot)$ is a function that returns the distance between two images measured in the feature space f, and R, and N are respectively the set of the relevant and non-relevant images.

5   Experimental Results

5.1   Datasets

Experiments have been carried out using the Caltech-256 dataset, from the California Institute of Technology[1], that consists of 30607 images subdivided into 257 semantic classes [8]. Five different features have been extracted, namely the *Tamura* features [16] (18 components), the *Scalable Color* (64 components), *Edge Histogram* (80 components), *Color Layout* descriptors (12 components) [1], and the *Color and Edge Directivity Descriptor* (*Cedd*, 144 components) [3]. The open source library LIRE (Lucene Image REtrieval) has been used for feature extraction [11].

5.2   Experimental Set-Up

In order to test the performances, 500 query images have been randomly extracted from the dataset, covering all the semantic classes. The top twenty best-scored images for each query are returned to the user. Relevance feedback is

---

[1] http://www.vision.caltech.edu/Image_Datasets/Caltech256/

performed by marking images belonging to the same class of the query as relevant, and all other images in the top twenty as non-relevant. Performances are evaluated in terms of retrieval precision, and recall.

In order to evaluate the improvement attained by the proposed method in the next sub-section we will show the results attained separately in each feature space, and the performance related to the six combination techniques described in Section 4.

5.3   Results

Figures 1(a) and 1(b) show the performance of the proposed dissimilarity space representation compared to the six combination techniques described in Section 4, and the performance attained separately in each feature space.

By inspecting the behavior of the precision reported in Fig. 1(a), it can be easily seen that the highest performances are provided by the proposed dissimilarity (DS) based technique, and by the MEAN of the relevance scores. The combination of the relevance scores by the MAX and RR Weight rules allows attaining higher precision results than those attained by four out of the five features considered. This result is quite reasonable as typically the goal of the combination is to avoid choosing the worst problem formulation. In addition, it can be seen that the weighted combination (RR Weight) provides a lower result compared to the arithmetic MEAN, thus confirming the difficulty in providing an effective estimation of the weights. Finally, the worst result is attained by the MIN rule, that represents the logical AND function. Thus, we can conclude that, at least for the considered data set, the fusion of information from multiple feature spaces is more effective than the selection of one feature space. In addition, the results attained by the proposed DS space and the MEAN rule confirm that an unweighted combination can be more effective than weighted combination or selection. If we consider the two techniques based on the combination of the distances, namely the SUM rule and the NBW rule, we can see that their performances are lower than those of the techniques based on the combination of scores.

If we consider the recall (Fig. 1(b)), we can see that MEAN rule, and the DS approaches are still the best technique. It is worth noting that all the combination techniques, except for the MIN, provide an improvement in recall with respect to the performance attained in the individual feature spaces. In particular the fusion techniques working at the distance level provided good results, quite close to those attained by the RR Weight rule.

The proposed approach based on dissimilarity spaces not only allows attaining good performances in terms of precision and recall, but also requires a low effort in terms of computational time if compared to other combination techniques (Fig. 2). This effect can be explained by considering that all the distances from each image to the query are computed only once, during the first retrieval iteration. All the following iterations can exploit this result, and all the computation are made in the low-dimensional dissimilarity space.

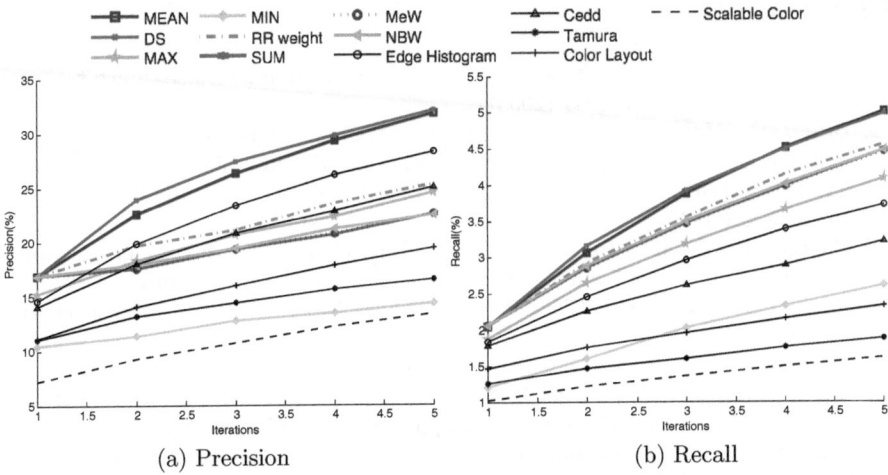

(a) Precision   (b) Recall

Fig. 1. Caltech-256 Dataset - Precision and Recall for 5 rounds of relevance feedback

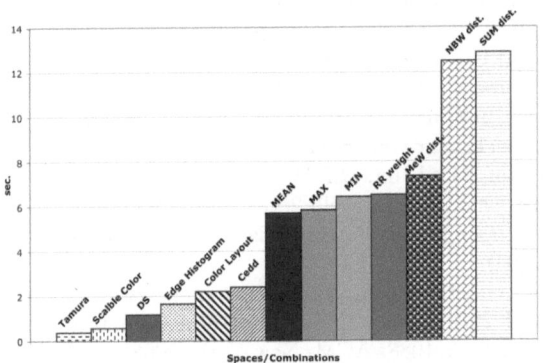

Fig. 2. Caltech-256 Dataset - Mean execution time for 1 round of relevance feedback

6   Conclusion

In this paper we proposed a technique that addresses the problem of the combination of multiple-features for image retrieval with relevance feedback. We showed
that a dissimilarity representation of images allows to combine nicely and effectively a number of feature spaces. In particular, reported results show that the
proposed technique allows outperforming other combination methods, both in
terms of performances and computational time. In addition, this method scales
well with the number of features, as the addition of one feature space adds one
component to the dissimilarity vector, and the distances in the original feature
spaces from the query needs to be computed only once.

References

1. Information technology - Multimedia content description interface - Part 3: Visual, ISO/IEC Std. 15938-3:2003 (2003)
2. Bruno, E., Moenne-Loccoz, N., Marchand-Maillet, S.: Learning user queries in multimodal dissimilarity spaces. In: Detyniecki, M., Jose, J.M., Nürnberger, A., van Rijsbergen, C.J. (eds.) AMR 2005. LNCS, vol. 3877, pp. 168–179. Springer, Heidelberg (2006)
3. Chatzichristofis, S.A., Boutalis, Y.S.: Cedd: Color and edge directivity descriptor: A compact descriptor for image indexing and retrieval. In: Gasteratos, A., Vincze, M., Tsotsos, J.K. (eds.) ICVS 2008. LNCS, vol. 5008, pp. 312–322. Springer, Heidelberg (2008)
4. Giacinto, G.: A nearest-neighbor approach to relevance feedback in content based image retrieval. In: CIVR 2007: Proceedings of the 6th ACM International Conference on Image and Video Retrieval, pp. 456–463. ACM, New York (2007)
5. Giacinto, G., Roli, F.: Dissimilarity representation of images for relevance feedback in content-based image retrieval. In: Perner, P., Rosenfeld, A. (eds.) MLDM 2003. LNCS, vol. 2734, pp. 202–214. Springer, Heidelberg (2003)
6. Giacinto, G., Roli, F.: Bayesian relevance feedback for content-based image retrieval. Pattern Recognition 37(7), 1499–1508 (2004)
7. Giacinto, G., Roli, F.: Nearest-prototype relevance feedback for content based image retrieval. In: ICPR (2), pp. 989–992 (2004)
8. Griffin, G., Holub, A., Perona, P.: Caltech-256 object category dataset. Tech. Rep. 7694, California Institute of Technology (2007), http://authors.library.caltech.edu/7694
9. Kittler, J., Hatef, M., Duin, R.P.W., Matas, J.: On combining classifiers. IEEE Trans. Pattern Anal. Mach. Intell. 20(3), 226–239 (1998)
10. Kuncheva, L.I.: Combining Pattern Classifiers: Methods and Algorithms. Wiley, Chichester (2004)
11. Lux, M., Chatzichristofis, S.A.: Lire: lucene image retrieval: an extensible java cbir library. In: MM 2008: Proceeding of the 16th ACM International Conference on Multimedia, pp. 1085–1088. ACM, New York (2008)
12. Nguyen, G.P., Worring, M., Smeulders, A.W.M.: Similarity learning via dissimilarity space in cbir. In: Wang, J.Z., Boujemaa, N., Chen, Y. (eds.) Multimedia Information Retrieval, pp. 107–116. ACM, New York (2006)
13. Pekalska, E., Duin, R.P.W.: The Dissimilarity Representation for Pattern Recognition: Foundations And Applications (Machine Perception and Artificial Intelligence). World Scientific Publishing Co., Inc., River Edge (2005)
14. Piras, L., Giacinto, G.: Neighborhood-based feature weighting for relevance feedback in content-based retrieval. In: WIAMIS, pp. 238–241. IEEE Computer Society, Los Alamitos (2009)
15. Rui, Y., Huang, T.S.: Relevance feedback techniques in image retrieval. In: Lew, M.S. (ed.) Principles of Visual Information Retrieval, pp. 219–258. Springer, London (2001)
16. Tamura, H., Mori, S., Yamawaki, T.: Textural features corresponding to visual perception. IEEE Trans. Systems, Man and Cybernetics 8(6), 460–473 (1978)
17. Zhang, L., Lin, F., Zhang, B.: Support vector machine learning for image retrieval. In: ICIP (2), pp. 721–724 (2001)

Discrete Point Based Signatures and Applications to Document Matching

Nemanja Spasojevic, Guillaume Poncin, and Dan Bloomberg

Google Inc., 1600 Amphitheatre Parkway, Mountain View, CA 94043, USA
sofra@google.com, gponcin@google.com, dbloomberg@google.com

Abstract. Document analysis often starts with robust signatures, for instance for document lookup from low-quality photographs, or similarity analysis between scanned books. Signatures based on OCR typically work well, but require good quality OCR, which is not always available and can be very costly. In this paper we describe a novel scheme for extracting discrete signatures from document images. It operates on points that describe the position of words, typically the centroid. Each point is extracted using one of several techniques and assigned a signature based on its relation to the nearest neighbors. We will discuss the benefits of this approach, and demonstrate its application to multiple problems including fast image similarity calculation and document lookup.

Keywords: image processing, feature extraction, image lookup.

1 Introduction

Over the past decade vast amounts of digitalized documents have become available. Projects like Google Books and Internet Archive have brought millions of books online, extremely diverse in form and content. Such a corpus requires fast and robust document image analysis. This starts with image morphology techniques, which are very effective for various image processing tasks such as extraction of word bounding boxes, de-skewing, connected component analysis, and page segmentation into text, graphics and pictures [1].

Image feature extraction is a well studied problem. Many techniques like SURF [2] and SIFT [3], [4], FIT [5] perform well at point matching across images, and image lookup from a database. However these techniques do not fare as well on repetitive patterns, such as text in document images. In addition, both SURF and SIFT extract thousands of key features per image, and features are matched by nearest neighbor search in the feature space which requires sophisticated indexing mechanisms. Unlike images of 3D objects, document images are projections of 2D image (paper) on the 3D scene (with significant warping caused by camera proximity and paper curvature). Locally Likely Arrangement Hashing (LLAH) [6] exploits this by making signatures from affine invariant features. This was shown to work well and achieves high precision in document page retrieval from a corpus of 10k pages. LLAH uses discrete features which are directly used for indexing, thus simplifying the process. Other retrieval techniques use word shape coding [7],

G. Maino and G.L. Foresti (Eds.): ICIAP 2011, Part I, LNCS 6978, pp. 149–158, 2011.

which should be robust to image deformation, but still suffer significant degradation in performance on non-synthetic document images.

In this paper we describe a novel method for extracting signatures based solely on the pixels of the page, and show how it can be applied to a range of problems in document matching. Our signatures are created from either centroids of the words, or centers of word bounding boxes. Because of this, we can use them to match documents for which we only have word bounding box information. The number of signatures extracted per page is the same as the number of words, which results in an order of magnitude less features than previous techniques.

We will demonstrate the use of these signatures in two applications: page similarity detection and image lookup from a database of indexed images. This work complements similar techniques based on OCR'd text to find similar pages or similar regions between two books, but works independently of language. In particular it can be applied to documents where OCR usually performs poorly (Chinese text, old text, etc.).

2   Algorithm Overview

In this section we cover point cloud extraction from the raw image and creation of signatures. We also describe two possible improvements: filtering high-risk signatures and superposition for ambiguous signatures.

2.1   Word Position Extraction

In order to perform signature extraction on a point cloud, the first step is to extract word centroids from the document image. For this, we use an open source image processing library named *Leptonica* [8]. We use a number of steps, including conversion of the raw image (Fig. 1a) to grayscale, background normalization (Fig. 1b), binarization (Fig. 1c), deskewing (Fig. 1d), and a series of morphological operations to extract the connected components representing words (Fig. 1e). The final step is to extract the centroids (Fig. 1f). This approach assumes that the document image has either vertical or horizontal text lines, which is typically the case in printed material and digital books (or we can correct for).

The benefit of operating on word positions is that it only requires the word bounding boxes, not the image. This way we can use, PDF or OCR word bounding box information as a source of points, which greatly reduces the cost of computing signatures when working with a large number of pages.

2.2   Signature Calculation from Point Cloud

An overview of signature extraction is shown in Algorithm 1. The basic idea is, for each point, to select $kNNCount$ nearest neighbors, sort them in radial distance order, and compute the angle between the selected and neighbor point. The angle is discretized based on how many bits we decide to use. Discretized angles (sub-signatures) are concatenated together in order of radial distance,

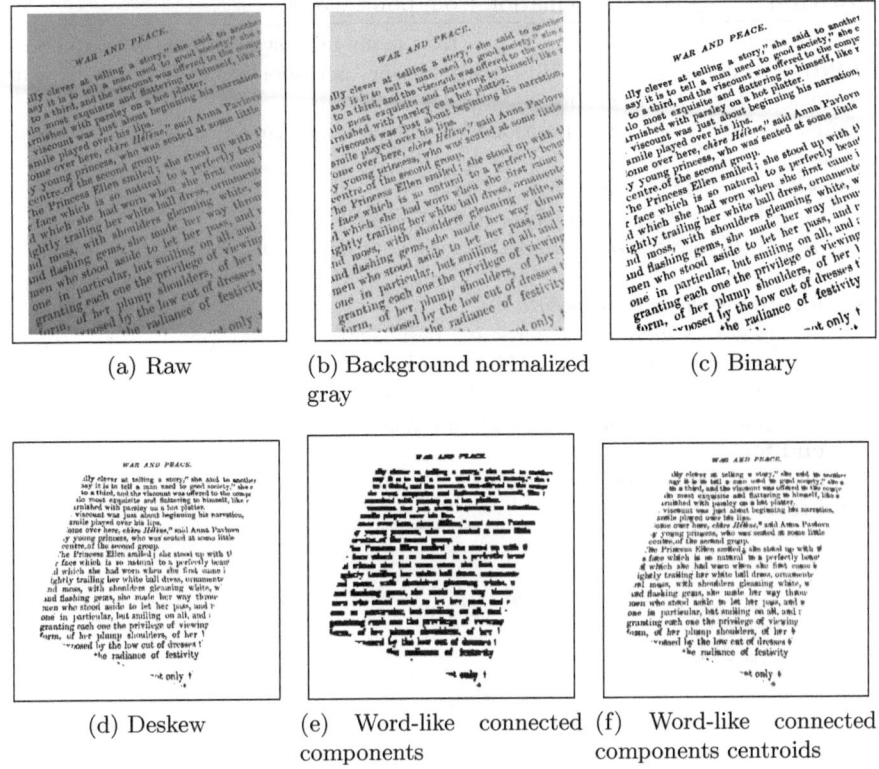

     (a) Raw   (b) Background normalized   (c) Binary

                                gray

    (d) Deskew   (e) Word-like connected (f) Word-like connected

                       components   components centroids

Fig. 1. Examples of image processing steps while extracting the word centroids

creating the final signature. An illustration of this process is shown on Fig. 2. In this paper a single angle is represented by 4 bits, i.e. using buckets of 22.5°.

Distortions such as skew are corrected for during word extraction. Other types of distortion like severe warping due to lens distortion or paper curvature are more challenging to deal with. However, the signature extraction algorithm achieves a reasonable degree of robustness to those. The main failure mode for matching signatures is actually word segmentation errors leading to changes in the neighborhood of a point. For example the merging of two words into one causes a word centroid to be shifted and another one to be missing. Flips in discretized values of the angle or reordering of the radial distances can also modify the signature. However in practice, such failures have limited impact on the result since they only affect small regions.

2.3   Signature Filtering Based on Estimated Risk

The idea is to only keep signatures that are stable with high confidence. We filter out signatures with high probability of bits shifting or flipping. Each signature S is composed of smaller sub-signatures $S = [s(0)][s(1)]..[s(N)]$. We consider

---

Algorithm 1. Signature calculation from point cloud

1: $kNNCount \leftarrow 8$ How many nearest neighbors we care about
2: $kBitPerAngle \leftarrow 4$ How many bits per neighbor to neighbor we dedicate
3: $kMask \leftarrow 1 << kBitPerAngle$
4: $points$ Word positions for a given image (image or word box based)
5: $signatures \leftarrow \emptyset$
6: **for all** $point \in points$ **do**
7:      $nn\_points \leftarrow NearestNeighbors(point, kNNCount)$
8:      $nn\_points \leftarrow SortByRadiusInIncreasingOrder(point, nn\_points)$
9:      $signature \leftarrow 0$
10:     **for all** $neighbor \in nn\_points$ **do**
11:         $alpha \leftarrow CalculateAngle(point, neighbor)$
12:         $alpha\_discrete \leftarrow floor(kMask \times alpha/2\_PI)$
13:         $signature \leftarrow signature << kBitPerAngle$
14:         $signature \leftarrow signature \mid alpha\_discrete$
15:     **end for**
16:     $signatures.push\_back(signature)$
17: **end for**

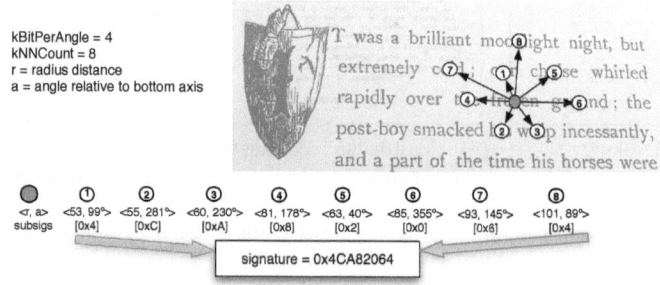

Fig. 2. Overview of signature creation from word centroids

signature variations that comes from slight shifts of word positions. Small shifts may lead to changes in discretized angle value, e.g. $s(0)$ flipping from 13 to 14 due to small word position shifts, or in the order of sub-signatures, e.g. $s(0)$ and $s(1)$ swapping as they had almost same radial distance. If we can estimate the confidence of a given signature, we can filter out weaker ones.

Let's consider the probability of a discretized angle flipping. If we have angle α and its discrete value a, where for example $a \leq \alpha < a + 1$ then we can say that distance of from the edge is $\epsilon = \mid \alpha - (a + 0.5) \mid$ assuming that $a + 0.5$ is edge of discretization. One can easily see that if $\epsilon = 0$ a random perturbation of points will lead to a flip in 50% of the cases. If we say that probability of flip is $p(\epsilon)$, then the probability that the entire signature changes due to at least one sub signature flipping can be expressed as:

$$P_{flip} = 1 - \prod_{i=1}^{N} (1 - p(\epsilon_i)) \qquad (1)$$

Similarly we can estimate the probability of two neighboring points swapping as $p(\delta r)$ where $\delta r = \mid r_1 - r_2 \mid / \left(\frac{1}{2}(\mid r_1 \mid + \mid r_2 \mid) \right)$ is the relative radial distance between consecutive neighbor points from the choosen point. The probability that the signature changes can be expressed as:

$$P_{swap} = 1 - \prod_{i=1}^{N-1} \left(1 - p\left(\frac{\mid r_i - r_{i+1} \mid}{\frac{1}{2}(\mid r_i \mid + \mid r_{i+1} \mid)} \right) \right) \tag{2}$$

Finally the chance of a signature changing due to swap or flip is $P_{flip\_or\_swap} = 1 - (1 - P_{flip}) \times (1 - P_{swap})$.

In this paper we naively modeled the probability distribution as $p(x, w) = 0.5 \times (w - x)$ if $x \in [0, w)$ and 0 otherwise. x is variable, while w is a threshold parameter. For angular discretization $w_{angle} = 0.05$, and for radius risk $w_{radius} = 0.01$. In experiments where we use signature filtering, signatures with $P_{flip\_or\_swap} > 0.6$ are filtered out.

2.4   Superposition of Ambiguous Signatures

Let us consider the problem of angle discretization. We often end up on one side or the other side an edge in the discretization function. One option is to use both values when composing the signature. We can consider a signature to be a superposition of states (angles), and by calculating the signature we project mixtures to their discrete values. But we can also create all possible projections; i.e. the set of all signatures. For example if the 1st and 3rd sub-signatures have two possible states then we have 4 possible signatures. For example $[\{s_1, s_1'\}][s_2][\{s_3, s_3'\}][s_4]$ would lead to $\{[s_1][s_2][s_3][s_4], [s_1][s_2][s_3'][s_4], [s_1'][s_2][s_3][s_4], [s_1'][s_2][s_3'][s_4]\}$. In the following, superposition was used only where indicated for angles within $\epsilon < 0.05$ of a discretization edge.

3   Evaluation on Synthetic Data

In this section we evaluate precision and recall of signatures for the task of matching two pages. The evaluation is done on synthetic data. All the signatures used were 32bit (generated based on 8 nearest neighbors). We start from an 'original' point cloud and a 'copy' point cloud derived from the original with some added distortions. All points are within $H = 1200$ pixels, $W = 1600$ pixels, and we use a fixed number of points per page for the original page ($N_{original} = 300$) (on average one point per 80×80 pixel square). The 'copy' page then has ($N_{copy} = N_{original} \times (1 - C_{drop})$) points where $N_{original} \times C_{drop}$ points are randomly be dropped to simulate word segmentation errors. Each copy point is also moved from its original position as follows:

$$(x_{copy}, y_{copy}) = (x_{original} + rand() \times C_{drift}, y_{original} + rand() \times C_{drift}) \tag{3}$$

where $rand()$ is random float from $[0, 1)$. In this way we try to emulate realistic differences that may occur between two sets of point clouds.

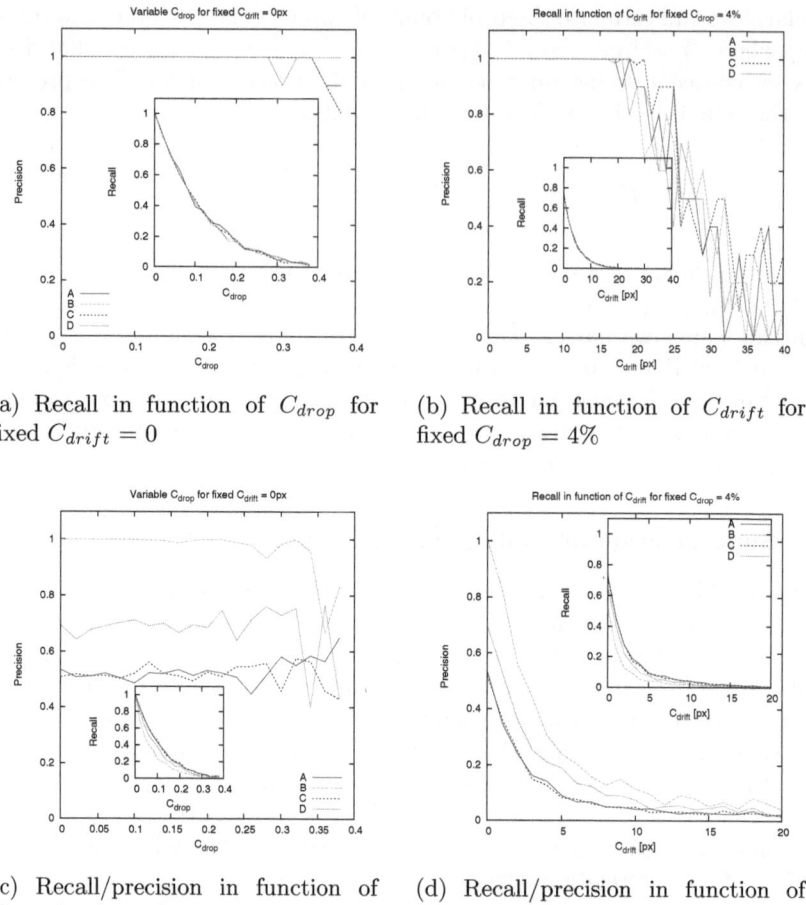

(a) Recall in function of C_{drop} for fixed $C_{drift} = 0$

(b) Recall in function of C_{drift} for fixed $C_{drop} = 4\%$

(c) Recall/precision in function of C_{drop} for fixed $C_{drift} = 0$

(d) Recall/precision in function of C_{drift} for fixed $C_{drop} = 4\%$

Fig. 3. Signature precision/recall evaluation for the random (3a, 3b) and grid (3c, 3c) point cloud distribution. Four scenarios: (A) original signatures, (B) risky signatures filtered, (C) superposition of ambiguous signatures, (D) only using unique signatures.

We experimented first with randomly distributed point clouds, and then with points aligned on a square grid (e.g. Chinese texts) with small x and y variations (less than 5 pixels). These two sets are called *random* and *grid* set.

Random Distribution: On Fig. 3a, 3b we see that precision for this set remains 100% even as the amount of drop increases. This is expected, because signatures in this case are fairly unique and can withstand some amount of perturbations. For variable drop at $C_{drop} = 10\%$ recall is 40% (Fig. 3a), while for variable drift (drop fixed at 4%) at $C_{drift} = 5\%$ pixels recall recall is 20% (Fig. 3b). In both cases recall shows a steep drop, but there is significant robustness to points being dropped and shifted around. Also it's interesting to note that there is practically no difference in results between scenarios A-D.

Grid Distribution: This is probably the hardest type of distribution for our algorithm. It is so regular that unrelated signatures are likely to collide since neighborhoods all look similar. This is verified in Fig. 3d, where starting precision is 50% (curve A). This means that many unrelated signatures match across two identical point clouds. Note that precision is constant as we randomly drop points, and it seems to perform best in the case where we filter out weak signatures, but also pretty well when we filter out signatures that occur multiple times on one page. However in Fig. 3d, 3c we see that increased precision when filtering comes at the expense of recall.

4   Document Image Similarity Application

A common application of analysis is the detection of similar pages. For instance, one may want to cluster duplicate pages when merging multiple scans of a document, or find corresponding pages between 2 different versions. This can be done based on OCR text, but OCR is an expensive operation, which does not work well on all scripts. Document digitization is typically done using sheetfed scanners or other techniques in which acquired images have little warping, moderate skew ($\pm 3°$), no scale difference, and small translation variation. These are optimal conditions for our signatures, although we have shown how we could correct for imperfect conditions. Jaccard J similarity is used to estimate document image similarity. The similarity between two pages is calculated as:

$$J_s(p_1, p_2) = \frac{|\, S(p_1) \cap S(p_2)\, |}{|\, S(p_1) \cup S(p_2)\, |} \tag{4}$$

where $S(p)$ is the set of extracted signatures from page p. Depending on image distortion and word segmentation discrepancy between pages, the number of matching signatures may be low compared to the total number of signatures. We can get better estimate using matching signatures to calculate an affine transform from one image to another, and then use that transform to align word bounding boxes between pages. We declare that boxes match when their centroid is close after transformation and they have roughly the same size. From the matching box count we calculate similarity J_b as:

$$J_b(p_1, p_2) = \frac{|\, MatchCount(p_1, p_2)\, |}{|\, BoxCount(p_1) + BoxCount(p_2) - MatchCount(p_1, p_2)\, |} \tag{5}$$

In Fig. 4 we show both signature matches and box matches. The two example pages we use are in Chinese and Arabic. Initially few signatures match (Fig. 4a, 4c), but the number improves significantly after alignment. Similarities were initially 19% for Chinese, and 5% for Arabic. After alignment and similarity recalculation, we measured 93% for Chinese and 37% for Arabic. Note that in Arabic we still fail to align many boxes due to inconsistent word segmentation, but we align enough to have confidence in the measurement. With Chinese, similarity is high because segmentation is consistent despite being wrong.

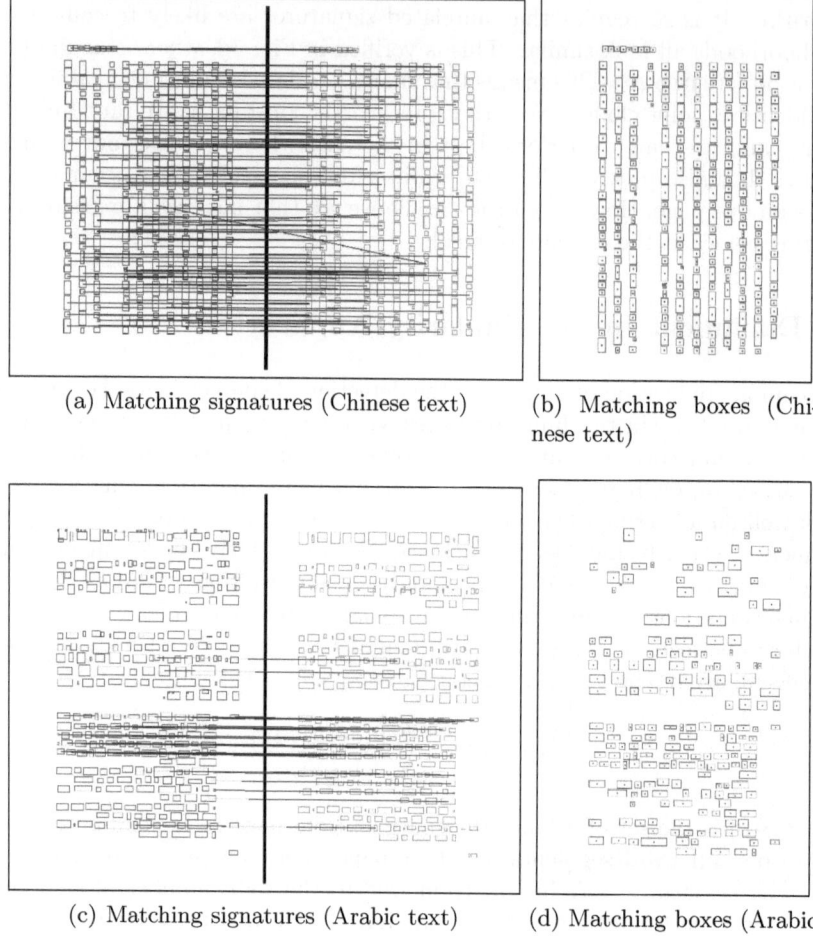

(a) Matching signatures (Chinese text)

(b) Matching boxes (Chinese text)

(c) Matching signatures (Arabic text)

(d) Matching boxes (Arabic text)

Fig. 4. Examples of matching signatures and aligned boxes in the page similarity calculation. Shown on example of Chinnese (4a, 4b), and Arabic (4c, 4d) scripts.

5   Document Image Lookup Application

Document retrieval based on a photograph of a page is another common application. For example, a user takes a photo with his mobile phone, sends it to a server, and the digital reference to the page is returned. We ran two experiments. The first with a small set of English language classics for a total of 4.1K pages. The second with a much wider variety of books, totalling 1M pages. In both cases, we queried for 120 photographs of book pages taken with a NexusOne Phone camera (5MP). Our server runs on a 2.2GHz PC with 8GB of RAM.

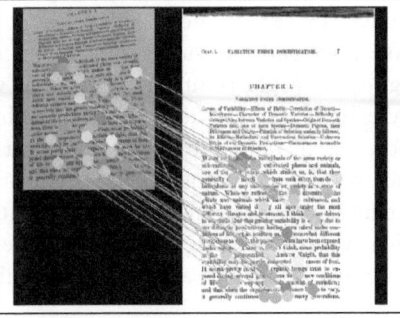
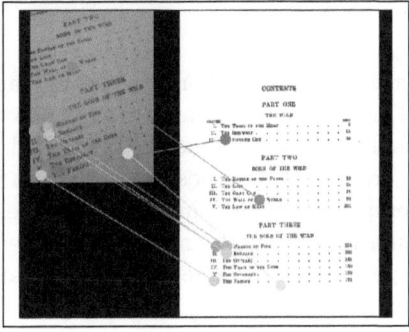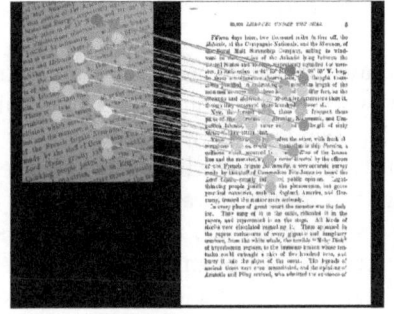

Fig. 5. Examples of image lookups

Table 1. Image lookup results

Index Size	Accuracy	Signature size [bits]
4.1K	0.966	16
4.1K	0.949	32
1M	0.871	32

The pipeline is straightforward. We built an index based on OCR bounding boxes since we had them available (instead of running the image processing). The signature calculation and indexing takes on average 8.6ms per page (for 32bit signatures, for the 4.1K set). The index maps the signature to a list of $< book, page, position >$ tuples. At query time, we first process the input image using Leptonica to extract word centroids from which we generate signatures. We look up each signature in the index and bucket hits per book page. Finally we pick the page with the largest number of matching signatures. We only need to look up around 200 signatures, so queries run very quickly.

Overall accuracy exceeds 95% on the 4.1K set, as shown in Table 1. We tried various signature sizes and found that they had little impact on the results themselves. The downside of small signatures is that lookup takes longer because we run into many more false positives. Lookup times are sub-millisecond, once we have the list of query signatures.

We then ran the same experiment on a set of 1M pages, which is orders of magnitude larger than other sets that we know of, e.g [6],[5]. Accuracy degraded only a little to 87%, demonstrating that this technique works well with tens of thousands of books worth of data in the index. The index is also compact: 1M pages resulted in about 386M signatures, allowing us to store the entire index in memory on a single machine ($< 4GB$). We filtered out 0.8% of signatures, which were shared by over 1000 pages, to prevent false positives. On average each query results in about 2000 candidate matches, most of which agree.

6   Conclusion

We have presented a method for extracting signatures from document pages, that works using bounding boxes found from either OCR, or image processing independent of either language or the age of the document. We showed the robustness of this technique on synthetic data, including post-processing to improve the results in various cases. We then presented two simple applications of these signatures: one related to page similarity measurement, and the other to the retrieval of documents from camera pictures on a large corpus. The latter was shown to be efficient on a large corpus of 1M pages.

References

1. Bloomberg, D., Vincent, L.: Document Image Analysis, Mathematical morphology: theory and applications, Najman L., Talbot H. (ed.), pp. 425–438 (2010)
2. Bay, H., Tuytelaars, T., Van Gool, L.: SURF: Speeded up robust features. In: Leonardis, A., Bischof, H., Pinz, A. (eds.) ECCV 2006. LNCS, vol. 3951, pp. 404–417. Springer, Heidelberg (2006)
3. Lowe, D.G.: Distinctive image features from scale-invariant keypoints. IJCV 60(2), 91–110 (2004)
4. Ke, Y., Sukthankar, R.: PCA-SIFT: A More Distinctive Representation for Local Image Descriptors. In: Proc. CVPR 2004, pp. 506–513 (2004)
5. Liu, Q., Yano, H., Kimber, D., Liao, C., Wilcox, L.: High accuracy and language independent document retrieval with a fast inv. t. In: Proc. ICME 2009, pp. 386–389 (2009)
6. Nakai, T., Kise, K., Iwamura, M.: Hashing with Local Combinations of Feature Points and Its App. In: Proc. CBDAR 2005, pp. 87–94 (2005)
7. Shijian, L., Linlin, L., Chew Lim, T.: Document Image Retrieval through Word Shape Coding. IEEE TPAMI 30(11), 1913–1918 (2008)
8. http://www.leptonica.org/

Robustness Evaluation of Biometric Systems under Spoof Attacks

Zahid Akhtar, Giorgio Fumera, Gian Luca Marcialis, and Fabio Roli

Dept. of Electrical and Electronic Eng., Univ. of Cagliari
Piazza d'Armi, 09123 Cagliari, Italy
{z.momin,fumera,marcialis,roli}@diee.unica.it
http://prag.diee.unica.it

Abstract. In spite of many advantages, multi-modal biometric recognition systems are vulnerable to spoof attacks, which can decrease their level of security. Thus, it is fundamental to understand and analyse the effects of spoof attacks and propose new methods to design robust systems against them. To this aim, we are developing a method based on *simulating* the fake score distributions of individual matchers, to evaluate the relative robustness of different score fusion rules. We model the score distribution of fake traits by assuming it lies between the one of genuine and impostor scores, and parametrize it by a measure of the relative distance to the latter, named *attack strength*. Different values of the attack strength account for the many different factors which can affect the distribution of fake scores. In this paper we present preliminary results aimed at evaluating the capability of our model to approximate realistic fake score distributions. To this aim we use a data set made up of faces and fingerprints, including realistic spoof attacks traits.

Keywords: Biometric systems, Performance evaluation, Spoof attacks, Adversarial pattern recognition.

1 Introduction

Biometrics are biological or behavioural characteristics that are unique for each individual. In order to combat growing security risks in information era, academies, governments and industries have largely encouraged research and adoption of biometric identification systems. The main advantage of biometric technologies compared to conventional identification methods is replacing "what you have" and "what you know" paradigms with "who you are" one, thus preventing identity fraud by using biometrics patterns that are claimed to be hard to forge.

However, several researches have shown that some biometrics, such as face and fingerprint, can be stolen, copied and replicated to attack biometric systems [1,2]. This attack is known as *spoof attack*, and also named as *direct attack*. It is carried out by presenting replicated biometric trait to the biometric sensor. "Liveness" testing (vitality detection) methods have been suggested among feasible counteractions against spoof attacks. Liveness testing, which aims to detect

G. Maino and G.L. Foresti (Eds.): ICIAP 2011, Part I, LNCS 6978, pp. 159–168, 2011.

whether the submitted biometric trait is live or artificial, is performed by either software module based on signal processing or hardware module embedded into the input device itself [2,3]. But, so far, the literature review states that no effective method exists yet. Moreover, the collateral effect when biometric systems are coupled with liveness detection methods is the increase of false rejection rate.

In our opinion, it is pivotal to develop also methods, beside liveness detection ones, to design secure biometric systems. A straightforward approach could be to fabricate fake traits to evaluate the security of the system under design. However, constructing reliable fake replicas and simulating all possible ways in which they can be realised, is impractical [1]. A potential alternative is to develop methods based on *simulating* the distribution of fake biometric traits. To the best of our knowledge, no systematic research effort has been carried out toward this direction yet. The only works which addressed this issue are [5,14,15], where the fake distribution is simulated by assuming that attacker is able to replicating exactly the targeted biometric (worst-case scenario): in other words, the fake score distribution coincides with that of genuine users.

Based on above motivation, we are currently developing a method for evaluating the robustness of multi-modal systems to spoof attacks, based on simulating the score distribution produced by fake traits at the matchers output, and then on evaluating the relative robustness of different score fusion rules. Due to the unknown impact of several factors, such as particular biometric trait being spoofed, forgery techniques and skills used by the attackers, etc., on position and shape of score distribution, we make substantive assumptions on the potential form and shape it can get. In particular, we argue that the fake score distribution generated by comparing fake replica of a given subject with the corresponding template of that subject, is between impostor and genuine distributions. On the basis of these considerations, as starting point of our research, we model fake scores as a combination of the genuine and impostor ones, on the basis of a single parameter, that we call "attack strength". This parameter controls the degree of similarity of the fake and genuine scores, with respect to the impostor scores. The attack strength quantifies the effect of several factors mentioned above, and allow to figure out more possible scenarios that the only worst-case one [5,14,15]. To evaluate the robustness of a given multi-modal system under spoof attacks using our method, the testing impostor scores of the matcher under attack have to be replaced with simulated fake scores generated as mentioned above. The system designer can also evaluate the robustness of the system by repeating the above procedure for different values of attack strength parameter. In this paper, we present preliminary results aimed at evaluating the capability of our model to approximate realistic fake score distributions.

Our model of the fake score distribution is presented in Sect. 2. In Sect. 3 we describe its preliminary experimental validation on two data sets of faces and fingerprints including real spoof attacks.

2   A Model of the Match Score Distribution Produced by Spoof Attacks

We denote the output score of a given biometric matcher as random variable s, and denote with G and I the event that the input biometric trait comes respectively from a genuine or an impostor user. The respective score distributions will be denoted as $p(s|G)$ and $p(s|I)$. In the standard design phase of a multi-modal biometric verification systems, the score of the individual matchers $s_1, s_2, \ldots$ are combined using some fusion rule, and a decision threshold t is set on the fused matching score $s_f = f(s_1, s_2, \ldots)$, so that a user is accepted as genuine if $s_f \geq t$, and is rejected as an impostor otherwise. The threshold t is usually set according to applications requirements, like a desired false acceptance rate (FAR) or genuine acceptance rate (GAR) value. This defines the so-called operational point of the biometric system. The FAR and GAR values are estimated from training data made up of a set G_{tr} of genuine scores and a set I_{tr} of impostor scores.

A straightforward way to analyse the performance of biometric system under spoof attacks is to fabricate fake biometric traits and present them to the system. However, this can be a lengthy and cumbersome task [4]. An alternative solution for multi-modal systems is to *simulate* the effects of spoof attacks on the matching score of the corresponding biometric trait. This is the approach followed in [5,14,15]. In these works, the robustness of multi-modal systems against spoof attacks was evaluated in a worst-case scenario, assuming that the matching score produced by a spoofed trait is identical to the score produced by the original trait of the corresponding genuine user. Accordingly, the score distribution of spoofed traits was assumed to be identical to the genuine score distribution.

However, when a fake trait is presented to the biometric sensor, many factors can influence the resulting output score distribution, such as the particular biometric trait spoofed, the forgery approach, the ability of the attacker in providing a "good" biometric trait of the targeted subject as model for his replica, the specific matching algorithm used by the system, the degree of "robustness" of the representation and matcher themselves to noisy patterns, etc. In practice it can be very difficult, if not impossible, to systematically construct fake biometric traits with different degrees of similarity to the original traits. Due to the current very little knowledge on how aforesaid factors affect the fake score distribution, we argue that the only feasible way is to simulate their effect.

A different scenario than the worst-case one considered in [5,14,15] could be modelled by considering a score distribution of fake traits lying between the genuine and impostor distributions. For example, in the case of fingerprint, its "similarity" to the impostors distribution will be caused by several factors as artefacts in the replica, the image distortion from the mould to the cast, the good/bad pressure of the attacker on the sensor surface when placing the spoofed fingerprint, whilst its "similarity" to the genuine users one is given by the fact that several important features, as the ridge texture and minutiae locations, will be the same of the correspondent subject. In absence of more specific information on the possible shapes that the fake score distribution may exhibit, we propose to simulate the one of any individual matcher as follows: denoting the event that the input biometric trait comes from a spoof attack as F, and the corresponding score distribution as $p(s|F)$, we replace each impostor score s_I with a fictitious score s_F given by

$$s_F = (1 - \alpha)s_I + \alpha s_G , \tag{1}$$

where s_G is a randomly drawn genuine score, and $\alpha \in [0, 1]$ is a parameter which controls the degree of similarity of the distribution of fake scores to the one of genuine scores. The resulting distribution of fictitious fake scores $p(s|F)$ is thus "intermediate" between the ones of $p(s|I)$ and $p(s|G)$. By using different values of α, one gets different possible distributions: the higher the α value, the closer $p(s|F)$ to the genuine score distribution $p(s|G)$. Accordingly, we name α "attack strength". This parameter, α, and related Eq. (1), are aimed not to model the physical fake generation process, but only its effect on the corresponding distribution $p(s|F)$, which depends on several causes like the ones mentioned above. In this paper, we want to investigate if the above model allows us to obtain a reasonable approximation of realistic fake score distributions.

Since the designer has no a priori information about the possible characteristics of the attacks the system may be subject to, he should consider several, hypothetical distributions corresponding to different α, and evaluate the robustness of the score fusion rules of interest against each of them.

Accordingly, Algorithm 1 details the proposed procedure. First, the decision threshold t on the combined score s_f has to be estimated from training data made up of a set of genuine and impostor scores, G_{tr} and I_{tr}, as described above. In the standard performance evaluation procedure, the performance is then evaluated on a distinct test set of genuine and impostor scores, denoted as G_{ts} and I_{ts}. To evaluate the performance under a spoof attack, we propose instead to replace the impostor scores I_{ts} corresponding to the matcher under attack with a set of fictitious fake scores F_{ts}, obtained from Eq. (1). This can be done several times, using different α values to evaluate the performance under spoof attacks of different strength.

Note that using Eq. (1), if the randomly chosen genuine score s_G is higher than the impostor score s_I, then the latter is replaced by a greater fictitious fake score s_F. Therefore, for any threshold value t the FAR evaluated under a simulated spoof attack is likely to be higher than the FAR evaluated in the standard way, without spoof attacks. The GAR remains unchanged instead, as spoof attacks do not affect genuine scores. Accordingly, as the value of α increases, the corresponding FAR is likely to increase from the values attained for $\alpha = 0$, corresponding to the absence of attacks, to the worst-case corresponding to $\alpha = 1$. Hence, the above procedure allows one to evaluate how the system's performance degrades for different potential fake score distributions characterised by a different attack strength. In particular, it can be useful to check the amount of the relative "shift" (the corresponding α value) of the impostor score distribution toward the genuine one, such that the system's performance (the FAR) drops below some given value. The more gracefully the performance degrades (namely, the higher the α value for which the FAR drops below some value of interest), the more robust a system is.

Algorithm 1. Procedure for evaluating the performance of a multi-modal biometric system under a simulated spoof attack

Inputs:

- A training set (G_{tr}, I_{tr}) and a testing set (G_{ts}, I_{ts}) made up of N vectors of matching scores coming from genuine and impostor users;
- α: the attack strength value for the matcher under attack.

Output: The system's performance under a simulated spoof attack with attack strength α.

1: Set the threshold t from training data (G_{tr}, I_{tr}), according to given performance requirements.
2: Replace the scores I_{ts} of the matcher under attack with a same number of fictitious fake scores F_{ts} generated by Eq. (1).
3: Evaluate the performance of the multi-modal system on the scores (G_{ts}, F_{ts})

3   Experimental Results

In this section, we report a preliminary validation of our model of the fake score distribution of a single matcher, using two data sets including realistic spoof attacks. More precisely, our aim is to investigate whether realistic fake score distributions can be reasonably approximated by our model, for some α values.

3.1   Data Sets

Since no biometric data sets including spoof attack samples are available publicly, we collected two sets of face and fingerprint images and created spoof attacks. The collected data set contains face and fingerprint images of 40 individuals, with 40 genuine samples and 40 fake samples (spoof attacks) per individual.

Face images were collected under different facial expressions and illumination. Spoofed face images were created with a "photo attack" [11]: we put in front of the camera the photo of each individual displayed on a laptop screen. For each individual, we created 40 spoofed images.

Fingerprint images were collected using Biometrika FX2000 optical sensor. Fake fingers were created by the consensual method with liquid silicon as carried out in [6,7,8]. We fabricated fake fingerprint using plasticine-like material as the mould while two-compound mixture of liquid silicon and a catalyst as cast. The main property of the material utilised as the cast is high flexibility silicon resin (SILGUM HF) with a very low linear shrinkage. Further details on fingerprint spoof production can be found in [12].

The fingerprint and the face recognition systems used in the experiments were implemented using the minutiae-based Neurotechnologs VeriFinger 6.0 and the elastic bunch graph matching (EBGM) [9], respectively.

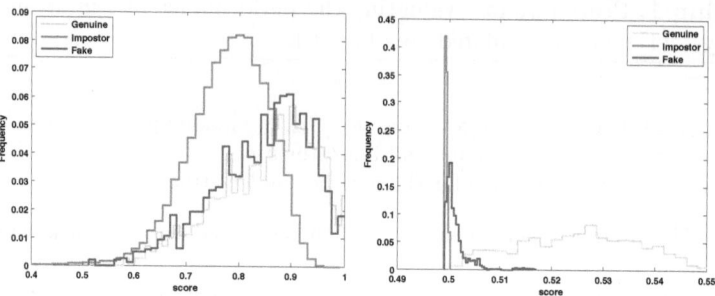

Fig. 1. Histograms of genuine, impostor and fake scores computed with the collected face (left) and fingerprint (right) image data sets

3.2   Results

In Fig. 1 the histograms of genuine, impostor and fake scores computed with the above data sets are shown. It is worth noting that these distributions exhibit two very different degrees of "attack strength": the fake score distribution of fingerprints is close to the impostor distribution, while the one of faces is much close to the genuine distribution. This provides a first, qualitative support to the assumption behind our model, namely that different, realistic fake score distributions can lie at different relative "distances" from the genuine and impostor ones.

To investigate whether the realistic fake scores distributions of Fig. 1 can be reasonably approximated by our model, for some α value, we evaluated the dissimilarity between them and the ones provided by our model, as a function of the attack strength α, and empirically computed the α value that minimised the dissimilarity between the two distributions. The fictitious fake scores were obtained as described in Algorithm 1. To assess the dissimilarity between the two distributions, we used the L1-norm Hellinger distance [13], also called Class Separation Statistic [10]. The L1-norm Hellinger distance between two probability distribution functions $f(x)$ and $g(x), x \in \mathcal{X}$ can be measured as:

$$\int |f(x) - g(x)| \mathrm{d}x.$$

Since this is a non-parametric class separation statistic, it can be used for all possible distributions.

The α values which minimise the dissimilarity between the fake score distribution obtained by our method and the real one is reported in Table 1. The corresponding distributions are depicted in Fig. 2.

Fig. 2 and Table 1 show that our approximation is rather good for the face data set. It is less good for the fingerprint data set instead, but it could be still acceptable to the aim of evaluating the relative robustness of different score fusion rules in a multi-modal system, which is the final aim of this model. Let us better explain this point. Obviously, in practice the designer of a biometric

Table 1. Minimum values of the Hellinger distance between the real distribution of fake scores and the one obtained by our model, as a function of α, for the face and fingerprint data sets. The corresponding α value is also shown.

Data set	Hellinger distance	α
Face	0.0939	0.9144
Fingerprint	0.4397	0.0522

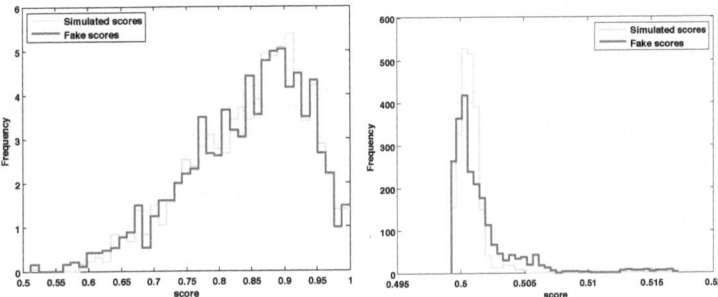

Fig. 2. Probability distributions of the scores of fake faces (left) and of fake fingerprints (right) obtained from our data sets (blue), and obtained by our method for fake score simulation (green), for the α value of Table 1

system can not not know in advance what shapes the fake score distributions will exhibit, if the system will be subject to a spoof attack. Accordingly, the robustness of a multi-modal system must be evaluated for several α values. What the above results show is a preliminary evidence that the simulated distributions one obtains using our model, for different α values, can actually give reasonable approximations of possible, realistic distributions.

For the sake of completeness, we also evaluated the accuracy of our model of fake score distribution in approximating the performance of the *individual* matcher under attack, for the α values of Table 1 that give the best approximation of the fake score distribution, although this is not the final aim of this model as explained above.

When the system is under a spoof attack, only the False Acceptance Rate (FAR) value changes, while the genuine acceptance rate (GAR) remains unchanged, since it does not depend on the matching scores of the impostors. To check the accuracy of the FAR approximated by our model, we compared the FAR of the mono-modal system attained under real spoof attacks with the FAR provided by our model, for all possible values of the threshold.

Fig. 3 shows the FAR as a function of the threshold for the uni-modal biometric system when no spoof attack is included in the data set (i.e., using only the genuine and impostor data; the "no attack" curve), under a real spoof attack against the face (fingerprint) matcher (using the fake biometric traits of

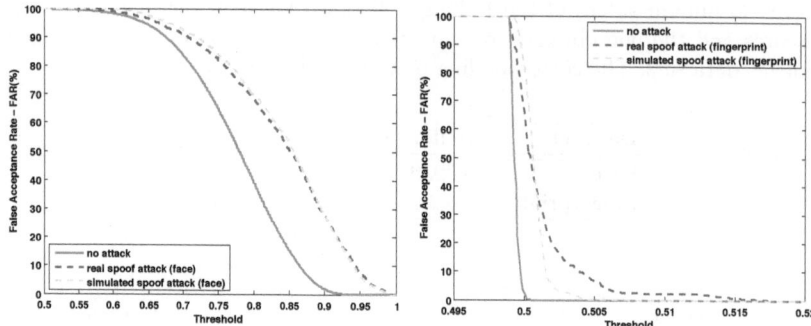

Fig. 3. FAR of the uni-modal biometric systems as a function of the threshold applied to the score, when the data set does not contain spoof attacks ("no attack" curve), under a real spoof attack against the face (left) or fingerprint (right) matcher ("real spoof attack" curve), and under a spoof attack simulated with our method ("simulated attack" curve)

our data set; the "real spoof attack" curve), and by a simulated spoof attack (using the fake scores provided by our method with the α values of Table 1; the "simulated attack" curve). It can be seen from Fig. 3 that our model provides a quite accurate approximation of the FAR in the case of face spoofing (Fig. 3, left): the maximum absolute difference between the real and the approximated FAR is 0.02. In the case of fingerprint spoofing (Fig. 3, right), our model over-estimates the FAR by an amount of up to 0.03 for threshold values lower than 0.502, while it underestimates the FAR up to a larger amount for threshold values greater than 0.502. This is due to the positive skewness of the real fake fingerprint scores, as can bee seen in Fig. 2. Note however that the threshold t corresponding to the zeroFAR operational point is 0.500, as can be seen from Fig. 1 (right). It is worth remarking that zeroFAR is the operational point such that the threshold leads to a zero FAR value on training data, and maximises the correspondent GAR value. Therefore, threshold values more than this one are out of the designer interest and can be neglected. This means that threshold values where the real FAR is underestimated by our model can be neglected as well, since they are localised for threshold values higher than 0.502.

Accordingly, let us focus in particular on high security operational points like the zeroFAR and 1% FAR, which are very crucial in order to assess the system robustness. The corresponding FAR attained by the fake score distribution in our data set ("Real FAR") and the approximated FAR using our model is reported in Table 2. We also report for comparison the approximated FAR obtained using the worst-case assumption of [5,14,15]. The reported results show that our method provides a good approximation of the performance of the two considered uni-modal systems under spoof attacks, at these operational points. The overestimation of the values for the fingerprint system is in some sense beneficial, since it puts the designer in the position to expect a performance decrease

Table 2. Comparison between the FAR attained at the zeroFAR and 1% FAR operational points by the uni-modal biometric system under a real spoof attack ("real FAR") and the FAR approximated by our model ("approximated FAR")

	Operational point	Real FAR	Approximated FAR (our model)	Approximated FAR (worst-case assumption)
Face	zeroFAR	0.048	0.042	0.114
System	1%FAR	0.235	0.233	0.243
Fingerprint	zeroFAR	0.506	0.625	0.948
System	1%FAR	0.600	0.808	0.951

higher than that occurring in the real case. In addition, it can be seen that our model is more flexible and appropriate for fake score distributions quite far from the worst-case one, as happens for fingerprints.

To sum up, our preliminary results provide some evidence that our model is able to reasonably approximate realistic distributions of the matching scores produced by spoof attacks.

4 Conclusions

Assessing the robustness of multi-modal biometric verification systems under spoof attacks is a crucial issue, do to the fact that replicating biometrics is a real menace. The state-of-the-art solves this problem by simulating the effect of a spoof attacks in terms of fake score distribution modelling, for each individual matcher. In particular, the fake score distribution is assumed to be coincident to the genuine users one, thus drawing a worst-case scenario.

However, a more realistic modelling should take into account a larger set of cases. Unfortunately, the approach of fabricating fake biometric traits to evaluate the performance of a biometric system under spoof attacks is impractical. Hence, we are developing a method for evaluating the robustness of multi-modal systems against spoof attacks, based on *simulating* the corresponding score distribution.

In this work we proposed a model of the fake score distribution that accounts for different possible realistic scenarios characterised by factors like different spoofing techniques, resulting in different degrees of similarity between the genuine and the fake score distribution. Such factors are summarised in our model in a single parameter associated to the degree of similarity of the fake score distribution to the genuine one, which is named accordingly "attack strength". A designer may use this method to generate several fake distributions for different α values, to analyse the robustness of the multi-modal system under design.

Preliminary experimental results provided some evidence that our model is capable to give reasonable approximations of realistic fake score distributions, and also to be a good alternative to the model based on the worst-case scenario adopted so far. Currently, we are working on constructing data sets containing spoofing attacks of different biometric traits, spoofing techniques, matchers, etc., to give a more extensive validation of our model, and to evaluate the effectiveness of our method for robustness evaluation of multi-modal systems under spoof attacks.

Acknowledgment. This work was partially supported by the TABULA RASA project, 7th Framework Research Programme of the European Union (EU), grant agreement number: 257289, and by the PRIN 2008 project "Biometric Guards - Electronic guards for protection and security of biometric systems" funded by the Italian Ministry of University and Scientific Research (MIUR).

References

1. Matsumoto, T., Matsumoto, H., Yamada, K., Hoshino, S.: Impact of artificial "gummy" fingers on fingerprint systems. In: Optical Security and Counterfeit Deterrence Techniques IV. Proc. of SPIE, vol. 4677, pp. 275–289 (2002)
2. Kim, Y., Na, J., Yoon, S., Yi, J.: Masked Fake Face Detection using Radiance Measurements. J. Opt. Soc. Am. - A 26(4), 760–766 (2009)
3. Kang, H., Lee, B., Kim, H., Shin, D., Kim, J.: A study on performance evaluation of the liveness detection for various fingerprint sensor modules. In: Proc. Seventh Int. Conf. on Know. Based Intel. Info. and Engg. Sys., pp. 1245–1253 (2003)
4. Marcialis, G.L., Lewicke, A., Tan, B., Coli, P., Grimberg, D., Congiu, A., Tidu, A., Roli, F., Schuckers, S.: First International Fingerprint Liveness Detection Competition. In: Proc. 14th Intl. Conf. on Image Analysis and Proc., pp. 12–23 (2009)
5. Rodrigues, R.N., Ling, L.L., Govindaraju, V.: Robustness of Multimodal Biometric Methods against Spoof Attacks. JVLC 20(3), 169–179 (2009)
6. Abhyankar, A., Schuckers, S.: Integrating a Wavelet Based Perspiration Liveness Check with Fingerprint Recognition. Patt. Rec. 42(3), 452–464 (2009)
7. Coli, P., Marcialis, G.L., Roli, F.: Fingerprint Silicon Replicas: Static and Dynamic Features for Vitality Detection using an Optical Capture Device. Int'l J. of Image and Graphics 8(4), 495–512 (2008)
8. Marcialis, G.L., Roli, F., Tidu, A.: Analysis of Fingerprint Pores for Vitality Detection. In: Proc. 12th Int'l Conf. on Pattern Recognition, pp. 1289-1292 (2010)
9. Bolme, D.S.: Elastic Bunch Graph Matching. Master's Thesis: Dept. of Comp. Science, Colorado State University (2003)
10. Jain, A.K., Prabhakar, S., Chen, S.: Combining Multiple Matchers for a High Security Fingerprint Verification System. PRL 20(11-13), 1371–1379 (1999)
11. Pan, G., Wu, Z., Sun, L.: Liveness detection for face recognition. In: Recent Advances in Face Recognition, pp. 236–252 (2008)
12. http://prag.diee.unica.it/LivDet09
13. LeCam, L.: Asymptotic Methods in Statistical Decision Theory. Springer, Heidelberg (1986)
14. Rodrigues, R.N., Kamat, N., Govindaraju, V.: Evaluation of Biometric Spoofing in a Multimodal System. In: Proc. Fourth IEEE Int. Conf. Biometrics: Theory Applications and Systems, pp. 1–5 (2010)
15. Johnson, P.A., Tan, B., Schuckers, S.: Multimodal Fusion Vulnerability to Non-Zero Effort (Spoof) Imposters. In: Proc. IEEE Workshop on Information Forensics and Security, pp. 1–5 (2010)

A Graph-Based Framework for Thermal Faceprint Characterization

Daniel Osaku[1,*], Aparecido Nilceu Marana[1], and João Paulo Papa[2,**]

Department of Computing, São Paulo State University - UNESP, Bauru, Brazil
dosaku@uol.com.br, {nilceu,papa}@fc.unesp.br

Abstract. Thermal faceprint has been paramount in the last years. Since we can handle with face recognition using images acquired in the infrared spectrum, an unique individual's signature can be obtained through the blood vessels network of the face. In this work, we propose a novel framework for thermal faceprint extraction using a collection of graph-based techniques, which were never used to this task up to date. A robust method of thermal face segmentation is also presented. The experiments, which were conducted over the UND Collection C dataset, have showed promising results.

Keywords: Faceprint, Image Foresting Transform, Optimum-Path Forest, Thermal Face Recognition.

1 Introduction

Biometric identification systems, which are based on physical, behavioral and physiological features of a given person, have been widely used in the last years as an alternative to increase the security level or replace the traditional identification systems, which are mainly based on possession or knowledge.

Although fingerprint recognition still remains the most used biometric technique, such approach is very sensitive to fingers' imperfections, which can be congenital or acquired over time. Other alternatives have been extensively pursued, such as iris and face recognition in visible spectrum. While the former has prohibitive costs, the latter is extremely dependent on the environment illumination and is not straightforward to distinguish twins or similar people.

Therefore, aiming to get deeper with face recognition, thermal face imagery has been used, since the temperature in the different regions of the human face allow its characterization. This thermal map is directly related to the blood vessels network of the face, which is unique for each individual [5].

Keeping up this in mind, several works have proposed new approaches for face recognition in infrared spectrum using physiological features. Akhloufi and Bendada [1], for instance, introduced the concept of *faceprint*, in which the

---

* Supported by Fapesp Grant # 2009/12437-9.
** Partially supported by CNPq Grant # 481556/2009-5 (ARPIS) and FAPESP Grant
# 2009/16206-1.

G. Maino and G.L. Foresti (Eds.): ICIAP 2011, Part I, LNCS 6978, pp. 169–177, 2011.

physiological features of the face are extracted by identifying the boundaries of isothermal regions. The final result looks like a traditional fingerprint, with the crests representing the vessels and the valleys the isothermal regions.

Buddharaju et al. [5] provided a comprehensive study on physiological-based face recognition. Their work proposed a methodology to segment the blood vessels network and extract the thermal minutia points in order to compare the faces of a given database. Later, Buddharaju and Pavlidis [3] proposed a new methodology to correct the problems of false acceptance ratio in their previous work, mainly because of the methodological weakness in the feature extraction and face matching algorithms. Thus, the main contributions of their work were twofold: (i) the first one is related with the blood vessels segmentation step, which was improved in order to remove false contours of the vascular network, and (ii) the second one concerns with the development of a novel face matching algorithm, which considers pose deformations and face expression changes.

Chel et al. [6] evaluated the application of PCA technique in infrared images, and also reported its impact over images acquired with different conditions of illumination, expression and appearance. Finally, Buddharaju et al. [4] proposed a feature-based approach that characterizes the shape of isothermal regions, for further face recognition.

This paper proposes a framework to compute the faceprint using a set of graph-based tools provided by the Image Foresting Transform (IFT) [9] and by Optimum Path Forest (OPF) [11]. While the former addresses image processing techniques, the latter handles unsupervised classification. To the best of our knowledge, this is the first time that IFT and OPF are used to this purpose. The remainder of this paper is organized as follows. Section 2 describes IFT and OPF. The proposed methodology is presented in Section 3. Finally, conclusions and future works are stated in Section 4.

2   Background Theory

In this section we introduce some basic concepts about the Image Foresting Transform and the Optimum-Path Forest.

2.1   Image Foresting Transform

The Image Foresting Transform proposed by Falcão et al. [9] is a graph-based tool to the design of image processing operators. Each pixel is modeled as a node, and a predefined adjacency relation originates a graph over the image. After that, the IFT algorithm begins a competition process between some key samples (seeds) in order to partition the graph into optimum-path trees (OPTs), which are rooted at the seeds.

The competition process is ruled by a path-cost function that needs to be constrained under some restrictions [9], and may simulate a region growing process in the graph. As one can see, by selecting a proper adjacency relation, seeds estimation methodology and path-cost function, one can design an image processing

operator based on the IFT paradigm. Actually, the IFT can be seen as a generalization of the Dijkstra's algorithm to compute shortest paths, in the sense that
IFT allows to use different path-cost functions and with multiple source nodes.

One can find several implementations of IFT-based segmentation. In this
work, we used the IFT-WT (IFT-Watershed) approach [9], which works similar
to the Watershed algorithm [12]. More details about IFT can be found in [9].

2.2   Optimum-Path Forest

Keeping in mind the idea of IFT, the Optimum-Path Forest is a framework to
the design of pattern classifiers based on discrete optimal partitions of the feature space. In this case, each dataset sample is represented by its corresponding
feature vector in a node, which originates a graph together with a predefined
adjacency relation. Both nodes and arcs can be weighted with density values
and the distance between feature vectors, respectively.

Given some key samples (prototypes), which work similar to the seeds in
the IFT algorithm, the OPF tries to partition the graph in OPTs, which may
represent clusters (unsupervised classification) or labeled trees (supervised classification). In this work, we applied the unsupervised OPF, also known as OPF
clustering, in order to group isothermal regions of the face.

In this case, the prototypes are chosen as the samples that fall in the regions
with highest density. The OPF clustering uses a k-nn adjacency, in which the
best value of k is chosen as the one the minimizes a minimum cut over the
graph [11], and it is bounded by k_{max}. This parameter is chosen by user, and
is responsible to allow a wider search range for OPF to compute the density of
each node, which is calculated over its k-neighborhood.

3   Graph-Based Framework

Facial images obtained through infrared devices contain thermal information
about the blood vessels network, which originates regions with different temperatures (Figure 3a). This thermal map produces an unique thermal faceprint for
each individual [1].

In this work, we propose a novel framework to obtain faceprints from thermal
images composed by four steps, as described by Figure 1. Each one of these
steps is modeled to be conducted with the graph-based framework described in
the previous section. These steps will be further detailed in the next sections.
In order to validate the proposed method, we used images obtained from UND
Collection C dataset [7].

3.1   Pre-processing

In infrared images, a set of possible face transformations (rotation, scaling and
translation) and sensor-dependent variations (e.g., automatic gain control calibration and bad sensor points) could undermine the recognition performance.

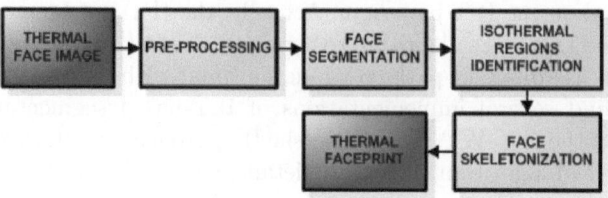

Fig. 1. General pipeline for the proposed methodology: the blue boxes denote the four main steps, while the red ones mean the input and output to the system

This impact can be minimized by performing some pre-processing operations, as follows:

1. Integer to float conversion to proceed with pixel-based operations at the image;
2. Pixel normalization to compensate brightness and contrast variations and
3. Histogram equalization in order to reduce image variation due to lighting and sensor differences.

3.2   Face Segmentation

The image segmentation concerns with to divide an image in regions that share certain features, aiming to separate the object of interest from its background. In our proposed approach, we divide the segmentation process in several steps in order to minimize errors in the face extraction procedure. These steps provide the basis for the proposed face extraction methodology. Figure 2 displays the detailed pipeline for the face segmentation schema adopted here.

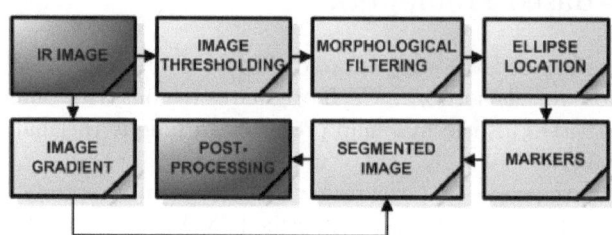

Fig. 2. Proposed face segmentation pipeline composed by five main steps: image thresholding, morphological filtering, ellipse location, markers finding and post-processing

Image thresholding. This first step attempts to separate the individual's face from its background. Given that thermal face images are characterized by high contrast between background and foreground, we applied the Otsu threshold [10] in order to obtain a binary image. Figures 3a and 3b display the original and thresholded images, respectively.

Morphological Filtering. Although the thresholding process may achieve good results, some images still could contain imperfections, such as face pixels disconnected from the foreground and holes inside the region of interest. In order to circumvent such problems, we performed morphological closing and opening operations at the images of the database, as shown in Figure 3c.

Ellipse Location. Given that a human face has quite similarity with the geometric figure defined by an ellipse, this step consists in finding the biggest ellipse contained at the image's background. Such task was accomplished by using an IFT-based method to find ellipses proposed by Andaló et al. [2]. The idea is to find the biggest ellipse within a homogeneous region with a center in a point that belongs to this region (Figure 3d).

Markers. The next step consists in to execute the segmentation process using the IFT-WT, which requires the use of both internal (foreground) and external (background) markers. In order to automatically find them, we propose here to use the ellipse found in the previous step as the basis to compute the internal and external markers. Since we have the ellipse (Section 3.2), we may execute erosion and dilation operations on that in order to find the internal and external ellipses (Figure 3e), that are used as internal and external markers, respectively. This eliminates the need of markers manually selected by user.

The segmentation is then performed using the markers at gradient image (Figure 3f). The result of the segmentation is shown in Figure 3g.

3.3 Post-processing

The last step consists in to post-process the segmented face, since some images still have some imperfections (Figure 3g). The idea is to obtain the binary mask of the segmented image and then run the ellipse location step again (Section 3.2). The final result is obtained by applying the ellipse's mask to the original image (Figure 3a). Figure 3h displays the resulting face extracted after the post-processing procedure.

3.4 Isothermal Regions Identification

This step comprises with the identification of the isothermal regions, that is, regions that have homogeneous temperature, which are related to the pixels's brightness. We propose to handle the isothermal regions identification using the OPF clustering algorithm introduced by Rocha et al [11]. For that, we used $k_{max} = 100$.

The value used for k_{max} was chosen after several experiments, and seemed to be the best option in our case. Low values of k_{max} lead us to an over clustering, whereas high values tend to merge regions. The reason for that relies on the fact that the best k value is chosen as the one that minimizes a minimum cut over the graph [11], and its search range is bounded by k_{max}.

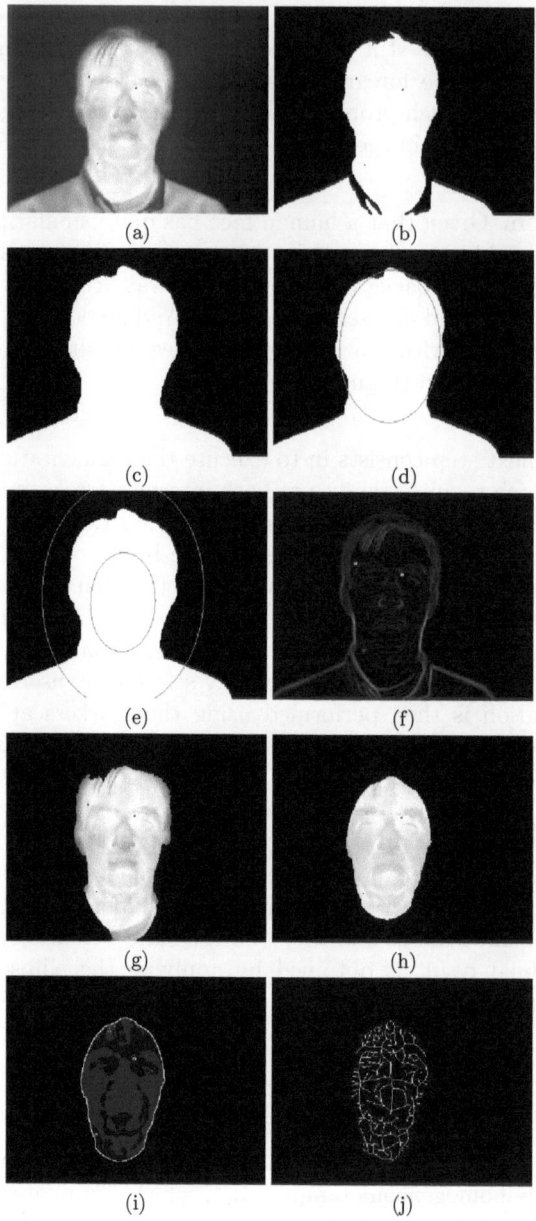

Fig. 3. Proposed methodology for faceprint extraction: (a) original thermal image in grayscale (8 bits/pixel), (b) thresholded image according to Section 3.2, (c) image after morphological filtering (Section 3.2), (d) ellipse location at face (Section 3.2), (e) internal and external markers defined by the eroded and dilated ellipses, respectively, (f) gradient image, (g) segmented image with some imperfections, (h) face extracted at the final of the segmentation process described in Figure 2, (i) face with isothermal regions grouped by OPF clustering and (j) faceprint obtained at the final of the proposed methodology depicted in Figure 1

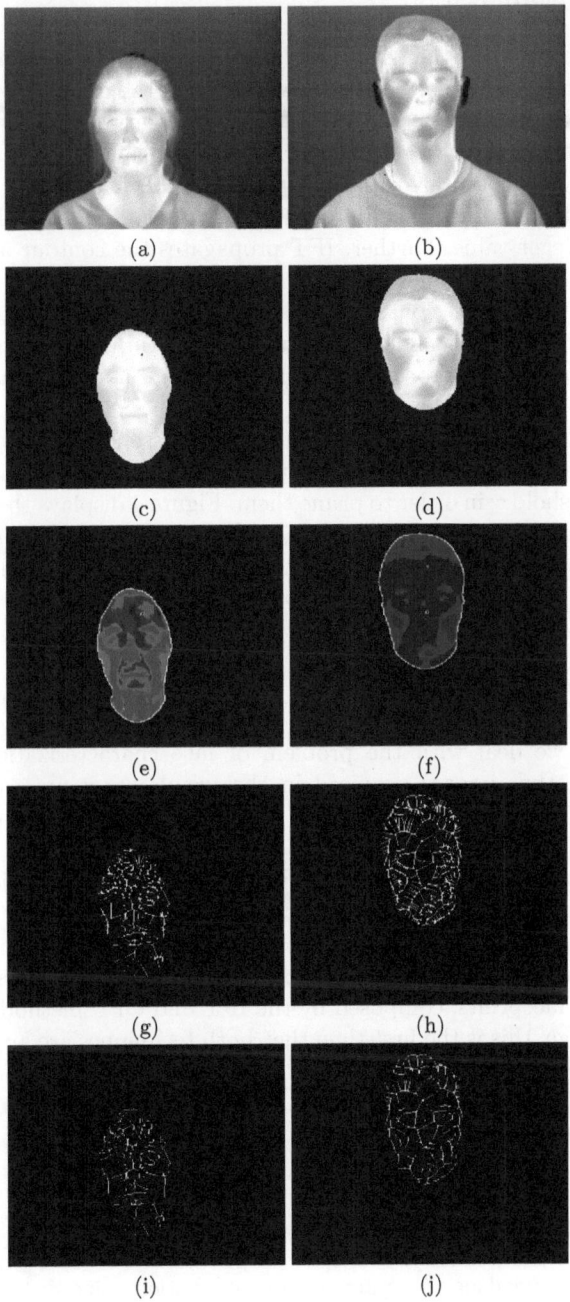

(a)   (b)

(c)   (d)

(e)   (f)

(g)   (h)

(i)   (j)

Fig. 4. Proposed methodology applied to two images of the UND Collection C dataset [7]: (a)-(b) infrared images, (c)-(d) face extracted according to Section 3.2, (e)-(f) isothermal regions identified (Section 3.4), and skeletons obtained with $\tau = 1$ in (g)-(h) and $\tau = 3$ in (i)-(j)

3.5   Face Skeletonization

The last step consists in the characterization of the isothermal regions through the face skeletonization. The edge representation of the objects by their internal and external skeletons has been studied for several years, since them are compact representations and allow to rebuild the object. In this work, we carried out this phase with an IFT-based algorithm proposed by Falcão et al. [8].

The main idea is to use each border pixel as a seed, and then label it with a consecutive integer value. Further, IFT propagates the contour and the pixel's labels, using a path-cost function based on the Euclidean distance. This process outputs three images: (i) contour label map, (ii) pixel label map and (iii) an image filtered with the Euclidean Distance Transform. By filtering the contour label map image with a procedure described in [8], one can obtain its internal skeletons.

However, the main drawback of such image representation concerns with the irrelevant branches produced during the image skeletonization. Thus, one need to consider a threshold γ in order to prune them. Figure 4 displays the methodology applied to different individuals of UND Collection C dataset. Figures 4g, 4h, 4i and 4j show the thermal faceprint obtained with different values of pruning thresholds.

4   Conclusions

In this paper we deal with the problem of face characterization in infrared imagery. Since that images acquired in the visible spectrum can be affected by distortions in illumination, thermal images may appear to overcome such problems.

Some recent works have proposed to build a faceprint of individuals, instead of carrying out with holistic methods for thermal face recognition. The faceprint is based on the blood vessels network of the face, which is unique for each individual, even for twins. Thus, we propose here a novel graph-based framework to obtain such faceprints, composed by the IFT and OPF methods. To the best of our knowledge, this is the first time that both techniques are applied to tackle this problem. We also presented a robust procedure to extract faces from thermal images. Nowadays, our ongoing research has been guided to extract information about faceprints in order to associated them to the individuals of the dataset.

References

1. Akhloufi, M., Bendada, A.: Infrared face recognition using distance transforms. In: Proceedings of the World Academy of Science, Engineering and Technology, vol. 30, pp. 160–163 (2008)
2. Andaló, F.A., Miranda, P.A.V., Torres, R.d.S., Falcão, A.X.: Shape feature extraction and description based on tensor scale. Pattern Recognition 43(1), 26–36 (2010)

3. Buddharaju, P., Pavlidis, I.T.: Physiological face recognition is coming of age. In: Proceedings of the Conference on Computer Vision and Pattern Recognition, pp. 128–135. IEEE Computer Society, Los Alamitos (2009)
4. Buddharaju, P., Pavlidis, I.T., Kakadiaris, I.A.: Face recognition in the thermal infrared spectrum. In: Proceedings of the Conference on Computer Vision and Pattern Recognition Workshop, vol. 8, pp. 167–191 (2004)
5. Buddharaju, P., Pavlidis, I.T., Tsiamyrtzis, P., Bazakos, M.: Physiology-based face recognition in the thermal infrared spectrum. IEEE Transactions on Pattern Analysis and Machine Intelligence 29(4), 613–626 (2007)
6. Chen, X., Flynn, P.J., Bowyer, K.W.: Ir and visible light face recognition. Computer Vision and Image Understanding 99(3), 332–358 (2005)
7. Chen, X., Flynn, P.J., Bowyer, K.W.: Ir and visible light face recognition. Computer Vision and Image Understanding 99(3), 332–358 (2005)
8. Falcão, A.X., da Costa, L.F., Cunha, B.S.: Multiscale skeletons by image foresting transform and its application to neuromorphometry. Pattern Recognition 35(7), 1571–1582 (2002)
9. Falcão, A.X., Stolfi, J., Lotufo, R.A.: The image foresting transform: Theory, algorithms, and applications. IEEE Transactions on Pattern Analysis and Machine Intelligence 26(1), 19–29 (2004)
10. Otsu, N.: A threshold selection method from gray-level histograms. IEEE Transactions on Systems, Man and Cybernetics 9(1), 62–66 (1979)
11. Rocha, L.M., Cappabianco, F.A.M., Falcão, A.X.: Data clustering as an optimum-path forest problem with applications in image analysis. International Journal of Imaging Systems and Technology 19(2), 50–68 (2009)
12. Vincent, L., Soille, P.: Watersheds in digital spaces: An efficient algorithm based on immersion simulations. IEEE Transactions on Pattern Analysis and Machine Intelligence 13(6), 583–598 (1991)

Reflection Removal for People Detection in Video Surveillance Applications*

Dajana Conte, Pasquale Foggia, Gennaro Percannella, Francesco Tufano,
and Mario Vento

Dipartimento di Ingegneria Elettronica e Ingegneria Informatica
Università di Salerno
Via Ponte don Melillo, I-84084 Fisciano (SA), Italy
{dconte,pfoggia,pergen,ftufano,mvento}@unisa.it

Abstract. In this paper we present a method removing reflection of people on shiny floors in the context of people detection for video analysis applications. The method exploits chromatic properties of the reflections and does not require a geometric model of the objects. An experimental evaluation of the proposed method, performed on a significant database containing several publicly available videos, demonstrates its effectiveness. The proposed technique also favorably compares with respect to other state of the art algorithms for reflection removal.

1 Introduction

Correct segmentation of foreground objects is important in video surveillance and other video analysis applications. In order to achieve an accurate segmentation, artifacts related to lighting issues such as shadows and reflections must be detected and properly removed. In fact, if a shadow or a reflection is mistakenly included as part of a detected foreground object, several problems may severely impact the accuracy of the subsequent phases of the application.

While many papers have been devoted to shadow removal [4,6], the problem of reflections has received comparatively much less attention; however, in some environments, reflections can be more likely than shadows, and usually they are harder to deal with. Examples are indoor scenes when the floor is smooth and shiny, or outdoor scenes in rainy weather conditions. Shadows and reflections differ under several respects; the most important differences are in position and color. The position of a shadow depends on the light sources, while reflections (assuming that the reflecting surface is a horizontal floor) are always located below the corresponding object. As regards the color, a shadow depends only on the color of the background and on the light sources (it has a darker shade of the same color of the background); on the other hand, the color of a reflection also depends on the color of the object. As a consequence of these differences, methods for shadow removal cannot be effectively applied for removing reflections.

One of the earliest work is the paper by Teschioni and Regazzoni [7], following an approach very similar to the techniques commonly used for shadow removal. In particular, a model of the color properties of a reflection is assumed; the pixels consistent with

---

* This research has been partially supported by A.I.Tech s.r.l., a spin-off company of the University of Salerno (www.aitech-solutions.eu).

G. Maino and G.L. Foresti (Eds.): ICIAP 2011, Part I, LNCS 6978, pp. 178–186, 2011.

this model are grouped using a region growing technique, and then discarded from the foreground. The method makes the assumption that the pixels of the foreground objects are significantly different (in the RGB space) from both the ones in the background and the ones in the reflections; when this assumption is not satisfied, it is likely that parts of the objects will be mistaken as reflections, even if their position would make this unplausible.

A completely different approach is proposed by Zhao and Nevatia in [8]. Their algorithm is based on the hypothesis that the foreground object is a person, and uses a geometrical model of a person to recognize those parts of the foreground that have to be labeled as reflections. Unfortunately this method does not work if the scene includes other kinds of objects, or even people carrying large objects such as backpacks, suitcases or umbrellas.

The recent paper by Karaman et al. [5] presents a more sophisticated method that takes into account both geometric and chromatic information to remove the reflections. The method is based on the "generate and test" approach, where for each detected foreground region several hypotheses are made on the vertical position of the object baseline. For each position, the algorithm generates a synthetic reflection by combining the pixels of the background and of the part of the region that lies above the baseline, adding a blur effect to take into account the imperfect smoothness of the floor surface. Then, the baseline for which the synthetic reflection is most similar to the observed one, is selected, and all the pixels below this baseline are removed from the foreground object. This method is fairly general and robust, since it does not require an a priori knowledge of the shape of the objects. On the other hand, the "generate and test" process is computationally expensive, because for each hypothesis an image has to be generated and matched with the observed region. Furthermore, the pixel combination and blurring require parameters depending on the characteristics of the floor, implicitly assuming that the floor smoothness and reflectivity are uniform.

In this paper we propose a reflection removal technique that is similarly based on the evaluation of multiple hypotheses for the object baseline. The proposed method does not make assumptions on the characteristics of the floor surface, and so can easily work with heterogeneous floors. Furthermore, it is extremely efficient because it does not involve the actual generation of a synthetic reflection, and the test phase exploits an incremental scheme of computation to evaluate each baseline very quickly.

2   The Proposed Method

We assume that our algorithm is applied to the output of a foreground detection system based on background subtraction. It does not require a specific background subtraction technique and can be used as a postprocessing phase of any existing foreground detection module.

We briefly recall that a foreground detection system compares the current frame to a background reference image (suitably created and updated), and finds the frame pixels whose color is significantly different from the corresponding background pixels, using some sort of thresholding technique. Such pixels are grouped into connected components called *foreground regions*. Our method assumes that each foreground region contains either a single object or a group of objects at the same distance from the camera

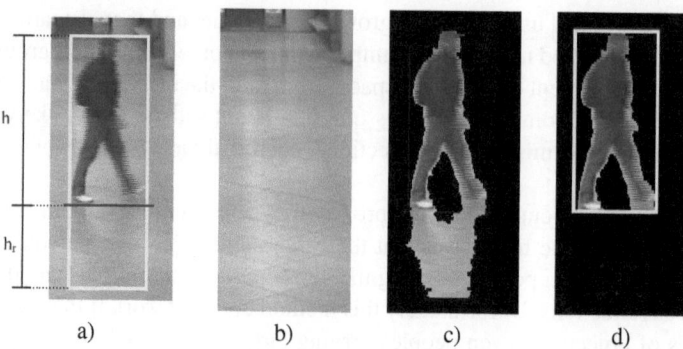

Fig. 1. a) a portion of the input image, containing a person whose height is h with its relative reflection h_r on the floor. The horizontal line represents the ideal cut separating the person from its background; b) the background reference image B(.); c) The foreground mask F(.) obtained by using a standard detection algorithm and d) the foreground mask after the removal of the reflection.

(e.g. a person with his/her luggage); hence the object can be separated from its reflection using a single horizontal line that we call the *cut line*. Note that the actual shape of the object does not need to be known in advance, so the method can be used even when the scene contains several kinds of objects. The method exploits the following property of the pixels belonging to a reflection: they are, on the average, much more similar in color to the background than the other foreground pixels are, although they are not so similar as to be considered part of the background. This happens because part of the color of the floor gets blended with the color of the reflected object to form the reflection color. Figure 1 presents an example of a person with a reflection on the floor, and the corresponding output of the foreground detection. The figure also shows the ideal cut line for this image, and the background reference image.

The proposed method, on the basis of these assumptions, determines the ideal cut line as the row of the foreground detected object that:

– minimizes the average difference in color between the detected object and the background for all the rows below it;
– on the contrary, maximizes the average difference in color between the detected object and the background for all the rows above it;

In order to quantitatively evaluate the difference in color, we introduce the following notations: $F(x, y)$ is the color of the pixel at position (x, y) in the foreground region, $B(x, y)$ the color of the corresponding pixel in the background image, and $r(k)$ the set of pixels belonging to the generic row k of the foreground region; we measure the average difference of color, along the row $r(k)$ of the detected foreground object, the following quantity:

$$d(k) = \frac{\sum\limits_{(x,y)\in r(k)} \|F(x, y) - B(x, y)\|}{|r(k)|} \tag{1}$$

where $\|.\|$ is the Euclidean norm in the color space, and $|.|$ is the cardinality of a set.

Figure 2c reports the graph representing $d(k)$ for any row k of the detected foreground image. The actual determination of the ideal cut line is obtained on the basis of the values of $d(k)$, by considering for each candidate cut line k, the difference $\Delta(k)$ between the integral of $d(i)$ for the set of the rows above k, and the one of $d(j)$ of the set of the rows below. By denoting with $R_a(k)$ and $R_b(k)$, respectively the set of the rows above and below k, we define:

$$\Delta(k) = \frac{1}{|R_a(k)|} \cdot \sum_{i \in R_a(k)} d(i) - \frac{1}{|R_b(k)|} \cdot \sum_{j \in R_b(k)} d(j) \qquad (2)$$

According to this definition, $\Delta(k)$ represents the difference between the average foreground–background dissimilarity above the candidate cut line k and the average dissimilarity below k. It is simple to verify that, if $d(i)$ is greater for the rows belonging to the object than for those of the reflection, starting from the top of the detected foreground object, $\Delta(k)$ increases, reaching its maximum in correspondence with the ideal cut line, and then it decreases as k approaches the bottom of the reflection. The ideal cut line σ can be consequently determined by searching for the relative maximum of $\Delta(k)$:

$$\sigma = k : \Delta(k) \geq \Delta(j), j \in [0, h + h_r] \qquad (3)$$

Figure 2d reports the graph representing $\Delta(k)$ for any row k of the detected foreground image.

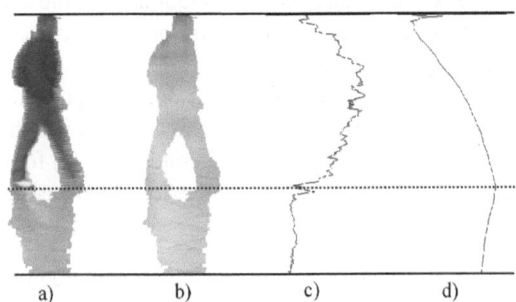

Fig. 2. a) a portion of an image and b) the corresponding background; c) the function $d(k)$ for each row of the image (k is on the vertical axis, the value of $d(k)$ on the horizontal one); d) the function $\quad(k)$ for each row of the image (k is on the vertical axis, the value of $\quad(k)$ on the horizontal one)

In real cases, the $\Delta(k)$ function is not as well-behaved as in the ideal case, showing a few spurious maxima in addition to the one corresponding to the ideal cut line. These spurious maxima are due to the effect of noise and to dishomogeneity in the color of the foreground object, which may be locally very similar to the background color. To filter out the spurious maxima, we have introduced the following criteria, based on geometrical and physical considerations:

a. the maximum is discarded if it is too isolated, i.e. the average value of $\Delta(k)$ in a neighborhood of the maximum differs from the maximum by more than a threshold (both the width of the neighborhood and the threshold are parameters of the algorithm); the rationale of this criterion is that an isolated maximum is more likely due to noise than to the underlying trend of the function;

b. the maximum is discarded if its position is below the middle of the detected foreground region; in fact, it is geometrically unlikely that a reflection is larger than the actual object, if the floor surface is horizontal and the object is not significantly inclined with respect to the vertical;

c. the maximum is discarded, if its value is negative; a negative value of $\Delta(k)$ would mean that the object is more similar to the background that its reflection, and this is incompatible with the assumptions of the method.

On the basis of these considerations, the method operates according to Algorithm 1.

From a computational complexity point of view, a naive computation of $\Delta(k)$ would require for each value of k the scanning of the whole detected region, in order to compute the average difference from the background above and below row k. Since this process would have to be repeated for each row, the resulting complexity would be $O(w \cdot h^2)$ (w and h are respectively the width and the height of the region).

The method described by Algorithm 1 computes the function much more efficiently, using two improvements with respect to a naive implementation:

- the algorithm keeps the values of $d(i)$ in a data structure, so that each $d(i)$ is only computed once; this would reduce the computational complexity to $O(w \cdot h + h^2)$, where the first term is due to the computation of $d(i)$ and the second term to the computation of $\Delta(k)$ given the $d(i)$ values;
- while iterating over the rows for computing $\Delta(k)$, the algorithm keeps in two variables the sum of the $d(i)$ above and below row k; these variables can be updated in $O(1)$ at each step, and avoid the need to iterate over $d(i)$ for computing the two sums of equation 2; hence the overall complexity is reduced to $O(w \cdot h + h) = O(w \cdot h)$.

Thus the proposed algorithm is very efficient even on large foreground regions, requiring a time that is negligible with respect to the overall processing of a frame.

3   Experimental Evaluation

Experiments were carried out using the object detection algorithm described in [3], characterized by a good trade-off between the detection performance and the computational complexity. The dataset used for the tests is composed by four real-world videos. All refer to indoor scenarios with reflecting floorings. The first and the second video sequences (hereinafter referred to as V1 and V2), belonging to the PETS2006 dataset [1], were taken in the hall of a railway station. Both videos show the same scene but from different view angles. The third video (hereinafter referred to as V3), was acquired by the authors nearby a subway platform. Finally, also the last video refers to a subway platform, and it belongs to the AVSS2007 dataset [2]. In all videos the reflections are

Algorithm 1. The pseudo-code of the algorithm.

{ Compute $d(.)$ and its sum }
$Sum \leftarrow 0$
for $i = 0$ to $height - 1$ **do**
  $d(i) \leftarrow \quad _{(x\ y)\in r(i)} \quad F(x,y) - B(x,y) \ /|r(i)|$
  $Sum \leftarrow Sum + d(i)$
end for

{ Compute $(.)$ }
$SumAbove \leftarrow d(0)$
$SumBelow \leftarrow Sum - d(0)$
for $row = 1$ to $height - 1$ **do**
  $(row) \leftarrow \dfrac{SumAbove}{row} - \dfrac{SumBelow}{height - row}$
  $SumAbove \leftarrow SumAbove + d(row)$
  $SumBelow \leftarrow SumBelow - d(row)$
end for

{Compute the best local maximum among the ones satisfying the criteria of feasibility}
$BestMax \leftarrow -1$
$BestCut \leftarrow -1$
for $row = height/2$ to $height - 1$ **do**
  if (row) is a local maximum AND $(row) > 0$ AND (row) is not isolated **then**
    if $(row) > BestMax$ **then**
      $BestMax \leftarrow$ (row)
      $BestCut \leftarrow row$
    end if
  end if
end for
if $BestMax > 0$ **then**
  RETURN $BestCut$
else
  RETURN Nothing
end if

mainly generated by persons in the scene. For each video, a ground truth has been produced by inspecting the objects detected at each frame and choosing by hand the most appropriate cutting line, on the basis of the visual appearance. Of course, the objects missed by the used detection algorithm, as well as the wrongly detected ones (those corresponding to partial detections of the persons) have been discarded: the method cannot recover from such errors due to the previous detection phase. Table 1 reports the main characteristics of the used video sequences, and for each of them the total number of considered objects. For experimental purposes, the detected objects have been classified into two classes, *reflected objects* and *unreflected objects*: we considered an object as affected by reflection when $h_r/(h_r + h) \geq 0.15$, i.e. when its reflection is at least 15% of its apparent height.

Table 1. Dataset main characteristics. All videos were acquired at 4CIF resolution and 25 fps.

ID	Dataset / Video sequence	Number of frames	Type of objects	Total objects
V1	PETS2006 / S1-T1-C (view 1)	3021	unreflected	117
			reflected	1528
V2	PETS2006 / S1-T1-C (view 3)	3021	unreflected	2375
			reflected	556
V3	sequence acquired by the authors	880	unreflected	214
			reflected	955
V4	AVSS2007 / AB_Easy	5474	unreflected	1206
			reflected	1326

In order to measure the effectiveness of the proposed system, we have to consider that there can be two kinds of errors:

- the algorithm fails to remove completely the reflection of a detected object;
- the algorithm remove completely the reflection, but also cuts away part of the object; we call this situation an *overcut*.

The following indices have been defined to provide a quantitative evaluation of the two errors for a single object i:

$$RH(i) = \frac{h_r(i)}{h_r(i) + h(i)}, \quad OE(i) = \frac{h_o(i)}{h(i)}$$

where $h_o(i)$ is the height of the portion of the object that is erroneously removed by the algorithm in case of an overcut error.

$RH(i)$ is the height of the reflection normalized on the total height of the i-th bounding box; if the algorithm manages to completely remove the reflection, $RH(i)$ should become 0. More generally, it is expected that the value of $RH(i)$ is reduced by the application of the algorithm. We call $RH(i)$ the *reflection error*.

$OE(i)$ is a measure of the overcut relative to the true height of the object; in the ideal case, if the algorithm does not cut away a part of the object, the value of $OE(i)$ after the removal is 0. We call $OE(i)$ the *overcut error*. Notice that, before the application of the algorithm, $OE(i) = 0$, since no part of the detected region has been removed.

It is evident that only one of the two indices can be greater than 0; in fact, RH is meaningful when part of the reflection still remains after the cut, while OE must be considered when the cut removes the whole reflection and (possibly) part of the actual object. If we denote with N and M the cardinality of the two disjoint sets of boxes on which the proposed cut line is respectively below or above the ideal cut line, then the performance of the system over all the objects can be expressed in terms of the following two indices:

$$MRH = \frac{1}{N} \cdot \sum_{i:RH(i)>0} RH(i)$$

$$MOE = \frac{1}{M} \cdot \sum_{i:OE(i)>0} OE(i)$$

which are the average reflection and overcut errors.

Table 2. Performance of the proposed reflection removal method

Video ID	Method	Type of objects	Reflection Error			Overcut Error
			MRH (before)	MRH (after)	%	MOE
V1	proposed method	unreflected	0.082	0.057	31.0%	0.031
		reflected	0.436	0.391	10.3%	0.030
	Karaman	unreflected	0.080	0.068	15.4%	0.071
		reflected	0.417	0.304	27.2%	0.027
V2	proposed method	unreflected	0.095	0.037	61.1%	0.026
		reflected	0.206	0.070	66.0%	0.057
	Karaman	unreflected	0.089	0.053	40.4%	0.026
		reflected	0.241	0.133	44.8%	0.106
V3	proposed method	unreflected	0.127	0.068	46.6%	0.064
		reflected	0.294	0.075	74.4%	0.039
	Karaman	unreflected	0.135	0.035	74.3%	0.063
		reflected	0.320	0.217	32.0%	0.220
V4	proposed method	unreflected	0.057	0.023	59.6%	0.040
		reflected	0.202	0.065	67.8%	0.031
	Karaman	unreflected	0.060	0.042	30.0%	0.103
		reflected	0.203	0.095	53.2%	0.063

The experimental results are reported in Table 2. For comparison, the table also reports the performance of another recent reflection removal algorithm by Karaman et al. [5] for which we have provided our own implementation. The motivation behind the choice of Karaman's algorithm is twofold: first, it is a very recent approach presented in the literature that was tested also on standard datasets, where it has shown a very interesting performance; second, our proposed approach is similar to Karaman's one as they are both based on the evaluation of multiple hypotheses for the object baseline, even if the two methods make different assumptions about the properties of the reflections. Table 2 reports in the fourth, fifth and sixth columns the reflection removal performance expressed in terms of the MRH index. This analysis was done considering only the objects with no overcut, comparing the reflection error before and after the reflection removal. The sixth column shows the relative improvement in the MRH index. Finally, the rightmost column of the table shows the MOE index.

Considering the non-overcut objects, the results show a consistent reduction of the reflection error, which is decreased in most cases by more than 50%. Notice that the error is reduced also for unreflected objects. As evident from the last column in Table 2, the algorithm does not introduce appreciable overcut errors: less than 5% in the average on both the unreflected and the reflected objects. It should be also considered that on reflected objects, an overcut means that the whole reflection has been eliminated; thus for the reflected samples reported in this table, the algorithm yields a significant improvement in the height estimation. In comparison with the algorithm by Karaman, the proposed method performs better on some video sequences (V2 and V4), and has similar results on the others (V1 and V3).

Finally, it is important to highlight that, as already anticipated in a previous section, the proposed reflection removal method has a negligible impact on the overall processing time. In fact, we have experimentally verified that the adoption of the reflection removal procedure produces an increase of 1.5% of the processing time with respect to the original foreground detection algorithm.

4   Conclusions

In this paper we have presented a novel algorithm for reflection removal, based on fairly general assumptions and computationally efficient.

The algorithm has been experimentally validated on a significant database of real videos, using quantitative measurements to assess its effectiveness. The experiments have shown that the proposed algorithm significantly reduce the error in the estimation of the actual height for objects with a reflection, while unreflected objects are left substantially unchanged. The method has been also experimentally compared with the algorithm by another recent approach for reflection removal by Karaman et al, showing in almost cases significant performance improvements.

As a future work, a more extensive experimentation will be performed, adding other algorithms to the comparison and enlarging the video database to provide a better characterization of the advantages of the proposed approach.

References

1. Dataset for PETS 2006, http://www.cvg.rdg.ac.uk/PETS2006/
2. i-Lids dataset for AVSS 2007,
   http://www.eecs.qmul.ac.uk/~andrea/avss2007_d.html
3. Conte, D., Foggia, P., Petretta, M., Tufano, F., Vento, M.: Evaluation and improvements of a real-time background subtraction method. In: Kamel, M.S., Campilho, A.C. (eds.) ICIAR 2005. LNCS, vol. 3656, pp. 1234–1241. Springer, Heidelberg (2005)
4. Horprasert, T., Harwood, D., Davis, L.: A statistical approach for real-time robust background subtraction and shadow detection (1999)
5. Karaman, M., Goldmann, L., Sikora, T.: Improving object segmentation by reflection detection and removal. In: Proc. of SPIE-IS&T Electronic Imaging (2009)
6. Shen, J.: Motion detection in color image sequence and shadow elimination. Visual Communications and Image Processing 5308, 731–740 (2004)
7. Teschioni, A., Regazzoni, C.S.: A robust method for reflection analysis in color image sequences. In: IX European Signal Processing Conference, Eusipco 1998 (1998)
8. Zhao, T., Nevatia, R.: Tracking multiple humans in complex situations. IEEE Trans. on Pattern Analysis and Machine Intelligence 26, 1208–1221 (2004)

The Active Sampling of Gaze-Shifts

Giuseppe Boccignone[1] and Mario Ferraro[2]

[1] Dipartimento di Scienze dell'Informazione Universitá di Milano
via Comelico 39/41, 20135 Milano, Italy
boccignone@dsi.unimi.it
[2] Dipartimento di Fisica Sperimentale, Universitá di Torino
via Pietro Giuria 1, 10125 Torino, Italy
ferraro@ph.unito.it

Abstract. The ability to predict, given an image or a video, where a human might fixate elements of a viewed scene has long been of interest in the vision community.

In this note we propose a different view of the gaze-shift mechanism as that of a motor system implementation of an active random sampling strategy that the Human Visual System has evolved in order to efficiently and effectively infer properties of the surrounding world. We show how it can be exploited to carry on an attentive analysis of dynamic scenes.

Keywords: active vision, visual attention, video analysis.

1 Introduction

Gaze shifts are eye movements that play an important role: the Human Visual System (HVS) achieves highest resolution in the fovea and the succession of rapid eye movements (saccades) compensates the loss of visual acuity in the periphery when looking at an object or a scene that spans more than several degrees in the observer's field of view. Thus, the brain directs saccades to actively reposition the center of gaze on circumscribed regions of interest, the so called "focus of attention" (FOA), to sample in detail the most relevant features of a scene, while spending only limited processing resources elsewhere. An average of three eye fixations per second generally occurs, intercalated by saccades, during which vision is suppressed. Frequent saccades, thus, avoid to build enduring and detailed models of the whole scene. Apparently evolution has achieved efficient eye movement strategies with minimal neural resources devoted to memory [13].

In order to take into account these issues, much research has been devoted specially in the fields of computational vision (see, for instance, [9,10,22]).

Interestingly enough, one point that is not addressed by most models is the "noisy", idiosyncratic variation of the random exploration exhibited by different observers when viewing the same scene, or even by the same subject along different trials [15]. Such variations speak of the stochastic nature of scanpaths. Indeed, at the most general level one can assume any scanpath to be the result of a random walk performed to visually sample the environment under the constraints of both the physical information provided by the stimuli (saliency or

G. Maino and G.L. Foresti (Eds.): ICIAP 2011, Part I, LNCS 6978, pp. 187–196, 2011.

conspicuity) and the internal state of the observer, shaped by cognitive (goals, task being involved) and emotional factors. Under this assumption, the very issue is how to model such "biased" random walk.

In a seminal paper [6], Brockmann and Geisel have shown that a visual system producing Lévy flights implements an efficient strategy of shifting gaze in a random visual environment than any strategy employing a typical scale in gaze shift magnitudes. Lévy flights provide a model of diffusion characterized by the occurrence of long jumps interleaved with local walk.

To fully exploit diffusion dynamics, in [3], a gaze-shift model (the Constrained Lévy Exploration, CLE) was proposed where the scanpath is guided by a Langevin equation,

$$\frac{d\mathbf{x}}{dt} = -U(\mathbf{x}) + \xi, \tag{1}$$

on a potential $U(\mathbf{x})$ modelled as a function of the saliency (landscape) and where the stochastic component ξ represents random vector sampled from a Lévy distribution (refer to [3] for a detailed discussion, and to [11,12] for application to robot vision relying on Stochastic Attention Selection mechanisms). The basic assumption was the "foraging metaphor", namely that Lévy-like diffusive property of scanpath behavior mirrors Lévy-like patterns of foraging behavior in many animal species [24]. In this perspective, the Lévy flight, as opposed, for instance, to Gaussian walk, is assumed to be essential for optimal search, where optimality is related to efficiency, that is the ratio of the number of sites visited to the total distance traversed by forager [24]. An example depicting the difference between Gaussian and Lévy walks is provided in Fig. 1. In [4] a new method, the Lévy Hybrid Monte Carlo (LHMC) algorithm was presented, in which gaze exploration is obtained as a sampling sequence generated via a dynamic Monte Carlo technique. Interestingly enough the method described in [3] can be recovered as a special case of [4] .

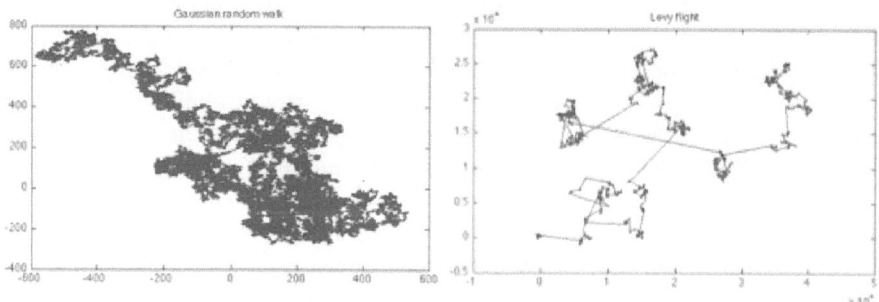

Fig. 1. Different walk patterns obtained through the method described in [7] (defined in the next section). Top: Gaussian walk (characteristic exponent or index of stability $\alpha = 2$); bottom: Lévy walk ($\alpha = 1.2$). Horizontal and vertical axes represent $(X, \ )$ coordinates; note the different scale at which the flight lenghts are obtained.

This perspective suggested a different and intriguing view of the saccadic mechanism: *that of a motor system implementation of an active random sampling strategy that the HVS has evolved in order to efficiently and effectively infer properties of the surrounding world.* In this note we further develop this view, showing how it can be exploited to carry on an attentive analysis of dynamic scenes.

2 Background

The fact that a visual system producing Lévy flights implements an efficient strategy of shifting gaze in a visual environment should not be surprising [6,3]. Indeed, movements of some biological organisms can be represented as a trajectory constructed as a Random Walk Model (a simplified version of Eq. 1),

$$\mathbf{x}_{t+1} = \mathbf{x}_t + \xi, \tag{2}$$

where ξ is a random vector representing the displacement $\mathbf{x}_t \to \mathbf{x}_{t+1}$ occurring at time t. In other terms, a discrete-time continuous-space Lévy flight is a Markovian Random Walk process controlled by the conditional probability $P(\mathbf{x}_{t+1}|\mathbf{x}_t)d\mathbf{x}_{t+1}d\mathbf{x}_t$ for the walker to be in the region $\mathbf{x}_{t+1}+d\mathbf{x}_{t+1}$ at time $t+1$, if he was in the region $\mathbf{x}_t + d\mathbf{x}_t$ at time t. We restrict ourselves to

$$P(\mathbf{x}_{t+1}|\mathbf{x}_t) \approx P(\mathbf{x}_{t+1} - \mathbf{x}_t) = P(\xi) \sim |\xi|^{-(1+\alpha)}, \tag{3}$$

for large $|\xi|$, namely the lenght of the single diplacement ξ. The nature of the walk is determined by the asymptotic properties of such conditional probability (see, for instance, [5]): for the characteristic exponent $\alpha \geq 2$ the usual random walk (brownian motion) occurs; if $\alpha < 2$, the distribution of length jump is "broad" and the so called Lévy flights or Lévy walks take place[1].

Intuitively, Lévy flights (walks) consist of many short flights and occasional long flights: sample trajectories of an object undergoing ordinary random walk and Lévy flights, are presented in Fig. 1.

Coming back to Eq. 2, one can simulate a Brownian walk (e.g. , $\alpha = 2$) by sampling ξ from a Normal distribution at each time step t; for what concerns Lévy motion, one could choose, for instance, to sample from a Cauchy distribution ($\alpha = 1$). However, this constrains to specific values of the characteristic exponent α, which might become a limit for the analysis of natural phenomena. At the most general level we can choose $P(\xi) = f(\alpha, \beta, \gamma, \delta)$ to be an alpha-stable distribution [14] of characteristic exponent (index of stability), skewness, scale and location parameters $\alpha, \beta, \gamma, \delta$, respectively. In general there exists no simple closed form for the probability density function of stable distributions: they

---

[1] The Lévy Flight and the Lévy Walk are trajectories constructed in the same way spatially but differ with respect to time. The Lévy Flight displaces in space with one unit of time being spent for each element between turning points. It could be considered a discrete version of the Lévy Walk. The Lévy Walk on the other hand is a continuous trajectory and has a velocity vector.

admit explicit representation of the density function only in the following cases: the Gaussian distribution $P(2, 0, \gamma, \delta)$, the Cauchy distribution $P(1, 0, \gamma, \delta)$ and the Lévy distribution $P(1/2, 1, \gamma, \delta)$. Nevertheless, there are efficient numerical methods to generate samples $\xi \sim f(\alpha, \beta, \gamma, \delta)$ ([8], cfr. the Simulation Section).

Examples of Lévy flights and walks have been found in many instance of animal mouvement such as spider monkey [16], albatrosses [23], jackals [1] and in general such dynamics have been found to be be essential for optimal exploration in random searches [24].

It is also becoming apparent that other complex models can be subsumed within a Lévy flight paradigm [18]. For example, in the composite Brownian walk (CBW), taking into account search patterns within a landscape of patchily distributed resources, a searcher moves in a straight-line between patches and adopts more localized Brownian movements within a patch. The search pattern is therefore adaptive because detection of a food item triggers switching from an extensive mode of searching for clusters of food to intensive within-cluster searching for individual prey items. Benhamou [2] showed that this CBW out-performs any Lévy-flight searching strategy with constant μ. The CBW can, however, be interpreted as an adaptive Lévy-flight searching pattern in which the inter-patch straight-line motions correspond to Lévy flights with $\mu \to 1$ and where the intra-patch motion corresponds to a Lévy flight with $\mu = 3$. This adaptive Lévy flight is an optimal strategy because $\mu \to 1$ is optimal for the location of randomly, sparsely distributed patches that once visited are depleted and because $\mu = 3$ flights are optimal for the location of densely but random distributed within-patch resources [17]. In complex environments a mixed strategy could be an advantage. For instance Reynolds and Frye [19], by analizing how fruit flies (*Drosophila melanogaster*) explore their landscape have reported that once the animals approach the odor, the freely roaming Lévy search strategy is abandoned in favour of a more localized (Brownian) flight pattern. Such mixed-type motion behavior is indeed the one taken into account in our active sampling strategy.

3   The Active Sampling Strategy

Denote $\mathbf{x}_{FOA}(t) = \mathbf{x}_t$ the oculomotor state which results, at time t, in a gaze fixation at a certain point of the world. The state of the world at time t is then observed as "filtered" by the attentional focus set $\mathbf{x}_t$, and producing an observation likelihood $P(w_t|\mathbf{x}_t)$.

The observation likelihood is obtained as follows. Given a fixation point at time t the current frame is blurred with respect to a Gaussian function centered at $\mathbf{x}_t$ in order to simulate foveation of the scene. Then, relying on motion features, a Bayesian surprise map $\mathcal{SM}(t) = \{\mathcal{SM}(t)_i\}_{i=1}^N$ at each site $i \in \Im$, $\Im$ being the spatial support of current frame (see Itti and Baldi [9] for detailed discussion). Based on such map we sample a subset $\Im_s$ of salient interest points $\{\mathcal{SM}(t)_s\}_{s=1}^{Ns}$, where $N_s = |\Im_s| < N$, through a weighted random sampling:

1. Sample:

$$s \sim Unif(\Omega). \tag{4}$$

2. Sample:

$$u_s \sim Unif(0,1). \tag{5}$$

3. Accept, s_t^s with probability $P_s = \frac{\mathcal{SM}(t)_s}{\mathcal{SM}(t)_{max}}$, if $u_s < P_s$.

Then, we subdivide the image into L windows and let $P(w_t^s)$ be the probability that the s-th point falls in the window w_t when $t \to \infty$, in other words $P(w_t^s)$ is the asymptotic probability distribution. Subregion partitioning of the image, which performs a coarse-graining of the saliency map, is justified by the fact that gaze-shift relevance is determined according to the clustering of salient points that occur in a certain region of the image, rather than by single points. Thus, the image is partitioned into Nw rectangular windows. For all Ns sampled points, each point occurring at site s, is assigned to the corresponding window, and probability $P(w_t = l|\mathbf{x}_t)$, $l = 1 \cdots Nw$ is empirically estimated as

$$P(w_t = l|\mathbf{x}_t) \simeq \frac{1}{Ns} \sum_{k=1}^{Ns} \chi_{k,l}, \tag{6}$$

where $\chi_{k,l} = 1$ if $\mathcal{SM}(t)_k \in w_t = l$ and 0 otherwise.

Under a sampled observation $w_t \sim P(w_t = l|\mathbf{x}_t)$, the observer has to determine the joint probability $P(\mathbf{x}_{t+1}, a_t|\mathbf{x}_t, w_t)$ of taking an action a_t and of achieving the next state $\mathbf{x}_{t+1}$. We can rewrite such joint probability as:

$$P(\mathbf{x}_{t+1}, a_t|\mathbf{x}_t, w_t) = P(\mathbf{x}_{t+1}|\mathbf{x}_t, a_t)P(a_t|\mathbf{x}_t, w_t). \tag{7}$$

The term $P(\mathbf{x}_{t+1}|\mathbf{x}_t, a_t)$ represents the dynamics, given action a_t, while $P(a_t|\mathbf{x}_t, w_t)$ is the probability of undertaking such action a_t given the current state of affairs $(\mathbf{x}_t, w_t)$.

We can represent the action as the pair $a_t = (z_t, \theta_t)$, where z_t is a discrete random variable $z_t = \{z_t = k\}_{k=1}^K$, K being the number of possible actions, $\theta_t = \{\theta_t = k\}_{k=1}^K$ representer the parameters related to the action. Under the hypotheses motivated in the previous section, $z_t = \{1, 2\}$, for foraging and exploratory behaviors, then

$$P(a_t|\mathbf{x}_t, w_t) = P(z_t, \theta_t|\mathbf{x}_t, w_t) = P(z_t|\theta_t)P(\theta_t|a_t, b_t), \tag{8}$$

where a_t, b_t are the hyperparameters for the prior $P(\theta_t|a_t, b_t)$ on parameters θ_t.

Since in our case the motor behavior is chosen among two possible kinds, $p(z = k|\theta) = \theta_k$ is a Binomial distribution whose conjugate prior is the Beta distribution, $p(\theta) = Beta(\theta; a_t, b_t)$.

Coming back to Eq.7, we write that oculomotor flight will be governed by a Binomial distribution,

$$P(\mathbf{x}_{t+1}|\mathbf{x}_t, a_t) = \sum_{z_t} [P(\mathbf{x}_{t+1}|\mathbf{x}_t, \eta)]^{z_t}, \tag{9}$$

where $P(\mathbf{x}_{t+1}|\mathbf{x}_t, z_t = k) = P(\mathbf{x}_{t+1}|\mathbf{x}_t, \eta_k)$ is the flight generated according to motor behavior $z_t = k$ and regulated by parameters η_k. Here $P(\mathbf{x}_{t+1}|\mathbf{x}_t, \eta_k)$, is the probabilistic representation of the jump

$$\mathbf{x}_{t+1} = \mathbf{x}_t + \xi_k, \tag{10}$$

where $\xi_k \sim f(\eta_k)$, $f(\eta_k) = f(\alpha_k, \beta_k, \gamma_k, \delta_k)$ being an alpha-stable distribution of characteristic exponent, skewness, scale and location parameters $\alpha_k, \beta_k, \gamma_k, \delta_k$, respectively.

The last thing we have to take into account are the hyperparameters a_t, b_t of the Beta distribution that govern the choice of motor behavior regime.

Following the discussion in the previous Section, here we assume that at each time t these are "tuned" as a function of the order/disorder of the scene: intuitively, a completely ordered or disordered scenario will lead to longer flights so as to gather more information, whilst at the edge of order/disorder enough information can be gathered via localized exploration. This insight can be formalized as follows.

Consider again the probability distribution $P(w_t = l|\mathbf{x}_t)$. The corresponding Boltzmann-Gibbs-Shannon entropy is $S = -k_B \sum_{i=1}^{N} P(w_t = i|\mathbf{x}_t) \log P(w_t = i|\mathbf{x}_t)$, where k_B is the Boltzmann's constant; in the sequel, since dealing with images we set $k_B = 1$. The supremum of S is obviously $S_{sup} = \ln N$ and it is associated to a completely unconstrained process, that is a process where $S =$const, since with reflecting boundary conditions the asymptotic distribution is uniform.

Following Shiner *et al.* [20] it is possible to define a disorder parameter Δ as $\Delta \equiv S/S_{sup}$ and an order parameter Ω as $\Omega = 1 - \Delta$ can be defined. Note that by virtue of our motion features and related Bayesian surprise, a disordered frame event will occur when either no motion (static scene) or extreme motion (many objects moving in the scene) is detected. Thus, the hyperparameter update can be written as

$$\begin{aligned} a_{(t)} &= a_{(0)} + f_a(\Omega_t), \\ b_{(t)} &= b_{(0)} + f_b(\Delta_t), \end{aligned} \tag{11}$$

and is simply computed as a counter of the events "ordered frame", "disordered frame", occurred up to time t, namely, $f_a(\Omega_t) = N_{\Omega_t}$ and $f_b(\Delta_t) = N_{\Delta_t}$.

4   Simulation

Simulation has been performed on the publicly available datasets, the *BEHAVE Interactions Test Case* (http://homepages.inf.ed.ac.uk/rbf/BEHAVE), the CAVIAR data set (http://homepages.inf.ed.ac.uk/rbf/CAVIAR/), the UCF Crowd Data set (http://server.cs.ucf.edu/ vision/) The first comprises videos of people acting out various interactions, under varying illumination

Fig. 2. Examples of videos used in the simulation. Left: video with no dynamic events happening from the CAVIAR data set; center: a video from the BEHAVE test set showing a few significative and temporally ordered dynamic events (one man cycling across the scene followed by a group of people interacting); right: a crowded scene from the UCF dataset.

conditions and spurious reflections due to camera fixed behind a window; the second include people walking alone, meeting with others, window shopping, entering and exiting shops; the third contains videos of crowds and other high density moving objects. The rationale behind the choice of these data sets stems from the possibility of dealing with scenes with a number of moving objects at a different level of complexity (in terms of order/disorder parameters as defined in the previous Section). At a glance, three representative examples are provided in Fig. 2.

Given a fixation point $\mathbf{x}_t$ at time t (the frame center is chosen for $t = 1$), the current RGB frame of the input sequence is blurred with respect to a Gaussian function centered at x_t and down-sampled via a 3-level Gaussian pyramid. At the pre-attentive stage, optical flow features $\mathbf{v}_{n,t}$ are estimated on the lowest level of the pyramid following [21].

The related motion energy map is the input to the Bayesian surprise step, where the surprise map $\mathcal{SM}(t)$ is computed [9]. Then, through weighted random sampling, the set of salient interest points $\{\mathcal{SM}(t)_s\}_{s=1}^{Ns}$ is generated and $P(w_t = l|\mathbf{x}_t)$ is estimated; for this purpose $Ns = 50$ interest points are sampled and we use $Nw = 16$ windows/bins, their size depending on the frame size $|\Im|$, to compute Eq.6.

At this point we can compute the order/disorder parameters, and in turn the hyperparameters of the Beta distribution are updated via Eq. 11. This is sufficient to set the bias of the "behavioral coin" (Eq.8) and the coin is tossed (Eq. 9). This allows to choose parameters $\alpha_k, \beta_k, \gamma_k, \delta_k$ and to sample the flight vector ξ. For the examples shown here the Brownian regime is characterized by $\alpha_1 = 2, \beta_1 = 0, \gamma_1 = |FOA|, \delta_1 = 0$ and the Lévy one by $\alpha_2 = 1.3, \beta_2 = 0, \gamma_2 = |FOA|, \delta_2 = 0$. Here $|FOA|$ indicates approximately the radius of a FOA, $|FOA| \approx 1/8 \min[w, h]$, w, h being the horizontal and vertical dimensions of the frame.

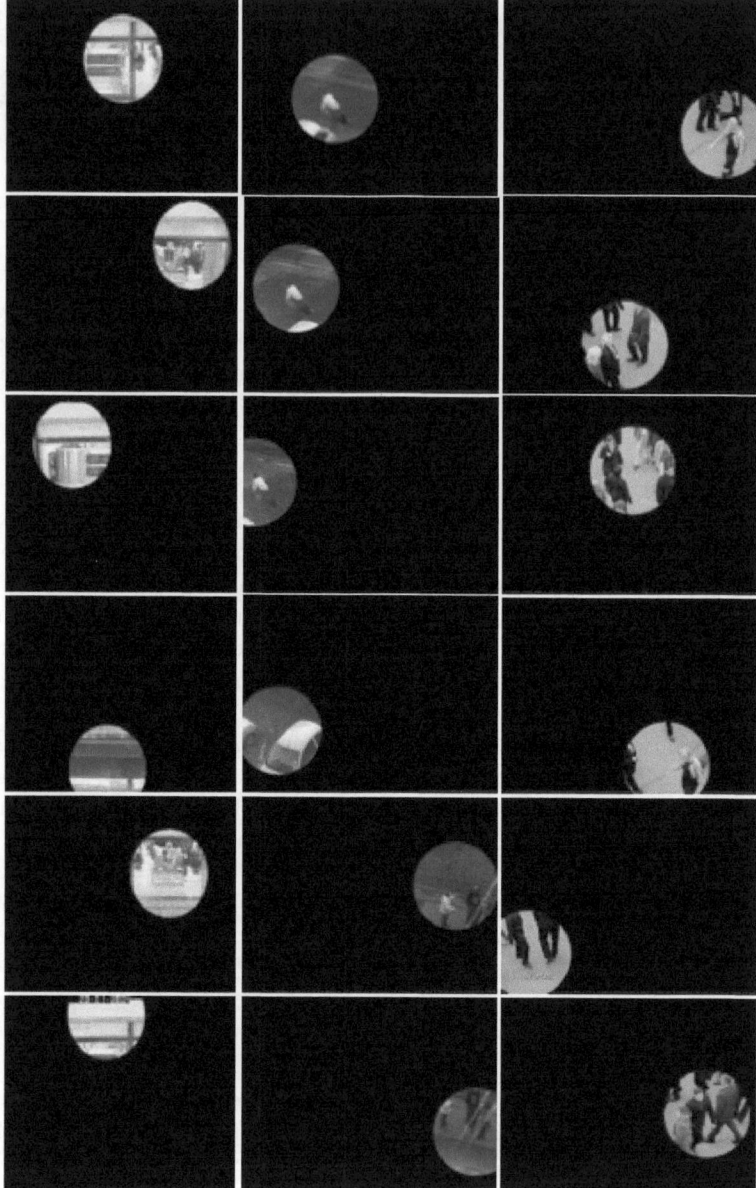

Fig. 3. An excerpt of typical results obtained along the simulation. Left and right columns shows that gaze samples the scene by principally resorting to a Lévy strategy as a result of no interesting dynamic events or too many dynamic events occurring in the scene. In the middle column a complex behavior is generated: a Brownian regime is maintained during an interesting event (resulting in bicyclist tracking), followed by a Lévy exploration to search for a new interesting event, ended by backing to a Brownian dynamics for sampling the new event of interest (people discussing).

Given the parameters, an alpha-stable random vector ξ can be sampled in several ways[8]. The one applied here [7] is the following.

1. Generate a value Z from a standard stable $f(\alpha, \beta, 0, 1)$:

$$V \sim U\left(-\frac{\pi}{2}, \frac{\pi}{2}\right) ; \tag{12}$$

$$W \sim \exp(1). \tag{13}$$

2. If $\alpha \neq 1$:

$$Z = S_{\alpha,\beta} \frac{\sin(\alpha(V + B_{\alpha,\beta}))}{\cos(V)^{1/\alpha}} \left(\frac{\cos(V - \alpha(V + B_{\alpha,\beta}))}{W}\right)^{\frac{1-\alpha}{\alpha}}, \tag{14}$$

where $S_{\alpha,\beta} = \frac{\arctan(\beta\tan(\pi\alpha/2))}{\alpha}$ and $B_{\alpha,\beta} = \left(1 + \beta^2\tan^2(\pi\alpha/2)\right)^{1/2\alpha}$.

3. If $\alpha = 1$, then

$$Z = \frac{2}{\pi}\left[\left(\frac{\pi}{2} + \beta V\right)\tan(V) - \beta\log\left(\frac{W\cos(V)}{\frac{\pi}{2} + \beta V}\right)\right]. \tag{15}$$

Once a value Z from a standard stable $f(\alpha, \beta, 0; 1)$ has been simulated, in order to obtain a value ξ from a stable distribution with scale parameter γ and location parameter δ, the following transformation is required: $\xi = Z + \delta$ if $\alpha \neq 1$; $\xi = \gamma Z + \frac{2}{\pi}\beta\gamma\log(\gamma) + \delta$, if $\alpha = 1$.

Eventually, the new FOA x_{t+1} is determined via Eq. 2.

An illustrative example, which is representative of results achieved on such data-set, is provided in Fig. 3, where the change of motor behavior regime is readily apparent as a function of the complexity of scene dynamics.

The system is currently implemented in plain MATLAB code, with no specific optimizations and running on a 2 GHz Intel Core Duo processor, 2 GB RAM, under Mac OS X 10.5.8. As regards actual performance, most of the execution time is spent to compute the saliency map, which takes an average elapsed time of 0.8 secs per frame, whilst only 0.1 sec per frame is devoted to the FOA sampling. Clearly, the speed-up in this phase is due to the fact that once the set of salient interest points $\{\mathcal{SM}(t)_s\}_{s=1}^{Ns}$ has been sampled, then subsequent computations only deal with $Ns = 50|$ points, a rather sparse representation of the original frame.

References

1. Atkinson, R.P.D., Rhodes, C.J., Macdonald, D.W., Anderson, R.M.: Scale-free dynamics in the movement patterns of jackals. Oikos 98(1), 134–140 (2002)
2. Benhamou, S.: How many animals really do the Lévy walk? Ecology 88(8), 1962–1969 (2007)
3. Boccignone, G., Ferraro, M.: Modelling gaze shift as a constrained random walk. Physica A: Statistical Mechanics and its Applications 331(1-2), 207–218 (2004)
4. Boccignone, G., Ferraro, M.: Gaze shifts as dynamical random sampling. In: Proceedings of 2nd European Workshop on Visual Information Processing (EUVIP 2010), pp. 29–34. IEEE Press, Los Alamitos (2010)

5. Bouchaud, J., Georges, A.: Anomalous diffusion in disordered media: statistical mechanisms, models and physical applications. Physics Reports 195, 127–293 (1990)
6. Brockmann, D., Geisel, T.: The ecology of gaze shifts. Neurocomputing 32(1), 643–650 (2000)
7. Chambers, J.M., Mallows, C.L., Stuck, B.W.: A method for simulating stable random variables. J. Am. Stat. Ass. 71(354), 340–344 (1976)
8. Fulger, D., Scalas, E., Germano, G.: Monte Carlo simulation of uncoupled continuous-time random walks yielding a stochastic solution of the space-time fractional diffusion equation. Physical Review E 77(2), 21122 (2008)
9. Itti, L., Baldi, P.: Bayesian surprise attracts human attention. Vision Research 49(10), 1295–1306 (2009)
10. Itti, L., Koch, C., Niebur, E.: A model of saliency-based visual attention for rapid scene analysis. IEEE Trans. Pattern Anal. Machine Intell. 20, 1254–1259 (1998)
11. Martinez, H., Lungarella, M., Pfeifer, R.: Stochastic Extension to the Attention-Selection System for the iCub. University of Zurich, Tech. Rep (2008)
12. Nagai, Y.: From bottom-up visual attention to robot action learning. In: Proceedings of 8 IEEE International Conference on Development and Learning, IEEE Press, Los Alamitos (2009)
13. Najemnik, J., Geisler, W.S.: Optimal eye movement strategies in visual search. Nature 434(7031), 387–391 (2005)
14. Nolan, J.P.: Stable Distributions - Models for Heavy Tailed Data. Birkhauser, Boston (2011)
15. Privitera, C.M., Stark, L.W.: Algorithms for defining visual regions-of-interest: Comparison with eye fixations. IEEE Trans. Pattern Anal. Machine Intell. 22(9), 970–982 (2000)
16. Ramos-Fernandez, G., Mateos, J.L., Miramontes, O., Cocho, G., Larralde, H., Ayala-Orozco, B.: Lévy walk patterns in the foraging movements of spider monkeys (Ateles geoffroyi). Behavioral Ecology and Sociobiology 55(3), 223–230 (2004)
17. Reynolds, A.: How many animals really do the Lévy walk? Comment. Ecology 89(8), 2347–2351 (2008)
18. Reynolds, A.M.: Optimal random Lévy-loop searching: New insights into the searching behaviours of central-place foragers. EPL (Europhysics Letters) 82, 20001 (2008)
19. Reynolds, A.M., Frye, M.A.: Free-flight odor tracking in Drosophila is consistent with an optimal intermittent scale-free search. PLoS One 2(4), 354 (2007)
20. Shiner, J.S., Davison, M., Landsberg, P.T.: Simple measure for complexity. Physical Review E 59(2), 1459–1464 (1999)
21. Simoncelli, E., Adelson, E., Heeger, D.: Probability distributions of optical flow. In: Proceedings of IEEE Computer Society Conference on Computer Vision and Pattern Recognition, CVPR 1991, pp. 310–315. IEEE, Los Alamitos (2002)
22. Torralba, A.: Contextual priming for object detection. Int. J. of Comp. Vis. 53, 153–167 (2003)
23. Viswanathan, G.M., Afanasyev, V., Buldyrev, S.V., Murphy, E.J., Prince, P.A., Stanley, H.E.: Lévy flight search patterns of wandering albatrosses. Nature 381(6581), 413–415 (1996)
24. Viswanathan, G.M., Raposo, E.P., da Luz, M.: Lévy flights and superdiffusion in the context of biological encounters and random searches. Physics of Life Rev. 5(3), 133–150 (2008)

SARC3D: A New 3D Body Model for People Tracking and Re-identification

Davide Baltieri, Roberto Vezzani, and Rita Cucchiara

Dipartimento di Ingegneria dell'Informazione - University of Modena and Reggio
Emilia, Via Vignolese, 905 - 41125 Modena - Italy
{davide.baltieri,roberto.vezzani,rita.cucchiara}@unimore.it

Abstract. We propose a new simplified 3D body model (called SARC3D)
for surveillance application, which can be created, updated and compared in real-time. People are detected and tracked in each calibrated
camera, with their silhouette, appearance, position and orientation extracted and used to place, scale and orientate a 3D body model. For each
vertex of the model a signature (color features, reliability and saliency) is
computed from 2D appearance images and exploited for matching. This
approach achieves robustness against partial occlusions, pose and viewpoint changes. The complete proposal and a full experimental evaluation
are presented, using a new benchmark suite and the PETS2009 dataset.

Keywords: 3D human model, People Re-identification.

1 Introduction and Related Work

People Re-identification is a fundamental task for the analysis of long-term activities and behaviors of specific people. Algorithms have to be robust in challenging situations, like widely varying camera viewpoints and orientations, varying
poses, rapid changes in part of clothes appearance, occlusions, and varying lighting conditions. Moreover, people re-identification requires elaborate methods in
order to cope with the widely varying degrees of freedom of a person's appearance. Various algorithms have been proposed in the past, based on the kind of
data available: a first category of person re-identification methods rely on biometric techniques, such as face [1] or gait, where high resolution or PTZ cameras
are required in this case. Other approaches suppose well constrained operative
conditions, calibrated cameras and precise knowledge of the scene geometry; the
problem is then simplified by adding spatial and/or temporal constraints, in
order to greatly reduce the candidate set [2,3,4]. Finally, most re-identification
methods purely rely on appearance-based features; a comparison and evaluation
of some of them is reported in [5,6]. For example, Farenzena et el [6] proposed
to divide the person appearance into five parts using rule based approaches to
detect head, torso, legs and image symmetries to split torso and leg regions
into left and right ones. For each region, a set of color and texture features are
collected for the matching step. Alahi *et al* [7] proposed a general framework
for simultaneous tracking and re-detection by means of a grid cascade of dense

G. Maino and G.L. Foresti (Eds.): ICIAP 2011, Part I, LNCS 6978, pp. 197–206, 2011.

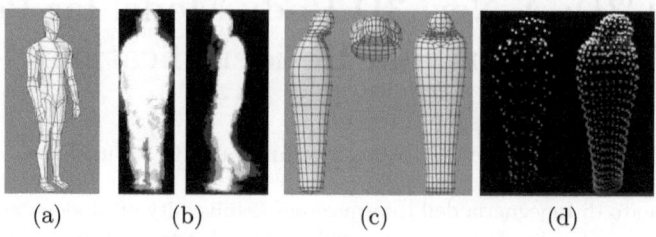

(a)   (b)   (c)   (d)

Fig. 1. (a) a human 3d model, (b) average silhouettes used for the model creation, (c) our simplified human model, (d) the vertices sampling used in our tests

region descriptors. Various descriptors have been evaluated, like SIFT, SURF and covariance matrices, and the latter are shown to outperform the formers. Finally, [8] proposed the concept of Panoramic Appearance Map to perform re-identification. This map is a compact signature of the appearance information of a person extracted from multiple cameras, and can be thought of as the projection of a person appearance on the surface of a cylinder.

In this paper we present the complete design of a method for people re-identification based on 3D body models. The adoption of 3D body models is new for re-identification, differently from other computer vision fields, such as motion capture and posture estimation [9,10]. The challenges connected with 3D models rely on the need for precise people detection, segmentation and estimation of the 3D orientation for a correct model to image alignment. However, our proposal has several benefits; first of all, we provide an approximate 3D body model with a single shape parameter, which can be learned and used for comparing people from very few images (even only one). Due to the precise 3D feature mapping, the comparison allows to look for details, and not only to global features, it also allows to cope with partial data and occluded views; finally, the main advantage of a 3D approach is to be intrinsically independent from point of views. Our approximate body model (called SARC3D) is not only fast, but also suitable for low resolution images as the ones typically acquired by surveillance cameras. A preliminary version of this work was presented in [11], while in this paper we present the complete proposal together with a comprehensive set of experiments, both on a new benchmark suite, available with the corresponding annotation, and on real scenarios using the PETS2009 dataset.

2 The SARC3D Model

First of all the generic body model of a person must be defined. Differently from motion capture or action recognition systems, we are not interested in the precise location and pose of each body part, but we need to correctly map and store the person's appearance. Instead of an articulated body model (as in fig. 1(a)), we propose a new monolithic 3d model, called "SARC3D". The model construction has been driven by real data: side, frontal and top views of generic people, which were extracted from various surveillance videos; thus, an average silhouette has been computed for each view, and has been used for the creation of a graphical 3d body model (see fig. 1(b)), producing a sarcophagus-like (fig. 1(c)) body hull. The final body model is a set of vertices regularly sampled from the sarcophagus surface. The number of sampled vertices could be selected accordingly to the required resolution. In our tests on real surveillance setups, we used from 153 to 628 vertices (fig. 1(d)). Other sampling densities of the same surface have been tested, but the selected ones outperformed the others on specificity and precision tests, and are a good trade-off between speed and efficacy. As a representative signature, we created an instance Γ^p of the generic model for each detected person p-th, characterized by a scale factor (to cope with different body builds) and relating appearance information (i.e., color and texture) for each vertex, defined as:

$$\Gamma^p = \{h^p, \{v_i^p\}\}, p \in [1 \ldots P], i \in [1 \ldots M] \tag{1}$$

where h^p is the person height, as extracted by the tracking module, and used as the scale factor for the 3D model; v_i^p is the vertex set; P is the number of people in the gallery and M is the number of vertices. For each vertex the following five features are computed and stored:

- $\boldsymbol{n}_i$: the normal vector of the 3D surface computed at the i-th vertex location; this feature is pre-computed during the sampling of the model surface;
- c_i: the average color;
- $\mathbf{H_i}$: a local HSV histogram describing the color appearance of the vertex neighbor; it is a normalized histogram with 8 bins for the *hue* channel and 4 bins for the *saturation* and *value* channels respectively;
- θ_i: the optical reliability value of the vertex, which takes into account how well and precisely the vertex color and histogram have been captured from the data.
- s_i: the saliency of the vertex, which indicates its uniqueness with respect to the other models; i.e., the saliency of a vertex will be higher in correspondence to a distinctive logo and lower on a common jeans patch.

2.1 Positioning and Orientation

The 3D placement of the model in the real scene is obtained from the output of a common 2D surveillance system working on a set of calibrated cameras. Assuming a vertical standing position of the person, the challenging problem to solve is the estimation of his horizontal orientation. To this aim, we consider that people move forward and thus we exploit the trajectory on the ground plane to give a first approximation using a sliding window approach. Given a detected person, we consider a window of N frames and the corresponding trajectory on the ground plane. A quadratic curve is then fitted on the trajectory and the fit score is used as orientation reliability. If it is above a predefined threshold, the final orientation is generated from the curve tangent.

In fig.2(a) and 2(b) a sample frame of the corresponding model placement and orientation is provided. In particular, the sample positions used for the curve fitting and orientation estimation are highlighted.

2.2   Model Creation

Given the 3D placement and orientation, the appearance part of the model can be recovered from the 2D frames. Projecting each vertex v_i to the camera image plane, the corresponding nearest pixel $x(v_i), y(v_i)$ is obtained. The vertex color c_i is initialized directly using the image pixel:

$$c_i = I\left(x(v_i), y(v_i)\right); \qquad (2)$$

where I is the analyzed frame, $x(v_i)$ and $y(v_i)$ are the image coordinates of the projection of the vertex v_i. The histogram $\mathbf{H_i}$ is computed on a squared image patch centered around $(x(v_i), y(v_i))$. The size of the patch is selected taking into account the sampling density of the 3D model surface and the medium size of the blobs items. In our experiments, we used 10x10 blocks. Finally, the optical reliability value is initialized as $\theta_i = \boldsymbol{n}_i \cdot \boldsymbol{p}$, where $\boldsymbol{p}$ is the normal to the image plane (equal to the inverted direction vector of the camera). The reason behind the adoption of the dot product is that data from front-viewed vertices and their surrounding surface are more reliable than that from lateral viewed vertices. The vertices belonging to the occluded side of the person are also projected onto the image, but their reliability has a negative value, due to the opposite directions of $\boldsymbol{n}_i$ and $\boldsymbol{p}$. Thus each vertex of the model is initialized even with a single image: from a real view if available or using a sort of symmetry-based hypothesis in absence of information. The vertices having no match with the current image (i.e. the vertices projected outside of the person silhouette) are also initialized with a copy of the feature vector of the nearest initialized vertex and their reliability values set to zero. By means of the reliability value, vertices directly seen at least once ($\theta > 0$), vertices initialized using a mirroring hypothesis ($-1 \leq \theta < 0$) and vertices initialized from its neighborhood ($\theta = 0$) are distinguishable. The described steps of the initialization phase are depicted in Fig. 2(c).

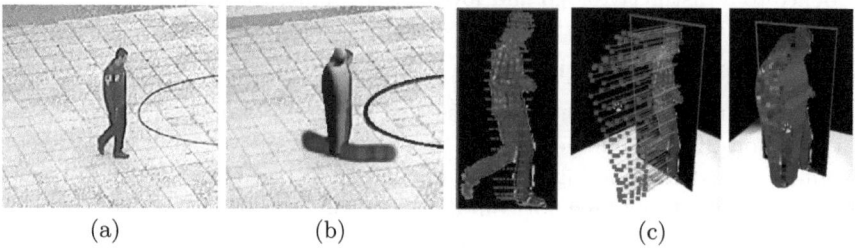

(a)   (b)   (c)

Fig. 2. (a) A frame from a video, (b) Automatic 3D positioning and orientation (c) Initialization of the 3D model of a person: model to image alignment, projection of the model vertex to the image plane, vertex initialization or update

If multiple cameras are available or if the short-term tracking system provides more detections for the same object, the 3D model could integrate all the available frames. For each of them, after the alignment step, a new feature vector is computed for each vertex successfully projected inside the silhouette of the person. The previously stored feature vector is then merged or overwritten with the new one, depending on the signs of the reliabilities. In particular, direct measures ($\theta > 0$) always overwrite forecasts ($\theta < 0$), otherwise they are merged, as in the following equations:

$$\hat{c}_i^p = \frac{\theta_i^p c_i^p + \theta_i^s c_i^s}{\theta_i^p + \theta_i^s}, \qquad \hat{\mathbf{H}}_i^p = \frac{\theta_i^p \mathbf{H}_i^p + \theta_i^s \mathbf{H}_i^s}{\theta_i^p + \theta_i^s}, \qquad \hat{\theta}_i^p = \frac{\theta_i^p + \theta_i^s}{2} \qquad (3)$$

The normal vector n_i^p does not change in the merging operation, since it is constantly obtained during the model generation, while the saliency s_i^p is recomputed after the merging. Figure 3 shows some sample models created from one or more images.

2.3   Occlusion Management and View Selection

Not all views should be used for the initialization and update of the model. Errors in the tracking step, noise and bad calibration could lead to degradation of the model. To this aim a rule based approach is proposed to select and exploit for the model initialization and update the best views only.

Occlusion Check: In addition to 2D occlusion detection algorithms [12], a computer graphic based generative approach is used: for the selected camera view and for each person p visible from that camera, a binary image mask $\hat{I}_p$ is rendered. Each time two model masks are overlapping or connected an occlusion is detected. To avoid false pixel to model assignments, both the occluding and the occluded models are not updated. A visual example of the 3D occlusion detection is shown in Fig. 4.

Model to Foreground Overlapping: The reliability of the model positioning could be evaluated considering the overlapping area between the 2D foreground mask and the rendered images $\hat{I}_p$. For each person, the overlapping score R_p is computed as the ratio between the number of foreground pixels that overlap with $\hat{I}_p$ with respect to the total number of silhouette pixels. If R_p is higher than a strong threshold (e.g., 95% in our experiments) the selected view is marked as good. Otherwise the alignment is not precise enough or the person is not assuming a standing position compliant with the sarcophagus model.

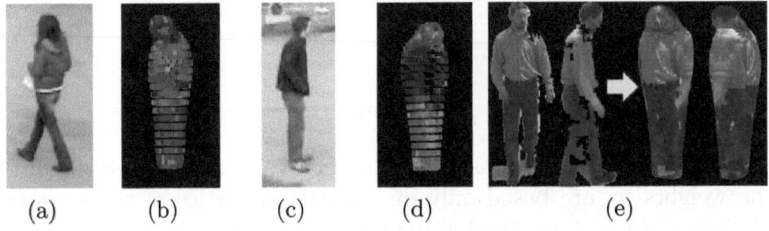

(a)   (b)   (c)   (d)   (e)

Fig. 3. Various models created with the corresponding source images

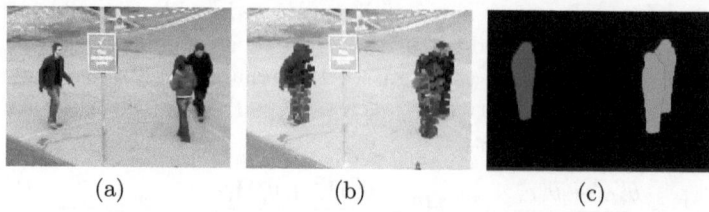

(a)	(b)	(c)

Fig. 4. Occlusion detection: (a) the input frame, (b) the aligned 3D models and (c) the mask generated by the rendering system. Since the blue and green objects are connected, the corresponding models are frozen and not updated during the occlusion.

Orientation Reliability: As mentioned before, the reliability of the orientation estimation could be evaluated considering the fitting score of the quadratic curve as described in Section 2.1. A similar check could be performed considering the sequence of the estimated orientations: if the distribution of the differences between consecutive orientations has a high variance, the trajectory is not stable and the orientation becomes unreliable.

If all the above conditions hold true, the estimated orientation and position is considered reliable and the selected view could be exploited to initialize (or update) the model.

3   People Re-identification

One of the main applications of the SARC3D model is people re-identification. The goal of the task is to find possible matches among couples of models from a given set of SARC3D items. First, we define the distance between two feature vectors. Using the optical reliability θ_i and the saliency s_i as weighting parameters, the Hellinger distance between histograms is used:

$$d(v_i^p, v_i^q) = d_{He}\left(\mathbf{H_i^p}, \mathbf{H_i^q}\right) = \quad 1 - \sqrt{\sum_{h,s,v} H_i^p(h,s,v) \cdot H_i^q(h,s,v)}. \tag{4}$$

The distance $D_H(\Gamma^p, \Gamma^q)$ between two models Γ^p and Γ^q is the weighted average of the vertex-wise distances, using the product of the reliabilities as weight.

$$D_H(\Gamma^p, \Gamma^t) = \frac{\sum_{i=1...M} \left(w_i \cdot d(v_i^p, v_i^q)\right)}{\sum_{i=1...M} (w_i)} \tag{5}$$

where

$$w_i = f(\theta_i^p) \cdot f(\theta_i^q) \tag{6}$$

This generic global distance assumes that each vertex has the same importance and the weights w_i are based only on optical properties of the projections or the reliability of the data. We believe that global features are useful to reduce the number of candidates or if the resolution is low. However, the final decision should be guided by original patterns and details, as humans normally do to recognize people without biometric information (e.g., a logo in a specific position of the shirt). To this aim we have enriched the vertex feature vector v_i^p with a saliency measure $s_i^p \in [0 \ldots 1]$. Given a set of body models, the saliency of each vertex is related to its minimum distance from all the corresponding vertices belonging to the other models:

$$s_i^p \propto \min_t \ d_H(\mathbf{H_i^p}, \mathbf{H_i^t}) \ + s_0, \qquad s_i^p = 1 \qquad (7)$$

where s_0 is a fixed parameter to give a minimum saliency to each vertex. If s is low, the vertex appearance is not distinctive; otherwise, the vertex has completely original properties and it could be used as a specific identifier of the person. The corresponding saliency-based distance D_S can be formulated based on new weights by substituting w_i' to w_i in eq.6.

$$w_i' = f(\theta_i^p) \cdot f(\theta_i^q) \cdot s_i^p \cdot s_i^q \qquad (8)$$

This saliency-based distance D_S cannot be used instead of Eq. 5, since it focuses on details discarding global information and then leading to macroscopic errors; the re-identification should be based on both global (D_H) and local (D_S) similarities. Thus, the final distance measure D_{HS} used for re-identification is the product of the two contributions $D_{HS} = D_H \cdot D_S$.

4 Experimental Results

Many experiments have been carried out on real videos and on our benchmarking suite. From its introduction, ViPER [5] is the reference dataset for re-identification problems. Unfortunately, it contains two images for each target only. Thus we propose a suitable benchmark dataset[1] with 50 people, consisting of short video clips captured with a calibrated camera. The annotated data set is composed by four views for each person, 200 snapshots in total. Some examples are shown in Fig. 5(a), where the output of the foreground segmentation is reported.

For each testing item we ranked the training gallery using the distance metrics defined in the previous section. The summarized performance results are reported using the cumulative matching characteristic (CMC) curve [5]. Each test was replicated exhaustively choosing different combinations of images.

In fig.6 we report the performance obtained using the proposed distances on SARC3D models. Two different images for each person were chosen as training set (i.e. for the model creation) while the remaining formed the test set. Each test was replicated six times using different split of the images into training and testing sets. For each testing image, the ranking score was obtained considering the model-to-model matching schema using the histogram based distance D_H and the saliency-based one D_{HS}. For comparison sake, we have evaluated the

---

[1] Available here: `http:\imagelab.ing.unimore.it\sarc3d`

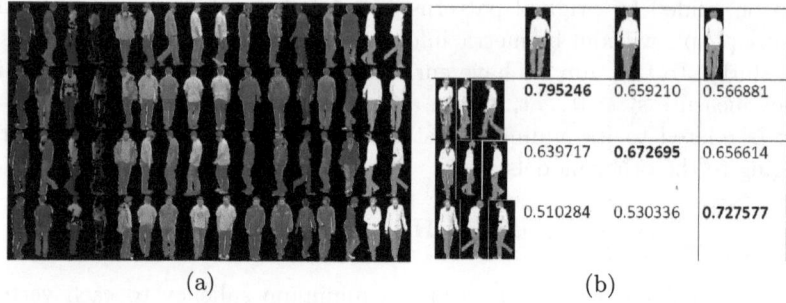

(a)   (b)

Fig. 5. (a) Selected images from the benchmarking suite, (b) distance matrix obtained from three very similar people: the three images used for the model creation (rows) and the test images (columns) are also shown

results obtained using the Euclidean distance between image pixels and the vertex mean color c_i instead of the histogram-based one of Eq. 4. In Fig. 6(b) some key values extracted from our experiments are reported, showing the performance improvements obtained using 3D-3D model matching, instead of 3D-2D measures and, in the last row, adopting the saliency measure.

In this paper we assumed to have a sufficiently accurate tracking system, which gives the 2D foreground images used for the model alignment. However, the proposed method is reliable and robust enough, even in case of approximated alignments. The use of a generic sarcophagus-like model and local color histograms instead of detailed 3D models and point-wise colors goes precisely in this direction and allows to cope with small alignment errors. In addition, the introduction of the normal vector to the 3D surface assign strong weights to the central points of the people and very low weights to lateral points, which are the most hit by misalignments.

In table 1 the system performance in presence of random perturbations of the correct alignment is reported; both errors on the ground plane localization and on the orientation have been introduced. The performance reported on the table

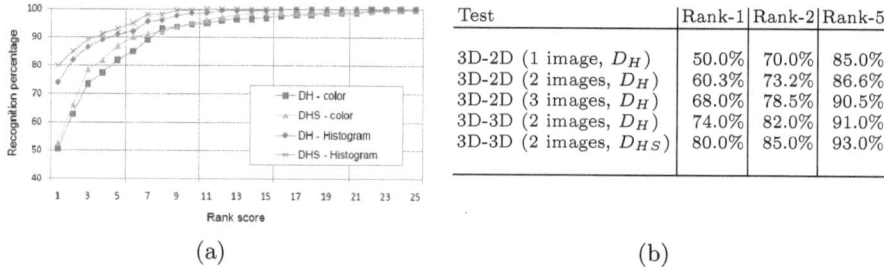

Test	Rank-1	Rank-2	Rank-5
3D-2D (1 image, D_H)	50.0%	70.0%	85.0%
3D-2D (2 images, D_H)	60.3%	73.2%	86.6%
3D-2D (3 images, D_H)	68.0%	78.5%	90.5%
3D-3D (2 images, D_H)	74.0%	82.0%	91.0%
3D-3D (2 images, D_{HS})	80.0%	85.0%	93.0%

(a)   (b)

Fig. 6. (a) Experimental results of the 2 image model-to-model test with the D_{HS} distance, (b) Performance evaluation of the system using 1,2 or 3 images to create the model. The last row shows the improvements using the saliency based distance.

Table 1. Performance evaluation of the system using random perturbation of the 3D model localization and orientation (3D-2D matching)

Rank	Correct Alignment	With noise on localization					With noise on orientation						
		2 px (15%)	4 px (30%)	6 px (45%)	12 px (75%)	16 px (90%)	5°	10°	15°	30°	40°	60°	90°
1	0.68	0.63	0.60	0.58	0.51	0.45	0.61	0.60	0.58	0.55	0.54	0.53	0.50
2	0.78	0.75	0.73	0.72	0.65	0.60	0.74	0.75	0.76	0.70	0.69	0.69	0.66
3	0.83	0.83	0.81	0.80	0.74	0.68	0.80	0.81	0.81	0.77	0.75	0.81	0.75
4	0.85	0.86	0.85	0.84	0.76	0.75	0.85	0.84	0.83	0.81	0.79	0.85	0.80
5	0.90	0.88	0.86	0.85	0.78	0.78	0.89	0.87	0.88	0.86	0.82	0.87	0.84
10	0.96	0.96	0.97	0.95	0.94	0.90	0.96	0.96	0.96	0.94	0.95	0.96	0.95

(a)   (b)   (c)   (d)

Fig. 7. (a) PETS dataset, sample frame from camera 1, (b,c,d) system output superimposed to camera 1, 2 and 3 frames

shows that our system is still reliable, even in the case of non precise model alignment and orientation, keeping good results with localization precision up to 6 pixels (45% overlap between the projected bounding box of the model and the image blob) and orientation up to 30 degrees.

We also tested the system on the PETS 2009 dataset [13] to evaluate the proposed method in real life conditions. The *City center* sequence, with three overlapping camera views, was selected. A 12.2m × 14.9m ROI was choosen, which is visible from all cameras. The proposed method was added on top of a previously developed tracking system [14], the goal of our method was to repair broken track and re-identify people that enter and exit the rectangular ROI. The D_H distance is used, together with the 153-vertices model, since the low resolution. Fig. 7 shows some sample frames from the system in action. The obtained precision and recall are 80.2% and 88.7

5   Conclusions

We proposed a new and effective method for people re-identification. Differently from currently available solutions we exploited a 3D body model to spatially localize the identifying patterns and colors on the model vertices. In this way, occlusion and view dependencies are intrinsically solved. Results both in real surveillance videos and in a proposed benchmark dataset are very promising. In this dataset, standard approaches based on 2D models fail, since the points of view are very different and the automatic segmentation is not precise enough. We believe that this new explored way based on 3D body models could be the starting point for future innovative solutions.

Acknowledgments. The work is currently under development as a doctorate project of the ICT School of the University of Modena and Reggio Emilia and within the project THIS (JLS/2009/CIPS/ AG/C1-028), with the support of the Prevention, Preparedness and Consequence Management of Terrorism and other Security-related Risks Programme European Commission - Directorate-General Justice, Freedom and Security.

References

1. Viswanathan, G.M., Raposo, E.P., da Luz, M.: Lévy flights and superdiffusion in the context of biological encounters and random searches. Physics of Life Rev 5(3), 133–150 (2008)
2. Javed, O., Shafique, K., Rasheed, Z., Shah, M.: Modeling inter-camera space-time and appearance relationships for tracking across non-overlapping views. Computer Vision and Image Understanding 109, 146–162 (2008)
3. Makris, D., Ellis, T., Black, J.: Bridging the gaps between cameras. In: Proc. of CVPR, pp. 205–210 (2004)
4. Vezzani, R., Baltieri, D., Cucchiara, R.: Pathnodes integration of standalone particle filters for people tracking on distributed surveillance systems. In: Proc. of ICIAP, Vietri sul Mare, Italy (2009)
5. Gray, D., Brennan, S., Tao, H.: Evaluating Appearance Models for Recognition, Reacquisition, and Tracking. In: Proc. of PETS 2007 (2007)
6. Farenzena, M., Bazzani, L., Perina, A., Murino, V., Cristani, M.: Person re-identification by symmetry-driven accumulation of local features. In: Proc. of CVPR, pp. 2360–2367 (2010)
7. Alahi, A., Vandergheynst, P., Bierlaire, M., Kunt, M.: Cascade of descriptors to detect and track objects across any network of cameras. Computer Vision and Image Understanding 114, 624–640 (2010)
8. Gandhi, T., Trivedi, M.: Panoramic Appearance Map (PAM) for Multi-camera Based Person Re-identification. In: Proc. of AVSS, pp. 78–78 (2006)
9. Andriluka, M., Roth, S., Schiele, B.: Monocular 3d pose estimation and tracking by detection. In: Proc. of CVPR, pp. 623–630 (2010)
10. Colombo, C., Del Bimbo, A., Valli, A.: A real-time full body tracking and humanoid animation system. Parallel Comput. 34, 718–726 (2008)
11. Baltieri, D., Vezzani, R., Cucchiara, R.: 3D body model construction and matching for real time people re-identification. In: Proc. of Eurographics Italian Chapter Conference (EG-IT 2010), Genova, Italy (2010)
12. Vezzani, R., Grana, C., Cucchiara, R.: Probabilistic people tracking with appearance models and occlusion classification: The ad-hoc system. Pattern Recognition Letters 32, 867–877 (2011)
13. PETS: Dataset - Performance Evaluation of Tracking and Surveillance (2009)
14. Vezzani, R., Cucchiara, R.: Event driven software architecture for multi-camera and distributed surveillance research systems. In: Proceedings of the First IEEE Workshop on Camera Networks - CVPRW, San Francisco, pp. 1–8 (2010)

Sorting Atomic Activities for Discovering Spatio-temporal Patterns in Dynamic Scenes

Gloria Zen[1], Elisa Ricci[2], Stefano Messelodi[3], and Nicu Sebe[1]

[1] Department of Information Engineering and Computer Science, Università di Trento, Italy
{gloria.zen,nicu.sebe}@disi.unitn.it
[2] Department of Electronic and Information Engineering, Università di Perugia, Italy
elisa.ricci@diei.unipg.it
[3] Fondazione Bruno Kessler, Trento, Italy
messelod@fbk.eu

Abstract. We present a novel non-object centric approach for discovering activity patterns in dynamic scenes. We build on previous works on video scene understanding. We first compute simple visual cues and individuate elementary activities. Then we divide the video into clips, compute clip histograms and cluster them to discover spatio-temporal patterns. A recently proposed clustering algorithm, which uses as objective function the Earth Mover's Distance (EMD), is adopted. In this way the similarity among elementary activities is taken into account. This paper presents three crucial improvements with respect to previous works: (i) we consider a variant of EMD with a robust ground distance, (ii) clips are represented with circular histograms and an optimal bin order, reflecting the atomic activities'similarity, is automatically computed, (iii) the temporal dynamics of elementary activities is considered when clustering clips. Experimental results on publicly available datasets show that our method compares favorably with state-of-the-art approaches.

Keywords: Dynamic scene understanding, Earth Mover's Distance, Linear Programming, Dynamic Time Warping, Traveling Salesman Problem.

1 Introduction

State-of-the-art video surveillance systems are currently based on an object centric perspective [1,2], *i.e.* first interesting objects in the scene are detected and classified according to their nature (*e.g.* pedestrians, cars), then tracking algorithms are used to follow their paths. However these approaches are suboptimal when monitoring wide and complex scenes and it is essential to take into account the presence of several occlusions and the spatial and temporal correlations between many objects. Therefore, unsupervised non-object centric approaches for dynamic scene understanding have gained popularity in the last few years [3,4,5,6]. These methods use low level features (such as position, size and motion of pixels or blobs) to individuate elementary activities. Then, by analyzing the co-occurrences of events, high level activity patterns are discovered. Most of the recent approaches for complex scene analysis are based on Probabilistic Topic Models [3,4,5]. It has been shown that these methods are particularly

G. Maino and G.L. Foresti (Eds.): ICIAP 2011, Part I, LNCS 6978, pp. 207–216, 2011.

effective to discover activity patterns and to model other interesting aspects such as the behaviors'correlation over time and space. However, since they rely on the traditional word-document paradigm for representing atomic activity distributions into clips, they discard any information about the similarity among elementary activities. To overcome this drawback recently Zen *et al.* [6] proposed a different approach. By optimizing a cross-bin distance function (*i.e.* EMD) rather than a bin-to-bin one, they showed that the problem of discovering high-level activity patterns in dynamic scenes can be modeled as a simple Linear Programming (LP) problem. To achieve scalability on large datasets they also propose a simplification of the optimization problem by establishing a words'order and considering only the similarity among adjacent words. However in [6] the order of atomic activities is chosen based on heuristics. In this paper, we propose a more rigorous strategy to sort atomic activities relying on the adaptation of the traveling salesman problem (TSP) to the task at hand. We also adopt a circular histogram representation for clips and optimize a variant of EMD with a robust ground distance. This allows us to improve the accuracy of clustering results at the expenses of a modest increase of the computational cost with respect to [6]. Here, as in many previous works [3,4,5,6], clips are represented by histograms. This has a beneficial effect in terms of filtering out noise. On the other hand any information about the temporal dynamics of atomic activities inside a clip is ignored. To compensate for this fact in this paper we propose to compute a dynamic time warping (DTW) similarity score between pairs of clips and to construct a nearest neighbor graph which is used to bias clips assignment toward appropriate clusters. The paper is organized as follows. In Section 2 the work in [6] is briefly summarized. Section 3 presents our approach for ordering atomic activities and the resulting LP problem. The proposed strategy for incorporating temporal information into the learning algorithm is also discussed. Experimental results are presented in Section 4. Finally, in Section 5 the conclusions are drawn.

2   Discovering Patterns with Earth Mover's Prototypes

The approach proposed in [6] is articulated in two phases. In the first phase simple motion features are extracted from the video and used to individuate elementary activities. In the second phase the video is divided into short clips and for each clip c an histogram h_c counting the occurrences of the elementary activities is computed. Then the clips are grouped according to their similarity and a small set of histogram prototypes representing typical activity patterns occurring in the scene are extracted. The Earth Mover's Prototypes learning algorithm in [6] amounts into solving the following optimization problem:

$$\min_{\boldsymbol{p}_i \geq 0, \, \sum p_i = 1} \sum_{i=1}^{N} \mathcal{D}_E(\boldsymbol{h}_i, \boldsymbol{p}_i) + \lambda \sum_{i \neq j} \eta_{ij} \max_{q=1...D} |p_i^q - p_j^q| \tag{1}$$

where $\{\boldsymbol{h}_1, \ldots, \boldsymbol{h}_N\}$, $\boldsymbol{h}_i \in \mathbb{R}^D$, are the original clip histograms, $\{\boldsymbol{p}_1, \ldots, \boldsymbol{p}_N\}$, $\boldsymbol{p}_i \in \mathbb{R}^D$, are the computed prototypes and $\mathcal{D}_E(\boldsymbol{h}, \boldsymbol{p})$ is the EMD [7]. The EMD among two normalized histograms (*i.e.* $\sum_q p^q = 1$ and $\sum_t h^t = 1$) is defined as:

$$\mathcal{D}_E(\boldsymbol{h}, \boldsymbol{p}) = \min_{f \ t \ 0} \sum_{q,t=1}^{D} d_{qt} f_{qt} \quad \text{s.t.} \quad \sum_{q=1}^{D} f_{qt} = h^t, \quad \sum_{t=1}^{D} f_{qt} = p^q \tag{2}$$

Solving (1) the prototypes $\boldsymbol{p}_i$ are computed in order to maximize their similarity with respect to the original histograms $\boldsymbol{h}_i$. At the same time the number of different prototypes is imposed to be small by minimizing their reciprocal differences. The relative importance of the two requirements is controlled by the positive coefficient λ. The coefficients $\eta_{ij} \in \{0, 1\}$ are fixed and are used to select the pairs of histograms which must be merged. In practice, by substituting the definition (2) into (1), the following LP is obtained:

$$\min_{p_i, f^i{}_t, \zeta_{ij} \ 0} \sum_{i=1}^{N} \sum_{q,t=1}^{D} d_{qt} f_{qt}^i + \lambda \sum_{i \neq j} \eta_{ij} \zeta_{ij} \tag{3}$$

$$\text{s.t.} \quad -\zeta_{ij} \leq p_i^q - p_j^q \leq \zeta_{ij}, \ \forall q, \forall i, j = 1 \dots N, \quad i \neq j$$

$$\sum_{q=1}^{D} f_{qt}^i = h_i^t, \ \forall t, \ \forall i = 1 \dots N \qquad \sum_{t=1}^{D} f_{qt}^i = p_i^q, \ \forall q, \forall i = 1 \dots N$$

where the slack variables ζ_{ij} are introduced.

A nice characteristic of (3) is that the ground distances d_{qt} can encode information about the similarity between atomic activities. However this flexibility comes at the expenses of a considerable computational cost. This cost is especially high due to the large number of flow variables f_{qt}^i, which is quadratic in the size of histograms D. Therefore, in order to speed up calculations, a modification of (3) which adopts EMD with L$_1$ distance over bins as ground distance (*i.e.* $d_{qt} = |q - t|$) is proposed in [6]. In this case, referred as EMD-L$_1$, the optimization problem simplifies and the number of flow variables reduces from $O(ND^2)$ to $O(ND)$ [8]. The idea is that similar atomic activities should correspond to neighboring bins in activity histograms. To this aim the atomic activities are sorted according to the associated location and motion information. However in [6] simple heuristics are used in this phase. In the following section we present a more rigorous approach to sort atomic activities.

3 Circular Earth Mover's Prototypes

In this Section we present the main phases of our approach, starting from low level features extraction to high-level activity patterns discovery by computing Circular Earth Mover's Prototypes.

3.1 Computing Atomic Activities

Given an input video, first low level features are extracted. We apply a background subtraction algorithm [9] to extract pixels of foreground. For these pixels we also compute the optical flow vectors using the Lucas-Kanade algorithm. By thresholding the magnitude of the optical flow vectors, foreground pixels are classified into static and

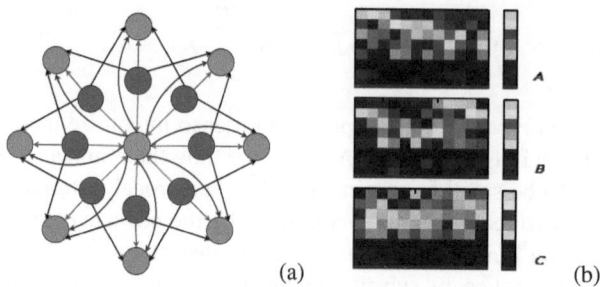

(a)   (b)

Fig. 1. (Best viewed in color.) **(a)** An example of the flow network associated to (4). Two circular histograms with $D = 8$ bins are compared. The green node is the transhipment vertex. Ingoing edge (green) cost is the threshold (2 in this case) and outgoing edge (blue) cost is 0. Red edges have cost 0. The black edges are 1-cost edges. **(b)** The importance of considering the temporal dynamics of atomic activities when comparing clips. For three clips (A,B,C) the sequences of frame histograms (left) and the associated average clip histograms (right) are shown. While all the average histograms are similar, only clip A and clip B should be assigned to the same cluster. In fact the temporal dynamics of atomic activities is very different in clip C.

moving pixels. Moving foreground pixels are also classified based on the direction of the optical flow vector. We consider 8 possible directions. Then we divide the scene into $p \times q$ patches. For each patch a descriptor vector $v = [x \; y \; f_g \; \bar{d}_{of} \; \bar{m}_{of}]$ is computed where (x, y) denotes the coordinates of the patch center in the image plane, f_g represents the percentage of foreground pixels and $\bar{d}_{of}$ and $\bar{m}_{of}$ are respectively the mode of the orientations distribution and the average magnitude of optical flow vectors. For patches of static pixels we set $\bar{d}_{of} = \bar{m}_{of} = 0$. Valid patches are only those with $f_g \geq T_{fg}$. Then the K-medoids algorithm is applied to the set of valid patch descriptors $w = [x \; y \; \bar{d}_{of} \; \bar{m}_{of}]$ and a dictionary of atomic activities is constructed.

3.2 Ordering Atomic Activities

The K-medoids algorithm provides a set of D centroids $c^d = [x^d \; y^d \; \bar{d}^d_{of} \; \bar{m}^d_{of}]$ representing typical elementary activities. However elementary activities are not independent and it is desirable to take into account their correlation when learning activity prototypes. To this aim in [6] the authors proposed to order atomic activities in a way that neighboring activities correspond to similar ones. While in [6] the order is determined based on simple heuristics, in this paper we propose to adopt a TSP strategy to sort atomic activities. The TSP [10] is a well known combinatorial optimization problem. The goal is to find the shortest closed tour connecting a given set of cities, subject to the constraint that each city must be visited only once. In this paper we model the task of computing the optimal order of atomic activities as the problem of computing the optimal city tour.

Formally, the TSP can be stated as follows. A distance matrix $\mathbf{D}$ with elements d_{qt}, $q, t = 1, \ldots, D$ and $d_{qq} = 0$ is given. The value d_{qt} represents the distance between the q-th and the t-th city. A city tour can be represented by a cyclic permutation π of

$\{1, 2, \ldots, D\}$ where $\pi(q)$ represents the city that follows the city q on the tour. The TSP is the problem of finding a permutation π that minimizes the length of the tour $\ell = \sum_{q=1}^{D} d_{q\pi(q)}$. In this paper we consider the symmetric TSP (i.e. $d_{qt} = d_{tq}$) and a metric as intercity distance, i.e. we set $d_{qt} = \mathcal{D}(c^q, c^t) = \overline{(x^q - x^t)^2 + (y^q - y^t)^2 +}$ $\gamma \overline{(\bar{d}^q_{of} - \bar{d}^t_{of})^2 + (\bar{m}^q_{of} - \bar{m}^t_{of})^2}$. The parameter γ is used to balance the importance of the position and the motion information.

The TSP can also be formulated as a graph theoretic problem. Given a complete graph $G = (V, E)$ the cities correspond to the node set $V = \{1, 2, \ldots, D\}$ and each edge $e_{qt} \in E$ has an associated weight d_{qt}. The TSP amounts to find a Hamiltonian cycle, i.e. a cycle which visits each node in the graph exactly once, with the least weight in the graph. For this task, the tour length of $(D - 1)!$ permutation vectors have to be compared. This results in a problem which is known to be NP-complete. However there are several heuristic algorithms for solving the symmetric TSP. In this paper we use a combination of the Christofides heuristic [11] for tour construction and simulated annealing for tour improvement.

3.3 Discovering Circular Earth Mover's Prototypes

To discover activity prototypes the video is divided into short clips and for each clip c a circular histogram h_c is created with bin orders obtained by solving the TSP. Finally, the clips are grouped according to their similarity and a set of circular histogram prototypes p_i is computed. They represent the salient activities occurring in the scene.

To this aim we solve (3). However, in this paper we use as ground distance a thresholded modulo L_1 distance [12] i.e. we set $d_{qt} = \min(\min(|q - t|, D - |q - t|), 2)$. The adoption of this ground distance with respect to the L_1 distance proposed in [6] allows us to deal in a principled way with circular histograms at the expenses of a modest increase of the computational cost. In fact, as shown [13], the adoption of a thresholded distance implies the introduction of a transhipment vertex, with slight increase of the number of flow variables. Moreover, it has been shown that saturated distances are beneficial in terms of accuracy results in several applications. With thresholded ground distance, the EMD (2) assumes the form:

$$\min_{f_{,+1}, f_{,-1}, f_{,D+1} \geq 0} \sum_{q=1}^{D} f_{q,q+1} + \sum_{q=1}^{D} f_{q,q-1} + 2 \sum_{q=1}^{D} f_{q,D+1} \qquad (4)$$

$$\text{s.t.} \quad f_{q,q+1} - f_{q+1,q} + f_{q,q-1} - f_{q-1,q} + f_{q,D+1} = h^q - p^q \quad \forall q$$

where the flow variables $f_{q,D+1}$ correspond to the links connecting sources to the transhipment vertex. Figure 1.a depicts the associated flow network. In practice with respect to (2), in (4) only flows between neighbor bins and flows between sources and the transhipment vertex are considered. The number of flow variables is still $O(D)$. Note also that since histograms are circular $q + 1 = 1$ if $q = D$ and $q - 1 = D$ if $q = 1$.

The proposed prototype learning algorithm is obtained substituting the EMD definition (4) into (1), *i.e.* solving:

$$\min \quad \sum_{i=1}^{N}\sum_{q=1}^{D} f_{q,q+1}^i + \sum_{i=1}^{N}\sum_{q=1}^{D} f_{q,q-1}^i + 2 \sum_{i=1}^{N}\sum_{q=1}^{D} f_{q,D+1}^i + \lambda \sum_{i\neq j} \eta_{ij}\zeta_{ij} \tag{5}$$

$$\text{s.t.} \quad -\zeta_{ij} \leq p_i^q - p_j^q \leq \zeta_{ij}, \ \forall q, \ \forall i,j, \ i \neq j$$

$$f_{q,q+1}^i - f_{q+1,q}^i + f_{q,q-1}^i - f_{q-1,q}^i + f_{q,D+1}^i = h_i^q - p_i^q, \ \forall q, i$$

$$p_i^q, f_{q,q+1}^i, f_{q,q-1}^i, f_{q,D+1}^i, \zeta_{ij} \geq 0$$

The resulting optimization problem is a LP with $n_{var} = 4ND + \frac{1}{2}N(N-1)$ variables
if we impose each prototype to be close to each other, *i.e.* $\eta_{ij} = 1 \ \forall i \neq j$. Therefore
the number of variables is slightly larger than in EMD-L$_1$ [6] where $n_{var} = 2N(D-
1) + ND + \frac{1}{2}N(N-1)$. However (5) allows us to deal with circular histograms and to
consider an optimal order of atomic activities. This provides more accurate clustering
results as shown in the experimental section.

3.4 Embedding Temporal Information into Clustering

A nice characteristic of (5) is that, thanks to the introduction of the binary coefficients
$\eta_{ij} \in \{0,1\}$, it is possible to select the pairs of histograms which must be merged.
Generally a comparison among all possible pairs $\{p_i, p_j\}$, $i \neq j$, is required, imposing
all prototypes to be close to each other. However by choosing only few $\eta_{ij} = 1$ it
is possible to embed into the clustering algorithm some a-priori knowledge about the
subset of histograms which must be fused. For example in [6], for each histogram h_i the
set of P nearest neighbors is identified and $\eta_{ij} = 1$ if h_j is a neighbor of h_i. This has
the effect of producing a biased clustering assignment which is imposed to reflect the
structure of the nearest neighbor graph. Moreover it greatly reduces the computational
cost of solving (2) since the number of slack variables (*i.e.* constraints) is limited.

While in [6] a simple Euclidean distance is adopted, in this paper we present a better
strategy to create a nearest neighbor graph. We propose to compute the distance among
clips by taking into account the temporal dynamics of atomic activities inside the clip.
More specifically for each clip c we consider not only the average histogram h_c but also
the sequence of histograms $H_c = \{h_c^1, \ldots h_c^M\}$ where h_c^i is the histogram of elementary
activities computed on the i-th frame. Then, to construct the nearest neighbor graph,
we propose to adopt a function which measures the distance between two histogram
sequences considering the match of their alignment. This allows us to account for small
shifts inside a clip and to consider two clips as similar only if the activity patterns
inside them have a similar temporal structure. This concept is exemplified in Fig.1.b.
In particular in this paper we consider the dynamic time warping and longest common
subsequence (LCSS) distances [14].

DTW. Given two clips H_a and H_b and the set $\mathcal{A}$ of all possible alignments ρ between
them, the DTW distance is defined as:

$$D_{DTW}(H_a, H_b) = \min_{\rho \in \mathcal{A}(H_a, H_b)} \sum_{i=1}^{\rho} \kappa(h_a^{\rho(i)}, h_b^{\rho(i)})$$

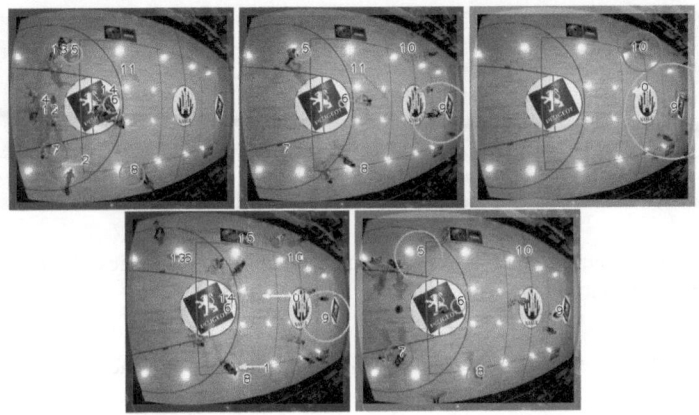

Fig. 2. APIDIS dataset: typical activities automatically discovered solving (5)

where $\kappa(\cdot)$ is the L_1 distance between histograms. Dynamic programming is used to compute the DTW distance, *i.e.* the optimal alignment between the two sequences of histograms.

LCSS. LCSS is also an alignment tool but is more robust to noise and outliers than DTW because not all points need to be matched. Instead of a one-to-one mapping between points, a point with no good match can be ignored to prevent unfair biasing. The LCSS distance is defined as:

$$D_{LCSS}(H_a, H_b) = 1 - \frac{LCSS(H_a, H_b)}{M}$$

As for DTW, dynamic programming can be used to compute LCSS, *i.e.* :

$$LCSS(H_a, H_b)=$$
$$= \begin{cases} 0, & m = 0 \mid n = 0 \\ 1 + LCSS(H_a^{m-1}, H_b^{n-1}), & \kappa(\mathbf{h}_a^m, \mathbf{h}_b^n) \leq \epsilon, \ |n - m| < \delta; \\ \max(LCSS(H_a^{m-1}, H_b^n), LCSS(H_a^m, H_b^{n-1})), & \text{otherwise} \end{cases}$$

4   Experimental Results

We tested the proposed approach on two publicly available datasets. Our method is fully implemented in C++ using the libraries OpenCV and GLPK 4.2.1 (GNU Linear Programming Kit) as the backend linear programming solver.

The first dataset is taken from **APIDIS**[1] and consists in a video sequence where players involved in a basketball match are depicted. A sequence of 3000 frames is chosen. The patch size is set to 16×16 pixels and each clip contains 60 frames, corresponding

---

[1] http://www.apidis.org/Dataset/

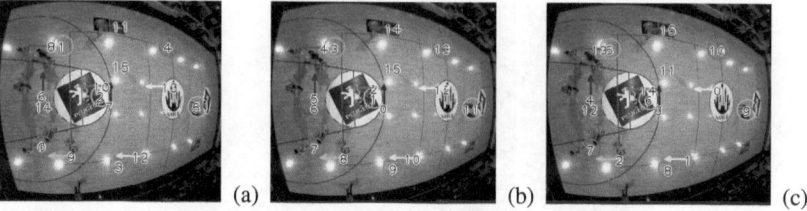

(a)   (b)   (c)

Fig. 3. APIDIS dataset: different orders of atomic activities. (a) Heuristics. (b) TSP (only position). (c) TSP (position and motion).

Table 1. APIDIS dataset: clustering accuracy (%) of the proposed approach with different orders of atomic activities

Random	Heuristics	TSP (only position)	TSP (position and motion)
78	80	80	**88**

to a time interval of about 3 sec. The number of atomic activities is fixed to 16. Solving (5) we automatically identify the five main activities occurring in the scene. Figure 2 shows a representative frame for each of them: (i) the blue team is attacking while the yellow team is on defence (green), (ii) the players are moving away from the yellow team's court side (blue), (iii) the blue team is on the defence (yellow), (iv) the players are moving back towards the yellow team's side (violet) and (v) the players of the blue team are shooting free throws (orange). Similar activities were also discovered in [6]. Table 1 shows some results of a quantitative evaluation of our method. Here the cluster assignments obtained solving (5) are compared with the ground truth build as in [6], based on the event annotation taken from the APIDIS website. The clustering performance corresponding to different ways of ordering atomic activities are compared. It is straightforward to observe that the TSP method outperforms other strategies. Moreover both motion and position information are crucial for obtaining accurate results (γ is set to 0.5). Figure 3 depicts an example of different orders of atomic activities. In this case the order based on the heuristics corresponds to sort activities considering first static ones ordered according to their position along the x axis, then those with motion different from zero.

The second dataset[2] [3] depicts a complex traffic scenes with cars moving in proximity of a **Roundabout**. It corresponds to a video of about 1 hour duration (93500 frames, 25 Hz framerate). In our experiment we consider only the first 30 minutes in order to compare quantitatively our results with those provided in [6,15]. In this case the patch size is set to 12×12 pixels while histograms of activities have 16 bins. Fig. 4 depicts an example of the typical activities discovered for this dataset, corresponding mainly to a horizontal and a vertical flow of vehicles. Table 2 shows the results of a quantitative evaluation. The proposed algorithm outperforms state-of-the-art approaches *e.g.* EMD-L_1 and L_1 methods [6] and PLSA and hierarchical PLSA [15]. The associated color

---

[2] http://www.eecs.qmul.ac.uk/ jianli/Roundabout.html

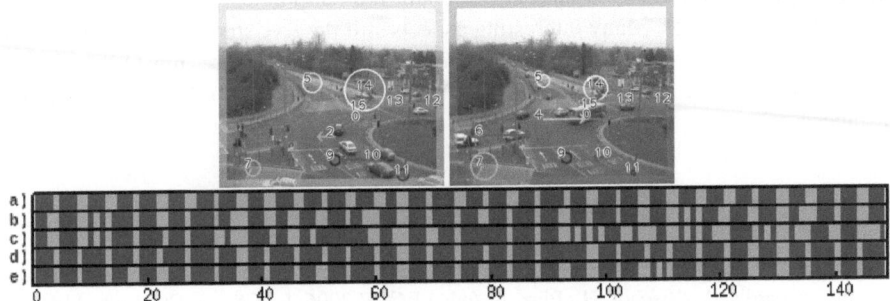

Fig. 4. Roundabout dataset. (top) Typical activities automatically discovered solving (5). (bottom) Temporal segmentation: (a) Ground truth, (b) Hierarchical PLSA [15], (c) Standard PLSA [15], (d) Our appraoch (5) - LCSS (TSP), (e) EMD-L$_1$ [6].

Table 2. Roundabout dataset: Clustering performance

method	accuracy (%)
L$_1$ [6]	86.4
EMD-L$_1$ (random) [6]	72.3
EMD-L$_1$ (heuristics) [6]	86.4
(5) - DTW (random)	81.63
(5) - LCSS (random)	83
(5) - DTW (TSP)	**87.75**
(5) - LCSS (TSP)	**87.75**
Standard PLSA [15]	84.46
Hierarchical PLSA [15]	72.30

bars depicting the results of temporal segmentation are shown in Fig. 4 (bottom). As observed for the APIDIS dataset, choosing a suitable order of atomic activities is crucial: using a random order the performance decrease significantly. Moreover a TSP strategy is also desirable with respect to an approach based on heuristics. Finally the adoption of DTW and LCSS distances for setting the coefficient η_{ij} further improves the clustering results. In fact, a better nearest neighbor graph drives the clustering algorithm towards more accurate solutions. Some videos showing the results of our experiments can be found at *http://tev.fbk.eu/people/ricci/iciap2011.html*.

5   Conclusions

We presented a novel method for discovering spatio-temporal patterns in dynamic scenes. Differently from most of the previous works on non-object centric dynamic scene analysis, our approach provides a principled way to deal with similarity of elementary activities while learning high-level activity prototypes. It relies on an automatical way to compute the optimal order of atomic activities and to an adaptation of the clustering algorithm in [6] to take into account the temporal dynamics of atomic activities inside the clips. Many interesting aspects still deserve study. For example more sophisticated mechanisms to filter out the noise of low level features must be exploited. On a theoretical side, we are currently investigating an approach for learning the ground distances.

References

1. Wang, X., Tieu, K., Grimson, W.E.L.: Learning semantic scene models by trajectory analysis. In: Leonardis, A., Bischof, H., Pinz, A. (eds.) ECCV 2006. LNCS, vol. 3953, pp. 110–123. Springer, Heidelberg (2006)
2. Zelniker, E.E., Gong, S.G., Xiang, T.: Global Abnormal Detection Using a Network of CCTV Cameras. In: Workshop on Visual Surveillance (2008)
3. Hospedales, T., Gong, S., Xiang, T.: A Markov Clustering Topic Model for Mining Behaviour in Video. In: IEEE International Conference on Computer Vision, ICCV (2009)
4. Varadarajan, J., Emonet, R., Odobez, J.-M.: Probabilistic Latent Sequential Motifs: Discovering temporal activity patterns in video scenes. In: British Machine Vision Conference, BMVC (2010)
5. Kuettel, D., Breitenstein, M.D., Van Gool, L., Ferrari, V.: What's going on? Discovering Spatio-Temporal Dependencies in Dynamic Scenes. In: IEEE Conference on Computer Vision and Pattern Recognition, CVPR (2010)
6. Zen, G., Ricci, E.: Earth Mover's Prototypes: a Convex Learning Approach for Discovering Activity Patterns in Dynamic Scenes. In: IEEE Conference on Computer Vision and Pattern Recognition, CVPR (2011)
7. Rubner, Y., Tomasi, C., Guibas, L.J.: The Earth Mover's Distance as a Metric for Image Retrieval. International Journal of Computer Vision (IJCV) 40(2), 99–121 (2000)
8. Ling, H., Okada, K.: An efficient earth mover's distance algorithm for robust histogram comparison. IEEE Transactions on Pattern Analysis and Machine Intelligence (PAMI) 29(5), 840–853 (2006)
9. Stauffer, C., Grimson, W.E.L.: Adaptive background mixture models for real-time tracking. In: IEEE Conference on Computer Vision and Pattern Recognition, CVPR (1999)
10. Lawler, E.L., Lenstra, J.K., Rinnooy Kan, A.H.G., Shmoys, D.B.: The Traveling Salesman Problem. Wiley, New York (1985)
11. Christofides, N.: Worst-case analysis of a new heuristic for the travelling salesman problem. Technical Report 388, Graduate School of Industrial Administration, Carnegie-Mellon University, Pittsburgh, PA (1976)
12. Pele, O., Werman, M.: A linear time histogram metric for improved SIFT matching. In: Forsyth, D., Torr, P., Zisserman, A. (eds.) ECCV 2008, Part III. LNCS, vol. 5304, pp. 495–508. Springer, Heidelberg (2008)
13. Pele, O., Werman, M.: Fast and robust earth mover's distances. In: IEEE International Conference on Computer Vision, ICCV (2009)
14. Rabiner, L., Juang, B.: Fundamentals of Speech Recognition. Prentice-Hall, Englewood Cliffs (1993)
15. Li, J., Gong, S., Xiang, T.: Global Behaviour Inference using Probabilistic Latent Semantic Analysis. In: British Machine Vision Conference, BMVC (2008)

Intelligent Overhead Sensor for Sliding Doors: A Stereo Based Method for Augmented Efficiency

Luca Bombini, Alberto Broggi, Michele Buzzoni, and Paolo Medici

VisLab – Dipartimento di Ingegneria dell'Informazione
Università degli Studi di Parma
{bombini,broggi,buzzoni,medici}@vislab.it
http://www.vislab.it

Abstract. This paper describes a method to detect and extract pedestrians trajectories in proximity of a sliding door access in order to automatically open the doors: if a pedestrian walks towards the door, the system opens the door. On the other hand if the pedestrian trajectory is parallel to the door, the system does not open. The sensor is able to self-adjust according to changes in weather conditions and environment. The robustness of this system is provided by a new method for disparity image extraction.

The rationale behind this work is that the device developed in this paper avoids unwanted openings in order to decrease needs for maintenance, and increase building efficiency in terms of temperature (i.e. heating and air conditioning). The algorithm has been tested in real conditions to measure its capabilities and estimate its performance.

Keywords: safety sensor, sliding doors, obstacle detection, pedestrian detection, trajectory planning, stereo vision.

1   Introduction

Current access control systems for automatic door control require a sensor able to detect a moving object or a pedestrian crossing the gate. This approach does not take into account the trajectory of pedestrians, and therefore can not estimate movements. As an example, if a pedestrian crosses the area in front of the door but does not want to cross the gate, the control access board detects his/her presence and anyway opens the door. In this case, the system is not efficient, since it leads to a waste of energy in terms of electricity, air conditioning, or heating and decrease the system lifetime with unnecessary open/close actions.

The sensor presented in this paper solves the problem of unwanted openings exploiting a stereo vision system based on a 3D reconstruction: CMOS cameras with *fish-eye* optics placed over the sliding doors are used to acquire stereo images.

A very general setup has been studied to fit different kinds of automatic door accesses. Another feature of this system is that it can be integrated or adapted to a wide range of existing doors and customer requirements.

G. Maino and G.L. Foresti (Eds.): ICIAP 2011, Part I, LNCS 6978, pp. 217–226, 2011.

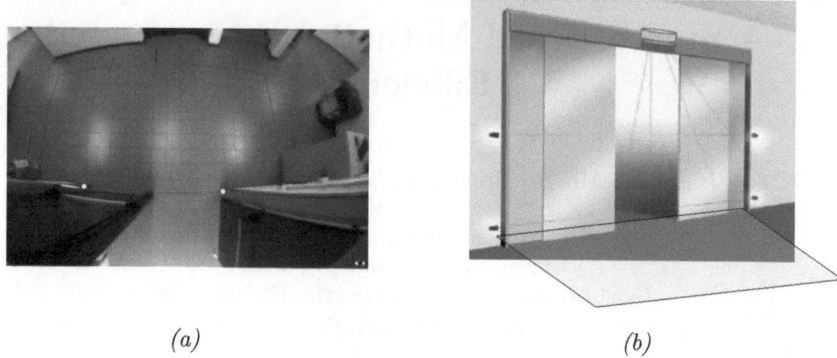

(a)   (b)

Fig. 1. *(a)* An image acquired by a single camera installed on an sliding doors access. *(b)* The cameras system (green) is placed on the door access and can replace usual sensor (red). The area monitored by the new vision-based system (green plane) is larger than the one covered by infrared sensors (red line) *(b)* and therefore allows the detection of pedestrian trajectories.

A pair of low-cost cameras are installed on the top of the door, overlooking the crossing area (fig. 1.*a*). The cameras are connected to an embedded computer that processes images using a stereo vision based algorithm. The computer is connected to the access control board and therefore can open or close the door as required.

This paper is organized as follows: in Section 2 the algorithm structured is sketched; Section 3 shows a the new method for the computation of the disparity image; results, final considerations, and ideas for future developments are then presented in Section 4.

2   The Algorithm

This section presents the overview and implementation of the algorithm used in the system. Figure 2 shows the flowchart of the man algorithm steps.

Given two stereo images, the direction of objects moving in the scene is detected and analyzed to trigger the doors opening.

The algorithm does not require calibration: the IPM images (Inverse Perspective Mapping) [1,2] of the scene are generated by a Look Up Table (LUT) computed off-line (see Section 3).

The images obtained so far are used to create the reference background thanks to a standard background algorithm generation using a circular buffer of the last N images [4]. The two background images are compared against input images performing a difference: binarized images (masks) are then generated after thresholding and dilating the difference images.

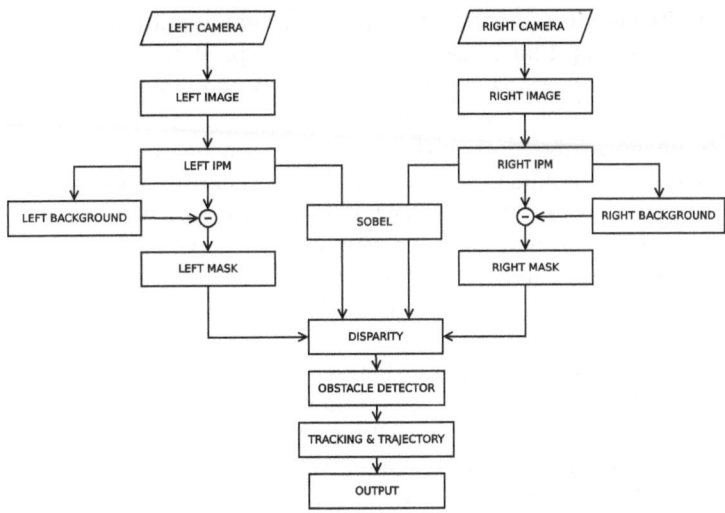

Fig. 2. Flowchart of the proposed algorithm

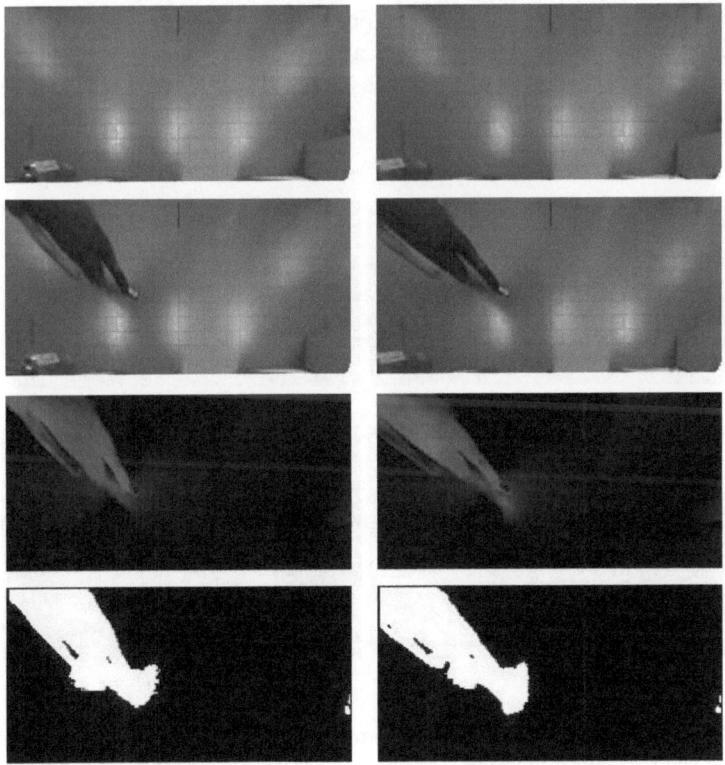

Fig. 3. From top: background, current, difference, and binarized images for left and right cameras

A Sobel filter is used to reduce the changing light conditions problem and to produce an edges map that is used to generate Disparity Space Image (DSI) [3]. Thanks to the preliminary use of the IPM, the resulting DSI has to be interpreted using a specific approach (see Section 3)

The DSI image is filtered using the two masks previously calculated, therefore considering only the points of the first mask and validating only the points that match in the other mask. This approach reduces false positives and errors in DSI computation, such as the ones due to repetitive patterns and reflections.

The obtained images are used by the obstacle detector to find each obstacle in the scene. In fact, the DSI image is clustered using the contiguity of points in the image and their location in the world as constraints (see Section 3). The result of this step is a list of blobs that belong to the obstacles in the framed scene (fig. 4.a). Each blob allows the computation of the 3D transformation related to the corresponding obstacle. And, specifically, the size and position with respect to the ground. The obstacles with a small size and an height lower than 20 cm are discarded. The detected size and position of the remaining blobs are then used to track obstacles. The trajectory of each obstacle is computed through a linear regression of the past positions (fig. 4.b). The system triggers the door opening (fig. 4.c) if the trajectory is aimed towards the door for two successive frames or if the object is close to the door.

For safety reasons a small proximity area near the door is considered as well; the door will be opened if any obstacle is detected inside this area, independently from its direction of movement.

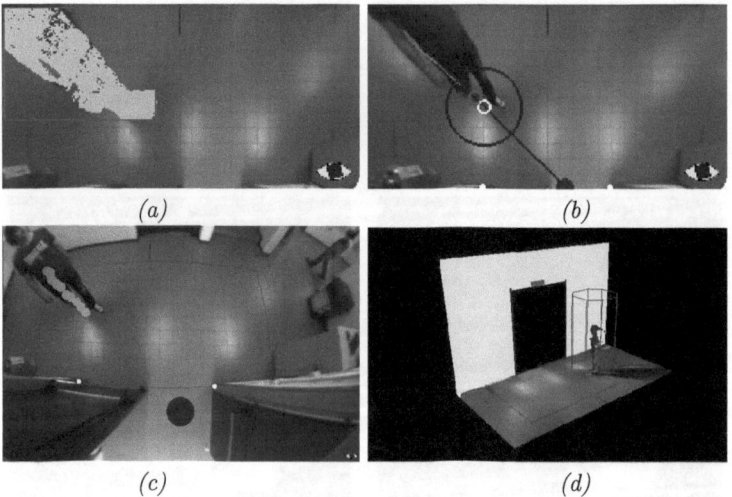

Fig. 4. (a) A cluster DSI Blob, (b) the computed trajectory crosses the doorstep (delimited by the two white markers, (c) output, (d) 3D-reconstruction

3   Disparity and DSI from Stereo IPM Transformation

In this section a novel method to compute the disparity on IPM images is presented. Initially, the formulas to generate the IPM images are analyzed, then the concept of disparity on these images to create the DSI will be introduced.

3.1   Inverse Perspective Mapping

The Inverse Perspective Mapping (IPM) is a geometric transformation that allows to remove the perspective effect from images acquired by a camera [1,2]. It resamples the source image mapping each pixel in a different position and creating a new image (IPM image) which represents an bird-eye view of the observed scene.

Unfortunately is not possible to correctly remap all pixels of the source image but only those that belong to a known plane (usually the ground, $Z = 0$). The points that do not belong to this plane are remapped in an incorrect mode that depends on their position in the world and the position of the cameras.

An off-line LUT is used for the generation of the IPM, that can be defined from the intrinsic and extrinsic parameters of the vision system or, as in this case, through a manual mapping of points of a calibration grid and a subsequent interpolation. The LUT allows to remove the perspective effect and the distortion of the source image all together [5,6,7].

A point not belonging to the $Z = 0$ plane is remapped on the IPM image in a specific way that can be understood by studying the process of generation of the IPM. The IPM image is the result of two perspective transformations: a direct one and a reverse one. The world points are mapped into a perspective image through a perspective transformation and then in an IPM image assuming all pixels belonging to the ground.

The equation of a perspective transformation can be expressed in three-dimensional homogeneous coordinates:

$$\begin{bmatrix} u \\ v \\ 1 \end{bmatrix} = \mathbf{AR} \left(\begin{bmatrix} X \\ Y \\ Z \end{bmatrix} - \begin{bmatrix} x_0 \\ y_0 \\ z_0 \end{bmatrix} \right) . \tag{1}$$

where (X, Y, Z) are the coordinates of the points in the world, (x_0, y_0, z_0) are the coordinates of the camera position in the world, and (u, v) are the coordinates in the perspective image. $\mathbf{A}$, and $\mathbf{R}$ are the matrices (3×3) of the intrinsic and rotation parameters, respectively.

The IPM transformation can be defined as:

$$\left(\begin{bmatrix} X' \\ Y' \\ Z' \end{bmatrix} - \begin{bmatrix} x_0 \\ y_0 \\ z_0 \end{bmatrix} \right) = \mathbf{R}^{-1}\mathbf{A}^{-1} \begin{bmatrix} u \\ v \\ 1 \end{bmatrix} . \tag{2}$$

With Z' the height from ground of the point to remap. Without additional information and assuming $Z' = 0$, equations (1) and (2) lead to:

$$\left(\begin{bmatrix} X' \\ Y' \\ 0 \end{bmatrix} - \begin{bmatrix} x_0 \\ y_0 \\ z_0 \end{bmatrix} \right) = \overbrace{\mathbf{R}^{-1}\mathbf{A}^{-1}\mathbf{A}\mathbf{R}}^{I} \left(\begin{bmatrix} X \\ Y \\ Z \end{bmatrix} - \begin{bmatrix} x_0 \\ y_0 \\ z_0 \end{bmatrix} \right) . \tag{3}$$

which, since the homogeneous coordinates are defined up to a proportionality factor λ, gives the final expression:

$$\begin{bmatrix} X' - x_0 \\ Y' - y_0 \\ -z_0 \end{bmatrix} = \begin{bmatrix} X - x_0 \\ Y - y_0 \\ Z - z_0 \end{bmatrix} \lambda . \tag{4}$$

λ can be obtained thanks to the $Z' = 0$ assumption:

$$\lambda = \frac{-z_0}{Z - z_0} . \tag{5}$$

and, can be substituted in eq.(4), allowing X', Y' computation:

$$X' = -z_0 \frac{X - x_0}{(Z - z_0)} + x_0 \quad Y' = -z_0 \frac{Y - y_0}{(Z - z_0)} + y_0 . \tag{6}$$

It can be noticed that equations (6) do not depend on the lens distortion, intrinsic parameters and orientation of the cameras, but only on their position in the world. Therefore, the IPM can be easily computed using a LUT and not requiring a complex calibration step.

From these results, it is evident that the same point in the world (X, Y, Z) is remapped on different IPM coordinates (X', Y') depending on the camera position in the world. Only if the world point belongs to the ground, $Z = 0$, the equations are reduced at $X' = X$ and $Y' = Y$.

Equations (6) can be merged together, to obtain the important result:

$$\frac{X' - x_0}{X - x_0} = \frac{Y' - y_0}{Y - y_0} . \tag{7}$$

A vertical obstacle features constant X and Y, therefore equation (7) can be written as:

$$X' - x_0 = m'(Y' - y_0) \tag{8}$$

where the line slope m' is defined as:

$$m' = \frac{X - x_0}{Y - y_0} . \tag{9}$$

All the world points that belong to a vertical obstacle are then remapped into IPM image on the line connecting the pin-hole and the the base position of the obstacle. This line depends on the coordinates (X, Y) of the world point and on camera location (x_0, y_0, z_0). Since the dependence from Z has been removed, eq. (7), the z_0 coordinate does not influence the line equation on which the vertical obstacles are mapped, but only their size in the IPM image.

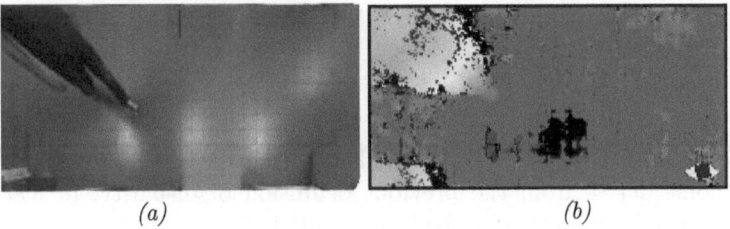

(a)   (b)

Fig. 5. *(a)* Right IPM images, *(b)* DSI

3.2   The Disparity and the 3D Reconstruction on IPM Images

Let us consider a system with two stereo cameras aligned on a line parallel to Y-axis, with the same height h and with baseline b. The right camera coordinates are $(x_0, -b_r, h)$, while the left ones are (x_0, b_l, h).

Using equations (6), the IPM coordinates of a generic world point $P(X, Y, Z)$ can be computed. The point is then remapped applying the IPM to the images captured by the cameras:

$$X'_r = X'_l = -h\frac{X - x_0}{Z - h} + x_0 . \tag{10}$$

$$Y'_r = -h\frac{Y + b_r}{Z - h} - b_r \quad Y'_l = -h\frac{Y - b_l}{Z - h} + b_l \tag{11}$$

Solving the coordinates Y'_r as a function of the coordinate Y'_l:

$$Y'_r = Y'_l - b\left(\frac{h}{Z - h} + 1\right) = Y'_l + b\frac{Z}{h - Z} . \tag{12}$$

and given that the coordinates X'_r and X'_l are the same for the two images, $(X'_r = X'_l)$, the concept of disparity is defined as:

The disparity of a world point with coordinate (X, Y, Z) projected onto two IPM images is equal to:

$$\Delta = Y'_r - Y'_l = b\frac{Z}{h - Z} . \tag{13}$$

This definition of disparity leads to the following considerations:

- Since in the IPM image only points of the world with a height less than the height of the camera, $Z < h$, can be remapped, the disparity is always positive.
- The disparity of points does not depend on coordinates (X, Y) of the points in the world, but only on their coordinate Z. This implies that all points placed at the same height Z in the world have the same disparity Δ in the IPM image. The ground, having no height ($Z = 0$), has zero disparity $\Delta = 0$. In addition, the disparity of features has an upper limit of h.

- The disparity proportionally depends on the distance between the two cameras (baseline).
- The disparity position does not depend on the x_0 coordinate of the cameras.

The knowledge of the disparity of an IPM image allows a three-dimensional reconstruction of the scene. The formulas to estimate the word coordinates of a IPM point derive from the previous definition of disparity; in fact, the Z coordinate can be directly obtained from eq.(13):

$$Z = h\left(1 - \frac{b}{\Delta + b}\right) = h\frac{\Delta}{\Delta + b} . \tag{14}$$

with h being the height from ground and b the cameras baseline.

Replacing this value for Z in eq.(6), it can be obtained:

$$X'_r = -h\frac{X - x_0}{\left(h\frac{\Delta}{\Delta + b} - h\right)} + x_0 \quad Y'_r = -h\frac{Y + b_r}{\left(h\frac{\Delta}{\Delta + b} - h\right)} - b_r . \tag{15}$$

Therefore allowing to compute the world coordinates (X, Y) as function of disparity:

$$X = X'_r\left(\frac{b}{\Delta + b}\right) + x_0\left(\frac{\Delta}{\Delta + b}\right) \quad Y = Y'_r\left(\frac{b}{\Delta + b}\right) - b_r\left(\frac{\Delta}{\Delta + b}\right) . \tag{16}$$

3.3   DSI on IPM Images

The Disparity Space Image computation is traditionally used on standard stereo images, but it can be efficiently used for IPM images too (see fig. 5.b). In fact the algorithm for the implementation of DSI [3] requires to search for each pixel of the right image the corresponding pixel on the left image (homologous pixel). In IPM images, the homologous pixels are searched for in the same row in each images. In fact, equation 13, shows that the coordinates X' of a world point only depends on its height and is the same in the two images, therefore easing DSI computation.

In addiction it can further noticed for DSI on IPM images that:

- The disparity of pixels is always positive.
- The ground has zero disparity.
- The disparity of vertical object increases with its height, featuring a gradient along the line defined by eq. (7).
- Given that the same obstacle is differently deformed by the IPM depending on camera position, the search of homologous points has to cope with differences when computing the matches.
- If the best correlation between two homologous point is under a given threshold, the DSI value is undefined, indicating that the pixel has not been found in the other image.
- The size of the IPM images is a parameter that highly influences the computation time of the DSI. More precisely the time needed for DSI computations is cubically proportional to the IPM image size.

4   Results and Conclusions

The system was tested in real conditions to measure its capabilities and evaluate its performance, using as computing architecture a SmallPC industrial computer with a processor Intel Core 2 Duo 2.5GHz and 2G DDR DRAM. The system was developed using Linux operating system and C++ programming language. Two cameras were installed over a public area door acquiring many image sequences. The format video was set to 752 × 480 pixel at 15 Hz, but it was reduced to 240 × 120 pixel in IPM images to reduce the computational load. The images have been analyzed in laboratory and the algorithm has been tested in a lot of situations.

The system has proven to work correctly and reach good results in different situations (fig. 6). The most critical situation that affects the system performance is the presence of an excessive number of obstacles in the scene. In such a case, the objects trajectory is not correctly detected, due to difficulties in tracking all objects, and therefore generating some false positives leading to unwanted openings. Anyway, the system always assures the door opening when a person really wants to cross the gate thanks to the implementation of the proximity area attention (fig. 6.c).

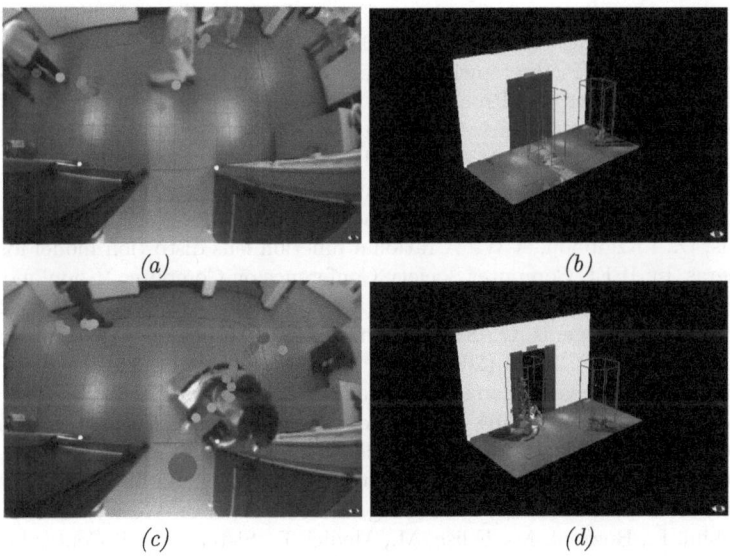

Fig. 6. *(a)* Algorithm output in a multi-pedestrian scenario, *(b)* 3D-reconstruction, *(c)* a pedestrian detected in proximity area, *(d)* 3D-reconstruction

The stereo system presented in this paper has a great advantage with respects to classic disparity obstacle detectors, in fact it uses a calibration based on a LUT. The LUT for the removal of the perspective effect and camera distortion is computed with a simple grid positioned in the area of interest during the first installation. This area is easily tunable acting on software parameters and therefore enabling the use of the system for different doors and environments. An automatic calibration application is currently under development: it will ease the installation and reduce the installation time, making easier the work of the installers.

A new installation and testing in a shopping center is scheduled to compare and tune the system performance against a classic infrared approach. An extension of the system for monitoring bus doors also for safety purposes and not only for optimization is currently under development.

A patent [10] was filed to cover this approach that might turn into a product.

Acknowledgements. The work has been supported by the Industria2015 AU-TOBUS Italian project.

References

1. Bertozzi, M., Broggi, A., Fascioli, A.: Stereo Inverse Perspective Mapping:Theory and Applications. Image and Vision Computing Journal 8(16), 585–590 (1998)
2. Bertozzi, M., Broggi, A., Fascioli, A.: An extension to the Inverse Perspective Mapping to handle non-flat roads. In: Procs. IEEE Intelligent Vehicles Symposium 1998, Stuttgart, Germany, pp. 305–310 (1998)
3. Felisa, M., Zani, P.: Incremental Disparity Space Image computation for automotive applications. In: Procs. IEEE/RSJ Intl. Conf. on Intelligent Robots and Systems, St.Louis, Missouri, USA (October 2009)
4. Benezeth, Y., Jodoin, P.M., Emile, B., Laurent, H., Rosenberger, C.: Review and evaluation of commonly-implemented background subtraction algorithms. In: ICPR 2008 19th International Conference on Pattern Recognition (December 2008)
5. Claus, D., Fitzgibbon, A.W.: A rational function lens distortion model for general cameras. In: IEEE Computer Society Conference on Computer Vision and Pattern Recognition (CVPR 2005), San Diego, Calif., USA, vol. 1, pp. 213–219 (2005)
6. Devernay, F., Faugeras, O.: Straight lines have to be straight. Machine Vision and Applications 13(1), 14–24 (2001)
7. Tsai, R.Y.: A versatile camera calibration technique for high-accuracy 3D machine vision metrology using off-the-shelf TV cameras and lenses. IEEE Journal of Robotics and Automation, 323–334 (1987)
8. Jahne, B.: Digital Image Processing, 5th edn. Springer, Berlin (2002)
9. Pratt, W.K.: Digital Image Processing, 3rd edn. Addison-Wesley, Milano (2001)
10. Bombini, L., Buzzoni, M., Felisa, M., Medici, P.: Sistema per il Controllo di Porte Automatiche (March 2010), CCIAA di Milano, Patent application MI2010A000460

Robust Stereoscopic Head Pose Estimation in Human-Computer Interaction and a Unified Evaluation Framework

Georg Layher[1], Hendrik Liebau[1], Robert Niese[2], Ayoub Al-Hamadi[2],
Bernd Michaelis[2], and Heiko Neumann[1]

[1] University of Ulm, Institute of Neural Information Processing
{georg.layher,hendrik.liebau,heiko.neumann}@uni-ulm.de
[2] Otto-von-Guericke University Magdeburg, IESK
{robert.niese,ayoub.al-hamadi,bernd.michaelis}@ovgu.de

Abstract. The automatic processing and estimation of view direction and head pose in interactive scenarios is an actively investigated research topic in the development of advanced human-computer or human-robot interfaces. Still, current state of the art approaches often make rigid assumptions concerning the scene illumination and viewing distance in order to achieve stable results. In addition, there is a lack of rigorous evaluation criteria to compare different computational vision approaches and to judge their flexibility. In this work, we make a step towards the employment of robust computational vision mechanisms to estimate the actor's head pose and thus the direction of his focus of attention. We propose a domain specific mechanism based on learning to estimate stereo correspondences of image pairs. Furthermore, in order to facilitate the evaluation of computational vision results, we present a data generation framework capable of image synthesis under controlled pose conditions using an arbitrary camera setup with a free number of cameras. We show some computational results of our proposed mechanism as well as an evaluation based on the available reference data.

Keywords: head pose estimation, stereo image processing, human computer interaction.

1 Introduction

Human-centered computing has paved the ground for the integration of new technologies into the development of natural and intuitive-to-use interfaces for advanced human-computer interaction (HCI). The visual modality is an important channel for non-verbal communication transmitting social signals which contain rich behavioral cues concerning attention, communicative initiative, empathy, etc. (Pentland 2007; Vinciarelli et al. 2008). Such cues are conveyed through various behavioral signals and combinations thereof which allow their automatic analysis by computers (Corso et al. 2008; Jacob 1996; Turk 2004). These social signals can be interpreted to reason about the user state or intention in a context-dependent fashion to launch differentiated communicative or social reactions in

G. Maino and G.L. Foresti (Eds.): ICIAP 2011, Part I, LNCS 6978, pp. 227–236, 2011.
© Springer-Verlag Berlin Heidelberg 2011

next generation HCI systems. Here we focus on one of the most meaningful cues concerning the user's focus of attention relative to the observer's (camera's) view direction which can be derived from an actor's head pose direction. Vision-based estimation of the human head pose received more and more attention over the last decade and a wealth of methods and mechanisms have been proposed. An overview is presented in (Murphy-Chutorian & Trivedi 2009). Still, the robustness of many of these approaches suffer from large errors and redundancies in the feature extraction and matching mechanisms, variations in illumination conditions, or size changes due to variable camera-actor distances. Here, we propose a new stereo matching approach that operates on features derived from hierarchical processing of images pairs in a biologically inspired architecture of static form processing. Stereo matching is subsequently driven by learned intermediate-level image features which allow robust matching and fast false targets reduction.

Considering the evaluation of methods, a large number of different approaches entail numerous different training and test data sets. For example, Murphy-Chutorian & Trivedi (2009) quote no less than 14 different databases. However, such data sets are often somehow adapted to support the particular research focus of a given approach which leads to restrictions in the variation of free parameters. This makes it virtually impossible to compare fundamentally different computational methods for estimating head poses using only one of the currently available datasets. In addition to our newly developed method for pose estimation, we propose a data generation framework that enables us to generate input for any kind of head pose estimation approach and thus allowing a comprehensive comparison.

The rest of the paper is structured as follows. In Section 2 we introduce a new stereo matching mechanism for robust head pose estimation. Section 3 details the suggested approach for the development of an evaluation framework. In Section 4 we demonstrate the functional significance of the proposed computational method in conjunction with the evaluation framework.

2   Stereo Head Pose Estimation

We propose a robust approach to estimate the 3-dimensional (3D) head pose from stereoscopic image pairs. The approach mainly consists of three processing stages. First, the images of a stereo pair are initially processed to detect and localize facial features of different complexity and at different image scales. The processing scheme employed here follows a biologically inspired model architecture of the cortical ventral pathway in the primate visual system (Riesenhuber & Poggio 1999). Second, the different features detected in the left and right image frames of a stereo pair are subsequently matched. Our approach is inspired by recent findings that visual features with intermediate size and complexity are superior in terms of their specificity and their misclassification probability (Ullman et al. 2002). Following this observation we propose making use of intermediate level features as learned by the previously mentioned processing hierarchy. Using such complex features reduces the numbers of false targets in the stereo correspondence problem and thus quickly resolves the matching ambiguities. At the same time the features used still allow spatial resolutions of sufficient detail. Correspondence is established by using a simple correlation-based matching approach. Third, the facial feature localization and the subsequent calculation of disparity values between each pair of corresponding features allow to reconstruct the 3D location of a patch of matched image regions. The estimated 3D surface information in world coordinates is then used to fit a so-called facial plane (defined by the eyes and the mouth) to estimate the pose angle. In the following, we briefly describe the object recognition framework used for the localization of the face and the enclosed features. Subsequently, we address the issue of how the positions of the features can be used for an efficient and robust estimation of the head pose.

2.1 Hierarchical Feature Extraction

For the detection of the head and the facial features, we implemented a slightly modified version of the biologically inspired object-recognition model proposed by Mutch & Lowe (2008). It is based on a model by Serre et al. (2005), which in turn extends the above-mentioned "HMAX" model proposed by Riesenhuber & Poggio (1999). The proposed model architecture utilizes alternating layers of so-called simple (S) and complex (C) cells. In a nutshell, simple cells perform a local linear filtering operation, i.e. a convolution, and combine spatially adjacent elements into a high-order feature. Complex cells, on the other hand, increase position and scale invariance by a nonlinear pooling operation. As sketched in Fig. 1, our model variant consists of five different layers of processing. The input layer transforms the original image (I_{in}) into a pyramidal representation of different spatial scales (I_{scale}) through proper low-pass filtering and downsampling,

$$I_{scale}(x,y) = (I_{in}(x,y) \star \frac{1}{2\pi\sigma^2} exp(-\frac{x^2+y^2}{2\sigma^2})) \cdot III_{\Delta x, \Delta y}(x,y) \qquad (1)$$

where $\star$ denotes the spatial convolution operator, σ denotes the width of the Gaussian low-pass filter (for reducing the signal frequency content), and $III(\cdot)$

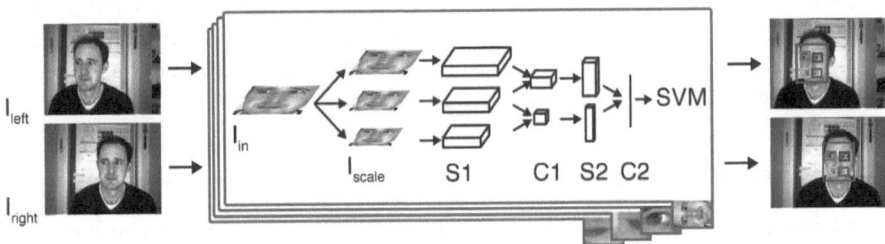

Fig. 1. In the first processing step, a face and its correspondent are detected in a stereo image pair. The same object-recognition framework is then used to localize both eyes and the mouth corners within the two facial regions. Over all, the framework has to be trained on four different kinds of features (faces, eyes, left and right mouth corners).

is the down-sampling function (Shah function, Bracewell 1978) with $\Delta x, \Delta y$ denoting the sampling width on a rectangular grid. Each scale representation is then convolved with 2D Gabor filters of four orientations. These filters resemble the shape of receptive field profiles of cells in the mammalian visual cortex (Hubel & Wiesel 1959). The Gabor filter kernels are generated by Gaussian window functions modulated by oriented sinusoidal wavefronts

$$g_{\sigma,\omega}(x,y) = exp(-\frac{x^2+y^2}{2\sigma^2}) \cdot f_{wave}(\omega_0{}^T \cdot (x,y)) \qquad (2)$$

with $f_{wave}(x) = \cos(x)$ or $f_{wave}(x) = \sin(x)$ (even and odd modulation functions), σ again denotes the width of the Gaussian, and $\omega_0 = (\omega_{x0}, \omega_{y0})$ denotes the modulating wave of a given frequency and orientation. Convolution of these kernels from the bank of filters with orientations and scales,

$$I_{\sigma,\omega}(x,y) = I_{scale}(x,y) \star g_{\sigma,\omega}(x,y) \qquad (3)$$

generates output activations in layer S1 leading to a 4D feature representation (position, scale and orientation). Unlike Serre et al. (2005) receptive fields of different sizes are modeled here through the image pyramid instead of Gabor filters tuned to different spatial frequencies. Consequently, the number of necessary convolution operations is reduced. Layer C1 cells pool the activities of S1 cells in the same orientation channel over a small local neighborhood. This leads to a locally increasing scale, position, and size invariance, while decreasing the spatial extent of the representation. The intermediate feature layer S2 performs a simple template matching of patches of C1 units and a number of learned prototypes utilizing a sliding window approach (following the suggestion by Mutch & Lowe 2008). In a nutshell, the learning algorithm selects the most descriptive and discriminative prototypes among an exhaustive number of C1 patches that were randomly sampled during the learning process. Disregarding the learning process the prototypes resemble spatial filters of higher feature complexity concerning their response selectivity. The correlation responses of the prototypes with the input patterns generate a feature vector of corresponding response amplitudes. In the last layer the S2 prototype responses are pooled over all positions and scales by choosing the maximum value for each prototype to yield a single feature vector as C2 response. These vectors finally serve as input to a linear support vector machine (SVM) that is used for the classification of faces, as well as the associated facial features.

2.2   Stereo Matching

Input images of a stereo pair are processed individually by using the hierarchical processing architecture sketched above. Using pairs of stereo images with slightly different views on the scene the next step is to automatically determine the matching points from one image with the corresponding points in the second image of the pair. The feature sets available from the preprocessing stage range from small localized features at the finest resolution to intermediate size patches

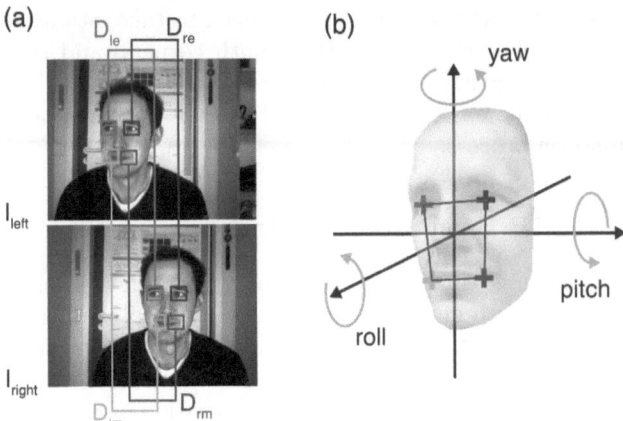

Fig. 2. Disparity values are calculated for all four facial features (a). The infered information about the 3D-positions of the features is used to determine the orientation of the facial plane (b).

up to the full image at the coarsest resolution level of the pyramid. Most traditional stereo matching approaches operate on fine resolutions utilizing certain similarity measures for initial correspondence findings. Pyramid matching approaches using a coarse-to-fine scheme make use of differently filtered versions of the input (compare, e.g., Szeliski 2011). Unlike these approaches, we make use of the findings of Ullman et al. (2002) that intermediate level image fragments are most distinctive in providing structural information about the presence of features. More precisely, a face fragment of intermediate size provides reliable indicators of face presence and locality. For the increase in reliability one has to pay the price that the likelihood of appearance of the same pattern in a new face image is low. A small fragment, corresponding to a local filter response, increases the likelihood but at the same time reduces the specificity (increase of uncertainty). This property is also reflected in the auto-correlation surface E_{AC} using a sum-of-squared-distance measure: For a small fragment of low complexity the correlation surface is noisy and rather unspecific while for intermediate complexity it becomes more pronounced and less error-prone due to the noise. The intermediate-size fragments still allow localization in the image such that, e.g., the left and right eye can be distinguished (results not shown). Considering the correspondence finding in stereo images these intermediate-level fragments help to improve the stability of finding candidate matches and reduce the false-targets uncertainty.

Our matching approach makes use of a simple correlation-based similarity measure using image fragments of intermediate complexity. The correlation measure is maximized starting with the reference template fragment and using a search window along epipolar lines in the left and right image (Barnard & Fischler 1982; Hannah 1988). It is important to stress, that we do not need to estimate a dense disparity map. Instead, we need to take into account only few facial features (namely the eyes and the mouth corners) and their stereo correspondence to subsequently estimate the spatial head orientation (see Fig. 2). This considerably reduces the computational costs of the correspondence finding and pose estimation process.

2.3   Estimating 3D Head Pose Orientation

Given the disparity values of the left and right eye and the corners of the mouth, as well as the focal length and the baseline of the stereo camera system, we are able to estimate a plane that fits the four spatial points of eye and mouth in 3D. The plane is fitted using a least-squares approach minimizing the distance of the projections of the four 3D coordinates. The orientation of the resulting plane is then used as an estimation of the head pose (see Fig. 2b).

3   Data Generation Framework

For the generation of image sequences as training and testing data we use a textured generic 3-D model. Based on three key points, i.e. left and right eye plus upper lip point, the generic model is adapted to a specific face. The key points are gathered from a frontal face image of a given subject using the Viola & Jones (2004) Haar-like feature detection algorithm with cascades provided by Santana (2005). Based on the actual inter-ocular distance (which must be provided) and the ratio of face width and height estimated from the detected image points, i.e. eye distance and distance between eye axis and mouth, the generic 3-D model is scaled. The texture is assigned to the frontal image of the face. The availability of textured 3-D models enables quick OpenGL rendering of images under controlled pose conditions (see Fig.3) using an arbitrary camera setup with a free number of cameras. This way image material can be generated for accuracy testing of pose estimation methods. Also training data can be created for holistic pose estimation approaches. An example for a two camera setup is given in Figure 3. For the parameterization of the virtual cameras we chose to emulate a real camera setup in our vision lab. This has two advantages; on the one hand we determined the camera viewing parameters through calibration of the real cameras. On the other hand, after simulation and accuracy testing, we can directly turn our method into real application, using the same camera setup.

4   Estimation Results

The proposed stereo head pose estimation approach was tested under two different conditions. At first we used a local head pose database in combination with the framework described in Section 3 to generate test data with known ground truth. In Fig. 3 an exemplary picture of the database, as well as the resulting stereo image pairs for different head poses are shown. The estimated

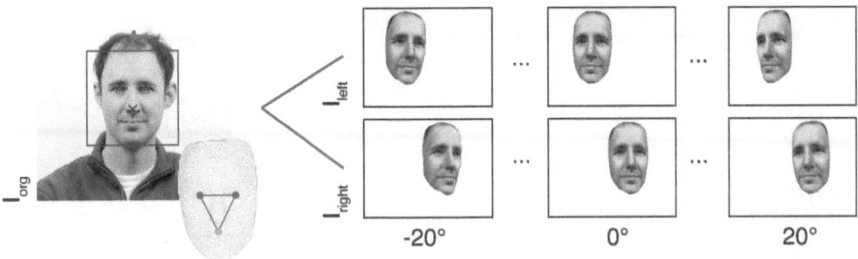

Fig. 3. A local database in which head poses were systematically w.r.t. yaw and pan angle served as input for the generation of artificial images of different head poses (as described in Section 3; the database is available via http://www.uni-ulm.de/in/neuroinformatik/mitarbeiter/g-layher; Weidenbacher et al. 2007). The virtual camera setup was adjusted to the specification of the Bumblebee®2 Stereo Vision Camera System (Point Grey Research Inc.) which was used in a real world testing scenario (Fig. 4b). The artificially generated stereo images were used to evaluate the head pose estimation approach as described in Section 2.

head pose is displayed in Fig. 4a. For the test data shown in Fig. 3, the estimation error increases for larger yaw angles, but never exceeds 3° over head poses in the range of $[-20°,20°]$. Secondly, we used a sequence of real world images to test the accuracy of the estimated head pose under unrestricted conditions. The actual head pose within the sequence was unknown, but the subject was told to systematically rotate his head from left to right. As shown in Fig. 4b, the estimated head pose reflects that fact. Note that the underlying classifiers used for the localization of the head and the facial features were all trained using the FERET Database (Phillips et al. 2000) and thus have no association to the test data at all. Fig. 4c shows the capability of the proposed approach under varying camera-actor distances.

5   Discussion and Conclusion

In this study we have demonstrated that online stereo head pose estimation can be achieved by utilizing features of intermediate complexity. Stereoscopic head pose estimation is simplified by matching only significant facial features, thus keeping the disparity map sparse. The use of intermediately complex templates for matching left and right image pairs allows to use a simple correlation-based criterion as similarity measure. The distinctiveness of the features used greatly reduces the matching ambiguities to improve the reliability of the correspondence estimation. While many previous approaches assume that the observer keeps his viewing distance in a restricted range, the approach here allows large changes in viewing distance (or scale) while at the same time continuously varying the view direction (yaw and pitch) and changing the facial mimics. The approach meets several design criteria that were suggested by Murphy-Chutorian & Trivedi

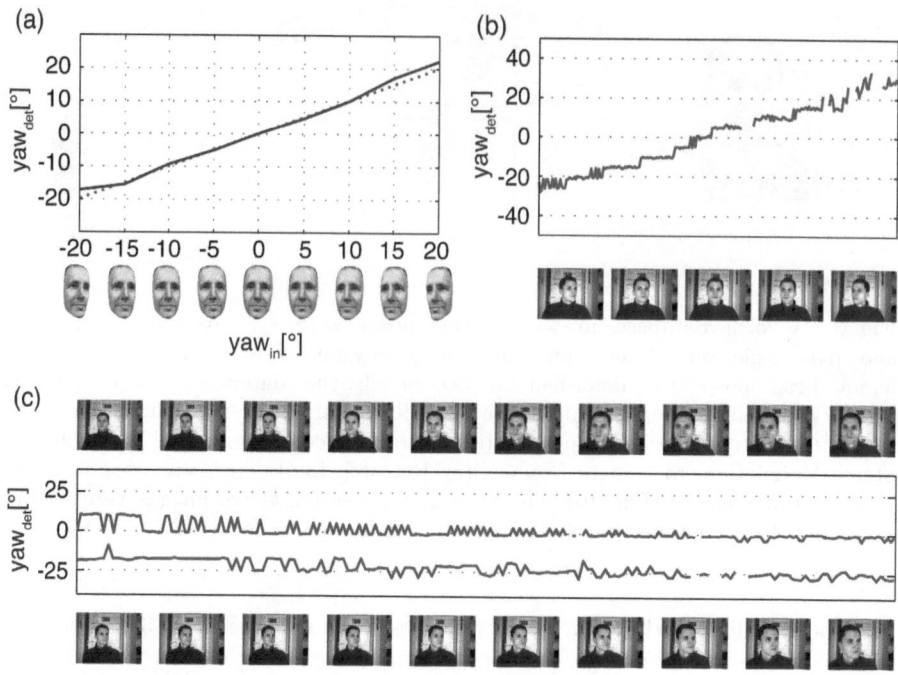

Fig. 4. Head pose estimation results. (a) An artificially generated sequence of stereo images (using the 3D face model desribed in Section 3) contains pictures of horizontal head poses in a range of −20° to +20°, in steps of 5° (yaw angle) used as input (ground truth). The resulting head pose estimates lead to an error that is less than 3° (with large yaw angles producing larger errors). (b) A sequence of real world stereo images was obtained using the Bumblebee®2 Camera System. The subject was instructed to rotate his head continuously from left to right, thus yielding different yaw angles for head poses. Even though no ground truth data is available, it can be seen that the estimated head pose follows the pose characteristic of the head. (c) The capability of the proposed approach under varying camera-actor distances was tested using two real world sequences with different but constant yaw angles (as shown above and below the plot). As can be seen, there is only a little change in the estimated head poses for differing distances. Note that the apparently large differences of the estimated head poses in the left half of the plot are caused by a mere difference of one pixel in the disparity values. This reflects the fact, that with an increasing camera-actor distance, the pixel resolution is more and more restricting the number of distinguishable head poses.

(2009), namely accuracy (achieving pose estimates with less than 3° mean absolute error for yaw/pitch angles below +/- 40°), autonomy of processing without manual initialization or feature localization, invariance against person identity and illumination, and independence of the spatial resolution and viewer distance. Apart from the technical approach of pose estimation, we also suggested a strategy to evaluate the method by using a textured 3D model to flexibly generate synthetic ground truth data for arbitrary camera views.

In all, we suggest that the approach makes a valuable contribution to build more flexible vision-based conversational systems in human-computer interaction and affective communication. For example, the approach presented here allows continuous and reliable head pose estimation and the recognition of head gestures analyzing social signals in communication (Morency & Darrell 2004; Vinciarelli et al. 2008; Pentland 2007). Other approaches also used stereo-based head pose estimation. For example, Voit et al. (2006) estimate head poses in meeting room scenarios using multiple cameras, while Morency et al. (2006) estimate pose information to detect conversational turns for an embodied conversational agent. In these scenarios the variation of viewer distance and thus image resolution can be assumed to change only within a limited range and that the luminance conditions can be controlled. Furthermore the precision of the focus of attention (as estimated from the head pose direction) was not evaluated in detail. Here, we suggest an approach that successfully handles these variable conditions and allows further applications in interactive scenarios. For example, we are interested to already set an interactive companion system into an alert condition when a person approaches. During the course of approaching the person could be analyzed in terms of communicative social signals and the vision-based system might properly react ahead of the first interaction. We will address this in future investigations.

Acknowledgments. This research has been supported by a Grant from the Transregional Collaborative Research Center SFB/TRR62 "Companion-Technology for Cognitive Technical Systems" funded by the German Research Foundation (DFG). Portions of the research in this paper use the FERET database of facial images collected under the FERET program, sponsored by the DOD Counterdrug Technology Development Program Office.

References

Pentland, A.: Social Signal Processing. IEEE Signal Processing Magazine 24(4), 108–111 (2007)

Vinciarelli, A., Pantic, M., Bourlard, H., Pentland, A.: Social Signals, their Function, and Automatic Analysis: a Survey. In: Proceedings of the 10th international conference on Multimodal interfaces (ICMI 2008), pp. 61–68. ACM, New York (2008)

Corso, J.J., Ye, G., Burscbka, D., Hager, G.D.: A Practical Paradigm and Platform for Video-Based Human-Computer Interaction. IEEE Computer 41(5), 48–55 (2008)

Jacob, R.: Human-Computer Interaction: Input Devices. ACM Computing Surveys 28(1), 177–179 (1996)

Turk, M.: Computer Vision in the Interface. Commun. ACM. 47(1), 60–67 (2004)

Murphy-Chutorian, E., Trivedi, M.M.: Head Pose Estimation in Computer Vision: A Survey. IEEE Trans. on Pattern Analysis and Machine Intelligence 31(4), 607–626 (2009)

Riesenhuber, M., Poggio, T.: Hierarchical Models of Object Recognition in Cortex. Nature Neuroscience 2(11), 1019–1025 (1999)

Ullman, S., Vidal-Naquet, M., Sali, E.: Visual Features of Intermediate Complexity and their Use in Classification. Nature Neuroscience 5(7), 682–687 (2002)

Mutch, J., Lowe, D.G.: Object Class Recognition and Localization Using Sparse Features with Limited Receptive Fields. Int. J. Comput. Vision 80(1), 45–57 (2008)

Serre, T., Wolf, L., Poggio, T.: Object Recognition with Features Inspired by Visual Cortex. In: Proceedings of the 2005 IEEE Computer Society Conference on Computer Vision and Pattern Recognition (CVPR 2005), pp. 994–1000. IEEE Computer Society, Washington (2005)

Bracewell, R.N.: The Fourier Transform and its Applications. McGraw-Hill, Columbus (1978)

Hubel, D.H., Wiesel, T.N.: Receptive Fields of Single Neurones in the Cat's Striate Cortex. Journal of Physiology 148(3), 574–591 (1959)

Szeliski, R.: Computer Vision - Algorithms and Applications. Springer, London (2011)

Barnard, S.T., Fischler, M.A.: Computational Stereo. ACM Computing Surveys (CSUR) 14(4), 553–572 (1982)

Hannah, M.J.: Digital Stereo Image Matching Technique. In: Proc. XVIth ISPRS Congress (Int'l Soc. for Photogrammtery and Remote Sensing), Commission III, Kyoto, Japan, vol. XXVII, Part B3, pp. 280–293 (1988)

Castrillón-Santana, M., Lorenzo-Navarro, J., Déniz-Suárez, O., Isern-González, J., Falcón-Martel, A.: Multiple Face Detection at Different Resolutions for Perceptual User Interfaces. In: Marques, J.S., Pérez de la Blanca, N., Pina, P. (eds.) IbPRIA 2005. LNCS, vol. 3522, pp. 445–452. Springer, Heidelberg (2005)

Viola, P., Jones, M.J.: Robust Real-Time Face Detection. Int. J. Comput. Vision 57(2), 137–154 (2004)

Phillips, P.J., Moon, H., Rizvi, S.A., Rauss, P.J.: The FERET Evaluation Methodology for Face-Recognition Algorithms. IEEE Trans. on Pattern Analysis and Machine Intelligence 22(1), 1090–1104 (2000)

Morency, L.-P., Darrell, T.: From Conversational Tooltips to Grounded Discourse: Head Pose Tracking in Interactive Dialog Systems. In: Proceedings of the 6th International Conference on Multimodal Interfaces, pp. 32–37. ACM, New York (2004)

Lades, M., Vorbrüggen, J.C., Buhmann, J., Lange, J., von der Malsburg, C., Wurtz, R.P., Konen, W.: Distortion Invariant Object Recognition in the Dynamic Link Architecture. IEEE Trans. on Computers 42(3), 300–311 (1993)

Voit, M., Nickel, K., Stiefelhagen, R.: A Bayesian Approach for Multi-View Head Pose Estimation. In: Proceedings of the 2006 IEEE International Conference Multisensor Fusion and Integration for Intelligent Systems (MFI 2006), pp. 31–34. IEEE Computer Society, Washington (2006)

Morency, L.-P., Christoudias, C.M., Darrell, T.: Recognizing Gaze Aversion Gestures in Embodied Conversational Discourse. In: Proceedings of the 8th International Conference on Multimodal Interfaces, ICMI 2006 (2006)

Weidenbacher, U., Layher, G., Strauss, P.-M., Neumann, H.: A Comprehensive Head Pose and Gaze Database. In: 3rd IET International Conference on Intelligent Environments, IE 2007 (2007)

Automatic Generation of Subject-Based Image Transitions

Edoardo Ardizzone, Roberto Gallea, Marco La Cascia, and Marco Morana

Università degli Studi di Palermo
{ardizzon,robertogallea,lacascia,marcomorana}@unipa.it

Abstract. This paper presents a novel approach for the automatic generation of image slideshows. Counter to standard cross-fading, the idea is to operate the image transitions keeping the subject focused in the intermediate frames by automatically identifying him/her and preserving face and facial features alignment. This is done by using a novel Active Shape Model and time-series Image Registration. The final result is an aesthetically appealing slideshow which emphasizes the subject. The results have been evaluated with a users' response survey. The outcomes show that the proposed slideshow concept is widely preferred by final users w.r.t. standard image transitions.

Keywords: Face processing, image morphing, image registration.

1 Introduction

In recent years, the diffusion of digital image acquisition devices (e.g., mobile phones, compact digital cameras, smartphones) encouraged people to take more and more pictures. However, the more pictures are stored the more users need automatic tools to manage them. In particular, personal photo libraries show peculiar characteristics as compared to generic image collection. Personal photos are usually captured on the occasion of real-life events (e.g., birthdays, weddings, trips), so that it is common the presence of people in most of the images. Moreover a relatively small number of different individuals (i.e., the family) is usually shown across the whole library and this allows to achieve reliable results with automatic approaches.

Many methods for photo collection management were proposed focusing on users' request for accessing a subset of stored data according to some particular picture properties.

In this work we address the scenario where the user wants to manage its own photo collection according to *who* is depicted in each photo. Starting from a collection of personal images, we propose a tool for the automatic slideshow of a sequence of pictures depicting the same individual. Firstly the whole collection is searched for faces, while face identities are assigned with an automatic approach [1]. Then each face is processed for automatically finding the position of some facial feature points. Finally, image sequences that contain the same person are animated by applying a morphing approach.

G. Maino and G.L. Foresti (Eds.): ICIAP 2011, Part I, LNCS 6978, pp. 237–246, 2011.
© Springer-Verlag Berlin Heidelberg 2011

The problem of the automatic morphing of pairs of digital images has been investigated since the early '90s. More recentlty, several techniques have been focused on face morphing. In [2] and [3] two systems are proposed for the automatic replacement of faces in photographs using Active Shape Models (ASMs). Both of them use standard ASMs on simple images, i.e., high-quality frontal faces, thus it seems that such approaches are not robust enough to be used with faces captured "in the wild". This is a strong limitation since even if a number of features detection techniques are available in literature, their performances signitificaly drop when such methods are applied on real-life images.

Some commercial systems, such as *Fantamorph* [4], allows users to create face animation. Although its latest version provides support for face and facial features detection, user intervention is often needed to aid the detection process and no face identification is available.

The paper is arranged as follows: a description of the system will be given in Sect. 2. The Sect. 2.1 will give an overview of the face processing techniques we developed, while the image alignment methods are described in Sect. 2.2. Experimental results will be shown and discussed in Sect. 3. Conclusions will follow in Sect. 4.

To better evaluate the results of the proposed work, sample videos of produced slideshows are available at `http://www.dinfo.unipa.it/~cvip/pps/`.

2   Methods

The whole system realizing the image animation can be subdivided into two main blocks (see Fig. 1): face processing and image alignment modules. The first is responsible of detecting and identifying the subject, the second operates the spatial transformation across the transitions.

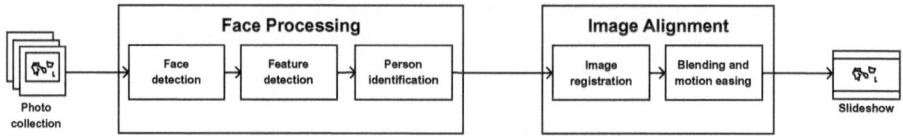

Fig. 1. Block diagram of the proposed system

2.1   Face Processing

Face processing is considered nowadays as one of the most important application of image analysis and understanding.

Face detection, i.e., determine if and where there is a face in the image, is usually the first step of face processing techniques. A number of challenges are associated with face detection due to several factors. For example, face appearance may heavily change according to the relative camera-face pose, to the presence of facial hair, or on the occurrence of lighting variations. All these factors are further stressed in real-life photo collections.

Many face detection techniques [5] have been proposed and even if face detection is not a solved problem, some methods have reached a certain level of maturity. In this work we used the state of the art approach to face detection, i.e., the framework proposed by Viola and Jones [6], due to its efficiency and classification rate.

When making the animation of two subsequent pictures, as much local information as possible is needed to preserve the appearance of face regions. Thus, the corners of the bounding box obtained from the face detection step are not sufficient and local facial features need to be detected.

Early approaches for facial feature detection focused on template matching to detect eyes and mouth [7], but these features are not suitable for noisy images such as real-life photos. More recent models, i.e., ASM, AFM, AAM, offer more robustness and reliability working on local *feature points* position. Active Shape Model (ASM) [8] extends Active Contour Model [9] using a flexible statistical model to find feature points position consistently with the training set. Active Appearance Model (AAM) [10] combines shapes with gray-level face appearance.

Due to their time efficiency, ASMs are one of the most used approach both for face detection and real time face tracking. However the efficiency of this approach is heavily dependent on the training data used to build the model, that in most cases consists of face databases (e.g., FERET, Color FERET, PIE, Yale, AR, ORL, BioID) acquired under constrained conditions. Such collections of faces looks very different from those acquired in everyday life, thus in the considered scenario this approach is unsatisfactory. For this reason we used personal collections both for training and testing. The training set is a collection of faces detected in a private personal collection, while the whole system has been tested on a publicly available dataset [11] enabling future comparison.

Once faces have been detected, we focused on using ASMs to find a predefined set of fiducial points in face images. Considering a training set of 400 size-normalized faces (200x200 pixels) detected in a private photo collection, we developed three models, shown in Fig. 2, using 45, 30 and 23 landmarks respectively. The first model, Fig. 2(a), is composed of five shapes, i.e., right eye, left eye, mouth, nose-eyebrows and face profile. This model is frequently used in ASMs-based works, however we performed some tests on real-life pictures noticing that in most cases, due to lighting variations, face profile and internal shapes (mouth and eyes) are misplaced while the nose-eyebrow contour appears as the most robust to pose changes. For this reason the first model has been refined as shown in Fig. 2(b). In this case we considered a single shape by removing the tip of the nose and linking mouth and eyes contours. However we discovered that the tip of the nose is fundamental for the right positioning of surrounding features, moreover the areas around the eyes are frequently noisy due to hard intensity changes caused by occlusions (e.g., glasses). Consequently, the contours of the eyes are not reliable enough.

Taking into account such considerations, our final model (Fig. 2(c)) consists of a single shape that follows the top contour of the eyebrows and the bottom contour of nose and mouth. A comparison of the feature detected with the three models is shown in Fig. 2.

Fig. 2. The three developed models. In leftmost column red points represent the first and last landmark for each shape. Other columns depict examples of features detection. Each row shows the shapes obtained by using the first, second and third face model respectively.

The 23 detected points are used as input of the morphing algorithm. These points, in conjunction with information about *who* is in the picture (i.e., the face identity) allow to perform the automatic animation of the photo sequence.

Organizing the collection based on *who* is in the photo generally requires much work from the user that, in the worst case, has to manually annotate all the photos in the collection. Here we used a data association framework for people re-identification [1] that takes advantage of an important constraint: a person can not be present two times in the same photo and if a face is associated to an identity, the remaining faces in the same photo must be associated to other identities. The problem is modeled as the search for probable associations between faces detected in subsequent photos using face and clothing descriptions. In particular, a two level architecture is adopted: at the first level, associations are computed within meaningful temporal windows (events); at the second level, the resulting clusters are re-processed to find associations across events.

The output of the re-identification process is a set of identities associated to each face detected in the collection.

2.2   Image Alignment and Photo Transitions

Given the correspondences between two consecutives images, the next task to be performed is the smooth transition bringing the first images onto the second, keeping the main object (i.e., the face of the considered person) aligned. Such problem can be regarded as a time-series registration task, or more specifically, a morphing problem. Some issues need to be considered for this purpose: choosing the registration function for the alignment, the easing function for the transition and the motion function for the feature points.

Registration Function. During the whole transition, in order to mantain spatial coherence, the feature points need to be kept in alignment. Such problem is defined image registration. It can be considered as the geometric function to apply to the image I (input image), to bring it in correspondence with the image R (reference image). Many approaches exist for such purpose, in this work we adopted two strategies. The first one is based on region-wise affine registration, the latter leverages onto thin-plate spline transforms.

In the case of *region-wise multi-affine transformation*, the main idea is to subdivide the images in sub-regions, each one defined by three feature points, resulting corresponding triangles are aligned using affine transformations. Points triangulation in I and R is obtained using the classic Delaunay triangulation.

Since the vertices of the obtained triangles tessellation and their respective displacements are known, for each triangle is possible to recover the mapping function for all the points belonging to it. Each 2d affine transformation requires six parameters, so writing down the transformation (both x and y coordinates) for three points, produces a system of six equations, sufficient for its complete determination, as stated in (1).

$$\begin{bmatrix} x \\ y \\ 1 \end{bmatrix} = \begin{bmatrix} a\ b\ c \\ d\ e\ f \\ 0\ 0\ 1 \end{bmatrix} \begin{bmatrix} x_0 \\ y_0 \\ 1 \end{bmatrix} = \begin{bmatrix} ax_0 + by_0 + c \\ dx_0 + ey_0 + f \\ 1 \end{bmatrix} \qquad \begin{cases} x_1 = ax_{0\ 1} + by_{0\ 1} + c \\ y_1 = dx_{0\ 1} + ey_{0\ 1} + f \\ x_2 = ax_{0\ 2} + by_{0\ 2} + c \\ y_2 = dx_{0\ 2} + ey_{0\ 2} + f \\ x_3 = ax_{0\ 3} + by_{0\ 3} + c \\ y_3 = dx_{0\ 3} + ey_{0\ 3} + f \end{cases} \quad (1)$$

Each transformation is recovered from a triangle pair correspondence, and the composition of all the transformations allows the full reconstruction of the image. In addition, to avoid crisp edge across triangles edges, the *fuzzy kernel regression* approach [12] was used for transformation smoothing.

The second approach is based on *thin-plate spline transformations* (TPS) [13]. The TPS is a parametric interpolation function which is defined by $D(K + 3)$ parameters, where D is the number of spatial dimensions of the datasets and K is the number of the given landmark points where the displacement values are known. The function is a composition of an affine part, defined by 3 parameters, and K radial basis functions, defined by an equal number of parameters. Its analytic form is defined as:

$$g(p) = ax + by + d + \sum_{i=1}^{k} \rho\left(\|\mathbf{p} - \mathbf{c}_i\|^2\right) w_i; \qquad \mathbf{p} = \begin{bmatrix} x_p \\ y_p \end{bmatrix}, \mathbf{c}_i = \begin{bmatrix} x_{c_i} \\ y_{c_i} \end{bmatrix} \quad (2)$$

where $\mathbf{p}$ is the input point, $\mathbf{c}_i$ are the landmark points and the radial basis function $\rho(r)$ is given by:

$$\rho(r) = \frac{1}{2}r^2 \log r \tag{3}$$

All of the TPS parameters are computed solving a linear system defined by a closed-form minimization of the bending energy functional. Such functional is given by:

$$E_{tps} = \sum_{i=1}^{k} \|y_i - g(\mathbf{p}_i)\| + \lambda \int \int \left[\left(\frac{\partial^2 g}{\partial x^2}\right)^2 + 2\left(\frac{\partial^2 g}{\partial xy}\right)^2 + \left(\frac{\partial^2 g}{\partial y^2}\right)^2 \right] dx dy. \tag{4}$$

The functional is composed by two terms: the data term and the regularization term. The former minimizes the difference between known and recovered displacements at landmark points, the latter minimizes the bending energy of the recovered function, i.e., maximises its smoothness and it is weighted by the parameter λ. As mentioned before, for this expression a closed-form analytical solution exists, from which is possible to recover all of the required spline function parameters. The main characteristic of this function is that it exhibits minimum curvature properties.

Notwithstanding the used deformation function, for the morphing animation purposes, at each time step both the two images A and B involved in the transition need to be registered to an intermediate image defined by the current frame considered, whose feature points lie onto the path connecting their corresponding feature points (Fig. 3).

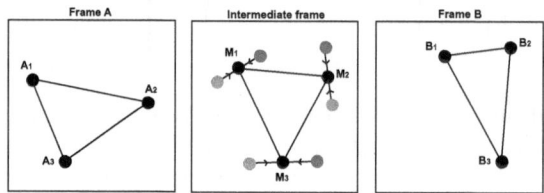

Fig. 3. When transitioning from frame A to frame B, both feature points of A and B, need to move to intermediate feature points to reconstruct intermediate image M

Easing Function. The visual transition between the two images is realized through a blending of the two images registered to the intermediate images. Such blending is produced as the weighted sum of the intensity level of the images, where the weighting factor is defined by a blending paramter b. Varying the parameter in function of the time during the animation, produces an easing visual effect. The easiest variation criterion is to adjust it linearly with the time variation. However, linear easing is equivalent to no easing at all, this is quite unaesthetic since the variation occurs instantly, resulting in a crisp bad-looking mechanical effect. For this reason, for determining $b(t)$, several non-linear easing functions based on Penner's formula [14] have been implemented.

Basically, three types of easing are used, ease-in (slow start with instantaeous stop), ease-out (instantaneous start and slow stop) and ease-in-out (slow start and stop), using four function shapes, quadratic, cubic, sinusoidal and exponential. In Fig.4 are reported the 12 resulting different easing functions.

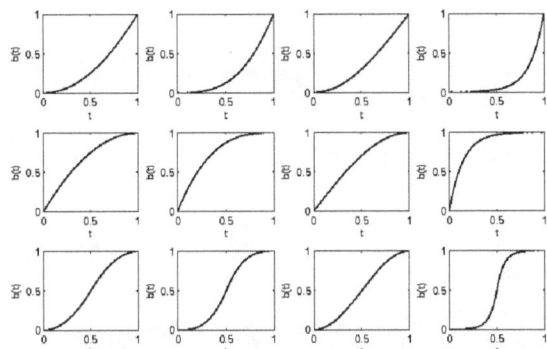

Fig. 4. Easing functions used for the determination of the blending parameter $b(t)$ and the position parameter $p(t)$. From left to right: quadratic, cubic, sinusoidal and exponential easing. From top to bottom: ease-in, ease-out, ease-in-out functions.

Motion Function. As for blending, an easing factor for the path followed by the feature points during the transition is introduced too. Such parameter $p(t)$, determines the equation of motion of the points as a combination of feature positions in image A and B. The same functions in Fig.4 are used for computing $p(t)$.

3   Results and Discussion

For the proposed approach, a quantitative evaluation is neither possible nor required, since the purpose of the project is to produce aesthetically appealing animations for subject-based slideshows. Thus, the evaluation was just qualitative. However, statistics of people opinions about the images were collected. In particular, we created a benchmark of videos and conducted a user study to compare our method to standard slideshows. Then, we present an analysis of users' responses, finding an agreement on the evaluation of the results.

3.1   Video Benchmarks

We created a benchmark of videos based on the images in the Gallagher Collection Person Dataset [11]. The videos were realized using 24 frames per second and 5 image transitions with a running time of 1 second each. For each test case two videos were generated, one using our slideshow approach, and one using standard cross-fading. In Fig.5 some frames of a video are shown. Top and bottom rows depict the videos generated with the proposed and the standard approach respectively. In particular frame 1, 7, 13, 19 and 24 are illustrated.

To better evaluate the results of the proposed work, sample videos of produced slideshows are available at http://www.dinfo.unipa.it/~cvip/pps/.

Fig. 5. Frame 1, 7, 13, 19 and 24 of an example of benchmark video for the proposed slideshow (top row), and the standard cross-fading slideshow (bottom row)

3.2   Users' Evaluation

In order to accomplish user evaluation, a user's reponse analysis system was developed, relying onto the web-based survey system provided by the *RetargetMe* framework [15]. Once the system is set up, the survey is taken by 113 people with different background in image analysis spanning from *no experience* to *advanced*. For each survey 10 test cases are submitted. For each test case two videos are presented and the person is requested to choose which one is the preferred. Alternatively only one video can be presented and the person is requested to give an absolute evaluation with a vote between 1 and 5.

In Fig. 6 the outcome of the survey is reported: averagely about the 80% of the people preferred the proposed version of the transition, while this percentage increase for people very familiar with image analysis and processing concepts. Fig.7 illustrates statistics of the votes given from an absolute point of view. The boxplot diagram shows that the mean vote for the proposed slideshow (≈ 4.33) is higher than the vote for the standard one (≈ 3.32). The motivation given when preferring the standard slideshow is generally related to the presence of

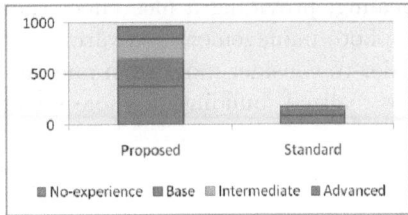

Background	People	Proposed	Standard
No-exp.	45	373 (82.89%)	77 (17.11%)
Base	33	269 (81.52%)	61 (18.48%)
Intermediate	23	203 (88.26%)	27 (11.74%)
Advanced	12	114 (95.00%)	6 (05.00%)
Total	113	959 (84.87%)	171 (15.13%)

Fig. 6. Statistics of users preferences about the proposed slideshow and the cross-fading slideshow. Stacked bars are used to express the different backgrounds of people who took the survey.

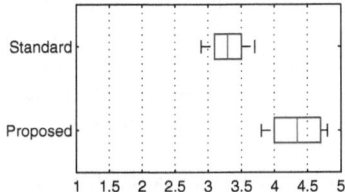

Fig. 7. Boxplot diagrams of absolute evaluation, expressing the votes assigned to the videos created with the proposed slideshow (bottom row), and the standard cross-fading effect (top row)

deformations in non-subject regions of the images, which are needed to preserve alignment of subject regions. However, this should not be considered a system fault, since this issue is inherent to the system design itself.

4   Conclusion and Future Works

A novel method for fully automatic subject-based slideshow generation was presented. The method is aimed to realize a cross-fade image transition which keep focused the person representing the main subject of the pictures. This is achieved by means of aligning and morphing the subject face (along facial features) while realizing the transition. This allows to keep the attention of the user onto the subject, attaining a pleasant and aesthetically attractive visual result. Given a set of images, the system recognizes and locates the subject, then performs a time-varying registration (i.e., a morphing) to mantain it aligned during the animation. Such alignment is realized using a deformation function based on triangles mesh with locally affine transformations or thin plate spline surfaces. In addition easing functions are used to give a more natural and pleasant aesthetic look to the transitions and object movement.

The system was evaluated using a web-based framework submitted to heterogeneous audience, which judged the videos according to the visual appealing, both through comparisons and absolute assessment. The survey response was that the proposed slideshows are more appealing, providing a nice effect that can be implemented as a feature in personal photo management software.

The project can be further extended in order to consider more than one person, in addition other warping effects can be realized, building a whole set of transition plugins.

References

1. Lo Presti, L., Morana, M., La Cascia, M.: A data association algorithm for people re-identification in photo sequences. In: 2010 IEEE International Symposium on Multimedia (ISM), pp. 318–323 (2010)
2. Min, F., Lu, T., Zhang, Y.: Automatic face replacement in photographs based on active shape models. In: Asia-Pacific Conference on Computational Intelligence and Industrial Applications, PACIIA 2009, vol. 1, pp. 170–173 (2009)
3. Terada, T., Fukui, T., Igarashi, T., Nakao, K., Kashimoto, A.: Yen-Wei Chen. Automatic facial image manipulation system and facial texture analysis. In: Fifth International Conference on Natural Computation, ICNC 2009, vol. 6, pp. 8–12 (2009)
4. Fantamorph. Abrosoft (2010), http://www.fantamorph.com
5. Yang, M.-H., Kriegman, D.J., Ahuja, N.: Detecting faces in images: a survey. IEEE Transactions on Pattern Analysis and Machine Intelligence 24(1), 34–58 (2002)
6. Viola, P., Jones, M.J.: Robust real-time face detection. Int. J. Comput. Vision 57(2), 137–154 (2004)
7. Yuille, A.L., Cohen, D.S., Hallinan, P.W.: Feature extraction from faces using deformable templates. In: IEEE Computer Society Conference on Computer Vision and Pattern Recognition, Proceedings CVPR 1989, pp. 104–109 (June 1989)
8. Cootes, T.F., Taylor, C.J., Cooper, D.H., Graham, J.: Active shape models—their training and application. Comput. Vis. Image Underst. 61(1), 38–59 (1995)
9. Kass, M., Witkin, A., Terzopoulos, D.: Snakes: Active contour models. International Journal of Computer Vision V1(4), 321–331 (1988)
10. Cootes, T.F., Edwards, G.J., Taylor, C.J.: Active appearance models. In: Burkhardt, H., Neumann, B. (eds.) ECCV 1998. LNCS, vol. 1407, p. 484. Springer, Heidelberg (1998)
11. Gallagher, A., Chen, T.: Clothing cosegmentation for recognizing people. In: Proc. CVPR (2008)
12. Gallea, R., Ardizzone, E., Gambino, O., Pirrone, R.: Multi-modal image registration using fuzzy kernel regression. In: ICIP 2009, pp. 193–196 (2009)
13. Bookstein, F.L.: Principal warps: Thin-plate splines and the decomposition of deformations. IEEE Transactions on Pattern Analysis and Machine Intelligence 11, 567–585 (1989)
14. Penner, R.: Programming Macromedia Flash MX. McGraw-Hill, New York (2002)
15. Rubinstein, M., Gutierrez, D., Sorkine, O., Shamir, A.: A comparative study of image retargeting. ACM Transactions on Graphics (SIGGRAPH) 29(5) (2010)

Learning Neighborhood Discriminative Manifolds for Video-Based Face Recognition

John See[1] and Mohammad Faizal Ahmad Fauzi[2]

[1] Faculty of Information Technology, Multimedia University,
Persiaran Multimedia, 63100 Cyberjaya, Selangor, Malaysia
[2] Faculty of Engineering, Multimedia University,
Persiaran Multimedia, 63100 Cyberjaya, Selangor, Malaysia
{johnsee,faizal1}@mmu.edu.my

Abstract. In this paper, we propose a new supervised Neighborhood Discriminative Manifold Projection (NDMP) method for feature extraction in video-based face recognition. The abundance of data in videos often result in highly nonlinear appearance manifolds. In order to extract good discriminative features, an optimal low-dimensional projection is learned from selected face exemplars by solving a constrained least-squares objective function based on both local neighborhood geometry and global manifold structure. The discriminative ability is enhanced through the use of intra-class and inter-class neighborhood information. Experimental results on standard video databases and comparisons with state-of-art methods demonstrate the capability of NDMP in achieving high recognition accuracy.

Keywords: Manifold learning, feature extraction, video-based face recognition.

1 Introduction

Recently, manifold learning has become an increasingly growing area of research in computer vision and pattern recognition. With the rapid development in imaging technology today, it plays an important role in many applications such as face recognition in video, human activity analysis and multimodal biometrics, where the abundance of data often demands better representation.

Typically, an image can be represented as a point in a high-dimensional image space. However, it is common presumption that the perceptually meaningful structure of the data lies on or near a low-dimensional manifold space [1]. The mapping between high- and low-dimensional spaces is accomplished through dimensionality reduction. This remains a challenging problem for face data in video, where large complex variations between face images can be better represented by extracting good features in the low-dimensional space.

In this paper, we propose a novel supervised manifold learning method called Neighborhood Discriminative Manifold Projection (NDMP) for feature extraction in video-based face recognition. NDMP builds a discriminative eigenspace

G. Maino and G.L. Foresti (Eds.): ICIAP 2011, Part I, LNCS 6978, pp. 247–256, 2011.

projection of the face manifold based on the intrinsic geometry of both intra-class and inter-class neighborhoods. For each training video, a set of face representative exemplars are automatically selected through clustering. With these exemplars, an optimal low-dimensional projection is learned by solving a constrained least-squares (quadratic) objective function using local neighborhood and global structural constraints. A compact generalized eigenvalue problem is formulated, where new face data can be linearly projected to the feature space. Finally, the test video sequences are classified using a probabilistic classifier.

2   Related Work

Classical linear dimensionality reduction methods such as Principal Component Analysis (PCA) [2], Multidimensional scaling (MDS) [3] and Linear Discriminant Analysis (LDA) [4] are the most popular techniques used in many applications. These linear methods are clearly effective in learning data in simple Euclidean structure. PCA learns a projection that maximizes its variance while MDS preserves pairwise distances between data points in the new projected space. With additional class information, LDA learns a linear projection that maximizes the ratio of the between-class scatter to the within-class scatter. The biggest drawback of these methods is that they fail to discover the intrinsic dimension of the image space due to its assumption of linear manifolds. Also, overfitting remains a common problem in these methods.

The emergence of nonlinear dimensionality reduction methods such as Locally Linear Embedding (LLE) [5] and Isomap [6] signalled the beginning of a new paradigm of manifold learning. These methods are able to discover the underlying high-dimensional nonlinear structure of the manifold in a lower dimensional space. LLE seeks to learn the global structure of a nonlinear manifold through a linear reconstruction that preserves its local neighborhood structure. Isomap is similar in nature, but geodesic distances between points are computed before reducing the dimensionality of space with MDS.

The main disadvantage of these methods is that they cannot deal with the *out-of-sample* problem – where new data points cannot be projected onto the embedded space. This limits their potential usage for classification and recognition tasks. Several works [7,8] reported good recognition rates in a video-based face recognition setting by using LLE to build a view-based low-dimensional embedding for exemplar selection, but stops short of utilizing it for feature representation.

In recent developments, various manifold learning methods such as Locality Preserving Projections (LPP) [9], Marginal Fisher Analysis (MFA) [10] and Neighborhood Preserving Embedding (NPE) [11] have been proposed to solve the out-of-sample problem. These methods resolve this limitation by deriving optimal linear approximations to the embedding using neighborhood information in the form of neighborhood adjacency graphs (with simple binary or heat kernel functions). While these methods can be performed in supervised mode with additional class information, their discriminative ability is only limited to neighborhood point relationships by graph embedding. It does not distinguished between neighborhood structures created by intra-class and inter-class groups of points within the same manifold, which can be very disparate in nature.

The major contribution of this paper centers upon the novel formulation of a neighborhood discriminative manifold learning method that exploits the intrinsic local geometry of both intra-class and inter-class neighborhoods while preserving its global manifold structure.

3   NDMP Algorithm

In this section, we give some motivations behind our work, followed by an overview of the Locally Linear Embedding (LLE) algorithm and a detailed description of the proposed NDMP method. Brief remarks on some closely-related techniques are also provided.

3.1   Motivations

The rich literature of manifold learning methods offers many potential avenues for improving existing shortcomings. With this, we want to highlight two main motivations behind our work.

1. While recent works [9,10,11] have addressed the out-of-sample problem with added discriminative properties, they do not discriminate between within-class and between-class neighborhood structures while keeping the global manifold fitted. Although similar to the NPE [11] in terms of its theoretical foundation (based on LLE [5]), our final objective function is different.
2. Most works reported impressive results on data sets with limited face variations such as ORL, Yale and UMIST datasets [12]. In video sequences, the large amount of face variations may prove to be a challenging task due to the difficulty in finding good meaningful features.

3.2   Locally Linear Embedding (LLE)

Assume $X = \{x_i \in \Re^D | i = 1, \ldots, N\}$ represent the input data in Euclidean space consisting of N face samples in D dimensions, belonging to one of C classes $\{c | c \in \{1, \ldots, C\}\}$. The LLE algorithm can be summarized into 3 main steps:

Step 1: For each point x_i, compute its k nearest neighbors.

Step 2: Compute the reconstruction weights W that best reconstruct each point x_i from its neighbors by minimizing the cost function

$$\epsilon_{rec}(W) = \sum_{i=1}^{N} \left\| x_i - \sum_{j=1}^{k} W_{ij} x_j \right\|^2 , \tag{1}$$

where W_{ij} is the reconstruction weight vector, subject to constraints $\sum_{i=1}^{N} W_{ij} = 1$ and $W_{ij} = 0$ if x_i and x_j are not neighbors.

Step 3: Compute the embedding coordinates y_i best reconstructed by the optimal reconstruction weight W by minimizing the cost function

$$\epsilon_{emb}(y) = \sum_{i=1}^{N} \left\| y_i - \sum_{j=1}^{k} W_{ij} y_j \right\|^2, \tag{2}$$

subject to constraints $\sum_{i=1}^{N} y_i = 1$ and $\sum_{i=1}^{N} y_i y_i^T / N = I$ where I is an identity matrix. The new coordinates in embedded space, $Y = \left\{ y_i \in \Re^d | i = 1, \ldots, N \right\}$ is a $d \times N$ matrix, where the dimensionality of the new embedded points, $d < D$.

3.3   Neighborhood Discriminative Manifold Projection (NDMP)

Considering the large amount of face variations in each training video sequence, we first apply LLE to reduce the dimensionality of the data. Then, faces in each training video sequence are grouped into clusters using hierarchical agglomerative clustering (HAC) [13]. For each cluster, the face that is nearest to the cluster mean is selected as an *exemplar*, or a representative image of the cluster. Similar to these approaches [7,8], the subject in each training video is represented by a set of M exemplars, which are automatically extracted from the video. Features will be extracted from the exemplar sets using the NDMP method. Finally, subjects in the test video sequences are identified using a probabilistic classifier.

In the proposed NDMP method, we first construct two sets of reconstruction weights – one each for the *intra-class* and *inter-class* neighborhoods, unlike the NPE which uses only one set of weights. The optimal projection is then formulated as a constrained minimization problem.

Construction of Intra-class and Inter-class Neighborhood Subsets. For clarity, the exemplars are also regarded as *data points* in most part of the paper. Let all data points in a local neighborhood Ψ comprise of two disjointed neighborhood subsets – *intra-class* subset $\{\psi_p | p = 1, \ldots, k\}$ and *inter-class* subset $\{\psi_q' | q = 1, \ldots, k'\}$. Compute the k nearest *intra-class* neighbors and k' nearest *inter-class* neighbors of each point x_i. A point x_j is an *intra-class* neighbor of point x_i if they belong to the same class ($c_i = c_j$). Similarly, point x_j is an *inter-class* neighbor of point x_i if they do not belong to the same class ($c_i \neq c_j$).

It is possible for a local neighborhood to have unequal number of intra-class and inter-class nearest neighbors. However, for the sake of uniformity of weight matrices, we fix $k = k'$. Due to class set limitation, the number of intra-class and inter-class neighbors is restricted to a maximum of $M - 1$, where M is the number of exemplars in a class.

Formulation of Neighborhood Reconstruction Weights. Based on Eq. (1), we formulate two reconstruction cost functions to obtain neighborhood reconstruction weights that best reconstruct each point x_i based on its type of neighbors. Both intra-class reconstruction weight matrix W^r and inter-class reconstruction weight matrix W^e can be computed by minimizing the respective cost functions:

$$\epsilon_{rec}(W^r) = \sum_{i=1}^{N} \left\| x_i - \sum_{j=1}^{k} W_{ij}^r x_j \right\|^2 , \tag{3}$$

$$\epsilon_{rec}(W^e) = \sum_{i=1}^{N} \left\| x_i - \sum_{j=1}^{k'} W_{ij}^e x_j \right\|^2 . \tag{4}$$

Both Eqs. (3) and (4) can be computed in closed form by determining the optimal weights through the local covariance matrix [5]. Alternatively, it can also be solved using a linear system of equations.

Formulation of Optimal Projection. In this step, an optimal projection is learned to enable new data points to be linearly embedded in the NDMP feature space. Typically, a linear subspace projection

$$Y = A^T X \tag{5}$$

maps the original data points X to the projected coordinates in embedded space Y by a linear transformation matrix $A = \{a_i \in \Re^D | i = 1, ..., d\}$.

Similar to NPE algorithm [11], the *intra-class* cost function can be formulated by expanding the least-squares term:

$$\begin{aligned}
\epsilon_{emb}^r(Y) &= \sum_{i=1}^{N} \left\| y_i - \sum_{j=1}^{k} W_{ij}^r y_j \right\|^2 \\
&= \left\| Y - \sum_{i=1}^{N}\sum_{j=1}^{k} W_{ij}^r y_j \right\|^2 \\
&= Tr\left(Y \left[\delta_{ij} - 2W^r + \|W^r\|^2 \right] Y^T \right) \\
&= Tr\left(Y(I - W^r)^T(I - W^r)Y^T \right) \\
&= Tr\left(Y M^r Y^T \right) ,
\end{aligned} \tag{6}$$

where $Tr\{.\}$ refers to the trace of the matrix and the orthogonal *intra-class* weight matrix

$$M^r = (I - W^r)^T(I - W^r) . \tag{7}$$

Likewise, the *inter-class* cost function and its orthogonal *inter-class* weight matrix is derived as

$$\epsilon_{emb}^e(Y) = Tr(Y M^e Y^T) , \tag{8}$$

$$\text{where } M^e = (I - W^e)^T(I - W^e) . \tag{9}$$

Substituting Eq. (5) to both cost functions yields:

$$\epsilon_{emb}^r(X) = Tr(A^T X M^r X^T A) \tag{10}$$
$$\epsilon_{emb}^e(X) = Tr(A^T X M^e X^T A) . \tag{11}$$

Motivated by Fisher's discrimination criterion [4], the objective function can be formulated to incorporate discriminative property that enables the compaction of intra-class neighborhood and the dispersion of inter-class neighborhood (see Fig. 1 for illustration). The intra-class cost function, ϵ_{emb}^r can be minimized so that the overall weighted pairwise distances between intra-class neighbors in embedded space are reduced.

Since the total sum of weights is subjected to $\sum_{i=1}^N W_{ij} = 1$ or $Tr(W) = I$, the inter-class cost function can be formulated as a *local constraint*,

$$A^T X M^e X^T A = I . \tag{12}$$

In order to maintain global rotational invariance within the embedding structure, another constraint is used for further optimization. From Step 3 of the LLE algorithm, rotational invariance is achieved by subjecting $\sum_{i=1}^N y_i y_i^T / N = I$ or $YY^T/N = I$, resulting in a *global constraint*,

$$A^T X X^T A = NI . \tag{13}$$

Modeling a constrained optimization problem involving two constraints results in the following Lagrangians:

$$L_\ell(\lambda_\ell, A) = A^T X M^r X^T A + \lambda_\ell(I - A^T X M^e X^T A) \tag{14}$$
$$L_g(\lambda_g, A) = A^T X M^r X^T A + \lambda_g(NI - A^T X X^T A) . \tag{15}$$

By forming two equations from their derivatives *w.r.t.* A and solving them simultaneously, it can be easily shown that $\lambda_\ell = \lambda_g = \lambda$ based on the constraints in Eqs. (12) and (13). Thus, the corresponding unified Lagrangian,

$$L'(\lambda, A) = A^T S A + \lambda((N+1)I - A^T C_\ell A - A^T C_g A) , \tag{16}$$

where $S = X M^r X^T$, $C_\ell = X M^e X^T$ and $C_g = X X^T$.

Setting its gradient to zero,

$$\frac{\partial L'}{\partial A} = 0 \Rightarrow 2SA - \lambda [2C_\ell A + 2C_g A] , \tag{17}$$

we can then rewrite it as a generalized eigenvalue problem,

$$SA = \lambda [C_\ell + C_g] A . \tag{18}$$

Note that matrices S, C_ℓ and C_g are all symmetric and semi-positive definite. The optimal embedding A is solved by taking d eigenvectors associated with the d smallest eigenvalues ($\lambda_1 < \ldots < \lambda_d$), where $d < D$.

In a generalized case, a constraint tuning parameter $\beta = \{\beta|\ 0 \leq \beta \leq 1\}$ can be introduced to allow both local and global constraints to be adjusted according to importance,

$$SA = \lambda [\beta C_\ell + (1 - \beta)C_g] A . \tag{19}$$

For equal contribution from both constraints, we can fix $\beta = 0.5$.

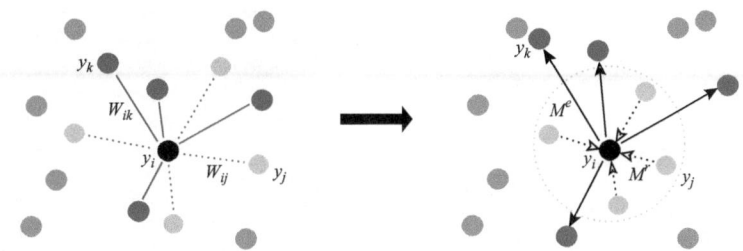

Fig. 1. A data point (black) is embedded in NDMP-space using weights from intra-class (blue) and inter-class (red) neighborhoods. Solving the constrained minimization problem leads to distinction between intra-class and inter-class structures.

4   Experimental Results

The proposed NDMP method was tested on two standard video face datasets: Honda/UCSD [14] and CMU MoBo [15] in order to ensure a comprehensive evaluation was conducted. The first dataset, Honda/UCSD, which was collected for video-based face recognition, consists of 59 video sequences of 20 different people (each person has at least 2 videos). Each video contains about 300-600 frames, comprising of large pose and expression variations and significant head rotations. The second dataset, CMU MoBo is a commonly used benchmark dataset for video-based face recognition consisting of 96 sequences of 24 different subjects (each person has 4 videos). Each video contains about 300 frames. For both datasets, faces were extracted using the Viola-Jones cascaded face detector [16], resized to grayscale images of 32×32 pixels, followed by histogram equalization to remove illumination effects. Some sample images are shown in Fig. 2.

For each subject, one video sequence is used for training, and the remaining video sequences for testing. To ensure extensive evaluation on the datasets, we construct our test set by randomly sampling 20 sub-sequences consisting of 100 frames from each test video sequence. The test sequences are evaluated using a probabilistic Bayes classifier where the class with the maximum posterior probability, $c^* = arg\ max\ P(c|x_1, \ldots, x_N)$ is the identified subject.

We use 7 exemplars for Honda/UCSD and 6 exemplars for MoBo[1], while intra-class and inter-class neighbors are fixed at $k = k' = M - 1$. The tuning parameter is set at $\beta = 0.75$ for all our experiments. It should be noted that the optimal number of feature dimensions for all methods were determined empirically through experiments.

---

[1] The number of exemplars selected from each video is heuristically determined from the residual error curve of clustering distance criterion [13].

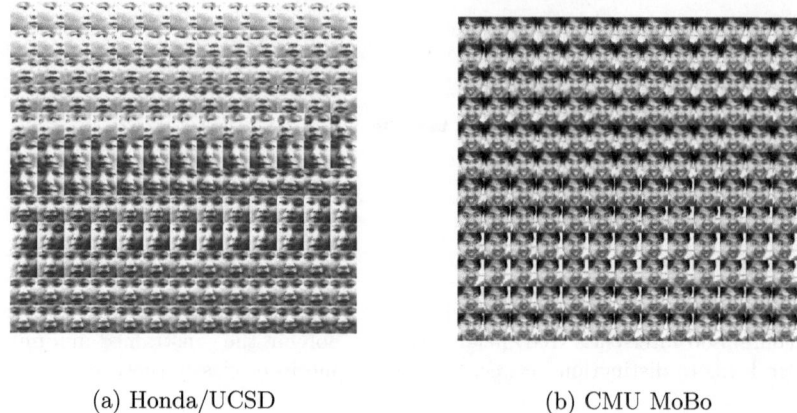

(a) Honda/UCSD   (b) CMU MoBo

Fig. 2. Sample images extracted from a video sequence of different datasets

4.1   Comparative Evaluation

The proposed NDMP method is compared against other classical (PCA, LDA) and recent state-of-art methods (LPP, MFA, NPE). The overall recognition performance on both Honda/UCSD and CMU MoBo datasets is summarized in Table 1, which shows that the NDMP method can outperform other methods in recognizing faces in video sequences. The strength and robustness of NDMP over the rest is more apparent in the Honda/UCSD data set, where video sequences possess a wide range of complex face poses and head rotations. Interestingly, LPP and MFA performed rather poorly in this dataset. The constraint tuning parameter, β in the generalized NDMP (see Eq. (19)) can be tuned to values between 0 and 1 to adjust the contribution of the local neighborhood constraint and global invariance constraint. From Fig. 3, we can observe that while both constraints seemed equally important, slightly better results can be expected by imposing more influence towards constraining the local neighborhood structure.

4.2   Rank-Based Identification Setting

To further evaluate the reliability of the NDMP method in a rank-based identification setting, we present its performance using a cumulative match curve (CMC). To accommodate this evaluation setting, we adopt a probabilistic voting strategy where the top n matches based on posterior probability scores are given a vote at each frame. The class with the majority vote is identified as the subject in the test sequence.

The CMCs of various methods evaluated on both datasets in Fig. 4 showed that NDMP consistently yielded better recognition rates throughout rank-n top matches for the Honda/UCSD. It also achieved a perfect recognition score (100%) for the CMU MoBo dataset with the top 3 matches. In contrast, global

Table 1. Recognition rates (%) of various manifold learning methods on the evaluated datasets

Datasets	Methods					
	PCA	LDA	LPP	MFA	NPE	NDMP
Honda/UCSD	60.7	68.9	56.8	57.4	71.7	86.9
CMU MoBo	86.6	92.6	89.3	91.4	96.3	97.7

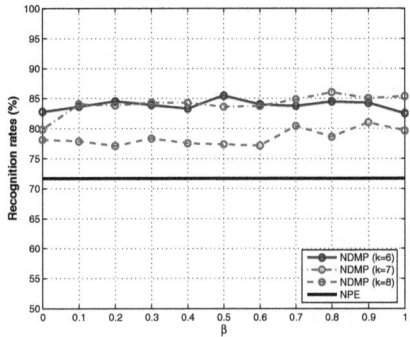

Fig. 3. Comparison of different values of β on the Honda/UCSD dataset

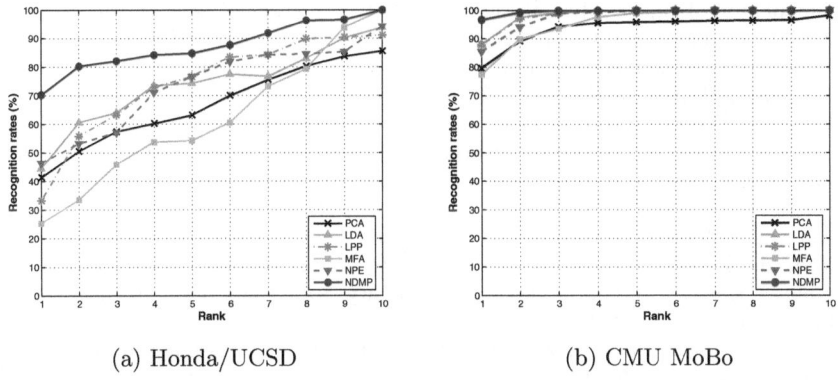

(a) Honda/UCSD   (b) CMU MoBo

Fig. 4. Comparison of cumulative match curves of various methods

methods such as PCA and LDA performed poorly due to their inability to learn the nonlinear manifold of appearance variations that is inherent in videos. It can be observed that the performance of other local neighborhood-preserving methods (MFA, LPP, NPE) tend to improve rapidly as the rank increases.

5   Conclusion

In this paper, we present a novel supervised manifold learning method called Neighborhood Discriminative Manifold Projection (NDMP) for video-based face recognition. The NDMP method constructs a discriminative eigenspace projection from nonlinear face manifolds based on both local neighborhood geometry and global manifold structure. Extensive experiments on standard video face datasets demonstrated the robustness and effectiveness of the proposed method compared to classical and recent state-of-art methods. In the future, the NDMP method can be further generalized to a nonlinear form in kernel feature space. Its potential usage in practical real-world applications is also worth exploring.

Acknowledgments. The authors wish to thank all who had provided invaluable suggestions throughout the development of this work. The authors also pay tribute to Sam Roweis, co-author of LLE, on his passing away in early 2010.

References

1. Brand, M.: Charting a manifold. In: Proc. of NIPS 15, pp. 961–968 (2003)
2. Turk, M., Pentland, A.: Eigenfaces for recognition. J. Cogn. Neurosc. 3(1), 71–86 (1991)
3. Cox, T.F., Cox, M.A.A.: Multidimensional Scaling. Chapman and Hall, Boca Raton (2001)
4. Belhumeur, P.N., Hespanha, J.P., Kriegman, D.: Eigenfaces vs Fisherfaces: Recognition using class specific linear projection. IEEE Trans. PAMI 19, 711–720 (1997)
5. Roweis, S.T., Saul, L.: Nonlinear dimensionality reduction by locally linear embedding. Science 290, 2323–2326 (2000)
6. Tenenbaum, J.B., de Silva, V., Langford, J.C.: A global geometric framework for nonlinear dimensionality reduction. Science 290, 2319–2323 (2000)
7. Hadid, A., Peitikäinen, M.: From still iamge to video-based face recognition: An experimental analysis. In: IEEE FG, pp. 813–818 (2004)
8. Fan, W., Wang, Y., Tan, T.: Video-based face recognition using bayesian inference model. In: Kanade, T., Jain, A., Ratha, N.K. (eds.) AVBPA 2005. LNCS, vol. 3546, pp. 122–130. Springer, Heidelberg (2005)
9. He, X.F., Niyogi, P.: Locality preserving projections. In: Proc. of NIPS 16, pp. 153–160 (2003)
10. Yan, S., Xu, D., Zhang, B., Zhang, H.J., Yang, Q., Lin, S.: Graph embedding: A general framework for dimensionality reduction. IEEE Transactions on PAMI 29(1), 40–51 (2007)
11. He, X., Cai, D., Yan, S., Zhang, H.J.: Neighborhood preserving embedding. In: IEEE ICCV, pp. 1208–1213 (2005)
12. Gross, R.: Face databases. In: Li, S.Z., Jain, A.K. (eds.) Handbook of Face Recognition, Springer, Heidelberg (2005)
13. Duda, R., Hart, P., Stork, D.: Pattern Classification. Wiley, Chichester (2000)
14. Lee, K.C., Ho, J., Yang, M.H., Kriegman, D.: Visual tracking and recognition using probabilistic appearance manifolds. CVIU 99(3), 303–331 (2005)
15. Gross, R., Shi, J.: The CMU Motion of Body (MoBo) Database. Technical Report CMU-RI-TR-01-18, Robotics Institute, CMU (2001)
16. Viola, P.: Jones. M.: Rapid object detection using a boosted cascade of simple features. In: CVPR, pp. 511–518 (2001)

A Novel Probabilistic Linear Subspace Approach for Face Applications

Ying Ying and Han Wang

School of Electrical and Electronic Engineering,
Nanyang Technological University,
50 Nanyang Avenue, Singapore, 639798
{yi0002ng,hw}@ntu.edu.sg

Abstract. Over the past several decades, pattern classification based on subspace methodology is one of the most attractive research topics in the field of computer vision. In this paper, a novel probabilistic linear subspace approach is proposed, which utilizes hybrid way to capture multi-dimensional data extracting maximum discriminative information and circumventing small eigenvalues by minimizing statistical dependence between components. During features extraction process, local region is emphasized for crucial patterns representation, and also statistic technique is used to regularize these unreliable information for both reducing computational cost and maintaining accuracy purposes. Our approach is validated with a high degree of accuracy with various face applications using challenging databases containing different variations.

Keywords: probabilistic analysis, linear subspace, face application.

1 Introduction

The ultimate goal of pattern recognition is to discriminate the class of observed objects with the minimum misclassification rate. Thus, in the discriminating process, a pattern recognition system intrinsically utilizes low dimensionality to represent the input data. Subspace analysis is a powerful tool of seeking low-dimensional manifolds which models continuous variations in patterns, and new image can be embedded into these manifolds for classification. Among the numerous techniques published in the past few years, ways directly solves the classification and clustering problems, such as sparse representation [1] and subspace arrangements [2]. Another category emphasizes on feature extraction, like Laplacian Eigenmaps (LE) [3], locality preserving projections (LPP) [4], and marginal Fisher analysis (MFA) [5]. The superiority of subspace method can be concluded as aiming at reducing the computational complexity of the classification with minimum loss of discriminative information [6]. This can be done by maximizing the information carried by the data in the extracted low-dimensional subspace, and as evidenced by the fact that vast majority of the proposed approaches are based on "most discriminative" criteria which extract maximum discriminative information in the form of reduced low-dimensional space from

G. Maino and G.L. Foresti (Eds.): ICIAP 2011, Part I, LNCS 6978, pp. 257–266, 2011.

large scale, such as LE, LPP, and MFA. While the other objectives are to circumvent the over-fitting problem of the classification and enhance the accuracy and robustness. Curse of dimensionality, small sample size, or noise removal effect are some problematic and harmful situations for robust classification. One way is to regularize these unreliable statistic or remove corresponding dimensions. Although various regularization techniques are proposed, they are supposed to be applied before dimensionality reduction because regularization in classification stage cannot recover the improperly removed dimensions in dimensionality stage. Based on the understanding of the roles of subspace method, we can find most top performers of the state-of-art subspace-based approaches adopt either various regularized discriminative analysis or two-stage approaches to realize superiorities and also boost the classification accuracy.

According the principles behind the subspace analysis, here, we propose a novel probabilistic approach for pattern classification. Basically, it adopts multiple linear subspace methods to capture multi-dimensional data which extracts maximum discriminative information and circumventing small eigenvalues by minimizing the statistical dependence between components. During the features extraction process, local region is emphasized for crucial patterns representation, and also statistic technique is used to regularize these unreliable information for both reducing computational cost and maintaining accuracy purposes. Our discussion in this paper is limited on face applications which involves two- and multi-class classification problems, and proposed approach can be also developed for other general pattern recognition tasks.

The remainder of this paper is structured as follows: Section 2 analyzes the proposed probabilistic linear subspace approach. Performances of different face applications and conclusion will be presented in section 3 and 4, respectively.

2   Probabilistic Linear Subspace Approach

As one of the most popular subspace methods, PCA yields the projection directions that maximize the data structure information in the principal space across all classes, and hence is optimal for data reconstruction. Although PCA-based classification approaches reduce computational complexity by capturing low-dimensional data, the dimensionality utilized is chosen according to the different targets which means the performance utilizing extracted feature cannot be guaranteed in case of the data with large deviation from training samples. Unfortunately, unwanted variations like lighting, facial expression, or other reality factors always exist. Some work suggests to discard several most significant principal components and better clustering of projected samples can be achieved. Yet, it is unlikely that these principal components correspond solely to variations, as a consequence, information that is useful for discrimination may also lose. On the other hand, PCA-feature only analyzes the first- and second-order statistics information capturing amplitude spectrum of images, since PCA seeks directions that are most independent from each other. High-order statistics containing phase spectrum hold important information which can be transformed be variation immunity. Thus, these information should be well modeled for boosting the classification accuracy and robustness.

Classification is to assign a given pattern to one of the categories. The minimum probability of misclassification is achieved by assigning pattern to class that has maximum probability after pattern has been observed, thus the Bayes theory can be utilized to evaluate the discriminant function,

$$f_i(x) = \ln P(x|C_i) + a_i \tag{1}$$

where x and a_i represent the input and the threshold for class i, respectively. For further analysis, data usually is modeled by an analytical form, such like multivariate Gaussian distribution. As the most natural distribution, it is an appropriate second-order statistics model for many situations [6]. After eigen-decomposition, Eq. 1 is simplified as,

$$f_i(x) = -\frac{1}{2}(x - \bar{X}_i)^T \Sigma_i^{-1}(x - \bar{X}_i) + b_i$$
$$= -\frac{1}{2}\sum_j \frac{(w_j - \bar{W}_j)^2}{\lambda_j} + b_i \tag{2}$$

where $\bar{X}_i$ and Σ_i are the mean and covariance distribution, and w_j and $\bar{W}_j$ are the projections of x and $\bar{X}_i$ on the orthonormal eigenvectors corresponding to the eigenvalue λ_j of Σ_i.

Moghaddam [7] and Jiang [8] suggested to separate the discriminant function into two parts and replace small eigenvalues by a constant,

$$f_i(x) = -\frac{1}{2}[\sum_{j=1}^{m} \frac{(w_j - \bar{W}_j)^2}{\lambda_j} + \sum_{j=m+1}^{n} \frac{(w_j - \bar{W}_j)^2}{\rho}] + b_i \tag{3}$$

where ρ is computed as a fixed percentage of corresponding eigenvalues or an upper bound of small eigenvalues. All these adding a constant to all eigenvalues or replacing the unreliable eigenvalues by a constant are reported with better classification performances. These PCA variants still adopt single model to extract discriminate feature and decreased large deviation effect caused by small eigenvalues. However these "unreliable" small eigenvalues actually do contain high-order statistic information which contributes for classification and cannot be simply replaced by a constant instead of completely modeling. One more issue is all these modeling is based on Gaussian distribution assumption, which can not completely describe the reality cases. Inspired from it, we propose our subspace-based two-stage model.

2.1 Proposed Approach

Considering the outstanding performance for both dimensionality reduction and discriminative information representation, PCA feature is competent for modeling the first- and second-order statistics. The rest small eigenvalues, as the second part decomposed in Eq. 3, can be described by independent component analysis (ICA) [9]. Although ICA is a generalization of linear subspace approach, unlike PCA, it searches for a transformation that minimizes the statistical dependence between components, and thus provides a good representation of data by virtue of exploiting the entire data space. However, ICA loses merit when dealing Gaussian distributed data [10] while showing the superiority in encoding non-Gaussian distribution. As the reasons stated for building up multivariate Gaussian distribution, the existence of non-Gaussian distribution cannot be ignored, such as geometrical variation, which here can be compensated by ICA.

Our strategy is to use PCA to isolate low-order statistical information modeling Gaussian distribution, simultaneously, ICA is applied to represent high-order data which is more superior to model non-Gaussian distribution. Thus, each reconstructed image $\hat{x}$ can be written as,

$$
\begin{aligned}
\hat{x} &= (\Phi_m \Phi_n) * (\Phi_m \Phi_n)^T * x \\
&= \Phi_m \Phi_m^T * x + \Phi_n \Phi_n^T * x \\
&= \Phi_m W_m + U_n (W_{invt}^T)^{-1} \Phi_n^T * x \\
&= \Phi_m W_m + U_n (\Phi_n W_{invt}^{-1})^T * x \\
&= \Phi_m W_m + U_n B_n
\end{aligned}
\tag{4}
$$

where Φ_m and Φ_n are the first m and residual n principle components in the eigenvector matrix Φ, respectively; W_m represents the projection in the PCA space; U_n and B_n denote the independent basis image and coefficient in the residual ICA space, respectively.

Generally, subspace methods encode the gray scale correlation among every pixel position statistically, and any variation can cause severe changes of information representation. However, pattern recognition based on local regions are reported having a demonstrably superior performance to those which exploit global information [10,11,12]. Firstly, as local region exhibiting less statistical complexity, method using local space will be more robust to illumination change. Also, local pattern might vary less under pose changes than global one. Lastly, local feature is more robust against partial occlusions since local region recognition is little affected. So Eq. 4 can be further rewritten according to the specific subregions defined containing important and meaningful local features,

$$
\begin{aligned}
\hat{x}_s &= (\Phi_s)_m (\Phi_s)_m^T * x_s + (\Phi_s)_n (\Phi_s)_n^T * x_s \\
&= (\Phi_s)_m (W_s)_m + (U_s)_n (B_s)_n
\end{aligned}
\tag{5}
$$

Having derived the principal and independent components, each local component is projected onto PCA and ICA spaces, respectively. Thus the statistic analysis is applied to regularize PCA projection weight W_m and ICA coefficient B_n. Because data usually has divergent and complicated distributions in reality case, instead of a single multivariate Gaussian model, a weighted mixture of multivariate Gaussian (GMM) distribution is adopted here to increase modeling precision and reduce misclassification rate. The likelihood probability of the jth local region x_s^j can be written as,

$$
\begin{aligned}
P(x_s^j|C_i) &= \frac{exp[-\frac{1}{2}(x_s^j - \bar{X}_s^j)^T(\Sigma_s^j)^{-1}(x_s^j - \bar{X}_s^j)]}{(2\pi)^{L/2}|\Sigma_s^j|^{1/2}} \\
&= \frac{exp[-\frac{1}{2}d(x_s^j)]}{(2\pi)^{L/2}|\Sigma_s^j|^{1/2}} \\
&= \frac{exp\{-\frac{1}{2}d[(x_s)_{PCA}^j] - \frac{1}{2}d[(x_s)_{ICA}^j]\}}{(2\pi)^{(m+n)/2}|(\Sigma_s)_m^j(\Sigma_s)_n^j|^{1/2}} \\
&= P((x_s)_{PCA}^j|C_i) * P((x_s)_{ICA}^j|C_i) \\
&= \sum_{k=1}^{a^j} P[(W_s)_k^j|\Omega_k^j] * \sum_{l=1}^{b^j} P[(B_s)_l^j|\Gamma_l^j]
\end{aligned} \tag{6}
$$

where a^j and b^j are the GMM cluster numbers for PCA and ICA spaces of class i, where $\sum a^j = 1$ and $\sum b^j = 1$; $d(x^j)$ is the corresponding Mahalanobis distance; $\Omega_k^j = (\pi_k^j, \mu_k^j, \Phi_k^j)$ and $\Gamma_l^j = (\varphi_l^j, \nu_l^j, \Psi_l^j)$ are the kth and lth Gaussian parameter sets containing likelihood probability weights π_k^j and φ_l^j, mean vectors μ_k^j and ν_l^j, covariance matrixes Φ_k^j and Ψ_l^j in PCA and ICA spaces, respectively. For estimating parameter sets Ω_k^j and Γ_l^j, the initial values of mean μ_k^j and ν_l^j, and covariance Φ_k^j and Ψ_l^j of cluster j in class i can be calculated by being partitioned by modified k-means clustering, which uses the likelihood probability of multivariate Gaussian model as the measure of the nearest neighbor rule. The corresponding initial weights a^j and b^j are defined as the ratio of subspace data number in cluster j to the total training number. Then the expectation maximization (EM) algorithm iteratively optimizes Ω_k^j and Γ_l^j to local maximum in the total likelihood of the training set. During the integration procedure [11], likelihood probability of the entire image x can be expressed as,

$$
P(x|C_i) = \prod_{j=1}^{Q} P(x_s^j|C_i) \tag{7}
$$

where Q is the subregion number. Therefore, the discriminant function Eq 1 becomes,

$$
f_i(x) = \ln[\prod_{j=1}^{Q} P(x_s^j|C_i)] + a_i = \sum_{j=1}^{Q} \ln[P(x_s^j|C_i)] + a_i \tag{8}
$$

and x is classified to the class with maximum probability.

The outline of the proposed probabilistic subspace approach for both training and testing procedures is shown in Fig. 1.

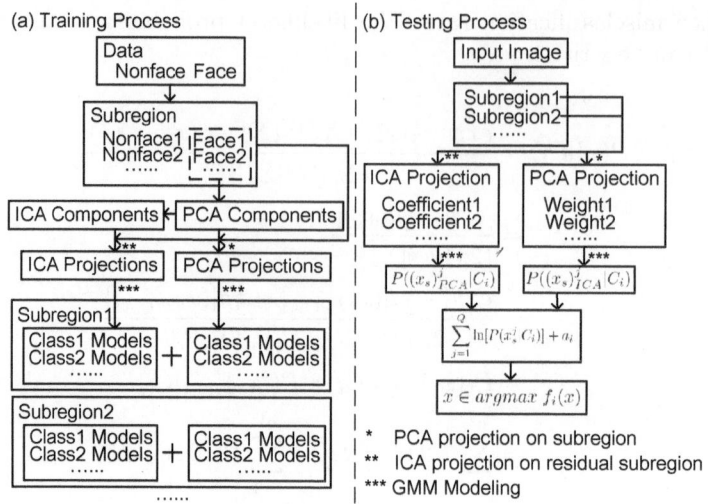

Fig. 1. Flowcharts of proposed approach: (a) training, and (b)testing process

3   Experiment

3.1   Multi-view Face Detection

Similar to the work [13], the experiment platform adopted here comprises preprocessing, detection, and postprocessing modules. After geometric normalization
and lighting correction utilized in the preprocessing, three view-based face detectors are trained by the proposed approach using canonical facial data, which continuously cover the whole out-of-plane rotation. Postprocessing including grouping, averaging, and filtering is used to solve multi-resolution issue, and gives
unambiguous location and scale of face without information loss.

Experiment data includes FacePix [14,15], PIE, Pointing04 [16], and CMU
profile [11] datasets as face images; Caltech background dataset, Caltech and
Pasadena Entrances 2000 dataset, Caltech Houses dataset and Fifteen Scene Categories dataset as nonface images. FacePix provides multi-view faces with continuous pose changing, PIE and Pointing04 have discrete pose changing faces, and
CMU datasets contain faces with unestimated poses under complex background.
Canonical images, subregions, PCA reconstructed subregions, and residual subregions defined for different face views are illustrated in Fig. 2. It can be seen
that reconstructed subregions are similar to their low-pass filtered equivalents
and the residual ones which are characterized by high-frequency components
are less sensitive to illumination variations, which is equivalent to the low- and
high-order statistics separation.

The proposed subspace model, designated as "PCA+ICA", is compared with
other two approaches, namely "PCA" and "ICA", in which PCA and ICA are
employed solely to model data respectively. The current "PCA+ICA" model

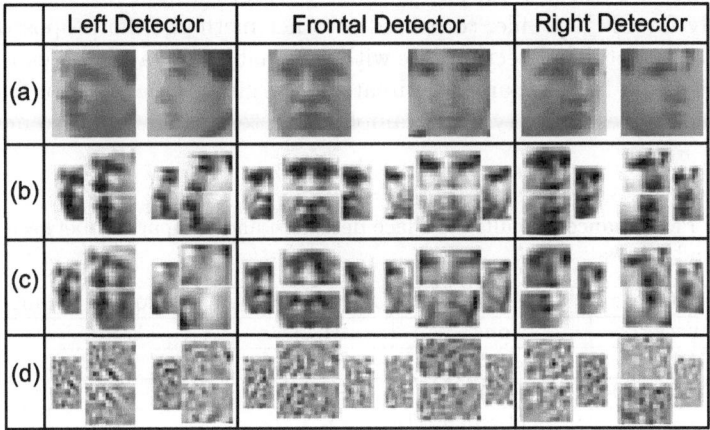

Fig. 2. (a) Original images; (b) subregions; (c) PCA reconstructed subregions; and (d) residual subregions

yields the best performance which demonstrates information distributed in the low- and high-frequency components are appropriately preserved by PCA and ICA, respectively, see Fig. 3a. Another experiment is conducted between current local subspace face detector, designated as "Local+PCA+ICA", and the "Global+PCA+ICA" detector in which the input is the entire patch. Fig. 3b plots the corresponding ROC curves, in which the "Local+PCA+ICA" yields better result than the global one.

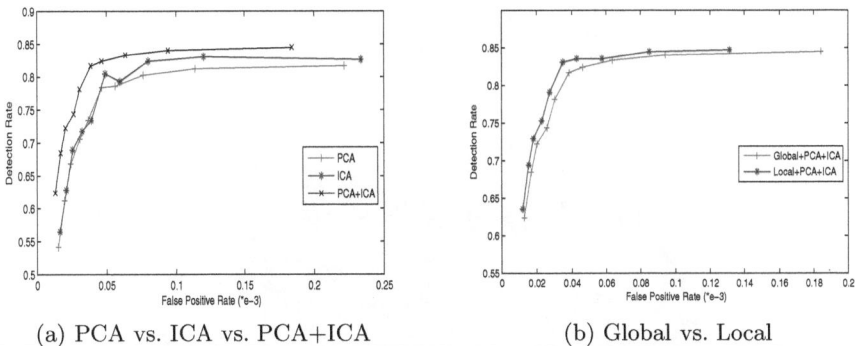

(a) PCA vs. ICA vs. PCA+ICA   (b) Global vs. Local

Fig. 3. ROC for multi-view face detection based on different approaches

Besides the evaluations conducted for showing the superiority of the proposed approach, quantitative experiment also is done and comparison with other outstanding methods are summarized in Table 1. Results using FacePix, PIE,

Pointing'04, and CMU profile data averagely achieve around 90% detection rates. Especially for CMU profile, compared to other methods, the proposed method obtains an acceptable detection rate with reasonable false alarm. Meanwhile, the pose of detected face is coarsely estimated according to the detector from which the patch survives. Some typical examples detected by the proposed method are shown in Fig. 4.

Table 1. Performance of multi-view face detection and comparison between proposed method and others

Database	Method	Detection Rate	False Positive Number
FacePix	our	94.13%	281
PIE	our	88.76%	1348
Pointing'04	our	91.52%	537
CMU profile	Schneiderman [11]	92.7%	700
	Jones [17]	83.1%	700
	Wu [18]	91.3%	415
	our	86.17%	715

(a) FacePix dataset

(b) PIE dataset

Tile: -30   Tile: -15   Tile: 0   Tile: +15   Tile: +30

(c) Pointing'04 dataset

(d) CMU profile dataset

Fig. 4. Results for multi-view face detection application tested on different databases; for CMU profile dataset, red, green, and blue rectangles present results detected by left, frontal, and right detectors, respectively

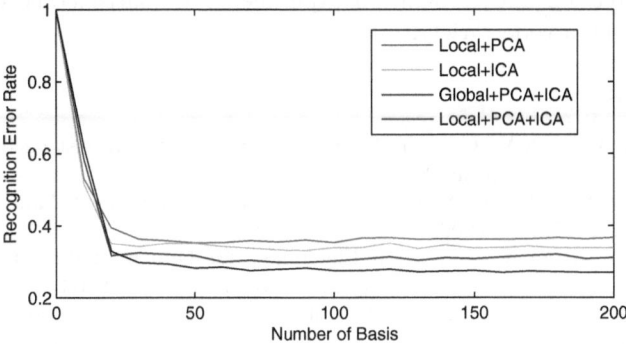

Fig. 5. Top recognition error rate for the number of selected basis components

3.2 Face Recognition

The proposed method is applied for face recognition and satisfied performance is achieved. The evaluation performance is conducted based on AR database [19], including frontal view faces with different facial expressions, illumination conditions, and partial occlusions. For each subject, 20 images are randomly chosen and treated as gallery, and the left 6 images are for testing. "Local+PCA+ICA", "Global+PCA+ICA", "Local+PCA", and "Local+ICA" curves are shown in Fig. 5, which clearly prove that our model has advantage over other ones.

The two- and multi-class pattern classification applications demonstrated above consistently show the classification ability of our probabilistic subspace approach. Based on the PCA work, the ICA model representing the residual high-order statistic information reveals the ability of describing non-Gaussian distribution. Especially with the local component emphasizing, the robustness and accuracy are both boosted since partial occlusion, illumination, geometry variations are alleviated to different extent.

4   Conclusion

This paper proposes a probabilistic two-stage linear subspace method for pattern classification. Besides the principal dimension statistics extracted by PCA, high-order information is analyzed by ICA model by projecting the corresponding coefficient to its feature space. Not only this, non-Gaussian distribution modeled by ICA feature shows the same importance as Gaussian assumption for pattern recognition. On the other hand, feature within subregions shows strong robustness against variations, e.g. lighting, geometry, and occlusion, since features are selectively emphasized. The weighted GMM probabilistic model also makes data representation more completely. ICA model, local feature, and probabilistic analysis all contribute to classification accuracy boosting, while being exempted from ICA-reconstruction helps to keep from the reconstruction cost. It can be concluded from experiments that the performances of our method in both face detection and recognition experiments are satisfactory with images characterized by a wide variety.

References

1. Wright, J., Yang, A.Y., Ganesh, A., Sastry, S.S., Ma, Y.: Robust Face Recognition Via Sparse Representation. TPAMI 31, 210–227 (2009)
2. Vidal, R., Ma, Y., Sastry, S.: Generalized Principal Component Analysis (GPCA). TPAMI 27, 1945–1959 (2005)
3. Belkin, M., Niyogi, P.: Laplacian Eigenmaps for Dimensionality Reduction and Data Representation. Neural Computation 15, 1373–1396 (2003)
4. He, X., Yan, S., Hu, Y., Niyogi, P., Zhang, H.J.: Face Recognition Using Faces. TPAMI 27, 328–340 (2005)
5. Yan, S., Xu, D., Zhang, B., Yang, Q., Zhang, H., Liu, S.: Graph Embedding and Extensions: A General Framework for Dimensionality Reduction. TPAMI 29, 40–51 (2007)
6. Jiang, X.D.: Linear Subspace Learning-Based Dimensionality Reduction. Signal Processing Magazine 28, 16 (2011)
7. Moghaddam, B.: Principal Manifolds and Probabilistic Subspace for Visual Recognition. TPAMI 24, 780–788 (2002)
8. Jiang, X.D., Mandal, B., Kot, A.C.: Enhanced Maximum Likelihood Face Recognition. Electronic Letter 42, 1089–1090 (2006)
9. Bartlett, M.S., Movellan, J.R., Sejnowski, T.J.: Face Recognition by Independent Component Analysis. Neural Networks 13, 1450–1464 (2002)
10. Kim, T.K., Kim, H., Hwang, W., Kittler, J.: Independent Component Analysis in a Local Facial Residue Space for Face Recognition. Pattern Recognition 37, 1873–1885 (2004)
11. Schneiderman, H., Kanade, T.: A Statistical Method for 3D Object Detection Applied to Faces and Cars. In: IEEE International Conference on Computer Vision and Pattern Recognition, pp. 746–751 (2000)
12. Heisele, B., Serre, T., Poggio, T.: A Component-Based Framework for Face Detection and Identification. IJCV 74, 167–181 (2007)
13. Ying, Y., Wang, H., Xu, J.: An Automatic System for Multi-View Face Detection and Pose Estimation. In: IEEE Internation Conference on Control, Automation, Robotics and Vision, pp. 1101–1108 (2010)
14. Black, J., Gargesha, M., Kahol, K., Kuchi, P., Panchanathan, S.: A Framework for Performance Evaluation of Face Recognition Algorithms. In: Internet Multimedia Systems II ITCOM (2002)
15. Little, G., Krishna, S., Black, J., Panchanathan, S.: A Methodology for Evaluating Robustness of Face Recognition Algorithms with Respect to Changes in Pose and Illumination Angle. In: ICASSP (2005)
16. Gourier, N., Hall, D., Crowley, J.L.: Estimating Face Orientation from Robust Detection of Salient Facial Features. In: International Workshop on Visual Observation of Deictic Gestures (2004)
17. Jones, M., Viola, P.: Fast Mulit-View Face Detection. Technical Report TR2003-96, Mitsubishi Electric Research Labs (2004)
18. Wu, B., Ai, H.Z., Huang, C., Lao, S.H.: Fast Rotation Invariant Multi-View Face Detection Based on Real Adaboost. In: IEEE 6th International Conference on Automatic Face and Gesture Recognition, pp. 79–84 (2004)
19. Martinez, A.M., Benavente, R.: The AR face database. CVC Tech. Report #24 (1998)

Refractive Index Estimation of Naturally Occurring Surfaces Using Photometric Stereo

Gule Saman and Edwin R. Hancock*

Department of Computer Science, University of York, YO10 5GH, UK

Abstract. This paper describes a novel approach to the computation of refractive index from polarisation information. Specifically, we use the refractive index measurements to gauge the quality of fruits and vegetables. We commence by using the method of moments to estimate the components of the polarisation image computed from intensity images acquired by employing multiple polariser angles. The method uses photometric stereo to estimate surface normals and then uses the estimates of surface normal, zenith angle and polarisation measurements to estimate the refractive index. The method is applied to surface inspection problems. Experiments on fruits and vegetables at different stages of decay illustrate the utility of the method in assessing surface quality.

Keywords: Refractive index estimation, Photometric stereo, Polarisation Information, Fresnel Theory.

1 Introduction

The physics of light has been widely exploited for surface inspection problems. Although the majority of methods make use of images in the visble, infra-red or ultraviolet ranges, there is a wealth of additional information that can be exploited including the pattern of light scattering, multispectral signatures and polarisation. In fact, the optical properties of surfaces prove to be particularly useful for assessing the quality of changes in naturally occurring surfaces. Here the scattering properties of visible light have been exploited for acquiring morphological information about the tissues. The two factors that effect the scattering of light from biological surfaces are, tissue morphology and biochemistry. Morphology affects the distribution of the scattered light, while the refractive index is determined by the biochemical composition of the tissue [1]. The refractive index is the ratio of the speed of light in vacuum to that in the material, and it hence determines the light transmission properties of a material. To understand the detailed propagation in biological tissue consisting of cells, the Mie theory has been used for approximating the scattering of light assuming that the cells are unifomrly sized homogeneous spheres [2]. A more complex approach is provied by the finite-difference time-domain (FDTD) model where Maxwell's equations are used for modelling the scattering of light from biological cells [3].

Refractive index therefore plays an important role in characterising the properties of biological tissue. One way of measuring refractive index is to turn to the Fresnel

---

* Edwin Hancock is supported by the EU FET project SIMBAD and by a Royal Society Wolfson Research Merit Award.

G. Maino and G.L. Foresti (Eds.): ICIAP 2011, Part I, LNCS 6978, pp. 267–275, 2011.

theorty and to use polarisation measurements. Polarisation information has been used for developing algorithms for a diverse set of problems in computer vision ranging from surface inspection to surface reconstruction. The Fresnel theory is used to determine the parallel and perpendicular components of the electric field for incident light in order to, model the transmission and reflection of these components [4]. The Fresnel theory of light has been used by Wolff [5] for developing a polarisation based method for identifying metal surfaces and dielectrics. It can be applied at a smooth boundary between two media as a quantitative measure of reflection and refraction of incident light [4,6]. Generally speaking, modeling dielectrics is straightforward as compared to modeling metals, since in the latter case the incident electromagnetic field induces surface currents for which the Fresnel theory alone is insufficient. When dealing with dielectrics, it is convenient to distinguish between the specular and diffuse polarisation. In specular polarisation, polarised incident light is reflected from the reflecting surface where the orientation of the surface determines the plane of polarisation. In diffuse polarisation, unpolarised incident light is subjected to subsurface scattering before being re-emitted hence spontaneously acquiring polarisation[5].

A polaroid filter can be used as an analyser for measuring the degree of polarisation and phase of both diffuse and specular polarisation. The refractive index and the degree of polarisation can be physically determined by the zenith angle between the remitted light and the surface where the phase angle is determined by the azimuth angle of the remitted light to the surface. The surface orientation of a reflecting surface can be determined for a constant refractive index and vice versa by using the Fresnel theory of light, which is dictated by the polarisation nature of the incident light and the geometry of the scattering process. Polarisation information proves to be very useful for determining the surface quality by using refractive index estimation or surface shape for surfaces of constant refractive index.

The aim of this paper is mainly the computation of the refractive index of a material by using the Fresnel theory. Firstly, the method of moments is used which was proposed in [7] for moments estimates from multiple polarisation images. Secondly, photometric stereo is used for images taken with three different light source directions in order to determine the surface normals and consequently the angle of incidence for the incident light. Finally, the Fresnel theory for diffuse reflectance is used for estimating the refractive index. We use samples of fruits and vegetables at different stages of decay and surface texture to explore how effectively can the method be used for revealing local variations in refractive index.

2   Polarisation Image

For computing the components of the polarisation image, we follow the method that was proposed in [7] for diffuse reflectance using robust moment estimators, where light undergoes subsurface reflections before being re-emitted. The Fresnel theory gives the relationship between the degree of polarisation and the angle of reflection of the reflected light.

2.1   Data Collection

We collect a succession of images of the subject with different orientations of the analyser polaroid for measuring the polarisation state using the geodesic light dome[8]. The object is placed in the center of the geodesic light dome for image acquisition. We used a Nikon D200 digital SLR camera, with fixed exposure and aperture settings for obtaining the data set. An unpolarised light source has been used for the experiments where the images have been captured with the analyser angle being changed by increments of $10°$ to give 19 images per object.

Since the light source locations have been carefully calibrated, the use of the geodesic light dome gives accurate measures for the angle of incidence of the incident light source as shown in Figure (1) for a wrinkled apple, Figure (2) for an orange and Figure (3) for a tomato. The experiments are conducted in a dark room with matte-black walls hence, leading to minimal refelctions from the surroundings.

2.2   Robust Moments Estimators for Polarisation Image

The conventional way of estimating the components of the polarisation image, i.e. mean intensity, I_0, degree of polarisation, ρ, and phase, ϕ, has been to use the least-squares fitting of 3 images [9]. As has been mentioned earlier, we use the robust moments estimators for computing the components of the polarisation image from a larger set of data [7]. The method is explained as follows:

If the angle of the analyser is taken as β_i, where the index of the analyser angle is i. At the pixel indexed p with the analyser angle indexed i, the predicted brightness is

$$I_p^i = \hat{I}_p\left\{1 + \rho_p\cos(2\beta_i - 2\phi_p)\right\}. \tag{1}$$

where $\hat{I}_p$, ρ_p and ϕ_p are the mean intensity, polarisation and phase at the pixel indexed p.

We take N equally spaced polarisation images, where the polariser angle index is $i = 1, 2, ..., N$. To compute the polarisation parameters, we commence by normalising the pixel brightness values. Therefore,

$$x_p^i = (I_p^I - \hat{x}_p)/\hat{x}_p, \tag{2}$$

where

$$\hat{x}_p = 1/N\sum_{i=1}^{N}x_p^i. \tag{3}$$

At the pixel p the normalised brightness has variance: $\sigma_p^2 = 1/N\sum_{i=1}^{N}(x_p^i - \hat{x}_p)^2$. The moments estimators of the three components of the polarisation image are as follows:

$$\hat{I}_p = 1/N\sum_{i=1}^{N}I_p^I, \ \rho_p = \sqrt{2/\pi}\sigma_p \tag{4}$$

and

$$\phi_p = 1/2\cos^{-1}(\langle\hat{x}_p\cos(2\beta_i)\rangle/\pi\rho_p). \tag{5}$$

3   Surface Normal Estimation

Photometric stereo has been used for estimating the angle of incidence for images acquired for three light source directions as has been proposed in [10] and further used in [11]. The mean-intensity component of the polarisation image has been used as input to the photometric stereo for computing the surface normals.

Let $S_m = (S_1|S_2|S_3)$ be the matrix with the three light source vectors as columns, N_p the surface normal at the pixel indexed p and $\hat{J}_p = (\hat{I}1_p, \hat{I}2_p, \hat{I}3_p)^T$ be the vector of the three mean brightness values recorded at the pixel indexed p with the three different light source directions. Under the assumption of Lambertian reflectance, at pixel p we have

$$\hat{J}_p = S_m N. \tag{6}$$

The surface normal can be calculated from the vector of brightness values J_p and the inverse of the source matrix S_m. The reflectance factor, R, is calculated by taking the magnitude of the right side of equation (7) because the surface normal, N, is assumed to have unit length

$$R_p = |[S_m]^{-1}\hat{J}_p|. \tag{7}$$

The unit normal vector is calculated as follows:

$$N = (1/R) * [S_m]^{-1}\hat{J}_p. \tag{8}$$

The images taken across the polarizer angles are reconstructed using the following equation:

$$J_p^i = S_m.N(1 + \rho_p cos(2\beta_i - 2\phi_p)). \tag{9}$$

The surface normal information is used to compute the angle of incidence of the incident light, which is given by the dot product of the source vector and the surface normal.

4   Refractive Index Estimation

Fresnel theory of light predicts that light incident on a surface is partially polarised and refracted while penetrating the surface. Scattering due to the structure of the reflecting surface depolarises the incident light. The remitted light is refracted into the air and hence is refracted and partially polarised in the process. From the Fresnel theory of light, the relationship between the degree of diffuse polarisation, the angle between the surface normal and the remitted light θ and the refractive index n is given as follows:

$$\rho = \frac{(n-\frac{1}{n})^2 \sin^2 \theta}{2+2n^2-(n+\frac{1}{n})^2 \sin^2 \theta+4 \cos \theta \sqrt{n^2-\sin^2 \theta}}. \tag{10}$$

If ρ and θ are known then the above equation can be solved for the refractive index n. The refractive index is given by the roots of quartic equation:

$$A^2n^4 + (2AC - 1)n^3 + (2AB + C^2 + \sin^2 \theta)n^2 + 2BCn + B^2 = 0. \tag{11}$$

where $A = ((1+\rho) \sin^2 \theta - \frac{2\rho}{4\rho \cos \theta})$, $B = (\frac{(1+\rho) \sin^2 \theta+\rho}{4\rho \cos \theta})$ and $C = -\frac{2((1-\rho) \sin^2 \theta+\rho)}{4\rho \cos \theta}$. In practice, we find that only one root falls within the physical range of refractive index encountered for biological tissue, i.e. $1 < n < 2.5$. For specular polarisation, there is a similar equation but its solution requires only the solution of a cubic equation in refractive index [12]. We have used the Newton-Raphson method to compute the roots of the equation (11), 10 iterations have been used to get the final value for the refractive index.

There are other methods that have been proposed for estimating the refractive index, which are as follows: multi-spectral polarisation imagery from a single viewpoint [13], the spectral dependence of the refractive index has been studied in [14], [15], [16]. Our method differs from these as we have used the Fresnel theory in conjunction with photometric stereo and estimates of the polarisation image from the robust moments estimator for estimating the refractive indices.

5   Experiments

As has been already mentioned the images were acquired in a dark room. The objects and the camera are positioned on the same axis while the LED (light sources) are positioned at different angles in the geodesic light dome designed by Cooper et al. for controlling the light sources while the camera operation is manual [8]. A linear polarising filter is placed in front of the camera lens where the orientation is changed manually. The experiments were carried out on a wrinkled apple, orange and tomato. The orange

Fig. 1. The scene for a wrinkled apple using three different light source directions

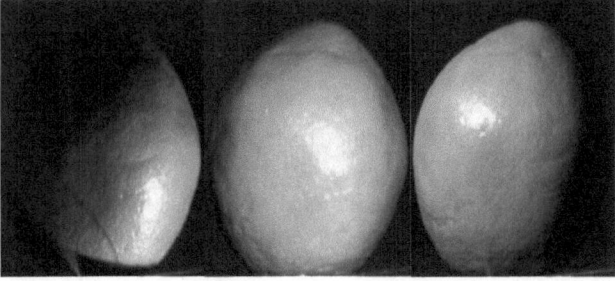

Fig. 2. The scene for an orange using three different light source directions

Fig. 3. The scene for a tomato using three different light source directions

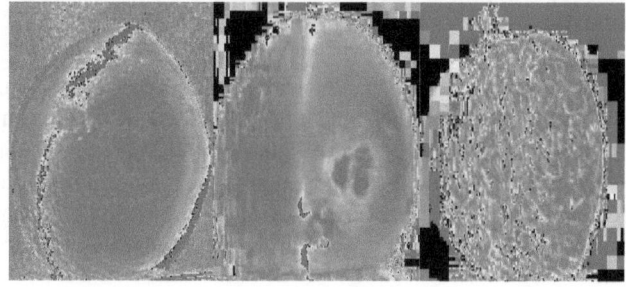

Fig. 4. The refractive index images for an orange, tomato and apple, respectively

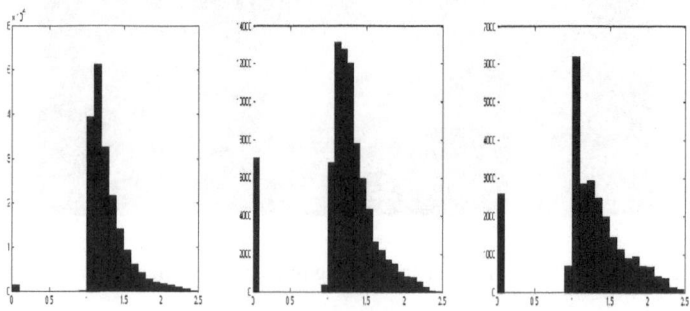

Fig. 5. The histograms for an orange, tomato and apple, respectively

has been chosen in order to test the method due to presence of natural indentations in the surface. We have placed lumps of blu-tac on the surface of the apple. The aim here is to detect whether the method can deal with local variations of shape and refractive index. For the three objects the values for the refractive indices fall in the range $1 < n < 2.5$. There are some outliers which have been filtered. These mainly correspond to non-physical values of the refractive index, which are less than unity. The refractive index images for the different objects are shown in Figure (4). From the images it is clear that

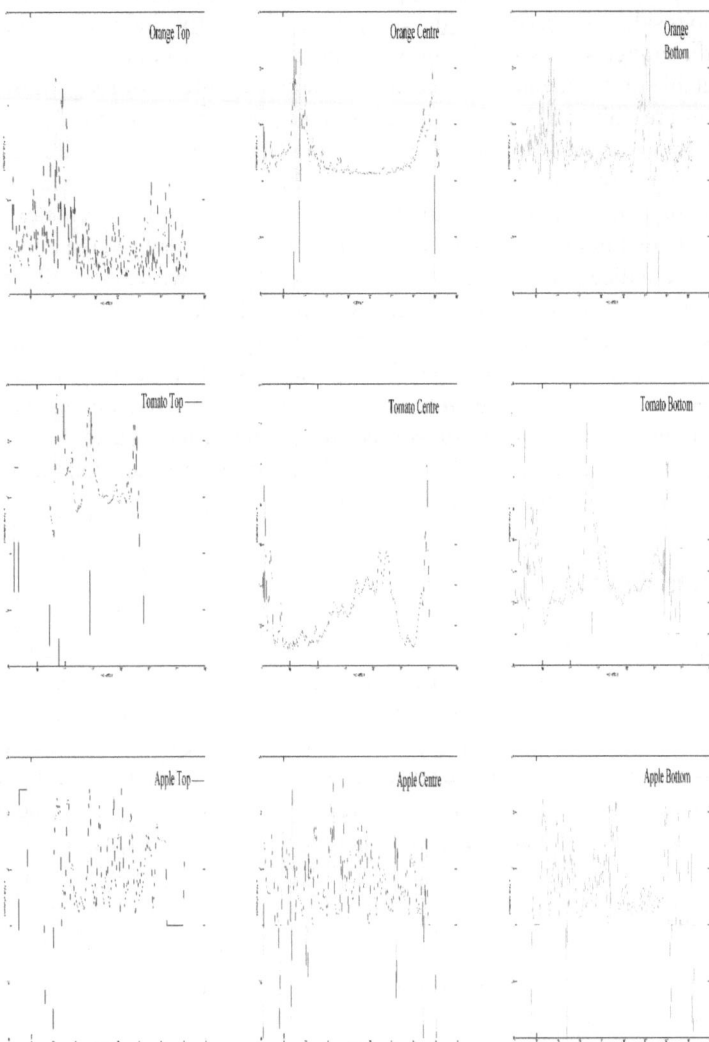

Fig. 6. [Top to Bottom]The top, centre and bottom profiles of an orange, tomato and apple, respectively

the blu-tac on the apple can be detected as both changes in shape and refractive index. Also, regions of specularity have refractive index $'0'$ since, we use the Fresnel equation for the diffuse reflectance. Figure (5) shows histograms of the refractive indices for the apple, tomato and an orange. The modal values are consistent with tabulated values for the refractive index, and the non physical outliers are well separated from the main distribution. Figure. (6) shows refractive index profiles across the physical top, centre and bottom for each object. For the apple and tomato, the profiles are flat, showing that there is no residual shape-bias in the estimation of refractive index. However, the profiles for the orange shows significant variation near the object boundary, and this may be attributable to its indented surface and the boundary effects of roughness. However, the profiles are quite stable in the centre of the object. Infact the profiles of all of the object, show more variation near the boundaries and also where there is a change of material (e.g. the blu-tac lumps). In the case of the apple where the surface is wrinkled due to loss of water content and rotting, there is variation in refractive index.

It is worth mentioning that there are potential sources of error in the computation of the refractive indices because the surface for the apple is wrinkled. This has resulted in inter-reflections and these in turn lead to unrealistic values. On the other hand, the tomato is smooth apart from where there is a change of material where the blue-tac lumps have been added. This change of composition is enhanced in the refractive index image. For the orange the dents in the skin appear as refractive index variations, probably due to inter-reflections and sub-surface reflections. There is also the possibility of error due to noise and camera jitter. Also, the degree of polarisation computations might not be accurate due to misalignment of the polariser angles.

6   Conclusions

In this paper we have exploited the information that is acquired from the Fresnel theory and polarisation information for refractive index estimation. Our approach has been to use information from the polarisation image computed using robust moments estimation and surface normal estimation using photometric stereo. These results are used as inputs to the Fresnel equation for diffuse reflectance, in order to compute the refractive index of the material. The computation is considered to be effective since the refractive indices vary with the change in material and do not exhibit significant shape bias.

References

1. Dunn, A., Richards-Kortum, R.: Three-Dimensional Computation of Light Scattering From Cells. IEEE Journal of Selected topics in Quantum Electronics 2(4) (1996)
2. Saidi, I., Jacques, S., Kittel, F.: Mie and Rayleigh modeling of visible-light scattering in neonatal skin. Applied Optics 34, 7410–7418 (1995)
3. Yee, K.: Numerical Solutions of initial boundary value problems involving Maxwell's equations in instotropic media. IEEE Transactions on Antennas Propagation AP-14, 302–307 (1966)
4. Hecht, E.: Optics, 4th edn. Addison-Wesley, Reading (2002)
5. Wolff, L.B., Boult, T.E.: Constraining Object Features using a Polarisation Reflectance Model. IEEE Transactions on Pattern Analysis and Machine Intelligence 13, 635–657 (1991)
6. Born, M., Wolf, E.: Principles of Optics, 7th edn. Cambridge University Press, Cambridge (1999)
7. Saman, G., Hancock, E.R.: Robust Computation of the Polarisation Image. In: International Conference on Pattern Recognition (2010)

8. Cooper, P., Thomas, M.: Geodesic Light Dome. Department of Computer Science, University of York, UK (March 2010),
   http://www-users.cs.york.ac.uk/-pcc/Circuits/dome
   (accessed on: September 10, 2010)
9. Atkinson, G., Hancock, E.R.: Robust estimation of reflectance functions from polarization. Springer, Heidelberg (2007)
10. Woodham, R.J.: Photometric method for determining surface orientation from multiple images. Optical Engineering 19(1) (1980)
11. Coleman, E.N., Jain, R.: Obtaining 3-Dimensional Shape of textured and Specular surfaces using four-source photometry. Computer Graphics and Image Processing 18(4), 309–328 (1982)
12. Atkinson, G., Hancock, E.R.: Recovery of Surface Orientation from Diffuse Polarization. IEEE Transactions on Image Processing 15(6) (2006)
13. Huynh, C.P., Robles-Kelly, A., Hancock, E.R.: Shape and Refractive Index Recovery from single-view polarisation images. In: IEEE Conference on Computer Vision and Pattern Recognition (2010)
14. Chang, H., Charalampopoulos, T.T.: Determination of the wavelength dependence of refractive indices of flame soot. Royal Society (1990)
15. Bashkatov, A.N., Genina, E.A., Kochubey, V.I., Tuchin, V.V.: Estimation of wavelength dependence of refractive index of collagen fibers of scleral tissue. In: Proceedings of SPIE, vol. 4162, p. 265 (2000)
16. Ding, H., Lu, J.Q., Wooden, W.A., Kragel, P.J., Hu, X.: Refractive indices of human skin tissues at eight wavelengths and estimated dispersion relations between 300 and 1600nm. Journal Physics in Medicine and Biology 51(6) (2006)

Synchronous Detection for
Robust 3-D Shape Measurement
against Interreflection and Subsurface Scattering

Tatsuhiko Furuse, Shinsaku Hiura*, and Kosuke Sato

Graduate School of Engineering Science, Osaka University
1-3 Machikaneyama, Toyonaka, Osaka 560-8531 Japan

Abstract. Indirect reflection component degrades the preciseness of 3-D measurement with structured light projection. In this paper, we propose a method to suppress the indirect reflection components by spatial synchronous detection of structured light modulated with MLS (Maximum Length Sequence, M-sequence). Our method exploits two properties of indirect components; one is the high spatial frequency component which is attenuated through the scattering of projected light, and the other is the geometric constraint between projected light and its corresponding pixel of camera.

Several experimental results of measuring translucent or concave objects show the advantage of our method.

Keywords: Subsurface Scattering, Interreflection, Shape Measurement.

1 Introduction

In this paper, we present a novel method to precisely measure the 3-D shape of object by suppressing subsurface scattering and interreflection which disturb the correct detection of structured light projected on the object.

In principle, triangulation method using structured light projection detects the point of reflection where the incoming light firstly reaches to the surface of the object. In other words, the method assumes that the directly illuminated part only is brightly shining, or otherwise dark. However, in general, the luminance of the scene consists of not only direct but also indirect reflection component (Figure 1). Direct reflection component is the ideal light for 3-D measurement because the line of sight intersects with the incoming light beam exactly on the object surface. On the other hand, indirect component does not directly correspond to the shape of the object and may cause errors.

Indirect component can be classified to subsurface scattering and interreflection. Subsurface scattering is the reflection observed as blurring of light on the surface. Incoming light penetrates into a translucent object, and scattered by particles such as pigments. Therefore, the light exits the surface at a point different from the entrance, and introduces error to measured shape as shown in Figure 1. Interreflection is a phenomenon that the light reflected on the object act as a second light source illuminating the other surface, and produces spurious shape apart from the true surface.

---

* Now at Graduate School of Information Sciences, Hiroshima City University.

G. Maino and G.L. Foresti (Eds.): ICIAP 2011, Part I, LNCS 6978, pp. 276–285, 2011.
© Springer-Verlag Berlin Heidelberg 2011

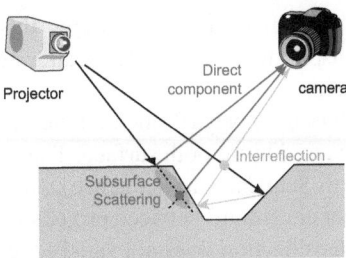

Fig. 1. Error caused by subsurface scattering and interreflection. Subsurface scattering introduces a displacement of measured shape (red dot). Interreflection also generates a spurious shape (green dot) apart from the true surface.

In general, more precisely we want to measure the detail of object, the more strong influence we have via subsurface scattering and interreflecion, even if the object seems opaque in macroscopic scale. For example, a material which attenuates the light by half in 1mm on its pathway does not affect to the measurement whose resolution is 10mm, but may have an influence for the case of requirement of 0.1mm resolution. Most non-metallic objects such as plastic, cloth, paper and wood have more or less subsurface scattering. Similarly, the intensity of interreflection gets higher if the distance between two interacting surfaces is small. Therefore, suppression of such indirect components is important for precise shape measurement in industry and digital archives.

1.1   Related Works

Subsurface scattering have been well addressed in computer graphics area. Jensen et al. [6] proposed a mathematical model of subsurface scattering to represent BSSRDF (Bidirectional Surface Scattering Reflectance Distribution Function) with parametric function, and many techniques to render photorealistic images of such objects have been proposed.

In recent years, 3-D measurement of non-lambertian object such as transparent, translucent or specular object have extensively attacked [4]. As described before, sub-surface scattering is the main issue to measure the translucent object using structured light. Godin et al. [2] analyzed the defocus and noise of penetrating light by projecting spot light onto a marble stone, because statues made of marble stones are need to be measured in digital archiving area [7] . Goesele et al. [3] measured the shape of translu-cent object by using laser light and HDR (High Dynamic Range) camera. However, it takes very long time to measure the whole object because it is based on spot light projection method.

Methods for suppressing the influence of indirect components have been proposed. T.Chen et al. [1] improved the phase shifting method by using polarization filter to sup-press indirect components. Contrary, Nayar et al. [10] proposed a method to separate direct and indirect reflection component by projecting high spatial frequency pattern onto the object. Since intensity via indirect component receives contributions from var-ious part of projected light, high spatial frequency components tend to be attenuated through the light transport of indirect reflection. Using this principle, Mukaigawa et al. [9] analyzed the light transport in homogeneous scattering media using similar high spatial frequency stripe pattern.

Our method is basically based on Nayar's method, but their method is not effective for specular interreflection caused by smooth surface. Therefore we extend the method with MLS (Maximum Length Sequence, M-sequence) which is used for checking the geometric constraint (epipolar constraint) between projector and camera. Generally, observed light via specular interreflection does not satisfy the geometric constraint, and synchronous detection effectively suppress the component.

Synchronous detection techniques are widely used in control engineering, communication technologies and weak signal detection. For example, CDMA (Code Division Multiple Access) uses pseudorandom sequence such as MLS to separate multiplexed signals or suppressing multi-path fading effect caused by reflected signals in temporal domain. The principle of proposed method is similar from the mathematical point of view, but ours is not temporal but spatial (or geometric).

2   Surpression of Indirect Component Using Spatially Modulated Light

In this section, we will describe the method to suppress the indirect component of projected light using spatially modulated light. Firstly we will start from an extension of slit light projection method, then apply the principle to the Gray-code projection method.

2.1   Spatially Modulated Slit Light

MLS (Maximum Length Sequence, M-sequence) is a pseudorandom binary sequence which has three advantages for our purpose. Firstly, MLS contains high frequency component which is more likely to be attenuated through indirect light transport. The second advantage is that MLS has high autocorrelation value only if the phase difference is zero. MLS is also easy to implement because it has only two binary values, therefore no photometric calibration is necessary for projector. To suppress the indirect component of projected light, we spatially modulate the light along the slit direction using MLS as shown in Figure 2. The intensity distribution of projected pattern $L(x_p, y_p, t)$ is denoted as

$$L(x_p, y_p, t) = \begin{cases} M((t - x_p) \mod T) & y_p = \bar{y}_p \\ 0 & y_p \neq \bar{y}_p \end{cases} \tag{1}$$

where T is the cycle of MLS function $M(t)$, and (x_p, y_p) is the coordinate on the projector image. The slit light is parallel to the axis x_p, and the position of the slit is determined by $\bar{y}_p$. The MLS pattern is shifted along the slit light for one cycle, therefore T different pattern is projected for each slit position.

The image of the scene is captured by the camera as shown in Figure 2. In prior to the measurement, the geometric relationships between projector and camera are calibrated using a reference object. This process is common to the calibration of quantitative 3-D measurement, therefore this process does not lose the feasibility at all. From the

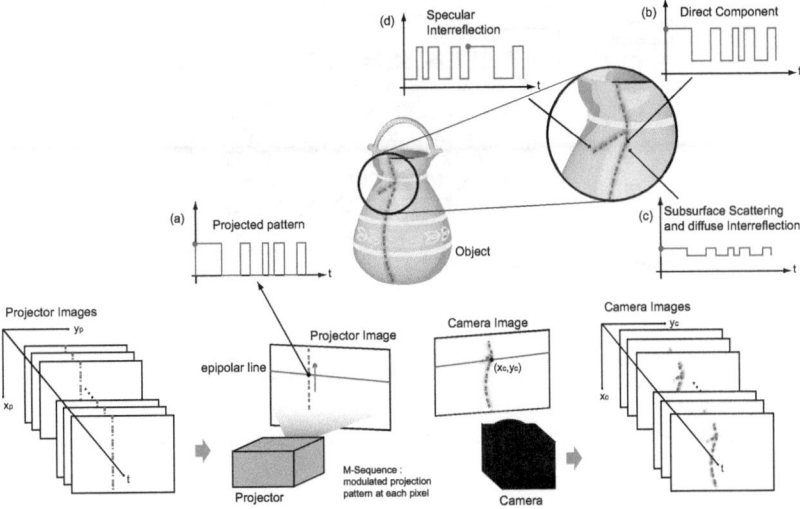

Fig. 2. Schematic illustration of proposed method. Slit light is modulated with MLS code, and captured by a camera. The amplitude of intensity by subsurface scattering is very low, and the phase of the intensity sequence by interreflection is shifted.

calibration parameters of camera and projector, we can derive the following epipolar constraint equation,

$$\left(x_p \; y_p \; 1 \right) F \begin{pmatrix} x_c \\ y_c \\ 1 \end{pmatrix} = 0 \tag{2}$$

where F is called fundamental matrix [8]. Using this equation, we can calculate the epipolar line on the projector image for each pixel (x_c, y_c) on camera image. The epipolar line has a intersection with the slit light, and we can decide the corresponding phase of MLS of projected light for each point on the image. More specifically, we can determine the corresponding coordinate $(x_p, \bar{y}_p)$ on the projector image by

$x_p = -\frac{e_2 \bar{y}_p + e_3}{e_1}$ where

$$\begin{pmatrix} e_1 \\ e_2 \\ e_3 \end{pmatrix} = F \begin{pmatrix} x_c \\ y_c \\ 1 \end{pmatrix}. \tag{3}$$

The process of synchronous detection of modulated slit light is illustrated in Figure 3. We simply calculate the correlation between binary value of projected MLS pattern and intensity sequence from the camera image, then use the correlation value for slit detection instead of the raw intensity on the image.

As described in the paper [10], high spatial component in projected light is more likely to be attenuated by the propagation of subsurface scattering or diffuse interreflection. However, specular interreflection preserves the structure of projected light, therefore, frequency analysis is not sufficient to suppress the effect. Fortunately, specular interreflection can be assumed as a light projected from the other point, therefore, the

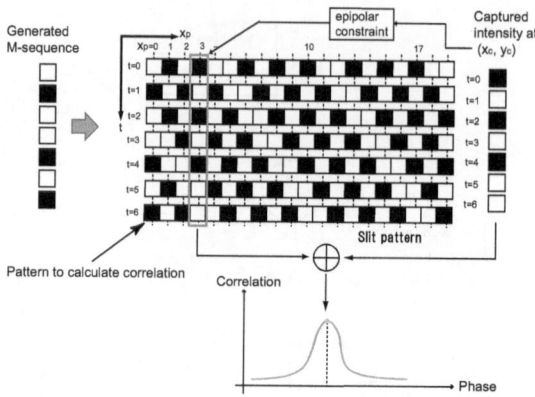

Fig. 3. Synchronous detection of captured intensity using geometric constraint

epipolar constraint is no longer satisfied. In the next section, we will show the faster method based on Gray code projection.

2.2 Modulation of Gray Code Pattern

As shown in the experimental results below, proposed method with synchronous detection using modulated slit light is effective for suppressing indirect component. However, it takes much time to measure the whole shape of the object. The number of projection pattern and corresponding image capture is $T \cdot N$ where T is the cycle of MLS and N is the number of slit planes which corresponds to the depth resolution. Actually in our experiment, we used parameters of $T = 31$ and $N = 1024$ which take around 18 minutes if the frame rate of projection and acquisition is 30 frames/sec.

Fortunately, several methods to accelerate the range measurement using structured light have been proposed, and the temporal space coding [5,11] is one of the most successful method using Gray-code. If we use complementary pattern projection[1], it only needs $2 \log_2 N$ patterns for measuring the whole object. Therefore we combine our synchronous detection framework with temporal space coding method.

The projection patterns which consist of Gray-code and MLS are illustrated in Figure 4. The geometric synchronous detection requires that the phase of MLS should be uniquely determined by the coordinate on the image plane (x_c, y_c). In the case of slit light projection method, the phase of MLS can be uniquely determined because the cross section of slit light and epipolar line is a single point. Contrary to the case of slit-code projection, the bright part in projection pattern has certain width, therefore we should align the lines of MLS code parallel to epipolar lines as shown in Figure 5. This arrangement make the phase corresponds to a epipolar line unique, therefore we can use the same synchronous detection method again.

---

[1] For each bit plane, positive and its inverted pattern are projected, and sign of subtraction of two images is used for binarized image.

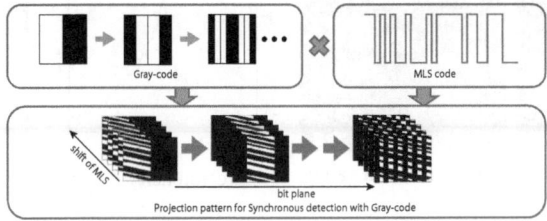

Fig. 4. Modulated Gray-code pattern for temporal space coding method with synchronous detection

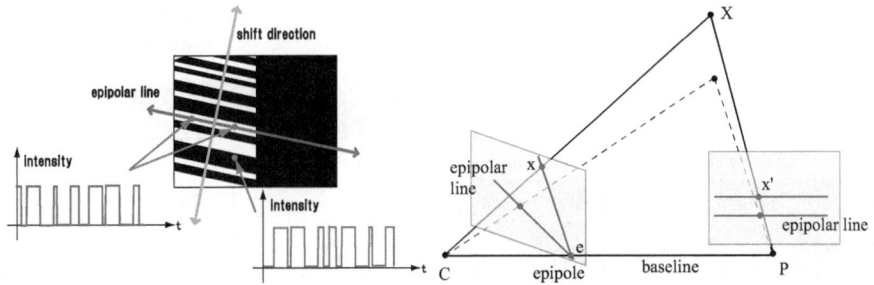

Fig. 5. Arrangement of MLS code in the projection pattern. Arrangement of camera and projector. If the image plane of the projector is parallel to the baseline, all epipolar line on the projector image is parallel.

IF the epipolar lines are not parallel each other, the density of the code is not homogeneous. However, the density of the pattern decides the performance of eliminating indirect component. Therefore, we arrange the image plane of the projector parallel to the baseline as shown in Figure 5. In this case, the all epipolar lines on the projector image are parallel. Contrary, the epipolar lines on the camera image are not necessary to be aligned parallel.

By using the temporal space coding method with synchronous detection, the total number of projected pattern is $2T \log_2 N$. The condition of $T = 31$ and $N = 1024$ takes only 21 seconds with frame rate of 30 frames/sec. This is not only 50 times faster than the case of slit light projection method with synchronous detection, but also even faster than simple slit light projection.

3   Experiments

The appearance of experimental setup is shown in Figure 7. We used a liquid crystal projector (EPSON EMP-1710, resolution : 1024×768pixel) with a convex lens which focal length is 330mm. The convex lens is used to adjust the focus on the object at short distance. The cycle of used MLS code is $T = 31$, and the calibration of the system is done in prior to the measurement using a reference object.

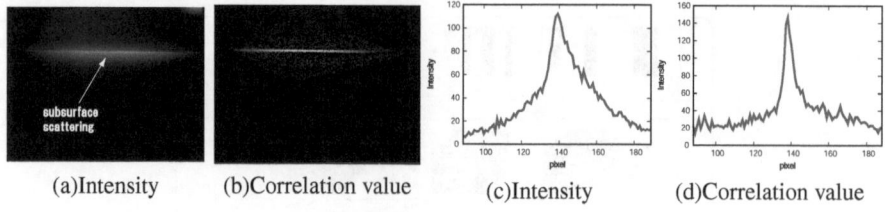

(a)Intensity   (b)Correlation value   (c)Intensity   (d)Correlation value

Fig. 6. Observed slit light under subsurface scattering

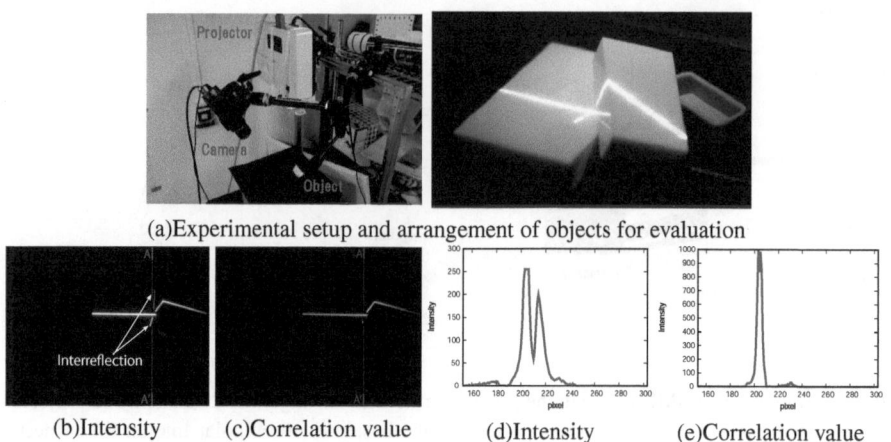

(a)Experimental setup and arrangement of objects for evaluation

(b)Intensity   (c)Correlation value   (d)Intensity   (e)Correlation value

Fig. 7. Observed slit light under interreflection

3.1 Suppression of Indirect Component

In prior to showing the performance of shape measurement, we will show the result of suppressing indirect component in this section. Figure 6 is a result of suppressing subsurface scattering using an opalescent acrylic plate. Figure 6(a) and (b) show the intensity distribution of normal and synchronous detection respectively. The intensity on a vertical line at the center of Figure (a) and (b) are shown in Figure 6(c) and (d) respectively. The half-value widths of slit light, 21.2 and 6.7 pixels respectively, show that the subsurface scattering is effectively suppressed by our method.

Figure 7 shows the result of suppressing specular interreflection. The object is a stack of zirconia ceramic blocks shown in Figure 7(a). Figure 7(d) and (e) are the intensity distribution on the line A-A' in Figure 7(b) and (c) respectively. False slit is well eliminated by synchronous detection though it is clearly visible in raw intensity image.

3.2 Shape Measurement with Slit Light Projection

The results of shape measurements are shown in the last 2 pages. Figure 8 is the result with suppressed interreflection. The object (small figure of a house, width is around 40mm) is measured from the front side of the house. Interreflection causes a false shape

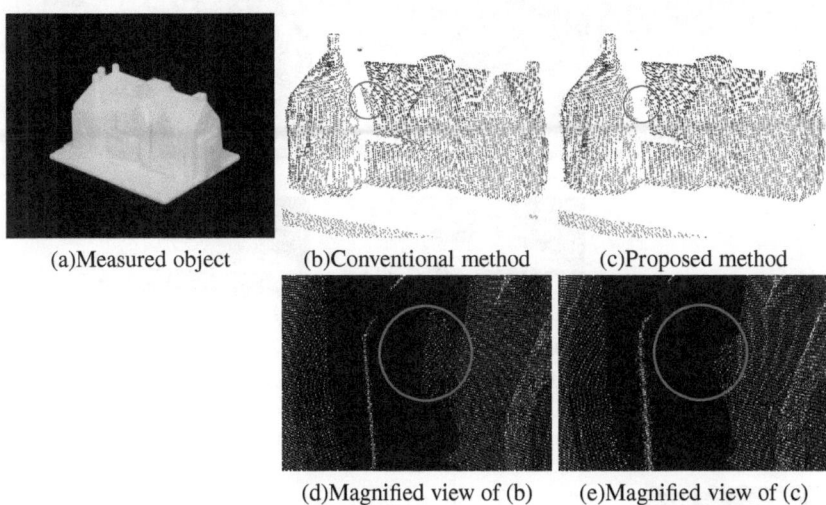

(a)Measured object　　(b)Conventional method　　(c)Proposed method

(d)Magnified view of (b)　　(e)Magnified view of (c)

Fig. 8. Experimental result 1 (slit light projection)

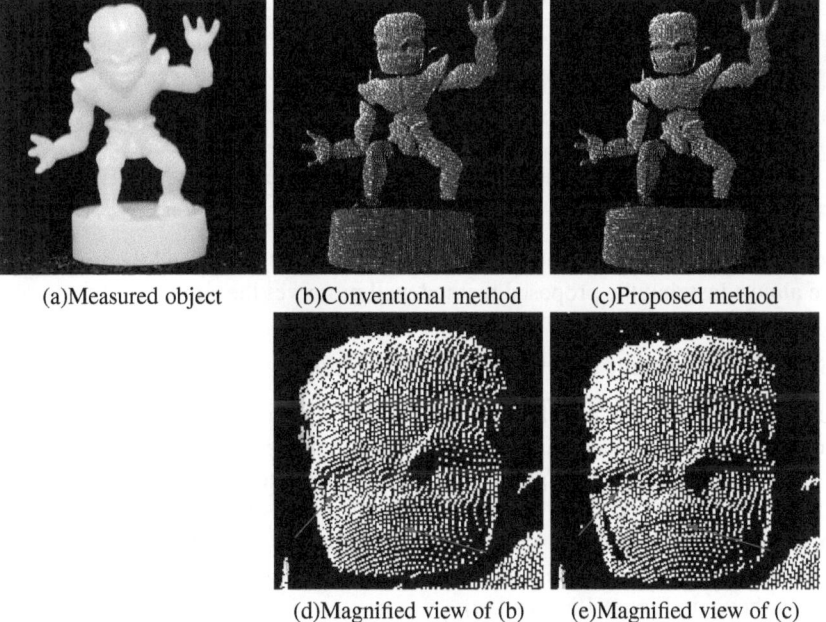

(a)Measured object　　(b)Conventional method　　(c)Proposed method

(d)Magnified view of (b)　　(e)Magnified view of (c)

Fig. 9. Experimental result 2 (slit light projection)

(indicated by a red circle) with conventional method, and it is well suppressed by the proposed method.

The result with translucent object with subsurface scattering is shown in Figure 9. Since the figure is so small (height is around 38mm), some details of the figure are lost

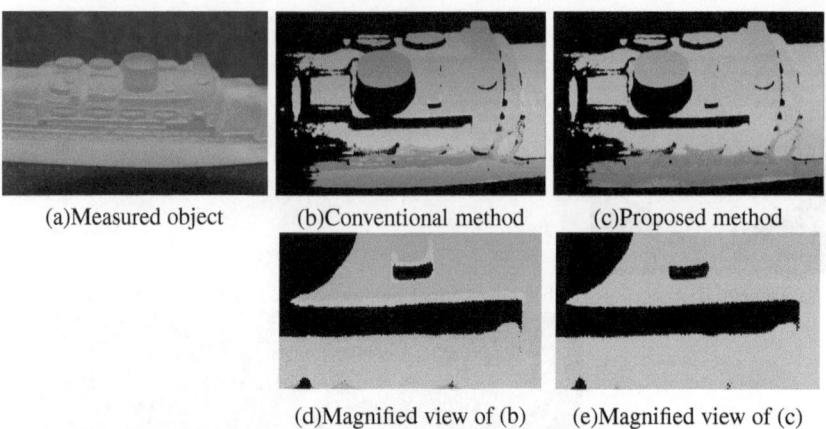

(a)Measured object   (b)Conventional method   (c)Proposed method

(d)Magnified view of (b)   (e)Magnified view of (c)

Fig. 10. Experimental result 1 (temporal space coding)

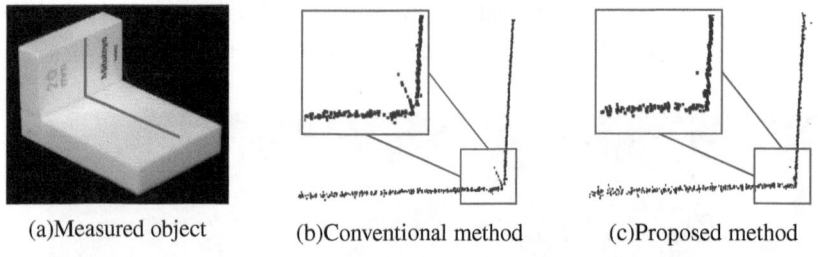

(a)Measured object   (b)Conventional method   (c)Proposed method

Fig. 11. Experimental result 2 (temporal space coding)

with conventional method. As shown in Figure 9(d), the edges around the mouse and eyes are almost lost, but the proposed method well preserves the edge.

3.3   Shape Measurement with Gray-Code Projection

Indirect component of structured light causes not only displacement of measured shape but also spurious shape by decoding error for temporal space coding method. Figure 10 shows the result of measuring a figure of a ship made of flosted glass. Figure 10(b) - (e) shows the color map of measured height of obliquely placed object. At the edges of flat part, false step shape is observed with conventional method shown in Figure 10(b) and (d). Contrary, the shape measured with proposed method has no false step shape around the edge as shown in Figure 10(c) and (e).

To show the result of suppressing specular interreflection, we used a stack of ceramic blocks again as shown in Figure 11. Figure 11(b) shows that conventional method produces a spurious shape around the concave edge of the object, and the shape of the edge is not acute but rounded. Contrary, with our proposed method, the shape of the object is properly measured as shown in Figure 11(c).

The main contribution of our method is to suppress the erroneous result caused by indirect components. The amount of error much depends on the shape and material of the object, and the other conditions such as the angle of incident light also heavily affect to the result. Therefore, it is not adequate to show a quantitive comparison between conventional and proposed method with measured shape. Instead, we showed the effect of our method with the difference of intensity distribution in section 3.1.

4   Conclusion

We proposed the 3-D shape measurement method robust against both interreflection and subsurface scattering. MLS pattern is very useful to combine two suppression principles based on geometric constraint and transfer characteristics in spatial frequency. Experimental results showed the practicality of our method for translucent or shiny object with fine details.

Since the false image caused by the interreflection sometimes has exactly same phase as the true one, multiple cameras or projectors should be effective to verify the true image using multiple epipolar lines. Synchronous detection technique will be also useful for disambiguating simultaneous projections from multiple projectors.

References

1. Chen, T., Lensch, H.P.A., Fuchs, C., Seidel, H.P.: Polarization and phase-shifting for 3d scanning of translucent objects. In: Proc. CVPR, pp. 1–8 (2007)
2. Godin, G., Rioux, M., Beraldin, J.A.: An assessment of laser range measurement on marble surfaces. In: Conference on Optical 3D Measurement Techniques, pp. 49–56 (2001)
3. Goesele, M., Lensch, H.P.A., Lang, J., Fuchs, C., Seidel, H.P.: Disco – acquisition of translucent object. In: Proc. SIGGRAPH, pp. 835–844 (2004)
4. Ihrke, I., Kutulakos, K.N., Lensch, H.P.A., Magnor, M., Heidrich, W.: State of the art in transparent and specular object reconstruction. In: Proc. Eurographics, pp. 1–22 (2008)
5. Inokuchi, S., Sato, K., Matsuda, F.: Range-imaging system for 3d object recognition. In: Proc. of 7th International Conference on Pattern Recognition, pp. 806–808 (1984)
6. Jensen, H.W., Marschner, S.R., Levoy, M., Hanrahan, P.: A practical model for subsurface light transport. In: Proc. SIGGRAPH, pp. 511–518 (2001)
7. Levoy, M., Pulli, K., Curless, B.: Digital michelangelo project: 3d scanning of large statues. In: Proc. SIGGRAPH, pp. 131–144 (2000)
8. Luong, Q.T., Faugeras, O.D.: The fundamental matrix: Theory, algorithms, and stability analysis. International Journal of Computer Vision 17(1), 43–75 (1996)
9. Mukaigawa, Y., Yagi, Y., Raskar, R.: Analysis of light transport in scattering media. In: Proc. CVPR, pp. 153–160 (2010)
10. Nayar, S.K., Krishnan, G., Grossberg, M., Raskar, R.: Fast separation of direct and global components of a scene using high frequency illumination. In: Proc. SIGGRAPH, pp. 9350–9943 (2006)
11. Sato, K., Inokuchi, S.: Range-imaging system utilizing nematic liquid crystal mask. In: Proc. of 1st International Conference on Computer Vision, pp. 657–661 (1987)

Unambiguous Photometric Stereo
Using Two Images

Roberto Mecca[1] and Jean-Denis Durou[2]

[1] Dipartimento di Matematica "G. Castelnuovo"
Sapienza - Universitá di Roma
roberto.mecca@mat.uniroma1.it
www.mat.uniroma1.it/~mecca
[2] IRIT - Université Paul Sabatier - Toulouse
Jean-Denis.Durou@irit.fr

Abstract. In the last years, the 3D reconstruction of surfaces which represent objects photographed by simple digital cameras has become more and more necessary to the scientific community. Through the most various mathematical and engineering methods, scientists continue to study the Shape-from-shading problem, using the photometric stereo technique which allows the use of several light sources, but keeps the camera at the same point of view. Several studies, through different advances on the problem, have checked that in the applications, the smallest number of photos that have to be considered is three. In this article we analyze the possibility to determine the objects' surface using two images only.

Keywords: Shape-from-shading, photometric stereo, normals integration, PDE numerical analysis, boundary conditions.

1 Introduction

Many articles have been written about the impossibility to solve the Shape-from-shading problem (SFS) considering only one picture [1], even if a recent perspective SFS model exploiting the attenuation of the lighting with respect to the distance to the light source has been shown to yield to a well-posed problem, if complemented by reasonable assumptions [2]. This impossibility, from the PDEs point of view, comes out from the difficulty we meet in differentiating the concave surfaces from the convex ones. The most natural way to solve the problem is to use more than one picture. First introduced by Woodham [3], photometric stereo (PS) consists in using several images which portray the object photographed always from the same point of view, but with different light source positions [4].

There are two main approaches to solve the SFS-PS problem. The first one aims at computing in each point the normal to the surface to be reconstructed. If the albedo is supposed to be known, this approach has the drawback to be well-posed only if a minimum of three images of a differentiable surface are used (we emphasize that the required regularity of the surface, in real applications,

G. Maino and G.L. Foresti (Eds.): ICIAP 2011, Part I, LNCS 6978, pp. 286–295, 2011.
© Springer-Verlag Berlin Heidelberg 2011

can be seen as a supplementary disadvantage). Its vantage is that, even if only two images are used, the number of solutions of the problem (for a differentiable surface) is a priori predictable through the study that we propose in this work.

The second approach, which is more recent, aims at solving the PDE model. If we still suppose the albedo to be known, it has the advantage of admitting only one solution even if only two images are used. It is also possible to approximate the solution even if the surface is Lipschitzian (that is, almost everywhere differentiable). The drawback is that it is well-posed only if we preliminary know the height of the surface on the boundary of the image (i.e. only if we know the boundary condition of the differential problem).

The main idea of this paper is to approximate the boundary condition by integrating the normal field only on the boundary of the image, and then to solve the PS problem anywhere else using the PDE approach. In Section 2, we recall the differential and non-differential formulations of SFS-PS. In Section 3, we show that in some points, two images are enough to deduce the normal univocally. Section 4 is dedicated to the tests and Section 5 to conclusion and perspectives.

2   Main Features of the Photometric Stereo Technique

2.1   Shape-from-Shading

We start by giving a brief outline of the SFS problem and introducing the basic assumptions. We attach to the camera a 3D Cartesian coordinate system xyz, such that xy coincides with the image plane and z with the optical axis. Under the assumption of orthographic projection, the visible part of the surface is a graph $z = u(x, y)$. For a Lambertian surface of uniform albedo equal to 1, lighted by a unique light source located at infinity in a direction indicated by the unitary vector $\omega = (\omega_1, \omega_2, \omega_3) = (\tilde{\omega}, \omega_3) \in \mathbb{R}^3$, the SFS problem can be modeled by the following "image irradiance equation" [5]:

$$n(x, y) \cdot \omega = I(x, y) \qquad \forall (x, y) \in \overline{\Omega} \qquad (1)$$

where $I(x, y)$ is the greylevel at the image point (x, y) and $n(x, y)$ is the unitary outgoing normal to the surface at the scene point $(x, y, u(x, y))$. The greylevel I, which is the datum in the model, is assumed to take real values in the interval $[0, 1]$. The height u, which is the unknown, has to be reconstructed on a compact domain $\overline{\Omega} = \Omega \cup \partial\Omega \subset \mathbb{R}^2$ called the "reconstruction domain". It does not explicitly appear in Eq. (1), but implicitly through the normal $n(x, y)$, since this vector can be written:

$$n(x, y) = \frac{1}{\sqrt{1 + |\nabla u(x, y)|^2}} [-\nabla u(x, y), 1]^{\top} \qquad (2)$$

Combining Eqs. (1) and (2), we arrive to the following differential formulation of the SFS problem:

$$\frac{-\nabla u(x, y) \cdot \tilde{\omega} + \omega_3}{\sqrt{1 + |\nabla u(x, y)|^2}} = I(x, y) \qquad \forall (x, y) \in \Omega \qquad (3)$$

which is a first order non-linear PDE of the Hamilton-Jacobi type. Eq. (3) with the add of a Dirichlet boundary condition $u(x,y) = g(x,y) \; \forall(x,y) \in \partial\Omega$, do not admit a unique solution if the brightness function I reaches its maximum i.e., if there are points $(x,y) \in \Omega$ such that $I(x,y) = 1$. In this case, we cannot distinguish whether a surface is concave or convex ("concave/convexity ambiguity", see [5]).

With the purpose to prove the existence of a unique solution, we increase the information about the surface considering the photometric stereo technique.

2.2 Photometric Stereo: Differential Approach

The first approach to PS is based on the differential formulation (3) of the SFS problem, that is, using two images we have:

$$\begin{cases} \dfrac{-\nabla u(x,y) \cdot \widetilde{\omega} + \omega_3}{\sqrt{1 + |\nabla u(x,y)|^2}} = I(x,y) & \text{a.e. } (x,y) \in \Omega \\[2mm] \dfrac{-\nabla u(x,y) \cdot \widetilde{\omega}' + \omega_3'}{\sqrt{1 + |\nabla u(x,y)|^2}} = I'(x,y) & \text{a.e. } (x,y) \in \Omega \\[2mm] u(x,y) = g(x,y) & \forall(x,y) \in \partial\Omega \end{cases} \qquad (4)$$

It is a PDEs non-linear system with the add of a Dirichlet boundary condition that admits a unique solution in the space of Lipschitzian functions. This means that, even if a surface is differentiable for almost every $(x,y) \in \overline{\Omega}$ it is possible to approximate it using a convergent numerical scheme [6]. The only drawback of this formulation concerns the boundary condition knowledge. In fact, beyond the image data (I, I') and the light vectors (ω, ω'), $g(x,y)$ (taken in the space of the Lipschitz functions $W^{1,\infty}(\partial\Omega)$) represents an additional information that we must know to make this approach work.

Let us explain how the differential approach works. To arrive to the final PDE formulation, we simplify the system (4) eliminating its non-linearity, supposing that $1 \geq I(x,y) > 0$ everywhere. That is, we consider for example the following equality from the first equation:

$$\sqrt{1 + |\nabla u(x,y)|^2} = \frac{-\nabla u(x,y) \cdot \widetilde{\omega} + \omega_3}{I(x,y)} \qquad (5)$$

and replacing (5) into the other equation we obtain a linear equation, $\forall(x,y) \in \Omega$:

$$[I'(x,y)\omega_1 - I(x,y)\omega_1']\frac{\partial u}{\partial x} + [I'(x,y)\omega_2 - I(x,y)\omega_2']\frac{\partial u}{\partial y} = I'(x,y)\omega_3 - I(x,y)\omega_3' \qquad (6)$$

Considering also the same boundary condition as that of (4), it is possible to arrive to the following linear problem:

$$\begin{cases} b(x,y) \cdot \nabla u(x,y) = f(x,y) & \text{a.e. } (x,y) \in \Omega \\ u(x,y) = g(x,y) & \forall(x,y) \in \partial\Omega \end{cases} \qquad (7)$$

where

$$\begin{cases} b(x,y) = [I'(x,y)\omega_1 - I(x,y)\omega_1', I'(x,y)\omega_2 - I(x,y)\omega_2']^\top \\ f(x,y) = I'(x,y)\omega_3 - I(x,y)\omega_3' \end{cases} \qquad (8)$$

With these elements it is possible to enunciate the following result [6]:

Theorem 1. *Let $b(x,y)$ and $f(x,y)$ be both bounded functions defined by (8), where I and I' are two greylevel functions such that $1 \geq I, I' > 0$, with a jump discontinuity on the piecewise regular curve $\gamma(s)$ and $g(x,y) \in W^{1,\infty}(\partial\Omega)$. If $\gamma(s)$ is not a characteristic curve of the problem (7) then it admits a unique Lipschitzian solution $u(x,y)$.*

2.3 Photometric Stereo: Non-differential Approach

The other approach is based on the local estimation of the outgoing unitary normal to the surface. With the same data as before, Eq. (1) gives the following non-linear system in the coordinates of the normal, in each point $(x,y) \in \overline{\Omega}$:

$$\begin{cases} n_1(x,y)^2 + n_2(x,y)^2 + n_3(x,y)^2 = 1 \\ \omega_1 n_1(x,y) + \omega_2 n_2(x,y) + \omega_3 n_3(x,y) = I(x,y) \\ \omega_1' n_1(x,y) + \omega_2' n_2(x,y) + \omega_3' n_3(x,y) = I'(x,y) \end{cases} \qquad (9)$$

This approach goes on with the integration of the normal field using Eq. (2), all over the domain $\overline{\Omega}$ [7]. Its drawback is that the non-linear system (9) has not a unique solution in general.

The purpose of our work is to find and, in particular, characterize the zones of the images where the solution to (9) is unique. This permits us to understand, before the integration of the gradient field, the number of possible surfaces approximated by this approach. With the aim to combine both approaches, we study in detail the problem (9) giving information about all possible solutions. We advance that there are two local solutions at the most, but the problem is that they can globally generate much more than two surfaces.

3 Photometric Stereo with 2 Images: Normal Uniqueness

3.1 General Study of the Problem

We now focus on the problem of normal estimation, emphasizing one more time that it is based on a local study of the images. For each pixel we want to estimate the unitary vector which represents the outgoing normal to the surface. We can determine the set of visible normals as the superior part of a sphere centered at the origin and with radius one, that is the Northern hemisphere of the so-called Gaussian sphere $\mathcal{S}$ (see Fig. 1). Referring to the 3D Cartesian coordinate system xyz, these vectors $n = [n_1, n_2, n_3]^\top$ are those such that $n_3 > 0$. We now consider the two light vectors ω and ω' and the sets of lighted normals, that is those such that $w \cdot n > 0$ and $w' \cdot n > 0$. On S, these sets are limited by the two planes π and π' passing through the origin and orthogonal to w and w'. Now, we can determine the set of possible normals for each twice-lighted point (x, y), which can be summed up in the following non-linear system:

$$\begin{cases} n_1(x, y)^2 + n_2(x, y)^2 + n_3(x, y)^2 = 1 \\ w_1 n_1(x, y) + w_2 n_2(x, y) + w_3 n_3(x, y) \geq 0 \\ w'_1 n_1(x, y) + w'_2 n_2(x, y) + w'_3 n_3(x, y) \geq 0 \\ n_3(x, y) \geq 0 \end{cases} \quad (10)$$

On the other hand, in each twice-lighted point (x, y), the linear system (9) usually admits two solutions. An important study is carried out on the straight line $\Delta = \pi' \cap \pi''$, which is supported by the vector $r = w \times w'$. This allows us to establish a direct connection between the solutions of (9). For each twice-lighted point (x, y) such that (9) admits two solutions $\hat{n}$ and $\hat{\hat{n}}$, the locations of these normals on S define a straight line which is parallel to Δ since, according to the Lambertian model, $\hat{n}$ and $\hat{\hat{n}}$ form the same angles with w and w' (see Fig. 1).

3.2   Normal Uniqueness Obtained by Visibility or by Coincidence

The first case in which normal uniqueness using two images we can be proved is explainable geometrically, considering the set obtained changing the sign of the last inequality of the system (10) (which represents the condition of visibility) and then projecting it, according to the direction r, on the other side of the sphere. Therefore, in order to determine these points in the images, the first step is to determine the set $\Omega_G \subset \overline{\Omega}$ of points (x, y) such that there exists a solution $\hat{n}$ to the following system:

$$\begin{cases} n_1(x, y)^2 + n_2(x, y)^2 + n_3(x, y)^2 = 1 \\ w_1 n_1(x, y) + w_2 n_2(x, y) + w_3 n_3(x, y) \geq 0 \\ w'_1 n_1(x, y) + w'_2 n_2(x, y) + w'_3 n_3(x, y) \geq 0 \\ n_3(x, y) < 0 \end{cases} \quad (11)$$

Clearly, these normals $\hat{n}$ cannot be considered as possible candidates for the normal field of the surface taken into consideration, because they are located on the non-visible part of S (see the yellow part S_Y of S in Fig. 1). The normals which could be candidate for the normal field are those in the set in biunique correspondence with S_Y with respect to the direction r (see the green part S_G of S in Fig. 1). Note that, if w or w' is vertical i.e., equal to $[0, 0, 1]^\top$, then r is horizontal and, therefore, the set Ω_G is empty.

The second way of obtaining normal uniqueness corresponds to the limit case where the two solutions $\hat{n}$ and $\hat{\hat{n}}$ of (9) coincide. The set $\Omega_R \subset \overline{\Omega}$ thus contains the points (x, y) where the normal is orthogonal to the direction r (see the red line S_R on S in Fig. 2), which is a geodesic line on S between two extreme points $\hat{P}$ and $\hat{P}'$.

Looking at Figs. 1 and 2, it is obvious that $S_R \cap S_G = \emptyset$ as soon as $w_3 > 0$ and $w'_3 > 0$, which implies that $\Omega_G \cap \Omega_R = \emptyset$ as well.

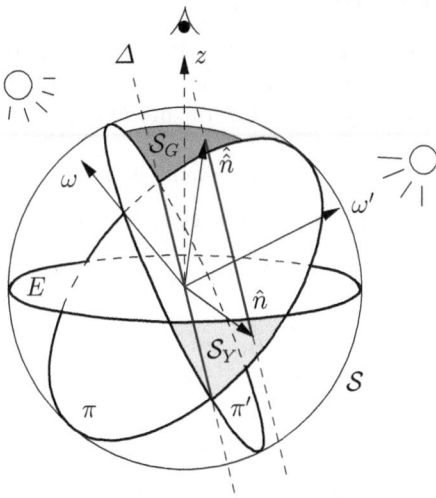

Fig. 1. The planes   and   ′ are orthogonal, respectively, to the light vectors   and
  ′. The intersection between   and   ′ is denoted as   . Each normal $\hat{\hat{n}}$ pointing to the
green area $\mathcal{S}_G$ is known without ambiguity, since the second possible normal $\hat{n}$ points
to the yellow area $\mathcal{S}$, which is a twice lighted but non-visible part of $\mathcal{S}$ (because it
lies under the equator E).

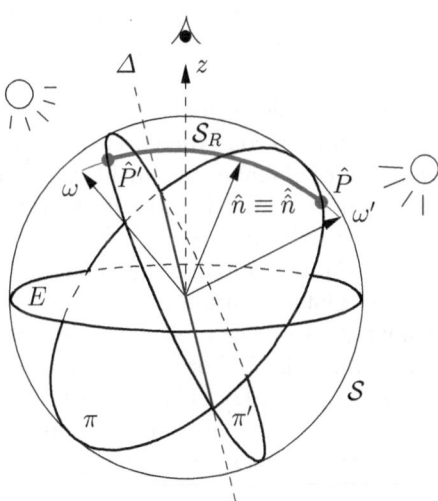

Fig. 2. The geodesic $\mathcal{S}_R$ in red is a part of the intersection between $\mathcal{S}$ and the plane
supported by   and   ′, limited by $\hat{P}$ and $\hat{P}'$. Each normal pointing to $\mathcal{S}_R$ is known
without ambiguity, since both normals $\hat{n}$ and $\hat{\hat{n}}$ coincide in this case.

3.3   Finding the Sets Ω_G and Ω_R

Here we describe how to find the sets Ω_G and Ω_R in the reconstruction domain $\overline{\Omega}$. For each normal $\tilde{n} = [\tilde{n}_1, \tilde{n}_2, \tilde{n}_3]^\top \in \mathcal{S}_G \cup \mathcal{S}_R$, we have to calculate the correspondent couple of greylevels $(\tilde{I}, \tilde{I}')$ using (1), i.e.:

$$\begin{cases} \tilde{I} = \omega_1 \tilde{n}_1 + \omega_2 \tilde{n}_2 + \omega_3 \tilde{n}_3 \\ \tilde{I}' = \omega_1' \tilde{n}_1 + \omega_2' \tilde{n}_2 + \omega_3' \tilde{n}_3 \end{cases} \tag{12}$$

and check if, for each pixel $(i, j) \in \overline{\Omega}$, the greylevels $(I_{i,j}, I_{i,j}')$ are such that:

$$\begin{cases} |I_{i,j} - \tilde{I}| < \varepsilon \\ |I_{i,j}' - \tilde{I}'| < \varepsilon \end{cases} \tag{13}$$

for a small fixed value of ε ($\varepsilon = 0.001$ is used in the tests).

Definition 2. *Given a pair of images, we call Ω_G^p and Ω_R^p the sets of pixels of $\overline{\Omega}$ which belong to Ω_G and Ω_R and are determined using the criterion (13).*

As we will see in the numerical tests, the sets Ω_G^p and Ω_R^p, depending on the shape of the surface, can be made of several disjoint parts, that is $\Omega_G^p = \Omega_G^p(1) \cup \ldots \cup \Omega_G^p(n_G)$ and $\Omega_R^p = \Omega_R^p(1) \cup \ldots \cup \Omega_R^p(n_R)$.

3.4   Predictability of the Number of Global Solutions

Let us suppose that the system (9) always admits two solutions $\hat{n}$ and $\hat{\hat{n}}$. In fact, (9) could have no solution in some points where the greylevels do not perfectly match the Lambertian model. Nevertheless, we know under this assumption that there exist either one or two possible normals in each twice-lighted point $(x, y) \in \overline{\Omega}$. If moreover the surface to be reconstructed is supposed to be differentiable everywhere, then the number of global normal fields is predictable. For example, if Ω_R^p is empty while Ω_G^p is not empty, then the normal field is unique, since all the normals point toward the twice-lighted part $\mathcal{S}_U$ of $\mathcal{S}$ which lies between $\mathcal{S}_G$ and $\mathcal{S}_R$ (see Figs. 1 and 2). More generally, this analysis of the problem allows us to predict the number of global solutions.

Another interesting advantage one can take from the study of the sets Ω_G^p and Ω_R^p is related to the PDE approach (7). For example, let us suppose that no pixel lying on the boundary $\partial\Omega$ belongs to Ω_R^p. This means that there are two different normal fields along the boundary i.e., two different values for ∇u according to (2), and finally two boundary conditions $u(x, y) = g(x, y)$ considering:

$$g(\beta(t)) = g_1(\beta(0)) + \int_0^t \nabla u(\beta(s))\beta'(s)ds \tag{14}$$

where $\beta(t)$ is a parametrization of $\partial\Omega$. Then, according to Theorem 1, we can approximate the height of the surface with two values in this case (supposing one more time that the surface is differentiable on the boundary $\partial\Omega$).

4   Numerical Tests

Here we present some numerical tests on the synthetic surfaces shown in Fig. 3: the surface on the left is differentiable everywhere, whereas that on the right is only Lipschitzian (differentiable almost everywhere).

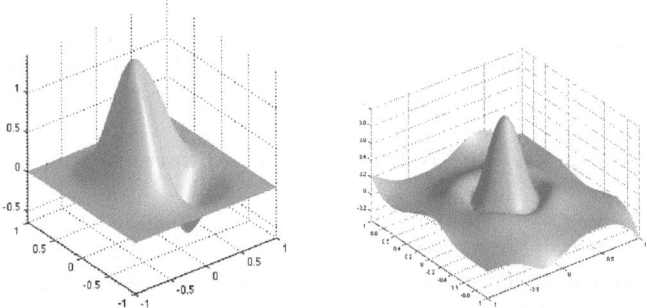

Fig. 3. Surfaces S_{Regular} (left) and $S_{\mathrm{Lipschitz}}$ (right) used in the numerical tests

4.1   Let's Count the Solutions!

A first example uses the pair of images of S_{Regular} shown in Fig. 4, over which the two sets Ω_G^p and Ω_R^p are superimposed. Below each image, the spherical coordinates (φ, θ) of the light vector, such that $\omega = (\sin\varphi\cos\theta, \sin\varphi\sin\theta, \cos\varphi)$, are given.

In order to understand how to count the solutions, let us introduce the sets S_U and S_D, which are the subsets of the Gaussian sphere lying, respectively, upon and below the geodesic S_R.

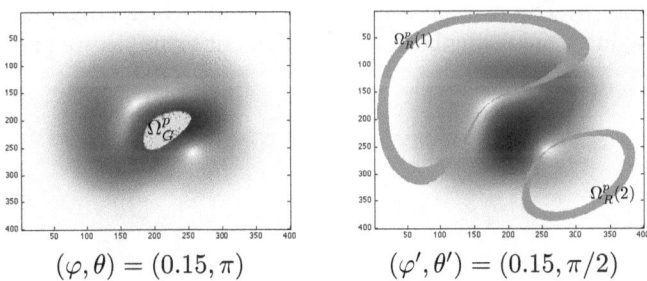

$$(\varphi, \theta) = (0.15, \pi) \qquad\qquad (\varphi', \theta') = (0.15, \pi/2)$$

Fig. 4. Pair of synthetic images of the surface S_{Regular} (400×400 pixels) used to count the solutions. Note that all the pixels are twice-lighted.

Now, taking into account that these sets can be mapped to the reconstruction domain $\overline{\Omega}$ (namely $\Omega_U^p = \Omega_U^p(1)\cup\ldots\cup\Omega_U^p(n_U)$ and $\Omega_D^p = \Sigma_D^p(1)\cup\ldots\cup\Omega_D^p(n_D)$), we show in Fig. 5 the four solutions deduced from the sets previously described.

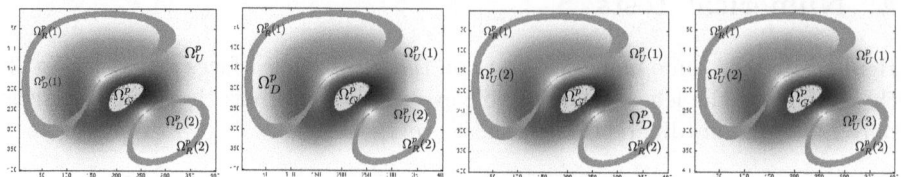

Fig. 5. All possible combinations that allow us to predict the number of solutions

4.2   Combining the Differential and Non-differential Approaches

A second example uses the pair of images of $S_{\text{Lipschitz}}$ shown in Fig. 6, over which the set of points where the surface is not differentiable is superimposed in blue. A problem that we want to avoid is the presence of pixels of Ω_R^p on the boundary $\partial\Omega$, since this would give rise to an ambiguity. In fact, in each pixel of Ω_R^p, the normal can cross the geodesic S_R, passing from S_D to S_U (or vice versa) or remain on the same side of S_R. With this aim we choose the light vectors ω and ω' very close to each other, in order to reduce the sizes of S_G and S_R, and therefore to reduce those of Ω_G^p and Ω_R^p. As in the first test, if some pixels of $\partial\Omega$ belong to Ω_R^p, then we are able to count the different boundary conditions. In our example, we find that $\Omega_G^p = \emptyset$ and $\Omega_R^p = \emptyset$. This means that there are only two possible boundary conditions, but only one is admissible for a correct approximation of the surface. The remaining question is thus: is it possible to find the correct boundary condition?

Once the two possible normal fields are computed along $\partial\Omega$, we can integrate them using a very simple method which consists in fixing the height in some reference pixel and then computing the height along $\partial\Omega$ as Strat did (see [8]). It is well-known that the integration of an irrotational field along a closed contour is equal to zero [7]. Therefore, it is possible to decide which boundary condition is the right one, just comparing both integrals. In our example, we find an $L^\infty(\partial\Omega)$

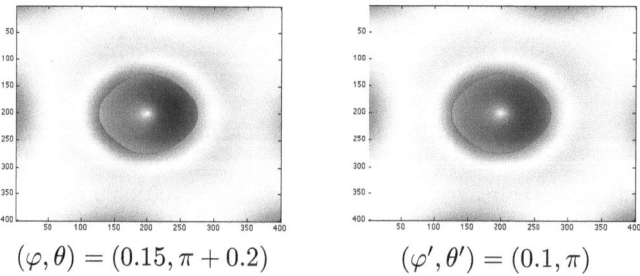

$$(\varphi, \theta) = (0.15, \pi + 0.2) \qquad (\varphi', \theta') = (0.1, \pi)$$

Fig. 6. Pair of synthetic images of the surface $S_{\text{Lipschitz}}$ (400×400 pixels) used for the approximation of the boundary condition. In blue is emphasized the set of points where the surface is not differentiable.

norm error on $\partial\Omega$ between the real and the predicted boundary conditions equal to 5.38×10^{-3} for S_{Regular} and to 3.325×10^{-2} for $S_{\text{Lipschitz}}$. Unfortunately, due to the lack of space, we cannot show here the whole 3D-reconstructions obtained using the PDE approach (7).

5 Conclusion and Perspectives

In this paper, we addressed photometric stereo using two images only. This particular situation is rarely studied because using more than three images usually suffices to render the problem well-posed. Nevertheless, there are at least two reasons which can validate our work. First, the situation where all the light vectors are coplanar is known to be ill-posed, and reduces to the case with two lights only. Note that this is exactly the case of an outdoors scene lighted by the Sun. Second, the "standard" PS technique supposes that the albedo of the scene is unknown, in order to linearize the problem, but this is not always necessary and can moreover give rise to inaccuracies in the reconstructions. From that point of view, our work is an interesting insight in the non-linear PS problem.

Our main result is to show that the number of solutions of PS using two images is predictable, thanks to particular points where the normal can be estimated without ambiguity. More tests have to be performed in order to more clearly show the accuracy of our approach, since this work was rather theoretical. In addition, it must be questionned to which extent our results could be useful to standard PS, since n images induce C_2^n pairs of images!

References

1. Durou, J.-D., Falcone, M., Sagona, M.: Numerical Methods for Shape-from-shading: A New Survey with Benchmarks. CVIU 109(1), 22–43 (2008)
2. Prados, E., Faugeras, O., Camilli, F.: Shape from Shading: a well-posed problem? Rapport de Recherche 5297, INRIA, Sophia Antipolis, France (August 2004)
3. Woodham, R.J.: Photometric Method for Determining Surface Orientation from Multiple Images. Optical Engineering 19(1), 139–144 (1980)
4. Kozera, R.: Existence and Uniqueness in Photometric Stereo. Appl. Math. Comp. 44(1), 1–103 (1991)
5. Horn, B.K.P., Brooks, M.J.: Shape from Shading. MIT Press, Cambridge (1989)
6. Mecca, R.: A constructive differential approach to photometric shape-from-shading (2010) (preprint)
7. Durou, J.-D., Aujol, J.-F., Courteille, F.: Integrating the Normal Field of a Surface in the Presence of Discontinuities. In: Cremers, D., Boykov, Y., Blake, A., Schmidt, F.R. (eds.) EMMCVPR 2009. LNCS, vol. 5681, pp. 261–273. Springer, Heidelberg (2009)
8. Horn, B.K.P., Brooks, M.J.: The Variational Approach to Shape From Shading. CVGIP 33(2), 174–208 (1986)

Von Kries Model under Planckian Illuminants

Michela Lecca and Stefano Messelodi

Fondazione Bruno Kessler - Center for Information Technology
via Sommarive 18, Povo, 38123 Trento, Italy
{lecca,messelod}@fbk.eu

Abstract. Planckian illuminants and von Kries diagonal model are commonly assumed by many computer vision algorithms for modeling the color variations between two images of a same scene captured under two different illuminants. Here we present a method to estimate a von Kries transform approximating a Planckian illuminant change and we show that the Planckian assumption constraints the von Kries coefficients to belong to a ruled surface, that depends on physical cues of the lights. Moreover, we provide an approximated parametric representation of such a surface, making evident the dependence of the von Kries transform on the light color temperature and on the intensity.

1 Introduction

A same scene captured under two different lights produces different colors. This is because the color of an image recorded by a camera depends on the spectral power of the light illuminating the scene, on the geometrical and physical conditions of the scene and on the characteristics of the device used for the acquisition. Understanding the relation between the colors of two scenes imaged under different light is a crucial task for its many applications, like image retrieval and indexing [18], image segmentation [16], and shadow removal [2].

The von Kries diagonal model and/or Planckian illuminants are two common assumptions for modeling the color variations due to light changes. In this paper we investigate the relation between these two hypotheses.

The von Kries model approximates an illuminant change by a linear diagonal map that rescales independently the camera responses. The diagonal elements of the matrix representing this map are named *von Kries coefficients* and they completely determine the color transform. Despite its simplicity, the von Kries model has been proved to be a good approximation for color variations due to a photometric change [4], [5], [1].

Planckian illuminants are lights whose spectral power can be expressed by Planck's law, i.e. they behave like a black body radiator. Many natural lights such as the sunlight and fluorescent lamps satisfy Planck's approximation. This hypothesis is used in many recent works for removing shadows from pictures [8], [7]. Changes of Planckian illuminants are commonly modeled by a linear map, called *Bradford transform* and defined on the XYZ color space [11], [20].

G. Maino and G.L. Foresti (Eds.): ICIAP 2011, Part I, LNCS 6978, pp. 296–305, 2011.

The contributions of our work are three:

1. given a Bradford transform τ relating two Planckian illuminants, we estimate its diagonal approximation $\mathcal{K}$ by a least square based method, that computes the von Kries coefficients of $\mathcal{K}$ by minimizing a L^1 distance from images re-illuminated by τ and $\mathcal{K}$; according to the work [4], we empirically show that a linear diagonal map suffices for describing a Planckian illuminant change;
2. we show that under the Planckian assumption the von Kries coefficients are not independent each to other but they form a ruled surface parametrized by the physical cues of the varying illuminants;
3. finally, we derive an approximated equation for the von Kries surface and we show how it can be used for computing the color temperature and intensity of an illuminant.

Our empirical analysis has been carried out on two public real-world datasets [15],[3] and on some synthetic data generated from them.

Section 2 describes the von Kries model; Section 3 defines the Planckian lights and illustrates the Bradford model. Section 4 reports our experimental analysis and conclusions.

2 The von Kries Model

Hereafter we assume that the response of a camera to the light reflected from a point x in a scene is coded in a triplet $\mathbf{p}^T(x) = (p_0(x), p_1(x), p_2(x))$, where

$$p_i(x) = \int_\Omega E(\lambda) S(\lambda, x) F_i(\lambda) \, d\lambda, \quad i = 0, 1, 2. \tag{1}$$

λ is the wavelength of the light illuminating the scene, E its spectral power distribution, S the reflectance of the surface to which the point belongs, and F_i is the spectral sensitivity function of the sensor. The integral ranges over the visible spectrum, i.e. $\Omega = [380, 780]$ nm.

The von Kries diagonal model approximates the spectral sensitivity of the camera sensor by a delta function, i.e. it assumes that each sensor responds only to a single wavelength of light: $F_i(\lambda) = \delta(\lambda - \lambda_i)$, for each $i = 0, 1, 2$. With this assumption, the responses $\mathbf{p}(x)$ and $\mathbf{p}(x)$ at x taken by the same camera under two different illuminants are linearly related by the following Equation:

$$(p_0(x), p_1(x), p_2(x)) = \left(\frac{E(\lambda_0)}{E'(\lambda_0)} p_0'(x), \frac{E(\lambda_1)}{E'(\lambda_1)} p_1'(x), \frac{E(\lambda_2)}{E'(\lambda_2)} p_2'(x) \right) \tag{2}$$

i.e. the von Kries diagonal model approximates the change of illuminant mapping $\mathbf{p}$ onto $\mathbf{p}'$ by a linear transform that rescales each channel independently.

In the following, for each $i = 0, 1, 2$, we set $\alpha_i := E(\lambda_i)[E'(\lambda_i)]^{-1}$ and we refer to the parameters α_0, α_1 and α_2 as the *von Kries coefficients*.

3   Planckian Illuminants and Bradford Transform

An illuminant is said *Planckian* if its spectral power is given by the Planck law

$$E(\lambda, T, I) = Ic_1\lambda^{-5}\left(e^{\frac{c_2}{\lambda T}} - 1\right)^{-1}. \tag{3}$$

In this Equation, variables λ, I and T denote respectively the wavelength, the intensity and the color temperature of the illuminant, while the terms c_1 and c_2 are constants, more precisely $c_1 = 3.74183 \cdot 10^{-16}$ W m^2 and $c_2 = 1.4388 \cdot 10^{-2}$ K m (W = Watt, m = meter, K = Degree Kelvin). The intensity of the light describes its brightness and the color temperature is a measurement in Degrees Kelvin of its hue. For instance, the color temperature of a candle flame ranges over [1850, 1930] K, while the sun light at sunrise or at sunset has a color temperature between 2000 and 3000 K.

The color of a Planckian light is often codified by the 2D vector χ of its chromaticities in the CIE XYZ color space. The computation of the color temperature of a light source is not easy because it differs from light to light upon the physical nature of the light. The chromaticities of the most Planckian illuminants have been tabulated empirically [10], [9]. Approximated formulas are also available [14].

A color change due to a variation of Planckian illuminants is modeled by the *Bradford transform* [11], [20]. This relates the XYZ coordinates $[X, Y, Z]$ and $[X', Y', Z']$ of the RGB responses $\mathbf{p}$ and $\mathbf{p}'$ by the linear transform

$$[X', Y', Z']^T = M\mathcal{D}M^{-1}[X, Y, Z]^T, \tag{4}$$

where M is the *Bradford matrix* and $\mathcal{D}$ is a diagonal matrix representing the relation between the colorimetric properties (color temperatures and intensities) of σ and σ'.

Bradford matrix has been obtained empirically from Lam's experiments described in [11]:

$$M = \begin{bmatrix} 0.8951 & 0.2664 & -0.1614 \\ -0.7502 & 1.7135 & 0.0367 \\ 0.0389 & -0.0685 & 1.0296 \end{bmatrix} \text{ and } \mathcal{D} = \frac{Y_\sigma}{Y_{\sigma'}} \begin{bmatrix} \frac{x}{y} \frac{y'}{x'} & 0 & 0 \\ 0 & 1 & 0 \\ 0 & 0 & \frac{1-x}{y} \frac{-y'}{1-x'} \frac{y'}{-y'} \end{bmatrix}$$

where $[x_\sigma, y_\sigma]$ and $[x_{\sigma'}, y_{\sigma'}]$ are the chromaticities of the color temperatures of σ and σ' respectively. Y_σ and $Y_{\sigma'}$ are the Y coordinates of the white reference of the illuminants σ and σ' respectively.

In the RGB space, the Bradford transform can be re-written as follows:

$$\mathbf{p}'^T = CM\mathcal{D}M^{-1}C^{-1}\mathbf{p}^T := B\mathbf{p}^T, \tag{5}$$

where C is the 3×3 matrix mapping the XYZ coordinates into the RGB coordinates and B indicates the product $CM\mathcal{D}M^{-1}C^{-1}$.

4   Von Kries Coefficients under Planckian Illuminants

Here we analyze the relation between the von Kries model and the Planckian assumption. The Section is organized as follows. In Subsection 4.1 we provide a method for approximating a Planckian illuminant change by a von Kries transform. In Subsection 4.2 we measure the goodness of the proposed approximation. In Subsection 4.3 we show how the von Kries coefficients are constrained by Planckian illuminants and how this fact can be used for determining the physical cues of the illuminant of an image. Finally, in Subsection 4.4, we summarize the results we obtained and we outlines our future work.

The experiments we describe here, have been carried out on the public databases Outex [15] and UEA dataset [3]. Both these databases consist of images taken under three different Planckian illuminants.

The Outex database collects different image sets for empirical evaluation of texture classification and segmentation algorithms. In this work we consider the test set named Outex_TC_00014, which consists of three sets of 1360 texture images viewed under the illuminants INCA, TL84 and HORIZON with color temperature 2856 K, 4100 K and 2300 K respectively. Figure 1 shows an image of Outex taken under these illuminants.

Fig. 1. Outex: a texture image taken under the Planckian illuminants INCA, TL84 and HORIZON (from left to right)

The UEA dataset comprises 28 design patterns, each one captured under 3 illuminants with 4 different cameras. The illuminants are indicated by Ill A (tungsten filament light, 2856 K), Ill D65 (simulated daylight, 6500 K), and TL84 (fluorescent tube, 4100 K). An example is shown in Figure 2.

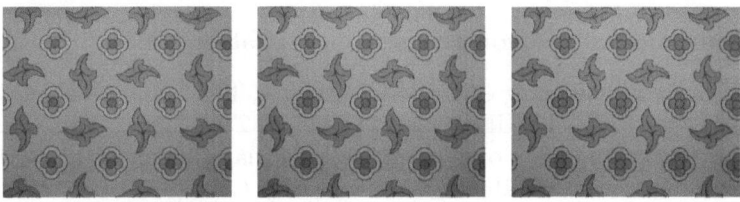

Fig. 2. UEA dataset: an image taken by Camera 1 under the Planckian illuminants Ill A, TL84 and Ill D65 (from left to right)

4.1 Estimating the von Kries Approximation

The algorithm we propose for estimating the von Kries approximation does not consider possible affine distortions between the re-illuminated images, like for instance change of scales or of in-plane orientation. For affine transformed images, other methods can be used, as for instance [12], [13].

Given an image $\mathcal{A}$ taken under a source illuminant σ and a Planckian illuminant change τ modeled by the Bradford transform (5), we define the von Kries approximation $\mathcal{K}$ of τ as the diagonal linear map whose coefficients α_0, α_1, α_2 minimize the following distance:

$$d(\alpha_0, \alpha_1, \alpha_2) := \sum_{x \in \mathcal{A}} \| (B - K)\mathbf{p}(x)^T \|^2 . \tag{6}$$

Here B and K are the matrices associated (with respect to the canonical basis of $\mathbf{R}^3$) to τ and $\mathcal{K}$ respectively. The ith diagonal element K_{ii} of K is the ith von Kries coefficient α_i. The sum is computed over all the pixels x of the image $\mathcal{A}$.

The ith von Kries coefficient minimizing (6) is thus given by

$$\alpha_i = \frac{\sum_x \sum_{j=0}^2 b_{ij} p_i(x) p_j(x)}{\sum_x p_i(x)^2} \tag{7}$$

Hereafter we indicate by $\tau(\mathcal{A})$ and $\mathcal{K}(\mathcal{A})$ the images obtained by re-illuminating $\mathcal{A}$ by τ and $\mathcal{K}$ respectively.

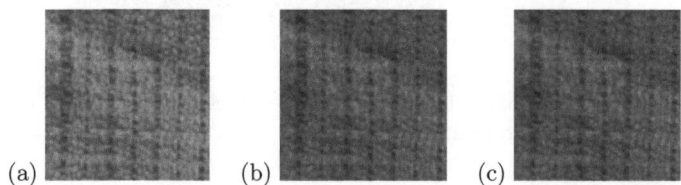

(a)   (b)   (c)

Fig. 3. Outex: The image (a) has been re-illuminated by a transform   mapping the color temperature of INCA (2856 K) onto 6500 K and rescaling its intensity by 0.5. The result is the picture (b). Image (c) is the image (a) remapped by our estimate $\mathcal{K}$. Images (b) and (c) look very similar.

4.2 Evaluation of the von Kries Approximation

For each source illuminant $\sigma \in \mathcal{I} = \{\text{INCA/ILL A, TL84, HORIZON/ILL D65}\}$ of the datasets Outex and UEA, we defined a set $\mathcal{T}$ of 40 Bradford transforms (Equation (5)), mapping σ onto an other Planckian illuminant σ', with color temperature $T_{\sigma'} = (2000 + 1000\,t)$K and intensity $I_{\sigma'} = (0.5 + 0.25\,m)$ with $t = 0, \ldots, 8$, $m = 0, \ldots, 4$. For the considered datasets, no information about the intensity of the source illuminants are provided. Therefore, each transform of $\mathcal{T}$ rescales the intensity I_σ of the source illuminant by 0.5, 0.75, 1.0, 1.25 and 1.5. Figure 3 shows an example of such a transform.

We re-illuminated each image of Outex and UEA captured under the illuminant σ by these transforms and for each of them we estimate the von Kries approximation.

First of all, we note that the von Kries coefficients α_0, α_1 and α_2 do not depend on the image $\mathcal{A}$ used in Equation (6). In fact, let $\alpha_i^{\mathcal{A}}$, $i = 0, 1, 2$, be the von Kries coefficients of the transform $\mathcal{K}$ approximating τ for the image $\mathcal{A}$. Indicated by N_{DB} the number of images in the database viewed under the illuminant σ, we computed the mean values

$$\alpha_i = \frac{1}{N_{DB}} \sum_{\mathcal{A}} \alpha_i^{\mathcal{A}}, \qquad i = 0, 1, 2. \tag{8}$$

and we observed that the standard deviations from these values are very small, being them 0.03 on average, that is the 3 % of the measure.

For each pair of images $(\tau(\mathcal{A}), \mathcal{K}(\mathcal{A}))$, we measure the goodness of the estimate of $\mathcal{K}$ on $\mathcal{A}$ by the L^1 RGB distance between the images $\tau(\mathcal{A})$ and $\mathcal{K}(\mathcal{A})$:

$$d_{\tau(\mathcal{A}), \mathcal{K}(\mathcal{A})} = \frac{1}{N} \sum_{x} |\mathbf{p}(x)^{\tau(\mathcal{A})} - \mathbf{p}(x)^{\mathcal{K}(\mathcal{A})}|, \tag{9}$$

where N is the number of pixels of $\mathcal{A}$. Distance (9) has been normalized to range over $[0, 1]$. Closer to zero $d_{\tau,\mathcal{K}}$ is, more accurate the von Kries approximation is. Then, we measure the accuracy of the approximation $\mathcal{K}$ by the mean value $d_{\tau,\mathcal{K}}$ of $d_{\tau(\mathcal{A}), \mathcal{K}(\mathcal{A})}$ averaged on the number of database images (1360 for Outex, 28 for each camera of UEA). To measure how much a transform τ modifies the image, we compute the RGB distance $d_{\mathcal{A},\mathcal{A}'}$: we expected that the mean value of $d_{\mathcal{A},\mathcal{A}'}$ is much greater than $d_{\tau,\mathcal{K}}$.

Figure 4 shows the results for Outex database, for which the accuracy $d_{\tau,\mathcal{K}}$ varies from 0.0 to 0.057 and is 0.022 on average. Similar results are obtained for UEA, where the mean value of $d_{\mathcal{A},\mathcal{A}'}$ varies from 0.002 to 0.052 and its average is 0.014. The value of the mean RGB distance $d_{\mathcal{A},\mathcal{A}'}$ ranges over $[0.005, 0.333]$ with average 0.177 for Outex, and over $[0.012, 0.284]$ with average 0.176 for UEA.

According to [4], we found that the von Kries model is a good approximation for a Planckian illuminant change. However, the approximation accuracy decreases by increasing the intensity and the color temperature of the illuminant, i.e. by increasing the number of saturated pixels, whose percentage is particularly high (greater than 50%) for the transforms with a color temperature gap of 4000 K and intensity 1.25 times greater than that of the original image. Figure 3 shows an example of color correction provided by our estimate. We got analogous accuracies for the estimate of the von Kries map that approximates the changes between two illuminants σ and σ' of $\mathcal{I}$ on the real-world images of Outex and UEA.

4.3 Constraints on the von Kries Coefficients

Our experiments show that the von Kries coefficients we estimated are not independent each to other. We claim that the Planckian assumption constraints the

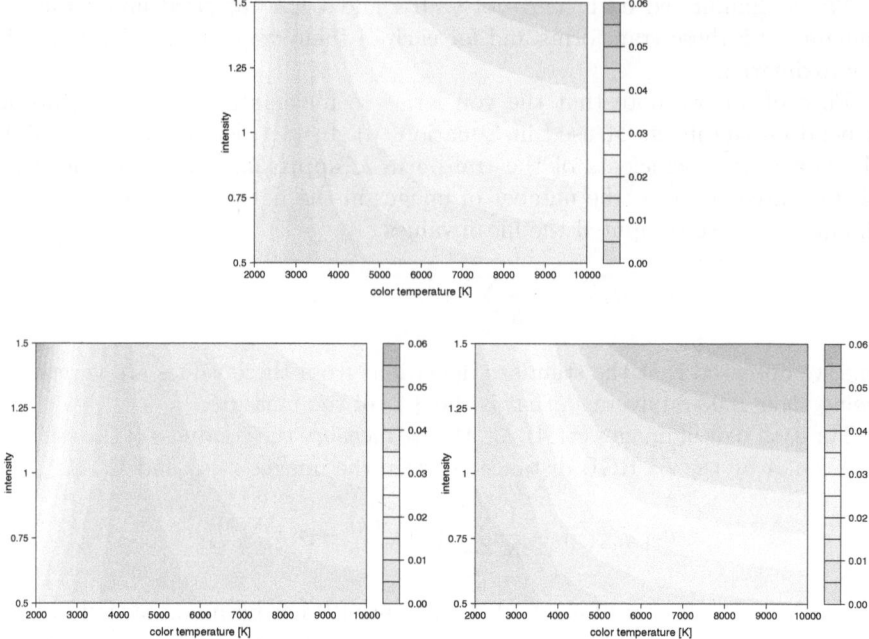

Fig. 4. Outex: INCA (top), TL84 (bottom, left), and HORIZON (bottom, right). L^1 RGB distance between the images re-illuminated by the Planckian model and the images re-illuminated by our von Kries approximation. The L^1 RGB distance varies from 0.0 to 0.06.

triplets $(\alpha_0, \alpha_1, \alpha_2)$ to belong to a ruled surface $\mathcal{S}$ (that we name *von Kries surface*) dependent on the color temperatures and on the variation of the illuminant intensities.

Each source illuminant σ with color temperature T_σ and intensity I_σ defines a von Kries surface $\mathcal{S}$ given by

$$\alpha_i = \alpha_i(T_{\sigma'}, I_{\sigma'}) \qquad i = 0, 1, 2, \tag{10}$$

where the α_i's are the von Kries coefficients of the map $\mathcal{K}$ that approximates the change moving σ onto a Planckian illuminant σ' with color temperature $T_{\sigma'}$ and intensity $I_{\sigma'}$.

From Equations (5) and (7), we have that each von Kries coefficient can be re-written as $\alpha_i = \frac{Y}{Y_\prime}\alpha_i^*(T_{\sigma'})$, where α_i^* is the ith von Kries coefficient of the linear map $\mathcal{K}$ that approximates the Bradford transform τ^* mapping T_σ onto $T_{\sigma'}$ and leaving unchanged the light intensity I_σ. By observing that $\frac{Y}{Y_\prime} = \frac{I}{I_\prime}$, we have that $\alpha_i = \frac{I}{I_\prime}\alpha_i^*(T_{\sigma'})$. Let K and B^* be the matrices representing $\mathcal{K}$ and τ^* in the canonical basis of $\mathbf{R}^3$. Since $\mathcal{K}$ is an approximation of τ^*, there exists a matrix H such that $K = HB$ and the difference $K - HB$ is close to the

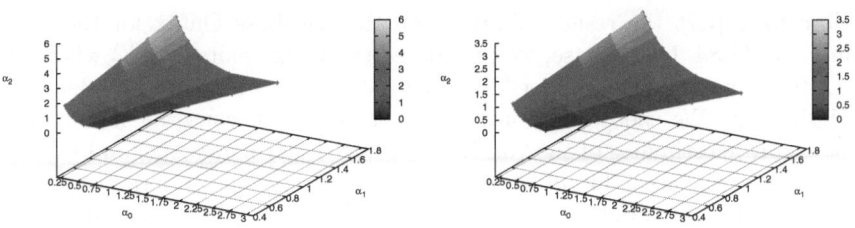

Fig. 5. UEA dataset: von Kries surfaces of cameras 1 (top) and 4 (bottom) for the illuminant σ = ILL A

identity matrix. Therefore, $\alpha_i^*(T_{\sigma'}) = \sum_{j=0}^{2} h_{ij} b_{ji}^*(T_{\sigma'})$, where $i = 0$, 1, 2, and h_{ij} is the ij-th element of H and b_{ji}^* is the ij-th element of B^*. We have that

$$\alpha_i(I_{\sigma'}, T_{\sigma'}) = \frac{I_\sigma}{I_{\sigma'}} \sum_{j=0}^{2} h_{ij} b_{ii}^*(T_{\sigma'}). \tag{11}$$

Equation (11) makes evident the relation between the von Kries coefficients and the photometric properties of the illuminants σ and σ': it describes a ruled surface depending on the intensity and on the color temperature of σ'. By varying discretely the intensity and the color temperature of the source illuminant σ, we obtain a sheaves of such surfaces.

As a consequence, we have that von Kries surface $\mathcal{S}$ is completely determined by the Equation of the 3D curve $\alpha^*(T_{\sigma'}) = (\alpha_0^*(T_{\sigma'}), \alpha_1^*(T_{\sigma'}), \alpha_2^*(T_{\sigma'}))$, because each point on $\mathcal{S}$ is a rescaled version of a point onto $\alpha^*(T_{\sigma'})$. Thus, the von Kries maps that approximates Planckian variations of the color temperature but not of the intensity suffices for determining the von Kries surface.

We remark that for each fixed source illuminant σ there exists a von Kries surface, that differs from device to device. This is because the values of the von Kries coefficients depend on the acquisition device (see for instance, Figure 5).

Finally, we briefly discuss a possible usage of the von Kries surfaces for recovering the color temperature and the intensity of an illuminant, a crucial task for many imaging applications [17], [19].

Let σ be a Planckian illuminant with known color temperature T_σ and intensity I_σ and let $\mathcal{S}$ the von Kries surface of a camera C computed with respect to the source illuminant σ. Let $\mathcal{A}$ and $\mathcal{A}'$ be two images of the same scene taken by C under the illuminants σ and σ' respectively. Of course we suppose σ and σ' to be Planckian. We employ the von Kries surface $\mathcal{S}$ to determine the color temperature $T_{\sigma'}$ and the intensity $I_{\sigma'}$ of σ' as follows: (1) we estimate the von Kries coefficients $\beta_0, \beta_1, \beta_2$ of the von Kries map relating $\mathcal{A}$ and $\mathcal{A}'$; (2) we compute the triplet $(\alpha_0, \alpha_1, \alpha_2)$ on the von Kries map that has the minimum Euclidean distance from $(\beta_0, \beta_1, \beta_2)$ and we return the color temperature and intensity correspondent to $(\alpha_0, \alpha_1, \alpha_2)$.

Here we report the results obtained on the database Outex for the source illuminant TL84. In this case, we considered the image pairs $(\mathcal{A}, \mathcal{A}')$ where $\mathcal{A}$ is an Outex picture imaged under TL84 and $\mathcal{A}'$ is the same picture under INCA or HORIZON. We choose this case among the others because for it we got the most accurate estimates of the von Kries coefficients. Figures 6(a) and (b) show the von Kries surface $\mathcal{S}$ for TL84 with the estimates of the von Kries coefficients of the transforms mapping (a) TL84 onto INCA and (b) TL84 onto HORIZON, respectively. The estimates we obtain determine a range of color temperatures and intensities. In particular, the color temperature of INCA varies over [3000, 4000] K, but the 70% about of the estimates are closer to 3000 K than to 4000 K. For HORIZON the color temperature ranges over [2000, 4000] K with the most part of the data (about the 90 %) in [2000 K, 3000 K]. Similarly, we obtained a variability range for the intensity, with the 99% of the estimates between 1.0 and 1.25 for INCA and between 0.75 and 1.0 for HORIZON.

The results can be further refined, by considering a finer von Kries surface and by restricting the search for the triplet realizing the minimum distance to the ranges found before. Nevertheless, in general, obtaining an accurate estimate of these photometric parameters is a hard problem [19], also when calibrated images are used [6].

4.4   Conclusions and Future Directions

In this work we show empirically that the von Kries model approximates well a change of Planck's illuminants. The main consequence is that under Planck's hypothesis, the von Kries coefficients are the points of a ruled surface, whose mathematical expression highlights their dependency on the physical properties of the light and on the camera cues. Here we used the von Kries surfaces for estimating the color temperature and intensity of the illuminant under which an image has been captured. Our future work will be focus on other possible applications of this results, as for instance the development of device-independent approaches for the estimation of scene illuminants.

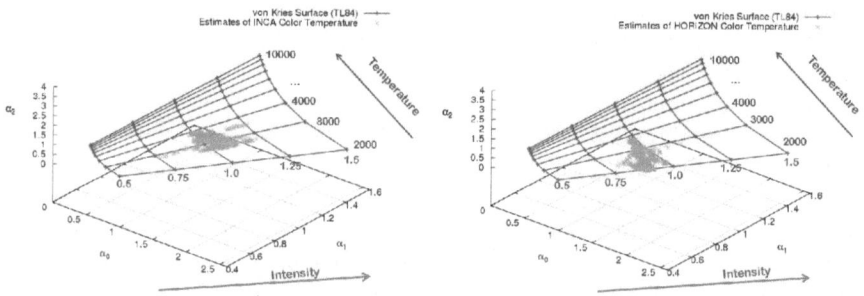

Fig. 6. Outex: color temperature and intensity estimates for illuminant INCA (on left) and HORIZON (on right) by using the von Kries surface with respect to TL84

References

1. Barnard, K., Ciurea, F., Funt, B.: Sensor sharpening for computational color constancy. J. Opt. Soc. Am. A 18, 2728–2743 (2001)
2. Finlayson, G.D., Hordley, S.D., Drew, M.S.: Removing shadows from images. In: Heyden, A., Sparr, G., Nielsen, M., Johansen, P. (eds.) ECCV 2002. LNCS, vol. 2353, pp. 129–132. Springer, Heidelberg (2002)
3. Finlayson, G., Schaefer, G., Tian, G.Y.: The UEA uncalibrated colour image database. Technical Report SYS-C00-07, School of Information Systems, University of East Anglia, Norwich, United Kingdom (2000)
4. Finlayson, G.D., Drew, M.S., Funt, B.V.: Diagonal transforms suffice for color constancy. In: Proc. of International Conference of Computer Vision (1993)
5. Finlayson, G.D., Drew, M.S., Funt, B.V.: Spectral sharpening: sensor transformations for improved color constancy. J. Opt. Soc. Am. A 11(5), 1553–1563 (1994)
6. Finlayson, G.D., Hordley, S.D., Hubel, P.M.: Color by correlation: a simple, unifying framework for color constancy. IEEE Transactions on Pattern Analysis and Machine Intelligence 23(11), 1209–1221 (2001)
7. Finlayson, G.D., Hordley, S.D., Lu, C., Drew, M.S.: On the removal of shadows from images. IEEE Trans. Pattern Anal. Mach. Intell. 28, 59–68 (2006)
8. He, Q., Chu, C.-h.H.: Intrinsic images by fisher linear discriminant. In: Bebis, G., Boyle, R., Parvin, B., Koracin, D., Paragios, N., Tanveer, S.-M., Ju, T., Liu, Z., Coquillart, S., Cruz-Neira, C., Müller, T., Malzbender, T. (eds.) ISVC 2007, Part II. LNCS, vol. 4842, pp. 349–356. Springer, Heidelberg (2007)
9. Judd, D.B., MacAdam, D.L., Wyszecki, G., Budde, H.W., Condit, H.R., Henderson, S.T., Simonds, J.L.: Spectral distribution of typical daylight as a function of correlated color temperature. J. Opt. Soc. Am. 54(8), 1031–1036 (1964)
10. Kelly, K.L.: Lines of constant correlated color temperature based on macadam's (u,) uniform chromaticity transformation of the cie diagram. J. Opt. Soc. Am. 53(8), 999–1002 (1963)
11. Lam, K.M.: Metamerism and colour constancy (1985)
12. Lecca, M., Messelodi, S.: Computing von kries illuminant changes by piecewise inversion of cumulative color histograms. ELCVIA 8(2) (2009)
13. Lecca, M., Messelodi, S.: Illuminant Change Estimation via Minimization of Color Histogram Divergence. In: Trémeau, A., Schettini, R., Tominaga, S. (eds.) CCIW 2009. LNCS, vol. 5646, pp. 41–50. Springer, Heidelberg (2009)
14. McCamy, C.S.: Correlated color temperature as an explicit function of chromaticity coordinates. Color Research & Application 17(2), 142–144 (1992)
15. Ojala, T., Topi, M., Pietikäinen, M., Viertola, J., Kyllönen, J., Huovinen, S.: Outex - new framework for empirical evaluation of texture analysis algorithms. In: Proc. of the 16 th ICPR 2002, vol. 1, p. 10701. IEEE Computer Society, Los Alamitos (2002)
16. Ozden, M., Polat, E.: A color image segmentation approach for content-based image retrieval. Pattern Recognition, 1318–1325 (2007)
17. Reinhard, E., Ashikhmin, M., Gooch, B., Shirley, P.: Color transfer between images. IEEE Comput. Graph. Appl. 21(5), 34–41 (2001)
18. Schettini, R., Ciocca, G., Zuffi, S.: A survey of methods for colour image indexing and retrieval in image databases. In: Color Imaging Science: Exploiting Digital Media (2001)
19. Tominaga, S., Ishida, A., Wandell, B.A.: Color temperature estimation of scene illumination by the sensor correlation method. Syst. Comput. Japan 38, 95–108 (2007)
20. Takahama, K., Nayatani, Y., Sobagaki, H.: Formulation of a nonlinear model of chromatic adaptation. Color Research & Application 6(3) (1981)

Colour Image Coding with Matching Pursuit in the Spatio-frequency Domain

Ryszard Maciol, Yuan Yuan, and Ian T. Nabney

School of Engineering and Aplied Science,
Aston University, Aston Triangle, Birmingham, B4 7ET, United Kingdom
{maciolrp,y.yuan1,i.t.nabney}@aston.ac.uk

Abstract. We present and evaluate a novel idea for scalable lossy colour image coding with Matching Pursuit (MP) performed in a transform domain. The idea is to exploit correlations in RGB colour space between image subbands after wavelet transformation rather than in the spatial domain. We propose a simple quantisation and coding scheme of colour MP decomposition based on Run Length Encoding (RLE) which can achieve comparable performance to JPEG 2000 even though the latter utilises careful data modelling at the coding stage. Thus, the obtained image representation has the potential to outperform JPEG 2000 with a more sophisticated coding algorithm.

Keywords: Colour image coding, Matching Pursuit, Wavelets, Run Length Encoding.

1 Introduction

1.1 Colour Image Coding

Due to the large size of raw image and video data files there is great demand for lossy compression methods. Most still image and video data are in colour and are represented for display in RGB colour space thus tripling the raw file size comparing to grayscale. Nevertheless, most of the research effort in algorithms for lossy colour image compression is focused on single-channel methods which are then extended to exploit inter-colour redundancies by applying decorrelating transforms. The current coding standard JPEG 2000 utilises the YC_bC_r transform which attempts to separate luminance (Y) from chrominance (C). Coarser quantisation of C channels improves coding performance without significant visual degradation [2]. JPEG 2000 also utilises the concept of transform coding using discrete wavelets and supports the generation of scalable bit-streams. Sparse approximation techniques that raised interest in the field of image and video compression in the late 90s [1,14] could potentially be the next step in scalable image coding. Moreover they provide new options to exploit inter-channel redundancies in colour images [5,9].

G. Maino and G.L. Foresti (Eds.): ICIAP 2011, Part I, LNCS 6978, pp. 306–317, 2011.
© Springer-Verlag Berlin Heidelberg 2011

Algorithm 1. Single channel Matching Pursuit [11].

Initialisation: $Rf_1 = f$.
for $n = 1$ to N **do**
    Find atom $g_n \in \mathcal{D}$ such that:
    $| \langle Rf_n, g_n \rangle | = \max_{g \in \mathcal{D}} (| \langle Rf_n, g \rangle |)$.
    Update residual:
    $Rf_{n+1} = Rf_n - \langle Rf_n, g_n \rangle g_n$.
end for

1.2 Matching Pursuit

Mallat and Zhang proposed in 1993 [11] a simple greedy technique to obtain a sparse approximation of a given signal f from a Hilbert space $\mathcal{H}$. The algorithm, called Matching Pursuit (MP), finds the approximation of f by a sum of N atoms g_{γ_n} selected from a dictionary $\mathcal{D}$:

$$f \approx \sum_{n=1}^{N} \langle Rf_n, g_{\gamma_n} \rangle g_{\gamma_n}. \tag{1}$$

The dictionary is a set of functions from $\mathcal{H}$ normalised to have unit norm. For any dictionary that spans $\mathcal{H}$ a decomposition given by Eq. 1 converges to f as $N \to \infty$ [11]. Full Search MP, used in image and video compression applications [3,4] for single channel signals, is summarised by Alg. 1. At each iteration the atom most correlated with the actual signal residual Rf_n is selected and removed from Rf_n.

1.3 Multi-channel Matching Pursuit

MP can be extended to decompose vector signals without losing the convergence property [9]. The atom that, according to some criterion, best matches all the components of the input signal is selected. Multi-channel MP for RGB images is summarised by Alg. 2.

Algorithm 2. Multi-channel Matching Pursuit for RGB images.

Initialisation: $Rf_1^r = f^r, Rf_1^g = f^g, Rf_1^b = f^b$.
for $n = 1$ to N **do**
    Find atom $g_n \in \mathcal{D}$ that maximises the L_2-norm:
    $g_n = \max_{g \in} \sqrt{\langle Rf_n^r, g \rangle^2 + \langle Rf_n^g, g \rangle^2 + \langle Rf_n^b, g \rangle^2}$.
    Update residuals:
    $Rf_{n+1}^r = Rf_n^r - \langle Rf_n^r, g_n \rangle g_n$.
    $Rf_{n+1}^g = Rf_n^g - \langle Rf_n^g, g_n \rangle g_n$.
    $Rf_{n+1}^b = Rf_n^b - \langle Rf_n^b, g_n \rangle g_n$.
end for

This algorithm was applied in image space (i. e. to raw RGB values) to colour image coding in [5]. The idea was to explore inter-channel correlations and dependencies of a typical image directly in RGB colour space. In the spatio-frequency domain the dependencies between corresponding subbands of R, G and B channels can be even stronger [6]. In this paper we explain the idea of MP performed in the transform domain and apply it for the first time to colour image coding. The first time MP was performed in the wavelet transform domain for grayscale image coding in [19] and for grayscale video coding in [18]. It was shown in [19] that MP with wavelets can achieve a coding performance comparable to JPEG 2000 for grayscale images. We extend the ideas from [19] to colour coding proposing a new method of coding coefficients.

The next section discusses details of our implementation of MP. Section 3 analyses MP performed in the transform domain. Section 4 describes quantisation and coding of the MP data into bit-stream. Section 5 presents coding results and compares performance with JPEG 2000. Finally, Section 6 concludes the paper and gives the ideas for the performance improvement.

2   Implementation of Matching Pursuit

The main shortcoming of MP is high computational complexity of encoder (atom finding process). On the other side decoding (composing an image back) requires just summing up the atoms which makes MP suitable for asymmetric application in which one encodes the stream once and decodes many times. We argue here that using short support separable filters and performing search in a transform domain keeps also the computational complexity of encoder tractable. Separability refers to the property that each 2D dictionary entry is a tensor product of two 1D vectors. The dictionary is fully specified by a typically small set of 1D filters (mother functions). This reduces number of multiply-accumulate operations when calculating convolutions [14].

The MP algorithm is implemented similarly to the *full 2D separable inner product search* from [20]. The maximal inner products and the corresponding atom indexes are stored for each location in the image. At each iteration, inner products have to be recomputed only on a sub-area of the image. For colour coding it has to be done for all channels and requires approximately three times more multiply-accumulate operations than for grayscale. The overall complexity of our MP implementation can be summarised as:

$$\tau^{(sep)} = \tau^{(sep)}_{init} + \sum_{n=1}^{N} \left(\tau^{(sep)}_{update_n} + \tau^{(sep)}_{search_n} \right). \tag{2}$$

If W denotes the maximum length of bases in the dictionary, S_x the width and S_y the height of the image then, following [20], the overall complexity can be estimated as:

$$\tau^{(sep)} = O^{init}(S_x S_y K^2 W) +$$
$$O^{update}(NK^2 W^3) + O^{search}(N S_x S_y). \tag{3}$$

Eq. 3 shows that the size of the dictionary and lengths of bases are critical for complexity of the general MP algorithm with maximum length of basis more important than the number of bases. Moreover, when MP is performed in a transform domain we typically have: $K^2W^3 > S_xS_y$. This implies that recalculation of the inner products is the most demanding part of the algorithm.

3 Matching Pursuit in Transform Domain

MP has been found to be useful for residual video coding [14]. For still image coding the use of non-separable filters of footprint up to quarter of the image size to represent image features at different scales and Fast Fourier Transform (FFT) has been proposed in [4]. The coding performance was comparable to JPEG 2000 at low bit rates. However, matching of long and non-separable filters makes the method from [4] computationally extremely demanding. For practical image coding, as concluded in Sec. 2, one should prefer a dictionary with short filters. When the filters are shorter than 64 samples the FFT slows down the calculation of convolutions. To preserve low complexity and a dictionary capable of capturing image features at different scales the use of the 2D Discrete Wavelet Transform (2D-DWT) has been proposed in [19]. MP decomposition was performed for wavelet subbands. Like the codec in [4] the method proposed in [19] is comparable in coding performance to the JPEG 2000 standard but additionally has a tractable computational complexity.

In this work we present and analyse in more detail the idea of performing MP in transform domain. Performing MP after transformation like DCT or DWT reduces complexity and improves coding performance. Let us start with the simple observation that applying an orthonormal linear transform T does not change the output of MP. If the transform T is linear and preserves inner product,

$$\langle f, g \rangle = \langle T\{f\}, T\{g\} \rangle \text{ for all } f, g \in \mathcal{H}, \tag{4}$$

then the MP decomposition of signal f (see Eq. 1) obtained in the transform domain is:

$$T\{f\} \approx \sum_{n=1}^{N} \langle T\{Rf_n\}, T\{g_{\gamma_n}\} \rangle T\{g_{\gamma_n}\}. \tag{5}$$

In practice it can be computationally easier to match filters in the spatio-frequency domain. To give an example for the discrete case, consider a dictionary entry with support W: $g(t) = 1/\sqrt{W}$ for $t = 1, 2, \ldots, W$. Its DCT or DFT is the Dirac delta $g(\omega) = 1$ with support 1. Performing an inner product with such a short signal requires only one multiplication. It is known that for transforms like DCT, DFT or DWT the filters applied locally in the transform domain correspond to some global structures in the image domain. Therefore MP can be more efficient when performed in the transform domain. In principle, a very similar idea was used indirectly in [4] where convolutions with filters in a dictionary were performed in the Fourier domain. In [19] filters designed for video coding in the image domain were applied to wavelet subbands after performing

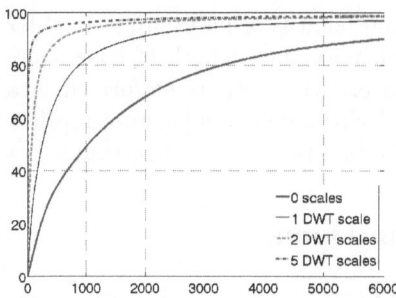

Fig. 1. Percentage of the image energy (y-axis) represented by a given number of atoms (x-axis) using different numbers of wavelet scales (grayscale Goldhill)

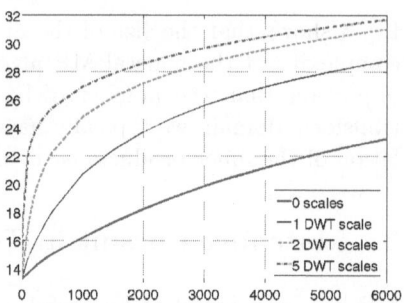

Fig. 2. PSNR performance in dB (y-axis) for a given number of atoms (x-axis) using different numbers of wavelet scales (grayscale Goldhill)

2D-DWT with CDF 9/7 filters from lossy mode of JPEG 2000. As the wavelet transform does not change a signal dimension, the overall size of a dictionary in the image domain remains the same. Thanks to the energy compaction property of DWT, the atoms found in the wavelet domain in initial iterations have high amplitudes. Hence, they contribute more to the whole image energy as shown in Fig. 1. In Fig. 2 we see corresponding values of PSNR. The dictionary applied for wavelet subbands is capable of giving a few orders of magnitude sparser representation than the same dictionary applied in the image domain. Moreover at initial steps of MP there are usually more atoms found in lower frequencies what gives a potential for more efficient coding. The dictionaries we use in this study for colour and grayscale coding were trained using Basis Picking method from [12] on colour and grayscale (i. e. luminance only) Goldhill image respectively. Both dictionaries contain 16 1D bases of maximal footprint 9.

4   Quantisation and Coding

4.1   Quantisation

For data compression applications MP decomposition has to be encoded into a bit-stream. The values $a_n = \langle Rf_n, g_{\gamma_n} \rangle$ have to be quantised (e. g. rounded) to the values A_n. Quantisation is performed inside the MP loop [14] with the aim of correcting the introduced quantisation error during later iterations. For the MP decomposition given by Eq. 1 the Parseval-like equality is satisfied [11]:

$$||f||^2 = \sum_{n=1}^{N} a_n^2 + ||Rf_{N+1}||^2. \tag{6}$$

Eq. 6 is a direct consequence of the update step from Alg. 1. If we replace a_n by A_n in the update step to reflect in-loop quantisation then, for real values, Eq. 6 will change to:

$$||f||^2 = \sum_{n=1}^{N} A_n(2a_n - A_n) + ||Rf_{N+1}||^2. \tag{7}$$

To preserve convergence of the algorithm the energy of residual Rf_n has to keep decreasing [11]. Therefore we may use any quantisation method for which $A_n(2a_n - A_n) > 0$ which is equivalent to a_n, A_n having the same sign and their absolute values to follow Eq. 8 [15].

$$0 < |A_n| < 2|a_n|. \tag{8}$$

Our grayscale implementation utilises Precision Limit Quantisation (PLQ) [13]. The original idea of PLQ is to keep only the most significant bit of a_n and some refinement bits governed by the parameter PL. Then the value $|a_n|$ is quantised to: $|A_n| = r2^k$, where k indicates bitplane and r refinement. The value of the parameter PL is taken to be $PL = 2$, as advised in [19], which in our case means that $r \in \{1.25, 1.75\}$.

The colour codec uses PLQ and Uniform Quantisation. The amplitude with maximal value over the three colour channels (a_n^{max}) is quantised using PLQ and serves as a base for grouping atoms. The atoms with the same quantised absolute value of maximal amplitude ($|A_n^{max}|$) compose one group. We record the channel c_n for which the maximal value occurred. The remaining two amplitudes for the other two colours are quantised using dead-zone uniform scalar quantisation with L bins [7]. The value of L has been experimentally chosen to be as low as $L = 2$ in order to maximally reduce the number of bits required. The two numbers d_n^1 and d_n^2, sent to the encoder, represent either dead-zone or the quantised amplitude with its sign.

4.2   Atom Encoding

After MP decomposition and quantisation, the data to be encoded form a matrix in which rows represent atoms. There are 8 columns containing the following variables for colour coding:

1 :	s_n,	sign of the maximal amplitude,	$s_n \in \{-1, 1\}$,
2 − 3 :	d_n^1, d_n^2,	quantised amplitude differences,	$d_n^* \in \{1, 2, \ldots, 2L + 1\}$,
4 :	c_n,	maximum amplitude colour channel,	$c_n \in \{1, 2, 3\}$,
5 :	w_n,	sub-band index,	$w_n \in \{1, 2, \ldots, 3S + 1\}$,
6 :	λ_n,	2D dictionary entry,	$\lambda_n \in \{1, 2, \ldots, 256\}$,
7 − 8 :	x_n, y_n,	atom location inside the sub-band w_n,	$x_n \in \{1, \ldots, W_{x_n}\}, y_n \in \{1, \ldots, W_{y_n}\}$.

For grayscale coding there are only 5 columns: s_n, w_n, λ_n, x_n and y_n from which only 3 are being reordered. Data from columns 1-6 (or 1-3 for grayscale) are encoded group by group using Alg. 3 based on Run Length Encoding (RLE). The rows are ordered in a lexicographical order recommended for databases indexes [8]. Encoding inside each group is done calling recursively column by column the Alg. 3. The coding performance depends on the column order (see Sec. 5).

Algorithm 3. One stage of encoding.

---

input: $\{v_s\}_{s=1\,2\,\ldots\,n}$ with $s < s'$ v_s $v_{s'}$
$s = 1$
while $s < n$ **do**
   if there are 2 times more symbols than alphabet entries remaining **then**
      encode all zero lengths (if any) and one non-zero run length l
      $s = s + l$
   else
      encode symbol v_s directly
      $s = s + 1$
   end if
end while

Therefore a fixed permutation of columns π is applied prior to the sorting. For each stage of encoding the input of Alg. 3 is the sorted sequence $\{v_s\}$ from an alphabet of the size determined by the column number. At each iteration a decision is made whether to encode the symbol v_s directly or to signal its run length. RLE is used when the run length of 2 or more symbols is expected. An expected run length is indicated by the ratio of the remaining symbols count to the size of the alphabet they can come from. The atom locations (the last two columns) are always encoded as the two raw values x_n and y_n from the ranges $1 \ldots W_{x_n}$ and $1 \ldots W_{y_n}$ respectively, where $W_{x_n} \times W_{y_n}$ is a dimension of the sub-band w_n. All the symbols are sent to the arithmetic coder [17] that uses models which assume uniform distributions for each column. Arithmetic coding allows, knowing the probability distribution of data, to achieve compression ratio close to a theoretical bound given by the Shannon's entropy [17]. Uniform distribution has the highest entropy among the discrete distributions. Therefore the results shown here can serve as the upper bound for the sizes of encoded streams. [1]

5   Coding Performance

As evaluation metric we use PSNR. For colour images it is averaged over RGB channels:

$$PSNR = 10 \log_{10} \left(\frac{3 \cdot 255^2}{MSE_r + MSE_g + MSE_b} \right), \qquad (9)$$

where MSE_r, MSE_g, MSE_b are mean squared errors calculated for R, G and B channels respectively. Although PSNR is known to correlate poorly with human visual perception, especially in the case of colour images, it measures the mathematical properties of the algorithms used. Comparisons with JPEG 2000 are done using the same wavelet filters and the same number of scales $S = 5$. For fair comparison of colour codecs an option of JPEG 2000 which minimises mean squared error (i. e *no_weights* switch for Kakadu implementation [16]) was used.

---

[1] More details about implementation of the whole coding system and its evaluation can be found in [10].

Table 1. Number of bits required for 6000 grayscale atoms for different column orders

image	the best order	the worst order	sub-optimal (π_g) (2,3,1)	$2\frac{worst-best}{worst+best}$
Barbara, 720 × 576	105699 (2,1,3)	115314 (1,3,2)	106577	8.70%
Goldhill, 720 × 576	102978 (2,3,1)	111453 (3,1,2)	102978	7.90%
Lena, 512 × 512	102321 (2,3,1)	110012 (3,1,2)	102321	7.24%
Lighthouse, 768 × 512	104441 (2,1,3)	112076 (1,3,2)	104905	7.05%
Parrots, 768 × 512	107218 (2,3,1)	113520 (3,1,2)	107218	5.71%
Peppers, 512 × 512	102222 (2,3,1)	108971 (3,1,2)	102222	6.39%

Table 2. Number of bits required for 6000 colour atoms for different column orders

image	the best order	the worst order	sub-optimal (π_c) (5,2,3,4,6,1)	$2\frac{worst-best}{worst+best}$
Barbara, 720 × 576	126937 (5,2,3,4,6,1)	148111 (6,1,4,5,3,2)	126937	15.40%
Goldhill, 720 × 576	124938 (5,3,2,4,6,1)	142009 (1,6,4,5,3,2)	124971	12.79%
Lena, 512 × 512	127077 (5,3,2,4,6,1)	142753 (6,1,4,5,2,3)	127113	11.62%
Lighthouse, 768 × 512	121129 (5,2,3,4,6,1)	147995 (1,6,4,5,3,2)	121129	19.97%
Parrots, 768 × 512	130512 (5,3,2,6,1,4)	145187 (6,4,1,5,3,2)	130739	10.65%
Peppers, 512 × 512	128686 (5,3,2,4,6,1)	138515 (6,1,4,5,2,3)	128803	7.36%

At first, a set of experiments for standard test images of different sizes has been done to find an optimal column order to apply Alg. 3. Each permutation was tried for 6 grayscale (see Tab. 1) and colour images (see Tab. 2). The differences in the size of a bit-stream for the different column orders are significant. For grayscale, where there are only 6 possible column permutations, the differences between maximum and minimum bit-stream sizes are less than 10% (Tab. 1). However, for colour, where we have 720 orders, the differences can be up to 20% (Tab. 2). In the proposed coding scheme the best or close to the best performance is achieved when atoms are sorted by wavelet scale first. Atom indexes and signs of the amplitudes are the last sorting criteria for both grayscale and colour. The column permutations that perform close to optimal for all tested images are: $\pi_g = (2, 3, 1)$ for grayscale and $\pi_c = (5, 2, 3, 4, 6, 1)$ for colours.

R-D performance plots are shown in Fig. 4 including PSNR results for a default mode of Kakadu. Fig. 3 presents a visual comparison example. In general both grayscale and colour codecs are comparable to JPEG 2000, often outperforming the latter at low bit rates. However for many standard test images like Parrots (Fig. 4c and 4f) or Lighthouse a coding performance is still significantly worse. On average (see Tab. 3) MP performs better than the standard at 0.1 bpp but for the higher rates the average performance is worse. We believe that modelling the distributions of wavelet scale indexes and run lengths for arithmetic coding could give a significant improvement. For example, as mentioned in Sec. 3, in initial iterations the atoms are more likely to be found in low frequencies.

(a) Original image   (b) MP, 6086 atoms, 29.95 dB

(c) J2K, default, 29.60 dB   (d) J2K, no-weights, 29.93 dB

Fig. 3. Visual comparison at 0.30 bpp against JPEG 2000 for Goldhill, 720 × 576

Table 3. Coding performance comparisons against JPEG 2000 at fixed bit-rates

grayscale image	0.1 bpp		0.3 bpp		0.5 bpp	
	J2K	MP-Gray	J2K	MP-Gray	J2K	MP-Gray
Barbara, 720 × 576	25.21	**26.02**	30.21	**30.44**	**33.35**	33.09
Goldhill, 720 × 576	28.90	**29.12**	32.30	**32.38**	**34.25**	34.11
Lena, 512 × 512	**29.90**	29.83	**34.94**	34.58	**37.32**	36.85
Lighthouse, 768 × 512	25.77	**25.81**	**29.57**	29.28	**32.11**	31.49
Parrots, 768 × 512	**33.62**	33.43	**39.04**	38.43	**41.61**	40.95
Peppers, 512 × 512	**29.66**	29.44	**34.16**	33.74	**35.84**	35.46
Average	28.84	**28.94**	**33.37**	33.14	**35.75**	35.33
colour image	J2K	MP-RGB	J2K	MP-RGB	J2K	MP-RGB
Barbara, 720 × 576	23.89	**24.23**	28.03	**28.04**	**30.33**	30.20
Goldhill, 720 × 576	**27.24**	27.22	29.93	**29.95**	**31.46**	31.38
Lena, 512 × 512	**27.68**	27.64	**31.31**	31.25	**32.97**	32.95
Lighthouse, 768 × 512	**25.18**	25.00	**28.68**	28.18	**30.95**	30.19
Parrots, 768 × 512	**30.72**	30.40	**35.92**	35.23	**38.54**	37.73
Peppers, 512 × 512	25.57	**25.80**	29.61	**29.68**	31.17	**31.24**
Average	26.71	**26.72**	**30.58**	30.39	**32.57**	32.28

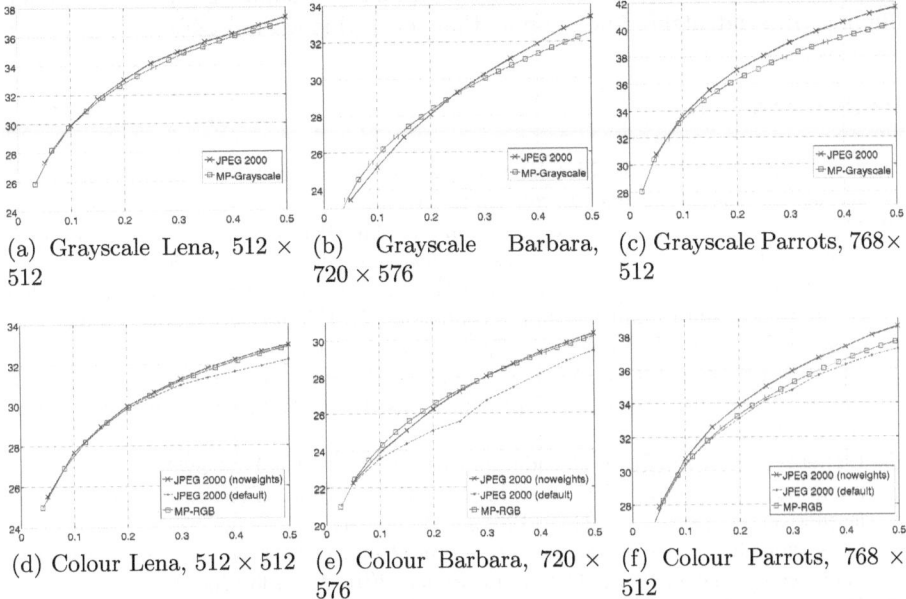

(a) Grayscale Lena, 512 × 512

(b) Grayscale Barbara, 720 × 576

(c) Grayscale Parrots, 768× 512

(d) Colour Lena, 512 × 512

(e) Colour Barbara, 720 × 576

(f) Colour Parrots, 768 × 512

Fig. 4. R-D performance comparisons between the proposed MP coding and Kakadu implementation of the JPEG 2000 standard (y-axis: PSNR [dB], x-axis: bit-rate [bpp])

Our current C++ implementation encodes 8000 colour atoms (this corresponds to a bit-rate of 0.40 bpp for Goldhill image) under Linux on PC with Intel Core 2 Duo in less than 0.2 s which is negligible comparing to finding these atoms by MP algorithm that takes around 80 seconds for the dictionary used in this work and images of dimension 720 × 576.

6   Conclusions

We have presented a novel approach of decomposing and encoding images that has shown comparable R-D performance to the current coding standard JPEG 2000 at low bit rates. Our idea of encoding atoms is especially promising for colour images. MP is performed after the discrete wavelet transform to reduce complexity and improve sparsity [19]. MP decomposition of an image is represented as a matrix of rows. These rows are sorted in lexicographical order after permutation of columns and encoded using run length and then arithmetic coding with simple data model that assumes equal probabilities of each type of symbol (each column). The optimal column orders were found for both grayscale and colour data. The open questions that are currently under our investigation include: more sophisticated data modelling for coding and finding the optimal dictionary.

Acknowledgement. Ryszard Maciol would like to thank Aston University for funding the studentship and support that made this work possible.

References

1. Bergeaud, F., Mallat, S.: Matching pursuit of images. In: Proc. International Conference on Image Processing, vol. 1, pp. 53–56 (1995)
2. Christopoulos, C., Skodras, A., Ebrahimi, T.: The JPEG 2000 still image compression standard. IEEE Signal Processing Magazine 18(5), 36–58 (2001)
3. Czerepinski, P., Davies, C., Canagarajah, N., Bull, D.: Matching pursuits video coding: Dictionaries and fast implementation. IEEE Transactions on Circuits and Systems for Video Technology 10(7), 1103–1115 (2000)
4. Figueras i Ventura, R.M., Vandergheynst, P., Frossard, P.: Low-rate and flexible image coding with redundant representations. IEEE Transactions on Image Processing 15(3), 726–739 (2006)
5. Figueras i Ventura, R.M., Vandergheynst, P., Frossard, P., Cavallaro, A.: Color image scalable coding with matching pursuit. In: Proc. IEEE International Conference on Acoustics, Speech and Signal Processing, vol. 3, pp. 53–56 (2004)
6. Gershikov, E., Lavi-Burlak, E., Porat, M.: Correlation-based approach to color image compression. Image Communications 22(9), 719–733 (2007)
7. Gray, A.M., Gersho, R.: Vector Quantization and Signal Compression, 5th edn. Kluwer Academic Publishers, Dordrecht (1992)
8. Lemire, D., Kaser, O.: Reordering columns for smaller indexes. Information Sciences 181(12), 2550–2570 (2011)
9. Lutoborski, A., Temlyakov, V.M.: Vector greedy algorithms. Journal of Complexity 19(4), 458–473 (2003)
10. Maciol, R., Yuan, Y., Nabney, I.T.: Grayscale and colour image codec based on matching pursuit in the spatio-frequency domain. Tech. rep., Aston University, http://eprints.aston.ac.uk/15194/ (2011)
11. Mallat, S.G., Zhang, Z.: Matching pursuits with time-frequency dictionaries. IEEE Transactions on Signal Processing 41(12), 3397–3415 (1993)
12. Monro, D.M.: Basis picking for matching pursuits image coding. In: Proc. International Conference on Image Processing, vol. 4, pp. 2495–2498 (2004)
13. Monro, D.M., Poh, W.: Improved coding of atoms in matching pursuits. In: Proc. International Conference on Image Processing, vol. 3, pp. 759–762 (2003)
14. Neff, R., Zakhor, A.: Very low bit-rate video coding based on matching pursuits. IEEE Transactions on Circuits and Systems for Video Technology 7(1), 158–171 (1997)
15. Neff, R., Zakhor, A.: Modulus quantization for matching-pursuit video coding. IEEE Transactions on Circuits and Systems for Video Technology 10(6), 895–912 (2000)
16. Taubman, D.: Kakadu JPEG 2000 implementation, http://www.kakadusoftware.com/

17. Witten, I.H., Neal, R.M., Cleary, J.G.: Arithmetic coding for data compression. Communications of the ACM 30(6), 520–540 (1987)
18. Yuan, Y., Monro, D.M.: 3D wavelet video coding with replicated matching pursuits. In: Proc. IEEE International Conference on Image Processing, vol. 1, pp. 69–72 (2005)
19. Yuan, Y., Monro, D.M.: Improved matching pursuits image coding. In: Proc. IEEE International Conference on Acoustics, Speech, and Signal Processing, vol. 2, pp. 201–204 (2005)
20. Yuan, Y., Evans, A.N., Monro, D.M.: Low complexity separable matching pursuits [video coding applications]. In: Proc. IEEE International Conference on Acoustics, Speech, and Signal Processing, vol. 3, pp. 725–728 (2004)

Color Line Detection

Vinciane Lacroix*

Signal and Image Centre Department, Royal Military Academy, 1000 Brussels, Belgium
Vinciane.Lacroix@elec.rma.ac.be

Abstract. Color line extraction is an important part of the segmentation process. The proposed method is the generalization of the Gradient Line Detector (GLD) to color images. The method relies on the computation of a color gradient field. Existing color gradient are not "oriented": the gradient vector direction is defined up to , and not up to 2 as it is for a grey-level image. An oriented color gradient which makes use of an ordering of colors is proposed. Although this ordering is arbitrary, the color gradient orientation changes from one to the other side of a line; this change is captured by the GLD. The oriented color gradient is derived from a generalization from scalar to vector: the components of the gradient are defined as a "signed" distance between weighted average colors, the sign being related to their respective order. An efficient averaging method inspired by the Gaussian gradient brings a scale parameter to the line detector. For the distance, the simplest choice is the Euclidean distance, but the best choice depends on the application. As for any feature extraction process, a post-processing is necessary: local maxima should be extracted and linked into curvilinear segments. Some preliminary results using the Euclidean distance are shown on a few images.

Keywords: color line, color edge, color ordering, Gaussian gradient.

1 Introduction

Feature extraction is important for Computer Vision. The argument to use color information for line detection is the same as for color edge detection: some linear features have a much better contrast in the colored than in the luminance image.

Color edge detection has largely been addressed and effective methods are available (see [1], [2] for an overview). Other features such as corners, interest points and lines have mainly been addressed on luminance images, although some of them include color or multi-band information [3], [4], [5]. Lines deserve specific attention because they are important cues that edge detectors often fail to detect. The methods proposed to detect color edges could be simply transposed to line detection, thus edge and lines (dis)similarities are worth summarizing. In grey-level, an edge *orientation* and *direction* are defined by an angle in the range $] - \pi \ \pi]$ and $] - \pi/2 \ \pi/2]$ respectively (see Figure 1), the orientation pointing towards *higher intensity* values. Although edges have an orientation, this information is hardly used to build contours. On the other hand, in grey-level images the orientation leads to two types of line: "dark" or "bright". Line detection in grey-level image thus usually involves ridge (bright) and/or valley (dark) detection. Note that "plateau-like" lines are ignored. Both edge and line detection involve

---

* This study is funded by the Belgian Ministry of Defense.

G. Maino and G.L. Foresti (Eds.): ICIAP 2011, Part I, LNCS 6978, pp. 318–326, 2011.

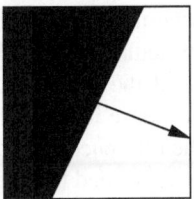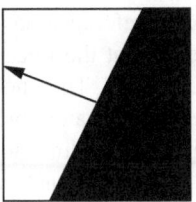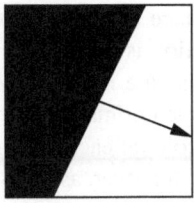

Fig. 1. The edge orientations in (a) and (b) are opposite while their directions shown in (c) and (d) are the same

a "resolution" or "scale" parameter. The resolution is linked to the minimum distance between the extracted features, but also to the level of noise in the input or smoothness desired for the output. The performance issues concerning edge and line detection are similar [6]: position precision, robustness to noise and computational complexity.

The proposed method for color line detection is made of three modules: the first one computes an "oriented" color gradient field including a smoothing parameter to cope with noise and an efficient implementation to address the computational complexity, the second produces a line strength and direction, and the last one — out of the scope of this article– extracts local maxima and links them into lines.

In Section 2 several strategies for color line detection are analyzed and our choice motivated. Section 3 presents an "oriented" color gradient field. Section 4 explains how the oriented color gradient is used to derive a line strength and direction. Results are shown and discussed in Section 5. Further discussion is provided in Section 6. Conclusions are summarized in Section 7.

2   Choice for a Color Line Detection Strategy

Our choice of strategy for color line detection is motivated by exploring several tracks and analyzing their drawbacks.

Color Lines from color edges: if edges may be detected, detecting lines involves only detecting edges at both sides of the line. In practice, two problems occur. First, the edge detection on very thin lines (one pixel wide) may often miss an edge at one side of the line, the edge model assuming some uniformity at each side of the edge. Second, as the aim of line detection is to locate the axis of the line, a necessary non-trivial process is required to match edges at each side of the line.

Color Line Template: masks similar to the generalized Sobel and Prewitt operator on a 3×3 window or on a 5×5 in color, could be designed and combined as they are for the GVDG [2]. Such an option would convey the same drawbacks as the ones they bring in color edge detection: their sensitivity to noise.

Transform color information to grey-level: a simple approach is to transform the multi-band information into a grey-level image and perform line detection on this image (on the first component of a PCA for example). In Remote Sensing where road detection on high resolution images may rely on line detection, the "spectral-angle difference" between the pixel radiance with respect to some reference has been used at this aim [5]. A black line detection is then performed so that the linear structures having the most similar signature as the reference are extracted. The choice of the reference is a critical issue: an almost optimal choice will produce a double response instead of a unique response if the color at each side of the line are closer to the reference than the color of the line itself. Moreover, as many as reference signatures are needed to detect all "types" of curvilinear structures.

Combining line detection in each plane: detecting lines in each band, then merging the information seems the most straightforward method; see Figure 2 for an example of edge and line norm fusion. Such an approach would however miss a yellow line at the interface of a red and green surface (see Figure 3 (left)) unless it is sought on the Value or Luminance plane, implying a transformation in another space. Which transformations, which planes to consider and how to combine the outputs? A generic but time consuming solution is proposed in [7].

Fig. 2. (a) Original Image (b) Edge Fusion: (R,G,B)= scaled Edge Norm in (R,G,B) (c) Line Fusion: scaled Line Norm in (R,G,B) with grey: no line, brighter: bright line, darker: dark line (GLD computation) [8]

Line detection exploiting local color variation: pixels located on the center of a line are characterized by low color variation along the line and a high variation in the perpendicular direction. If the line lies on a uniform background the color variation seen as a vector in the 3D color space at each side of the line will have opposite orientation; in the case of a colored line at the interface of two other colors, the orientation will not be opposite, but will vary significantly. Exploring the color variation in a 8-neighborhood will thus give an indication of the presence of a line. Color variation is captured by the color gradient field, but existing color gradient [1], [2], [9], [10] are not "oriented" as they cannot distinguish the passage from a color c_1 to a color c_2, from its symmetrical (see Figure 3 (right)); we thus introduce an "oriented" color gradient in the next section.

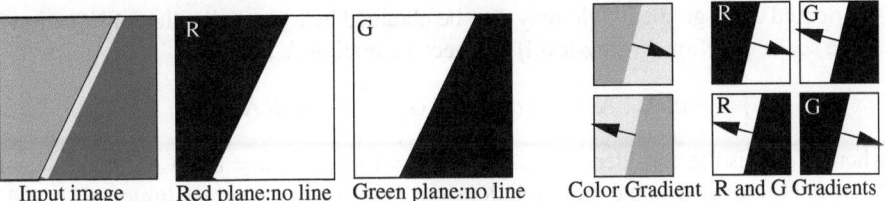

| Input image | Red plane:no line | Green plane:no line | Color Gradient | R and G Gradients |

Fig. 3. Left: input image on which marginal line output fusion fails; right: an oriented color gradient on two input images and their gradients on Red and Green planes

3 Oriented Color Gradient Field

A color image is seen as a mapping from $Z^2 \to Z^3$ where each point $p = (i, j)$ of the plane is mapped on a three dimensional vector $\mathbf{c_p} = (r_p, g_p, b_p)$ where $r_p, g_p,$ and b_p represent the red, green and blue values at the coordinates (i, j). The vector may also be seen as a point (in a three-dimensional space), namely, a "color" point.

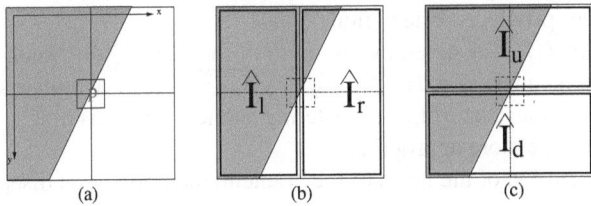

Fig. 4. (a) local window at p (b) x component of the gradient: $\hat{I}_r - \hat{I}_l$ (c) y component of the gradient: $\hat{I}_d - \hat{I}_u$

In grey-level images, the intensity variation is captured by $\mathbf{G} = (G_x, G_y)$, the Gradient of the Intensity: at each point, the gradient direction, its norm and its orientation provide the direction of the greatest intensity variation, the amount of variation and the direction of higher intensities respectively; it is the perfect candidate to describe an edge as displayed in Figure 1 (a–b). Each component of the gradient may be seen as the difference between two average intensities [11] or as the **Euclidean distance** between two averages intensities multiplied by a sign representing their respective mathematical order, as seen in Figure 4 where $\hat{I}_r, \hat{I}_l, \hat{I}_u, \hat{I}_d$ represent the average of the intensity in the zones at the right, left, up and down of the pixel respectively. Thus,

$$G_x(i, j) = s \, d(\hat{I}_r, \hat{I}_l) \equiv s \, d_{rl} \quad \text{and} \quad G_y(i, j) = s \, d(\hat{I}_d, \hat{I}_u) \equiv s \, d_{du} \tag{1}$$

where $d(a, b)$ is the Euclidean distance and $s = 1$ if $a > b$, $s = 0$ if $a = b$ and $s = -1$ if $a < b$. The norm of the gradient may then be rewritten as:

$$N(i, j) = \sqrt{d(\hat{I}_r, \hat{I}_l)^2 + d(\hat{I}_d, \hat{I}_u)^2} \equiv \sqrt{d_{rl}^2 + d_{du}^2}$$

An oriented color gradient field may thus be obtained by a generalization of Equation 3 from a scalar function I (grey-level) to a vector function $\mathbf{A}$ (color):

$$G_x(i,j) = s\, d(\hat{\mathbf{A}}_{\mathbf{r}}, \hat{\mathbf{A}}_{\mathbf{l}}) \equiv s\, d_{rl} \text{ and } G_y(i,j) = s\, d(\hat{\mathbf{A}}_{\mathbf{d}}, \hat{\mathbf{A}}_{\mathbf{u}}) \equiv s\, d_{du} \qquad (2)$$

where $d(a,b)$ is the considered distance and $s = 1$ if $a > b$, $s = 0$ if $a = b$ and $s = -1$ if $a < b$, according to some order relationship. This order is arbitrary (unique if $n = 1$) but all orders will generate similar gradient direction up to π, as the same rule applies to both directions. The lexicographical order is the most natural one. Given a vector $\mathbf{a} = (a_1, \ldots, a_n)$ and a vector $\mathbf{b} = (b_1, \ldots, b_n)$,

> $\mathbf{a} = \mathbf{b}$ iff $a_i = b_i$ for $i = 1, \ldots, n$.
> $\mathbf{a} < \mathbf{b}$ iff $a_1 < b_1$ or $a_i = b_i$ for $i = 1, \ldots, j - 1 < n$ and $a_j < b_j$
> $\mathbf{a} > \mathbf{b}$ iff $a_1 > b_1$ or $a_i = b_i$ for $i = 1, \ldots, j - 1 < n$ and $a_j > b_j$

Other ordering can be found in literature: color edge detectors based on vector order statistics require an ordering scheme [1]; morphological color edge detectors [10] are a special case of such detectors. However, to our knowledge, none of them uses an *absolute* ordering: in [10], the concept of "extremum" exists but there is no minimum nor maximum; in more general vector order statistics schemes, the concept of "rank" holds, but without privileged orientation.

If the intensity $I(i,j)$ is given by the norm of the vector $\|\mathbf{A}\|$, such as in the RGB representation of an image (i.e. $I(i,j) = \sqrt{A_R^2 + A_G^2 + A_B^2}$), a multi-spectral or hyperspectral representation, performing an ordering on the norm basis first, and then on each coordinates would be more appropriate.

In the generalization of the norm of the gradient, the orientation disappears:

$$N(i,j) = \sqrt{d(\hat{\mathbf{A}}_{\mathbf{r}}, \hat{\mathbf{A}}_{\mathbf{l}})^2 + d(\hat{\mathbf{A}}_{\mathbf{d}}, \hat{\mathbf{A}}_{\mathbf{u}})^2} \equiv \sqrt{d_{rl}^2 + d_{du}^2} \qquad (3)$$

If d is the Euclidean distance, it is easily seen that N in Equation 3 is the square root of the marginal squared gradients (i.e. gradient in red, green and blue planes) as already proposed in literature (see [2]). If color perception is an issue, the average intensities $\hat{I}_r, \hat{I}_l, \hat{I}_u, \hat{I}_d$ may be converted in the CIE-L*a*b* space (referred as CIE-Lab in the following) and, when the Euclidean distance is small, other distances such as CIE1994, CIE2000, or CMC could be used [12]. Although there is still no consensus on the best perceptual distance to use [13], it is recognized that the Euclidean distance in CIE-Lab is *not correct* for small distances (in [3] *the contrary* is assumed). Indeed, similar colors do not hold in spheres in the CIE-Lab space, but rather in ellipsoids, which shapes depend on the color position in the space and on the observer [14].

A two-dimensional color gradient field is thus available on each pixel p, its norm being defined by Equation 2, its direction (including an orientation) defined by the gradient vector in Equation 3. For an efficient computation of the averages, we recommend the weighting factors involved in the Gaussian gradient computation: on each color plane, for the x component of the gradient, a smoothing in the y direction using a Gaussian is performed first (details on the size of the window and the best way to compute the coefficients are given in [8]), then, instead of convolving the resulting image with the derivative of the same Gaussian along x direction as it done for the Gaussian gradient, the derivative mask is divided in a left and right mask, in order to compute a left and right average with all positive coefficients, as the "difference" between the averages is taken out of the computation. The average vector at left and at right may then be obtained by collecting the averages on each color plane. The y component is computed similarly. The scale parameter of the line detector is thus related to the σ of the Gaussian used in this averaging process.

4   Gradient Color Line Detection

The grey-level Gradient Line Detector [8] (GLD) exploits the gradient orientation change at each side of the line axis. At each pixel, pairs of opposite pixels in the 8-neighborhood (see Figure 5) producing a negative dot product — the guarantee of an orientation change— are considered, and the square root of the maximum absolute value is taken as line strength. The projection of each gradient along the line joining the pairs enables to distinguish dark lines (lines darker than the background) from bright ones; the method as such cannot detect "plateau" lines.

The oriented color gradient is thus used as input of the GLD, from which "bright" lines or "dark" lines are extracted. In the color context, "bright" and "dark" have to be interpreted with respect to the color order introduced: a line will be detected as bright if its color $c_l > c_b$, where c_l and c_b are the color of line and the background respectively. More precisely, at each pixel p, the oriented color gradient is computed according to Equation 2. In the 8-neighborhood of the pixel the 4 pairs of symmetrical pixels are considered (see Figure 5). In order to have a line at p, the projection of the oriented color gradient along the line joining some of such pairs should be of different sign. Thus let $\mathbf{d} = (d_x, d_y)$ be the vector joining p to q, where (q, r) is the considered pair. Compute $P_q = G_x(q)d_x + G_y(q)d_y$ and $P_r = -G_x(r)d_x - G_y(r)d_y$. Then if P_q and P_r have the same sign, compute D, the dot product of $\mathbf{G}(\mathbf{q})$ and $\mathbf{G}(\mathbf{r})$:

$$D = G_x(q)G_x(r) + G_y(q)G_y(r)$$

D should be negative, as the orientation of the gradient should change at each side of the line. If P_q and P_r are positive (negative), pixel p is a candidate for a dark (bright) line. Thus if $P_q < 0$ and $P_r < 0$, $D_{bright} = |D|$ else $D_{bright} = 0$; if $P_q > 0$ and $P_r > 0$, $D_{dark} = |D|$, else $D_{dark} = 0$. A pixel may be both candidate for a bright

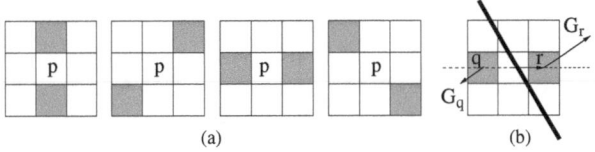

Fig. 5. (a) pixels pairs at p (b) gradient vectors configuration for a dark line at p

line in some direction (i.e. for some pair (q, r)) and dark in another one. Compute D_{bright} and D_{dark} for each of the 4 pairs and compute respectively L_b and L_d, the maximum of both values. The "bright" and "dark" line strength at p is then $\sqrt{L_b}$ and $\sqrt{L_d}$ respectively. For display purpose, it is convenient to define the "line strength" output at p as $L(p) = \sqrt{L_b}$ if $L_b > L_d$ or $L(p) = -\sqrt{L_d}$ if $L_d \geq L_b$. The direction of the line is given by the direction of the difference of the gradient vectors of the pairs providing the corresponding line strength.

5   Results

The color line filter is characterized by 3 parameters: the smoothing factor σ, the type of distance, and the ordering scheme. It has been applied to two images shown in Figure 6 with slightly different parameters. For the upper image $\sigma = 1$ while for the lower image a value of $\sigma = 0.5$ was necessary to separate some lines. This value is convenient for the upper image: note how the noise in the orange line is smooth out. On the lower image, despite the low value of σ, some lines cannot be separated so that a more precise line

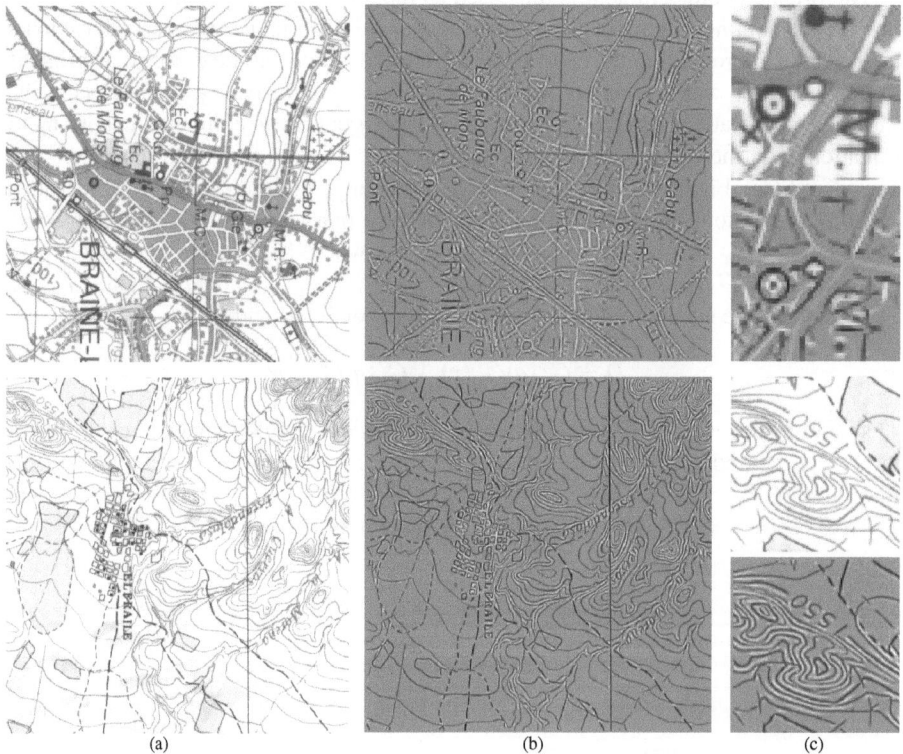

(a)   (b)   (c)

Fig. 6. (a) 512×512 color images; (b) Color line filter outputs : grey:0, whiter: "bright" line output, darker: "dark" line output. Filter parameters up: $(1, d_E, O_{rgb})$ down: $(0.5, d_E, O_{rgb})$ (c) zooms on part of the image.

detector should be used at some places. All other parameters are similar: the Euclidean distance (d_E) and the lexicographical order (O_{rgb}) were used. The curvilinear structures are well detected in both images.

6 Discussion

A potential undesirable effect of the introduction of an order in the color space is the instability of the edge orientation at the interface of some colors that have a similar color component although their distance is large. Consider the example shown in Figure 7, where one side has the uniform color a, and the other side includes some noise: one pixel has a color b, another one a color b', with b very close to b' and a far away from b. If the red components of the noisy pixels on the red axis are respectively just lower and greater than the component of a and if the ordering is made on the red component first, then the edge orientation will be inverted as shown in (b). Note that if another ordering is used (for example starting by the green component first) as in (c), the instability vanishes.

A solution might thus be to detect these specific cases i.e. when $\hat{A}_r$ and $\hat{A}_l$ or $\hat{A}_u$ and $\hat{A}_d$ are distant while sharing a similar color component, and use a different ordering scheme for the latter. Not more than three ordering scheme are necessary. Of course, the meaning of "dark" and "bright" will change accordingly.

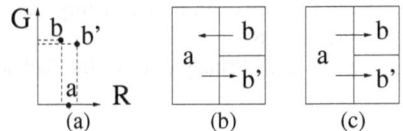

(a)   (b)   (c)

Fig. 7. Gradient orientation instability: (a) 3 colors in Color Space having 0 blue components (b) Oriented Gradient using O_{rgb} (c) Oriented Gradient using O_{grb}

7 Conclusions

The introduction of an arbitrary order in the color space enables to produce an oriented color gradient which in turn may be used for color line detection. The proposed color gradient field includes a scale parameter and an efficient implementation based on the Gaussian gradient is proposed. The color gradient computation enables to use other color distances at the price of an additional cost in computer time. The line detection computation is straightforward as it involves computing a few dot products at each pixel. "Bright" and "dark" line strength output are produced, they should be interpreted with respect to the color ordering introduced. The results on some images are encouraging.

References

[1] Zhu, S.-Y., et al.: A Comprehensive Analysis of Edge Detection in Color Image Processing. Optical Engineering 38(4), 612–625 (1999)

[2] Smolka, B., et al.: Noise Reduction and Edge Detection in Color Images. In: Smolka, B., et al. (eds.) Color Image Processing Methods and Applications, ch. 4, pp. 75–102. CRC Press, Boca Raton (2007)

[3] Ruzon, Tomasi: Edge, Junction, and Corner Detection Using Color Distributions. TPAMI 23(11), 1281–1295 (2001)

[4] Gevers, T., Van de Weijer, J., Stokman, H.: Color feature detection. In: Color Image Processing Methods and Applications, ch. 9, pp. 75–102. CRC Press, Boca Raton (2007)

[5] Christophe, E., Inglada, J.: Robust Road Extraction for High Resolution Satellite Images. In: ICIP (5), pp. 437–440 (2007)

[6] Canny, J.: A computational approach to edge detection. TPAMI 8(6), 679–697 (1986)

[7] Stokman, H., et al.: Selection and Fusion of Colour Models for Image Feature Detection. TPAMI 29(3), 371–381 (2007)

[8] Lacroix, V., Acheroy, M.: Feature-Extraction Using the Constrained Gradient. ISPRS J. of Photogram. and RS 53(2), 85–94 (1998)

[9] Di Zenzo, S.: A Note on the Gradient of a Multi-Image. CVGIP 33(1), 116–125 (1986)

[10] Evans, Liu: A Morphological Gradient Approach to Color Edge Detection. IEEE Transactions on Image Processing 15(6), 1454–1463 (2006)

[11] Lacroix, V.: A Three-Module Strategy for Edge Detection. TPAMI 10(6), 803–810 (1988)

[12] Ohta, N., Robertson, A.R.: Colorimetry Fundamentals and Applications. John Wiley & Sons, Ltd., Chichester (2005)

[13] Kuehni, R.: Color difference formulas: An unsatisfactory state of affairs. Color Research & Application 33(4), 324–326 (2008)

[14] Berns, R.: Billmeyer and Saltzman's Principles of Color Technology. John Wiley & Sons, Chichester (2000)

A New Perception-Based Segmentation Approach Using Combinatorial Pyramids

Esther Antúnez, Rebeca Marfil, and Antonio Bandera

Grupo ISIS, Dpto. Tecnología Electrónica, ESTI Telecomunicación,
Universidad de Málaga
Campus de Teatinos, 29071-Málaga, Spain
{eantunez,rebeca,ajbandera}@uma.es

Abstract. This paper presents a bottom-up approach for perceptual segmentation of natural images. The segmentation algorithm consists of two consecutive stages: firstly, the input image is partitioned into a set of blobs of uniform colour (pre-segmentation stage) and then, using a more complex distance which integrates edge and region descriptors, these blobs are hierarchically merged (perceptual grouping). Both stages are addressed using the Combinatorial Pyramid, a hierarchical structure which can correctly encode relationships among image regions at upper levels. Thus, unlike other methods, the topology of the image is preserved. The performance of the proposed approach has been initially evaluated with respect to groundtruth segmentation data using the Berkeley Segmentation Dataset and Benchmark. Although additional descriptors must be added to deal with textured surfaces, experimental results reveal that the proposed perceptual grouping provides satisfactory scores.

Keywords: perceptual grouping, irregular pyramids, combinatorial pyramids.

1 Introduction

Image segmentation is the process of decomposing an image into a set of regions which have some similar visual characteristics. These visual characteristics can be based on pixel properties as color, brightness or intensity or on other more general properties as texture or motion. Segmentation in regions may be achieved using pyramidal methods that provide hierarchical partitions of the original image. These pyramidal structures help in reducing the computational load associated to the segmentation process and allows to have a same object in different levels of representation. Basically, a pyramid represents an image at different resolution levels. Each pyramid level is recursively obtained by processing its underlying level. In this hierarchy, the bottom level contains the image to be processed. The main advantage of the pyramidal structure is that the parent-child relationships defined between nodes in adjacent levels can be used to reduce the time required to analyze an image. Besides, among the inherent properties of pyramids are [3]: reduction of noise and computational cost, resolution independent processing, processing with local and global features within the

G. Maino and G.L. Foresti (Eds.): ICIAP 2011, Part I, LNCS 6978, pp. 327–336, 2011.

same frame. Moreover, irregular pyramids adapt their structure to the data. A detailed explanation of pyramidal structures can be found in [12]. Combinatorial Pyramids are irregular pyramids in which each level of the pyramid is defined by a combinatorial map. A combinatorial map is a mathematical model describing the subdivision of a space. It encodes all the vertices which compound this subdivision and all the incidence and adjacency relationships among them. In this way, the topology of the space is fully described.

On the other hand, natural images are generally composed of physically disjoint objects whose associated groups of image pixels may not be visually uniform. Hence, it is very difficult to formulate a priori what should be recovered as a region from an image or to separate complex objects from a natural scene [10]. To achieve this goal several authors have proposed generic segmentation methods called 'perceptual segmentations', which try to divide the input image in a manner similar to human beings. Therefore, perceptual grouping can be defined as the process which allows to organize low-level image features into higher level relational structures. Handling such high-level features instead of image pixels offers several advantages such as the reduction of computational complexity of further processes. It also provides an intermediate level of description (shape, spatial relationships) for data, which is more suitable for object recognition tasks [16].

As the process to group pixels into higher level structures can be computationally complex, perceptual segmentation approaches typically combine a pre-segmentation stage with a subsequent perceptual grouping stage [1]. The pre-segmentation stage conducts the low-level definition of segmentation as a process of grouping pixels into homogeneous clusters, meanwhile the perceptual grouping stage performs a domain-independent grouping which is mainly based on properties such as the proximity, similarity, closure or continuity. It must be noted that the aim of these approaches is providing a mid-level segmentation which is more coherent with the human-based image decomposition. That is, it could be usual that the final regions obtained by these bottom-up approaches do not always correspond to the natural image objects [8,13].

This paper presents a hierarchical perceptual segmentation approach which accomplishes these two aforementioned stages. The pre-segmentation stage uses a colour-based distance to divide the image into a set of regions whole spatial distribution is physically representative of the image content. The aim of this stage is to represent the image by means of a set of blobs (superpixels) whose number will be commonly very much less than the original number of image pixels. Besides, these blobs will preserve the image geometric structure as each significant feature contain at least one blob. Next, the perceptual grouping stage groups this set of homogeneous blobs into a smaller set of regions taking into account not only the internal visual coherence of the obtained regions but also the external relationships among them. Both stages are addressed using the Combinatorial Pyramid. It can be noted that this framework is closely related to the previous works of Arbeláez and Cohen [1,2], Huart and Bertolino [8] and Marfil and Bandera [11]. In all these proposals, a pre-segmentation stage precedes the perceptual grouping stage: Arbeláez and Cohen propose to employ the extrema mosaic technique [2], Huart and Bertolino use the Localized Pyramid [8] and Marfil and Bandera employ the Bounded Irregular Pyramid (BIP) [11]. The result of this first grouping is considered in all these works as a graph, and the perceptual grouping is then achieved by means of a hierarchical process whose aim is to reduce the number of vertices of this graph. Vertices of the uppermost level will define a partition of the input image into a set of perceptually relevant regions. Different metrics and strategies have been proposed to address this second stage, but all of the previously proposed methods rely on the use of a simple graph (i.e., a region adjacency graph (RAG)) to represent each level of the hierarchy. RAGs have two main drawbacks for image processing tasks: (i) they do not permit to know if two adjacent regions have one or more common boundaries, and (ii) they do not allow to differentiate an adjacency relationship between two regions from an inclusion relationship. That is, the use of this graph encoding avoids that the topology will be preserved at upper levels of the hierarchies. Taking into account that objects are not only characterized by features or parts, but also by the spatial relationships among these features or parts [15], this limitation constitutes a severe disadvantage. Instead of simple graphs, each level of the hierarchy could be represented using a dual graph. Dual graphs preserve the topology information at upper levels representing each level of the pyramid as a dual pair of graphs and computing contraction and removal operations within them [9]. Thus, they solve the drawbacks of the RAG approach. The problem of this structure is the high increase of memory requirements and execution times since two data structures need now to be stored and processed. Combinatorial maps can be seen as an efficient representation of dual graphs in which the orientation of edges around the graph vertices is explicitly encoded using only one structure. Thus, the use of this structure reduces the memory requirements and execution times.

The rest of the paper is organized as follows: Section 2 describes the proposed approach. It briefly explains the main aspects of the pre-segmentation and perceptual grouping processes which are achieved using the Combinatorial Pyramid. Experimental results revealing the efficiency of the proposed method are presented in Section 3. Finally, the paper concludes along with discussions and future work in Section 4.

2   Segmentation Algorithm

As we aforementioned, the perceptual segmentation algorithm is divided in two stages: pre-segmentation and perceptual grouping stages. Moreover, in both stages the combinatorial map is employed as the data structure to represent each level of the pyramid. Combinatorial maps define a general framework, which allows to encode any subdivision of nD topological spaces orientable or non-orientable with or without boundaries. Formally speaking, a n-dimensional combinatorial map is described as a $(n+1)$-tuple $M = (D, \beta_1, \beta_2, ..., \beta_n)$ such that D is the set of abstract elements called *darts*, β_1 is a permutation on D

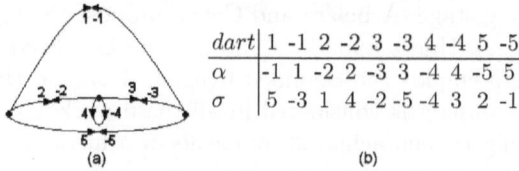

Fig. 1. a) Example of combinatorial map; and b) values of α and   for the combinatorial map in a)

and the other β_i are involutions on D. An involution is a permutation whose cycle has the length of two or less.

In the case of 2D, combinatorial maps may be defined with the triplet $G = (D, \alpha, \sigma)$, where D is the set of darts, σ is a permutation in D encoding the set of darts encountered when turning (counter) clockwise around a vertex, and α is an involution in D connecting two darts belonging to the same edge:

$$\forall d \in D, \alpha^2(d) = d \tag{1}$$

Fig. 1.a) shows an example of combinatorial map. In Fig. 1.b) the values of α and σ for such a combinatorial map can be found. In our approach, counter-clockwise orientation (ccw) for σ is chosen.

The symbols $\sigma^*(d)$ and $\alpha^*(d)$ stand, respectively, the σ and α orbits of the dart d. The orbit of a permutation is obtained applying successively such a permutation over the element that is defined. In this case, the orbit σ^* encodes the set of darts encountered when turning counter-clockwise around the vertex encoded by the dart d. The orbit α^* encode the darts that belong to the same edge. Therefore, the orbits of σ encode the vertices of the graph and the orbits of α define the edges of the graph. In the example of Fig. 1, $\alpha^*(1) = \{1, -1\}$ and $\sigma^*(1) = \{1, 5, 2\}$.

Given a combinatorial map, its dual is defined by $\bar{G} = (D, \varphi, \alpha)$ with $\varphi = \sigma \circ \alpha$. The orbits of φ encode the faces of the combinatorial map. Thus, the orbit φ^* can be seen as the set of darts obtained when turning-clockwise a face of the map. In the example of Fig. 1, $\phi^*(1) = \{1, -3, -2\}$.

Thus, 2D combinatorial maps encode a subdivision of a 2D space into vertices $(V = \sigma^*(D))$, edges $(E = \alpha^*(D))$ and faces$(F = \varphi^*(D))$.

When a combinatorial map is built from an image, the vertices of such a map G could be used to represent the pixels (regions) of the image. Then, in its dual $\bar{G}$, instead of vertices, faces are used to represent pixels (regions). Both maps store the same information and there is not so much difference in working with G or $\bar{G}$. However, as the base entity of the combinatorial map is the dart, it is not possible that this map contains only one vertex and no edges. Therefore, if we choose to work with G, and taking into account that the map could be composed by an unique region, it is necessary to add special darts to represent the infinite region which surrounds the image (the background). Adding these darts, it is avoided that the map will contain only one vertex. On the other hand, when $\bar{G}$ is chosen, the background also exists but there is no need to add special darts to represent it. In this case, a map with only one region (face) would be made out of two darts related by α and σ.

In our case, the base level of the pyramid will be a combinatorial map where each face represent a pixel of the image as an homogeneous region. The combinatorial pyramid is build reducing this initial combinatorial map successively by a sequence of contraction or removal operations [5,9].

In the following subsections, the application of the Combinatorial Pyramid to the pre-segmentation and perceptual grouping stages is explained in detail.

2.1 Pre-segmentation Stage

Let $G_0 = (D_0, \sigma_0, \alpha_0)$ be a given attributed combinatorial map with the vertex set $V_0 = \sigma^*(D)$, the edge set $E_0 = \alpha^*(D)$ and face set $F_0 = \varphi^*(D)$ on the base level (level 0) of the pyramid. In the same way, the combinatorial map on level k of the pyramid is denoted by $G_k = (D_k, \sigma_k, \alpha_k)$. As we aforementioned, each face of the base level represent a pixel of the image. Thus, faces are attributed with the colour of the corresponding pixel. The colour space used in our approach is the HSV space. The edges of the map are also attributed with the difference of colour of the regions separated by each edge. The hierarchy of graphs is built using the algorithm proposed by Haxhimusa et al [7,6], which is based on a spanning tree of the initial graph obtained using the algorithm of Borůvka [4]. Building the spanning tree allows to find the region borders quickly and effortlessly based on local differences in a color space. For each face $f \in F_k$ Borůvka's algorithm marks the edge $e \in E_k$ with the smallest attribute value to be removed. Now, unlike [7,6], two regions (faces) are merged if the difference of colour between them is smaller than a given threshold U_p. That is, the attribute of each edge marked to be removed for the Borůvka's algorithm is compared with the threshold U_p and if its value is smaller, that edge if added to a removal kernel $(RK_{k,k+1})$. In a second step, hanging edges are removed. Finally, a contraction kernel $(CK_{k,k+1})$ is applied to remove parallel edges, obtaining the new level of the pyramid. After a contraction step, the attributes of the surviving edges have to be updated with the colour distance of the faces that the new edge separates. This process is iteratively repeated until no more removal/contraction operation is possible. This stage results in an over-segmentation of the image into a set of regions with homogeneous colour. These homogeneous regions will be the input of the perceptual grouping stage.

2.2 Perceptual Grouping Stage

After the pre-segmentation stage, the perceptual grouping stage aims for simplifying the content of the obtained colour-based image partition. To achieve an efficient grouping process, the Combinatorial Pyramid ensures that two constraints are respected: (i) although all groupings are tested, only the best groupings are locally retained; and (ii) all the groupings are spread on the image so that no part of the image is advantaged. To join pre-segmentation and perceptual grouping stages, the last level of the Combinatorial Pyramid associated to the pre-segmentation stage will constitute the first level of the pyramid associated to the perceptual grouping stage. Next, successive levels will be built using the decimation scheme described in Section 2.1. However, in order to accomplish the perceptual grouping process, a distance which integrates boundary and region descriptors has been defined as a criteria to merge two faces of the combinatorial map.

The distance has two main components: the colour contrast between image blobs and the boundaries of the original image computed using the Canny detector. In order to speed up the process, a global contrast measure is used instead of a local one. It allows to work with the faces of the current working level, increasing the computational speed. This contrast measure is complemented with internal region properties and with attributes of the boundary shared by both regions. The distance between two regions (faces) $\mathbf{f}_i \in F_k$ and $\mathbf{f}_j \in F_k$, $\psi^{\alpha,\beta}(\mathbf{f}_i, \mathbf{f}_j)$, is defined as

$$\psi^{\alpha,\beta}(\mathbf{f}_i, \mathbf{f}_j) = \frac{d(\mathbf{f}_i, \mathbf{f}_j) \cdot b_{\mathbf{f}_i}}{\alpha \cdot (c_{\mathbf{f}_i \mathbf{f}_j}) + (\beta \cdot (b_{\mathbf{f}_i \mathbf{f}_j} - c_{\mathbf{f}_i \mathbf{f}_j}))} \tag{2}$$

where $d(\mathbf{f}_i, \mathbf{f}_j)$ is the colour distance between $\mathbf{f}_i$ and $\mathbf{f}_j$. $b_{\mathbf{f}_i}$ is the perimeter of $\mathbf{f}_i$, $b_{\mathbf{f}_i \mathbf{f}_j}$ is the number of pixels in the common boundary between $\mathbf{f}_i$ and $\mathbf{f}_j$ and $c_{\mathbf{f}_i \mathbf{f}_j}$ is the set of pixels in the common boundary which corresponds to pixels of the boundary detected by the Canny detector. α and β are two constant values used to control the influence of the Canny boundaries in the grouping process. Two regions (faces) will be merged if that distance, $\psi^{\alpha,\beta}(\cdot, \cdot)$, is smaller than a given threshold U_s. It must be noted that the distance $\psi^{\alpha,\beta}(\cdot, \cdot)$ between two regions (faces) is proportional to its colour distance. However, it must be also noted that this distance decreases if the most of the boundary pixels of one of the regions is in contact with the boundary pixels of the other one. Besides, the distance value will decrease if these shared boundary pixels are not detected by the Canny detector.

3   Experimental Results

In order to evaluate the performance of the proposed colour image segmentation approach, the Berkeley Segmentation Dataset and Benchmark (BSDB) has been employed[1] [14]. In this dataset, the ground-truth data is provided by a large database of natural images, manually segmented by human subjects. The methodology for evaluating the performance of segmentation techniques is based in the comparison of machine detected boundaries with respect to human-marked boundaries using the *Precision-Recall framework* [13]. This technique considers two quality measures: precision and recall. The *precision* (P) is defined as the fraction of boundary detections that are true positives rather than false positives. Thus, it quantifies the amount of noise in the output of the boundaries detector approach. On the other hand, the *recall* (R) is defined by the fraction of true positives that are detected rather than missed. Then, it quantifies

---

[1]   http://www.cs.berkeley.edu/projects/vision/grouping/segbench/

the amount of ground truth detected. Measuring these descriptors over a set of images for different thresholds of the approach provides a parametric Precision-Recall curve. The F-measure combines these two quality measures into a single one. It is defined as their harmonic mean:

$$F(P,R) = \frac{2PR}{P+R} \tag{3}$$

Then, the maximal F-measure on the curve is used as a summary statistic for the quality of the detector on the set of images. The current public version of the data set is divided in a training set of 200 images and a test set of 100 images. In order to ensure the integrity of the evaluation, only the images and segmentation results from the training set can be accessed during the optimization phase. In our case, these images have been employed to choose the parameters of the

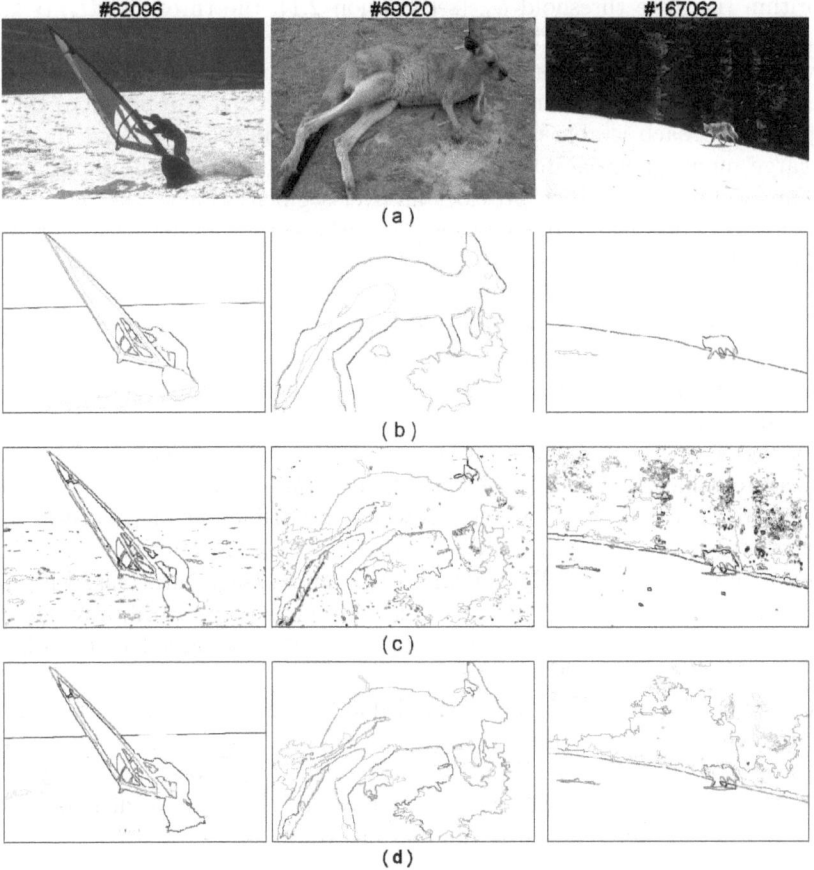

Fig. 2. a) Original images; b) boundaries of human segmentations; c) boundaries of pre-segmentation images; and d) boundaries of the regions obtained after the perceptual grouping

Table 1. Values of F for the images in Figure 2

	#62096	#69020	#167062
NoPG	0.85	0.63	0.41
PG	0.95	0.77	0.73

Table 2. Required time for each image in seconds

	#62096	#69020	#167062
Pre − segmentation	41.2	39.5	41.9
Perceptual Grouping	34.7	27.8	163.9
Total time	75.9	67.3	205.8

algorithm (i.e., the threshold U_p (see Section 2.1), the threshold U_s, α and β (see Section 2.2)). The optimal training parameters have been chosen. Fig. 2 shows the set of boundaries obtained in different segmentations of the original images as well as the ones marked by human subjects. It can be noted that the proposed approach is able to group perceptually important regions in spite of the large intensity variability presented on several areas of the input images. The pre-segmentation stage provides an over-segmentation of the image which overcomes the problem of noisy pixels [11], although bigger details are preserved in the final segmentation results.

The F-measure associated to each image in Fig. 2 can be seen in the Table 1. This Table shows the F-measure for the perceptual grouping stage (PG) as well as for the pre-segmentation stage ($NoPG$). These values of F reflect that adding a perceptual grouping stage improve significantly the results obtained in the segmentation.

On the other hand, Fig. 3 shows several images which have associated a low F-measure value. The main problems of the proposed approach are due to its inability to deal with textured regions which are defined at high natural scales. Thus, the tiger or the leopard in Fig. 3 are divided into a set of different regions. These regions do not usually appear in the human segmentations. The maximal F-measure obtained from the whole test set is 0.65. To improve it, other descriptors, such as the region area or shape, must be added to the distance $\psi^{\alpha,\beta}(\cdot,\cdot)$.

Regarding the execution times, the Table 2 summarize the processing time required for each of the images in Fig. 2. These times correspond to run the algorithm in a 1.60GHz Pentium PC, i.e. a sequential processor. It have to be noted that the proposed algorithm is mainly designed for parallel computing. Thus, it will run more efficiently in a parallel computer.

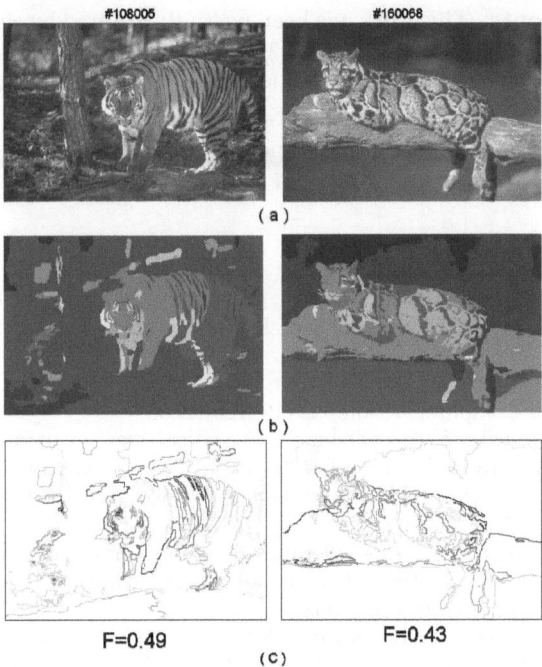

Fig. 3. a) Original images; b) example of segmented image; and c) set of obtained boundaries

4   Conclusions and Future Work

This paper presents a new perception-based segmentation approach which consists of two stages: a pre-segmentation stage and a perceptual grouping stage. In our proposal, both stages are conducted in the framework of a hierarchy of successively reduced combinatorial maps. The pre-segmentation is achieved using a color-based distance and it provides a mid-level representation which is more effective than the pixel-based representation of the original image. The combinatorial map which constitutes the top level of the hierarchy defined by the pre-segmentation stage is the first level of the hierarchy associated to the perceptual grouping stage. This second stage employs a distance which is also based on the colour difference between regions, but it includes information of the boundary of each region, and information provided by the Canny detector. Thus, this approach provides an efficient perceptual segmentation of the input image where the topological relationships among the regions are preserved.

Future work will be focused on adding other descriptors to the distance $\psi^{\alpha,\beta}(\cdot,\cdot)$, studying its repercussion in the efficiency of the method. Besides, it is necessary that the perceptual grouping stage also takes into account a texture measure defined at different natural scales to characterize the image pixels. This texture information could be locally estimated at the higher levels of the hierarchy.

Acknowledgments. This work has been partially granted by the Spanish Junta de Andalucía under project P07-TIC-03106 and by the Ministerio de Ciencia e Innovación (MICINN) and FEDER funds under projects no. TIN2008-06196 and AT2009-0026. This work extends the graph pyramid segmentation method proposed by Yll Haxhimusa, Adrian Ion and Walter Kropastch, Vienna University of Technology, Pattern Recognition and Image Processing Group, Austria [6].

References

1. Arbeláez, P.: Boundary extraction in natural images using ultrametric contour maps. In: Proc. 5th IEEE Workshop Perceptual Org. in Computer Vision, pp. 182–189 (2006)
2. Arbeláez, P., Cohen, L.: A metric approach to vector-valued image segmentation. Int. Journal of Computer Vision 69, 119–126 (2006)
3. Bister, M., Cornelis, J., Rosenfeld, A.: A critical view of pyramid segmentation algorithms. Pattern Recongition Letters 11(9), 605–617 (1990)
4. Borůvka, O.: O jistém problému minimálnim. Práce Mor. Přírodvěd. Spol. v Brně (Acta Societ. Scienc. Natur. Moravicae) 3(3), 37–58 (1926)
5. Brun, L., Kropatsch, W.: Introduction to combinatorial pyramids. In: Bertrand, G., Imiya, A., Klette, R. (eds.) Digital and Image Geometry. LNCS, vol. 2243, pp. 108–128. Springer, Heidelberg (2002)
6. Haxhimusa, Y., Ion, A., Kropatsch, W.G.: Evaluating hierarchical graph-based segmentation. In: Tang, Y.Y., et al. (eds.) Proceedings of 18th International Conference on Pattern Recognition (ICPR), Hong Kong, China, vol. 2, pp. 195–198. IEEE Computer Society, Los Alamitos (2006)
7. Haxhimusa, Y., Kropatsch, W.G.: Segmentation graph hierarchies. In: Fred, A., Caelli, T.M., Duin, R.P.W., Campilho, A.C., de Ridder, D. (eds.) SSPR&SPR 2004. LNCS, vol. 3138, pp. 343–351. Springer, Heidelberg (2004)
8. Huart, J., Bertolino, P.: Similarity-based and perception-based image segmentation. In: Proc. IEEE Int. Conf. on Image Processing, vol. 3, pp. 1148–1151 (2005)
9. Kropatsch, W.: Building irregular pyramids by dual graph contraction. IEEE Proc. Vision, Image and Signal Processing 142(6), 366–374 (1995)
10. Lau, H., Levine, M.: Finding a small number of regions in an image using low-level features. Pattern Recognition 35, 2323–2339 (2002)
11. Marfil, R., Bandera, A.: Comparison of perceptual grouping criteria within an integrated hierarchical framework. In: Torsello, A., Escolano, F., Brun, L. (eds.) GbRPR 2009. LNCS, vol. 5534, pp. 366–375. Springer, Heidelberg (2009)
12. Marfil, R., Molina-Tanco, L., Bandera, A., Rodríguez, J.A., Sandoval Hernández, F.: Pyramid segmentation algorithms revisited. Pattern Recognition 39(8), 1430–1451 (2006)
13. Martin, D., Fowlkes, C., Malik, J.: Learning to detect natural image boundaries using brightness, color, and texture cues. IEEE Trans. on Pattern Analysis Machine Intell. 26(1), 1–20 (2004)
14. Martin, D., Fowlkes, C., Tal, D., Malik, J.: A database of human segmented natural images and its application to evaluating segmentation algorithms and measuring ecological statistics. In: Proc. Int. Conf. Computer Vision (2001)
15. Pham, T., Smeulders, A.: Learning spatial relations in object recognition. Pattern Recognition Letters 27, 1673–1684 (2006)
16. Zlatoff, N., Tellez, B., Baskurt, A.: Combining local belief from low-level primitives for perceptual grouping. Pattern Recognition 41, 1215–1229 (2008)

Automatic Color Detection of Archaeological Pottery with Munsell System

Filippo Stanco[1], Davide Tanasi[2], Arcangelo Bruna[1], and Valentina Maugeri

[1] Dipartimento di Matematica e Informatica, Università di Catania,
viale A. Doria, 6 - 95125 Catania, Italy
{fstanco,bruna}@dmi.unict.it
[2] Arcadia University, Mediterranean Center for Arts and Sciences,
Palazzo Ardizzone, via Roma, 124 – 96100 Siracusa, Italy
dtanasi@mediterranencenter.it

Abstract. A main issue in the archaeological research is the identification of colored surfaces and soils through the application of Munsell system. This method widely used also in other fields, like geology and anthropology, is based on the subjective matching between the real color and its standardized version on Munsell chart. For preventing many possible errors caused by the subjectivity of the system itself, in this paper an automatic method of color detection on selected regions of digital images of archaeological pottery is presented.

Keywords: Archeological artifacts, color matching, Munsell color space.

1   Introduction

"What is color? It's a sensation, like hunger or fatigue, that exists only in our minds. Like hunger and fatigue, it's caused by external factors that we can measure and quantify, but measuring those external factors is no more measuring color than counting calorie intake tells us how hungry we are, or measuring exercise tells us how fatigued" [4].

Every observer perceives color differently. A major obstacle encountered when comparing colors is the choice of descriptive words. Color also varies in its appearance due to changes in the light source and the distance of the light source. The color identification as any other cognitive process can also be seriously influenced by cultural and linguistic background as well as psychological state [2]. Furthermore, it must also be taken into account that colors can only be described unequivocally as long as all the interlocutors can actually see them. If, however, one scholar receives the information exclusively from the oral or written reports of one of the others, she or he must try to picture a particular color without having perceived it herself or himself. The mental image thus created will thereby only in the rarest cases correspond to the visual impression which the other person was stimulated to communicate.

G. Maino and G.L. Foresti (Eds.): ICIAP 2011, Part I, LNCS 6978, pp. 337–346, 2011.
© Springer-Verlag Berlin Heidelberg 2011

Since color can only inadequately be described by verbal means, nowadays whenever one wants to make unequivocal systematically, constructed color chart are used.

At the beginning of the 20th century, Albert H. Munsell [5] brought clarity to color communication by establishing an orderly system for accurately identifying every color that exists. The Munsell color system is a way of precisely specifying colors and showing the relationship among colors. Every color has three qualities or attributes: hue, value and chroma. Munsell established numerical scales with visually uniform steps for each of these attributes.

Hue is that attribute of a color by which we distinguish red from green, blue from yellow, and so on. There is a natural order of hues: red, yellow, green, blue, purple. Then five intermediate hues were inserted: yellow-red, green-yellow, blue-green, purple-blue and red-purple, making ten hues in all. Paints of adjacent colors in this series can be mixed to obtain a continuous variation from one color to the other. For simplicity, the initials as symbols to designate the ten hue sectors are used: R, YR, Y, GY, G, BG, B, PB, P and RP.

Value indicates the lightness of a color. The scale of value ranges from 0 for pure black to 10 for pure white. Black, white and the grays between them are called "neutral colors", because they have no hue like the other "chromatic colors", that have it.

Chroma is the degree of departure of a color from the neutral color of the same value. The scale starts at zero, for neutral colors, but there is no arbitrary end to the scale, as new pigments gradually become available.

The Munsell color-order system has gained international acceptance. It is recognized in standard Z138.2 of the American National Standards Institute; Japanese Industrial Standard for Color, JIS Z 872; the German Standard Color System, DIN 6164; and several British national standards.

The reliability of Munsell's color scheme has been recently stressed by specific neurobiological researches which demonstrated how that system has successfully standardized color in order to match the reflectance spectra of Munsell's color chips with the sensitivity of the cells in the lateral geniculate nucleus (LGN cells), responsible for color identification. This statement makes Munsell charts appropriate for almost all jobs that require manual color identification by human agent [3].

In archaeology Munsell charts are widely used as the standard for color identification of soil profiles, organic materials, rock materials, colored glasses, metals, paintings, textiles and mainly pottery.

For which regards the interpretation of pottery the precise color identification of such parts like clay body, treated surfaces, core, and outer layers like slip and painting, it is fundamental for defining its stylistic and technical features.

As a coding framework, the charts both mediate perceptual access to the colored object being classified, and provide a color reference standard. This tool does not stand alone as a self-explicating artifact; instead its proper use is embedded within a set of systematic work practices, varying from community to community. As demonstrated in application in fields of archaeology, anthropology, these practices can contribute to misclassification of colors [2]. In fact, Munsell notations are not always unequivocal and the limits of their use are well known since decades [1].

Besides the above mentioned cultural, linguistic and psychological background, several other factors can misled the observer in the task of color identification of pottery. The most common are surface homogeneity of the material, state of color surface, color type, test condition, accuracy of assertion, color blindness, quality and type of the Munsell charts.

While Munsell system is ideally shaped for smoothed surfaces no displaying disturbing textures, the pottery surfaces are just in rarest cases homogenous both in relation in their color and their texture, often altered by cracks and superficial voids. Decorative techniques aimed to smooth, coat or glaze can also modify the real chromatic value of the surface. Some kind of patina and incrustation can cause misinterpretation of the color as well as artificial light sources, different than natural daylight must be avoided. Finally, an additional human error can be determined by the inaccuracy caused by tasks involving thousands of checks and by problems coming from quite common deficiencies in color perception [1]. In this perspective, the development of an automatic system for classification of colors in archaeology, and in particular in the field of pottery research, must be considered crucial for providing a solution to all the above mentioned problems.

In this paper, we propose an algorithm to extract an objective Munsell definition of colored selected regions of digital image. The method corrects the illumination defects in the picture in order to create the ideal illumination that permits one to extract the color information.

The rest of the paper is organized as follows. In Section 2 the proposed technique is described; the next Section reports a series of experiments devoted to assess the effectiveness of the method. Finally, some conclusions are given together with a few hints for the future work.

2 Proposed System

The proposed system is a semi-automatic algorithm aiming to find the best match between a user selected color in an archeological sherd with a color in the Munsell charts [5]. Focusing on a particular color in the sherd, the system must provide the color in the Munsell table that best matches it. There are several problems to overcome: first of all, the acquisition process is not usually obtained in good illumination conditions. Pictures are often acquired in artificially illuminated rooms, with uncontrolled light sources. It means that the color correction of the camera is not always able to compensate correctly for the illuminant. This problem, known as "white balance", is a main issue to deal with [6]. Secondly, the patch that the user is asked to point should be representative of the region. Noises (especially for low illuminated acquisitions) and dirty spots can make really difficult this process. Lastly, also the matching is not a minor problem, since the colors cannot be simply represented in the Munsell table; hence a different data space must be used. We define a system and a pipeline to overcome all these problems. A database has also been set up to make the tests and it is available in [7] to let the reader use and/or extend it for further research.

The proposed pipeline (Figure 1) is composed by a color correction module, a patch extraction, and a color matching. In the next sub-sections each block is analyzed in detail.

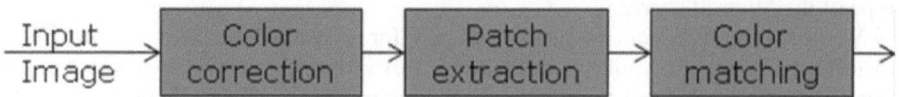

Fig. 1. Block scheme of the proposed algorithm pipeline

2.1 Color Correction

In the color correction module, the image is compensated for the illuminant. This problem is known as "white balance" and there are lots of algorithms in literature to reduce the problem in a fully automatic way [6, 8, 9, 10]. Unfortunately, a zero failure algorithm does not exist, since the white balance is an ill-posed problem and all the methods available are based on assumptions. Whenever, when the assumptions are not verified, the algorithm fails [11]. Moreover there is another problem: the pictures are obtained from a camera and the white balance is already applied (like other algorithms, e.g., color matrix, gamma correction, etc.). It may produce problems in color reproduction. In order to control these problems, we started taking pictures with a color checker chart acquired in the same image: first to obtain the best correction (to validate all the other steps of the algorithm); and second to create a ground truth to validate further methodologies.

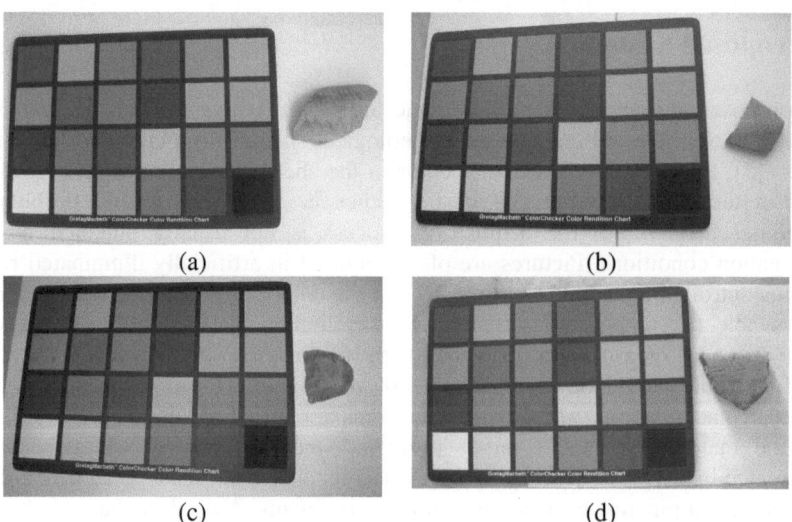

Fig. 2. Examples from the dataset with different illuminants; the effect of the illumination conditions is evident

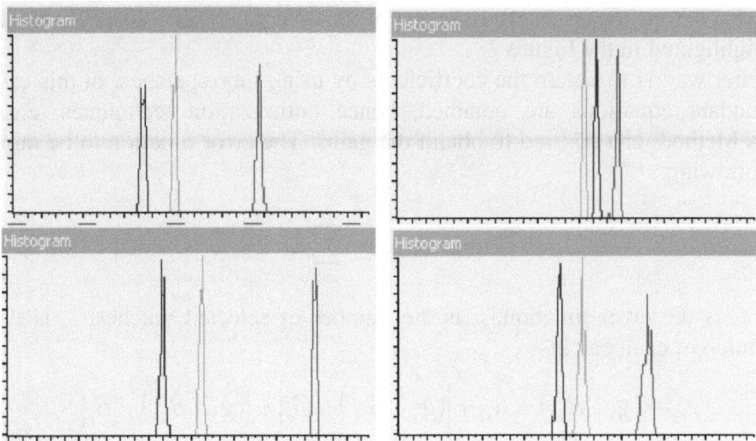

Fig. 3. Histogram of the Macbeth chart '*light skin*' patch in the four images in Figure 2. The RGB mean values are, respectively (from top to bottom, from left to right): (177-116-93), (157,143,133), (217,135,110), (181,133,116).

In the Figure 2 some pictures of the dataset are shown. They were acquired in different illumination conditions. In Figure 3 the related histograms of the '*light skin*' patch (the second patch of the Macbeth chart) are shown. It is evident the effect of the illuminant on the color rendition: without any post-processing correction, the color matching is impossible.

The algorithm proposed supposes that images are acquired with the Macbeth chart and the correction is performed compensating some patches of the chart.

It is supposed to compensate for the illuminant according to the von Kries–Ives adaptation [12], i.e., the correction can be obtained by multiplying every color component with an amplification coefficient:

$$\begin{bmatrix} R_{out} \\ G_{out} \\ B_{out} \end{bmatrix} = \begin{bmatrix} g_r & 0 & 0 \\ 0 & g_g & 0 \\ 0 & 0 & g_b \end{bmatrix} \cdot \begin{bmatrix} R_{in} \\ G_{in} \\ B_{in} \end{bmatrix} \qquad (1)$$

Where *(Rin,Gin,Bin)* is the original triplet, *(Rout,Gout,Bout)* is the corrected value and *(gr,gg,gb)* are the gains.

The weights can be found in different ways. Taking into account the Macbeth chart, the weights could be retrieved by constraining one of the patch to be same (e.g., in the *sRGB* color space) to the real value (since all the patches are completely characterized).

$$\begin{bmatrix} g_r \\ g_g \\ g_b \end{bmatrix} = \begin{bmatrix} R_t / R_{in} \\ G_t / G_{in} \\ B_t / B_{in} \end{bmatrix} \qquad (2)$$

Where (R_t, G_t, B_t) is the target sRGB triplet of the used patch (201,201,201) for the grey patch highlighted in the Figure 2.

A better way is to obtain the coefficients by using more patches. In this case a set of redundant equations are obtained, hence optimization techniques, e.g. Least Squares Method, can be used to obtain the gains. The error function to be minimized is the following:

$$E = \sum_{i=1}^{p} e_i^2 \tag{3}$$

Where E is the error function, p is the number of selected patches; e_i is the error contribution of each patch:

$$e_i = \left|(g_r \cdot R_{in}) - R_t\right| + \left|(g_g \cdot G_{in}) - G_t\right| + \left|(g_b \cdot B_{in}) - B_t\right| \tag{4}$$

This formula provided good results in terms of visual quality. Other error measures can be used, e.g., in a more perceptually uniform color space.

In our system we started using the six gray patches in the bottom of the chart. Of course, in order to reduce the noise, the patch color is obtained as mean of a patch crop. The entire process is shown in Figures 4 and 5.

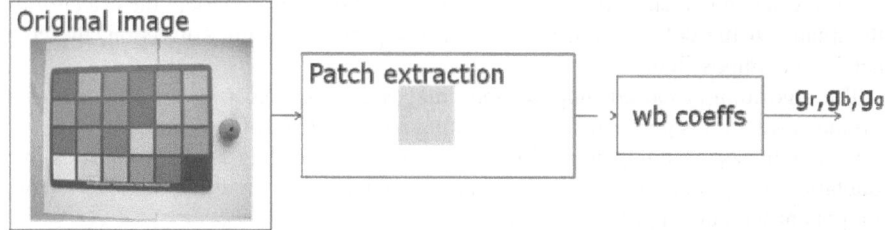

Fig. 4. Color correction module using one neutral patch

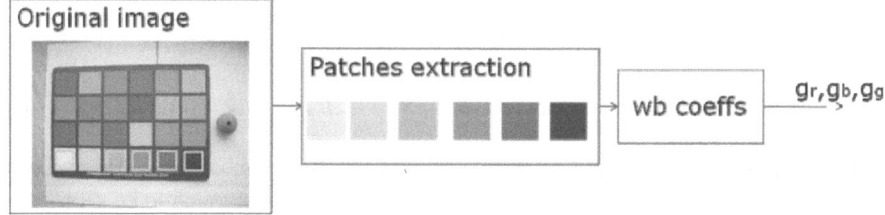

Fig. 5. Color correction module using six neutral patches

In the '*Patch extraction*', user has to select the patches and the system retrieves the mean value of the patch. In the '*wb coeffs*' block the system computes the gains according to the formulas shown above.

2.2 Patch Extraction

After the color correction, the user has also to choose the color to be matched in the Munsell table. A 'point and click' is the best user friendly way to do it. In order to reduce difficulties due to noises or scratches in the archeological finds, when the user points over a colored surface, a homogeneous patch is shown. The color of the patch is obtained, for the generic pointed pixel at position (p, q), as median of a square window:

$$C = median\left(C_{i,j} / i, j = p - n, ..., p + n\right);$$ (5)

Where $C=R$, G, or B; $n=10$ in the actual implementation. The use of the median instead of the mean value allows reducing the influences of impulsive noises and scratches in the patch extraction. In the Figure 6 is shown a snapshot of the software.

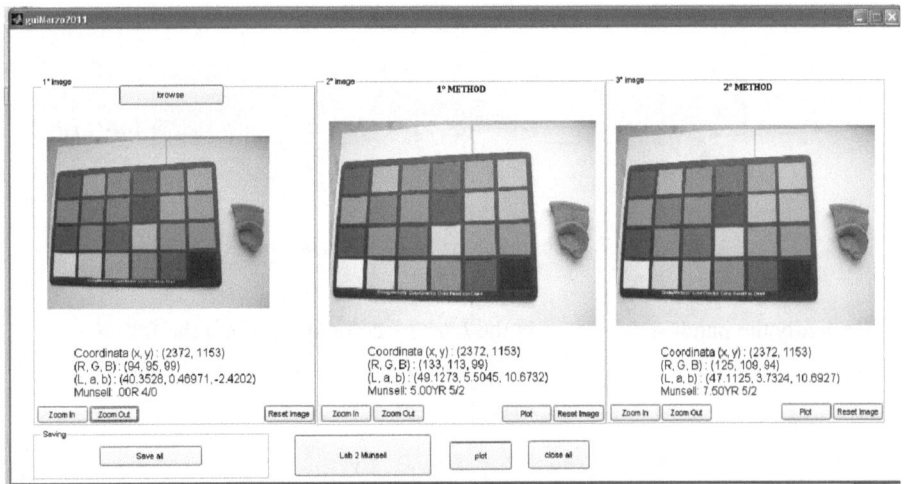

Fig. 6. A snapshot of our system

2.3 Color Matching

The color matching block aims to obtain the color in the Munsell table most similar to the patch chosen by the user. Each color is represented by three components: hue (H), value (V) and chroma (C). The representation is "H V/C". As example, 10YR 6/4 means that the hue is 10YR, i.e., a combination of yellow and red (note that the defined colors are R=Red, Y=Yellow, G=Green, B=Blue, and P=Purple); the value is 6 and the chroma is 4.

The main problem is that there is no direct formula to convert from Munsell patches to a representation in other color spaces. It also means that there is no way to work directly in the Munsell space, since it is quite difficult to define a distance measure to find the best fit. The earliest Munsell based difference formulae is Nickerson's "index of fading" [13] defined as:

$$\Delta E = \frac{2}{5}C\Delta H + 6\Delta V + 3\Delta C \qquad (6)$$

Where H, V, and C are Munsell coordinates.

It is a very old measure and authors decided to use also another measure using a perceptive uniform color representation. The DeltaE94 in the $L^*a^*b^*$ color space has been chosen. It is approximately *perceptually uniform*, i.e. a change of the same amount in a color value produces a change of about the same visual importance.

All the patches in the Munsell table were represented in the Lab color space. The block based scheme of the color matching phase is shown in Figure 7.

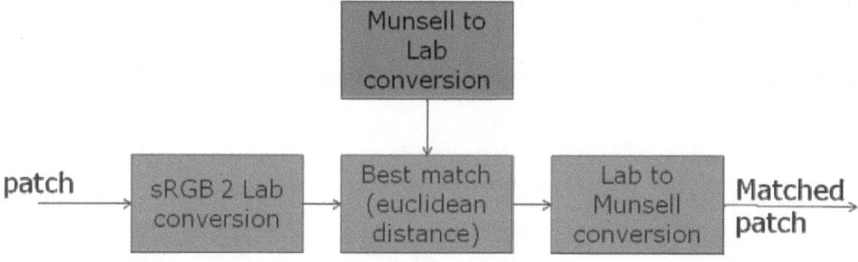

Fig. 7. Color matching block based scheme

Basically the patch is converted in the $L^*a^*b^*$ color space using the following:

$$L^* = 116 f\left(\frac{Y}{Y_n}\right) - 16$$

$$a^* = 500\left[f\left(\frac{X}{X_n}\right) - f\left(\frac{Y}{Y_n}\right)\right] \qquad (7)$$

$$b^* = 200\left[f\left(\frac{Y}{Y_n}\right) - f\left(\frac{Z}{Z_n}\right)\right]$$

All the patches of the Munsell table are also considered in the same color space. The best matching is performed using the minimum Euclidean distance between the patch and all the Munsell colors. The matched color is hence converted in the Munsell space and is provided to the user.

3 Experimental Results

First of all, we have tested that the proposed methods work correctly. To test this, we have acquired the image of the Munsell charts with the Macbeth color checker (Figure 8).

In this way, the algorithm of color correction works well if the single patch has the correct color as shown in the Munsell table, in 90% of the experiments.

In order to test our proposed method, in Table 1 some results using the image of archaeological sherds [7] are presented. The second column shows the color suggested by the archaeologist. This attributed color is very subjective. The third column reports the more representative color in the sherd computed in the input image without any corrections. In the fourth and fifth columns the results of our techniques are shown.

Even if they are different from the human suggestion, they are very close to this. Hence, this means that the system works in the right direction.

We have observed that experimental results are close to the archaeologist suggestions with a success rate of 85% when the images are compensated instead of the original 73% for the uncorrected images.

Fig. 8. Example of image used to validate the *color matching* block

Table 1. Some examples of color measures. The subjective archaeologist suggested color is compared with the algorithm results.

Image	Human identification	Without correction	I method	II method
9570	7.5 YR 7/4	10 YR 6/4	7.5 YR 6/4	5 YR 7/4
9579	10 R 7/6	5 YR 6/6	2.5YR 7/4	5 YR 7/3
9584	5 YR 6/2	10 YR 4/2	7.5 YR 6/2	5 YR 7/4
9591	5 YR 6/1	7.5 YR 5/2	10 YR 5/1	7.5 YR 5/2

4 Conclusions

A complete system to define the predominant color in archaeological sherds has been presented. It is an attempt to automate a manual methodology usually used by archeologists based on visual inspection and color matching of sherds (with Munsell table). Also a database has been created and it is available in [7].

The system aims to detect, starting from a single photo acquired with a common digital still camera, the real color of a patch pointed by the user and to retrieve the patch (in Munsell coordinates) with the best match. Color accuracy is important, but it cannot be ensured by the camera due to the critical illumination condition usually where the images are captured. The paper shows all the criticalities of the problem and proposes a methodology to overcome such problems. Particular attention has been used to select the proper color space and perceptive distance measures. Next steps will be focused on increase the reliability of the color correction, e.g., by using all the patches of the Macbeth chart or increasing the color accuracy for the patches nearer to the color of the find.

Next steps will be focused on increase the reliability of the color correction, e.g., by using a more perceptually uniform color space for the error measure shown in equation (4), or by using all the patches of the Macbeth chart or increasing the color accuracy for the patches nearer to the color of the find.

References

[1] Gerharz, R.R., Lantermann, R., Spennemann, D.R.: Munsell Color Charts: A Necessity for Archaeologists? Australian Historical Archaeology 6, 88–95 (1988)
[2] Goodwin, C.: Practices of Color Classification. Mind, Culture and Activity 7(1-2), 19–36 (2000)
[3] Conway, B.R., Livingstone, M.S.: A different point of hue. Proceedings of the National Academy of Sciences of the United States of America 102(31), 10761–10762 (2005)
[4] Blatner, D., Chavez, C., Fraser, B.: Real World Adobe Photoshop CS3. Peachpit press, Berkeley (2008)
[5] Munsell, A.H.: The Atlas of the Munsell Color System, Boston (1915)
[6] Bianco, S., Ciocca, G., Cusano, C., Schettini, R.: Improving Color Constancy Using Indoor–Outdoor Image Classification. IEEE Trans. on Image Processing 17(12) (2008)
[7] Database used in the experiments,
    http://www.archeomatica.unict.it/sherds&Macbeth.rar
[8] Hordely, S.D.: Scene illuminant estimation: Past, present, and future. Color Res. Appl. 31(4), 303–314 (2006)
[9] van de Weijer, J., Gevers, T., Gijsenij, A.: Edge-based Color Constancy. IEEE Trans. on Image Processing 16(9), 2207–2214 (2007)
[10] Gasparini, F., Schettini, R., Naccari, F., Bruna, A.R.: Multidomain pixel analysis for illuminant estimation. In: Proceedings of SPIE Electronic Imaging 2006 - Digital Photography II - 2006, San Josè (2006)
[11] Funt, B., Barnard, K., Martin, L.: Is machine colour constancy good enough? In: Burkhardt, H.-J., Neumann, B. (eds.) ECCV 1998. LNCS, vol. 1406, pp. 445–459. Springer, Heidelberg (1998)
[12] Von Kries, J.: Chromatic adaptation. Festchrift der Albrecht-Ludwigs-Universitat (1902) [Translation: MacAdam, D.L.: Colorimetry-Fundamentals. SPIE Milestone Series, vol. MS 77 (1993)]
[13] Nickerson, D.: The specification of color tolerance. Tex. Res. 6, 505–514 (1936)

Image Retrieval Based on Gaussian Mixture Approach to Color Localization

Maria Luszczkiewicz-Piatek[1] and Bogdan Smolka[2]

[1] University of Lodz, Faculty of Mathematics and Computer Science,
Department of Applied Computer Science, Banacha 22, 90-238 Lodz, Poland
mluszczkiewicz@math.uni.lodz.pl
[2] Silesian University of Technology, Department of Automatic Control,
Akademicka 16 Str, 44-100 Gliwice, Poland
bogdan.smolka@polsl.pl

Abstract. The paper focuses on the possibilities of color image retrieval of the images sharing the similar location of particular color or set of colors present in the depicted scene. The main idea of the proposed solution is based on treating image as a multispectral object, where each of its spectral channels shows locations of pixels of 11 basis colors within the image. Thus, each of the analyzed images has associated signature, which is constructed on the basis of the mixture approximation of its spectral components. The ability of determining of highly similar images, in terms of one or more basic colors, reveals that the proposed method provides useful and efficient tool for robust to impulse distortions image retrieval.

Keywords: color image retrieval, color composition, Gaussian mixture.

1 Introduction

The rapid growth in number of images acquired and available via World Wide Web creates a need for constant improvement of existing methods of efficient management and analysis of the vast amount of data.

Therefore, analyzing this huge amount of visual information for retrieval purposes, especially when a very specific retrieval criterion is set, still remains a challenging task and therefore there are many attempts to address it [1,2].

One of the fields intensively explored in this area is image retrieval, providing tools and methods addressing users needs concerning finding the images which are the retrieval of images which are considered similar taking into the account the spatial location of the colors within the depicted scene.

Spatial organization of colors has been recently explored in form of spatial statistics between color pixels, such as correlograms [3] or some filter outputs [4,5,6,7]. Related approaches use points of interest similarly to classic object recognition methods [8] and many other retrieval methods rely on segmentation as a basis for image indexing [9,10,11,12]. Mutual arrangements of regions in images are also the basis of the retrieval, however the representation of the relationship can be non-symbolic [13].

Thus, the problem addressed in this paper is as follows. Given a color image (query), the user expects to be provided with a set of candidate images sharing the same color arrangement as the query image, i.e. the same chosen color should be present in the

G. Maino and G.L. Foresti (Eds.): ICIAP 2011, Part I, LNCS 6978, pp. 347–355, 2011.

same location within the query image and the candidate images. Moreover, the user can decide whether the color similarity is analyzed in terms of the occurrence of one or more basic colors.

Although, many approaches explored the idea of the image similarity, expressed by color localization within the scene depicted by the image, the solution proposed in this paper provides a method which is robust to color outliers i.e. isolated pixels or small groups of pixels of color significantly different than that of their neighborhood. Such an approach provides a possibility of the construction of a compact structure of the image signature, which is, in case of the analyzed solution, the set of Gaussian Mixture Model (GMM) parameters, [14,15].

Therefore, the color image is represented as multispectral image formed of 11 binary images, representing the occurrence of certain base colors in particular locations. Each spectral image is then approximated by a *Gaussian Mixture Model*.

This approach offers the advantage of robustness to some pixel distortions, e.g. impulse noise or compression artifacts. This is obtained due to smoothing properties of Gaussian Mixture modeling. Moreover, when the model complexity is not unreasonably excessive, it can be assured that all redundant information is not modeled, i.e. is not introduced into retrieval and comparison process.

The analysis of the parameters of the evaluated models for the images for the images from the database of Wang [9] (of 1000 color images) enables to indicate the most similar pictures, sharing the same colors in the same spatial locations within the image.

The paper is organized as follows. Firstly, color categorization and the construction the spectral images are presented. Secondly, the modeling of the binary spectral images by mixtures of Gaussians is presented. Then, the evaluated results are shown and discussed. Finally, conclusions along with the remarks on future work are drawn.

2   Color Categorization and Spectral Image Formation

The first problem related with the color similarity among the image regions is bounded with the definition of the color similarity i.e. whether two colors are enough perceptually similar for human spectator to be considered as being the same. This problem is addressed by the approach based on the FEED[16] color categorization technique. This empirical categorization model enables to express each of HSI colors as one of 11 basic colors: red, green, blue, yellow, brown, orange, violet, pink, black, white or gray. This color space segmentation method is based on the Fast Exact Euclidean Distance (FEED) transform on the color markers placed on 2D projections of HSI color space. These color markers were a result of the empirical experiments, where human subjects were asked to categorize color stimuli into mentioned the 11 color categories.

Thus, the first step of the proposed solution is the categorization of the colors present in the given image. Therefore, each possible image is represented in terms of maximum 11 colors, as shown in Fig 1.

Having the image palette decreased, the analyzed image can be seen as a multispectral object i.e. each spectral image, related to the one of the basic colors, is the binary image representing the spatial occurrence of pixels of this color. Moreover, each spectral image indicates the set of color pixels, which were recognized by the FEED model as as belonging to the same color class.

Fig. 1. Exemplary results (right) of the original color image (left) categorization based on FEED technique for 11 base colors evaluated for 3 images of the database of Wang [9]

3   Mixture Modeling of Spectral Data

The next step of the proposed technique is the modeling of each of the spectral images using the Gaussian Mixture Model [14,15] and Expectation-Maximization algorithm (EM), [17].

In order to apply the proposed technique, each of binary spectral images are treated as histogram $\Phi(x, y)$, where x and y denotes the image size. This histogram is subjected then to mixture modeling. This approach describes color regions (in specified channel), present in depicted scene, using combination of 'blobs' (Gaussians).

However, to apply the proposed technique, and thus to assure the comparability of the evaluated results, the multispectral images, representing images stored in the analyzed database, should be resized using a method which does not introduce any new colors into a depicted scene during the resizing process.

For the need of this work the used Gaussian Mixture Model is obtained running 75 times the EM procedure for 7 model components. The choice of the model complexity and number of iterations is in general a trade-off between accuracy of the evaluated model and the obtained visual similarity, when some data should be excluded from further consideration. However, in the methodology proposed in this work, the main reasoning for such a choice of modeling parameters is the fact that the exact modeling of spectral data is not desirable, because of the fact that there can be data which should not be incorporated in further analysis, e.g. noise or small insignificant homogenous color regions, which do not convey any useful information. Thus, any possibly irrelevant data will be not considered, [18,19,20,21].

At this stage of the algorithm each image can be seen as set of spectral images, which are modeled using Gaussian Mixture Model, forming it signature, as shown in Fig 2 .

The evaluation of the GMM of the spectral image is illustrated by Fig. 3. Each spectral image, which is binary image conveying information about the spatial location of certain color, is approximated by GMM, reflecting the main structure of the spectral image content.

4   Image Retrieval

Having the spectral images modeled, the user can decide which color or colors reflects the similarity among the images. Therefore, the set of colors which will be taken into

Fig. 2. The construction of the set of GMM models which is a basis for the proposed image retrieval. The original image (left) was subjected to color categorization technique based on FEED model (middle) and each of the spectral images corresponding to 11 basic colors are modeled using GMM approach (right).

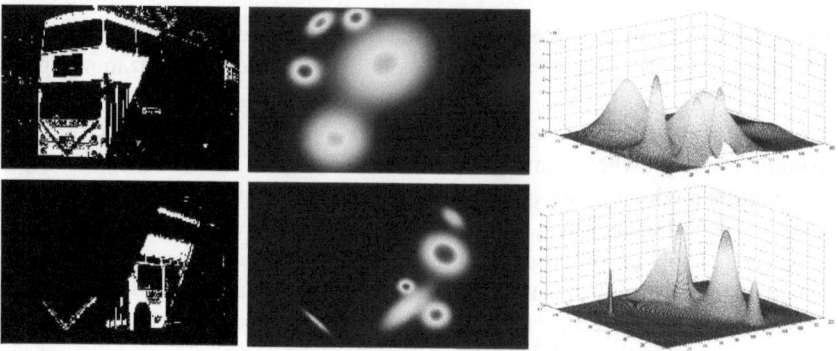

Fig. 3. The evaluation of the mixture model of spectral image corresponding to a chosen color (left) for original image shown in Fig. 2. The 2D (middle) and 3D (right) visualization of GMM are the basis for the retrieval.

account should be chosen. The mixture models of spectral images, corresponding to the user choice, are compared to GMM of respective spectral images of a given query.

For the user choice of colors, the index η, representing each database image, is constructed in relation to query image. The evaluated index also depends on the number of chosen colors, according to:

$$\eta_i = \sum_{k=1}^{m} d(Q_k, I_{i,k}) \tag{1}$$

where m denotes the number of spectral images (colors) taken into account during retrieval process, Q and I represent query and analyzed image from the database, indexed by i respectively. The distance between the spectral images Q_k and $I_{i,k}$ can be defined as Minkowski distance (e.g. L_1) or Earth Mover's Distance (EMD), [11]. The EMD is based on the assumption that one of the histograms reflects "hills" and the second represents "holes" in the ground of a histogram. The measured distance is defined as a minimum amount of work needed to transform one histogram into the other using a "soil" of the first histogram. As this method operates on signatures and their weights using GMM, we assigned as signature values the *mean* of each component and for the *signature weight* the weighting coefficient of each Gaussian in the model.

In details, let us assume that a user is interested in retrieval of images similar to a given query, having regions of the perceptually similar colors in corresponding locations. All images are subjected to the color categorization, mixture modeling, and similarity-to-query index construction. The exact value of the image index is bounded to the number of chosen colors, in terms of which, images are being compared. For each pair of the corresponding spectral images (of the query and the analyzed images) their mutual similarity is evaluated. Knowing that each spectral image is represented by a mixture model, the similarity evaluation resolves to the calculation of the similarity between evaluated GMM's for a corresponding color. The similarity indices for each spectral level are then combined to create the overall similarity-to-query index η for each image in the analyzed database. On the basis of those indexes the candidate images the most similar to given query are selected.

Let us note that when only a small set of basis color are chosen, the retrieved images will be similar only in terms of those colors, that can lead to the perceptual sensation that candidate images are in fact dissimilar when the overall image composition is taken into account. Moreover, not only the the undesirable data (as noise) can be omitted in the modeling process, but also the relatively small regions, in comparison to overall image size, can be not included into the mixture model. The ratio of data which is modeled to overall image data depends on the complexity of the used mixture model. Thus, increasing the model complexity, on the one hand, leads to more accurate reflection of the given data, but on the other hand, there is a possibility that redundant data will be taken into consideration.

5   Retrieval Results

The presented experiments were evaluated in two steps. Firstly, the color similarity was tested for a single color, i.e. images were retrieved only on the basis of the similarity of one chosen spectral image corresponding to a single chosen color (Fig. 4). The second part of the experiment was concentrated on the retrieval based on the combination of colors (see Figs. 5 and 6).

Let us note, that when the retrieval process is based only on a single color there is a chance that the overall similarity sensation can be different that those suggested by single color comparisons. However, these experiments prove the efficiency of the proposed methodology.

Moreover, the evaluated results for single color comparison for L_1 and EMD measures produce still accurate but slightly different sets of candidate images, because the first of the similarity measures does not take into account any, even slight spatial shift, of the location of the color regions in spectral image. The second of the similarity measures EMD, due to its inherent properties, allows small changes in color spatial location to claim similarity. However, this phenomenon does not occurs when multiple colors are analyzed simultaneously (as illustrated by Fig. 5 and 6), because when the numerous criteria are applied, only a smaller subset of images are capable to satisfy those conditions. In case of analyzed databases, chosen for the experiments described in this paper,

Red
(L₁D)

Red
(EMD)

White
(L₁D)

White
(EMD)

Green
(L₁D)

Green
(EMD)

Pink
(L₁D)

Pink
(EMD)

Yellow
(L₁D)

Yellow
(EMD)

Fig. 4. The results of the retrieval evaluated on the basis of the mixture approximation of spectral images corresponding to the single color chosen by user, evaluated for the database of Wang [9]. For the similarity measure for each color the L_1 or EMD methods we used.

the possible differences in evaluated results are unnoticeable. When the database of significantly larger size would be taken into account, these differences would be visible.

There is open question which of the similarity measure is more accurate. To address this problem, also the other aspects should be considered, such as the fact that for the EMD similarity measure the evaluation of the image signature, which need to be stored for further analysis, is reduced to set of GMM parameters. In case of pixel by pixel comparisons (as for L_1 measure), the completely evaluated model of 11 mixtures should be stored, which not always is acceptable.

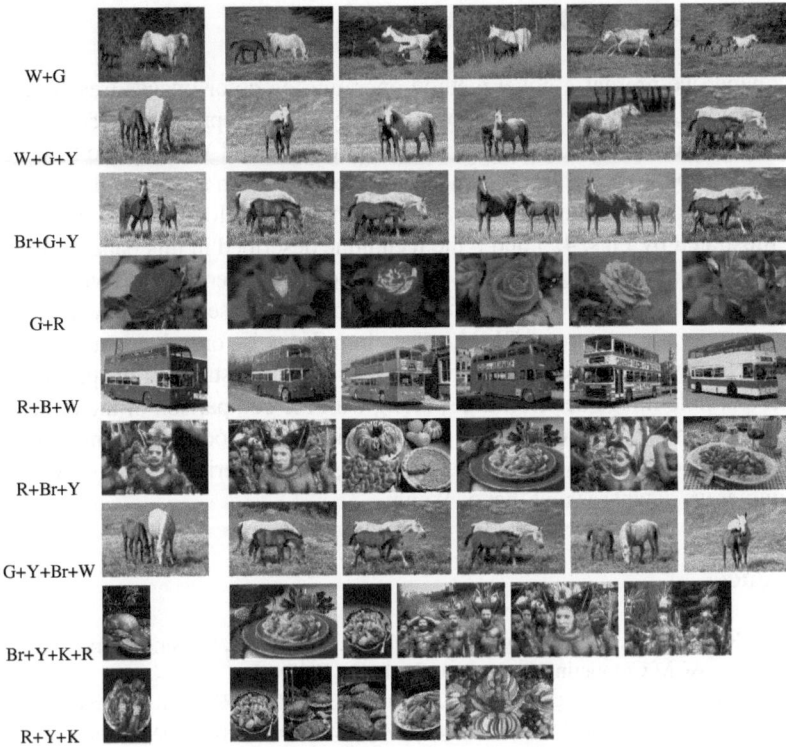

Fig. 5. The results of the retrieval evaluated on the basis of the mixture approximations of a set spectral images corresponding to the colors chosen by user. The experiments tested the efficiency of proposed method for the combination of the following colors, where: white, green, yellow, brown, red, blue and black are denoted by letters W, G, Y, Br, R, B and K.

Fig. 6. The results of the retrieval evaluated on the basis of the proposed method for the images of database of Webmuseum (http://www.ibiblio.org/wm/). The experiments tested the efficiency of proposed method for the combination of the following colors, where: yellow, blue, red, black, green and orange are denoted by letters Y,B,R,K,G and Z. The results evaluated for L_1 and EMD similarity measures provide the equivalent results.

6   Conclusions and Future Work

In this paper we present a novel method operating on color images treated as multi-spectral images, combined of the set of the binary images representing each of 11 basic colors. The experiments evaluated on images of database of Wang tested the effectiveness of the proposed method when only one color was taken as the similarity criterion, as long as when image similarity was based on the set of colors.

As illustrated by Figs. 4, 5 and 6 the technique described in this paper, utilizing the Gaussian mixture modeling approach to spatial location of color pixel within the color image, offers satisfactory retrieval results. Moreover, proposed technique is robust, due to the approximating nature of GMM, to any impulse noise or redundant information.

Future work will be based on the exploration of the robustness of the proposed technique to possible image deteriorations, along with the comparison of the effectiveness of the retrieval process in comparison to other existing methods, taking into account the spatial location of the analyzed colors. Moreover, these experiment will be evaluated on more numerous databases that that of Wang.

References

1. Datta, R., Joshi, D., Li, J., Wang, J.Z.: Image Retrieval: Ideas, Influences, and Trends of the New Age. ACM Computing Surveys 40(2), 1–60 (2008)
2. Zhou, X.S., Rui, Y., Huang, T.S.: Exploration of Visual Data. Kluwer, Dordrecht (2003)
3. Huang, J., et al.: Spatial Color Indexing and Applications. International Journal of Computer Vision 35(3), 245–268 (1999)
4. Pass, G., Zabih, R.: Comparing images using joint histograms. Journal of Multimedia Systems 7(3), 234–240 (1999)
5. Ciocca, G., Schettini, L., Cinque, L.: Image Indexing and Retrieval Using Spatial Chromatic Histograms and Signatures. In: Proc. of CGIV, pp. 255–258 (2002)
6. Lambert, P., Harvey, N., Grecu, H.: Image Retrieval Using Spatial Chromatic Histograms. In: Proc. of CGIV, pp. 343–347 (2004)
7. Hartut, T., Gousseau, Y., Schmitt, F.: Adaptive Image Retrieval Based on the Spatial Organization of Colors. Computer Vision and Image Understanding 112, 101–113 (2008)
8. Heidemann, G.: Combining Spatial and Colour Information For Content Based Image Retrieval. Computer Vision and Image Understanding 94, 234–270 (2004)
9. Wang, J.Z., Li, J., Wiederhold, G.: SIMPLIcity: Semantics-Sensitive Integrated Matching for Picture Libraries. IEEE Trans. Patt. Anal. Mach. Intel. 9, 947–963 (2001)
10. Rugna, J.D., Konik, H.: Color Coarse Segmentation and Regions Selection for Similar Images Retrieval. In: Proc. of CGIV, pp. 241–244 (2002)
11. Dvir, G., Greenspan, H., Rubner, Y.: Context-Based Image Modelling. In: Proc. of ICPR, pp. 162–165 (2002)
12. Jing, F., Li, M., Zhang, H.J.: An Effective Region-Based Image Retrieval Framework. IEEE Trans. on Image Processing 13(5), 699–709 (2004)
13. Berretti, A., Del Bimbo, E.: Weighted Walktroughs Between Extended Entities for Retrieval by Spatial Arrangement. IEEE Trans. on Multimedia 3(1), 52–70 (2002)
14. McLachlan, G., Peel, D.: Finite Mixtures Models. John Wiley & Sons, Chichester (2000)
15. Bilmes, J.: A Gentle Tutorial on the EM Algorithm and its Application to Parameter Estimation for Gaussian Mixture and Hidden Markov Models. University of Berkeley, ICSI-TR-97-021 (1997)

16. Van den Broek, E.L., Schouten, T.E., Kisters, P.M.F.: Modeling human color categorization. Pattern Recogn. Lett. 29(8), 1136–1144 (2008)
17. Dempster, A., Laird, N., Rubin, D.: Maximum Likelihood from incomplete data. J. Royal Stat. Soc. 39B, 1–38 (1977)
18. Luszczkiewicz, M., Smolka, B.: Gaussian Mixture Model Based Retrieval Technique for Lossy Compressed Color Images. In: Kamel, M.S., Campilho, A. (eds.) ICIAR 2007. LNCS, vol. 4633, pp. 662–673. Springer, Heidelberg (2007)
19. Luszczkiewicz, M., Smolka, B.: A Robust Indexing and Retrieval Method for Lossy Compressed Color Images. In: Proc. of IEEE International Symposium on Image and Signal, Processing and Analysis, pp. 304–309 (2007)
20. Luszczkiewicz, M., Smolka, B.: Spatial Color Distribution Based Indexing and Retrieval Scheme. In: Advances in Soft Computing, vol. 59, pp. 419–427 (2009)
21. Luszczkiewicz, M., Smolka, B.: Application of Bilateral Filtering and Gaussian Mixture Modeling for the Retrieval of Paintings. In: Proc. of PODKA, vol. 3, pp. 77–80 (2009)
22. Rubner, Y., Tomasi, C., Guibas, L.J.: The Earth Mover's Distance as a Metric for Image Retrieval. International Journal of Computer Vision 40(2), 99–121 (2000)

A Method for Data Extraction from Video Sequences for Automatic Identification of Football Players Based on Their Numbers

Dariusz Frejlichowski

West Pomeranian University of Technology, Szczecin,
Faculty of Computer Science and Information Technology,
Zolnierska 52, 71-210, Szczecin, Poland
dfrejlichowski@wi.zut.edu.pl

Abstract. In the paper the first stage of the approach for automatic identification of football players is presented. It is based on the numbers placed on their shirts. The method works with video frames extracted during a television sport broadcast. The element of the system described in this paper is devoted to the localisation of the numbers and their extraction for future recognition. It is simple, yet efficient and it is based on the use of appropriate ranges in various colour spaces. Four colour spaces were experimentally evaluated for this purpose. Thanks to this, the ranges could be established for particular kits. Firstly, the part of an image with a shirt was localised, and later, within this area, a number was found.

Keywords: players identification, feature localisation, colour spaces.

1 Introduction

Automatic identification of football players (or other sport disciplines) during a TV broadcast is a difficult and challenging task. However, the possible benefits arising from the use of a system realising such a function, especially taking into account the popularity of this sport, are providing researchers with constant work in this field. The aim is the identification of a sportsman (in real time) during a TV transmission, performed automatically, without (or in small level, semi-automatically) human interaction.

Many approaches, and even complete systems have been proposed so far (e.g. MATRIS [1], TRACAB [2], MELISA [3]), but in many cases they are not completely automatic or are limited to the problem of tracking. The most common approach is the use of a human operator of the broadcast, supported by a computer application. Usually a computer is only used to make the broadcast more attractive. In case of automatic systems employing image analysis and recognition algorithms and based on video sequence (or single frames extracted from it), two approaches are especially interesting. The first one is more popular and has

G. Maino and G.L. Foresti (Eds.): ICIAP 2011, Part I, LNCS 6978, pp. 356–364, 2011.

been explored for a longer period of time. It is based on the extraction and identification of numbers placed on the back of players' shirts. The second approach makes use of the most popular biometric feature — the face.

An example of the first approach is a method presented in [4], where the tracking of players and identification based on numbers was performed. The method used watersheds and Region Adjacency Graphs for feature localisation and extraction, and n-tuple, Hidden Markov Models and Neural Networks for classification. In [5] a simple descriptor was used for finding a white region on red-black stripes. The candidate regions were merged by means of graphs. The extracted numbers were identified with the use of the Principal Component Analysis (PCA).

An example of the second approach was presented in [6]. The identification described there was based on the players' faces. Firstly, scene change detection was carried out by means of the image histogram. Secondly, close-up detection was performed, which made the appearance of a face in a scene more probable. The next stage was the face localisation through searching for skin colour pixels in YCbCr and HSV spaces, followed by template matching. The identification stage involved Discrete Fourier Transform for feature extraction and PCA+LDA (Linear Discriminant Analysis) for feature reduction. This approach is strongly similar to [7], where some general biometric problems were explored and solved.

Sometimes, only the location of players is explored, e.g. for that purpose in [8] so-called mosaic images were used, and in [9] approximation of players' location (points in image) according to the play field model.

It is worth to mention that not only players are localised (and sometimes — tracked or identified). In [10] particular events (e.g. shot on goal, penalty, free kick) were detected, basing on fast camera movement (with the use of MPEG vectors). Similarly, in [11] 'highlights' were detected with the use of template matching, finding of play field zones and Finite State Machines. In [12] only the ball was detected by means of Support Vector Machines. Unfortunately, this approach was developed for traditional black and white football, therefore it is less practical nowadays.

The decision to grant the hosting of the 2012 UEFA European Football Championship to Poland and Ukraine has resulted in the significant increase of interest in those countries in the automatic analysis of video sequences coming from football matches. Identification of players in order to make the broadcast more attractive is the first obvious application. However, some other areas can be easily pointed out, e.g. analysis of the crowd connected to required standards of safety and security measures or an automatic analysis of exposure or visibility time of the advertising banners.

In this paper a method for automatic localisation of football players' kits and numbers is presented. It is the first stage of the algorithm for automatic identification of players basing on their numbers. The approach was tested on video sequences recorded during the World Cup in Germany in 2006 and partially presented in [13]. In the previous publication the entire proposed approach was outlined, but emphasis was placed on the localisation of shirts. However, it only constitutes the first part of the process. Since the goal is the automatic identification of players, the subsequent stage is more important, namely the extraction of the numbers placed on them. Hence, it constitutes the main topic of this paper — the approach and experimental results for this stage are presented. The extraction of a number from a shirt is described in Section 3. Earlier, in Section 2 the localisation of a shirt is briefly recalled.

2   Localisation of Players Based on Information about Their Kits Colour

The first step in the proposed approach is the localisation of pixels in a frame that belongs to a player on the play field. It is carried out by using information about the colours of sport kits assigned to teams. It is known that according to certain sports laws the kits of the two playing teams (and referees) have to be clearly different. Therefore, all football teams (not only national) have to use two (so called "home" and "away" colours), and sometimes even three kits. After the localisation of a shirt we are looking for numbers placed on it. In both cases (shirts and numbers) the appropriate ranges of values in colour spaces were experimentally established. Such an approach was applied with success in face localisation based on the skin colour ([7]).

A football player's kit is composed of three elements: a shirt, shorts (goalkeepers are allowed to wear tracksuit bottoms instead) and socks. In the proposed approach the first from the listed elements is especially important, because it constitutes the largest part of a kit and, what is more important, the numbers to be recognised are placed on it. However, if the colour of shorts and socks is the same as the colour of the shirts, they will be localised as well. It is not a problem, because the only difference is the larger area to explore in the next stage.

Four colour models were tested, namely: RGB, HSV, YCbCr and CIE L*a*b. Amongst them the best results were obtained when using YCbCr and HSV. It was caused by the characteristics of these colour spaces. Most important is the presence of a component more general from the other in particular model, e.g. Hue in HSV and luma in YUV. In the second case blue-difference and red-difference chroma components were even more helpful.

The process of establishing the appropriate ranges within the particular colour model components was based on the exploration of several dozens of shots for each colour of a kit (this number was different and depended on the 'uniqueness' of a colour) taken from various football matches. In some cases the localisation was successful in various spaces (for example the yellow colour was easy to extract in each explored model). If so, the one which separated a shirt from the rest of an image best was chosen. The achieved ranges for each colour class appearing during the World Cup 2006 are provided in Table 1 ([13]). For the purpose of the experiments the image sequences recorded from analogue TV during the FIFA World Cup 2006 in Germany (AVI, 720 × 576, interlaced video, MPEG-2) have been used. Some examples are provided in Fig. 1.

Fig. 1. Examples of the experimental data — frames for one class (white numbers on blue shirts, [13])

Table 1. The experimentally established ranges within the colour models for the localisation of players' shirts

	white	red	yellow	white-red	white-blue	orange	green	blue	maroon	navy-blue
Y	>90	—	—	<85 or >140	—	—	—	—	<85	—
Cb	120-150	90-130	30-80	90-150	120-145	—	95-120	135-190	115-135	—
Cr	—	>145	130-170	<125 or >160	110-135	—	100-122	95-125	135-165	—
H	—	—	—	—	—	0.045-0.1	—	—	—	0.4-0.7
S	—	—	—	—	—	>0.7	—	—	—	0.22-0.8
V	—	—	—	—	—	—	—	—	—	—

In the achieved image we have to reject some small undesirable objects that can appear as a result of various image or video distortions. In the proposed approach this was performed on the basis of the analysis of the boundary length of an object. The performed experiments gave the appropriate threshold — the length of the boundary has to exceed 200 points.

Unfortunately, in some cases the abovementioned method did not prove successful. This concerns the situations in which some other visible objects with the same colour as the requested shirt (e.g. banners) can be extracted as well (see Fig. 1). Hence, in order to remove them, it was assumed (basing on the experiments) that the product of both sizes of a rectangle enclosing the candidate object has to be higher than 0.65 and lower than 1.75 ([13]).

Fig. 2. The result of the localisation of a shirt

3 Extraction of a Number Placed on a Shirt

After performing the localisation of a shirt — the second problem concerned localisation of numbers placed on them. Unfortunately, the ranges established in the previous stage could not be applied again. This problem was mainly caused by the influence of the colour of the shirt on the colour of the number, e.g. blue digits enclosed by a red colour should be localised using other ranges than by white ones. Hence, new ranges in colour spaces were experimentally established. They are provided in Table 2 ([13]).

Table 2. The experimentally established ranges within the colour models for the localisation of numbers within the previously extracted frame area belonging to a shirt

	white (red)	white (blue)	white (green)	yellow	black (blue-white)	black (orange)	green	blue (white-blue)	blue (white)
Y	>110	>130	>95	—	45-130	105-130	—	—	—
Cb	125-155	125-160	115-145	65-105	120-140	125-175	60-95	—	—
Cr	—	—	—	135-155	—	—	110-140	—	—
H	—	—	—	—	—	—	—	0.45-0.6	0.3-0.8
S	—	—	—	—	—	—	—	—	0.22-0.57
V	—	—	—	—	—	—	—	0.19-0.6	0.3-0.75

The extracted numbers can be recognised with the help of shape descriptors. In case of international competitions employing national teams, such as The World Cup, The Confederations Cup, continental tournaments governed by particular football associations (e.g. The UEFA European Football Championship, The AFC Asian Cup, The Africa Cup of Nations, Copa America) usually the

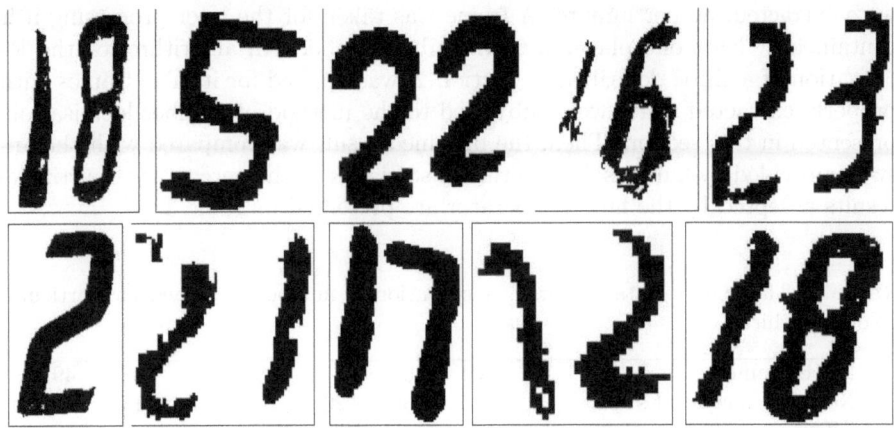

Fig. 3. Examples of numbers on sports shirts extracted by means of the approach described in the paper

problem of recognition of numbers can be made easier. It results from the strict regulations assumed, e.g. the limitations for possible numbers — from 1 to 23. It is different than in the case of football clubs, where those numbers are more inconstant.

The recognition of particular numbers can be more facilitated when taking into consideration only the players present on the play field at the moment. That gives at the most eleven possible numbers for one team. Moreover, the whole numbers can be treated as shapes for recognition, not the single digits. The small number of template classes enables it. However, in this case we have to combine the digits for two-digit numbers. It is simple, because if there are two 'candidates for digits' close to each other we can assume that they can produce a number. In the experiment this task was carried out through calculating the centroids of localised objects within the area of a shirt. If they were close in vertical and horizontal directions and the 'candidates' were sufficiently large, they were merged.

The last step in the approach is the rejection of small localised objects. Usually those are some undesirable objects that should not be taken into consideration. On the other hand, if those objects are numbers and are small it is still less probable to perform successful recognition using them. Therefore, they can be rejected without any influence on the overall efficiency of the approach. Some examples of the extracted using the described approach numbers are provided in Fig. 2. As can be seen the shape of extracted numbers can be strongly distorted, e.g. by creases in the shirts or imaging and weather conditions. It will strongly hamper the later recognition of numbers.

In order to estimate the efficiency of the proposed approach the experiment was performed, using 20 video sequences, lasting from 2 to 3 minutes, from various matches played during the 2006 World Cup. For each of them single frames were extracted, 30 per minute. A frame was taken for the later processing if it contained at least one player with a visible number. An algorithm for the localisation of a shirt, described in Section 2, was applied for it. The frames with properly extracted shirts were subjected to the method of number localisation, presented in this section. Then, the obtained result was compared with the answers provided by humans. The obtained statistics — the percentage of artificial results accepted by the human operator are presented in Table 3.

Table 3. The experimental results of localisation of numbers obtained for particular colours of shirts

No. of frames	52	63	31	67	47	19	29	38	49
No. of correct results	33	42	25	36	20	15	12	13	21
Efficiency	64%	67%	81%	54%	43%	79%	41%	34%	63%

The obtained results are far from ideal. The average efficiency was equal to 58% (227 from 395 investigated frames were properly processed). However, the very difficult nature of the investigated problem has to be emphasised. Apart from the mentioned strong distortions of the numbers there are many other problems to consider. Some of them were obvious even before starting the experiments. First of all, not every camera shot can be explored. For instance, it is useless to work with shots taken from long distance. In that case the player's number is not visible. Secondly, the grass's green colour is especially difficult in the method. During the World Cup 2006 three teams, namely Togo, Ireland and Ivory Coast, had green shirts. It is obvious that in some cases (e.g. depending on lighting conditions) the result of the extraction will cover a larger area, including some parts of the play field as well. The kit of a goalkeeper constitutes another problem. Usually it strongly differs from the original national kit. The same problem is related to a referee, but he does not lie in the area of interest as far as the proposed application is concerned. The very difficult problem is caused by some distortions of clothes, due to rain, dirt, tearing, etc. Those can significantly change the overall colour of a shirt.

Another interesting issue is the use of numbers strictly designed and prepared for a team. They are designed by the sports apparel manufacturers for particular national teams (sometimes more than one) and are used for some time (usually, at least one year) by them. This is helpful in the preparation of the templates for a particular football match.

4   Conclusions

In the paper an approach for the extraction of the numbers placed on sport shirts during the TV broadcast was proposed. This is the first step in the automatic system for identification of players basing on their numbers and is composed of two main stages: localisation of kits and later — numbers within the previously selected image area. The research was limited to the football matches. The experimentally established ranges in various colour spaces were used for the localisation.

Future works will be concentrated on the second stage — the recognition of the extracted numbers. Several shape description algorithms will be used for this purpose and the template matching approach will be applied. Firstly, the algorithms based on the transformation of points from Cartesian to polar coordinates will be investigated, since they obtained very promising results in the problem of shape recognition (e.g. [14], [15]).

Acknowledgments. The author of this paper wishes to thank gratefully MSc W. Batniczak for significant help in developing and exploring the presented approach.

References

1. Chandaria, J., Thomas, G., Bartczak, B., Koeser, K., et al.: Real-Time Camera Tracking in the MATRIS Project. In: Proc. of the International Broadcasting Convention, IBC, Amsterdam, the Netherlands, September 7-11 (2006)
2. TRACAB homepage, http://www.tracab.com
3. Papaioannou, E., Karpouzis, K., De Cuetos, P., Karagiannis, V., et al.: MELISA — A Distributed Multimedia System for Multi-Platform Interactive Sports Content Broadcasting. In: Proc. of the 30th EUROMICRO Conference, Rennes, France, September 1-3 (2004)
4. Andrade, E.L., Khan, E., Woods, J.C., Ghanbari, M.: Player Identification in Iinteractive Sport Scenes Using Region Space Analysis, Prior Information and Number recognition. In: Proc. of the International Conference on Visual Information Engineering, pp. 57–60 (July 2003)
5. Ghanbari, M., Andrade, E.L., Woods, J.C.: Outlier Removal for Player Identification in Interactive Sport Scenes Using Region Analysis. In: Proc. of the International Workshop on Image Analysis for Multimedia Interactive Services, WIAMIS 2004, Lisbon, Portugal (2004)
6. Frejlichowski, D., Wierzba, P.: A Face-Based Automatic Identification of Football Players During a Sport Television Broadcast. Polish Journal of Environmental Studies 17(4C), 406–409 (2008)
7. Kukharev, G., Kuzminski, A.: Biometric Techniques. Part 1. In: Face Recognition Methods (in Polish). WIPS Press, Szczecin (2003)
8. Yow, D., Yeo, B.-L., Yeung, M., Liu, B.: Analysis and presentation of Soccer highlights from digital video. In: Li, S., Teoh, E.-K., Mital, D., Wang, H. (eds.) ACCV 1995. LNCS, vol. 1035, pp. 499–503. Springer, Heidelberg (1996)
9. Choi, S., Seo, Y., Kim, H., Hong, K.-S.: Where Are the Ball and Players?: Soccer Game Analysis with Color-based Tracking and Image Mosaick. In: Del Bimbo, A. (ed.) ICIAP 1997. LNCS, vol. 1311, pp. 196–203. Springer, Heidelberg (1997)
10. Leonardi, R., Migliorati, P.: Semantic Indexing of Multimedia Documents. Trans. on IEEE Multimedia 9(2), 44–51 (2002)
11. Assfalg, J., Bertini, M., Colombo, C., del Bimbo, A., Nunziati, W.: Semantic Annotation of Soccer Videos: Automatic Highlights Identification. Computer Vision and Image Understanding 92, 285–305 (2003)

12. Ancona, N., Cicirelli, G., Stella, E., Distante, A.: Ball Detection in Static Images with Support Vector Machines for Classification. Image and Vision Computing 21, 675–692 (2003)
13. Frejlichowski, D.: Image Segmentation for the Needs of the Automatic Identification of Players Basing on Kits During the Sport Television Broadcast (in Polish). Methods of Applied Computer Science (2), 45–54 (2007)
14. Frejlichowski, D.: An Algorithm for Binary Contour Objects Representation and Recognition. In: Campilho, A., Kamel, M.S. (eds.) ICIAR 2008. LNCS, vol. 5112, pp. 537–546. Springer, Heidelberg (2008)
15. Frejlichowski, D.: Analysis of Four Polar Shape Descriptors Properties in an Exemplary Application. In: Bolc, L., Tadeusiewicz, R., Chmielewski, L.J., Wojciechowski, K. (eds.) ICCVG 2010, Part I. LNCS, vol. 6374, pp. 376–383. Springer, Heidelberg (2010)

Real-Time Hand Gesture Recognition
Using a Color Glove

Luigi Lamberti[1] and Francesco Camastra[2,*]

[1] Istituto Tecnico Industriale "Enrico Medi",
via Buongiovanni 84, 80046 San Giorgio a Cremano, Italy
luigi.lamberti@istruzione.it
[2] Department of Applied Science, University of Naples Parthenope,
Centro Direzionale Isola C4, 80143 Naples, Italy
francesco.camastra@uniparthenope.it

Abstract. This paper presents a real-time hand gesture recognizer based on a color glove. The recognizer is formed by three modules. The first module, fed by the frame acquired by a webcam, identifies the hand image in the scene. The second module, a feature extractor, represents the image by a nine-dimensional feature vector. The third module, the classifier, is performed by means of *Learning Vector Quantization*. The recognizer, tested on a dataset of 907 hand gestures, has shown very high recognition rate.

1 Introduction

Gesture is one of the means that humans use to send informations. According to Kendon's gesture continuum [1] the information amount conveyed by gesture increases when the information quantity sent by the human voice decreases. Moreover, the hand gesture for some people, e.g. the disabled people, is one of main means, sometimes the most relevant, to send information.

The aim of this work is the development of a real-time hand gesture recognizer that can also run on devices that have moderate computational resources, e.g. netbooks. The real-time requirement is motivated by associating to the gesture recognition the carrying out of an action, for instance the opening of a multimedia presentation, the starting of an internet browser and other similar actions. The latter requirement, namely the recognizer can run on a netbook, is desirable in order that the system can be used extensively in classrooms of schools as tool for teachers or disabled students.

The paper presents a real-time hand gesture recognition system based on a color glove that can work on a netbook. The system is formed by three modules. The first module, fed by the frame acquired by a webcam, identifies the hand image in the scene. The second module, a feature extractor, represents the image by means of a nine-dimensional feature vector. The third module, the classifier, is performed by *Learning Vector Quantization*.

---

* Corresponding author.

G. Maino and G.L. Foresti (Eds.): ICIAP 2011, Part I, LNCS 6978, pp. 365–373, 2011.

The paper is organized as follows: Section 2 describes the approach used; Section 3 gives an account of the segmentation module; the feature extraction process is discussed in Section 4; a review of Learning Vector Quantization is provided in Section 5; Section 6 reports some experimental results; in Section 7 some conclusions are drawn.

2   The Approach

Several approaches were proposed for gesture recognition [2]. Our approach was inspired by Virtual Reality [3] applications where the movements of the hands of people are tracked asking them to wear data gloves [4]. A data glove is a particular glove that has fiber-optic sensors inside that allow the track the movement of the fingers of hand. Our approach is similar to the Virtual Reality's one. We ask the person, whose gesture has to be recognized, to wear a glove or more precisely, a color glove. A color glove was recently used by Wang and Popovic [5] for the real-time hand tracking. Their color glove was formed by patches of several different colors[1]. In our system we use a wool glove[2] where three different colors are used for the parts of the glove corresponding to the palm and the fingers, whereas the rest of glove is black. One color is used to dye the palm, the remaining two to color differently adjacent fingers, as shown in figure 1. We

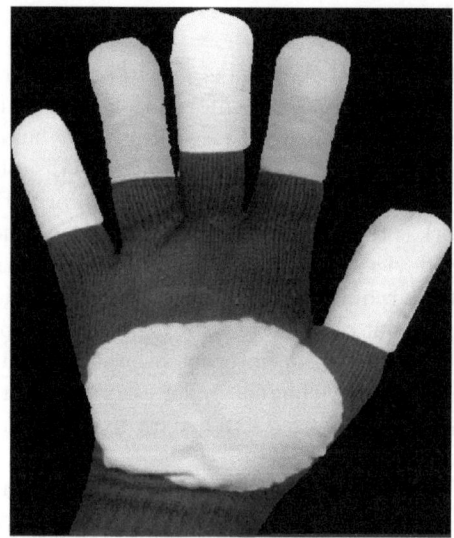

Fig. 1. The Color Glove used in our approach

---

[1] In the figure reported in [5] the glove seems to have at least seven different colors.

[2] Wool is not obviously compulsory, clearly cotton or other fabrics can be used.

have chosen to color the palm by magenta and the fingers by cyan and yellow. Further investigations seem to show that the abovementioned choice does not affect remarkably the performances of the recognizer.

Finally, we conclude the section with a cost analysis. In terms of costs, our glove compares favourably with data gloves. Our glove costs some euro, whereas the cost of effective data gloves can exceed several hundreds euro.

3   Segmentation Module

The gesture recognizer has three modules. The first one is the *segmentation* module. The module receives as input the RGB color frame acquired by the webcam and performs the segmentation process identifying the hand image. The segmentation process can be divided in four steps. The first step consists in representing the frame in *Hue-Saturation-Intensity (HSI)* color space [6]. We tested experimentally several color spaces, i.e. RGB, HSI, CIE XYZ, L*ab, L*uv and some others [7]. We chose HSI since it was the most suitable color space to be used in our segmentation process. Several algorithms were proposed [8] to segment color images. Our choice was to use the least expensive computationally segmentation strategy, i.e. a thresholding-based method. During the second step, the pixels of the image are divided in seven categories: "Cyan Pixels" (C), "Probable Cyan Pixels" (PC), "Yellow Pixels" (Y), "Probable Yellow Pixels" (PY), "Magenta Pixels" (M), "Probable Magenta Pixels" (PM), "Black Pixels" (B). A pixel, represented as a triple P=(H,S,I), is classified as follows:

$$\left\{ \begin{array}{ll} P \in C & \text{if } H \in [\Theta_1, \Theta_2] \wedge S > \Theta_3 \wedge I > \Theta_4 \\ P \in PC & \text{if } H \in [\Theta_{1r}, \Theta_{2r}] \wedge S > \Theta_{3r} \wedge I > \Theta_{4r} \\ P \in Y & \text{if } H \in [\Theta_5, \Theta_6] \wedge S > \Theta_7 \wedge I > \Theta_8 \\ P \in PY & \text{if } H \in [\Theta_{5r}, \Theta_{6r}] \wedge S > \Theta_{7r} \wedge I > \Theta_{8r} \\ P \in M & \text{if } H \in [\Theta_9, \Theta_{10}] \wedge S > \Theta_{11} \wedge I > \Theta_{12} \\ P \in PM & \text{if } H \in [\Theta_{9r}, \Theta_{10r}] \wedge S > \Theta_{11r} \wedge I > \Theta_{12r} \\ P \in B & \text{otherwise} \end{array} \right\}, \tag{1}$$

(a)   (b)

Fig. 2. The original image (a). The image after the segmentation process (b).

where Θ_{ir} is a relaxed value of the respective threshold Θ_i and $\Theta_i, (i = 1, \ldots, 12)$ are thresholds that were set up in a proper way.

In the third step only the pixels belonging to PC, PY and PM categories are considered. Given a pixel P and denoting with $N(P)$ its neighborhood, using the 8-connectivity [6], the following rules are applied:

$$\left\{\begin{array}{l} \text{If } P \in PC \wedge \exists_{Q \in N(P)} Q \in C \text{ then } P \in C \text{ else } P \in B \\ \text{If } P \in PY \wedge \exists_{Q \in N(P)} Q \in Y \text{ then } P \in Y \text{ else } P \in B \\ \text{If } P \in PM \wedge \exists_{Q \in N(P)} Q \in M \text{ then } P \in M \text{ else } P \in B \end{array}\right\}, \qquad (2)$$

In a nutshell, pixels belonging to PC, PY and PM categories are upgraded respectively to C,Y and M, respectively if in their neighborhood exists at least one pixel belonging to the respective superior class. The remaining pixels are degraded to black pixels. At the end of this phase only four categories, i.e. C, Y, M, and B, remain.

In the last step the connected components for the color pixels, i.e. the one belonging to the Cyan, Yellow and Magenta categories, are computed. Finally, each connected component is undergone to a *morphological opening* followed by a *morphological closure* [6]. In both steps the structuring element is a circle of radius of three pixels.

4   Feature Extraction

After the segmentation process, the image of the hand is represented by a vector of nine numerical features. The feature extraction process has the following steps. The first step consists in individuating the region formed by magenta pixels, that corrisponds to the palm of the hand. Then it is computed the centroid and the major axis of the region. In the second step the five centroids of yellow and cyan regions, corresponding to the fingers are individuated. Then, for each of the five regions, the angle $\theta_i(i = 1, \ldots, 5)$ between the main axe of the palm and the line connecting the centroids of the palm and the finger, is computed (see figure 3). In the last step the hand image is represented by a vector of nine normalized numerical features. As shown in figure 3, the feature vector is formed by nine numerical values that represent five distances $d_i(i = 1, \ldots, 5)$ and four angles $\beta_i(i = 1, \ldots, 4)$, respectively. Each distance measures the Euclidean distance between the centroid of the palm and the respective finger. The four angles between the fingers are easily computed by subtraction having computed before the angles θ_i. The extracted features are invariant, by construction, w.r.t. rotation and translation in the plane of the image.

Finally, all features are normalized. The distances are normalized, dividing them by the maximum value that they can assume [9]. The angles are normalized, dividing them by $\frac{\pi}{2}$ radians, assuming that it is the maximum angle that can be measured by the fingers.

In the description above we have implicitly supposed that exists only one region for the palm and five fingers. If the regions for the palm and the finger are not unique, the system uses a different strategy depending on it is already

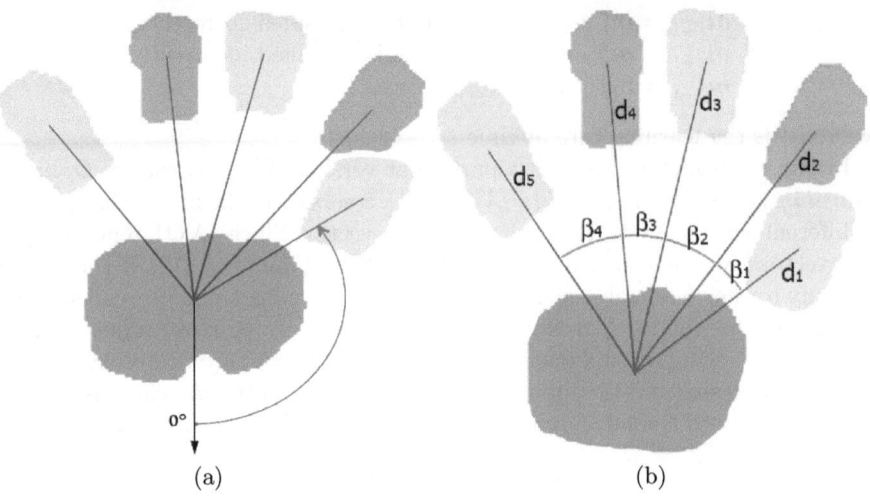

(a)   (b)

Fig. 3. (a) The angles   between the main axe of the palm and the line connecting the centroids of the palm and each finger, are computed. (b) The feature vector is formed by five distances $d_i(i = 1, \ldots, 5)$ and four angles $\beta_i(i = 1, \ldots, 4)$, obtained by subtraction, from angles   .

trained. If the system is not trained yet, i.e. it is in training the system takes for the palm and for each finger the largest region, in terms of area. If the system is trained it selects for each finger up to the top three largest regions, if they exist; whereas for the palm up the top two largest regions are picked. The chosen regions are combined in all possible ways yielding different possible hypotheses for the hand. Finally, the system selects the hypothesis whose feature vector is evaluated with the highest score by the classifier.

5   Learning Vector Quantization

We use *Learning Vector Quantization* (*LVQ*) as classifier since LVQ requires moderate computational resources compared with other machine learning classifiers (e.g. SVM [10]). We pass to describe LVQ. We first fix the notation; let $\mathcal{D} = \{x_i\}_{i=1}^{\ell}$ be a data set with $x_i \in \mathbb{R}^N$. We call *codebook* the set $W = \{w_k\}_{k=1}^{K}$ with $w_k \in \mathbb{R}^N$ and $K \ll \ell$. Vector quantization aims to yield codebooks that represent as much as possible the input data D. LVQ is a supervised version of vector quantization and generates codebook vectors (*codevectors*) to produce *near-optimal decision boundaries* [11]. LVQ consists of the application of a few different learning techniques, namely LVQ1, LVQ2 and LVQ3. LVQ1 uses for classification the nearest-neighbour decision rule; it chooses the class of the nearest codebook vector. LVQ1 learning is performed in the following way: if m_t^c [3] is the nearest codevector to the input vector x, then

---

[3] m_t^c stands for the value of m^c at time t.

$$m_{t+1}^c = m_t^c + \alpha_t[x - m_t^c] \text{ if } x \text{ is classified correctly}$$
$$m_{t+1}^c = m_t^c - \alpha_t[x - m_t^c] \text{ if } x \text{ is classified incorrectly} \qquad (3)$$
$$m_{t+1}^i = m_t^i \qquad\qquad i \neq c$$

where α_t is the learning rate at time t.

In our experiments, we used a particular version of LVQ1, that is *Optimized Learning Vector Quantization (OLVQ1)* [11], a version of the model that provides a different learning rate for each codebook vector. Since LVQ1 tends to push codevectors away from the decision surfaces of the *Bayes rule* [12], it is necessary to apply to the codebook generated a successive learning technique called LVQ2. LVQ2 tries harder to approximate the Bayes rule by pairwise adjustments of codevectors belonging to adjacent classes. If m^s and m^p are nearest neighbours of different classes and the input vector x, belonging to the m^s class, is closer to m^p and falls into a zone of values called *window* [4], the following rule is applied:

$$m_{t+1}^s = m_t^s + \alpha_t[x - m_t^s]$$
$$m_{t+1}^p = m_t^p - \alpha_t[x - m_t^p] \qquad (4)$$

It can be shown [13] that the LVQ2 rule produces an instable dynamics. To prevent this behavior as far as possible, the window w within the adaptation rule takes place must be chosen carefully.

In order to overcome the LVQ2 stability problems, Kohonen proposed a further algorithm (LVQ3). If m^i and m^j are the two closest codevectors to input x and x falls in the window, the following rule is applied[5]:

$$\left\{ \begin{array}{ll} m_{t+1}^i = m_t^i & \text{if } C(m^i) \neq C(x) \wedge C(m^j) \neq C(x) \\ m_{t+1}^j = m_t^j & \text{if } C(m^i) \neq C(x) \wedge C(m^j) \neq C(x) \\ m_{t+1}^i = m_t^i - \alpha_t[x_t - m_t^i] & \text{if } C(m^i) \neq C(x) \wedge C(m^j) = C(x) \\ m_{t+1}^j = m_t^j + \alpha_t[x_t - m_t^j] & \text{if } C(m^i) \neq C(x) \wedge C(m^j) = C(x) \\ m_{t+1}^i = m_t^i + \alpha_t[x_t - m_t^i] & \text{if } C(m^i) = C(x) \wedge C(m^j) \neq C(x) \\ m_{t+1}^j = m_t^j - \alpha_t[x_t - m_t^j] & \text{if } C(m^i) = C(x) \wedge C(m^j) \neq C(x) \\ m_{t+1}^i = m_t^i + \epsilon\alpha_t[x_t - m_t^i] & \text{if } C(m^i) = C(m^j) = C(x) \\ m_{t+1}^j = m_t^j + \epsilon\alpha_t[x_t - m_t^j] & \text{if } C(m^i) = C(m^j) = C(x) \end{array} \right\}, \qquad (5)$$

where $\epsilon \in [0, 1]$ is a fixed parameter.

6   Experimental Result

To validate the recognizer we selected 13 gestures, invariant by rotation and translation. We associated to each gesture a symbol, a letter or a digit, as shown in figure 4. We collected a database of 1541 gestures, performed by people of different gender and physique. The database was splitted with a random process into training and test set containing respectively 634 and 907 gesture. The number of classes used in the experiments was 13, namely the number of the different gestures in our database. In our experiments the three learning techniques, i.e.

---

[4] The window is defined around the midplane of m^s and m^p.
[5] $C(q)$ stands for the class of q.

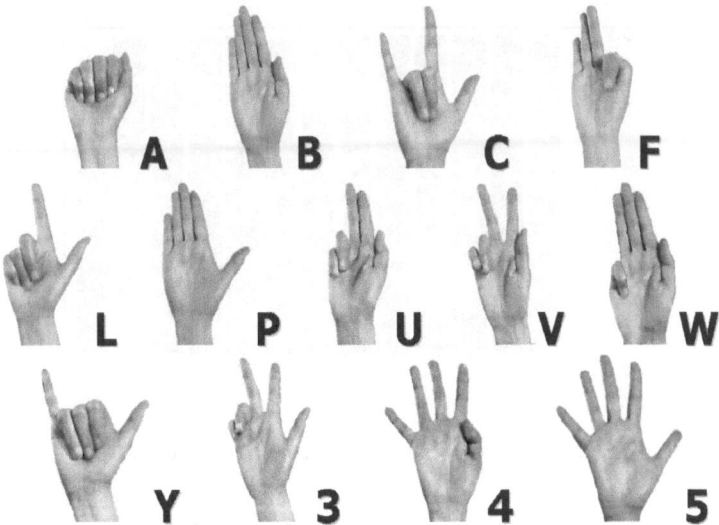

Fig. 4. Gestures represented in the database

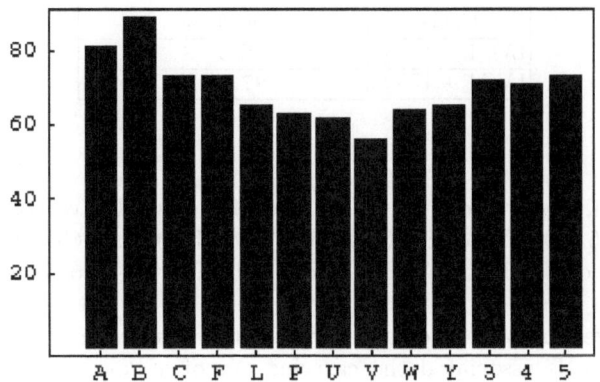

Fig. 5. Gesture distribution in the test set

LVQ1, LVQ2 and LVQ3, were applied. We trained several LVQ nets by specifying different combinations of learning parameters, i.e. different learning rates for LVQ1, LVQ2, LVQ3 and various total number of codevectors. The best LVQ net was selected by means of *crossvalidation* [14]. LVQ trials were performed using *LVQ-pak* [15] software package. Figure 5 shows the gesture distribution in the test set. In Table 1, for different classifiers, the performances on the test set, measured in terms of recognition rate in absence of rejection, are reported.

Our best result in terms of recognition rate is *97.79 %*. The recognition rates for each class are shown in figure 6.

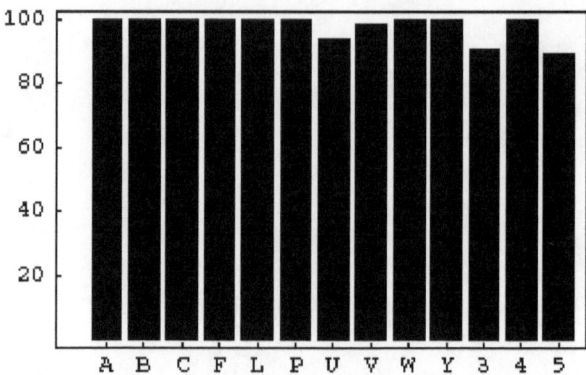

Fig. 6. Recognition rates, for each class, in the test set

Table 1. Recognition rates on the test set, in absence of rejection, for several LVQ classifiers

Algorithm	Correct Classification Rate
knn	85.67 %
LVQ1	96.36 %
LVQ1 + LVQ2	97.57 %
LVQ1 + LVQ3	**97.79** %

The system, implemented in C++ under Windows XP Microsoft and .NET Framework 3.5 on a Netbook with 32bit "Atom N280" 1.66 Ghz, Front Side Bus a 667 Mhz and 1 GB di RAM, requires 140 CPU msec to recognize a single gesture.

Finally, a version of the gesture recognition system where the recognition of a given gesture is associated the carrying out of a Powerpoint[6] command is actually in use at Istituto Tecnico Industriale "Enrico Medi" helping the first author in his teaching activity.

7   Conclusions

In this paper we have described real-time hand gesture recognition system based a color Glove. The system is formed by three modules. The first module identifies the hand image in the scene. The second module performs the feature extraction and represents the image by a nine-dimensional feature vector. The third module, the classifier, is performed by means of Learning Vector Quantization. The recognition system, tested on a dataset of 907 hand gestures, has shown a

---

[6] Powerpoint, .NET and Windows XP are registered trademarks by Microsoft Corp.

recognition rate close to 98%. The system implemented on a netbook requires an average time of 140 CPU msec to recognize a hand gesture. A version of the system where the recognition of a given gesture is associated the carrying out of a multimedia presentation command is actually used by the first author in his teaching duties. Since our system compares favourably, in terms of costs, with data glove, in the next future we plan to investigate its usage in virtual reality applications.

Acknowledgments. First the author wish to thank the anonymous reviewer for his valuable comments. The work was developed by L. Lamberti, under the supervision of F. Camastra, as final dissertation project for B. Sc. in Computer Science at University of Naples Parthenope.

References

1. Kendon, A.: How gestures can become like words. In: Crosscultural Perspectives in Nonverbal Communication, Toronto, Hogrefe, pp. 131–141 (1988)
2. Mitra, S., Acharya, T.: Gesture recognition: A survey. IEEE Transactions on Systems, Man and Cybernetics, Part C: Applications and Reviews 37(3), 311–324 (2007)
3. Burdea, G.C., Coiffet, P.: Virtual Reality Technology. John-Wiley & Sons, New York (2003)
4. Dipietro, L., Sabatini, A.M., Dario, P.: A survey of glove-based systems and their applications. IEEE Transactions on Systems, Man and Cybernetics 38(4), 461–482 (2008)
5. Wang, R.Y., Popovic, J.: Real-time hand-tracking with a color glove. ACM Transactions on Graphics 28(3), 461–482 (2009)
6. Gonzales, R.C., Woods, R.E.: Digital Image Processing. Prentice-Hall, Upper Saddle River (2002)
7. DelBimbo, A.: Visual Information Processing. Morgan Kaufmann Publishers, San Francisco (1999)
8. Cheng, H.D., Jiang, X.H., Sun, Y., Wang, J.: Color image segmentation: advances and prospects. Pattern Recognition 34(12), 2259–2281 (2001)
9. Lamberti, L.: Handy: Riconoscimento di semplici gesti mediante webcam. B. Sc. Dissertation (in Italian), University of Naples "Parthenope" (2010)
10. Shawe-Taylor, J., Cristianini, N.: Kernels Methods for Pattern Analysis. Cambridge University Press, Cambridge (2004)
11. Kohonen, T.: Self-Organizing Maps. Springer, Berlin (1997)
12. Duda, R.O., Hart, P.E., Stork, D.G.: Pattern Classification. John-Wiley & Sons, New York (2001)
13. Kohonen, T.: Learning vector quantization. In: The Handbook of Brain Theory and Neural Networks, pp. 537–540. MIT Press, Cambridge (1995)
14. Stone, M.: Cross-validatory choice and assessment of statistical prediction. Journal of the Royal Statistical Society 36(1), 111–147 (1974)
15. Kohonen, T., Hynninen, J., Kangas, J., Laaksonen, J., Torkkola, K.: Lvq-pak: The learning vector quantization program package. Technical Report A30, Helsinki University of Technology, Laboratory of Computer and Information Science (1996)

Improving 3D Reconstruction for Digital Art Preservation

Jurandir Santos Junior, Olga Bellon, Luciano Silva, and Alexandre Vrubel

Department of Informatics, Universidade Federal do Parana,
IMAGO Research Group, Curitiba, Brazil
{jurandiro,olga,luciano,alexandrev}@inf.ufpr.br
http://www.imago.ufpr.br

Abstract. Achieving a high fidelity triangle mesh from 3D digital reconstructions is still a challenge, mainly due to the harmful effects of outliers in the range data. In this work, we discuss these artifacts and suggest improvements for two widely used volumetric integration techniques: VRIP and Consensus Surfaces (CS). A novel contribution is a hybrid approach, named IMAGO Volumetric Integration Algorithm (IVIA), which combines strengths from both VRIP and CS while adds new ideas that greatly improve the detection and elimination of artifacts. We show that IVIA leads to superior results when applied in different scenarios. In addition, IVIA cooperates with the hole filling process, improving the overall quality of the generated 3D models. We also compare IVIA to Poisson Surface Reconstruction, a state-of-the-art method with good reconstruction results and high performance both in terms of memory usage and processing time.

Keywords: range data, 3D reconstruction, cultural heritage.

1 Introduction

The preservation of natural and cutural assets is one of the most challenging applications for digital reconstruction. It requires high fidelity results at the cost of high-resolution range acquisition devices, along with a set of algorithms capable of dealing with noise and other artifacts present in range data.

Several stages compose a complete 3D reconstruction pipeline, [1], [2], [3]. In this work, we focus on mesh integration within the pipeline proposed in [3]. Several approaches perform this task; we categorize them as presented in [1]. *Delaunay-based methods* [10] work on point clouds, but they are usually sensitive to noise and outliers since they interpolate the data points. *Surface-based methods* [11], [12] create surfaces directly, but some of them can catastrophically fail when applied on high curvature regions [6]. Also, topologically incorrect solutions can occur due to outliers in the input views.

Parametric surfaces methods [13], [14], [15] deform an initial approximation of the object through the use of external forces and internal reactions and constraints. Kazhdan *et al.* [9] calculate locally supported basis functions by solving

G. Maino and G.L. Foresti (Eds.): ICIAP 2011, Part I, LNCS 6978, pp. 374–383, 2011.
© Springer-Verlag Berlin Heidelberg 2011

sparse linear systems that represent a Poisson problem. A parallel and an out-of-core implementations of [9] are proposed in [16], [17]. One limitation of such algorithms is finding a balance between over smoothing effect and non-elimination of noisy surfaces. These methods may also fill holes incorrectly.

Volumetric methods [22], [23] create an implicit volumetric representation of the final model, which is also called a SDF (*Signed Distance Field*). The object surface is defined by the isosurface at distance value 0, that is usually extracted with the MC (*Marching Cubes*) [18]. Curless and Levoy [6] calculate the distance function according to the line-of-sight of the scanner and uses weights to assess the reliability of each measurement. Wheeler *et al.* [7] discard outliers through the calculation of a consensus surface. Volumetric methods use all available information and ensure the generation of manifold topologies [6]. The main limitation is their performance, both in terms of memory usage and processing time.

All integration approaches have their limitations. We have chosen volumetric methods because they impose fewer restrictions to the reconstructed objects; offer an easy way to change the precision of the output; can easily support the space carving technique [6], and can work in the presence of noisy input data. We implemented, tested and modified VRIP [6], [4] and Consensus Surface (CS) [7], [5]. So, in the attempt to solve their limitations, we developed a novel hybrid integration method, named IMAGO Volumetric Integration Algorithm (IVIA), which improves the fidelity of the generated 3D models. The first pass of our method approximate normals and eliminate outliers based on a modified VRIP. Then, a second pass reconstruct a bias-free surface using CS with our suggested improvements, and fully cooperates with the hole filling algorithm of [8].

IVIA was developed to handle with artifacts present on range data, such as noise, occlusions and false data. We show how it overcomes these problems and generates more accurate results. We also compare it to the Poisson Surface Reconstruction (PSR) [9]. Although PSR is not a volumetric approach, it is considered the state-of-the-art method for 3D reconstruction. Also, PSR presents superior results when compared to other methods and good performance both in terms of memory usage and processing time.

The remainder of this paper is organized as follows: first we brief analyze the artifacts present on range data in section 2. Then, in section 3 we suggest improvements for VRIP [6] and CS [7]. In section 4, we present our novel hybrid volumetric method, named IVIA. We discuss our results compared to VRIP, CS and PSR in section 5, followed by conclusions in section 6.

2   Range Data Analysis

To better understand the challenges of building a high quality integration algorithm, we analyzed the types of artifacts that appear in the input range data. In this section, we discussed data captured with a laser triangulation 3D scanner (*e.g.* Vivid 910 from Konica Minolta).

This analysis is important because those artifacts are complex and harder to automatically detect and discard. For instance, noise on the object surface is one of the most common artifact on range data. Even for very smooth objects, the captured data is usually rough. Another common artifact is data returned by the scanner that do not exist in the object. They usually appear as small groups of triangles in regions that should be empty. It is hard to eliminate this type of artifact without leading to the elimination of valid data. In [6] the authors proposed the space carving technique that was used to fill holes. However, they do not eliminate outliers as we do in our proposed approach.

Deformed surfaces is a more serious and relatively common artifact of difficult detection because they consist of large areas, usually connected to valid data, but completely wrong compared to the original object. Automatic detection and elimination of such artifacts is particularly difficult. Our IVIA approach successfully discards most of these artifacts, as discussed in section 5.

Another problem is due to the triangulation of the point cloud captured by the scanner. That introduces false internal and external silhouettes surfaces. Our solution is to check the angle between the face normal and the line-of-sight from the scanner position to the center of the face. If this angle is above some threshold, the face should be discarded. By using a threshold of 75° we achieved good results in all of our experiments.

Impossible to be captured data is a problem caused due to occlusions. For many applications, such holes are unacceptable and hole filling techniques complement the captured data [8]. The more information we can extract from the captured data, the better the completed surface will be. While other integration techniques usually discard helpful data that could be used in the hole filling process, IVIA was designed to fully cooperate with the hole filling stage.

3   VRIP, CS and PSR Approaches

As our hybrid method was developed to overcome the limitations of the VRIP and the CS, to better understand it, we assess them and suggest some improvements. As we compared our IVIA to PSR we shortly discuss this approach too.

3.1   Volumetric Range Image Processing (VRIP)

Like all volumetric integration methods, the goal of VRIP (Curless and Levoy [6]) is to calculate the signed distance from each voxel of the volume to the integrated surface. VRIP performs fast and allows incremental addition of views. Besides, it generates a smoother integrated surface, reduces noise and does not require all views being loaded in memory simultaneously.

The main limitation of VRIP is not comprise any specific process to discard outliers. It is even difficult to reduce their weight as each view is processed individually. It also integrates metrics from different viewpoints, resulting the SDF to be non-uniform. In addition, a flaw causes artifacts in corners and thin surfaces, as pointed by the authors [6]. As the algorithm combines positive and negative values in order to find the average surface, it creates creases on thin surfaces where one side would interfere with the other, increasing the thickness.

To reduce both the influence of outliers and the flaw near corners and thin surfaces, we propose a new weight attenuation curve according to the value of the signed distance. Instead of using a linear reduction from half of the range, as originally suggested in [6], we adopted a non-symmetric curve, giving more weight to measurements we are sure of (*i.e.* outside voxels). Although better visual results are obtained, this "trick" is not yet a definitive solution for a high fidelity reconstruction; however, it will be useful to our IVIA approach later.

3.2   Consensus Surface (CS)

Wheeler *et al.* [7] proposed to use consensual distances averaged to create the signed distance to the surface from the current voxel. Therefore, in consensual regions from several views, CS eliminates outliers and provides a relatively smooth integrated surface. However, it performs slowly, their magnitude and sign of the distance may be incorrectly calculated around the borders of views and when it automatically fills holes, the results may present inconsistencies.

The majority of the visible artifacts on CS result from incorrect signs being calculated for the distance values, as pointed by Sagawa *et al.* [20], [21]. However, they assumed that the consensus criterion successfully removes all outliers. We noticed in our experiments that even if the SDF signs are repaired, their magnitudes can still be wrong, and we cannot guarantee that the reconstruction will have high fidelity.

Trying to address the main limitations of CS, we propose modifications on it. First, we choose to use equal sized voxels to achieve the highest possible precision. As CS calculates distance measurements incorrectly when the nearest point in the view is on a mesh border, we choose to discard all of them. After the integration, we also perform an additional data validation by discarding voxels whose distances are not compatible with their neighbors. This process removes any eventual incorrect remaining data after the previous stages. Our modified CS still bears some limitations of the original algorithm such as: difficulty in choosing thresholds, the necessity to have all views loaded in memory simultaneously and outliers may appear on the integrated results. Our new IVIA algorithm combines the strengths of both the VRIP and our modified CS to overcome these problems aiming the generation of more accurate results.

3.3   Poisson Surface Reconstruction (PSR)

Kazhdan *et al.* [9] express surface reconstruction as the solution to a Poisson equation. Their idea is to compute a function χ that has value one inside the surface and zero outside (*i.e.* indicator function), and by extracting an appropriated isosurface they reconstruct the object surface. The oriented point set can be viewed as samples of the gradient from that function. They solve the indicator function by finding the scalar function whose gradient best matches the vector field $\vec{V}$, a variational problem of a standard Poisson equation.

PSR performs good both in terms of memory usage and processing time, and has the benefit to increase the resolution by increasing the octree depth. However, whenever resolution is increased the time performance becomes lower. It is also difficult finding a balance between over smoothing effect and non-elimination of noisy surfaces (see section 5). Besides, PSR does not ensure the generation of manifold topologies or when it creates manifold meshes it may occurs that outliers as vertices appear on the final model (see section 5). In additional, PSR does not use all available information as well (*e.g.* line-of-sight information) what may cause incorrectly connections of some regions, a problem pointed out by the authors [9].

4   The Proposed IVIA Approach

The greatest challenge to develop a high quality reconstruction is detecting and discarding several types of artifacts that usually appear on range data. We developed the a novel approach, named IMAGO Volumetric Integration Algorithm (IVIA), that automatically detects and eliminates most of them. We perform the integration in two passes. The first one creates a volumetric representation of the integrated model by using our modified version of the VRIP algorithm; the second one uses the representation created to detect and discard outliers during the generation of the final volumetric integration.

The goal of the first pass is to build an approximated volumetric representation through the modified VRIP presented in section 3.1. Besides, a binary volume *empty* is created to represent the space carving operation [6]. For each voxel there is a corresponding bit in *empty*; if the bit is set, the corresponding voxel is considered empty (*i.e.* outside the object). As the distance calculated is over the line-of-sight of the scanner, any negative values are considered "empty" and the corresponding bit of *empty* is set. It is important to notice that we use our modified distance weight curve, to help eliminate surface outliers.

After all views are processed, we apply a 3D mathematical morphology erosion operation on the binary volume *empty*. This is necessary for two reasons: first, to prevent any incorrect measurement "dig holes" in the object. Though unusual, we already noticed this type of error on some range images. The other reason is that *empty* is obtained through the union of empty spaces of all views. Therefore, it tends to represent the lowest measurement for each surface point, and not an average measurement. When reducing the empty space by a distance $+D_{max}$ we set aside a space near the surface to integrate the measurements from several views, and at the same time we keep an empty space representation to eliminate outliers far from the surface. Several artifacts presented in section 2 are eliminated by using *empty*.

Finishing the first pass, we smoothen the values of the volumetric representation, using a 3 x 3 x 3 filter with larger weights to central voxels on the mask (*i.e.* 2D smoothing filter). This smoothing completes individual voxels with plausible values and attenuate noise and surface outliers. This attenuation is important because this volumetric representation will be used to estimate the integrated object normals in the second pass. After that, the volumetric representation created is saved.

The second pass of the algorithm does the definitive integration, discarding outliers. This pass possesses elements from CS, like the Euclidian distance calculation. We used them to prevent the biased surfaces generated when VRIP distances are used. Besides, Euclidian distances improve the hole filling algorithm we used [8]. Only voxels near the view surfaces and not set in *empty* are evaluated. To find these voxels, we loop on each vertex of the view marking the voxels near them. This second pass has linear complexity on the number of views, instead of original CS, which is quadratic.

We calculate an estimated normal n for the current voxel v from the volumetric representation generated in the first pass. This normal will be used to validate the view data, discarding incorrect measurements and outliers. The normal n is calculated from the (x, y, z) gradients of volumetric representation. If the voxel does not have valid neighbors, n cannot be estimated. There are two possibilities in this case: if the boolean parameter $flagDiscardNoNormal$ is true, the voxel is skipped; otherwise, this voxel will not be validated by the normal n, which can lead to outliers being accepted on this voxel. The rare cases that require $flagDiscardNoNormal$ to be false are when there are very thin surfaces compared to the voxel size. In these cases, if $flagDiscardNoNormal$ is true, holes are usually created in the reconstructed surface.

Next, the nearest point p', its normal n', weight w and signed distance d are evaluated. These values are calculated as in CS by using a *kd-tree* of view i to find the nearest vertex. The search for the nearest point is done on the faces incident on this vertex by calculating point to triangle distances. This process also tells us if the nearest point belongs to the border of view i. If p' is located on a border, this measurement is discarded to avoid the problem discussed on section 3.2. Measurements larger than $+D_{max}$ are also discarded. Finally, if an estimated normal n was calculated, the angle between n and n' is calculated. If this angle is larger than the threshold *consensusAngle*, the measurement is discarded. In our experiments, a value from 30° to 45° for the threshold *consensusAngle* returned good results. This procedure solves the flaw of VRIP presented in section 3.1. Besides, most of the deformed surfaces presented in section 2 are eliminated.

Following, the basic reliability w of the measurement (which depends on the angle between the scanner line-of-sight and the normal n') is altered by other 3 factors. The first one is the border weight w_{border}, where measurements near the borders of the view have lower weights than interior measurements. This weight is used to discard p' if it is too small. The second factor is w_{angle}, due to the angle between n and n'. This factor, ranging from 0.0 to 1.0, therefore $1.0 \geq (n.n') \geq \cos(consensusAngle)$. The third factor $w_{distance}$ varies according to the signed distance d, like the weight curves from VRIP. Our IVIA approach uses a new weight curve as described in section 3.1. This curve allows a smooth and unbiased integration of the distance values. Finally, we integrate the measurements of all views through a weighted sum using the VRIP formula [6].

After the integration is completed, we have a last error elimination step. Neighboring voxels should have similar distance measurements because of the use of Euclidian distances. Therefore, two neighbor voxels should have a maximum distance difference equivalent to the distance between the voxel centers. Using 26-neighborhood, the distances between voxel centers can be $voxelSize$, $\sqrt{2} * voxelSize$ or $\sqrt{3} * voxelSize$. However, we cannot be so restrictive, because the integration slightly violates this condition, due to the several weights used. Nevertheless, neighboring voxels with excessively different values should be discarded to avoid the generation of bad surfaces on the final model. Therefore, we use a threshold $compatibleFactor$ (usually 1.5), which multiplied by the voxel center distance gives us the maximum allowed difference between neighboring voxels. An important detail is that this elimination cannot be done in a single pass, because a "wrong" voxel ends up spoiling its neighbors, since their difference would be too large. Because of that, we first gather all the "suspect" voxels, and sort them (using $bucketsort$) by the number of wrong neighbors (which can be 26 at maximum). Next, we eliminate the "most suspect" candidate, updating the suspect list to ignore neighborhood to voxels already eliminated. Therefore, the elimination is more selective, discarding only the really incompatible voxels. These voxels are tagged as unknown (distance value of $+D_{max}$ and weight 0).

Our IVIA algorithm deals with noise and outliers in several ways. With space carving, outliers far from the surface are eliminated. Both outliers near the surface and measurements out of consensus with the first pass are also eliminated. Measurements have their reliability assessed through several parameters to help decrease the influence of any outliers that had not been removed previously. Finally, after integration we eliminate the remaining incompatible data.

5   Experimental Results

We performed several experiments to assess the effectiveness of our method for applications that demand high fidelity 3D reconstructions, such as digital preservation of natural and cultural assets. The dataset used in the experiments is composed of art objects (from the Metropolitan Museum of Curitiba), fossils (from the Natural Science Museum of UFPR), insects (from the Biological Collections of UFPR), Baroque masterpieces (Aleijadinho's sculptures) and personal objects. The results show how our method deals with difficult situations.

The ability of IVIA to detect and discard outliers can be seen in Fig. 1. The captured views have several artifacts because the statue material (marble) is not optically cooperative. Fig. 1b shows false data and deformed surfaces. In Fig. 1c we show the influence of the space carving on the view data, and the successful detection and discarding of artifacts. Fig. 1a shows result without hole filling.

In Fig. 2 we show a comparison of results from the three volumetric algorithms when applied in a difficult case. IVIA generated the most reliable result. VRIP and our modified CS returned outliers on the final result, even after a postprocessing step that discarded disconnected geometries. We disabled this post-processing on IVIA to show its robustness in removing artifacts. The result from our modified CS (see section 3.2) is similar to the ones from IVIA; however, the result from the original CS has lots of outliers on the final result. VRIP result is good, except from the connected outliers and the biased surfaces, as well as the artifacts near corners and thin surfaces. In our IVIA result, the surface is as smooth as the VRIP one, but without the problems previously mentioned. The price we pay is slightly larger holes on regions of unreliable data.

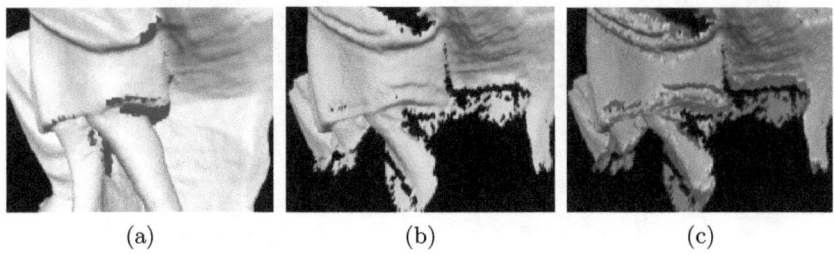

(a)   (b)   (c)

Fig. 1. Outlier removal of the IVIA: (a) statue integrated with IVIA; (b) one of the views, with several types of bad data; (c) space carving technique where red/green as influenced regions by the w_{angle}, and purple as discarded regions

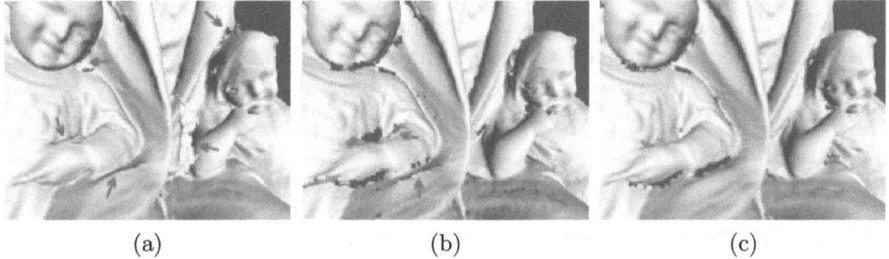

(a)   (b)   (c)

Fig. 2. Comparision of 3 integrations: (a) VRIP; (b) our modified CS; (c) IVIA. Several artifacts can be seen in (a) and (b) (arrows). In (c) IVIA eliminates almost all incorrect data and kept the surface smooth.

As mentioned IVIA was developed to cooperate with the holefilling algorithm of [8]. For that, IVIA performs a more effective outlier removal, uses Euclidian distances and space carving technique, producing more accurate results. In Fig. 3 we activate holefilling aiming to compare our results to PSR (original authors'). The IVIA's reconstructed model of Aleijadinho's Prophet Joel is showed in Fig 3a. Fig 3b shows the legs of Joel in details, where we can see that PSR reconstructed model (Fig 3b down) is over smoothed compared to our IVIA (Fig 3b up), we also noticed in the highlighted rectangle area that PSR creates outliers as vertices (see section 3.3). In Fig 3c we can see two detailed regions, Joel's hat (Fig 3c up) and Joel's vestment (3c down). We also tested IVIA on objects with smaller data sizes and two of them are shown in Fig 3d.

The computer used in all experiments was a 2.20GHz Core2 Duo PC, with 2 GB of RAM. Duck model had 67,772 faces, Wolf model had 206,627 faces, the marble statue (Fig 2) had approximately 1,300,300 faces and Joel model had 3,487,652 faces. To reconstruct Joel PSR took 3002s, VRIP took 2718s

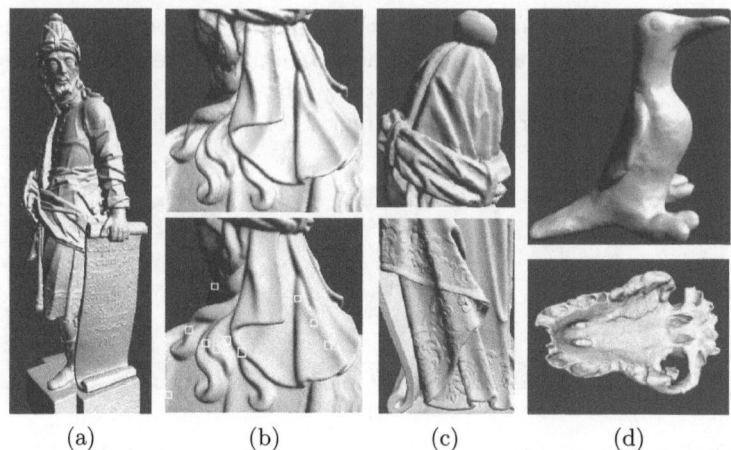

(a)   (b)   (c)   (d)

Fig. 3. Reconstructed models: (a) Prophet Joel; (b) comparison between IVIA and Poisson reconstructions of Joel's legs; (c) details of IVIA reconstruction of Joel's hat (up) and vestment (down); (d) Duck and Wolf models

and IVIA 11265s. However, the time performance of IVIA, can still be further optimized using 3D scan conversion on the space carving stage, the stage when the algorithm spends most of its execution time.

6   Final Remarks

Despite the good experimental results, IVIA may still be improved. Its use of VRIP on the first pass can lead to outliers being accepted later, since outliers can survive the first pass. What we did was to reduce the outlier influence by mixing them with good measurements. However, in very noisy regions, this methodology may fail. One possibility is to use feedback and more passes to improve outlier detection and removal between passes. An alternative is to combine parametric surfaces [14] with a volumetric representation and our outlier detection and removal techniques. Our work shows that more research is needed to guarantee high quality digital reconstruction results. We showed that real input range data has complex types of artifacts, and that widely used volumetric integration methods may still have major limitations. We proposed enhancements to both VRIP and CS, besides presenting our novel IVIA algorithm, which achieved precise reconstruction results even when compared to PSR.

Acknowledgments. The authors would like to thanks to CNPq, CAPES, UNESCO and IPHAN for supporting this research.

References

1. Bernardini, F., Rushmeier, H.: The 3D model acquisition pipeline. Computer Graphics Forum 21(2), 149–172 (2002)
2. Ikeuchi, K., Oishi, T., et al.: The Great Buddha project: Digitally archiving, restoring, and analyzing cultural heritage objects. IJCV 75, 189–208 (2007)

3. Vrubel, A., Bellon, O., Silva, L.: A 3D reconstruction pipeline for digital preservation of natural and cutural assets. In: Proc. of CVPR, pp. 2687–2694 (2009)
4. Levoy, M., Pulli, K., et al.: The Digital Michelangelo project: 3D scanning of large statues. In: SIGGRAPH, pp. 131–144 (2000)
5. Miyazaki, D., Oishi, T., Nishikawa, T., Sagawa, R., Nishino, K., Tomomatsu, T., Takase, Y., Ikeuchi, K.: The Great Buddha project: Modelling cultural heritage through observation. In: Proc. of VSMM, pp. 138–145 (2002).
6. Curless, B., Levoy, M.: A volumetric method for building complex models from range images. In: Proc. SIGGRAPH, pp. 303–312 (1996)
7. Wheeler, M.D., Sato, Y., Ikeuchi, K.: Consensus surfaces for modeling 3D objects from multiple range image. In: Proc. of ICCV, pp. 917–924 (1998)
8. Davis, J., Marschner, S.R., Garr, M., Levoy, M.: Filling holes in complex surfaces using volumetric diffusion. In: Proc. of 3DPVT, pp. 42–438 (2002)
9. Kazhdan, M., Bolitho, M., Hoppe, H.: Poisson surface reconstruction. In: Proc. of SIGGRAPH, pp. 61–70 (2006)
10. Edelsbrunner, H.: Shape reconstruction with Delaunay complex. In: Proc. of Latin Amer. Symp. Theoretical Informatics, pp. 119–132 (1998)
11. Turk, G., Levoy, M.: Zippered polygon meshes from range images. In: Proc. of SIGGRAPH, pp. 311–318 (1994)
12. Bernardini, F., Mittleman, J., Rushmeier, H., Silva, C., Taubin, G.: The ball-pivoting algorithm for surface reconstruction. In: IEEE TVCG, pp. 349–359 (1999)
13. Sharf, A., Lewiner, T., Shamir, A., Kobbelt, L., Cohen-Or, D.: Competing fronts for coarse-to-fine surface reconstruction. In: Computer Graphics, pp. 389–398 (2006)
14. Ohtake, Y., Belyaev, A., Alexa, M., Turk, G., Seidel, H.P.: Multi-level partition of unity implicits. ACM Transactions on Graphics 22, 463–470 (2006)
15. Fleishman, S., Cohen-Or, D., Silva, C.T.: Robust moving least-squares fitting with sharp features. ACM Transactions on Graphics 24, 544–552 (2005)
16. Bolitho, M., Kazhdan, M., Burns, R., Hoppe, H.: Parallel poisson surface reconstruction. In: Bebis, G., Boyle, R., Parvin, B., Koracin, D., Kuno, Y., Wang, J., Wang, J.-X., Wang, J., Pajarola, R., Lindstrom, P., Hinkenjann, A., Encarnação, M.L., Silva, C.T., Coming, D. (eds.) ISVC 2009, Part I. LNCS, vol. 5875, pp. 678–689. Springer, Heidelberg (2009)
17. Bolitho, M., Kazhdan, M., Burns, R., Hoppe, H.: Multilevel Streaming for Out-of-Core Surface Reconstruction. In: Proc. Eurographics (2007)
18. Lorensen, W.E., Cline, H.E.: Marching cubes: A high resolution 3D surface construction algorithm. In: Proc. of SIGGRAPH, pp. 163–169 (1997)
19. Hilton, A., Stoddart, A.J., Illingworth, J., Windeatt, T.: Reliable surface reconstructiuon from multiple range images. In: Buxton, B.F., Cipolla, R. (eds.) ECCV 1996. LNCS, vol. 1064, pp. 117–126. Springer, Heidelberg (1996)
20. Sagawa, R., Nishino, K., Ikeuchi, K.: Robust and adaptive integration of multiple range images with photometric attributes. In: Proc. of CVPR, pp. II:172–II:179 (2001)
21. Sagawa, R., Ikeuchi, K.: Taking consensus of signed distance field for complementing unobservable surface. In: Proc. of 3DIM, pp. 410–417 (2003)
22. Masuda, T.: Object shape modelling from multiple range images by matching signed distance fields. In: Proc. of 3DPVT, pp. 439–448 (2002)
23. Rocchini, C., Cignoni, P., Ganovelli, F., Montani, C., Pingi, P., Scopigno, R.: The marching intersections algorithm for merging range images. Visual Computer 20(2-3), 149–164 (2004)

Exploring Cascade Classifiers for Detecting Clusters of Microcalcifications

Claudio Marrocco, Mario Molinara, and Francesco Tortorella

DAEIMI, University of Cassino
Cassino (FR), Italy
{c.marrocco,m.molinara,tortorella}@unicas.it

Abstract. The conventional approach to the detection of microcalcifications on mammographies is to employ a sliding window technique. This consists in applying a classifier function to all the subwindows contained in an image and taking each local maximum of the classifier as a possible position of a microcalcification. Although effective such an approach suffers from the high computational burden due to the huge number of subwindows contained in an image. The aim of this paper is to experimentally verify if such problem can be alleviated by a detection system which employs a cascade-based localization coupled with a clustering algorithm which exploits both the spatial coordinates of the localized regions and a confidence degree estimated on them by the final stage of the cascade. The first results obtained on a publicly available set of mammograms show that the method is promising and has large possibility of improvement.

Keywords: Microcalcifications, mammography, computer aided detection, Adaboost, cascade of classifiers, clustering.

1 Introduction

Mammography is a radiological screening technique which makes it possible to detect lesions in the breast using low doses of radiation. At present, it represents the only not invasive diagnostic technique allowing the diagnosis of a breast cancer at a very early stage, when it is still possible to successfully attack the disease with a suitable therapy. For this reason, programs of wide mass screening via mammography for the female population at risk have been carried out in many countries. A particularly meaningful visual clue of breast cancer is the presence of clusters of microcalcifications (MC), tiny granule-like deposits of calcium that appear on the mammogram as small bright spots. Their size ranges from about 0.1 mm to 0.7 mm, while their shape is sometimes irregular. Isolated MCs are not, in most cases, clinically significant. However, the low quality of mammograms and the intrinsic difficulty in detecting likely cancer signs make the analysis particularly fatiguing, especially in a mass screening where a high number of mammograms must be examined by a radiologist in a day. In this case, a Computer Aided Detection (*CADe*) could be very useful to the

G. Maino and G.L. Foresti (Eds.): ICIAP 2011, Part I, LNCS 6978, pp. 384–392, 2011.

radiologist both for prompting suspect cases and for helping in the diagnostic decision as a "second reading". The goal is twofold: to improve both the sensitivity of the diagnosis, i.e. the accuracy in recognizing all the actual clusters and its specificity, i.e. the ability to avoid erroneous detections. The approach followed by the traditional CADe systems [1,2] entails a MC localization phase after that a successive clustering phase the localized regions are clustered with very simple rules based exclusively on their proximity, in order to individuate those clusters that are worth to be prompted. A widespread approach for accomplishing the localization step entails the application of statistical techniques directly to the image data [3]. To this aim, several statistical classifiers have been applied such as artificial neural network [4], support vector machine [1] or relevance vector machine [5]. All such methods rely on the sliding window technique that accomplishes the localization task by applying a classifier function on all the subwindows contained in an image and taking each local maximum of the classifier output as a pointer to a possible microcalcification. Although effective, the problem of such approaches is the computational burden involved by the huge number of subwindows typically contained in a whole image on which the classifier function (typically not very simple) should be applied.

In the Computer Vision field a very similar problem is the detection of human faces on images. In this framework, the approach presented by Viola and Jones in [6] has been particularly successful and nowadays it represents a reference solution for face detection tasks because of its fast execution and good performance. The most prominent characteristic of this algorithm is that it is still based on the sliding window technique, but the classifier function is structured as a "cascade" of simple classifiers. This allows background regions of the image to be quickly discarded while spending more computation on promising regions.

In this paper we present a method employing the cascade-based approach for detecting clustered microcalcifications. The difference with the original Viola-Jones approach is that the cascade does not make the final decision, but merely localizes on the mammogram the regions of interest (ROIs) most probable to contain a MC. The decision on the single ROI is postponed until the clustering phase, i.e. the confidence degree estimated for each ROI by the cascade is considered as an input feature for the clustering algorithm together with the spatial coordinates of the ROIs. In this way, we finally obtain a reliable partition of the ROIs in clusters of microcalcifications.

We performed several experiments on a publicly available set of mammograms and, according to the first results, the method is promising and shows large possibility of improvement.

2   The Cascade-Based Localization Method

The idea is to employ a *cascade of classifiers*, i.e., a sequence of base classifiers, to build a detector with both high specificity and sensitivity and low computational complexity. To this aim, the features employed should be both effective and simple to evaluate.

In particular, the input pattern to the detector is made of a set of rectangular features evaluated on the sliding subwindow. Figure 1 shows the features we have used for our implementation. In addition to the features employed in the original method [6] (fig. 1 a-e), we have added a further feature (fig. 1 f) more suitable for the shape of the microcalcifications. The value for each feature is calculated by subtracting the sum of the white rectangles from the sum of the black rectangles. Each of the feature types are scaled and shifted across all possible combinations on the subwindow. Since the subwindows we consider are 12x12, there are about 11000 possible features. For each of them, the evaluation can be performed very quickly thanks to a particular intermediate representation of the image, the *integral image* [6].

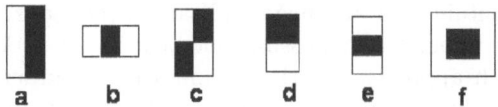

Fig. 1. The features used

Each stage of the cascade is built by means of a modified Adaboost learning algorithm that embeds a feature selection mechanism. In few words, Adaboost builds a "strong classifier" as linear combination $H(\mathbf{x}) = \sum_{t=1}^{T} \alpha_t h_t(\mathbf{x})$ of "weak classifiers" $h_t(\mathbf{x})$. In the Viola-Jones approach, the weak classifier is a simple decision stump using a single feature; the peculiarity is that the features are chosen on the basis of their performance. In other words, each new weak classifier involves finding the best performing feature on all the samples of the training set: in this way, the best T features are selected and the decision about the nature of a subwindow (MC or background) can be rapidly taken.

If a given subwindow is recognized as a MC, it is passed on to the next stage, otherwise discard it is immediately discarded (see fig. 2). In this way, the majority of subwindows containing easily detectable background are rejected by the early stages of the detector, while the most MC-like regions go through the entire cascade. This aims at reducing the number of false positives produced by the detector.

Accordingly, the classifier $H_i(\mathbf{x})$ at a given stage is built using the samples passed through all the previous stages. While the performance desired for the whole detector is very high, the learning target for each base classifier is reasonable since each stage should guarantee a high true positive t_i rate and a sufficiently low false positive rate f_i. The performance of the whole detector with K stages will thus be given by $TPR = \prod_{i=1}^{K} t_i$ and $FPR = \prod_{i=1}^{K} f_i$. A new stage is added to the cascade until the required performance is reached. As an example, to build a detector having $TPR = 0.990$ and $FPR = 0.001$ with base

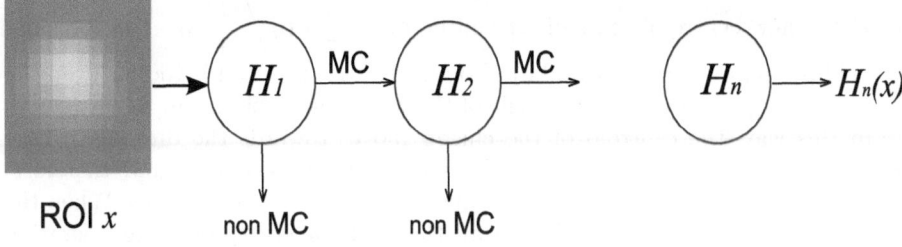

Fig. 2. The structure of the cascade

classifiers with $t_i = 0.999$ and $f_i = 0.5$, 10 stages will be needed. To this purpose, the training phase for the cascade requires a set of samples to train the single stages and a separated validation set for tuning each stage on the desired t_i and f_i. This is simply made by estimating on the validation set a threshold γ_i to be imposed on the value of $H_i(\mathbf{x})$.

It is worth noting that the cascade we employ is different from the scheme adopted in [6] which, in the final stage, provides a final decision about the ROI. In our approach, instead, the final stage produces a soft output which is a confidence degree about the presence of a MC in the ROI. In this way, the decision on the single ROI is postponed until the clustering phase so as to use for the final decision about the presence of a cluster both the estimated confidence degree and the spatial coordinates of the ROIs.

3   The Clustering Algorithm

Clustering is a well known topic in the image processing field. Roughly speaking, the goal of clustering is to achieve the best partition over a set of objects in terms of similarity. In particular, we consider a *sequential* clustering. Such kind of algorithm constructs the clusters on the basis of the order in which the points are brought and thus the final result is, usually, dependent on such order. In our system the ROIs are considered according to their confidence degrees (the largest value first) so that the clustering starts with the regions most likely to be microcalcifications. To this aim we have devised the *Moving Leader Clustering* (MLC) algorithm, which is a variation of the *leader follower clustering* [7]. It assumes as the centroid of the cluster the weighted centroid of mass of the ROIs belonging to the cluster. The centroid has to be calculated each time a new region is added to the cluster. The weight used for the ROIs is the respective confidence degree.

More formally, let us consider $n - 1$ ROIs $\mathbf{p}_1, \ldots, \mathbf{p}_{n-1}$ with $n > 2$ and confidence degree respectively $s_1, \ldots s_{n-1}$ grouped in a cluster C_1; its centroid is given by $\mathbf{c}_1 = \sum_1^{n-1} s_i \mathbf{p}_i \Big/ \sum_1^{n-1} s_i$. Let us call $\mathbf{p}_n$ the next ROI to be considered by the algorithm and let s_n be its confidence degree. It will be added to C_1 if its distance with respect to the centroid c_1 is less than a threshold R; in this case, the new centre of mass of C_1 will be $\mathbf{c}_1 = \sum_1^n s_i \mathbf{p}_i \Big/ \sum_1^{n-1} s_i$. A new cluster C_2 will be created with centroid $\mathbf{c}_1 = \mathbf{p}_n$ in the case the ROI is too far from the centroid of C_1. An example of result of the algorithm is shown in fig. 3.

In this way, the centroid of the cluster moves towards the direction where the points are more dense and with higher confidence degree and therefore, where there is a higher probability of finding new microcalcifications. When the clustering algorithm has considered all the ROIs, a post-processing operation is performed aimed at better grouping the ROIs. This consists in merging two clusters that share at least one ROI (see fig. 4). This approach has been called *Moving Leader Clustering with Merge* (MLCwM) [8] and avoids an excessive number of clusters that does not correspond to the real distributions.

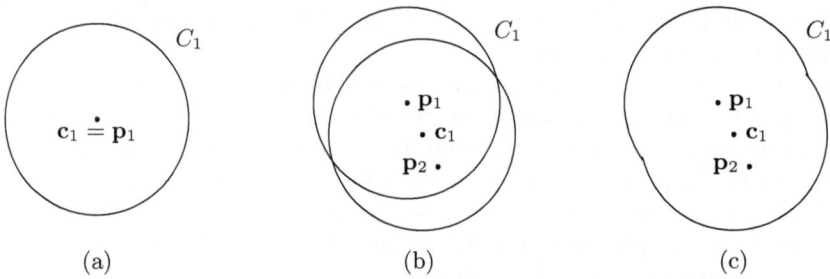

(a)   (b)   (c)

Fig. 3. The ROI $\mathbf{p}_1$ with highest confidence degree is the centroid $\mathbf{c}_1$ of a new cluster C_1 (a), when $\mathbf{p}_2$ is added to the cluster, the centroid moves accordingly (b) and the shape of the cluster is the union of the two circles (c)

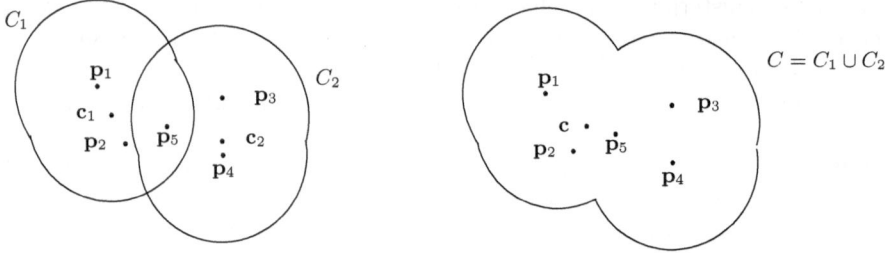

Fig. 4. When two different clusters share a ROI, they are merged and the shape of the cluster is the union of the two shapes

4   Experimental Results

The method has been tested on 145 images extracted from a publicly available database, the Digital Database for Screening Mammography (DDSM) [9]. The employed images are relative to 58 malignant cases and contain 119 clusters and 2849 microcalcifications.

In order to build a training set for the cascade classifier, several regions corresponding to positive and negative classes have been extracted from 100 images. In particular, the positive set consisted of 186 hand labeled microcalcifications scaled and aligned to a base resolution of 12 by 12 pixels while the negative set consisted of randomly extracted images with the same base resolution. Figure 5 shows an example of such regions for both positive and negative classes. The data set so obtained has then been equally parted into a training and a validation set respectively used to train each stage of the cascade and to verify its performance. During the test phase the remaining 45 images (containing 812 microcalcifications) have been used to evaluate the performance of the proposed approach.

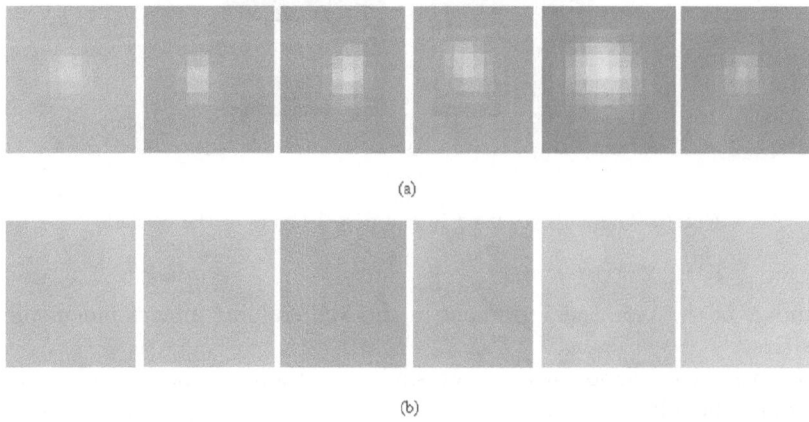

(a)

(b)

Fig. 5. An example of images used as training set: (a) positive class (microcalcifications), (b) negative class (non microcalcifications)

The cascade detector has been buit by choosing a target FPR equal to 0.001 with $t_i = 0.99$ and $f_i = 0.5$. The model employed almost 6000 regions for the negative classes in training set. For the validation set, instead, the number of negative elements in the validation set was increased till almost 230000 with the aim to increase the specificity of the detector, i.e., to reduce the number of false positive at the output of the cascade classifier. The cascade we obtained consisted of 12 layers using respectively 3, 2, 2, 50, 11, 18, 7, 50, 14, 21 and 24 features. The features used in the first three layers of the cascade are shown in fig. 6. Differently from what expected, the feature shown in fig. 1.f is never used in the first three layer of the cascade, i.e., it is less discriminant than the other features in the detection of the MC regions. This is probably due to the different size of the MC that varies from image to image (as shown by the images of the positive class reported in fig. 5). The other features, instead, are able to take

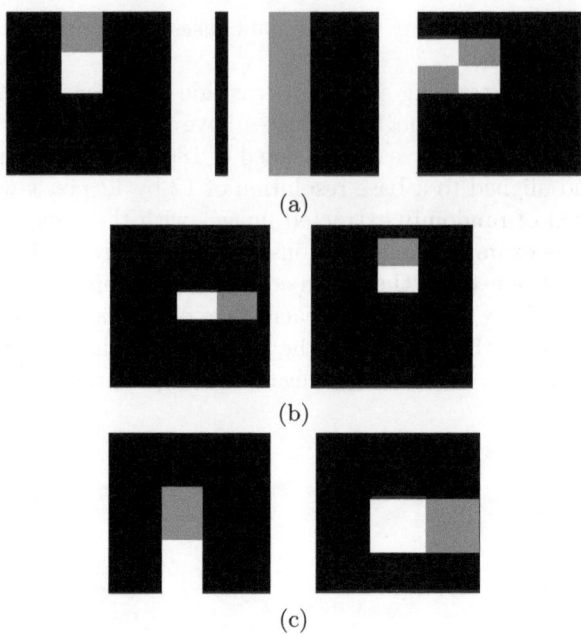

Fig. 6. The features used in the first 3 layers of the cascade

advantage of the brightness gradient in the MC training images independently of its size.

To have a comparison with other similar methods we have implemented a model of a monolithic AdaBoost classifier trained using a pixel-based approach with a 12x12 sliding window. In such case a CART decision tree with maximum depth equal to 3 and decision stumps as nodes functions has been used as weak learner. The number of boosting steps has been chosen equal to 10 while in this case the employed training set is the union of the starting training and validation set used for the cascade model.

The performance of the different proposed models have been evaluated in terms of Free-response Receiver Operating Characteristics (FROC) curve that plots the True Positive Rate (TPR), i.e., the number of regions correctly detected in the whole test set, versus the False Positive per image, i.e., the number of regions per image incorrectly classified as microcalcifications, when varying an opportune threshold. The results of the comparison are reported in fig. 7 where a threshold has been varied on the confidence degree obtained with the considered classifiers.

From this figure we can evince that the cascade approach is definitely superior to the pixel-based Adaboost in terms of both the performance measures. To remark this superiority we can note that the cascade reaches the 100% of detected clusters for 5.68 false positive per image while it is at 96% for 0.68 false positive per image. AdaBoost, instead, never reaches this performance in terms of TPR.

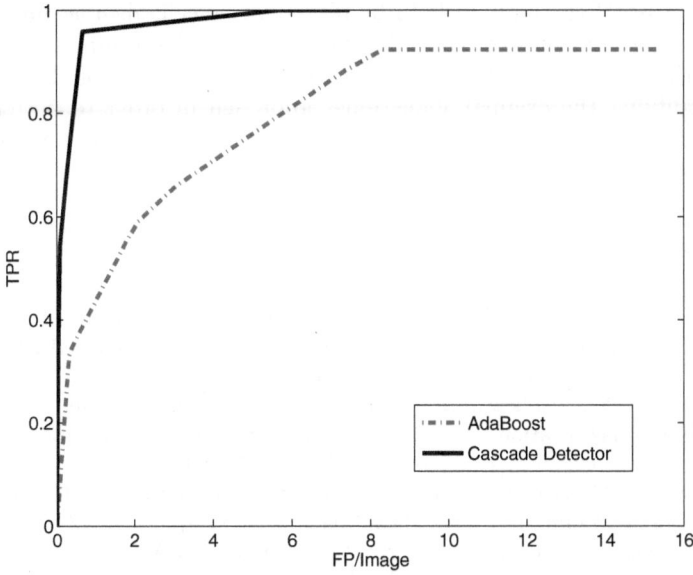

Fig. 7. The FROC curves obtained with the two compared approaches

It is also worth noting that another important advantage of the cascade approach is the considerable reduction of the elaboration time that is in terms of 64% with respect to the pixel-based AdaBoost detector.

5   Discussion and Conclusions

In this paper we have verified if the Viola-Jones approach can be profitably used in the case of detection of microcalcifications. Accordingly, we have modified this method in order to obtain as output a confidence degree that can be successively used in the clustering phase. From the first experiments, we can conclude that such approach gives good performance in both true positive rate and false positive per image. A closer look to the algorithm leads us to believe that there is still room for a significant improvement of the method performance. In fact, a critical point is the weighted accuracy used as performance measure to build the weak learners in each stage. This is not a good choice because the problem at hand is highly skewed (the number of negative samples is much greater than the positive ones) and thus a more appropriate performance index should be used (e.g. the area under the ROC curve). Another point to be considered is that, in the training phase, the boosting approach tries to improve its performance focusing on the most difficult samples and this implies a significant possibility of overfitting when dealing with very difficult samples [10]. In our case this is not very critical for the negative samples since the required FPR for each stage is not so demanding, but it remains a problem in presence of difficult positive samples because of the high TPR to be reached. A mechanism is thus required to moderate the influence of particularly difficult positive samples which could appear during the construction of the stage. The next steps of this work will aim at modifying the original Viola-Jones approach in order to better fit the requirements for MC detection. In particular, we will focus on new features and a different tuning method for the base classifiers.

References

1. El-Naqa, I., Yang, Y., Wernick, M.N., Galatsanos, N.P., Nishikawa, R.M.: A support vector machine approach for detection of microcalcifications. IEEE Transactions on Medical Imaging 21(12), 1552–1563 (2002)
2. Wei, L., Yang, Y., Nishikawa, R.M., Jiang, Y.: A study on several machine-learning methods for classification of malignant and benign clustered microcalcifications. IEEE Transactions on Medical Imaging, 24(3), 371–380 (2005)
3. Nishikawa, R.: Current status and future directions of computer-aided diagnosis in mammography. Computerized Medical Imaging and Graphics 31, 1357–1376 (2007)
4. Wu, Y., Giger, M.L., Doi, K., Vyborny, C.J., Schmidt, R.A., Metz, C.E.: Artificial neural networks in mammography: Application to decision making in the diagnosis of breast cancer. Radiology 187, 81–87 (1993)
5. Wei, L., Yang, Y., Nishikawa, R.M., Wernick, M.N., Edwards, A.: Relevance vector machine for automatic detection of clustered microcalcifications. IEEE Transactions on Medical Imaging 24(10), 1278–1285 (2005)
6. Viola, P., Jones, M.: Robust real-time face detection. International Journal of Computer Vision 57, 137–154 (2004)
7. Duda, R.O., Hart, P.E., Stork, D.G.: Pattern Classification, 2nd edn. John Wiley & Sons, Chichester (2001)
8. Marrocco, C., Molinara, M., Tortorella, F.: Algorithms for detecting clusters of microcalcifications in mammograms. In: Roli, F., Vitulano, S. (eds.) ICIAP 2005. LNCS, vol. 3617, pp. 884–891. Springer, Heidelberg (2005)
9. Kopans, D., Moore, R., Heath, M., Bowyer, K., Philip Kegelmeyer, W.: The digital database for screening mammography. In: Yaffe, M.J. (ed.) Proc. 5th Int. Workshop on Digital Mammography, pp. 212–218 (2001)
10. Ratsch, G., Onoda, T., Muller, K.R.: Soft margins for adaboost. Machine Learning 42(3), 287–320 (2001)

A Method for Scribe Distinction in Medieval Manuscripts Using Page Layout Features

Claudio De Stefano[1], Francesco Fontanella[1],
Marilena Maniaci[2], and Alessandra Scotto di Freca[1]

[1] Dipartimento di Automazione, Elettromagnetismo, Ingegneria dell'Informazione e
Matematica Industriale
Via G. Di Biasio, 43 – 03043 Cassino (FR) – Italy
[2] Dipartimento di Filologia e Storia
Via Zamosch, 43 – 03043 Cassino (FR) – Italy
University of Cassino
{destefano,fontanella,mmaniaci,a.scotto}@unicas.it

Abstract. In the framework of Palaeography, the use of digital image
processing techniques has received increasing attention in recent years,
resulting in a new research field commonly denoted as "digital palaeography". In such a field, a key role is played by both pattern recognition
and feature extraction methods, which provide quantitative arguments
for supporting expert deductions. In this paper, we present a pattern
recognition system which tries to solve a typical palaeographic problem: to distinguish the different scribes who have worked together to the
transcription of a single medieval book. In the specific case of a high standardized book typology (the so called Latin "Giant Bible"), we wished
to verify if the extraction of certain specifically devised features, concerning the layout of the page, allowed to obtain satisfactory results. To
this aim, we have also performed a statistical analysis of the considered
features in order to characterize their discriminant power. The experiments, performed on a large dataset of digital images from the so called
"Avila Bible" - a giant Latin copy of the whole Bible produced during
the XII century between Italy and Spain - confirmed the effectiveness of
the proposed method.

1   Introduction

In the context of palaeographic studies, there has been in the last years a growing scientific interest in the use of computer-based techniques of analysis, whose
aim is that of providing new and more objective ways of characterizing medieval
handwritings and distinguishing between scribal hands [3,4]. The application of
such techniques, originally developed in the field of forensic analysis, originated a
new research field generally known as *digital palaeography*. At a simpler level, the
digital approach can be used to replace qualitative measurements with quantitative ones, for instance for the evaluation of parameters such as the angle and the
width of strokes, or the comparison among digital examples of letter-forms. In
the above mentioned cases, technology is employed to perform "traditional" observations more rapidly and systematically than in the past. In contrast to this,

G. Maino and G.L. Foresti (Eds.): ICIAP 2011, Part I, LNCS 6978, pp. 393–402, 2011.
© Springer-Verlag Berlin Heidelberg 2011

there are some entirely new approaches emerged in the last few years, which have been made possible by the combination of powerful computers and high-quality digital images. These new approaches include the development of systems for supporting the experts' decisions during the analysis of ancient handwritings.

All the above approaches require the gathering of information from selected corpora of pre-processed digital images, aimed at the generation of quantitative measurements: these will contribute to the creation of a statistical profile of each sample, to be used for finding similarities and differences between writing styles and individual hands. In the treatment of the graphic sequence, possible methodologies range from the automatic recognition and characterization of single words and signs, to the reduction of the *ductus* to its basic profile, to the extraction of a more global set of "texture" features, depending on the detection of recurrent forms on the surface of the page. However promising, all these approaches haven't yet produced widely accepted results, both because of the immaturity in the use of these new technologies, and of the lack of real interdisciplinary research: palaeographers often missing a proper understanding of rather complex image analysis procedures, and scientists being unaware of the specificity of medieval writing and tending to extrapolate software and methods already developed for modern writings. However, the results of *digital palaeography* look promising and ought to be further developed. This is particularly true for what concerns "extra-graphic" features, such as those related to the layout of the page, which are more easily extracted and quantified. For instance, in case of highly standardized handwriting and book typologies, the comparison of some basic layout features, regarding the organization of the page and its exploitation by the scribe, may give precious clues for distinguishing very similar hands even without recourse to palaeographical analysis.

Moving from these considerations, we propose a pattern recognition system for distinguishing the different scribes who have worked together to the transcription of a single medieval book. The proposed system considers a set of features typically used by palaeographers, which are directly derived from the analysis of the page layout. The classification is performed by using a standard Multi Layer Perceptron (MLP) network, trained with the Back Propagation algorithm [10]. We have chosen MLP classifiers for two main reasons: on the one hand MLP classifiers are very simple, quite effective and exhibit a good generalization capability; on the other hand the main goal of our study is not that of building a top performing recognition system, but rather to verify that the use of page layout features allows obtaining satisfactory results. Finally, we have also performed a statistical analysis of the considered features in order to characterize the discriminating power of each of them. The results reported in Section 4 confirmed that the proposed method allowed us to select the feature subset which maximizes classification results.

A particularly favorable situation to test the effectiveness of this approach is represented by the so-called "Giant Bibles", a hundred or more of serially produced Latin manuscripts each containing the whole sacred text in a single volume of very large size (up to 600 x 400 mm and over). The Bibles originated

in Central Italy (initially in Rome) in the mid-11th century, as part of the po-
litical program of the "Gregorian Reform", dealing with the moral integrity and
independence of the clergy and its relation to the Holy Roman Emperor. Very
similar in shape, material features, decoration and script, the Bibles were produced by groups of several scribes, organizing their common work according to
criteria which still have to be deeply understood. The distinction among their
hands often requires very long and patient palaeographical comparisons.

In this context, we have used for our experiments the specimen known as
"Avila Bible", which was written in Italy by at least nine scribes within the third
decade of the 12h century and soon sent (for unknown reasons) to Spain, where
its text and decoration were completed by local scribes; in a third phase (during
the 15th century) additions were made by another copyist, in order to adapt the
textual sequence to new liturgical needs [8]. The Bible offers an "anthology" of
contemporary and not contemporary scribal hands, thus representing a severe
test for evaluating the effectiveness and the potentialities of our approach to the
distinction of scribal hands.

The remainder of the paper is organized as follows: Section 2 presents the
architecture of the system, Section 3 illustrates the method used for feature
analysis, while in Section 4 the experimental results are illustrated and discussed.
Finally, Section 5 is devoted to some conclusions.

2   The System Architecture

The proposed system receives as input RGB images of single pages of the
manuscript to be processed, and performs for each page the following steps:
pre-processing, segmentation, feature extraction, and *scribe distinction*.

In the pre-processing step noisy pixels, such as those corresponding to stains
or holes onto the page or those included in the frame of the image, are detected
and removed. Red out-scaling capital letters are also removed since they might
be all written by a single scribe, specialized for this task. Finally, the RGB image
is transformed into a grey level one and then in a binary black and white image.

In the segmentation step, columns and rows in each page are detected. The
Bible we have studied is a two column manuscript, with each column composed
by a slightly variable number of rows. The detection of both columns and rows is
performed by computing pixel projection histograms on the horizontal and the
vertical axis, respectively.

The feature extraction step is the most relevant and original part of our work
and it has been developed in collaboration with experts in palaeography and
following the suggestion reported in [11]. We have considered three main sets of
features, mainly concerning the layout of the page. The first set relates to properties of the whole page and includes the upper margin and the lower margin
of the page and the intercolumnar distance. Such features are not very distinctive for an individual copyist, but they may be very useful to highlight chronological and/or typological differences. The second set of features concerns the
columns: we have considered the number of rows in the column and the column

cubiculii. Dixeruntq; ei ferui fiu . Ecce au

Fig. 1. The number of peaks in the horizontal projection histogram of a row

exploitation coefficient [2]. The exploitation coefficient is a measure of how much the column is filled with ink, and is computed as:

$$exploitation\ coefficient = \frac{N_{BP}(C)}{N_P(C)} \qquad (1)$$

where the functions $N_{BP}(C)$ and $N_P(C)$ return the number of black pixels and the total number of pixels in the column C, respectively. Both features vary according to different factors, among which the expertise of the writer. In the case of very standardized handwritings, such as the "carolingian minuscule" shown by the Bible of Avila, the regularity in the values assumed by such features may be considered as a measure of the skill of the writer and may be very helpful for scribe distinction. The third set of features characterizes the rows, and includes the following features: weight, modular ratio, interlinear spacing, modular ratio/interlinear spacing ratio and peaks. The weight is the analogous of the exploitation coefficient applied to rows, i.e. it is a measure of how much a row is filled with ink. It is computed as in (1) but considering row pixels instead of column pixels. The modular ratio is a typical palaeographic feature, which estimates the dimension of handwriting characters. According to our definition, this feature is computed for each row measuring the height of the "centre zone" of the words in that row. Once the centre zone has been estimated, the interlinear spacing is the distance in pixels between two rows. Modular ratio, interlinear spacing and modular ratio/interlinear spacing ratio characterize not only the way of writing of a single scribe, but may also hint to geographical and/or chronological distinctions. In [8], for instance, the distance among layout lines in rows and the dimension of letters significantly differentiate Spanish and Italian minuscule. Highly discriminating features, such as the inter-character space and the number of characters in a row, imply the very difficult task of extracting the single characters contained in each word, which is far to be solved in the general case. Therefore, we have chosen to estimate the number of characters in a row by counting the number of peaks in the horizontal projection histogram of that row (see Fig. 1). The whole set of considered features is summarized in Table 1 reporting, for each of them, the associated identification number.

The last block performs the recognition task, which has the effect of identifying the rows in each page written by the same copyist. In our study, we have assumed that the manuscript has been produced by N different copyists, previously identified through the traditional palaeographical analysis. We have also assumed that each single pattern to be classified is formed by a group of M consecutive rows, described by using the previously defined features. More specifically, patterns belonging to the same page share the same features of both

Table 1. The considered features and the corresponding identification number (id)

1	intercolumnar distance	6	modular ratio
2	upper margin	7	interlinear spacing
3	lower margin	8	weight
4	exploitation	9	peak number
5	row number	10	modular ratio/interlinear spacing

the first and the second set, while feature values of the third set are averaged over the M rows forming each group. Summarizing, each pattern is represented by a feature vector containing 10 values. Finally, each pattern is attributed to one of the N copyist by using a Neural Network classifier: in particular we used a MLP trained with the Back Propagation algorithm [10].

3   Features Analysis

In order to identify the set of features having the highest discriminant power, we have used five standard *univariate* measures. Each of them ranks the available features according to a measure which evaluates the effectiveness in discriminating samples belonging to different classes. The final ranking of all the features has been obtained by using the Borda Count rule. According to such a rule, a feature receives a certain number of points corresponding to the position in which it has been ranked by each univariate measure. In our study, we have considered the following univariate measures: Chi-square [7], Relief [6], Gain ratio, Information Gain and Symmetrical uncertainty [5].

The *Chi-Square* measure estimates feature merit by using a discretization algorithm: if a feature can be discretized to a single value, then it can safely be removed from the data. The discretization algorithm, adopts a supervised heuristic method based on the χ^2 statistic. The range of values of each feature is initially discretized by considering a certain number of intervals (heuristically determined). Then, the χ^2 statistic is used to determine whether the relative frequencies of the classes in adjacent intervals are similar enough to justify the merging of such intervals. The formula for computing the χ^2 value for two adjacent intervals is the following:

$$\chi^2 = \sum_{i=1}^{2} \sum_{j=1}^{C} \frac{(A_{ij} - E_{ij})^2}{Eij} \tag{2}$$

where C is the number of classes, A_{ij} is the number of instances of the j-th class in the i-th interval and E_{ij} is the expected frequency of A_{ij} given by the formula $E_{ij} = R_i C_j / N_T$ where R_i is the number of instances in the i-th interval and C_j and N_T are the number of instances of the j-th class and total number of instances, respectively, in both intervals. The extent of the merging process is controlled by a threshold, whose value represent the maximum admissible difference among the occurrence frequencies of the samples in adjacent intervals. The value of this threshold has been heuristically set during preliminary experiments.

The second considered measure is the Relief, which uses instance based learning to assign a relevance weight to each feature. The assigned weights reflects the feature ability to distinguish among the different classes at hand. The algorithm works by randomly sampling instances from the training data. For each sampled instance, the nearest instance of the same class (nearest hit) and different class (nearest miss) are found. A feature weight is updated according to how well its values distinguish the sampled instance from its nearest hit and nearest miss. A feature will receive a high weight if it differentiates between instances from different classes and has the same value for instances of the same class.

Before introducing the last three considered univariate measures, let us briefly recall the well known information-theory concept of entropy. Given a discrete variable X, which can assume the values $\{x_1, x_2, \ldots, x_n\}$, its entropy $H(X)$ is defined as:

$$H(X) = -\sum_{i=1}^{n} p(x_i) \log_2(x_i) \tag{3}$$

where $p(x_i)$ is the probability mass function of the value x_i. The quantity $H(X)$ represent an estimate of the uncertainty of the random variable X. The entropy concept can be used to define the conditional entropy of two random variables X and Y taking values x_i and y_j respectively, as:

$$H(X|Y) = -\sum_{i,j} p(x_i, y_j) \log \frac{p(y_j)}{p(x_i, y_j)} \tag{4}$$

where $p(x_i, y_j)$ is the probability that $X = x_i$ and $Y = y_j$. The quantity in (4) represents the amount of randomness in the random variable X when the value of Y is known.

The above defined quantities can be used to estimate the usefulness of a feature X to predict the class C of unknown samples. More specifically, such quantities can be used to define the *information gain* (I_G) concept [9]:

$$I_G = H(C) - H(C|X) \tag{5}$$

I_G represents the amount by which the entropy of C decreases when X is given, and reflects additional information about C provided by the feature X.

The last three considered univariate measures uses the information gain defined in (5). The first one is the information Gain itself. The second one, called *Gain Ratio* (I_R), is defined as the ratio between the information gain and the entropy of the feature X to be evaluated:

$$I_R = \frac{I_G}{H(X)}. \tag{6}$$

Finally, the third univariate measure taken into account, called *Symmetrical Uncertainty* (I_S), compensates for information gain bias toward attributes with more values and normalizes its value to the range $[0, 1]$:

$$I_S = 2.0 \times \frac{I_G}{H(C) + H(X)} \tag{7}$$

4   Experimental Results

As anticipated in the Introduction, we have tested our system on a large dataset of digital images obtained from a giant Latin copy of the whole Bible, called "Avila Bible". The palaeographic analysis of such a manuscript has individuated the presence of 13 scribal hands. Since the rubricated letters might be all the work of a single scribe, they have been removed during the pre-processing step; we have therefore considered only 12 copyists to be identified. The pages written by each copyist are not equally numerous and there are cases in which parts of the same page are written by different copyists.

The aim of the classification step is that of associating each pattern, corresponding to a group of M consecutive rows, to one of the $N = 12$ copyists: in our experiments we have assumed $M = 4$, thus obtaining a database of 20867 samples extracted from the set of the 800 pages which are in two column format (the total number of pages in the Bible is 870). The database has been normalized, by using the Z-normalization method[1], and divided in two subsets: the first one, containing 10430 samples, has been used as training set for the neural network classifier, while the second one, containing the remaining 10437 samples, has been used for testing the system. For each class, the samples have been randomly extracted from the database in such a way to ensure that, approximately, each class has the same number of samples in both training and test set. Preliminary experiments have been performed for setting MLP parameter values: in particular, we have obtained the best results with 100 hidden neurons and 1000 learning cycles.

The accuracy achieved by using the whole set of features on training and test set, averaged over 20 runs, is 95.57% and 92.46% respectively. These results are very interesting since they have been obtained by considering only page layout features, without using more complex information relative to the shape of each sign: such information would be typically analyzed by palaeographers, but the process for automatically extracting them from the original images is very complex and not easy to generalize.

Table 2 reports the recognition rates obtained on the test set for each of the 12 copyists, together with the corresponding number of samples. The third and the fourth row of the table respectively report the average recognition rate and the variance obtained for each scribe over the 20 runs. Similarly, the fifth and the sixth row report the best and the worst recognition rate, respectively, over the 20 runs. The data in the table show that the worst performance is obtained for copyists represented by a reduced number of samples (the number of samples for each copyist is reported in the second row): in these cases, in fact, it is difficult to adequately train the MLP classifier. This happens, for instance, for the copyists B and W. In particular, the copyist B, for which only 5 samples are included in the training set, is completely confused with other copyists represented by a

---

[1] Note that the Z-normalization transforms the distribution of the original data into standard normal distribution (the mean is 0.0 and standard deviation is 1.0).

Table 2. Test accuracies obtained for each scribe

Scribe	A	B	C	D	E	F	G	H	I	W		
Samples	4286	5	103	352	1095	1961	446	519	831	44	522	266
Av. Acc.	97.10	24.00	65.60	83.50	93.00	90.40	85.90	83.40	94.80	49.90	93.70	79.90
Variance	0.01	16.71	0.86	0.61	0.03	0.04	0.14	0.05	0.00	5.46	0.03	0.03
max Acc.	99.00	100.00	87.00	98.00	96.00	94.00	91.00	86.00	95.00	100.00	96.00	84.00
min Acc.	95.00	0.00	57.00	69.00	90.00	88.00	82.00	80.00	94.00	27.00	90.00	78.00

higher number of samples. On the contrary, the best performance is obtained for the copyist A, which has the highest number of samples in the training set.

Further experiments have been performed in order to evaluate the discriminant power of each feature and to find the feature subset which maximizes classification results. As discussed in Section 3, we have considered five univariate measures, each providing a ranked list of the considered features. The results of this analysis are reported in Table 3, which shows the ranking relative to each measures. Although the different measures produced quite different results, they give a good insight about the best and worst features.

In order to compute the overall ranking of the features, we used the Borda count rule [1]. According to such rule, the overall score of each feature is obtained by using the formula:

$$Os_i = \sum_{j=1}^{5} 10 - \text{pos}_{ij} \tag{8}$$

where Os_i is the overall score of the i-th feature, while pos_{ij} is the position of the i-th feature in the j-ranking. Table 4 displays the overall ranking of the features obtained by using the Borda count rule.

Figure 2 shows the plot of the test set accuracy as a function of the number of features: for each number of features n_i, the first n_i features in the overall ranking have been considered, and 20 runs have been performed. Note that the most right bar refers to the results obtained considering the whole feature set. In the plot, for each number of features, the first and the third bar report the worst and the best recognition rate, respectively, while the second one reports the average

Table 3. Feature ranking according to the five considered measures. For each row, the most left numeric value indicates the best feature, while the most right value denotes the worst one.

Measure	Ranking
Chi Squared (C_S)	4 3 2 1 5 9 7 6 10 8
Relief (R_F)	5 4 1 9 3 7 6 10 8 2
Gain Ratio (I_R)	4 5 1 3 2 9 7 6 10 8
Information Gain (I_G)	4 3 2 1 5 9 7 6 10 8
Symmetrical Uncertainty (I_S)	4 3 5 1 2 9 7 6 10 8

Table 4. Overall ranking of the features

id	feature	score
4	exploitation	44
3	lower margin	35
5	row number	34
1	intercolumnar distance	32
2	upper margin	24
9	peak number	22
7	interlinear spacing	16
8	weight	16
6	modular ratio	11
10	modular ratio/interlinear spacing	6

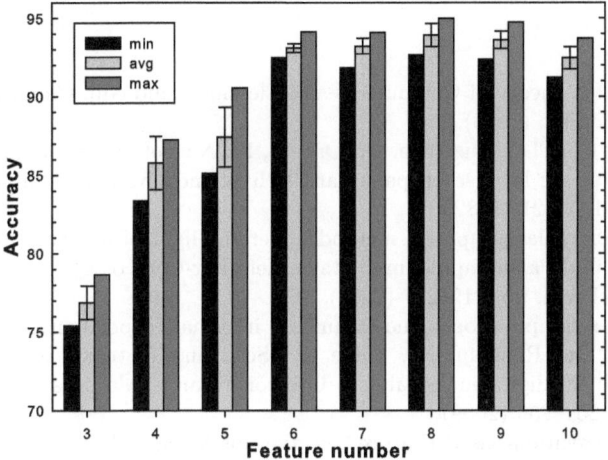

Fig. 2. Test set accuracy (averaged over 20 runs) vs feature number

recognition rate together with the corresponding variance. The results obtained by using one feature and two features have been omitted because they are too low (less than 60%). The data in the plot show that satisfactory results can be obtained considering at least the first six features in the overall ranking, while the best result has been obtained by using the first 8 features (95%), i.e. discarding the features representing the modular ratio and the "modular ratio/interlinear spacing". It is worth noting that while the performance difference between the use of eight and nine features is very small, the performance difference between the use of nine and ten features is more relevant. This means that the feature "modular ratio/interlinear spacing" is misleading.

5 Conclusion

We presented a novel approach for automatic scribe identification in medieval manuscripts. The task has been accomplished considering features suggested by paleograpy experts, which are directly derived from the analysis of the page layout. The experimental investigation has regarded two main aspects. The first one was intended to test the effectiveness of the considered features in discriminating different scribes. The second one was aimed at characterizing the discriminant power of each feature in order to find the best feature subset. The experimental results confirmed the effectiveness of the proposed approach.

Future work will include exploiting the information about the classification reliability. Such kind of information would allow palaeographers to find further confirmation of their hypothesis and to concentrate their attention on those sections of the manuscript which have not been reliably classified.

References

1. Black, D.: The Theory of Committees and Elections, 2nd edn. Cambridge University Press, London (1963)
2. Bozzolo, C., Coq, D., Muzerelle, D., Ornato, E.: Noir et blanc. premiers résultats d'une enquête sur la mise en page dans le livre médiéval. In: Il libro e il testo, Urbino, pp. 195–221 (1982)
3. Ciula, A.: The palaeographical method under the light of a digital approach. In: Kodikologie und Paläographie im digitalen Zeitalter-Codicology and Palaeography in the Digital Age, pp. 219–237 (2009)
4. Gurrado, M.: "graphoshop", uno strumento informatico per l'analisi paleografica quantitativa. In: Rehbein, M., Sahle, P., Schaßan, T. (eds.) Kodikologie und Paläographie im digitalen Zeitalter / Codicology and Palaeography in the Digital Age, pp. 251–259 (2009)
5. Hall, M.: Correlation-based Feature Selection for Machine Learning. Ph.D. thesis, University of Waikato (1999)
6. Kononenko, I.: Estimating Attributes: Analysis and Extensions of RELIEF. In: European Conference on Machine Learning, pp. 171–182 (1994)
7. Liu, H., Setiono, R.: Chi2: Feature Selection and Discretization of Numeric Attributes. In: ICTAI, pp. 88–91. IEEE Computer Society, Los Alamitos (1995)
8. Maniaci, M., Ornato, G.: Prime considerazioni sulla genesi e la storia della bibbia di avila. Miscellanea F. Magistrale, Spoleto (2010) (in press)
9. Quinlan, J.R.: C4.5: Programs for Machine Learning. Morgan Kaufmann Series in Machine Learning. Morgan Kaufmann, San Francisco (1993)
10. Rumelhart, D.E., Hinton, G.E., Williams, R.J.: Learning representations by back-propagating errors. Nature 323(9), 533–536 (1986)
11. Stokes, P.: Computer-aided palaeography, present and future. In: Rehbein, M., Sahle, P., Schaßan, T. (eds.) Kodikologie und Paläographie im digitalen Zeitalter / Codicology and Palaeography in the Digital Age. pp. 309–338 (2009)

Registration Parameter Spaces for Molecular Electron Tomography Images

Lennart Svensson*, Anders Brun, Ingela Nyström, and Ida-Maria Sintorn

Centre for Image Analysis,
Swedish University of Agricultural Sciences and
Uppsala University, Sweden
{lennart,anders,ingela,ida}@cb.uu.se
http://www.cb.uu.se

Abstract. We describe a methodology for exploring the parameter spaces of rigid-body registrations in 3-D. It serves as a tool for guiding and assisting a user in an interactive registration process. An exhaustive search is performed over all positions and rotations of the template, resulting in a 6-D volume, or fitness landscape. This is explored by the user, who selects and views suitable 3-D projections of the data, visualized using volume rendering. The 3-D projections demonstrated here are the maximum and average intensity projections of the rotation parameters and a projection of the rotation parameter for fixed translation parameters. This allows the user to jointly visualize projections of the parameter space, the local behaviour of the similarity score, and the corresponding registration of the two volumes in 3-D space for a chosen point in the parameter space. The procedure is intended to be used with haptic exploration and interaction. We demonstrate the methodology on a synthetic test case and on real molecular electron tomography data using normalized cross correlation as similarity score.

Keywords: volumetric registration, template matching, normalized cross correlation, molecular electron tomography.

1 Introduction

Optimization in high dimensional spaces is a difficult problem for any complex function. For simpler functions, e.g. monotonically increasing, exploitation schemes based on gradients can yield good results, but when more emphasis needs to be put on exploration due to local optima, the process can become too slow or the methods still easily get stuck in local optima. Both rigid and non-rigid registration are problems where the search space often is high dimensional. Here we describe a methodology for how rigid body registration parameter spaces can be presented and investigated to let the user guide the optimization process and simultaneously analyze the registration result.

---

* Corresponding author.

G. Maino and G.L. Foresti (Eds.): ICIAP 2011, Part I, LNCS 6978, pp. 403–412, 2011.

The application in this paper concerns Molecular Electron Tomography (MET) data, which is cryo electron microscopy combined with 3-D reconstruction techniques to create detailed tomograms of biological samples at the nanometer scale. MET allows for studying the structure, flexibility and interactions of molecules and macromolecules in solution (*in vitro*) as well as in tissue samples (*in situ*). A characteristic feature of MET volumes is however a very low signal to noise ratio. Registration is a very important process when analyzing this kind of data as it is very difficult to visually identify the objects or regions of interest in the very cluttered and complex volume images, and it is also difficult to analyze interactions and relations between different objects only based on vision.

Registration is usually performed with crystallographic data. The software Situs published in 1999 [12] has become a popular tool for this. It originally featured rigid body registration based on cross correlation. Shortly after, non-rigid registration was introduced by, e.g., Wriggers [10], who used a vector quantization registration scheme using restraints between vectors as a model for the structure flexibility. Birmanns introduced the use of haptics [2] to explore this landscape interactively and our approach is along the same lines.

The fitness landscape of rigid registration is defined by a scalar similarity function defined in the 6-D parameter space. We introduce projections of the fitness landscape into 3-D *score volumes*, visualized using a volume renderer, which allows the user to explore and navigate this high-dimensional space in search for relevant matches. This process is explained in Figure 1. It allows the user to simultaneously explore both the ordinary 3-D spatial domain, in which the registration is performed, and projections of the 6-D parameter space of rigid registration. In our application domain, the visualization and analysis of MET biological data, this methodology will be used in conjunction with haptic rendering and interaction, where a "3-D pen" with force feedback will be used to position and rotate the template image in the search image.

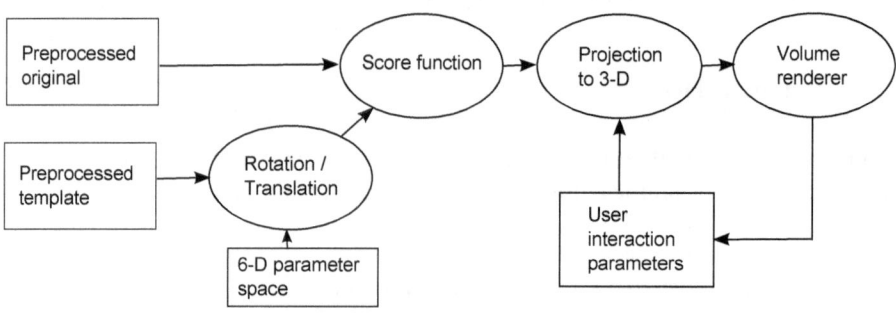

Fig. 1. Flowchart for the methodology

2   Methodology

Rigid-body registration involves six degrees of freedom: three degrees for translation, $\mathbf{t} \in \mathbf{R}^3$, and three for rotation, $\mathbf{A} \in SO(3)$. The special orthogonal group $SO(3)$ is the set of all rotation matrices.

To compute the total score volume, $s : \mathbb{R}^3 \times SO(3) \rightarrow \mathbb{R}$, we start by preprocessing the image data. Voxel values below a user set level, corresponding to the background noise level, are removed. An exhaustive registration search is then performed. We match the template with the original image volume using rotations and translations of the template evenly distributed in the image volume (see Sections 2.1 and 2.2).

The similarity function s, defined in the total parameter space, cannot be visualized in one 3-D view. For this reason, some kind of dimensionality reduction is required. We have taken two approaches in this paper to collapse this function which both provide 3-D volumetric projections suitable for exploring the parameter space of rigid registration:

- Mapping $s : \mathbb{R}^3 \times SO(3) \rightarrow \mathbb{R}$ to $f : \mathbb{R}^3 \rightarrow \mathbb{R}$, by summarizing the information over the rotation subspace (see Section 2.3). The result is a rectangular 3-D score volume, with the same dimensions as the original image, which provides an overview of the fitness landscape.
- Mapping $s : \mathbb{R}^3 \times SO(3) \rightarrow \mathbb{R}$ to $g : SO(3) \rightarrow \mathbb{R}$, by keeping a particular translation fixed (see Section 2.4). The result is a scalar function defined in a ball in 3-D, which provides a view of the fitness landscape of all rotations when a particular translation is fixed.

2.1   Distribution of Rotation Angles

Uniform sampling of the set of all rotations, $SO(3)$, is a non-trivial problem, which has been addressed in many publications, see e.g., [13]. Random sampling of $SO(3)$ is easier to facilitate, since rotations may be described by the set of unit quaternions (with antipodal points identified). Thus, random sampling of $SO(3)$ is equivalent to random sampling of the unit sphere in $\mathbb{R}^4$. However, random sampling introduces noise in the sampling, which makes the representation inefficient. To fix this, a relaxation procedure was devised to make the sampling more uniform.

1. Randomly sample N points on a unit 3-sphere in $\mathbb{R}^4$, store as rows in a $N \times 4$ matrix $\tilde{\mathbf{X}}^{(0)}$.
2. Double the dataset by mirroring, $\mathbf{X}^{(0)} = [\tilde{\mathbf{X}}^{(0)}; -\tilde{\mathbf{X}}^{(0)}]$.
3. Compute all pairwise Euclidean distances, store in a $2N \times 2N$ matrix $\mathbf{D}$.
4. Compute a weighted Laplacian matrix, $\mathbf{W}$, with coefficients: $W_{ij} = -1/(D_{ij} + 1)^{\alpha}$, when $i \neq j$, and $W_{ii} = -\sum_{k \neq i} W_{ik}$.
5. Relax points $\mathbf{X}^{(m+1)} = (\mathbf{I} + \mathbf{W})\mathbf{X}^{(m)}$.
6. Normalize each row in $\mathbf{X}$ to unit length.
7. If not converged, go to 3.
8. Remove the mirrored half of the points.

Running this scheme, it was observed that the average distance from a point to its closest neighbor increased with around 80%, for $N = 1000$, $\alpha = 20$ after 50 relaxation iterations. This indicates that the sampling becomes more uniform and thus also more efficient. By the use of a mirrored set of points, we ensured that the sampling was uniformly sampled from the set of *antipodally identified* points on the sphere in $\mathbb{R}^4$.

2.2 Similarity Measure

Choosing an appropriate scoring function is to a large extent a trade-off between accuracy and implementation speed. A thorough investigation of this subject was done by Wriggers [11]. Mutual information is an alternative similarity measure that was not covered by Wriggers. It is an established scoring function for many (medical) imaging modalities and has been found to work well also for this kind of data [9]. However, the extra generality of this measure, such as handling non-linear differences, was not needed in this case. Here, normalized cross-correlation [3] was chosen:

$$s(\mathbf{u}) = \frac{\sum_{\mathbf{x}} (I(\mathbf{x}) - \bar{I}_\mathbf{u})(T(\mathbf{x} - \mathbf{u}) - \bar{T})}{\left(\sum_{\mathbf{x}} (I(\mathbf{x}) - \bar{I}_\mathbf{u})^2 \sum_{\mathbf{x}} (T(\mathbf{x} - \mathbf{u}) - \bar{T})^2 \right)^{1/2}} \tag{1}$$

where $\bar{T}$ is the mean of the template and $\bar{I}_\mathbf{u}$ is the mean of the region under the template. The computation of normalized correlation was performed in the Fourier domain for computational efficiency [5].

2.3 Summarizing over the Rotation Subspace

We use the p-norm to summarize a scalar function over $SO(3)$. For any translation $\mathbf{t} \in \mathbb{R}^3$, we define

$$f(\mathbf{t}) = \left(\sum_{\mathbf{A} \in SO(3)} s(\mathbf{t}, \mathbf{A})^p \right)^{1/p}, \tag{2}$$

where $p \in [1, \infty[$. Cases of particular interest are:

- maximum projection, $p = \infty$, and
- average projection, $p = 1$.

To obtain a strong response with the average projection, the object needs to be rotationally symmetric, which is quite often the case for biological molecules. The combined analysis of the average and maximum projections is useful for identifying good registration positions for symmetric objects. They will have strong responses at good registration positions in both projections. It is similarly useful for discarding false positions for non symmetrical objects since, for good/correct positions, they will only have a strong response in the maximum projection. An alternative to simultaneously analyzing two projection volumes is to look at one projection using a different p-norm, which corresponds to mixing the maximum and average projection.

For analyzing and interpreting the score volumes the position of the center point of the template need to be considered. For symmetrical objects to generate strong and focused responses in the score volumes the rotation center of the template needs to be positioned midway on the symmetry axis/axes. If not, it will lead to multiple or smeared out and less pronounced maxima around the correct positions in the score volumes. If, for example, an object is relatively rotationally symmetric around one axis and the rotation center is positioned on that symmetry axis but not midway it will give rise to double maxima in the translation score volumes. Picture a vertical stick with the center point on the stick but somewhat offset from half of the length of the stick. One maximum will then correspond to the "correct" match and one to the upside down match.

2.4   Fixing the Translation in $\mathbb{R}^3$

For a particular translation $\mathbf{t} \in \mathbb{R}^3$, selected by the user, we define

$$g(\mathbf{A}) = s(\mathbf{t}, \mathbf{A}) \tag{3}$$

Via Euler's theorem, $g(\mathbf{A})$ may then be visualized as a scalar function defined in a ball with radius π in $\mathbb{R}^3$. In this ball, the identity rotation is placed at the center. Points within the ball specify an additional rotation around the vector from the center to the point itself, where the euclidean norm of the vector specifies the rotation angle.

Euler's theorem states that any rotation in 3-D can be described by a rotation around a particular axis, i.e., parameterized by the set of unit vectors, $||\hat{\mathbf{n}}|| = 1$, and a rotation angle, $\alpha \in (0, \pi)$. The mapping $h : SO(3) \rightarrow \mathbb{R}^3$, $\mathbf{y} = h(\hat{\mathbf{n}}, \alpha) = \alpha \hat{\mathbf{n}}$ provides a convenient coordinate system for visualizing the set of all rotations as a volume. In this mapping, every rotation is mapped to a point inside a ball with radius π. In order to place the maximum of $g(\mathbf{A})$ in the center of the sphere, $\hat{g}(\mathbf{A}) = g(\mathbf{A}\mathbf{A}_{max}^{-1})$ is visualized where $\mathbf{A}_{max}$ is the rotation with the maximum value of $g(\mathbf{A})$.

3   Experiments

The proposed methodology is illustrated on three different types of volumes and templates: a synthetic volume, a MET volume of antibodies in solution, and a MET volume of *in situ* skin cells. For each image type the original volume, translation score volumes, a rotation score volume, and registrations corresponding to different optima in the parameter spaces are shown.

3.1   Synthetic Data Set

For a first assessment, a synthetic volume image was generated, which is shown
in Figure 2 (a). It consists of one structure with three perpendicular arms of
different lengths, and is not intended to be a replica of real data, but a suitable
test structure to highlight features of the methodology. It is not symmetrical
around any axis and the inner parts of the structure have higher intensity. In the
search, we used a spatially identical template with linearly mapped intensities.

Figure 2 (b)–(d) show three different translation score volumes corresponding
to $p = 1$ (average projection), $p = 8$, and $p = \infty$ (maximum projection). For
this artificial example with flat background and only one object all three projections clearly show a maximum for the correct translation parameters and lower
scores (intensities) for positions when the template partly overlaps the object.
Figure 2 (e) shows the rotation projection for translation parameters interactively picked from the maximum in the $p = \infty$ translation projection. The center in that particular volume has the highest matching value and corresponds to
the correct registration shown in Figure 2 (f). The two second highest optima in
Figure 2 (e), correspond to the two rotations of the template where the directions
of all the arms coincide with the directions of the object arms. The registration
corresponding to one such rotation is shown in Figure 2 (g). The six slightly
lower optima (two on vertical "line" and four "dots") correspond to rotations
of the template where only two of the arms overlap with the object arms. One
such registration is shown in Figure 2 (h).

3.2   IgG Data Set

The second test case, is illustrated in Figure 3. It consists of a part of a volume
of proteins (antibodies) in solution, where a single Immunoglobulin G, IgG,
antibody (seen in the center of Figure 3 (a)), has been identified [8]. The IgG
molecule has three subgroups connected at one joint. In the test volume, the
three parts and the center joint are roughly spanning a plane. The antibody
template was created from the protein's atom positions deposited in the protein
data bank (PDB) [1,6,4]. A volume image where intensity represents density is
constructed by placing a Gauss-kernel, with standard deviation 1 nm, at each
atom position weighted by the atom mass, giving a resolution of 2 nm. The
intensity in each voxel in the image is generated by adding contributions from
all Gauss-kernels in the vicinity of that voxel [7].

Figure 3 (b)-(d) show the translation score volumes for, $p = 1$ (average projection), $p = 8$, and $p = \infty$ (maximum projection). For this image with lots of
small objects the benefit of studying different projections is especially large. The
correct position is not a pronounced maximum in the average projection but it
is in the $p = 8$ and maximum projections. In fact it is the only pronounced
optimum in the maximum projection score volume.

Figure 3 (e) shows the rotation score volume for a set of translational parameters interactively picked from the maximum in Figure 3 (d). In the rotational
subspace for the best translational position, Figure 3 (e), different local maxima

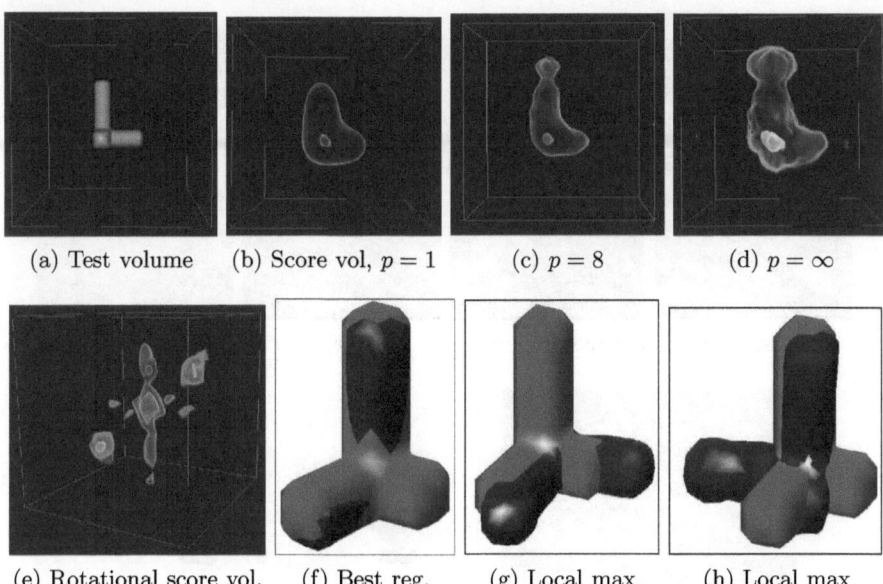

(a) Test volume   (b) Score vol, $p = 1$ (c) $p = 8$ (d) $p = \infty$

(e) Rotational score vol.   (f) Best reg.   (g) Local max   (h) Local max

Fig. 2. The synthetic test volume (a), translation score volumes (b)–(d), rotation score volume (e), and different registration results with the image volume structure in red and search template in blue (f)–(h). The score volumes have been rendered using an isosurface for low scores seen in low intensity, and a higher intensity isosurface for higher scores.

are visible that correspond to different orientations yielding a fairly good match. The arms and stem of the IgG are similar so if they are matched to each other in the wrong way, a high score is still acquired. Antipodal points (on the opposite side of the ball surface) refer to the same orientation. For a tri-symmetrical planar object, a total of eleven pronounced local maxima (central maximum plus five × 2 antipodal maxima) would be seen in the rotation score volume. In this case with IgG which is relatively symmetrical and relatively planar only nine pronounced local maxima (central maximum plus four × two antipodal maxima) are seen with the isosurface rendering levels chosen in Figure 3 (e).

One maxima is connected to two other maxima through a path of slightly higher intensities in the rotation score volume. Although not visible in this rendering such connecting paths can be seen for all the maxima. These paths of slightly higher match scores correspond to rotating the template while one component is fixed and overlapping one of the object components. Figure 3 (f) shows the registration corresponding to the center maximum in (e) and Fig. 3 (g) shows the registration corresponding to one of the other matches seen as non-central local maxima in (e).

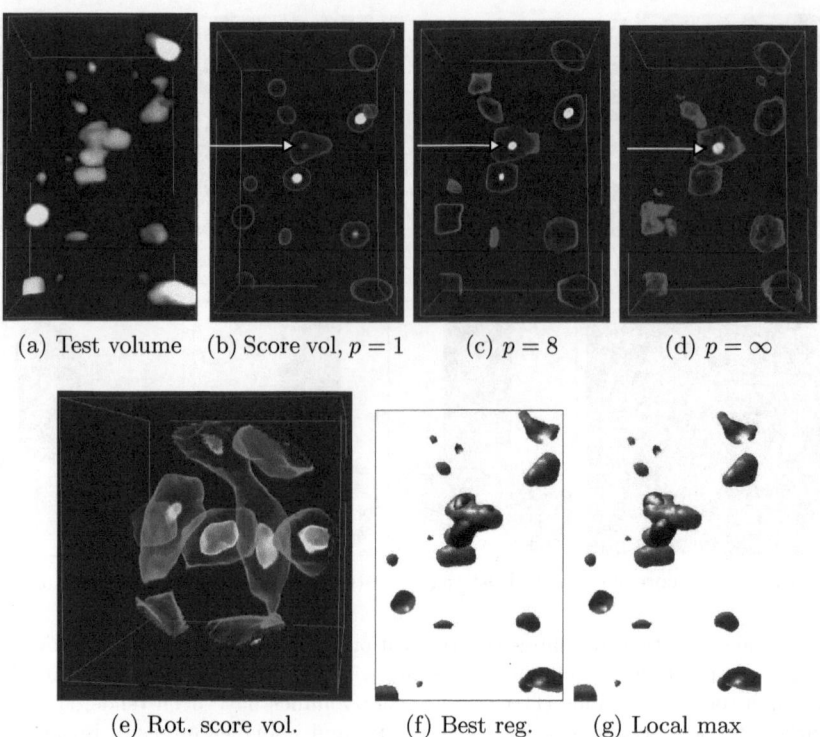

(a) Test volume   (b) Score vol, $p = 1$ (c) $p = 8$ (d) $p = \infty$

(e) Rot. score vol.   (f) Best reg.   (g) Local max

Fig. 3. The IgG test volume (a), translation score volumes with (b) – (d), rotation score volume (e), and different registration results (f)–(g). Since the IgG is rotationally asymmetric, the average projection yields a weak response for the correct position (marked with arrow), but as p is increased towards maximum projection, the correct position becomes clearer.

3.3   Desmosome Data Set

The third investigated data set, which can be seen in Figure 4, concerns a MET volume from a skin sample. The structures of interest in these volumes are chains of two corneodesmosin antibodies linked to a very bright and spherical gold particle. In Figure 4 (a) four such chains are visible as bright and slightly elongated structures. The template used is one of the chains cut from the volume and the task is to identify the three other chains.

The antibody chains in this volume are elongated structures, which are relatively rotationally symmetric around one axis. This implies mainly two things when interpreting the score volumes. Firstly that significant responses for the correct positions in average translation score volume would be expected. Secondly that, due to imperfectly positioned rotation point, the local maxima might be rather blurry in the average projection and multiple local maxima might be found for the correct position in all score volumes but especially in the maximum projection score volume. These phenomena can indeed be seen in the score

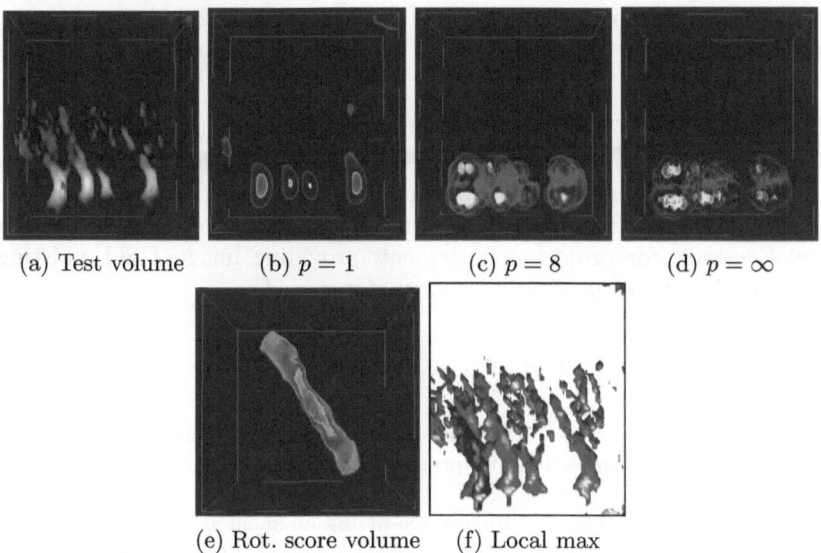

(a) Test volume   (b) $p = 1$ (c) $p = 8$ (d) $p = \infty$

(e) Rot. score volume   (f) Local max

Fig. 4. The desmosome test volume (a). Rotation score volume for a correct translation position (b). Translation score volumes (c) – (e) One registration result (f).

volumes. For the maximum projection case, Figure 4 (d), double maxima can be found for each antibody chain, one correct above the gold particle and one below corresponding to the "upside down" match. The average projection, Figure 4 (b), gives a strong but rather blurred response, centered at the gold particle, which is a displacement to the correct position. The rotation score volume for the correct translation position for one of the three antibody chains searched for has a clear cylinder shape, corresponding to rotations around the chain axis.

4   Conclusion and Future Work

We have presented a methodology for semi-automatic registration of MET data. The maximum projection has proven to be the overall best technique to summarize the non-spatial dimensions in the parameter space. The exploration of the rotational subspace, keeping the translation fixed, has also proven to be a valuable tool to understand the nature of a particular local maximum of the score function in the 6-D parameter space. In particular it reveals the maxima caused by symmetries, which are common in biological molecules.

In future work, we will include haptic rendering to give the user even stronger cues on the local fitting of the template volume. The projection from 6-D to 3-D, through p-norms, looks different depending on the choice of this center. Different ways of automatically determining this center is a possible future topic. We also aim to explore more ways to summarize the 6-D parameter space. In particular, the p-norms used in this paper does not take the *shape* of the local maxima and the local fitness landscape into account when it summarizes the score function over all rotations.

Acknowledgment. This work is part of the ProViz project[1] and funded through the *Visualization Program* by the Knowledge Foundation, Vårdal Foundation, the Foundation for Strategic Research, VINNOVA, and Invest in Sweden Agency.

Thanks to: Stina Svensson, Sara Sandin, Aurelie Laloef, Lars Norlén, and Daniel Evestedt, for project management, providing images, and assisting in implementation of the pre-integrated ray-caster used.

References

1. Berman, H., Westbrook, J., Feng, Z., Gilliland, G., Bhat, T., Weissig, H., Shindyalova, I., Bourne, P.: The protein data bank. Nucleic Acids Research 28, 235–242 (2000)
2. Birmanns, S., Wriggers, W.: Interactive fitting augmented by force-feedback and virtual reality. Journal of Structural Biology 144(1-2), 123–131 (2003)
3. Gonzalez, R.C., Woods, R.E.: Digital Image Processing, 3rd edn., vol. ch. 12. Prentice-Hall, Inc., Upper Saddle River (2006)
4. Harris, L.J., Larson, S.B., Hasel, K.W., McPherson, A.: Refined structure of an intact IgG2a monoclonal antibody. Biochemistry 36(7), 1581–1597 (1997)
5. Lewis, J.P.: Fast normalized cross-correlation (1995)
6. http://www.pdg.org/ (visited December 7, 2010)
7. Pittet, J.J., Henn, C., Engel, A., Heymann, J.B.: Visualizing 3D data obtained from microscopy on the internet. Journal of Structural Biology 125, 123–132 (1999)
8. Sandin, S., Öfverstedt, L.G., Wikström, A.C., Wrange, O., Skoglund, U.: Structure and flexibility of individual immunoglobulin G molecules in solution. Structure 12, 409–415 (2004)
9. Telenczuk, B., Ledesma-Carbato, M., Velazquez-Muriel, J., Sorzano, C., Carazo, J.M., Santos, A.: Molecular image registration using mutual information and differential evolution optimization. In: 3rd IEEE International Symposium on Biomedical Imaging: Nano to Macro, pp. 844–847 (2006)
10. Wriggers, W., Birmanns, S.: Using situs for flexible and rigid-body fitting of multiresolution single-molecule data. Journal of Structural Biology 133(2-3), 193–202 (2001)
11. Wriggers, W., Chacon, P.: Modeling tricks and fitting techniques for multiresolution structures. Structure 9, 779–788 (2001)
12. Wriggers, W., Milligan, R.A., McCammon, J.A.: Situs: A package for docking crystal structures into low-resolution maps from electron microscopy. Journal of Structural Biology 125(2-3), 185–195 (1999)
13. Yershova, A., Jain, S., Lavalle, S.M., Mitchell, J.C.: Generating uniform incremental grids on SO(3) using the Hopf fibration. Int. J. Rob. Res. 29, 801–812 (2010)

---

[1] http://www.cb.uu.se/research/proviz/

A Multiple Kernel Learning Algorithm for Cell Nucleus Classification of Renal Cell Carcinoma

Peter Schüffler[1,*,**], Aydın Ulaş[2,**],
Umberto Castellani[2], and Vittorio Murino[2,3]

[1] ETH Zürich, Department of Computer Science, Zürich, Switzerland
[2] University of Verona, Department of Computer Science, Verona, Italy
[3] Istituto Italiano di Tecnologia (IIT), Genova, Italy

Abstract. We consider a Multiple Kernel Learning (MKL) framework for nuclei classification in tissue microarray images of renal cell carcinoma. Several features are extracted from the automatically segmented nuclei and MKL is applied for classification. We compare our results with an incremental version of MKL, support vector machines with single kernel (SVM) and voting. We demonstrate that MKL inherently combines information from different input spaces and creates statistically significantly more accurate classifiers than SVMs and voting for renal cell carcinoma detection.

Keywords: MKL, renal cell carcinoma, SVM.

1 Introduction

Cancer tissue analysis consists of several consecutive estimation and classification steps which require intensive labor practice. The tissue microarray (TMA) technology enables studies associating molecular changes with clinical endpoints [7]. In this technique, $0.6mm$ tissue cylinders are extracted from primary tumor blocks of hundreds of different patients, and are subsequently embedded into a recipient paraffin block. Such array blocks can then be used for simultaneous analysis of primary tumors on DNA, RNA, and protein level.

In this work, we consider the computer based classification of tissue from renal cell carcinoma (RCC) after such a workflow has been applied. The tissue has been transferred to an array and stained to make the morphology of cells and cell nuclei visible. Current image analysis software for TMAs requires extensive user interaction to properly identify cell populations on the TMA images, to select regions of interest for scoring, to optimize analysis parameters and to organize the resulting raw data. Because of these drawbacks, pathologists typically collect tissue microarray data by manually assigning a composite staining score for each spot. Such manual scoring can result in serious inconsistencies between data collected during different microscopy sessions. Manual scoring also introduces a significant bottleneck that limits the use of tissue microarrays in high-throughput analysis.

---

* Corresponding author.
** Equal contributors.

G. Maino and G.L. Foresti (Eds.): ICIAP 2011, Part I, LNCS 6978, pp. 413–422, 2011.

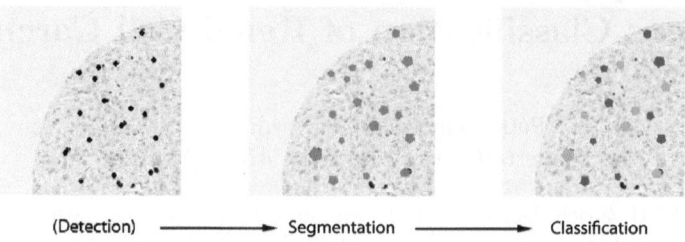

Fig. 1. One keypoint in the automatic TMA analysis for renal cell carcinoma is the nucleus classification. Nuclei are eosin stained and visible in the TMA image as dark blue spots. We want to simulate the classification of cell nuclei into cancerous or benign, which is recently done by trained pathologists by eye. The automatic approach comprises nucleus detection on the image, the segmentation of the nuclei and the classification, all based on training data labeled by two human experts.

The manual rating and assessment of TMAs under the microscope by pathologists is quite inconsistent due to the high variability of cancerous tissue and the subjective experience of humans, as shown in [4]. Therefore, decisions for grading and/or cancer therapy might be inconsistent among pathologists. With this work, we want to contribute to a more generalized and reproducible system that automatically processes TMA images and thus helps pathologists in their daily work.

For various classification tasks, SVM formulations involve using one data set and maximizing the margin between different classes. This poses a restriction on some problems, where different data representations are used. Combining the contribution of different properties is important in discriminating between cancerous and healthy cells. Multiple Kernel Learning (MKL) is a recent and promising paradigm, where the decisions of multiple kernels are combined to achieve better accuracies [1]. The advantage of this idea is to be able to utilize data from multiple sources. In MKL, multiple kernels are combined (see Section 3) globally. We also compare this idea with the usual classifier combination where outputs of multiple classifiers are combined [8,10].

In previous work, an automated pipeline of TMA processing was already proposed, concentrating on the investigation of various image features and associated kernels on the performance of a support vector machine classifier for cancerous cells [13]. In this work, we follow this workflow and extend the nucleus classification (Figure 1) by using MKL that combines information from multiple sources (in our case different representations). By considering different types of features, we show in Section 4 the importance of using shape features; our results show that MKL reaches significantly better accuracies than SVM and voting (VOTE) using the combination of multiple kernels.

Our contribution is to show how information from different representations can make this classification task easier: the MKL algorithm inherently combines data from different representations to get better classification accuracies. Instead of combining outputs of multiple classifiers, MKL uses an optimization procedure where data from all sources are seen during training and optimization is done accordingly. Our experiments demonstrate that although it is more costly to use MKL, the increase in accuracy is worth its cost.

The paper is organized as follows: in Section 2, we introduce the data set used in this study. We explain the methods applied in Section 3, and show our experiments in Section 4. We conclude in Section 5.

2   Data Set

2.1   Tissue Micro Arrays

Small round tissue spots of cancerous tissue are attached to TMA glass plate. The diameter of the spots is 1mm and the thickness corresponds to one cell layer. Eosin staining made the morphological structure of the cells visible, so that cell nuclei appear bluish in the TMAs. Immunohistochemical staining for the proliferation protein MIB-1 (Ki-67 antigen) makes nuclei in cell division status appear brown.

For computer processing, the TMA slides were scanned with a magnification of 40x, resulting in a per pixel resolution of $0.23\mu m$. The final spots of single patients are separately extracted as three channel color images of size 3000x3000px.

In this study, we used the top left quarter of eight tissue spots from eight patients. Therefore, each image shows a quarter of the whole spot, i.e. 100-200 cells per image (see Figure 2).

For training our models, the TMA images were independently labeled by two pathologists [4]. From such eight labeled TMA images, we extracted 1633 nuclei-patches of size 80x80 pixels. Each patch shows a cell nucleus in the center (see Figure 3). 1273 (78 %) from the nuclei form our datase, where the two pathologists agree on the label: 891 (70 %) benign and 382 (30 %) malignant nuclei.

2.2   Image Normalization and Patching

The eight images were adjusted in contrast to minimize illumination variances among the scans. To classify the nuclei individually, we extracted patches from the whole image such that each 80x80px patch has one nucleus in the center (see Figure 3). The locations of the nuclei were known from the labels of the pathologists. Both procedures drastically improved the following segmentation of cell nuclei.

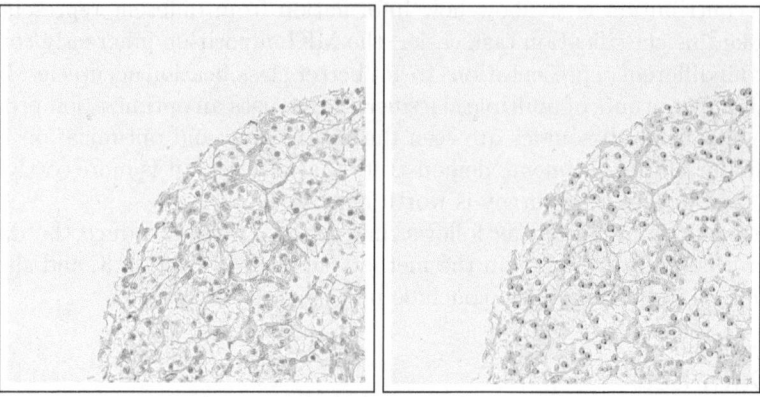

Fig. 2. Left: One 1500x1500px quadrant of a TMA spot from a RCC patient. **Right:** A pathologist exhaustively labeled all cell nuclei and classified them into malignant (black) and benign (red).

2.3   Segmentation

The segmentation of cell nuclei was performed with graphcut [3]. The gray intensities were used as unary potentials. The binary potentials were linearly weighted based on their distance to the center to prefer roundish objects lying in the center of the patch (see Figure 3). The contour of the segmented object was used to calculate several shape features as described in the following section.

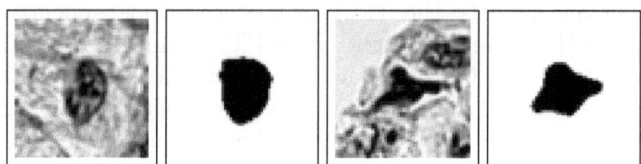

Fig. 3. Two examples of nucleus segmentation. The original 80x80 pixel patch are shown, each with the corresponding nucleus shape found with graphcut.

2.4   Feature Extraction

For training and testing the various classifiers we extracted several histogram-like features from the patches (see Table 1).

Table 1. Features extracted from patch images for training and testing. Except the PROP feature, all features are histograms normalized to sum up to one.

Shortcut	Feature Description
ALL	**Patch Intensity**: A 16-bin histogram of gray scaled patch
FG	**Foreground Intensity**: A 16-bin histogram of nucleus
BG	**Background Intensity**: A 16-bin histogram of background
LBP	**Local Binary Patterns**: This local feature has been shown to bring considerable performance in face recognition tasks. It benefits from the fact that it is illumination invariant.
COL	**Color feature**: The only feature comprising color information. The colored patch (RGB) is rescaled to size 5x5. The 3x25 channel intensities are then concatenated to feature vector of size 75.
FCC	**Freeman Chain Code**: The FCC describes the nucleus' boundary as a string of numbers from 1 to 8, representing the direction of the boundary line at that point ([6]). The boundary is discretized by subsampling with grid size 2. For rotational invariance, the first difference of the FCC with minimum magnitude is used. The FCC is represented in a 8-bin histogram.
SIG	**1D-signature**: Lines are considered from the object center to each boundary pixel. The angles between these lines form the signature of the shape ([6]). As feature, a 16-bin histogram of the signature is generated.
PHOG	**Pyramid histograms of oriented gradients**: PHOGs are calculated over a level 2 pyramid on the gray-scaled patches ([2]).
PROP	**Shape descriptors** derived from MATLAB's **regionprops** function: `Area BoundingBox(3:4)`, `MajorAxisLength`, `MinorAxisLength`, `ConvexArea`, `Eccentricity`, `EquivDiameter`, `Solidity`, `Extent`, `Perimeter`, `MeanIntensity`, `MinIntensity`, `MaxIntensity`;

3 Methodology

In this section, we summarize the MKL framework behind our experiments. The main idea behind support vector machines [15] is to transform the input feature space to another space (possibly with a greater dimension) where the classes are linearly separable. After training, the discriminant function of SVM becomes $f(x) = \langle w, \Phi(x) \rangle + b$, where w are the weights, b is the threshold and $\Phi(x)$ is the mapping function. Using dual formulation and the kernels one does not have to define this mapping function $\Phi(x)$ explicitly and the discriminant becomes as in (1) where $K(x_i, x)$ is the kernel.

$$f(x) = \sum_{i=1}^{N} \alpha_i y_i K(x, x_i) + b .\tag{1}$$

Using SVM with a single kernel would restrict us to use one feature set (or a concatenation of all feature sets) and complicates the possibility to exploit the information coming from different sources. As in classifier combination [8], we can combine multiple kernels using different feature sets and use this information to come up with more accurate classifiers [10]. The simplest way for this is to use an unweighted sum of kernel functions [11]. Lanckriet et al. [9] have formulated this semidefinite programming problem which allows finding the combination weights and support vector coefficients together. Bach et al. [1] reformulated the problem and proposed an efficient algorithm using sequential minimal optimization (SMO). Using Bach's formulation, with P kernels, the discriminant function becomes as in (2) where m indexes the kernel:

$$f(\boldsymbol{x}) = \sum_{m=1}^{P} \eta_m \sum_{i=1}^{N} \alpha_i y_i K_m(\boldsymbol{x}, \boldsymbol{x}_i) + b \,. \tag{2}$$

This allows us to combine different kernels in different feature spaces and this is the formulation we apply in this work. Here, kernels are combined globally, namely the kernels are assigned the same weights for the whole input space.

It has been shown by many researchers that using a subset of given classification algorithms increases accuracy rather than using all the classifiers [12,14]. Keeping this in mind, we apply the same idea to incrementally adding kernels to the MKL framework and compare the results.

The incremental algorithm works as follows: It starts with the most accurate kernel (classifier) on the validation folds (leave-the-other-fold-out), and adds kernels (classifiers) to the combination one by one. This procedure continues until all kernels (classifiers) are used or the average validation accuracy does not increase [14]. The algorithm starts with $E^0 \leftarrow \emptyset$, then at each step t, all the kernels (classifiers) $M_j \notin E^{(t-1)}$ are combined with $E^{(t-1)}$ to form S_j^t ($S_j^t = E^{(t-1)} \cup M_j$). We select S_{j*}^t which is the ensemble with the highest accuracy. If accuracy of S_{j*}^t is higher than $E^{(t-1)}$, we set $E^t \leftarrow S_{j*}^t$ and continue, else the algorithm stops and returns $E^{(t-1)}$.

4   Experiments

4.1   Experiment Setup

The data of 1273 nuclei samples is divided into ten folds (with stratification). We then train support vector machines (*sv1, sv2, svg*, see below) and MKL using these folds. We also combine the support vector machines using voting and report average accuracies using 10-fold CV. For the Gaussian kernel, σ is chosen using a rule of thumb: $\sqrt{D}$ where D is the number of features of the data representation. We compare our results using 10-fold CV t-test at $p = 0.05$. In the incremental learning part, we apply leave-the-other-fold-out cross validation (used for validation) to estimate which kernel and classifier should be added.

As a summary, we have 9 representations (ALL, BG, COL, FCC, FG, LBP, PHOG, SIG and PROP), three different kernels (linear kernel: *sv1*, polynomial kernel with degree 2: *sv2*, and Gaussian kernel: *svg*), and two combination algorithms (MKL, VOTE).

The SVM accuracies with each individual kernel are reported in Table 2. The best accuracy using a single SVM is 76.9 %. For most representations (except PHOG and COL), the accuracies of different kernels are comparable.

Table 2. Single support vector accuracies ($\in$ std) in %

	svl	sv2	svg
ALL	70.0$\in$0.2	71.6$\in$2.9	72.0$\in$3.2
BG	70.0$\in$0.2	71.2$\in$2.6	68.9$\in$2.3
COL	70.1$\in$0.2	63.6$\in$3.5	66.2$\in$2.3
FCC	70.0$\in$0.2	70.0$\in$0.2	67.4$\in$1.6
FG	70.0$\in$0.2	70.0$\in$3.2	70.5$\in$3.5
LBP	70.0$\in$0.2	66.9$\in$3.0	68.7$\in$4.4
PHOG	76.5$\in$3.7	72.0$\in$3.3	**76.9$\pm$3.6**
SIG	70.0$\in$0.2	68.6$\in$2.5	66.6$\in$2.6
PROP	75.7$\in$2.3	75.6$\in$2.6	74.1$\in$1.8

Next, we use the same kernel and combine all the feature sets we extracted. As shown in Table 3 (top), we can achieve an accuracy of 81.3 % using the linear kernel, by combining all representations. This shows that the combination of information from multiple sources might be important and, by using MKL, the accuracy can be increased around 5 %. We observe from the table also that when we use all kernels with *sv2*, we have a decrease in accuracy compared to the single best support vector machine. This is analogous to combining all classifiers in classifier combination. If one has relatively inaccurate classifiers, combining all may decrease accuracy. Instead, it might be better to choose a subset. This also shows that medically, all the information is complementary and should be used to achieve better accuracy. In Figure 4, we plotted the weights of MKL when we use the linear kernel. As expected, the two best representations PHOG and PROP have high weights. But the representation LBP that has very low accuracy when considered as a single classifier increases the accuracy when considered in combination. This shows that when considering combinations, even a representation which is not very accurate alone may contribute to the combination accuracy. From this, we also deduce that these three features are useful in discriminating between healthy and cancerous cells and we may focus our attention on these properties.

On the bottom part of Table 3, the results using the incremental algorithm are shown. We can see that we do not have an increase in accuracy compared to the best single support vector machine. In fact, the incremental algorithm cannot find a second complementary kernel which will increase accuracy when added to the single best. In principle, we expect the incremental algorithm to have better accuracies than combining all classifiers. We see this behavior for *sv2*. When we consider *svl*, combining all kernels seems to be better than the subset selection strategy. This might partially result from the fact that the incremental algorithm could not find a complementary kernel, and partially from the optimization formulation of MKL. In the incremental search, we discard kernels which do not improve the overall accuracy. On the other hand, in MKL, every kernel is given a weight and all kernels contribute to the solution of the problem. From this, we can say that it is better to use MKL instead of combining outputs of

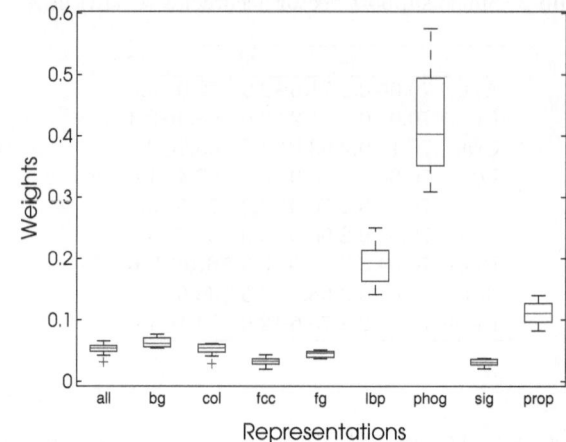

Fig. 4. Combination weights in MKL using the linear kernel

support vector machines using voting. We can also see the support of this claim in Table 3. When we use voting, combining all classifiers is always worse than the single best and always worse than MKL because the optimization procedure does not see the data, but only combines outputs of all classifiers. On the other hand, when we apply the incremental paradigm on classifier combination, we achieve better results than MKL because there are complementary classifiers which increase the accuracy.

Table 3. MKL accuracies (in %). Top: accuracy ($\in$ std) of combining all kernels. Bottom: accuracies calculated using the incremental algorithm, the number of kernels/classifiers selected.

	svl	sv2	svg
MKL	**81.3±3.6**	72.0∈3.3	76.9∈3.6
VOTE	70.0∈0.2	71.3∈1.7	72.4∈1.2
MKL		76.9∈3.6, 1	
VOTE		78.9∈2.5, 4	

4.2   Discussion

We have seen that MKL performs better than VOTE and SVMs with single kernel, when all kernels are combined. This is because the optimization procedure takes into account all data and gives weights to all kernels, so it can use all representations. On the other hand, when we apply the incremental algorithm, classifier combination achieves better accuracies than combining all classifiers.

MKL combines the underlying feature sets to make a better combination. In this work, we used three different kernels and two combination schemes to see how the change of each parameter effects the classification accuracy. We see that, when we use single support vector machines, all the kernels have comparable accuracies. The importance of each kernel function increases when the combination is considered, and combining outputs is less effective than combining the kernels themselves using optimization.

Also, we have seen that when we use the multiple kernel learning algorithm, we gain 5 % in accuracy compared to SVMs with single kernel. Combining all kernels here comes with a drawback. We have to use all kernels and extract all the features when we have to use this model but the increase in accuracy might be worth the cost. We see that when we use the incremental algorithm, we cannot add any kernels, so we are stuck in a local minimum. When we combine classifiers on the other hand, the incremental algorithm achieves more accurate results. Nevertheless, the best results are obtained when we use all representations using *svl* and this accuracy is the best result we have reached so far.

5   Conclusion

In this paper, we propose the use of the multiple kernel learning paradigm for the classification of nuclei in TMA images of renal clear cell carcinoma. We used support vector machines extensively through different feature sets in our previous work. This study extends those works by using several feature sets in a multiple kernel learning paradigm and compares the results with single support vector machines and combining outputs of support vector machines using voting.

We have seen that MKL performs better than SVMs and VOTE in most of the experiments. MKL exploits the underlying contribution of each feature set and heterogeneity of the problem, and by using multiple kernels, achieves better results than single kernels and voting of classifiers.

In this work, we used image based feature sets for creating multiple features. In a further application of this scenario, the use of other modalities or other features (e.g. SIFT) extracted from these images, as well as the incorporation of complementary information of different modalities to achieve better classification accuracy is possible. The incremental algorithm as implemented in this scenario does not work as well as combining all kernels using MKL. As a future work, we would like to implement other heuristics (decremental search, two step look-ahead incremental search, floating search etc.) so that we can achieve better accuracies without imposing too much cost on the system and using only a few kernel combinations. We also would like to apply a local multiple kernel combination framework [5] which is analogous to classifier selection in ensemble framework where the combination also depends on the input which puts forward the inherent localities of the data sets and automatically divides the data set into subsets within the optimization procedure.

Acknowledgements. We thank Dr. Cheng Soon Ong very much for helpful discussions and Dr. Mehmet Gönen for the implementation of the MKL algorithm. Also, we acknowledge financial support from the FET programme within the EU FP7, under the SIMBAD project (contract 213250).

References

1. Bach, F.R., Lanckriet, G.R.G., Jordan, M.I.: Multiple kernel learning, conic duality, and the smo algorithm. In: Proceedings of the Twenty-First International Conference on Machine Learning, ICML 2004, Banff, Alberta, Canada, pp. 41–48 (2004)
2. Bosch, A., Zisserman, A., Munoz, X.: Representing shape with a spatial pyramid kernel. In: CIVR 2007: Proceedings of the 6th ACM International Conference on Image and Video Retrieval, pp. 401–408. ACM, New York (2007)
3. Boykov, Y., Veksler, O., Zabih, R.: Efficient approximate energy minimization via graph cuts. IEEE Transactions on Pattern Analysis and Machine Intelligence 20(12), 1222–1239 (2001)
4. Fuchs, T.J., Wild, P.J., Moch, H., Buhmann, J.M.: Computational pathology analysis of tissue microarrays predicts survival of renal clear cell carcinoma patients. In: Metaxas, D., Axel, L., Fichtinger, G., Székely, G. (eds.) MICCAI 2008, Part II. LNCS, vol. 5242, pp. 1–8. Springer, Heidelberg (2008)
5. Gönen, M., Alpaydın, E.: Localized multiple kernel learning. In: Proceedings of the International Conference on Machine Learning, ICML 2008, pp. 352–359 (2008)
6. Gonzalez, R.C., Woods, R.E., Eddins, S.L.: Digital image processing using matlab, 993475 (2003)
7. Kononen, J., Bubendorf, L., et al.: Tissue microarrays for high-throughput molecular profiling of tumor specimens. Nat. Med. 4(7), 844–847 (1998)
8. Kuncheva, L.I.: Combining pattern classifiers: methods and algorithms. Wiley Interscience, Hoboken (2004)
9. Lanckriet, G.R.G., Cristianini, N., Bartlett, P., Ghaoui, L.E., Jordan, M.I.: Learning the kernel matrix with semidefinite programming. Journal of Machine Learning Research 5, 27–72 (2004)
10. Lee, W.-J., Verzakov, S., Duin, R.P.W.: Kernel combination versus classifier combination. In: Haindl, M., Kittler, J., Roli, F. (eds.) MCS 2007. LNCS, vol. 4472, pp. 22–31. Springer, Heidelberg (2007)
11. Moguerza, J.M., Muñoz, A., de Diego, I.M.n.: Improving support vector classification via the combination of multiple sources of information. In: Fred, A., Caelli, T.M., Duin, R.P.W., Campilho, A.C., de Ridder, D. (eds.) SSPR&SPR 2004. LNCS, vol. 3138, pp. 592–600. Springer, Heidelberg (2004)
12. Ruta, D., Gabrys, B.: Classifier selection for majority voting. Information Fusion 6(1), 63–81 (2005)
13. Schüffler, P.J., Fuchs, T.J., Ong, C.S., Roth, V., Buhmann, J.M.: Computational TMA analysis and cell nucleus classification of renal cell carcinoma. In: Goesele, M., Roth, S., Kuijper, A., Schiele, B., Schindler, K. (eds.) Pattern Recognition. LNCS, vol. 6376, pp. 202–211. Springer, Heidelberg (2010)
14. Ulaş, A., Semerci, M., Yıldız, O.T., Alpaydın, E.: Incremental construction of classifier and discriminant ensembles. Information Sciences 179(9), 1298–1318 (2009)
15. Vapnik, V.N.: Statistical learning theory. John Wiley and Sons, Chichester (1998)

Nano-imaging and Its Applications to Biomedicine

Elisabetta Canetta[1,*] and Ashok K. Adya[2]

[1] Cardiff University, School of Biomedical Sciences, Museum Avenue, PO Box 911,
Cardiff, CF10 3US, Wales, UK
[2] University of Abertay Dundee, Division of Biotechnology and Forensic Sciences,
Bell Street, Dundee, DD1 1HG, Scotland, UK
CanettaE@cardiff.ac.uk, ashok.adya@abertay.ac.uk

Abstract. Nanotechnology tools, such as Atomic Force Microscopy (AFM), are now becoming widely used in life sciences and biomedicine. AFM is a versatile technique that allows studying at the nanoscale the morphological, dynamic, and mechanical properties of biological samples, such as living cells, biomolecules, and tissues in their native state under physiological conditions. In this article, an overview of the principles of AFM will be first presented and this will be followed by discussion of some of our own recent work on the applications of AFM imaging to biomedicine.

Keywords: Atomic Force Microscopy, AFM, AFM imaging, elastic images, elastic modulus, biomedicine, cells, yeasts, biomolecules.

1 Introduction

Atomic force microscopy (AFM) was invented by Binning, Quate and Gerber in 1986 [1] and since then its use as a surface characterization tool has increased dramatically. In particular, in the last ten years AFM has become one of the most powerful nanotools in biology [2,3]. This success is due to the ability of the AFM to allow studying the structure, function, properties, and interaction of biological samples in their native state under physiological buffer conditions. Moreover, the excellent signal-to-noise ratio of the AFM allows obtaining images with sub-nanometer (< 1 nm) resolution which permits to investigate the ultrastructure of a broad range of samples ranging from cells (micrometer sized) to DNA and proteins (a few nanometers in size) [4]. Another advantage of AFM lies in minimal sample preparation requirements that preserve the real features of the specimens without damaging them or introducing any artifacts, thereby making AFM a very attractive and versatile nano-imaging tool in biological sciences [5].

The ability of AFM to visualize nanoscale topography and morphology of tissues, cells, and biomolecules permits a direct observation of the fine structural details of, e.g. cell surfaces, surface proteins, and DNA coiling. Processing of AFM images permits one to measure at the nanoscale, the surface roughness, texture, dimensions, and volumes of biological systems. This capability along with the ability of AFM to

---

* Corresponding author.

G. Maino and G.L. Foresti (Eds.): ICIAP 2011, Part I, LNCS 6978, pp. 423–432, 2011.
© Springer-Verlag Berlin Heidelberg 2011

detect pico-Newton force interactions between cells (cell-cell interaction), proteins (protein-protein interaction), a cell and a functionalized substrate (cell-substrate interaction), or a receptor and a ligand (lock-and-key interaction) has opened up completely new areas for investigating the mechanical and microrheological properties of biological samples, such as their adhesion, elasticity (property of a material which causes it to be restored to its original shape after applying a mechanical stress), viscosity (resistance of a fluid to deform under shear stress), and viscoelasticity (when materials change with time they respond to stress as if they are a combination of elastic solids and viscous fluids) [3].

In this paper, some of the relevant results by AFM obtained by the authors in the biomedical field will be presented. The computational processing of AFM images to obtain multiparameter analysis of cells and biomolecule functions will also be discussed.

2 Principles of AFM

The operating principle of an AFM is based upon scanning a fine tipped probe (called cantilever) just above a sample surface (Fig. 1) and monitoring the interaction force between the probe and the surface [3]. The cantilever obeys the Hooke's law, $F = k \cdot x$, where F is the force on the cantilever, k its spring constant, and x its deflection. This relationship permits one to obtain the tip-sample interaction force, provided that the spring constant of the cantilever is known and its deflection measured. To measure the cantilever deflection a laser beam is reflected off the back side of a reflective (gold- or aluminium-coated) cantilever towards a position sensitive detector which can measure both normal bending and torsion of the cantilever, corresponding to normal and lateral forces.

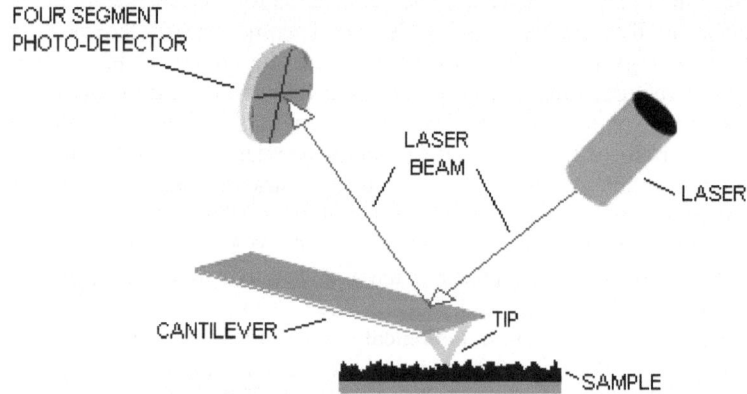

Fig. 1. Sketch of the operational principle of AFM

2.1 Application Modes of AFM

Imaging. During scanning of the sample surface, the deflection of the cantilever is measured and used as an input signal for the feedback circuitry to maintain constant the tip-sample distance. By recording the displacements of the tip along the z-axis it becomes possible to build the *topographic* (height) image of the sample surface. A variety of operational modes, such as contact mode, and intermittent-contact tapping mode are available to image a sample surface [3].

Force Spectroscopy. The main outcome of a force spectroscopy experiment is the force-distance (F-d) curve, which shows the force experienced by the cantilever both when the AFM probe is brought in contact with the sample surface (trace/approach) and separated from it (retrace/retract) (Fig. 2). Analysis of the F-d curves with suitable theoretical models can give information on the elastic and adhesive properties of the sample. In particular, the maximum adhesive force of the sample to the tip (F_{max}), the energy of adhesion (W_{adh}), and the Young's modulus (E) can be measured (Fig. 2).

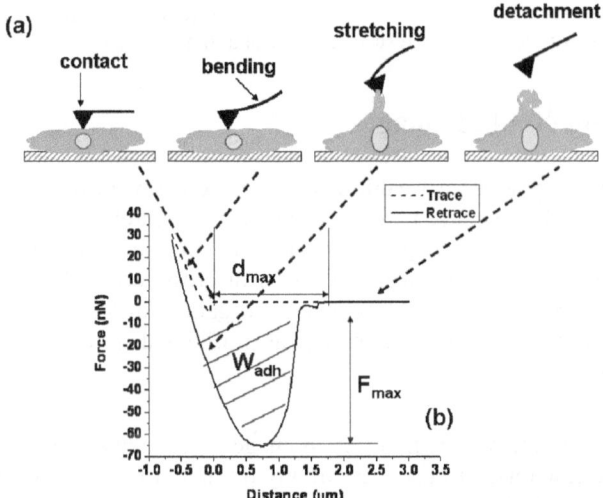

Fig. 2. (a) Sketch of the stages of a force spectroscopy experiment. Contact between the AFM probe and the sample surface and cantilever bending with AFM probe indentation occurs during the trace (approach) cycle of the AFM experiment. On the retrace (retract) cycle, the sample surface is stretched until the point of detachment. An example of a typical experimental F-d curve for a trace and retrace cycle is also presented; the stages are identified by the arrows. The meanings of F_{max}, d_{max}, and W_{adh} are also identified.

Force Mapping. It combines force measurements and imaging [6,7]. A force mapping data set contains an array of force curves and a sample image as well. It can be used to obtain a 2D map of the nanomechanical properties (e.g. elasticity) of the sample surface (Fig. 3). It is worth noting that force spectroscopy and force mapping can be done only in contact mode.

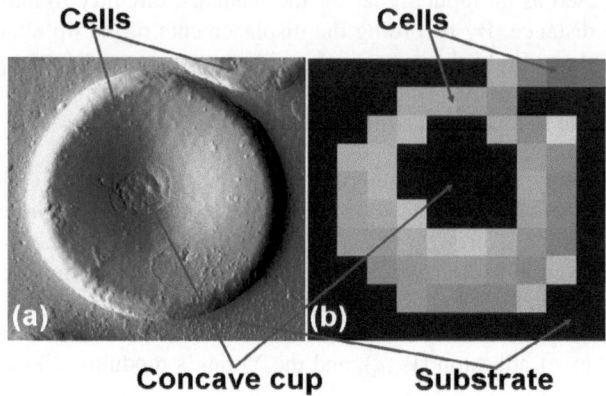

Fig. 3. AFM force mapping on a red blood cell (RBC). (a) AFM image of RBC (typical biconcave shape) on glass. (b) AFM elastic image of the RBC shown in (a). The difference in elasticity of the substrate and the cell surface allow obtaining an elastic map of the sample.

3 Processing of AFM Images

AFM experiments generate a wealth of data very quickly. In order to extract useful information from the data, each AFM image has to be carefully processed to remove sample tilt, scanner and probe geometry artifacts, and obtain correct image contrast and scaling. In addition, statistical evaluation of cross-sectional and surface texture analyses can be performed on AFM images to gain knowledge on the dimensions, surface roughness and texture, topographical profile and nanomechanical properties of the samples.

Image Enhancement. It is the first step in image processing and it mainly consists in: (i) leveling the raw image to take care of any tilt and bow that may be introduced to the image by the AFM scanner. This is achieved via background subtraction such as line-by-line or plane leveling; (ii) sharpening the raw image with high-pass filters to enhance the finest details of the sample surface topography; and (iii) smoothing the raw image to remove any background noise from the image. Here we wish to add a word of caution for the novice to be extremely careful because image enhancement, if not done properly, can introduce additional artifacts in the image.

Section Analysis. AFM images can be analysed by section analysis to determine the dimensions (e.g. length, width), and peak-to-valley heights of the sample (Fig. 4). In AFM images, horizontal dimensions of objects are usually greatly overestimated due to the well-known effect of tip convolution. Therefore, statistical methods should be used to correct the tip-convolution effects in AFM images in order to measure the real lateral dimension of the sample [8].

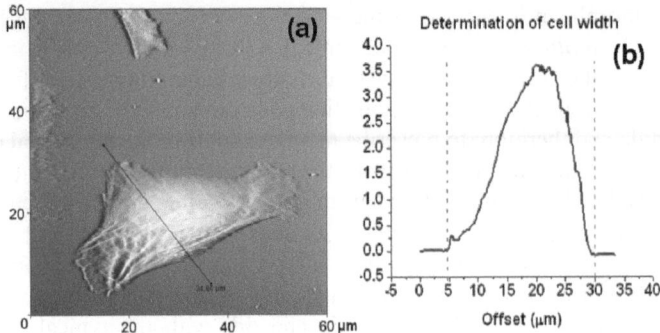

Fig. 4. (a) Section analysis on the AFM image of a living human urothelial cell (HUC) under physiological conditions, and (b) measure of the peak-to-valley height of the cell from cross-section analysis

Roughness Analysis. The roughness of the surface of a sample can be analysed by measuring the root mean square roughness, R_{rms}, on a height image, defined as the standard deviation from the mean data plane of the h (height) values of the AFM images within a selected region on the cell surface,

$$R_{rms} = \sqrt{\frac{\sum_{i=1}^{N}\left(h_i - \overline{h}\right)^2}{N}} .$$

(1)

In Eq. (1), h_i is the current height value, $\overline{h}$, the height of the mean data plane, and N, the number of points within the selected region of a given area. Roughness analysis is usually carried out on raw AFM images, i.e., images that are neither flattened nor elaborated with any filter (e.g., low pass or high pass filter).

4   Applications of AFM to Imaging of Biological Systems

Over the past 10 years, we have extensively used AFM in both imaging and force spectroscopy modes to investigate the ultrastructural and nanomechanical properties of different biological systems [9-11]. Some of our recent results on cells and biomolecules will be presented in the next sections.

4.1   Discrimination between *Candida albicans* and *Candida dubliniensis* Cells under Normal and Pathological Conditions

Candida albicans is a diploid fungus and a causal agent of opportunistic oral and genital infections in humans. *Candida dubliniensis* is commonly isolated from oral cavities and it is very closely related to *C. albicans* because both these yeasts are dimorphic. Dimorphism is defined as the switching between two cell-types. Both *C. albicans* and *C. dubliniensis* when they infect host tissues, switch from the usual unicellular yeast-like form into an invasive, multicellular filamentous form. Due to their dimorphic nature it is sometimes difficult for clinicians to distinguish between the two Candida and therefore to prescribe the more appropriate antifungal drug.

We used AFM imaging to investigate the ultrastructures of *C. albicans* and *C. dubliniensis* (isolated and sub-cultured from patients) after growing them in normal (30°C) and pathogenic (39°C) conditions. As shown in Figs. 5a and 5b, at 30°C *C. albicans* and *C. dubliniensis* cells remained non-pathogenic and they grouped together in colonies. However, increase in temperature led to the appearance of pathogenic cells for both strains of Candida (Figs. 5c and 5d) with the typical formation of hyphae.

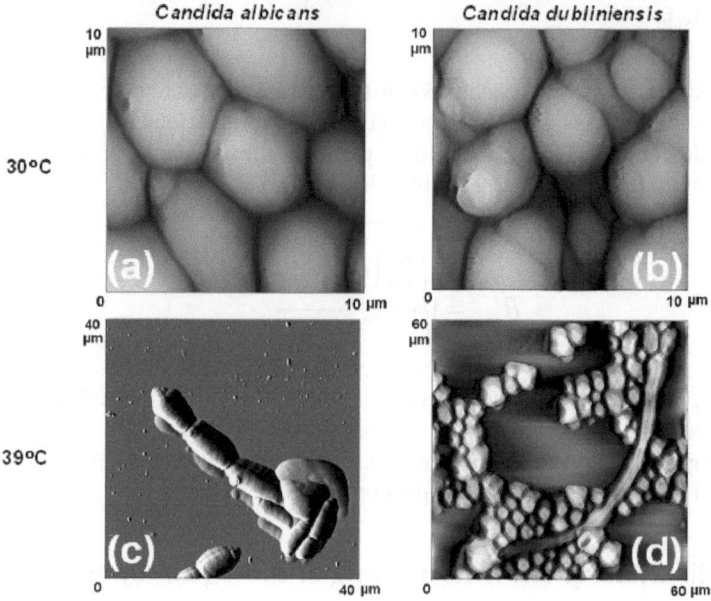

Fig. 5. AFM images of living *C. albicans* (Left) and *C. dubliniensis* (Right) yeast cells under physiological conditions, grown in non-pathogenic conditions at 30°C (upper row) and in pathogenic conditions at 39°C (bottom row). Height ranges in µm: (a) 0 - 4.3, (b) 0 - 3.2, (c) 0 - 5.0, (d) 0 - 4.3

Section analysis on the AFM images of the two types of Candida showed that for both *C. albicans* and *C. dubliniensis*, non-pathogenic cells were smaller than pathogenic cells. This difference could probably be due to the beginning of the process of extrusion of hyphae from the pathogenic yeast cells that resulted in an increase in their sizes. The main difference between *C. albicans* and *C. dubliniensis* revealed by the section analysis concerned their sizes. *C. dubliniensis* cells were found to be smaller than *C. albicans*. Moreover, the morphological structures of the hyphae of these two strains of Candida were quite different. In pathogenic *C. albicans* cells the hyphae were short $(21.2 \pm 3.1)\mu m$, and they were divided in segments with well defined secta formed between the segments. On the contrary, in pathogenic *C. dubliniensis* the hyphae were long $(52.1 \pm 8.4)\mu m$, and the secta dividing the segments were not as evident as in *C. albicans*.

4.2 Determination of Elasticity of Human Embryonic Stem Cells

Human embryonic stem cells (hESCs) are pluripotent, and they represent a promising source of cells for regenerative medicine. In a real clinical setting, the patient will receive mature, differentiated cells after they have been subjected to an appropriate differentiation protocol. However, most protocols result in a mixture of cells enriched for the desired population, but containing subpopulations of other cell types.

Separation and purification methods scalable to quantities of cells suitable for human cell replacement therapy are a major problem for stem cell therapy. Extensive biochemical and biomechanical characterisation of hESCs would be of great utility since such features might provide options for cell separation without requiring the use of antibodies as specific lineage markers. Cell elasticity is an important biomechanical parameter, which is primarily determined by the presence, number and distribution of specific organelles (e.g. nucleus, mitochondria) and the character and organisation of cytoskeletal elements (e.g. microfilaments, microtubules and intermediate filaments). It is well known that cell elasticity varies with cell function. Particular elasticities can also be associated with specific phenotypes. We employed AFM to determine the Young's (or elastic) modulus of hESCs. Single cell surfaces were mapped by performing AFM mapping experiments which allowed local variations due to cell structure to be identified.

AFM images showed the presence of two different phenotypes in the hESCs cell lines: (i) small and round (Fig. 6a) and (ii) large and spread (Fig. 6b). This difference could suggest that one phenotype corresponded to differentiated hESCs while the other to undifferentiated hESCs, but which one corresponded to what phenotype? A possible answer to this question was given by the Young's modulus values obtained from the AFM force mapping experiments. It was observed that the elasticity of hESCs varied at different points on the cell surface, reflecting their heterogeneous nature. In addition, variations in elasticity between cells that were larger than the variations on the individual cells were also observed. Such large variation in elasticity within hESCs of a particular cell line when compared to the variation across a single cell suggested significant structural differences between these cells. In particular, it was found that cells corresponding to phenotype (i) had higher Young's moduli, (8.16 ± 6.18) kPa, and thus they were stiff. However, cells corresponding to phenotype (ii) had lower Young's moduli, (0.0485 ± 0.0239) kPa, indicating that these cells were softer. These results seemed to suggest that stiffer cells might be differentiated while the softer ones might be undifferentiated, indicating a potential strategy for separation of differentiated and undifferentiated embryonic stem cells.

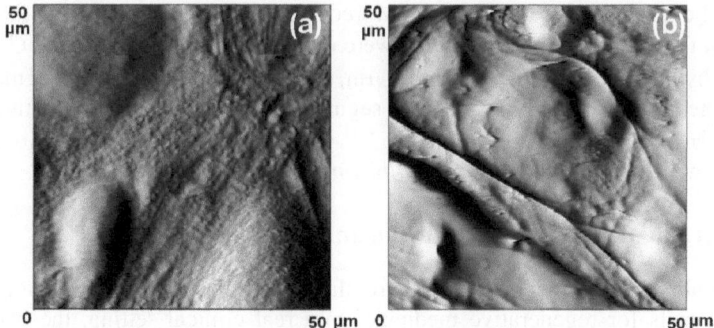

Fig. 6. AFM images of living hESCs in physiological conditions of probably (a) undifferentiated (0 - 4.9μm height range) and (b) differentiated (0 – 2.4μm height range) hESC cells

4.3 The Effects of Ionising Radiation on the Nanostructure of Pericardium Tissues

The pericardium is a protective membrane that surrounds the heart and it is composed of collagen and elastin fibres. Its functions are related to the mechanical and structural properties of the heart. The effects of ionising radiation, with respect to cardiac doses received in breast radiotherapy, on the nanostructure of fibrous pericardium tissue were investigated by AFM imaging. 40Gy is the dose limit placed on 100% of the heart volume in a clinical radiotherapy department, and it is considered to be the threshold at which pericarditis occurs. Therefore, a range of doses lower than 40Gy, as well as one of 80Gy was chosen to examine any structural changes of the pericardium under extreme conditions.

The control sample clearly showed (Fig. 7) that collagen fibrils bunched together and run parallel to one another. Fibrils had a distinct banding pattern, consisting of peaks and grooves, known as D-staggered configuration. Although, the AFM images do not seem to highlight a huge difference in the nanostructure of fibrils after exposure to ionising radiation, the section analysis carried out on the images showed that mean fibril width increases with radiation dose, with 80Gy having the largest mean fibril thickness by far, i.e., (111.2 ± 4.3) nm, compared with (88.1 ± 2.8) nm for the control tissue. Quite surprisingly, no threshold at 40Gy was found. The higher the radiation dose, the larger the amount of fibres that became affected and the more swollen the affected fibres became, ultimately increasing the mean collagen fibre width for that sample. In addition, the D-period of the fibrils in both the un-irradiated and irradiated tissues was found to be remarkably consistent over the range of doses, indicating that the banding period, and hence axial molecular arrangement, was unaffected by irradiation.

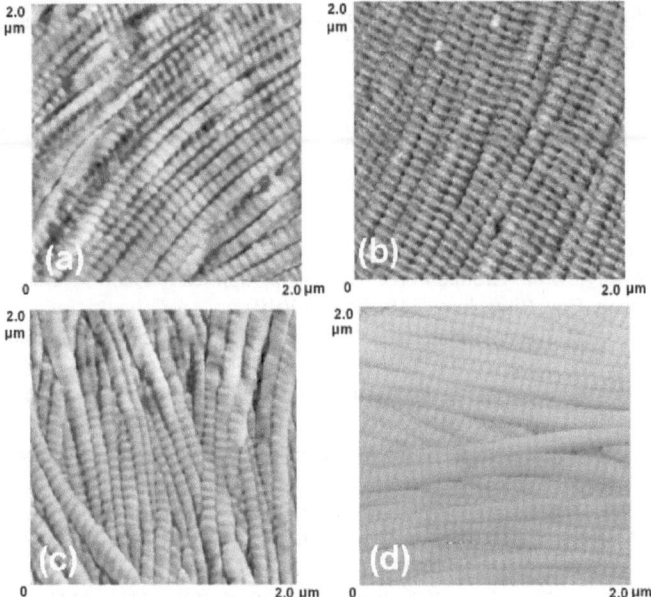

Fig. 7. AFM images of fibrous pericardial tissues exposed to (a) 0 Gy radiation (0-400nm); (b) 5 Gy radiation (0-450nm); (c) 40 Gy radiation (0-525nm); (d) 80 Gy radiation (0- 650nm) with – height ranges of AFM images given within brackets

These results showed that ionising radiation caused the collagen fibrils to increase in diameter, with the number of fibrils affected and the extent of swelling increasing with the radiation dose. Moreover, because one of the main mechanisms by which radiation interacts with proteins is through the formation of cross-links, it is very likely that the increased fibril width observed in the AFM images of pericardium tissue was due to the formation of cross-links within the collagen fibrils. The formation of cross-linking in fibrils could possibly account for the fact that the D-period value was retained for most of the irradiated samples, indicating that when cross-links were formed in fibrils, the molecules followed the regular arrangement of the natural collagen.

5 Conclusions

AFM has emerged as a very powerful technique in biomedicine for investigating the ultrastructural and nanomechanical properties of cells and biomolecules. This success is mainly due to the ability of AFM to provide high-resolution images without causing any damage during sample preparation and scanning. In addition, the multivariate array of image processing tools allow making direct measurements of the dimensions, profiles, surface roughness, volumes, and nanomechanics (e.g. elasticity and adhesion) of the samples. Recent developments in the technology to produce higher scan rates (e.g. collection of images in less than 1s) have pushed forward the design of control software [12] with new and very user-friendly user interfaces. These new developments of AFM will certainly boost even further its use in clinical settings as a tool for disease screening and diagnosis.

References

1. Binning, G., Quate, C.F., Gerber, C.: Atomic Force Microscope. Phys. Rev. Lett. 56, 930–933 (1986)
2. Ikai, A.: A Review on: Atomic Force Microscopy applied to Nano-Mechanics of the Cell. Adv. Biochem. Eng. Biotechnol. 119, 47–61 (2010)
3. Canetta, E., Adya, A.K.: Atomic Force Microscopy: Applications to Nanobiotechnology. J. Indian Chem. Soc. 82, 1147–1172 (2005)
4. Francis, L.W., Lewis, P.D., Wright, C.J., Conlan, R.S.: Atomic Force Microscopy comes of age. Biol. Cell, 133–143 (2010)
5. Fletcher, D.A., Mullins, R.D.: Cell Mechanics and Cytoskeleton. Nature 463, 485–492 (2010)
6. Haga, H., Sasaki, S., Kawabata, K., Ito, E., Ushiki, T., Sambongi, T.: Elasticity Mapping of Living Fibroblasts by AFM and Immunofluorescence Observation of the Cytoskeleton. Ultramicroscopy 82, 253–258 (2000)
7. Jung, Y.J., Park, Y.S., Yoon, K.J., Kong, Y.Y., Park, J.W., Nam, H.G.: Molecule-Level Imaging of Pax6 mRNA Distribution in Mouse Embryonic Neocortex by Molecular Interaction Force Microscopy. Nucl. Acids Res. 37, e10 (2009)
8. Schiffmann, K., Fryda, M., Goerigk, G., Lauer, R., Hinze, P.: Correction of STM Tip Convolution Effects in Particle Size and Distance Determination of Metal-C:H Films. Fresenius J. Anal. Chem. 358, 341–344 (1997)
9. Canetta, E., Walker, G.M., Adya, A.: Nanoscopic Morphological Changes in Yeast Cell Surfaces Caused by Oxidative Stress: An Atomic Force Microscopic Study. J. Microbiol. Biotechn. 19, 547–555 (2009)
10. Krysmann, M.J., Funari, S., Canetta, E., Hamley, I.W.: The Effect of PEG Crystallization on the Morphology of PEG-peptide Block Copolymers Containing Amyloid β Peptide Fragments. Macromol. Chem. Physic. 209, 883–889 (2008)
11. Canetta, E., Duperray, A., Leyrat, A., Verdier, C.: Measuring Cell Viscoelastic Properties Using a Force-Spectrometer: Influence of Protein Cytoplasm Interactions. Biorheology 42, 321–333 (2005)
12. Carberry, D.M., Picco, L., Dunton, P.G., Miles, M.J.: Mappifn Real-Time Images of High-Speed AFM Using Multitouch Control. Nanotechnology 20, 434018–434023 (2009)

IDEA: Intrinsic Dimension Estimation Algorithm

Alessandro Rozza, Gabriele Lombardi, Marco Rosa,
Elena Casiraghi, and Paola Campadelli

Dipartimento di Scienze dell'Informazione, Università degli Studi di Milano,
Via Comelico 39-41, 20135 Milano, Italy
rozza@dsi.unimi.it
http://security.dico.unimi.it/ fox721/

Abstract. The high dimensionality of some real life signals makes the usage of the most common signal processing and pattern recognition methods unfeasible. For this reason, in literature a great deal of research work has been devoted to the development of algorithms performing dimensionality reduction. To this aim, an useful help could be provided by the estimation of the intrinsic dimensionality of a given dataset, that is the minimum number of parameters needed to capture, and describe, all the information carried by the data. Although many techniques have been proposed, most of them fail in case of noisy data or when the intrinsic dimensionality is too high. In this paper we propose a local intrinsic dimension estimator exploiting the statistical properties of data neighborhoods. The algorithm evaluation on both synthetic and real datasets, and the comparison with state of the art algorithms, proves that the proposed technique is promising.

Keywords: Intrinsic dimension estimation, feature reduction, manifold learning.

1 Introduction

The high dimensionality of some real life signals, such as images, genome sequences, or EEG data, makes the usage of the most common signal processing and pattern recognition methods unfeasible. Nevertheless, many of these signals can be fully characterized by few degrees of freedom, represented by low dimensional feature vectors. In this case the feature vectors can be viewed as points constrained to lie on a low-dimensional manifold embedded in a higher dimensional space. The dimensionality of these manifolds is generally referred as *intrinsic dimensionality*. In more general terms, according to [8], a dataset is said to have intrinsic dimensionality equal to d $(1 \leq d \leq D)$ if its elements lie entirely within a d-dimensional subspace of $\Re^D$.

Intrinsic dimension estimation is important to discover structures, to perform dimensionality reduction and classification tasks. For this reason, in literature many techniques that estimate the intrinsic dimensionality have been proposed, which can be divided into two main groups: *global* approaches exploit the properties of the whole dataset to estimate the intrinsic dimensionality, whilst *local* approaches analyze the local behavior in the data neighborhoods.

G. Maino and G.L. Foresti (Eds.): ICIAP 2011, Part I, LNCS 6978, pp. 433–442, 2011.

The most cited example of *global* method is the Principal Component Analysis (PCA, [12]). Since it is the easiest way to reduce dimensionality, provided that a linear dependence exists, it is often used to pre-process the data. To this aim, PCA projects the points on the directions of their maximum variance, estimated by performing the eigen-decomposition on the data covariance matrix. Exploiting PCA, d can be estimated by counting the number of normalized eigenvalues that are higher than the threshold value ρ. Another interesting global approach is described in [3]; the authors exploit entropic graphs, such as the geodesic minimal spanning tree (GMST [2]) or the kNN-graph (kNNG), to estimate both the intrinsic dimensionality of a manifold, and the intrinsic entropy of the manifold random samples. Finally, we recall a well-known global intrinsic dimensionality estimator, which is the packing number method [13]. Unfortunately, global approaches often fail dealing with non-linearly embedded manifolds or noisy data, and they are usually computationally too expensive on high dimensional datasets.

Local intrinsic dimensionality estimators, such as LLE [17], TVF [16], and Hessian eigenmaps [5], are based on properties related to neighboring points in the given dataset. Most of these techniques consider hyperspheres with sufficiently small radius and centered on the dataset points, and they estimate some statistics by considering the neighboring points, included into the hypersphere; these statistics are expressed as functions of the intrinsic dimension of the manifold from which the points have been drawn. One of these techniques is the Correlation Dimension (CD) estimator [9]; it is based on the assumption that the volume of a d-dimensional set scales with its size r as r^d, which implies that also the number of samples covered by a hypersphere with radius r grows proportionally to r^d. Another well known technique is the Maximum Likelihood Estimator (MLE) [15] (and its regularization [10]), that applies the principle of maximum likelihood to the distances between close neighbors, and derive the estimator by a Poisson process approximation. To our knowledge, local estimators are more robust to noisy data than global ones, but most of them generally underestimate d when its value is sufficiently high. To address this problem few techniques have been proposed, among which the method described in [1] introduces a correction of the estimated intrinsic dimension based on the estimation of the errors obtained on synthetically produced datasets of known dimensionality.

In this work we present a local intrinsic dimension estimator, called IDEA, exploiting the statistical properties of manifold neighborhoods. Moreover, we compare our technique with state of the art algorithms. In Section 2 we describe our theoretical results, our base consistent estimator, and an asymptotic correction technique; in Section 3 experimental settings and results are reported; in Section 4 conclusions and future works are presented.

2   The Algorithm

In this section we present our theoretical results (see Section 2.1), our base algorithm (see Section 2.2), and an asymptotic estimation correction technique (see Section 2.3).

2.1   Theoretical Results

Suppose to have a manifold $\mathcal{M} \equiv \mathcal{B}(0,1)$, where $\mathcal{B}(0,1)$ is a d-dimensional centered open ball with unitary radius, and choose ψ as the identity map. To estimate the dimensionality d of $\mathcal{B}(0,1)$ we need to identify a measurable characteristic of the hypersphere depending only on d. To achieve this goal, we consider that a d dimensional vector randomly sampled from a d dimensional hypersphere according to the uniform probability density function (pdf), can be generated by drawing a point $\hat{z}$ from a standard normal distribution $\mathcal{N}(\cdot|0,1)$ and by scaling its norm (see Section 3.29 of [7]):

$$z = \frac{u^{\frac{1}{d}}}{\|\hat{z}\|}\hat{z}, \qquad \hat{z} \sim \mathcal{N}(\cdot|0,1) \tag{1}$$

where u is a random sample drawn from the uniform distribution $U(0,1)$.

Notice that, since u is uniformly distributed, the quantities $1 - u^{1/d}$ are distributed according to the beta pdf $\beta_{1,d}$ with expectation $\mathbb{E}_{u \sim U(0,1)}\left[1 - u^{1/d}\right] = \frac{1}{1+d}$. Therefore, the intrinsic dimensionality of the hypersphere is computed as:

$$\mathbb{E}_{z \sim \mathbf{B}_d}[1 - \|z\|] = \mathbb{E}_{z \sim \mathbf{B}_d}\left[1 - u^{\frac{1}{d}}\right] = \frac{1}{1+d} \Rightarrow d = \frac{\mathbb{E}_{z \sim \mathbf{B}_d}[\|z\|]}{1 - \mathbb{E}_{z \sim \mathbf{B}_d}[\|z\|]} \tag{2}$$

where $\mathbf{B}_d$ is the uniform pdf in the unit d-dimensional sphere. Notice that, embedding the hypersphere in a higher dimensional space $\Re^D$ by means of a map ψ that applies only a rotation, does not change this result.

More generally, we now consider points uniformly drawn from a d-dimensional manifold $\mathcal{M} \equiv \Re^d$ embedded in $\Re^D$ through a smooth map $\psi : \mathcal{M} \to \Re^D$; under these assumptions their norms may be not distributed as $u^{\frac{1}{d}}$. Nevertheless, being ψ a smooth map, close neighbors of $\mathcal{M}$ are mapped to close neighbors of $\Re^D$. Moreover, choosing a d-dimensional open ball $\mathcal{B}_d(c,\epsilon)$ with center $c \in \mathcal{M}$ and radius $\epsilon > 0$, as long as ψ preserves distances in $\mathcal{B}_d$, then for z uniformly drawn from $\mathcal{B}_d$, the distances $\frac{1}{\epsilon}\|\psi(c) - \psi(z)\| = \frac{1}{\epsilon}\|c - z\|$ are distributed as $u^{\frac{1}{d}}$, so that the result reported in Equation (2) is still valid and we obtain:

$$d = \frac{\mathbb{E}_{z \sim \mathbf{B}_d(c,\epsilon)}\left[\frac{1}{\epsilon}\|\psi(c) - \psi(z)\|\right]}{1 - \mathbb{E}_{z \sim \mathbf{B}_d(c,\epsilon)}\left[\frac{1}{\epsilon}\|\psi(c) - \psi(z)\|\right]} \tag{3}$$

where, $\mathbf{B}_d(c,\epsilon)$ is the uniform distribution in the ball $\mathcal{B}_d(c,\epsilon)$.

To further generalize our theoretical results, we consider a locally isometric smooth map $\psi : \mathcal{M} \to \Re^D$, and samples drawn from $\mathcal{M} \equiv \Re^d$ by means of a non-uniform smooth pdf $f : \mathcal{M} \to \Re^+$. Notice that, being ψ a local isometry, it induces a distance function $d_\psi(\cdot,\cdot)$ representing the metric on $\psi(\mathcal{M})$. Under these assumptions Equations (2,3) do not hold. However, without loss of generality, we consider $c = 0_d \in \Re^d$ and $\psi(c) = 0_D \in \Re^D$, and we show that any smooth pdf f is locally uniform where the probability is not zero. To this aim, assuming $f(0_d) > 0$ and $z \in \Re^d$, we denote with f_ϵ the pdf obtained by setting $f_\epsilon(z) = 0$ when $\|z\| > 1$, and $f_\epsilon(z) \propto f(\epsilon z)$ when $\|z\| \leq 1$. More precisely, denoting with $\chi_{\mathcal{B}_d(0,1)}$ the indicator function on the ball $\mathcal{B}_d(0,1)$, we obtain:

$$f_\epsilon(z) = \frac{f(\epsilon z)\chi_{\mathcal{B}_d(0,1)}(z)}{\int_{t \in \mathcal{B}_d(0,1)} f(\epsilon t)dt} \tag{4}$$

Theorem 1. *Given $\{\epsilon_i\} \to 0^+$, Equation (4) describes a sequence of **pdf** having the unit d-dimensional ball as support; such sequence converges uniformly to the uniform distribution $\mathbf{B}_d$ in the ball $\mathcal{B}_d(\mathbf{0}, 1)$.*

Proof. Evaluating the limit for $\epsilon \to 0^+$ of the distance between f_ϵ and $\mathbf{B}_d$ in the supremum norm we get:

$$\lim_{\to 0^+} \left\| f(z) - \mathbf{B}_d(z) \right\|_{\sup} = \lim_{\to 0^+} \left\| \frac{f(z) \; \mathcal{B}_d(0\;1)}{\int_{\mathcal{B}_d(0\;1)} f(t)dt} - \frac{\mathcal{B}_d(0\;1)}{\int_{\mathcal{B}_d(0\;1)} dt} \right\|_{\sup}$$

$$\{just\ notation\} = \lim_{\to 0^+} \left\| \frac{f(z)}{\int_{\mathcal{B}_d(0\;1)} f(t)dt} - \frac{1}{\int_{\mathcal{B}_d(0\;1)} dt} \right\|_{\sup \mathcal{B}_d(0\;1)}$$

$$\left\{ setting\ V = \int_{\mathcal{B}_d(0\;1)} dt \right\} = \lim_{\to 0^+} \left\| \frac{Vf(z) - \int_{\mathcal{B}_d(0\;1)} f(t)dt}{V \int_{\mathcal{B}_d(0\;1)} f(t)dt} \right\|_{\sup \mathcal{B}_d(0\;1)}$$

$$\left\{ 0 < \lim_{\to 0^+} V \int_{\mathcal{B}_d(0\;1)} f(t)dt < \right\} = \lim_{\to 0^+} \left\| Vf(z) - \int_{\mathcal{B}_d(0\;1)} f(t)dt \right\|_{\sup \mathcal{B}_d(0\;1)}$$

Defining:

$$min(\epsilon) = \min_{\mathcal{B}_d(\mathbf{0},1)} f(\epsilon z) \qquad max(\epsilon) = \max_{\mathcal{B}_d(\mathbf{0},1)} f(\epsilon z)$$

and noting that $min(\epsilon) > 0$ definitely since $f(\mathbf{0}_d) > 0$, we have:

$$V \cdot min(\epsilon) \leq \qquad Vf(\epsilon z) \qquad \leq V \cdot max(\epsilon)$$
$$V \cdot min(\epsilon) \leq \int_{\mathcal{B}_d(\mathbf{0},1)} f(\epsilon t)dt \leq V \cdot max(\epsilon)$$

thus their difference is bounded by $V(max(\epsilon) - min(\epsilon)) \xrightarrow[\epsilon \to 0^+]{} 0^+$.   □

Theorem 1 proves that the convergence of f_ϵ to $\mathbf{B}_d$ is uniform, so that in the limit ($\epsilon \to 0^+$) Equation (2) holds both for d-dimensional nonlinear manifolds embedded in $\Re^D$, and for points drawn by means of a non-uniform density function f. More precisely, for the smoothness and for the local isometry of ψ:

$$\mathbb{E}_{z \sim f} \left[d_\psi(\psi(z), \psi(\mathbf{0}_d)) \right] = \mathbb{E}_{z \sim f} \left[\|z\| \right] \xrightarrow[\epsilon \to 0^+]{} \mathbb{E}_{z \sim \mathbf{B}_d} \left[\|z\| \right] = m \qquad (5)$$

2.2   The Base Algorithm

Consider a d-dimensional manifold $\mathcal{M} \equiv \Re^d$ non-linearly embedded in $\Re^D$ through a smooth locally isometric map $\psi : \mathcal{M} \to \Re^D$. Given a sample set $\mathbf{X}_N = \{x_i\}_{i=1}^N = \{\psi(z_i)\}_{i=1}^N \subset \Re^D$, where $z_i \in \Re^d$ are independent identically distributed points drawn from $\mathcal{M}$ according to a smooth **pdf** $f : \mathcal{M} \to \Re^+$, our aim is to exploit our theoretical results to estimate the intrinsic dimensionality of $\mathcal{M}$ by means of the points in the set $\mathbf{X}_N$.

More precisely, the expectation of distances $\frac{1}{\epsilon} d_\psi(\psi(c), x)$ for infinitesimal balls $\mathcal{B}_D(\psi(c), \epsilon)$ with $c \in \mathcal{M}$ must be estimated. To this aim, for each point

$x_i \in X_N$ we find the set of $k+1$ $(1 \le k \le N-1)$ nearest neighbors $\hat{X}_{k+1}^N = \hat{X}_{k+1}^N(x_i) = \{x_j\}_{j=1}^{k+1} \subset X_N$. Call $\hat{x} = \hat{x}_{k+1}^N(x_i) \in \hat{X}_{k+1}^N$ the most distant point from x_i, and denote $X_k^N = X_k^N(x_i) = \hat{X}_{k+1}^N \backslash \{\hat{x}\}$. Notice that, when x_i is fixed, almost surely (a.s.) we have $\|x - x_i\| < \|\hat{x} - x_i\|$ $\forall x \in X_k^N$; therefore, we can consider points in X_k^N as drawn from the open ball $\mathcal{B}_D(x_i, \|\hat{x} - x_i\|)$. Exploiting this fact, in order to estimate the intrinsic dimension d of $\mathcal{M}$, we estimate the expectation of distances as follows:

$$m \simeq \frac{1}{k} \sum_{x \in X_k^N} \frac{\|x_i - x\|}{\|\hat{x} - x\|}$$

Note that m depends only upon the intrinsic dimensionality d of $\mathcal{M}$ and does not depend on the chosen center x_i.

Corollary 1. *Given two sequences $\{k_j\}$ and $\{N_j\}$ such that for $j \to +\infty$:*

$$k_j \to +\infty, \qquad N_j \to +\infty, \qquad \frac{k_j}{N_j} \to 0 \qquad (6)$$

We have the limit:

$$\lim_{j \to +\infty} \frac{1}{k_j} \sum_{x \in X_{k_j}^{N_j}} \frac{\|x_i - x\|}{\|\hat{x} - x\|} = m \qquad a.s. \qquad (7)$$

Proof. Considering the sequences $\{k_j\}$ and $\{N_j\}$, the conditions reported in Equation (6) ensure that $\epsilon = \|\hat{x} - x_i\| \to 0^+$ when $j \to +\infty$[1]. Theorem 4 in [4] ensures that geodetic distances in the infinitesimal ball converge to Euclidean distances with probability 1; furthermore, the sample mean is an unbiased estimator for the expectation (law of large numbers); moreover, Theorem 1 guarantees that the underlying pdf converges to the uniform one. We have:

$$\lim_{j \to +\infty} \frac{1}{k_j} \sum_{x \in X_{k_j}^{N_j}} \frac{\|x_i - x\|}{\|\hat{x} - x\|} = \lim_{j \to +\infty} \frac{1}{k_j \epsilon} \sum_{x \in X_{k_j}^{N_j}} d_\psi(x_i, x) + o(\epsilon) =$$

$$\lim_{\epsilon \to 0^+} \mathbb{E}_{z \sim f_\epsilon} \left[\frac{1}{\epsilon} d_\psi(\psi(z), \psi(0_d)) \right] = \lim_{\epsilon \to 0^+} \mathbb{E}_{z \sim \mathbf{B}_d(0, \epsilon)} \left[\frac{1}{\epsilon} d_\psi(\psi(z), \psi(0_d)) \right] =$$

$$\lim_{\epsilon \to 0^+} \frac{1}{\epsilon} \epsilon \, \mathbb{E}_{z \sim \mathbf{B}_d(0, 1)} [d_\psi(\psi(z), \psi(0_d))] = \mathbb{E}_{z \sim \mathbf{B}_d(0, 1)} [\|z\|] = m \qquad (8)$$

$\square$

By employing Equation (2) and Corollary 1, we get a consistent estimator $\hat{d}$ for the intrinsic dimensionality d of $\mathcal{M}$ as follows:

$$m \simeq \hat{m} = \frac{1}{Nk} \sum_{i=1}^N \sum_{x \in X_k^N} \frac{\|x_i - x\|}{\|\hat{x} - x\|}$$

$$d = \frac{m}{1-m} \simeq \frac{\hat{m}}{1-\hat{m}} = \hat{d} \qquad (9)$$

---

[1] See proof of Theorem 4 in [4] where k must be substituted by $o(n)$.

2.3   Asymptotic Correction

Although the algorithm described in Section 2.2 proposes a consistent estimator of the intrinsic dimension, when the dimensionality is too high, the number of sample points becomes insufficient to compute an acceptable estimation. This is due to the fact that, as shown in [6], the number of sample points required to perform dimensionality estimation with acceptable results, grows exponentially with the value of the intrinsic dimensionality ("curse of dimensionality").

In literature, some methods have been proposed to reduce this effect. For instance, in [1] the authors propose a correction of the estimated intrinsic dimension based on the estimation of the errors obtained on synthetically produced datasets of known dimensionality (hypercubes). Another interesting approach has been proposed in [3], where the authors propose a non parametric least square strategy based on re-sampling from the point population $\boldsymbol{X}_N$.

Similarly, in our work we propose a method that allows to study the asymptotic behavior described by the available data. To this aim, we adopt a Monte Carlo approach performing R runs of the algorithm reported in Section 2.2. We extract from the given dataset $\boldsymbol{X}_N$ random subsets $\boldsymbol{\mathcal{R}}_{r=1}^R$ with different cardinalities $\boldsymbol{R}_{r=1}^R$. The cardinalities $\boldsymbol{R}_r$ are randomly generated by means of the binomial distribution $Binom(N, p)$, where the value of p spans a fixed range[2]. The intrinsic dimensionality, estimated during each run, becomes a sample from a "trend curve"; moreover, for each subsample we choose the kNN parameter $k_r = \lceil k\sqrt{p} \rceil$, trying to emulate a sequence $\{k_r\}$ such that $k_r \to +\infty$, $\boldsymbol{R}_r \to +\infty$, and $\frac{k_r}{\boldsymbol{R}_r} \to 0$, thus fulfilling the conditions reported in Equation (6).

We noticed that, when the base algorithm proposed in Section 2.2 underestimates the intrinsic dimensionality, its application to point subsets $\boldsymbol{\mathcal{R}}_{r=1}^R$ with increasing cardinality produces increasing estimations of the intrinsic dimension $\hat{d} = \hat{d}(\boldsymbol{R}_r)$. As demonstrated in Section 2.2, these estimates converge to the real intrinsic dimensionality for $j \to +\infty$ (see conditions reported in Equation (6)). Our assumption, based on this empirical observation, is that the function $\hat{d}(N)$ has a horizontal asymptote. Therefore, we fit the pairs $\left(\log(\boldsymbol{R}_r), \hat{d}(\boldsymbol{R}_r) \right)$ by means of the parametric function[3] g described below:

$$\hat{d}(\boldsymbol{R}_r) \simeq g(\boldsymbol{R}_r) = a_0 - \frac{a_1}{\log_2(\frac{\boldsymbol{R}_r}{a_2} + a_3)} \tag{10}$$

where $\{a_i\}_{i=0}^3$ are fitting parameters controlling translation and scaling on both axes; their values are computed by a non-linear least squares fitting algorithm. Notice that, since $\lim_{\boldsymbol{R}_r \to +\infty} g(\boldsymbol{R}_r) = a_0$ then the asymptote of Equation (10) is $\hat{d} = a_0$. Moreover, the derivate $g' = D[g(\boldsymbol{R}_r)]$ shows that the parameter a_1 controls the increasing/decreasing behavior of the function g. For these reasons, when the estimated parameter $a_1 > 0$ (increasing function), we use the

---

[2] In our tests $p \in \{0.1, \quad , 0.9\}$.
[3] The choice of using 2 as the log base does not affect the results, being the change of base just a change of scale in the y axis.

Table 1. Brief description of the synthetic and real datasets, where d is the intrinsic dimension, and D is the embedding space dimension. In the second column of the synthetic data, the number in the subscript refers to the dataset name used by the generator proposed in [11].

Dataset	Name	d	D	Description
Syntethic	$\mathcal{M}_1$	10	11	Uniformly sampled sphere linearly embedded.
	$\mathcal{M}_2$	3	5	Affine space.
	$\mathcal{M}_3$	4	6	Concentrated figure, easy to confuse with a $3d$ one.
	$\mathcal{M}_4$	4	8	Non-linear manifold.
	$\mathcal{M}_5$	2	3	2-d Helix
	$\mathcal{M}_6$	6	36	Non-linear manifold.
	$\mathcal{M}_7$	2	3	Swiss-Roll.
	$\mathcal{M}_8$	12	72	Non-linear manifold.
	$\mathcal{M}_9$	20	20	Affine space.
	$\mathcal{M}_{10a}$	10	11	Uniformly sampled hypercube.
	$\mathcal{M}_{10b}$	17	18	Uniformly sampled hypercube.
	$\mathcal{M}_{10c}$	24	25	Uniformly sampled hypercube.
	$\mathcal{M}_{11}$	2	3	Möebius band 10-times twisted.
	$\mathcal{M}_{12}$	20	20	Isotropic multivariate Gaussian.
	$\mathcal{M}_{13}$	1	13	Curve.
Real	$\mathcal{M}_{Faces}$	3	4096	`ISOMAP` face dataset.
	$\mathcal{M}_{MNIST1}$	$8-11$	784	`MNIST` database (digit 1).
	$\mathcal{M}_{MNIST3}$	$12-14$	784	`MNIST` database (digit 3).

parameter a_0 as the final estimate for d; otherwise, we use the estimation obtained by the base algorithm applied to the whole dataset.

To obtain a stable estimation of the intrinsic dimension we execute the asymptotic correction algorithm 20 times and we average the obtained results.

3   Algorithm Evaluation

In this section we describe the datasets employed in our experiments (see Section 3.1), we summarize the adopted experimental settings (see Section 3.2), and we report the achieved results (see Section 3.3).

3.1   Dataset Description

To evaluate our algorithm, we have performed experiments on both synthetic and real datasets (see Table 1). To generate the synthetic datasets we have employed the tool proposed in [11]. The real datasets are the `ISOMAP` face database [18] and the `MNIST` database [14]. The `ISOMAP` face dataset consists in 698 gray-level images of size 64×64 depicting the face of a sculpture. This dataset has three degrees of freedom: two for the pose and one for the lighting direction. The `MNIST` database consists in 70000 gray-level images of size 28×28 of handwritten digits; in our tests we have used the 6742 training points representing

Table 2. Parameter settings for the different estimators: k represents the number of neighbors,   the edge weighting factor for kNN, M the number of Least Square (LS) runs, and N the number of resampling trials per LS iterations

Method	Synthetic	Real
PCA	$Threshold = 0.025$	$Threshold = 0.0025$
CD	$None$	$None$
MLE	$k_1 = 6 \ k_2 = 20$	$k_1 = 3 \ k_2 = 8$
kNNG$_1$	$k_1 = 6, k_2 = 20, \ = 1, M = 1, N = 10$	$k_1 = 3, k_2 = 8, \ = 1, M = 1, N = 10$
kNNG$_2$	$k_1 = 6, k_2 = 20, \ = 1, M = 10, N = 1$	$k_1 = 3, k_2 = 8, \ = 1, M = 10, N = 1$
IDEA	$k = 20$	$k = 8$

the digit 1 and the 6131 training points representing the digit 3. The intrinsic dimension of the datasets extracted from the MNIST database is not actually known, but some works have proposed similar estimations [11,4] for the different digits. More precisely, these works report an estimation in the range $d \in \{8..11\}$ for digit 1, and between 12 and 14 for the digit 3.

3.2 Experimental Setting

We have compared our method with well-known global (PCA, kNNG) and local (CD, MLE) intrinsic dimension estimators. For kNNG and MLE[4], we have used the authors' Matlab implementation, and we have employed the version in the toolbox of dimensionality reduction for the other algorithms[5].

We have used the synthetic dataset generator [11] to create 20 instances of each dataset reported in Table 1, each of which was composed by 2000 randomly sampled points. To execute multiple tests on $\mathcal{M}_{MNIST1}$ and $\mathcal{M}_{MNIST3}$, we have extracted 5 random subsets per dataset containing 2000 points each. To obtain a stable estimation, for each technique we have averaged the results achieved on the different subsets. Table 2 summarizes the employed configuration parameters. To relax the dependency of kNNG from the selection of k, we performed multiple runs with $k_1 \leq k \leq k_2$ and we averaged the achieved results.

3.3 Experimental Results

In this subsection results achieved on both real and synthetic datasets are reported. On synthetic datasets (see Table 3) our method has computed good approximations both on low and high intrinsic dimensional datasets, achieving results always comparable with those that better approximate the real intrinsic dimension. We further note that IDEA has generally outperformed global methods on strongly non-linearly embedded manifolds obtaining good estimations on

---

[4] http://www.eecs.umich.edu/ hero/IntrinsicDim/,
   http://www.stat.lsa.umich.edu/ elevina/mledim.m
[5] http://cseweb.ucsd.edu/ lvdmaaten/dr/download.php

Table 3. Results achieved on synthetic datasets by global and local approaches. In bold the best approximations achieved both by global and by local techniques.

Dataset	Int. Dim	PCA	kNNG$_1$	kNNG$_2$	CD	MLE	IDEA
$\mathcal{M}_{13}$	1	4	**1.00**	1.01	1.07	**1.00**	1.12
$\mathcal{M}_{5}$	2	3	1.96	**2.00**	**1.98**	1.96	2.07
$\mathcal{M}_{7}$	2	3	1.93	**1.98**	1.94	**1.97**	**1.97**
$\mathcal{M}_{11}$	2	3	1.96	**2.01**	2.23	2.30	**1.98**
$\mathcal{M}_{2}$	3	**3**	2.85	2.93	2.88	2.87	**3.07**
$\mathcal{M}_{3}$	4	**4**	3.80	4.22	3.16	3.82	**4.01**
$\mathcal{M}_{4}$	4	8	4.08	**4.06**	3.85	**3.98**	4.07
$\mathcal{M}_{6}$	6	12	**6.53**	13.99	**5.91**	6.45	6.79
$\mathcal{M}_{1}$	10	11	9.07	**9.39**	9.09	9.06	**10.35**
$\mathcal{M}_{10a}$	10	**10**	8.35	9.00	8.04	8.22	**10.07**
$\mathcal{M}_{8}$	12	24	**14.19**	8.29	10.91	**13.69**	14.45
$\mathcal{M}_{10b}$	17	**17**	12.85	15.58	12.09	12.77	**16.78**
$\mathcal{M}_{9}$	20	**20**	14.87	17.07	13.60	14.54	**16.81**
$\mathcal{M}_{12}$	20	**20**	16.50	14.58	11.24	15.67	**21.08**
$\mathcal{M}_{10c}$	24	**24**	17.26	23.68	15.48	16.80	**23.94**

Table 4. Results achieved on real datasets by global and local approaches. In bold the best approximations achieved both by global and by local techniques.

Dataset	Int. Dim	PCA	kNNG$_1$	kNNG$_2$	CD	MLE	IDEA
$\mathcal{M}_{Faces}$	3	21	**3.60**	4.32	**3.37**	4.05	3.73
$\mathcal{M}_{MNIST1}$	8-11	11.80	10.37	**9.58**	6.96	**10.29**	11.06
$\mathcal{M}_{MNIST3}$	12-14	9.80	37.70	**16.16**	10.16	15.67	**14.98**

high intrinsic dimensional datasets where local techniques have usually under-estimated the correct value. These behaviors have confirmed the quality of the asymptotic correction algorithm.

In Table 4 the results achieved on real datasets have been summarized; notice that, using real data that are often noisy, IDEA has obtained either the best approximation of the intrinsic dimension or stable results comparable with those achieved by the best performing technique.

4   Conclusions and Future Works

In this work we have presented a consistent local intrinsic dimension estimator that exploits the statistical properties of manifold neighborhoods. Moreover, we have proposed an asymptotic correction of our algorithm to reduce the under-estimate behavior that affects most of the local intrinsic dimension estimation methods. The promising results achieved, and the comparison with state of the art techniques, have confirmed the quality of the proposed technique. Moreover, in our tests, only IDEA has obtained good estimations both on low and high intrinsic dimensional datasets, and both on linear and non-linear embeddings.

In future works we want to focus on the asymptotic correction algorithm to improve the theoretical background about the fitting function.

References

1. Camastra, F., Vinciarelli, A.: Estimating the intrinsic dimension of data with a fractal-based method. IEEE Trans. PAMI 24, 1404–1407 (2002)
2. Costa, J.A., Hero, A.O.: Geodesic entropic graphs for dimension and entropy estimation in manifold learning. IEEE Trans. on SP 52(8), 2210–2221 (2004)
3. Costa, J.A., Hero, A.O.: Learning intrinsic dimension and entropy of high-dimensional shape spaces. In: EUSIPCO (2004)
4. Costa, J.A., Hero, A.O.: Learning intrinsic dimension and entropy of shapes. In: Krim, H., Yezzi, T. (eds.) Statistics and Analysis of Shapes. Birkhäuser, Basel (2005)
5. Donoho, D.L., Grimes, C.: Hessian Eigenmaps: New Locally Linear Embedding Techniques For High-Dimensional Data. Technical report (July 2003)
6. Eckmann, J.P., Ruelle, D.: Fundamental limitations for estimating dimensions and lyapunov exponents in dynamical systems. Physica D 56(2-3), 185–187 (1992)
7. Fishman, G.S.: Monte Carlo: Concepts, Algorithms, and Applications. Springer Series in Operations Research. Springer, New York (1996)
8. Fukunaga, K.: Intrinsic Dimensionality Extraction. Classification. In: Pattern Recognition and Reduction of Dimensionality (1982)
9. Grassberger, P., Procaccia, I.: Measuring the strangeness of strange attractors. Physica D: Nonlinear Phenomena 9, 189 (1983)
10. Gupta, M.D., Huang, T.: Regularized maximum likelihood for intrinsic dimension estimation. In: UAI (2010)
11. Hein, M.: Intrinsic dimensionality estimation of submanifolds in euclidean space. In: ICML, pp. 289–296 (2005)
12. Jollife, I.T.: Principal Component Analysis. Springer Series in Statistics. Springer, New York (1986)
13. Kégl, B.: Intrinsic dimension estimation using packing numbers. In: Proc. of NIPS, pp. 681–688 (2002)
14. LeCun, Y., Cortes, C.: Gradient-Based Learning Applied to Document Recognition. Proceedings of the IEEE 86(11), 2278–2324 (1998)
15. Levina, E., Bickel, P.J.: Maximum likelihood estimation of intrinsic dimension. In: NIPS, vol. 1, pp. 777–784 (2005)
16. Mordohai, P., Medioni, G.: Dimensionality estimation, manifold learning and function approximation using tensor voting. J. Mach. Learn. Res. 11, 411–450 (2010)
17. Roweis, S.T., Saul, L.K.: Nonlinear Dimensionality Reduction by Locally Linear Embedding. Science 290, 2323–2326 (2000)
18. Tenenbaum, J.B., Silva, V., Langford, J.C.: A Global Geometric Framework for Nonlinear Dimensionality Reduction. Science 290, 2319–2323 (2000)

Optimal Decision Trees Generation from *OR*-Decision Tables

Costantino Grana, Manuela Montangero,
Daniele Borghesani, and Rita Cucchiara

Dipartimento di Ingegneria dell'Informazione
Università degli Studi di Modena e Reggio Emilia
costantino.grana@unimore.it, manuela.montangero@unimore.it,
daniele.borghesani@unimore.it, rita.cucchiara@unimore.it

Abstract. In this paper we present a novel dynamic programming algorithm to synthesize an optimal decision tree from *OR*-decision tables, an extension of standard decision tables, which allow to choose between several alternative actions in the same rule. Experiments are reported, showing the computational time improvements over state of the art implementations of connected components labeling, using this modelling technique.

Keywords: Decision tables, Decision trees, Connected components labeling, Image processing.

1 Introduction

A decision table is a tabular form that presents a set of conditions which must be tested and a list of corresponding actions to be performed. Each combination of condition entries (*condition outcomes*) is paired to an *action entry*. In the action entries, a column is marked, for example with a "1", to specify whether the corresponding action is to be performed. The interpretation of a decision table is straightforward: all actions marked with 1s in the action entries vector should be performed if the corresponding outcome is obtained when testing the conditions [8]. In general, decision tables are used to describe the behavior of a system whose state can be represented as a vector, i.e. the outcome of testing certain conditions. Given a particular state, the system evolves by performing a set of actions that depend on the given state of the system. The state is described by a particular rule, the action by the corresponding row of the table.

Even if decision tables are easy to read and understand, their straightforward use might not be efficient. In fact, decision tables require all conditions to be tested in order to select the corresponding actions to be executed, and testing the conditions of the decision table has a cost which is related to the number of conditions and to the computational cost of each test. There are a number of cases in which not all conditions must be tested in order to decide which action has to be performed, because the specific values assumed by a subset of conditions might be enough to pick a decision. This observation suggests that

G. Maino and G.L. Foresti (Eds.): ICIAP 2011, Part I, LNCS 6978, pp. 443–452, 2011.

the order with which the conditions are verified impacts on the number of tests required, thus on the total cost of testing. Moreover, after selecting the first condition to be tested, the second condition to test might vary according to the outcome of the first test. What is thus obtained is a *decision tree*, a tree that specifies the order in which to test the conditions, given the outcome of previous tests. Changing the order in which conditions are tested might lead to more or less tests to be performed, and hence to a higher or lower execution cost. The *optimal decision tree* is the one that requires, on average, the minimum cost when deciding which actions to execute.

In a large class of image processing algorithms the output value for each image pixel is obtained from the value of the pixel itself and some of its neighbors. We can refer to these as *local* or *neighborhood* algorithms. In particular, for binary images, we can model local algorithms by means of a limited entry decision table, in which the pixels values are conditions to be tested and the output is chosen by the action corresponding to the conditions outcome.

In this paper we will address the recently introduced *OR*-decision tables [2], providing a dynamic programming algorithm to derive optimal decision trees for such decision tables. The algorithm is applied to the currently fastest connected components labeling algorithm, showing how the proposed optimal synthesis improves, in terms of efficiency, the previous approaches.

2   Decision Tables and Decision Trees

Given a set of conditions and the corresponding actions to be performed according to the outcome of these testing conditions, we can arrange them in a tabular form $\mathcal{T}$ called a *decision table*: each row corresponds to a particular outcome for the conditions and is called *rule*, each column corresponds to a particular action to be performed. A given entry $\mathcal{T}[i, j]$ of the table is set to one if the action corresponding to column j must be performed given the outcome of the testing condition as in row i; the entry is set to zero otherwise. Different rules might have different probability to occur and testing conditions might be more o less expensive to test. The order in which conditions are tested might be influent or not.

There are different kind of decision tables, according to the system they describe: a table is said to be a *limited entry* decision table [6] if the outcome of the testing conditions is binary; it is called *extended entry* otherwise. A table in which there is a row for every possible rule is said to be an *expanded* decision table, and *compressed* otherwise. We will call a decision table an *AND*-decision table if **all** the actions in a row must be executed when the corresponding rule occurs, instead we will call it an *OR*-decision table if **any** of the actions in a row might be executed when the corresponding rule occurs. In particular, *OR*-decision tables were firstly introduced by Grana *et al.* [2].

There are two main approaches to derive decision trees from decision tables: a top-down approach, in which a rule of the table is selected to be the root of the tree and the subtrees are build by recursively solving the problem on the two portions of the table that are obtained by removing the rule that has been placed in the root; a bottom-up approach, in which rules are grouped together if the associated actions can be decided by testing a common subset of conditions. We adopt the second approach, mainly because it has been shown to be able to handle tables having a larger number of conditions (hence, rules).

Schumacher *et al.* [8] proposed a bottom-up Dynamic Programming technique which guarantees to find the optimal decision tree given an expanded limited entry decision table, in which each row contains only one non-zero value. This strategy can be extended also to limited entry AND-decision tables. Lew [5] gives a Dynamic Programming approach for the case of extended entry and/or compressed AND-decision tables. In this paper, we extend Schumacher's approach to OR-decision tables.

2.1   Preliminaries and Notation

An OR-decision table is described by the following sets and parameters:

- $C = \{c_1, \ldots, c_L\}$ boolean conditions; the cost of testing condition c_i is given by $w_i > 0$;
- $R = \{r_1, \ldots, r_{2^L}\}$ rules; each rule is a boolean vector of L elements, element i corresponding to the outcome of condition c_i; the probability that rule r occurs is given by $p_r \geq 0$;
- $A = \{a_1, \ldots, a_M\}$ actions; rule r_i is associated to a non empty subset $A^{(i)} \subseteq A$ to be executed when the outcome of conditions c_j identifies rule r_i.

We wish to determine an efficient way to test conditions $c_1, \ldots, c_L$ in order to decide *as soon as possible* which action should be executed according to the values of conditions c_j. In particular, we wish to determine in which order conditions c_i have to be checked, so that the minimum number of tests are performed. Such information are given in the form of a tree, called decision tree, and here we will give an algorithm to find the optimal one, intuitively the one that stores these information in the most succinte way.

In the following we will call set $K \subseteq R$ a *k-cube* if it is a subset of rules in which the value of $L - k$ conditions is fixed. It will be represented as a L-vector of k dashes $(-)$ and $L - k$ values 0's and 1's. The set of positions containing dashes will be denoted as D_K. We associate to cube K a set of rules, denoted by A_K, that contains the intersection of the sets of actions associated to the rules in K (might be an empty set). The occurrence probability of the k-cube K is the probability P_K of any rule in K to occur, *i.e.* $P_K = \sum_{r \in K} p_r$. Finally, we will denote with $\mathcal{K}_k$ the set of the k-cubes, for $k = 0, \ldots, L$.

A *Decision Tree* for K is a binary tree T with the following properties:

1. Each leaf ℓ corresponds to a k-cube of rules, denoted by K_ℓ, that is a subset of K. Each leaf ℓ is associated to the set of actions A_{K_ℓ} associated to cube K_ℓ. Each internal node is labeled with a testing condition $c_i \in C$ such that there is a dash at position i in the vector representation of K. Left (resp. right) outgoing edges are labeled with 0 (resp. 1).

2. Two distinct nodes on the same root-leaf path can not be labeled with the same testing condition. Root-leaf paths univocally identify, by means of node and edges labels, the set of rules associated to leaves: conditions labeling nodes on the path must be set to the value of the label on the corresponding outgoing edges, the remaining conditions are set to a dash.

The cost of making a specific decision, is the cost of testing the conditions on a root-leaf path, in order to execute one of the actions associated to the leaf. On average, the cost of making decisions is given by the sum of the root-leaf paths weighted by the probabilty that the rules associated to leaves occur; *i.e.* the average cost of a decision tree is a measure of the cost of testing the conditions that we need to check in order to decide which actions to take when rules occur, weighted by the probability that rules occur.

We formally define the notions of cost and gain of decision trees on cubes, that will be used in the following.

Definition 1 (Cost and Gain of a Decision Tree). *Given a k-cube K, with dashes in positions D_K, and a decision tree T for K, cost and gain of T are defined in the following way:*

$$cost(T) = \sum_{\ell \in \mathcal{L}} \left(P_\ell \sum_{c_i \in path(\ell)} w_i \right), \tag{1}$$

where $\mathcal{L}$ is the set of leaves of the tree and the w_is are the costs of the testing conditions on the root-leaf path leading to $\ell \in \mathcal{L}$, denoted by $path(\ell)$;

$$gain(T) = P_K \sum_{i \in D_K} w_i - cost(T), \tag{2}$$

where $P_K \sum_{i \in D_K} w_i$ is the maximum possible cost for a decision tree for K.

An Optimal Decision Tree for k-cube K is a decision tree for the cube with minimum cost (might not be unique) or, equivalently, with maximum gain.

Observe that, when the probabilities of the rules in the leaves of the tree sum up to one, the cost defined in equation (1) is exaclty the quantity that we wish to minimize in order to find a decision tree of minimum average cost for the L-cube that describes a given *OR*-Decision table.

A simple algorithm to derive a decision tree for a k-cube K works recursively in the following way: select a condition index i that is set to a dash and make the root of the tree a node labeled with condition c_i. Partition the cube K into two cubes $K_{i,0}$ and $K_{i,1}$ such that condition c_i is set to zero in $K_{i,0}$ and to one in $K_{i,1}$. Recursively build decision trees for the two cubes of the partition, then make them the left and right children of the root, respectively. Recursion stops when all the rules in the cube have at least one associated action in common.

The cost of the outcoming tree is strongly affected by the order used to select the index that determines the cube partition. In the next section we give a dynamic programming algorithm that determines a selection order that produces a tree with maximum gain.

2.2   Dynamic Programming Algorithm

An optimal decision tree can be computed using a generalization of the Dynamic Programming strategy introduced by [8], with a bottom-up approach: staring from 0-cubes and for increasing dimension of cubes, the algorithm computes the gain of all possible trees for all cubes and keeps track of only the ones having maximum gain. The trick that allows to choose the best action to execute is to keep track of the intersection of actions sets of all the rules in the k-cube. It is possible to formally prove both the correctness and optimality of the algorithm.

```
 1: for K ∈ R do                           ▷ Inizialization of 0-cubes in R ∈ 𝒦₀
 2:     Gain*_K ← 0
 3:     A_K ← set of actions associated to rule in K
 4: end for
 5: for n ∈ [1, L] do                      ▷ for all possible cube dimentions > 0
 6:     for K ∈ 𝒦_n do                     ▷ for all possible cubes with n dashes
                                     ▷ compute current cube probability and set of actions
 7:         P_K ← P_{K_{j,0}} + P_{K_{j,1}}             ▷ where j is any index in D_K
 8:         A_K ← A_{K_{j,0}} ∩ A_{K_{j,1}}
                            ▷ compute gains obtained by ignoring one condition at the time
 9:         for i ∈ D_K do                  ▷ for all positions set to a dash
10:             if A_K ≠ ∅ then
11:                 Gain_K(i) ← w_i P_K + Gain*_{K_{i,0}} + Gain*_{K_{i,1}}
12:             else
13:                 Gain_K(i) ← Gain*_{K_{i,0}} + Gain*_{K_{i,1}}
14:             end if
15:         end for
                                     ▷ keep the best gain and its index
16:         i*_K ← arg max_{i∈D_K} Gain_K(i)
17:         Gain*_K ← Gain_K(i*_K)
18:     end for
19: end for
20: BuildTree(K ∈ 𝒦_L)                     ▷ Recursively build tree on entire set of rules

21: procedure BuildTree(K)
22:     if A_K ≠ ∅ then
23:         CreateLeaf(A_K)
24:     else
25:         left ← BuildTree(K_{i*_K,0})
26:         right ← BuildTree(K_{i*_K,1})
27:         CreateNode(c_{i*_K}, left, right)
28:     end if
29: end procedure
```

Figure 1 reports two examples of the algorithm steps with $L = 3$. **order** is the number of dashes in the cubes under consideration; the portion of a table characterized by the same order referes to one run of the **for** cycle in lines 5 - 19.

(a)

order	a b c	A	P	G1	G2	G3
0	0 0 0	1	0,125			
	0 0 1	3	0,125			
	0 1 0	1,3	0,125			
	0 1 1	3	0,125			
	1 0 0	2	0,125			
	1 0 1	3	0,125			
	1 1 0	2,4	0,125			
	1 1 1	4	0,125			
1	- 0 0	0	0,250	0,000		
	- 0 1	3	0,250	0,250		
	- 1 0	0	0,250	0,000		
	- 1 1	0	0,250	0,000		
	0 - 0	1	0,250	0,250		
	0 - 1	3	0,250	0,250		
	1 - 0	2	0,250	0,250		
	1 - 1	0	0,250	0,000		
	0 0 -	0	0,250	0,000		
	0 1 -	3	0,250	0,250		
	1 0 -	0	0,250	0,000		
	1 1 -	4	0,250	0,250		
2	- - 0	0	0,500	0,500	0,000	
	- - 1	0	0,500	0,250	0,250	
	- 0 -	0	0,500	0,000	0,250	
	- 1 -	0	0,500	0,500	0,000	
	0 - -	0	0,500	0,250	0,500	
	1 - -	0	0,500	0,250	0,250	
3	- - -	0	1,000	0,750	0,750	0,750

(b)

order	a b c	A	P	G1	G2	G3
0	0 0 0	1	0,100			
	0 0 1	3	0,100			
	0 1 0	1,3	0,100			
	0 1 1	3	0,100			
	1 0 0	2	0,100			
	1 0 1	3	0,200			
	1 1 0	2,4	0,100			
	1 1 1	4	0,200			
1	- 0 0	0	0,200	0,000		
	- 0 1	3	0,300	0,300		
	- 1 0	0	0,200	0,000		
	- 1 1	0	0,300	0,000		
	0 - 0	1	0,200	0,200		
	0 - 1	3	0,200	0,200		
	1 - 0	2	0,200	0,200		
	1 - 1	0	0,400	0,000		
	0 0 -	0	0,200	0,000		
	0 1 -	3	0,200	0,200		
	1 0 -	0	0,300	0,000		
	1 1 -	4	0,300	0,300		
2	- - 0	0	0,500	0,400	0,000	
	- - 1	0	0,500	0,200	0,300	
	- 0 -	0	0,400	0,000	0,300	
	- 1 -	0	0,600	0,500	0,000	
	0 - -	0	0,400	0,200	0,400	
	1 - -	0	0,600	0,300	0,200	
3	- - -	0	1,000	0,700	0,800	0,700

Fig. 1. Tabular format listing of the algorithm steps

a,b and **c** are the testing conditions, A (resp. P) the set of actions (resp. probability occurrence) associated to the cube identified by the values assumed by conditions on a particular row of the table. **Gi**, with $i = 1, 2, 3$, are the gains computed by the successive runs of the for cycle in lines 9 - 15. The highlighted values are those leading to the optimal solution (line 17). In case (a), a set of actions and alternatives with uniform probabilities is shown. The action sets get progressively reduced up to empty set, while the probabilities increase up to unity. While the n-cube order increases, different dashes can be selected (thus different conditions may be chosen) and the corresponding gain is computed. In this particular case, any choice of the first condition to be checked is equivalent (we conventionally choose the leftmost one). When the probability distribution of the rules changes, as in case (b), the gains change accordingly and, in the depicted case, a single tree (rooted with condition b) is the optimal one.

2.3   Computational Time

The total number of cubes that are analyzed by the algorithm is 3^L, one for all possible words of length L on the three letter alphabet $\{0, 1, -\}$. For each cube K of dimension n it computes: (1) the intersection of the actions associated to the cubes in one partition (line 8); this task that can be accomplished, in the

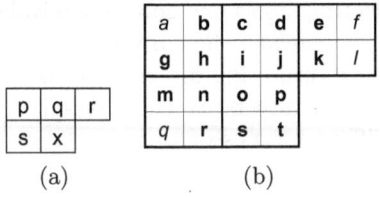

Fig. 2. The standard mask used for raster scan connected components labeling since original's Rosenfeld approach (a), and the 2×2 block based mask used in our novel approach. In both masks, non-bold letters identify pixels which are not employed in the algorithm.

worst case, in time linear with the number of actions. (2) The gain of $n \leq L$ trees, one for each dash (lines 9 - 15), each in constant time. Hence, as the recursive cosntruction of the tree adds only a non dominant additive cost, the computational time of the algortihm is upper bounded by:

$$3^L \cdot (|A| + L) \leq (2^2)^L \cdot (|A| + L) \in O(|R|^2 \cdot \max\{|A|, |C|\}), \qquad (3)$$

where C is the set of conditions (and $|C| = L$), R the set of rules (and $|R| = 2^L$), and A is the set of actions.

3 Optimizing Connected Components Labeling

A Connected Components Labeling algorithm assigns a unique identifier (an integer value, namely *label*) to every connected component of the image, in order to give the possibility to refer to it in the next processing steps. Usually, labeling algorithms deal with binary images.

The majority of images are stored in raster scan order, so the most common technique for connected components labeling applies sequential local operations in that order, as firstly introduced by Rosenfeld *et al.* [7]. This is classically performed in 3 steps, described in the following.

During the first step, each pixel label is evaluated locally by only looking at the labels of its already processed neighbors. When using 8-connectivity, these pixels belong to the scanning mask shown in Fig. 2(a). During the scanning procedure, pixels belonging to the same connected component can be assigned different (*provisional*) labels, which at some point will be marked as equivalent. In the second step, all the equivalent provisional labels must be merged into a single class. Modern algorithms process equivalences as soon as they are discovered (online equivalent labels resolution, as in [1]). Most of the recent optimizations techniques aim at increasing the efficiency of this step. Once the equivalences have been eventually solved, in the third step a second pass over the image is performed in order to assign to each foreground pixel the representative label of its equivalence class. Usually, the class representative is unique and it is set to be the minimum label value in the class.

The procedure of collecting labels and solving equivalences may be described by a *command execution metaphor*: the current and neighboring pixels provide a binary command word, interpreting foreground pixels as 1s and background pixels as 0s. A different action must be executed based on the command received. We may identify four different types of actions: *no action* is performed if the current pixel does not belong to the foreground, a *new label* is created when the neighborhood is only composed of background pixels, an *assign* action gives the current pixel the label of a neighbor when no conflict occurs (either only one pixel is foreground or all pixels share the same label), and finally a *merge* action is performed to solve an equivalence between two or more classes and a representative is assigned to the current pixel. The relation between the commands and the corresponding actions may be conveniently described by means of an *OR*-decision table.

As shown by Grana *et al.* [2], we can notice that, in algorithms with online equivalences resolution, already processed 8-connected foreground pixels cannot have different labels. This allows to remove merge operations between 8-connected pixels, substituting them with assignments of either of the involved pixels labels (many equivalent actions are possible, leading to the same result). It is also possible to enlarge the neighborhood exploration window, with the aim to speed up the connected components labeling process. The key idea comes from two very straightforward observations: when using 8-connection, the pixels of a 2×2 square are all connected to each other and a 2×2 square is the largest set of pixels in which this property holds. This implies that all foreground pixels in a the block will share the same label at the end of the computation. For this reason we propose to scan the image moving on a 2×2 pixel grid applying, instead of the classical neighborhood of Fig. 2(a), an extended mask of five 2×2 blocks, as shown in Fig. 2(b).

Employing all necessary pixels in the new mask of Fig. 2(b), we deal with $L = 16$ pixels (thus conditions), for a total amount of 2^{16} possible combinations. Grana *et al.* [2] employed the original Schumacher's algorithm, which required the conversion of the *OR*-decision table to a single entry decision table. This was performed with an heuristic technique, which led to producing a decision tree containing 210 nodes sparse over 14 levels, assuming all patterns occurred with the same probability and unitary cost for testing conditions. By using the algorithm proposed in this work, under the same assumptions, we obtain a much more compressed tree with 136 nodes sparse over 14 levels: the complexity in terms of levels is the same, but the code footprint is much lighter.

To push the algorithm performances to its limits, we further propose to add an occurrence probability for each pattern (p_r), which can be computed offline as a preprocessing stage on a subset of the dataset to be used, or the whole of it. The subset used for the probability computation obviously affects the algorithm performance (since we obtain a more or less optimal decision tree given those values), but the idea we want to carry out is that this optimization can be considered the upper bound of the performance we can obtain with this method.

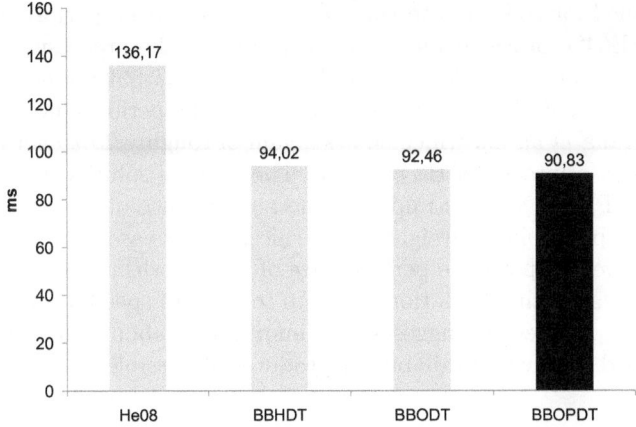

Fig. 3. The direct comparison between the He's approach (*He08*) with the three evolutions of block based decision tree approach, from the initial proposal with heuristic selection between alternative rules (*BBHDT*), further improved with the optimal decision tree generation (*BBOUDT*) and finally enhanced with a probabilistic weight of the rules (*BBOPDT*)

4   Experimental Results

In order to propose a valuable comparison with the state of the art, we used a dataset of Otsu-binarized versions of 615 high resolution document images. This dataset gives us the possibility to test the connected components labeling capabilities with very complex patterns at different sizes, with an average resolution of 10.4 megapixels and 35,359 labels, providing a challenging dataset which heavily stresses the algorithms. We performed a comparison between the following approaches:

- He's approach (*He07*), which highlights the benefits of the Union-Find algorithm for labels resolution implemented with the set of three arrays as in [3] and with the use of a decision tree to optimize the memory access with the mask of Fig. 2(a).
- The block based approach with decision tree generated with heuristic selection between alternatives as previously proposed by Grana *et al.* [2] (*BB-HDT*)
- The block based approach with *optimal* decision tree generated with the procedure as proposed in this work, *assuming uniform distribution of patterns* (BBOUDT)
- The block based approach with *optimal* decision tree generated with the procedure weighted with the pattern probabilities (*BBOPDT*)

For each of these algorithms, the median time over five runs is kept in order to remove possible outliers due to other tasks performed by the operating system. All algorithms of course produced the same labeling on all images, and a unitary cost is assumed for condition testing. The tests have been performed on a Intel Core 2 Duo E6420 processor, using a single core for the processing. The code is written in C++ and compiled on Windows 7 using Visual Studio 2008.

As reported in Fig. 3, we confirm the significant performance speedup presented by Grana *et al.* [2], which shows a gain of roughly 29% over the previous state-of-the-art approach of He *et al.* [4]. The optimal solution proposed in this work (BBODT) just slightly improves the performance of the algorithm. With the use of the probabilistic weight of the rules, in this case computed on the entire dataset, we can push the performance of the algorithm to its upper bound, showing that the optimal solution gains up to 3.4% of speedup over the original proposal. This last result, suggests that information about pattern occurrences should be used whenever available, or produced if possible.

5   Conclusions

In this paper we proposed a dynamic programming algorithm to generate an optimal decision tree from *OR*-decision tables. This decision tree represents the optimal arrangement of conditions to verify, and for this reason provides the fastest processing code to solve the neighborhood problem. The experimental section evidence how our approach can lead to faster results than other techniques proposed in literature. This suggests how this methodology can be considered a general modeling approach for many local image processing problems.

References

1. Di Stefano, L., Bulgarelli, A.: A simple and efficient connected components labeling algorithm. In: International Conference on Image Analysis and Processing, pp. 322–327 (1999)
2. Grana, C., Borghesani, D., Cucchiara, R.: Optimized Block-based Connected Components Labeling with Decision Trees. IEEE Transactions on Image Processing 19(6) (June 2010)
3. He, L., Chao, Y., Suzuki, K.: A linear-time two-scan labeling algorithm. In: International Conference on Image Processing, vol. 5, pp. 241–244 (2007)
4. He, L., Chao, Y., Suzuki, K., Wu, K.: Fast connected-component labeling. Pattern Recognition 42(9), 1977–1987 (2008)
5. Lew, A.: Optimal conversion of extended-entry decision tables with general cost criteria. Commun. ACM 21(4), 269–279 (1978)
6. Reinwald, L.T., Soland, R.M.: Conversion of Limited-Entry Decision Tables to Optimal Computer Programs I: Minimum Average Processing Time. J. ACM 13(3), 339–358 (1966)
7. Rosenfeld, A., Pfaltz, J.L.: Sequential operations in digital picture processing. J. ACM 13(4), 471–494 (1966)
8. Schumacher, H., Sevcik, K.C.: The Synthetic Approach to Decision Table Conversion. Commun. ACM 19(6), 343–351 (1976)

Efficient Computation of Convolution of Huge Images

David Svoboda

Centre for Biomedical Image Analysis
Faculty of Informatics, Masaryk University, Brno, Czech Republic
svoboda@fi.muni.cz

Abstract. In image processing, convolution is a frequently used operation. It is an important tool for performing basic image enhancement as well as sophisticated analysis. Naturally, due to its necessity and still continually increasing size of processed image data there is a great demand for its efficient implementation. The fact is that the slowest algorithms (that cannot be practically used) implementing the convolution are capable of handling the data of arbitrary dimension and size. On the other hand, the fastest algorithms have huge memory requirements and hence impose image size limits. Regarding the convolution of huge images, which might be the subtask of some more sophisticated algorithm, fast and correct solution is essential. In this paper, we propose a fast algorithm implementing exact computation of the shift invariant convolution over huge multi-dimensional image data.

Keywords: Convolution, Fast Fourier Transform, Divide-et-Impera.

1 Introduction

Convolution is a very important mathematical tool in the field of image processing. It is employed in edge detection [1], correlation [2], optical flow [3], deconvolution [4], simulation [5], etc. Each time the convolution is called plenty of primitive instructions (addition and multiplication) have to be computed. As the amount of processed image data still raises, there is considerable request for fast manipulation with huge image data. This means that the current image processing algorithms have to be capable of handling large blocks of image data in short time. As a lot of image processing methods is based on convolution, we will focus on this mathematical tool and its modifications that bring some acceleration or memory saving.

The basic convolution algorithm traces the individual pixel positions in the input image. In each position, it evaluates inner product of current pixel neighbourhood and flipped kernel. Although the time complexity of the algorithms based on this approach is polynomial [6] this solution is very slow. This is true namely for large kernels. There exist some improvements that guarantee lower complexity, however always with some limitations.

G. Maino and G.L. Foresti (Eds.): ICIAP 2011, Part I, LNCS 6978, pp. 453–462, 2011.

Separable convolution. The higher dimensional convolution with so called *separable* [7] kernels can be simply decomposed into several lower dimensional (cheaper) convolutions. Gaussian, DoG, and Sobel [1] are the representatives of such group of kernels. However, the deconvolution or template matching algorithms based on correlation methods [2] typically use general kernels, which cannot be characterized by special properties like separability.

Recursive filtering. The convolution is a process where the inner product, whose size corresponds to kernel size, is computed again and again in each individual pixel position. One of the vectors, that enter this operation, is always the same. Hence, we can evaluate the whole inner product only in one position while the neighbouring position can be computed as a slightly modified difference with respect to the first position. Analogously, the same is valid for all the following positions. The computation of the convolution using this approach is called *recursive filtering* [7]. Also this method has its drawbacks. The conversion of general convolution kernel into its recursive version is a nontrivial task. Moreover, the recursive filtering often suffers from inaccuracy and instability [8].

Fast convolution. While the convolution in time domain performs an inner product in each pixel position, all we have to do in Fourier domain is point-wise multiplication. Due to this convolution property we can evaluate the convolution in time $O(N \log N)$. This approach is known as a *fast convolution* [9]. Another advantage is that no restrictions are imposed on the kernel. Unfortunately, the excessive space requirements make this approach not very popular.

In this paper, we designed a fast algorithm capable of performing the convolution over huge image data. We considered both the image and the kernel of large size and the dimensions up to 3D. For the simplicity, all the statements will be explained for 1D space. The extension to higher dimensions will be either straightforward or we will explicitly focus on it. We did not impose any restrictions on the convolution kernel, i.e. the kernel neither was separable nor could be simply represented as a recursive filter.

2   Problem Analysis

In the following, we will focus on the efficient implementation that does not impose any special conditions on convolution kernels. Hence, the fast convolution will be under the scope. We will provide the analysis of time and space complexity. Regarding the former one we will focus on the number of complex additions and multiplications needed for the computation of studied algorithms.

Utilizing the convolution theorem and the fast Fourier transform the 1D convolution of two signals f and g requires

$$(M+N)\left[\frac{9}{2}\log_2(M+N)+1\right] \tag{1}$$

steps. Here $\texttt{size}(f) = M$, $\texttt{size}(g) = N$, and the term $(M + N)$ means that the processed image f was zero padded[1] to prevent the overlap effect caused by circular convolution. The kernel was modified in the same way. Another advantage of using Fourier transform stems from its separability. When convolving two 3D images f^{3d} and g^{3d}, where $\texttt{size}(f^{3d}) = M \times M \times M$ and $\texttt{size}(g^{3d}) = N \times N \times N$, we need only

$$(M+N)^3 \left[\frac{9}{2} \log_2 (M+N)^3 + 1 \right] \tag{2}$$

steps in total. Up to now, this method seemed to be optimal. Before we proceed, let us have a look at the space complexity of this approach.

If we do not take into account buffers for the input/output images and serialize both Fourier transforms, we need space for two equally aligned Fourier images and some negligible Fourier transform workspace. In total, it is

$$(M + N) \cdot C \tag{3}$$

bytes, where $(M+N)$ is a size of one padded image and C is a constant dependent on the required algorithm precision (single, double or long double). If the double precision is required, for example, then $C = 2 \cdot sizeof(double)$, which corresponds to two Fourier images used by real-valued FFT. In 3D case, when $\texttt{size}(f^{3d}) = M \times M \times M$ and $\texttt{size}(g^{3d}) = N \times N \times N$ the space needed by the aligned image data is proportionally higher: $(M + N)^3 \cdot C$ bytes.

For better insight, let us consider the complex simulation process, in which the convolution of two relatively small images 500×500×224 voxels and 128×128×100 voxels is called. In this example, the filtered image was a fraction of some larger 3D microscopic image and the kernel was an empirically measured point spread function (PSF). The convolution was one of image processing algorithm called from our simulation toolbox[2]. When this convolution was performed in double precision on Intel Xeon QuadCore 2.83 GHz computer it took 41 seconds using fast convolution while it lasted cca for 4 days when asked for the computation based on the basic approach. Although the FFT based approach seemed to be very promising, it disappointed regarding the memory requirements. Here are the reasons:

- For the computation of fast convolution, both the image and the kernel must be aligned (typically padded) to the same size.
- If not padded with sufficiently large zero area, the result may differ from that of basic approach due to the periodicity of discrete FT.
- In order to minimize the numerical inaccuracy, FFT over large memory blocks must be computed at least in double precision. In the example above, 2 GB of physical memory was required.

It is clear that using this method the memory overhead is very high. In the following, we will try to reduce the memory requirements while keeping the efficiency of the whole convolution process.

---

[1] The size of padded image should be exactly $(M + N - 1)$. For the sake of simplicity, we reduced this term to $(M + N)$ as we suppose $M \gg 1$ and $N \gg 1$.

[2] CytoPacq – a simulation toolbox: http://cbia.fi.muni.cz/simulator/

3   Method

Keeping in mind that due to the lack of available memory, direct computation of fast convolution is not realizable using common computers we will try to split the whole task into several subtasks. This means that the image and kernel will be split into smaller pieces, so called *tiles* that need not be of the same size.

3.1   Image Tiling

Splitting the convolved image f into smaller tiles $f_1, f_2, \ldots, f_m$, then performing m smaller convolutions $f_i \otimes g, i = \{1, \ldots, m\}$ and finally merging the results together with discarding the overlaps is a well known algorithm in digital signal processing. The implementation is commonly known as the *overlap-save method* [9].

Let us inspect the memory requirements for this approach. Again, let $\texttt{size}(f) = M$ and $\texttt{size}(g) = N$. As the filtered image f is split into m pieces, the respective memory requirements are lowered to

$$\left(\frac{M}{m} + N\right) \cdot C \tag{4}$$

bytes (compare to Eq. (3)). Hence, a slight improvement was reached. Concerning the time complexity, the fast convolution of two 1D signals of length M and N required $(M+N)(\frac{9}{2}\log_2(M+N) + 1)$ steps. If the image is split into m tiles, we need to perform

$$(M+mN)\left[\frac{9}{2}\log_2\left(\frac{M}{m}+N\right) + 1\right] \tag{5}$$

steps in total. If there is no division ($m=1$) we get the time complexity of the fast approach. If the division is total ($m=M$) we get even worse complexity than the basic convolution has. The higher the level of splitting is required the worse the complexity is. Therefore, we can conclude that splitting only the image into tiles does not help.

3.2   Kernel Tiling

From the previous text, we recognize that splitting only the image f might be inefficient. It may even happen that the kernel g is so large that splitting of only the image f does not reduce the memory requirements sufficiently. As the convolution belongs to commutative operators one could recommend swapping the image and the kernel. This may help, namely when the image is small and the kernel is very large. As soon as the image and the kernel are swapped, we can simply apply the overlap-save method. However, this approach fails when both the image and the kernel are too large.

Let us decompose the kernel g as well. Keeping in mind that the image f has already been decomposed into m tiles, we can focus on the manipulation with just one image tile $f_i, i \in \{1, \ldots, m\}$. For the evaluation of convolution of the

```
1:  (f, g) ← (input image, kernel)
2:  f → f₁, f₂, ..., fₘ {split 'f' into tiles according to overlap-save scheme}
3:  g → g₁, g₂, ..., gₙ {split 'g' into tiles according to overlap-add scheme}
4:  h ← 0 {create the output image 'h' and fill it with zeros}
5:  for i = 1 to m do
6:     for j = 1 to n do
7:        hᵢⱼ ← convolve(fᵢ, gⱼ)
              {use fast convolution}
8:        hᵢⱼ ← discard_overruns(hᵢⱼ)
              {discard hᵢⱼ overruns following overlap-save output rules}
9:        h ← h + shift(hᵢⱼ)
              {add hᵢⱼ to h following overlap-add output rules}
10:    end for
11: end for
12: Output ← h
```

Algorithm 1. Divide-et-impera approach applied to the convolution over large image data.

selected image tile f_i and the large kernel g we will employ so called *overlap-add method* [9]. This method splits the kernel g into n pieces, then it performs n smaller convolutions $f_i \otimes g_j, j = \{1, \ldots, n\}$, and finally it adds the partial solutions together preserving the appropriate overruns. This way, we obtain the Algorithm 1. For the detailed description of this algorithm see Figure 1.

3.3 Efficiency

Let us suppose the image f is split into m tiles and kernel g is decomposed into n tiles. The time complexity of the fast convolution $f_i \otimes g_j$ is

$$\left(\frac{M}{m} + \frac{N}{n}\right)\left[\frac{9}{2}\log_2\left(\frac{M}{m} + \frac{N}{n}\right) + 1\right]. \tag{6}$$

We have m image tiles and n kernel tiles. In order to perform the complete convolution $f \otimes g$ we have to perform $m \times n$ convolutions (see the nested loops in Algorithm 1) of the individual image and kernel tiles. In total, we have to complete

$$(Mn + Nm)\left[\frac{9}{2}\log_2\left(\frac{M}{m} + \frac{N}{n}\right) + 1\right] \tag{7}$$

steps. One can clearly see that without any division ($m = n = 1$) we get the complexity of fast convolution, i.e. the class $O((M+N)\log(M+N))$. For total division ($m = M$ and $n = N$) we obtain basic convolution, i.e. the complexity class $O(MN)$. Concerning the space occupied by our convolution algorithm, we need

$$\left(\frac{M}{m} + \frac{N}{n}\right) \cdot C \tag{8}$$

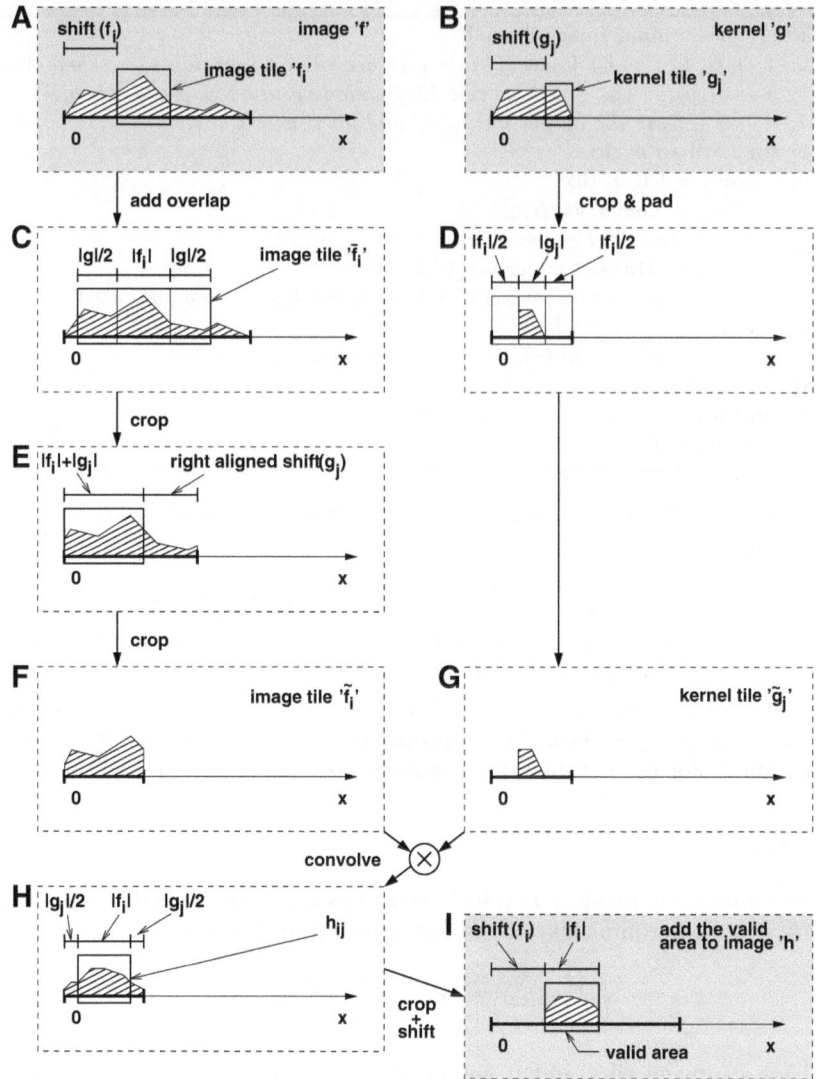

Fig. 1. Convolution of image tile f_i and kernel tile g_j: (A,B) The tiles f_i and g_j are selected. Remember the shift of these tiles with respect to the image they belong to. These shifts will be used later. (C) The tile f_i is extended and cropped. In this way, we obtain $\overline{f}_i$. (D) The tile g_j is cropped (redundant zeros are removed) and padded with zeros in order to get $\widetilde{g}_j$. (E) The area of size $\mathtt{size}(f_i) + \mathtt{size}(g_j)$ with the distance $\mathtt{shift}(g_j)$ from the right border is cropped. In this way, we get $\widehat{f}_i$. (F,G) The image buffers are aligned to the same size. They are ready for convolution. (H) The convolution is performed. (I) The solution h_{ij} is cut out and shifted to the correct position. Finally, it is added to the output image h. Take note of the light-gray background of some frames. In this way, we distinguish between input/output images (gray frames) and intermediate results (white frames).

bytes, where C is again the precision dependent constant and m, n are the levels of division of image f and kernel g, respectively.

All the previous statements are related only to 1D signal. Provided both image and kernel are D-dimensional cubes and the tiling proces is regular, we can combine Eq. (2) and Eq. (7) in order to get:

$$(Mn+Nm)^D \left[\frac{9}{2} \log_2 \left(\frac{M}{m} + \frac{N}{n} \right)^D + 1 \right] \tag{9}$$

This statement can be further generalized, i.e. the image and the kernel do not have be in the shape of cube and the tiling does not have to be identical in all the axes. It can be simply derived from the separability of multidimensional Fourier transform, which guarantees that the time complexity of the higher dimensional Fourier transform depends on the amount of processed voxels only. There is no difference in the time complexity if the higher-dimensional image is elongated or in the shape of cube.

4   Results

The proposed algorithm was compared to other freely available implementations of fast convolution – see Figure 2. For the computation of Fourier transform all three packages used FFTW library [10]. It can be clearly seen that the speed of our new approach is comparable with those implemented in the most common image processing toolkits and mostly it outperforms them. Unlike other toolkits, if the computer has less memory than required our new approach does not fail. It splits the task into subtasks and delegates the computing to the loop that successively executes the individual subtasks.

In the previous section, an algorithm of splitting the image f into m tiles and the kernel g into n tiles was proposed. Now we will answer the question regarding the optimal way of splitting the image and the kernel. We still assume that $\texttt{size}(f) = M$ and $\texttt{size}(g) = N$. As M and N are constants let us focus only on the relationship between m and n. Let us define the following substitution: $x = \frac{M}{m}$ and $y = \frac{N}{n}$. Here x and y stand for the sizes of the image and the kernel tiles, respectively. Applying this substitution to Eq. (7) and simplifying, we get

$$MN \left(\frac{1}{x} + \frac{1}{y} \right) \left[\frac{9}{2} \log_2(x + y) + 1 \right] \tag{10}$$

The plot of this function is depicted in Figure 3. The minimum of this function is reached if and only if $x=y$ and both variables x and y are maximized, i.e. the image and the kernel tiles should be of the same size and they should be as large as possible. In order to reach the optimal solution, the size of the tile should be the power of small primes [11]. In this sense, it is recommended to fulfill both criteria put on the tile size: the maximality (as stated above) and the capability of simple decomposition into small primes.

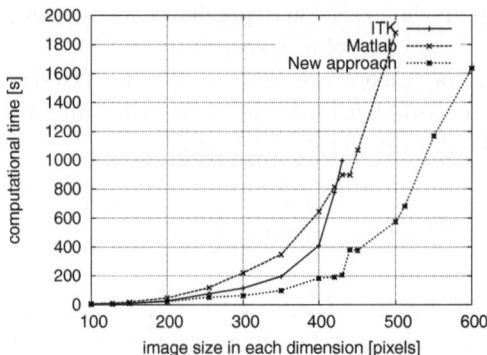

Fig. 2. A graph offering the comparison of the most common implementations of convolution and the new approach. Evaluated over two 3D images of identical size on Intel Xeon QuadCore 2.83 GHz computer with 32 GB RAM. Take note, that ITK and Matlab plots finish earlier as the computation for the images of large dimensions failed due to the lack of memory.

5   Application

Convolution is a core part of many image processing algorithms. In our group we employ the convolution in the process of simulation of fluorescence microscopy images [12]. The simulation process usually consists of three consecutive stages: generation of digital phantoms [13], simulation of optical system, and simulation of acquisition device. We use the convolution in the first two stages:

5.1   Generation of Digital Phantoms

A digital phantom is an image of an estimated model of studied object. As soon as the model is properly defined, a generation of the phantom is straightforward. However, if we want to generate more phantoms we have to solve the problem of collisions as it is reasonable to require the generation of non-overlapping objects only. For this purpose we utilized a correlation. Here, the input image is usually a large image with previously generated phantoms inside and the kernel is a smaller image with a newly generated phantom that should be added to the input image. The output correlated image contains zeros in the positions, where the new phantom will not overlap any of the already existing objects. As the only difference between correlation and convolution stems in the flipping of convolution kernel, the algorithm responsible for the correct placement of newly generated phantoms was substituted with the convolution with flipped kernel. The volume of the image to be correlated was set to the size $55 \times 55 \times 12$ microns which corresponded to $1024 \times 1024 \times 224$ voxels for the selected optical system[3].

---

[3] Optical system configuration: microscope Zeiss 200M, objective Zeiss Plan-Apochromat ($100\times/1.40$ Oil), confocal unit Yokogawa CSU-10, and camera Andor iXon DU888E (EM CCD 1024×1024 with pixel size 13 microns)

The kernel size corresponded to the bounding box volume of simulated object. In our case its was in average $10 \times 10 \times 10$ microns ($= 187 \times 187 \times 187$ voxels). In this example, the correlation memory requirements rose to 9.6 GB.

As long as we generated large complex cell structures we avoided the lack of memory because the computer we used was equipped with 32 GB RAM. The problem came out when the multiple spots within a large area were under the scope. Due to low default image resolution, the size of each spot was almost the same as one voxel. However, while the spot is expected to be of spherical shape, the voxel is always rectangular. The remedy was to sufficiently up-sample the image and the kernel. When we asked for n-tuple subpixel precision the size of both image and kernel rose n-times along each axis! The excessive memory requirements made the standard fast convolution approach unusable. For this purpose we used our new approach that can handle the data of any size.

5.2   Simulation of Optical System

The simulation of optical system is a next step that directly follows up on the generation of digital phantoms. The optical system is characterized by its PSF, that describes how the infinitely small impulse is transformed when passing through this system. This transform is usually modelled as a convolution, where the image to be convolved is the image containing digital phantoms and the convolution kernel is represented by the PSF. In this task we used the same optical system configuration and the same input image as in the example above. The kernel size, which was an empirically measured PSF (prepared by SVI Huygens® Pro software), was fixed to the size of $128 \times 128 \times 100$ voxels. The memory requirements for the convolution started at about 6.8 GB and they accordingly rose when asked for higher level of subpixel precision. Therefore, we again used our new approach capable of working with large image data.

6   Conclusion

We designed a fast algorithm for the computation of the convolution. While all the current methods impose some restrictions on the input data (size of the data, separable kernel, size of kernel and image should be powers of small primes), we do not require any. Prior to the computation, our algorithm verifies the amount of available memory. If there is not enough memory the principle of *divide-et-impera* is applied. Therefore, the unsolvable huge problem is split into several smaller subproblems. In our approach we split the input data (the image and the kernel). The efficiency of the computation highly depends on the level of division. If no division is realized, the time complexity is equal to the complexity of the fast convolution. On the other hand, if we were forced to split the image and the kernel into individual pixels the time complexity of the algorithm would belong to the same class as the basic approach. Following the graph in Fig. 3 one can clearly see that all the other cases belong to the complexity classes in between. In practice, we apply only a small level of division which leads nearly to the optimal

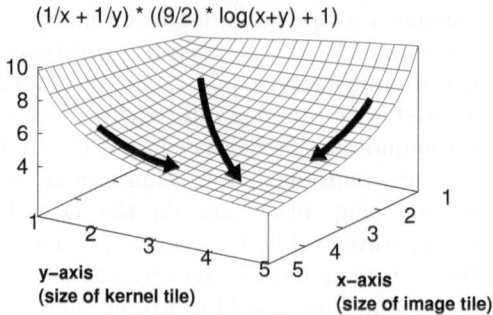

Fig. 3. A graph of a function that represents the time complexity of tiled convolution. The evident minimum occurs in the location, where both variables (size of the tiles) are maximized and equal at the same time.

solution (fast convolution). The implementation of this convolution algorithm is part of image processing library which is under GNU GPL and is available at: http://cbia.fi.muni.cz/projects/i3dlibs.html. The algorithm has been successfully used in the simulation toolbox developed in our group.

Acknowledgment. This work was supported by the Ministry of Education of the Czech Rep. (Projects No. MSM0021622419, No. LC535 and No. 2B06052).

References

1. Parker, J.R.: Algorithms for image processing and computer vision. Wiley, Chichester (1996)
2. Gonzalez, R.C., Woods, R.E.: Digital Image Processing. Prentice-Hall, Englewood Cliffs (2002)
3. Fleet, D.J., Jepson, A.D.: Computation of component image velocity from local phase information. Int. J. Comput. Vision 5(1), 77–104 (1990)
4. Verveer, P.J.: Computational and optical methods for improving resolution and signal quality in fluorescence microscopy, PhD Thesis. Delft Technical Univ. (1998)
5. Jensen, J.A., Munk, P.: Computer phantoms for simulating ultrasound B-mode and cfm images. Acoustical Imaging 23, 75–80 (1997)
6. Pratt, W.K.: Digital Image Processing. Wiley, Chichester (1991)
7. Jähne, B.: Digital Image Processing, 6th edn. Springer, Heidelberg (1997)
8. Smith, S.W.: Digital Signal Processing. Newnes (2002)
9. Jan, J.: Digital Signal Filtering, Analysis and Restoration. Telecommunications Series. INSPEC, Inc. (2000) ISBN: 0852967608
10. Frigo, M., Johnson, S.G.: The design and implementation of FFTW3. Proceedings of the IEEE 93(2), 216–231 (2005); Special issue on Program Generation, Optimization, and Platform Adaptation
11. Heideman, M., Johnson, D., Burrus, C.: Gauss and the history of the fast fourier transform. IEEE ASSP Magazine 1(4), 14–21 (1984) ISSN: 0740-7467
12. Svoboda, D., Kozubek, M., Stejskal, S.: Generation of digital phantoms of cell nuclei and simulation of image formation in 3d image cytometry. Cytometry Part A 75A(6), 494–509 (2009)
13. Rexilius, J., Hahn, H.K., Bourquain, H., Peitgen, H.-O.: Ground Truth in MS Lesion Volumetry – A Phantom Study. In: Ellis, R.E., Peters, T.M. (eds.) MICCAI 2003. LNCS, vol. 2879, pp. 546–553. Springer, Heidelberg (2003)

Half Ellipse Detection

Nikolai Sergeev and Stephan Tschechne

Institute for Neural Information Processing, University of Ulm,
James-Franck-Ring, 89069 Ulm, Germany
{nikolai.sergeev,stephan.tschechne}@uni-ulm.de

Abstract. This paper presents an algorithm of half ellipse detection
from color images. Additionally the algorithm detects two color average
values along the both sides of a half ellipse. In contrast to standard
methods the new one finds not only parameters of the entire ellipse but
also the end points of a half ellipse. The paper introduces a new way
of edge and line detection. The new detector of edges in color images
was designed to extract color on the both sides of an edge. The new line
detector is designed to optimize the detection of endpoints of a line.

Keywords: Edge detection, line detection, ellipse detection.

1 Introduction

In the literature there can be found several methods of ellipse detection e.g.
[Tsuji and Matsumoto, 1978, H.K. Yuen, 1988]. Most of them use Hough transformation [Hough, 1962].

The new object recognition system introduced in [Sergeev and Palm, 2011]
uses half ellipses. In other words not only ellipse parameters but also end points
of an ellipse segment are needed. For that reason a new half ellipse detection
algorithm was developed.

The algorithm consists of four steps: edge detection, line detection, combination of lines to chains, detection of half ellipses in line chains. These will be
consecutively described in the paper.

The first step is one of the best discussed topics in the literature. Unfortunately such edge detectors as [Canny, 1986, Prewitt, 1970] work with grayscale
images and do not extract color on the both sides of an edge. On this account a
new edge detector was developed.

Hough transformation is normally used to detect lines. As the original Hough
transformation is not quite suitable to locate endpoints of a line a new algorithm
was designed.

2 Edge Detection

At first the new concept of an edge will be introduced. A pixel corner will
be denoted as a vertex. An edge is a pair of neighbored vertexes. There are
horizontal and vertical edges.

G. Maino and G.L. Foresti (Eds.): ICIAP 2011, Part I, LNCS 6978, pp. 463–472, 2011.

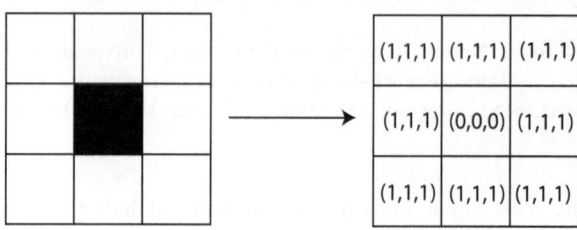

Fig. 1. A 3×3 pixel grid. One black pixel with RGB values $(0,0,0)$ at the center. Other pixels are white and have RGB value $(1,1,1)$.

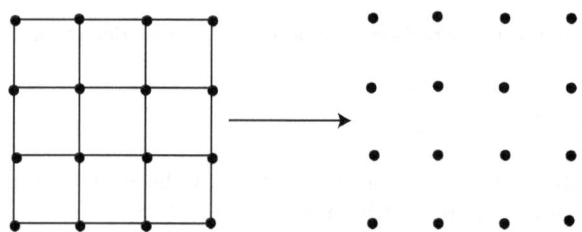

Fig. 2. The set of vertexes of the pixel grid

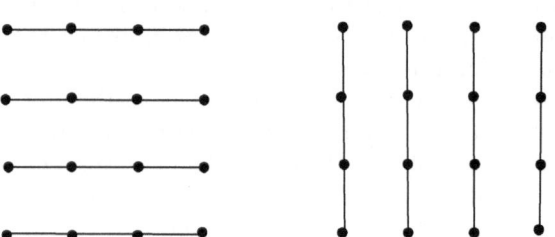

Fig. 3. Horizontal and vertical edges

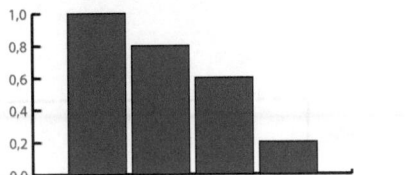 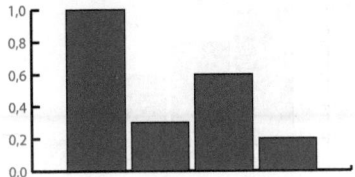

Fig. 4. Good(left) and bad(right) transition of the red RGB component in a line

Now it will be described how vertical contrast edges are detected. The horizontal ones are getting detected in exactly the same way after having transposed the pixel grid. Each line is getting processed separately. A line of pixels is a sequence of RGB vectors $(p_i)_{i \in \{1,\ldots,n\}} \subseteq \mathbb{R}^3$. For $1 \leq i < j \leq n$ contrast intensity $CI(i,j)$ is defined as

$$CI(i,j) = \frac{\|p_j - p_i\|}{j - i + 2} - \left(\sum_{k=i}^{j-1} \|p_{k+1} - p_k\| - \|p_j - p_i\| \right). \tag{1}$$

The first term $\frac{\|p_j - p_i\|}{j - i + 2}$ says the contrast intensity is the higher the bigger the difference and shorter transition is. Number 2 or generally $j - i + 2$ is chosen heuristically. This choice is due to adapt the edge detection to human visual perception. The second term $\sum_{k=i}^{j-1} \|p_{k+1} - p_k\| - \|p_j - p_i\|$ is due to diminish the contrast intensity in case of a "bad" transition as showed in Fig. 4. In case of a "good" transition the term is equal zero otherwise it is positive.

The first step of edge detection is to find $1 \leq i < j \leq n$ pairs with

$$i \leq k < l \leq j \Rightarrow CI(k,l) \leq CI(i,j) \tag{2}$$

In other words contrast intensity of (i,j)-transition should be bigger than contrast intensity of any inner transition. An (i,j)-transition satisfying 2 will be denoted as (i,j)-true transition. In the second step such (i,j)-true transitions have to be excluded for which a (k,l)-transition exists with

$$k < i < j \leq l \vee k \leq i < j < l. \tag{3}$$

In the next step a central edge has to be extracted from an (i,j)-transition as Fig.5 shows. In the last step each side of the extracted edge gets a RGB value. An edge is defined as an ordered pair of vertexes. Having the first and the second vertex an edge gets the left and the right side. As Fig.6 shows the left side of the edge gets the RGB value of the left outer pixel of a transition. The right side - that of the right outer one.

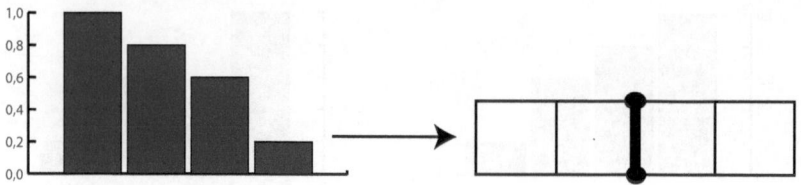

Fig. 5. An edge at the center of a transition

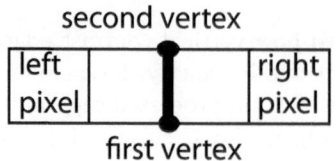

Fig. 6. Color extraction for an edge

3 Line Detection

At first a heuristic way to check if three discrete points "build" a line will be introduced. For the well known Manhattan norm $\|\cdot\|_M$ defined as

$$\|a\|_M = \sum_{i=1}^{2} |a_i| \tag{4}$$

three vertexes a, b, c build a line or $line(a, b, c) = 1$ if

$$\|c - a\|_M = \|c - b\|_M + \|b - a\|_M \tag{5}$$

and

$$\left| \frac{\|b - a\|_M}{\|c - a\|_M} |c.x - a.x| - |b.x - a.x| \right| \leq 1. \tag{6}$$

The last term can be formulated as integer multiplication

$$\|c.x - a.x\| \|b - a\|_M - |b.x - a.x| \|c - a\|_M | \leq \|c - a\|_M. \tag{7}$$

A line is defined as an ordered sequence of edges $(v_1^i, v_2^i)_{i \in \{1, \dots, n\}} \in \prod_{i \in \{1, \dots, n\}} \mathbb{R}^2 \times \mathbb{R}^2$. To checks if point p lies on a line the algorithm takes the first point of the line $first$ and tests for every vertex v_j^i of every edge if $line(first, v_j^i, p) = 1$. To check if an edge (p, q) lies on a line $(v_1^i, v_2^i)_{i \in \{1, \dots, n\}}$ both vertexes p, q of the edge get tested if they lie on the line. Finally $line(begin, p, q) = 1$ has to be true.

The process of line detection is the process of line proceeding. A typical situation is: A line is given. It has to be checked if there is an edge in direct neighborhood of the last point of the line which lies on the line in the sense defined above. If so a new edge gets added to the line. Initially a line consists of a single edge. The purpose is to proceed it maximally.

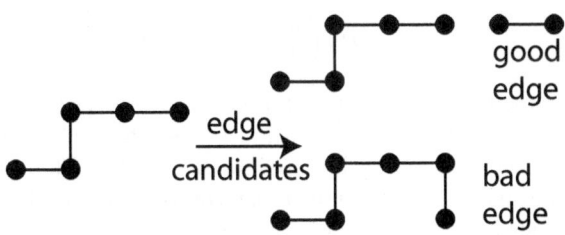

Fig. 7. Two edge candidates to proceed line. The bad one offends the first condition (5) of the line test.

4   Making Line Chains

Last section delivers a set of lines. The purpose of this section is to combine them to chains. Surely there are a lot of possible ways. To avoid the unnecessary details the most trivial one will be described.

To initialize a chain one line has to be selected randomly. The chain now consists of this one line. One of the endpoints of the initial line gets chosen arbitrarily. Let it be denoted as p. Than a set of all lines with one endpoint in some neighborhood of p has to be built. If the set is not empty a line $\tilde{l}$ has to be selected with for example the angle between l and $\tilde{l}$ closest to 180°. The selected line $\tilde{l}$ should be added to the chain and the procedure should be repeated for the remaining endpoint of $\tilde{l}$. The procedure must be iterated until the chain can not be proceeded any more. The entire process has to be replicated for the remaining end point of the initial line l.

One line can be part of only one chain. After having been chosen a line gets tagged as a used one and can not be used any more. It is due to reduce the number of produced chains and therefore run time.

The next chain should be initialized with a line untagged as a used one.

5   Definition, Representation and Detection of a Half Ellipse with Color

5.1   Definition of a Half Ellipse

For $\mathbb{C} = \mathbb{R}^2$ and $P(\mathbb{C})$ standing for power set of $\mathbb{C}$ a half ellipse is defined as a pair $(e, B) \in \mathbb{C}^2 \times P(\mathbb{C})$ with $e_1 \neq e_2$ for which $(a, b, t_0, \delta) \in [0, \infty) \times [0, \infty) \times \mathbb{R} \times \{-1, 1\}$ as well as $(c, \beta) \in \mathbb{C} \times \mathbb{R}$ exist so that

$$B = \left\{ T_c R_\beta \begin{pmatrix} a \cos t \\ b \sin t \end{pmatrix} \middle| t \in \begin{matrix} [t_0, t_0 + \delta\pi] \\ \cup \\ [t_0 + \delta\pi, t_0] \end{matrix} \right\} \tag{8}$$

and

$$e_1 = T_c R_\beta \begin{pmatrix} a \cos t_0 \\ b \sin t_0 \end{pmatrix}, \tag{9}$$

$$e_2 = T_c R_\beta \begin{pmatrix} a \cos(t_0 + \pi) \\ b \sin(t_0 + \pi) \end{pmatrix}. \tag{10}$$

T_c, R_β stand for translation, scaling and rotation respectively. The set of half ellipses will be denoted with HE. In other words a half ellipse consists of end-points $e_1, e_2 \in \mathbb{C}$ and of a set of bow points $B \in P(\mathbb{C})$. There are mainly two reasons to use half ellipses. An affine transformation $A \neq 0$ always maps a half ellipse onto another half ellipse. The second reason is the variety of half ellipses as Figure 8 shows.

Fig. 8. Examples of half ellipses

5.2   Rotation, Translation and Scaling Invariant Representation of a Half Ellipse

Now a unique rotation, translation and scaling invariant representation of a half ellipse will be introduced. At first two preliminary definitions are needed. For $x, y \in \mathbb{C}$ with $x \neq y$ the function $F^1_{x,y}$ is defined as

$$F^1_{x,y} : \begin{cases} \mathbb{C} \to \mathbb{C} \\ z \mapsto \frac{z-x}{y-x} \end{cases} . \tag{11}$$

Figure 9 shows the geometric meaning of the mapping. $F^1_{x,y}$ is an affine transformation with $F^1_{x,y}(x) = 0$ and $F^1_{x,y}(y) = 1$. The second function $F^2 : HE \to \mathbb{C}$ is defined as

$$F^2(e, B) = \begin{pmatrix} \max_{x \in B} \left| \left(F^1_{\frac{e_1+e_2}{2}, e_1}(x) \right)_1 \right| \\ \max_{x \in B} \left| \left(F^1_{\frac{e_1+e_2}{2}, e_1}(x) \right)_2 \right| \end{pmatrix}. \tag{12}$$

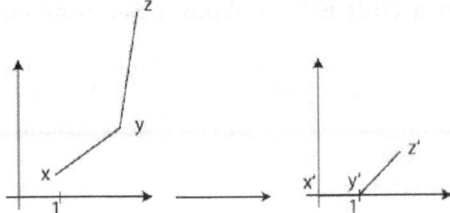

Fig. 9. Geometric meaning of $F^1_{x,y}(z)$

Finally the invariant representation $F^3 : HE \to \mathbb{C}$ is defined in such a way that for always existent $x, y \in B$ with

$$F^2(e, B) = \begin{pmatrix} \left| \left(F^1_{\frac{e_1+e_2}{2},\, e_1}(x) \right)_1 \right| \\ \left| \left(F^1_{\frac{e_1+e_2}{2},\, e_1}(y) \right)_2 \right| \end{pmatrix} \tag{13}$$

$$F^3(e, B) = \begin{pmatrix} z - SIGNUM(z) \\ \left(F^1_{\frac{e_1+e_2}{2},\, e_1}(y) \right)_2 \end{pmatrix} \tag{14}$$

with

$$z = \left(F^1_{\frac{e_1+e_2}{2},\, e_1}(x) \right)_1 . \tag{15}$$

For (e, B) in Figure 10

$$F^3(e, B) = \begin{pmatrix} M_1 - 1 \\ M_2 \end{pmatrix} . \tag{16}$$

It can be shown that for each $x \in \mathbb{C}$ a half ellipse $(e, B) \in HE$ exists with $F^3(e, B) = x$. Additionally for two half ellipses $(e, B), (\tilde{e}, \tilde{B}) \in HE$ with $F^3(e, B) = F^3(\tilde{e}, \tilde{B})$ it can be shown that they can be transformed in each other through translation, rotation and scaling. On the other side $F^3(e, B)$ is invariant to translation, rotation and scaling of (e, B).

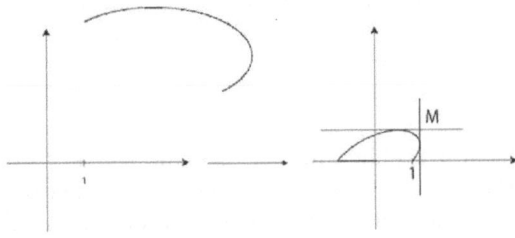

Fig. 10. Representation of a half ellipse

5.3   Extraction of a Half Ellipse from a Sequence of Lines

This section shows how to check whether a chain of lines is a half ellipse. The chain should be interpreted as a pair $(e, B) \in \mathbb{C}^2 \times P(\mathbb{C})$. e_1, e_2 are the endpoints of the line chain. $B \subseteq \mathbb{C}$ consists of all points of the chain.

Now $(a, b, t_0, \delta) \in [0, \infty) \times [0, \infty) \times \mathbb{R} \times \{-1, 1\}$ and $(c, \beta) \in \mathbb{C} \times \mathbb{R}$ must be found for which the corresponding half ellipse $(e, \tilde{B})$ would have $F^3(e, \tilde{B}) = F^3(e, B)$ At first $F^2(e, B) = M \in \mathbb{C}$ has to be determined. It is trivial to calculate as $x \in B$ just have to be inserted in $F^1_{\frac{e_1 + e_2}{2}, \, e_1}(\cdot)$. For

$$c = \sqrt{\frac{(M_1^2 + M_2^2) + \sqrt{(M_1^2 + M_2^2)^2 - 4M_2^2}}{2}} \tag{17}$$

and

$$d = \sqrt{\frac{(M_1^2 + M_2^2) - \sqrt{(M_1^2 + M_2^2)^2 - 4M_2^2}}{2}} \tag{18}$$

it can be shown that a can be chosen as $\frac{\|e_1 - e_2\|}{2} c$, b as $\frac{\|e_1 - e_2\|}{2} d$. It is known that

$$\cos^2 t_0 = \frac{1 - d^2}{c^2 - d^2}. \tag{19}$$

For $t = \arctan(\sin t_0 / \cos t_0)$ set $t_0 = -t$ if $F^3_{1/2}(e, B) \geq 0$ or $F^3_{1/2}(e, B) \leq 0$. Otherwise set $t_0 = t$. If $F^3_2(e, B) \geq 0$ than $\delta = 1$ otherwise $\delta = -1$.

The system says (e, B) is a half ellipse if all $x \in B$ lay in some ε-neighborhood of $(\tilde{e}, \tilde{B})$ with respect to maximum norm, which can be analytically determined as (a, b, t_0, δ) are known and only intersection points of four lines with the half ellipse are to be found Fig.11.

Fig. 11. Checking if a point is in an ε-neighborhood of a half ellipse with respect to maximum norm

5.4   Color Extraction

Similarly to an edge a half ellipse has the first and the last point. Hence it also has the left and the right site. A half ellipse is a sequence of lines. A line is a sequence of edges. The color of the right side of a half ellipse is set as the arithmetic mean of the right side RGB values of all edges belonging to the half ellipse. The color of the left site is set correspondingly.

6   Experimental Results

This paper describes the most important components of a half ellipse detector. Some details had to be omitted to spare space. The experimental results described below are valid for one implementation of the algorithm The half ellipse detector was developed for a new object recognition system. To evaluate the system the well known database COIL-100 (Columbia Object Image Library) was used. The data set is described in [Nene et al., 1996]. It contains 7200 color images of 100 3D objects. One image is taken per 5° of rotation.

The computer used in the experiments has a processor Intel(R) Core(TM)2 Duo CPU P8600 @2.40 GHz 2.40 GHz and 4.00 GB RAM. The system is implemented in Java.

There were made 2 experiments with slightly different parameter settings. In the first experiment 18 views(1 per 20°) were used to learn each object. The remaining 5400 images were analyzed. A recognition rate of 99.2% was reached. In the second experiment 8 views(1 per 45°) were used to learn an object. The other 6400 were analyzed. A recognition rate of 96.3% was reached.

Average time demand to extract half ellipses from one COIL-100 image is 95 milliseconds.

The Table 1 compares the system with alternative methods. It is based on the results described in [Obdrzalek and Matas, 2011],[Yang et al., 2000] and [Caputo et al., 2000].

Table 1. Comparison with alternative results

Method	18 views	8 views
LAFs	99.9%	99.4%
Sergeev and Palm	99.2%	96.3%
SNoW / edges	94.1%	89.2%
SNoW / intensity	92.3%	85.1%
Linear SVM	91.3%	84.8%
Spin-Glass MRF	96.8%	88.2%
Nearest Neighbor	87.5%	79.5%

7   Conclusion and Future Work

There have been mainly two reasons to develop the algorithm described in this paper. First of all the endpoints of half ellipses are very important for the object representation used in the new system [Sergeev and Palm, 2011]. Secondly the color on the both sides of a half ellipse has to be extracted.

This paper does not deliver a complete description of a half ellipse detector. As the space is limited it only contains a terse depiction of the principle elements of the algorithm. Nevertheless the remaining trivial details can be elaborated easily.

The running time as well as the quality of the detector still have to be improved significantly. The are probably a variety of equivalent heuristic means to upgrade the quality. At the same time only one way of running time optimization appears to be realistic: parallelization.

References

[Canny, 1986] Canny, J.: A computational approach to edge detection. IEEE Transactions on Pattern Analysis and Machine Intelligence (1986)

[Caputo et al., 2000] Caputo, B., Hornegger, J., Paulus, D., Niemann, H.: A spin-glass markov random field for 3d object recognition. In: NIPS 2000 (2000)

[H.K. Yuen, 1988] Yuen, H.K., Illingworth, J.: Ellipse detection using the hough transformation. In: 4. Alvey Vision Conference (1988)

[Hough, 1962] Hough, P.V.C.: Method and Means of Recognising Complex Patterns. US Patent 3069654 (1962)

[Nene et al., 1996] Nene, S.A., Nayar, S.K., Murase, H.: Columbia Object Image Library, COIL-100 (1996)

[Obdrzalek and Matas, 2011] Obdrzalek, S., Matas, J.: Object recognition using local affine frames. In: BMVC (2011)

[Prewitt, 1970] Prewitt, J.: Picture Processing and Psychopictorics, ch. (75-149). Academic Press, London (1970)

[Sergeev and Palm, 2011] Sergeev, N., Palm, G.: A new object recognition system. In: VISAPP 2011 (2011)

[Tsuji and Matsumoto, 1978] Tsuji, S., Matsumoto, F.: Detection of ellipses by a modified hough transform. IEEE Trans. Comput. (1978)

[Yang et al., 2000] Yang, M.-H., Roth, D., Ahuja, N.: Learning to recognize 3D objects with sNoW. In: Vernon, D. (ed.) ECCV 2000. LNCS, vol. 1842, pp. 439–454. Springer, Heidelberg (2000)

A Robust Forensic Hash Component
for Image Alignment

Sebastiano Battiato, Giovanni Maria Farinella, Enrico Messina,
and Giovanni Puglisi

Image Processing Laboratory
Department of Mathematics and Computer Science
University of Catania
Viale A. Doria 6 - 95125 Catania, Italia
{battiato,gfarinella,emessina,puglisi}@dmi.unict.it
http://iplab.dmi.unict.it

Abstract. The distribution of digital images with the classic and newest technologies available on Internet (e.g., emails, social networks, digital repositories) has induced a growing interest on systems able to protect the visual content against malicious manipulations that could be performed during their transmission. One of the main problems addressed in this context is the authentication of the image received in a communication. This task is usually performed by localizing the regions of the image which have been tampered. To this aim the received image should be first registered with the one at the sender by exploiting the information provided by a specific component of the forensic hash associated with the image. In this paper we propose a robust alignment method which makes use of an image hash component based on the Bag of Visual Words paradigm. The proposed signature is attached to the image before transmission and then analyzed at destination to recover the geometric transformations which have been applied to the received image. The estimator is based on a voting procedure in the parameter space of the geometric model used to recover the transformation occurred to the received image. Experiments show that the proposed approach obtains good margin in terms of performances with respect to state-of-the art methods.

Keywords: Image forensics, Forensic hash, Bag of Visual Word, Tampering, Geometric transformations, Image validation and authentication.

1   Introduction and Motivations

The growing demand of techniques useful to protect digital visual data against malicious manipulations is induced by different episodes that make questionable the use of visual content as evidence material [1]. Methods useful to establish the validity and authenticity of a received image are needed in the context of Internet communications. To this aim different solutions have been recently proposed in literature [2,3,4,5,6]. Most of them share the same basic scheme: i) a hash code based on the visual content is attached to the image to be sent; ii) the hash is analyzed at destination to verify the reliability of the received image.

G. Maino and G.L. Foresti (Eds.): ICIAP 2011, Part I, LNCS 6978, pp. 473–483, 2011.

An image hash is a distinctive signature which represents the visual content of the image in a compact way (usually just few bytes). The image hash should be robust against allowed operations and at the same time it should differ from the one computed on a different/tampered image. Image hashing techniques are considered extremely useful to validate the authenticity of an image received through the Internet. Although the importance of the binary decision task related to the image authentication, this is not always sufficient. In the application context of Forensic Science is fundamental to provide scientific evidences through the history of the possible manipulations applied to the original image to obtain the one under analysis. In many cases, the source image is unknown, and, as in the application context of this paper, all the information about the manipulation of the image should be recovered through the short image hash signature, making more challenging the final task. The list of manipulations provides to the end user the information needed to decide whether the image can be trusted or not.

In order to perform tampering localization[1], the receiver should be able to filter out all the geometric transformations (e.g., rotation, scaling) added to the tampered image by aligning the received image with the one at the sender [6]. The alignment should be done in a semi-blind way: at destination one can use only the received image and the image hash to deal with the alignment problem since the reference image is not available. The challenging task of recovering the geometric transformations occurred on a received image from its signature motivates this paper.

Despite different robust alignment techniques have been proposed by computer vision researchers [7], these techniques are unsuitable in the context of forensic hashing, since a fundamental requirement is that the image signature should be as "compact" as possible to reduce the overhead of the network communications. To fit the underlying requirements, authors of [5] have proposed to exploit information extracted through Radon transform and scale space theory in order to estimate the parameters of the geometric transformations. To make more robust the alignment phase with respect to manipulations such as cropping and tampering, an image hash based on robust invariant features has been proposed in [6]. The latter technique extended the idea previously proposed in [4] by employing the Bag of Visual Words (BOVW) model to represent the features to be used as image hash. The exploitation of the BOVW representation is useful to reduce the space needed for the image signature, by maintaining the performances of the alignment component.

Building on the technique described in [6], we propose a new method to detect the geometric manipulations occurred on an image starting from the hash computed on the original one. Differently than [6], we exploit replicated visual words and a voting procedure in the parameter space of the transformation model employed to establish the geometric parameters (i.e., rotation, scale, translation). As pointed out by the experimental results, the proposed approach obtains the

---

[1] Tampering localization is the process of localizing the regions of the image that have been manipulated for malicious purposes to change the semantic meaning of the visual message.

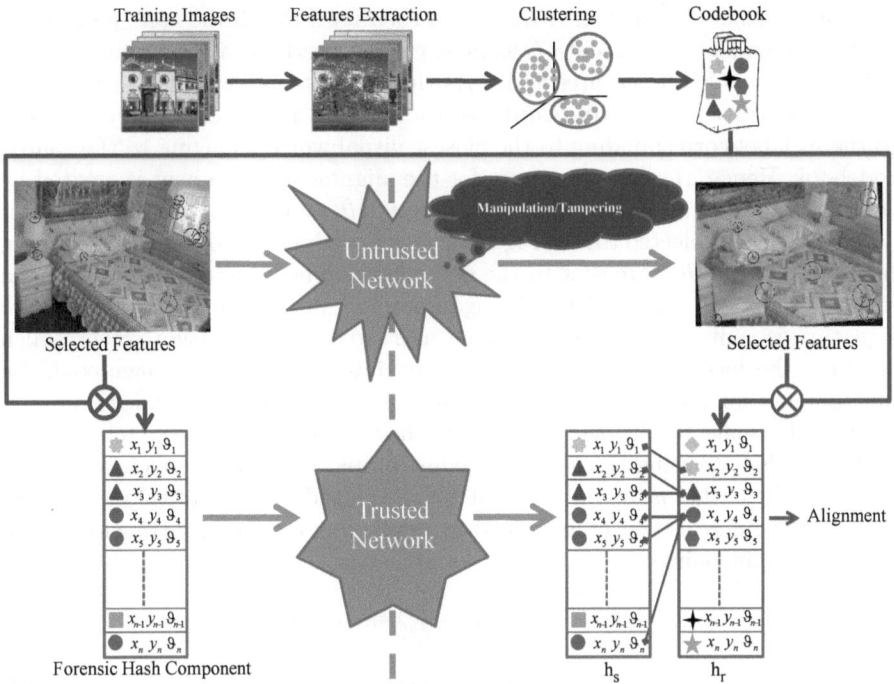

Fig. 1. Schema of the proposed approach.

best results with a significant margin in terms of estimation accuracy with respect to the approach proposed in [6].

The remainder of the paper is organized as follows: Section 2 presents the proposed framework. Section 3 reports the experiments and discusses the results. Finally, Section 4 concludes the paper with avenues for further research.

2   Proposed Approach

As stated in the previous section, one of the common steps of tampering detection systems is the alignment of the received image. Image registration is crucial since all the other tasks (e.g., tampering localization) usually assume that the received image is aligned with the original one, and hence could fail if the registration is not properly done. Classical registration approaches [7] cannot be directly employed in the considered context due the limited information that can be used (i.e., original image is not available and the image hash should be as short as possible).

The schema of the overall system is shown in Fig. 1. As in [6], we adopt a Bag of Visual Words based representations [8] to reduce the dimensionality of the feature descriptors to be used as hash component for the alignment. A codebook is generated by clustering the set of SIFT [9] extracted on training images. The pre-computed codebook is shared between sender and receiver. It should be noted that the codebook is built only once, and then used for all the communications between a sender and a receiver (i.e., no extra overhead for each communication). Sender extracts SIFT features and sorts them in descending order with respect to their contrast values. Afterward, the top n SIFT are selected and associated to the id label corresponding to the closest visual word belonging to the shared codebook. Hence, the final signature for the alignment component is created by considering the id label, the dominant direction θ, and the keypoint coordinates x and y for each selected SIFT (Fig. 1). The source image and the corresponding hash component (h_s) are sent to the destination. The system assumes that the image is sent over a network consisting of possibly untrusted nodes, whereas the signature is sent upon request through a trusted authentication server which encrypts the hash in order to guarantee its integrity [3]. The image could be manipulated for malicious purposes during the untrusted communication.

Once the image reaches the destination, the receiver generates the related hash signature (h_r) by using the same procedure employed by the sender. Then, the entries of the hashes h_s and h_r are matched by considering the id values (see Fig. 1). The alignment is hence performed by employing a similarity transformation of the keypoint pairs corresponding to matched hashes entries:

$$x_r = x_s \lambda \cos \alpha - y_s \lambda \sin \alpha + T_x \qquad (1)$$

$$y_r = x_s \lambda \sin \alpha + y_s \lambda \cos \alpha + T_y \qquad (2)$$

The above transformation is used to model the geometrical manipulations which have been done on the source image during the untrusted communication. The model assumes that a point (x_s, y_s) in the source image I_s is transformed in a point (x_r, y_r) in the image I_r at destination with a combination of rotation (α), scaling (λ) and translation (T_x, T_y). The aim of the alignment phase is the estimation of the quadruple $(\widehat{\lambda}, \widehat{\alpha}, \widehat{T_x}, \widehat{T_y})$ by exploiting the correspondences $((x_s, y_s), (x_r, y_r))$ related to matchings between h_s and h_r. We propose to use a cascade approach to perform the parameter estimation. First the estimation of $(\widehat{\alpha}, \widehat{T_x}, \widehat{T_y})$ is accomplished through a voting procedure in the parameter space (α, T_x, T_y). Such procedure is performed after filtering outlier matchings by taking into account the differences between dominant orientations of matched entries. Then the scaling parameter $\widehat{\lambda}$ is estimated by considering the parameters $(\widehat{\alpha}, \widehat{T_x}, \widehat{T_y})$ which have been previously estimated on the reliable information obtained through the filtering. The proposed method is detailed in the following.

Moving T_x and T_y on the left side and making the ratio of (1) and (2) the following equation is obtained:

$$\frac{x_r - T_x}{y_r - T_y} = \frac{x_s \cos \alpha - y_s \sin \alpha}{x_s \sin \alpha + y_s \cos \alpha} \qquad (3)$$

Solving (3) with respect to T_x we get the formula to be used in the voting procedure:

$$T_x = \left(\frac{x_s \cos \alpha - y_s \sin \alpha}{x_s \sin \alpha + y_s \cos \alpha} \right) (T_y - y_r) + x_r \qquad (4)$$

Each pair of coordinates (x_s, y_s) and (x_r, y_r) in (4) represents a line in the parameter space (α, T_x, T_y). An initial estimation of $(\widehat{\alpha}, \widehat{T_x}, \widehat{T_y})$ is obtained by considering the densest bin of a 3D histogram in the quantized parameter space. This means that the initial estimation of $(\widehat{\alpha}, \widehat{T_x}, \widehat{T_y})$ is accomplished in correspondence of the maximum number of intersections between lines generated by matched keypoints (Fig. 2).

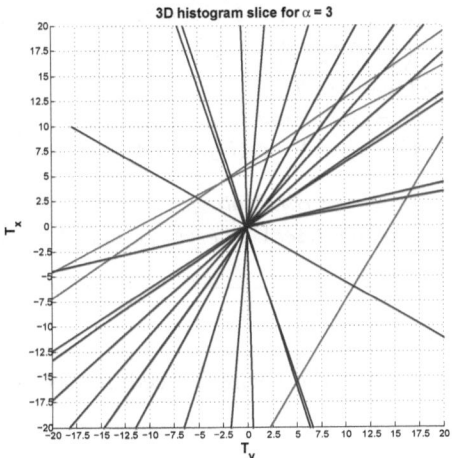

3D histogram slice for $\alpha = 3$

Fig. 2. A slices of the 3D histogram in correspondence of $\alpha = 3$, obtained considering an image manipulated with parameters $(\lambda, \alpha, T_x, T_y) = (1, 3, 0, 0)$. For a fixed rotational angle $\overline{\alpha}$, each pair of coordinates (x_s, y_s) and (x_r, y_r) votes for a line in the quantized 2D parameter space (T_x, T_y). Lines corresponding to inliers (blue) intersect in the bin $(T_x, T_y) = (0, 0)$, whereas the remaining lines (red) are related to outliers.

As said before, to discard outliers (i.e., wrong matchings) the information coming from the dominant directions (θ) of the SIFT are used during the voting procedure. In particular $\Delta\theta = \theta_r - \theta_s$ is a rough estimation of the rotational angle α. Hence, for each fixed triplet $(\overline{\alpha}, \overline{T_x}, \overline{T_y})$ of the quantized parameter space, the voting procedure considers only the matchings between h_s and h_r such that $|\Delta\theta - \overline{\alpha}| < t_\alpha$. The threshold value t_α is chosen to consider only matchings with a rough estimation $\Delta\theta$ which is closer to the considered $\overline{\alpha}$ (e.g., consider just matchings with a small initial error of ± 3.5 degree). The proposed approach is summarized in Algorithm 1.

The proposed method gives an estimation of rotation angle $\widehat{\alpha}$, and translation vector $(\widehat{T_x}, \widehat{T_y})$ by taking into account the quantized values used to build the 3D histogram into the parameter space. To refine the estimation we can use the m matchings which have been generated by the m lines which intersect in the selected bin. Specifically, for each pair $(x_{s,i}, y_{s,i})$, $(x_{r,i}, y_{r,i})$ corresponding to the selected bin, we consider $(\widehat{T_{x,i}}, \widehat{T_{y,i}}) = \left(\left(\frac{x_s \cos\overline{\alpha} - y_s \sin\overline{\alpha}}{x_s \sin\overline{\alpha} + y_s \cos\overline{\alpha}} \right) (\overline{T_y} - y_r) + x_r, \overline{T_y} \right),$

Algorithm 1. Parameters estimation through voting procedure.

Input: The set M of matching pairs $((x_s, y_s), (x_r, y_r))$
Output: The estimated parameter $(\widehat{a}, \widehat{T_x}, \widehat{T_y})$
begin
    $Initialize\ Votes(i, j, k) := 0\ \forall i, j, k;$
    for $\overline{\alpha} = -180, -179, \ldots, 0 \ldots, 179, 180$ **do**
        $V_\alpha = \{((x_s, y_s), (x_r, y_r)) \mid |(\theta_r - \theta_s) - \overline{\alpha}| < t_\alpha\};$
        foreach $((x_s, y_s), (x_r, y_r)) \in V_\alpha$ **do**
            for $\overline{T_y} = min_{T_y}, min_{T_y} - 1, \ldots, max_{T_y} - 1, max_{T_y}$ **do**
                $T_x := \left(\frac{x_s \cos \overline{\alpha} - y_s \sin \overline{\alpha}}{x_s \sin \overline{\alpha} + y_s \cos \overline{\alpha}} \right) (\overline{T_y} - y_r) + x_r;$
                $\overline{T_x} := Quantize(T_x);$
                $(i, j, k) := QuantizedValuesToBin(\overline{\alpha}, \overline{T_x}, \overline{T_y});$
                $Votes(i, j, k) := Votes(i, j, k) + 1;$

    $(i_{max}, j_{max}, k_{max}) = SelectBin(Votes);$
    $(\widehat{a}, \widehat{T_x}, \widehat{T_y}) := BinToQuantizedValues(i_{max}, j_{max}, k_{max});$
end

---

with $(\overline{\alpha}, \overline{T_y})$ obtained through the voting procedure (see Algorithm 1), and use the equations (5) and (6).

$$x_{r,i} = x_{s,i} \lambda_i \cos \alpha_i - y_{s,i} \lambda_i \sin \alpha_i + \widehat{T_{x,i}} \tag{5}$$

$$y_{r,i} = x_{s,i} \lambda_i \sin \alpha_i + y_{s,i} \lambda_i \cos \alpha_i + \widehat{T_{y,i}} \tag{6}$$

Solving (5) and (6) with respect to $a_i = \lambda_i \cos \alpha_i$ and $b_i = \lambda_i \sin \alpha_i$ we obtain

$$\widehat{a}_i = \frac{y_{r,i} y_{s,i} + x_{r,i} x_{s,i} - x_{s,i} \widehat{T_{x,i}} - y_{s,i} \widehat{T_{y,i}}}{x_{s,i}^2 + y_{s,i}^2} \tag{7}$$

$$\widehat{b}_i = \frac{x_{s,i} y_{r,i} - x_{r,i} y_{s,i} + y_{s,i} \widehat{T_{x,i}} - x_{s,i} \widehat{T_{y,i}}}{x_{s,i}^2 + y_{s,i}^2} \tag{8}$$

Since the ratio $\widehat{b}_i / \widehat{a}_i$ is by definition equals to $\tan \alpha_i$, we can estimate $\widehat{\alpha}_i$ with the following formula:

$$\widehat{\alpha}_i = \arctan \left(\frac{x_{s,i} y_{r,i} - x_{r,i} y_{s,i} + y_{s,i} \widehat{T_{x,i}} - x_{s,i} \widehat{T_{y,i}}}{y_{r,i} y_{s,i} + x_{r,i} x_{s,i} - x_{s,i} \widehat{T_{x,i}} - y_{s,i} \widehat{T_{y,i}}} \right) \tag{9}$$

Once $\widehat{\alpha}_i$ is obtained, the following equation derived from (5) and (6) is used to estimate $\widehat{\lambda}_i$

$$\widehat{\lambda}_i = \frac{1}{2} \left(\frac{x_{r,i} - \widehat{T_{x,i}}}{x_{s,i} \cos \widehat{\alpha}_i - y_{s,i} \sin \widehat{\alpha}_i} + \frac{y_{r,i} - \widehat{T_{y,i}}}{x_{s,i} \sin \widehat{\alpha}_i + y_{s,i} \cos \widehat{\alpha}_i} \right) \tag{10}$$

The above method produce a quadruple $(\widehat{\lambda}_i, \widehat{\alpha}_i, \widehat{T_{x,i}}, \widehat{T_{y,i}})$ for each matching pair $(x_{s,i}, y_{s,i})$, $(x_{r,i}, y_{r,i})$ corresponding to the bin selected with the voting procedure. The final transformation parameters $(\widehat{\lambda}, \widehat{\alpha}, \widehat{T_x}, \widehat{T_y})$ to be used for the alignment are computed by averaging over all the m produced quadruple:

$$\widehat{T_x} = \frac{1}{m}\sum_i^m \widehat{T_{x,i}} \quad \widehat{T_y} = \frac{1}{m}\sum_{i=1}^m \widehat{T_{y,i}} \quad \widehat{\alpha} = \frac{1}{m}\sum_{i=1}^m \widehat{\alpha}_i \quad \widehat{\lambda} = \frac{1}{m}\sum_{i=1}^m \widehat{\lambda}_i \quad (11)$$

It should be noted that some id values may appear more than once in h_s and/or in h_r. For example, it is possible that a selected SIFT has no unique dominant direction [9]; in this case the different directions are coupled with the same descriptor, and hence will be considered more than once by the selection process which generates many instance of the same id with different dominant directions.

As experimentally demonstrated in the next section, by retaining the replicated visual words the accuracy of the estimation increases, and the number of "unmatched" images decreases (i.e., image pairs that the algorithm is not able to process because there are no matchings between h_s and h_r). Differently than [6], the described approach considers all the possible matchings in order to preserve the useful information. The correct matchings are hence retained but other wrong pairs could be generated. Since the noise introduced by considering correct and incorrect pairs can badly influence the final estimation results, the presence of possible wrong matchings should be considered during the estimation process. The approach described in this paper deals with the problem of wrong matchings combining in cascade a filtering strategy based on the SIFT dominant direction (θ) with a robust estimator based on a voting strategy on the parameters' space. In this way the information of spatial position of keypoints and their dominant orientations are jointly considered. The scale factor is estimated only at the end of the cascade on reliable information. As shown in the experiments reported in the following section, replicated matchings help to better estimate the rotational parameter, whereas the introduced cascade approach allows robustness in estimating the scale factor.

3   Experimental Results

This section reports a number of experiments on which the proposed approach has been tested and compared with respect to [6]. The tests have been performed considering a subset of the fifteen scene category benchmark dataset [10]. The training set used in the experiments is built through a random selection of 150 images from the scene dataset. Specifically, ten images have been randomly sampled from each scene category. The test set consists of 5250 images generated through the application of different transformations on the training images[2]. Accordingly with [6], the following image manipulations have been applied (Tab. 1):

---

[2] Training and test sets used for the experiments are available at
http://iplab.dmi.unict.it/download/ICIAP_2011/Dataset.rar

Table 1. Image transformations

Operations	Parameters
Rotation (α)	3, 5, 10, 30, 45 degrees
Scaling (σ)	factor= 0.5, 0.7, 0.9, 1.2, 1.5
Cropping	19%, 28%, 36%, of entire image
Tampering	block size 50x50
Compression	JPEG Q=10
Various combinations of above operations	

cropping, rotation, scaling, tampering, JPEG compression. The considered transformations are typically available on image manipulation software. Tampering has been performed through the swapping of blocks (50 × 50) between two images randomly selected from the training set. Images obtained through various combinations of the basic transformations have been also included to make more challenging the task to be addressed. Taking into account the different parameter settings, for each training image there are 35 corresponding manipulated images into the test set.

To demonstrate the effectiveness of the proposed approach, and to highlight the contribution of the replicated visual words during the estimation, the comparative tests have been performed by considering our method and the approach proposed in [6]. Although Lu et al. [6] claim that further refinements are performed using the points that occur more than once, actually they do not provide any implementation detail. In our test we have hence considered two versions of [6] with and without replicated matchings. The approach proposed in [6] has been reimplemented. The Ransac thresholds used in [6] to perform the geometric parameter estimation have been set to 3.5 degrees for the rotational model and to 0.025 for the scaling one. These thresholds values have been obtained through data analysis (inliers and outliers distributions). In order to perform a fair comparison, the threshold t_α used in our approach to filter the correspondences (see Section 2) has been set with the same value of the threshold employed by Ransac to estimate the rotational parameter in [6]. The value T_y needed to evaluate (4) has been quantized considering a step of 2.5 pixels (see Fig. 2). Finally, a codebook with 1000 visual words has been employed to compare the different approaches. The codebook has been learned through k-means clustering on the overall SIFT descriptors extracted on training images.

First, let us examine the typically cases in which the considered approaches are not able to work. Two cases can be distinguished: i) no matchings are found between the hash built at the sender (h_s) and the one computed by the receiver (h_r); ii) all the matchings are replicated. The first problem can be mitigated considering a higher number of features (SIFT). The second one is solved only allowing the replicated matchings (see Section 2). As reported in Tab. 2, by increasing the number of SIFT there is a decreasing of the number of unmatched images for both approaches. In all cases the percentage of images on which our algorithm is not able to work is lower than the one of [6].

Tab. 3 shows the results obtained in terms of rotational and scale estimation through mean error. To properly compare the methods, the results have been

Table 2. Comparison with respect to unmatched images

	Unmatched Images			
Number of SIFT	15	30	45	60
Lu et al. [6]	11.73%	3.60%	1.64%	0.91%
Proposed approach	4.44%	1.26%	0.57%	0.46%

Table 3. Average rotational and scaling error

	Mean Error α				Mean Error σ			
Number of SIFT	15	30	45	60	15	30	45	60
Lu et al. [6]	12.8135	13.7127	13.2921	13.5840	0.1133	0.1082	0.1086	0.1124
Lu et al. [6] with replicated matchings	6.7000	4.1444	3.3647	2.8677	0.1522	0.1783	0.1981	0.2169
Proposed approach	2.2747	1.2987	0.6514	0.5413	0.0710	0.0393	0.0230	0.0183

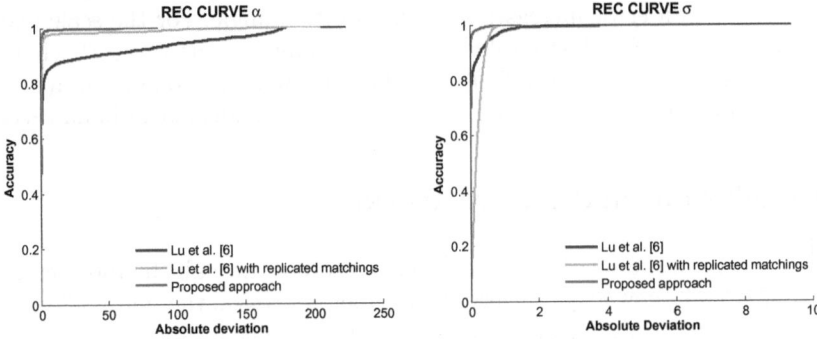

Fig. 3. REC curves comparison

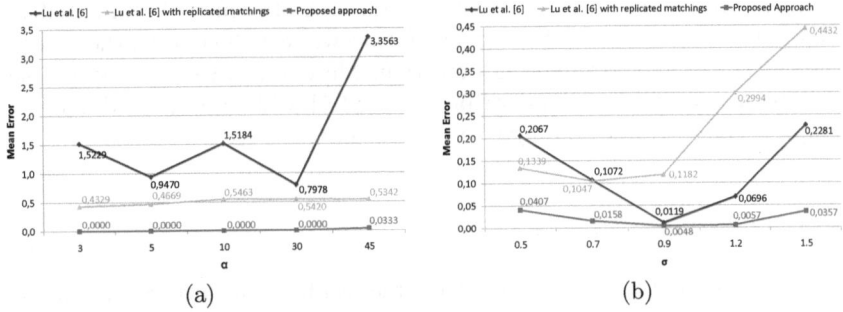

Fig. 4. Comparison on single transformation (60 SIFT). (a) Average rotation error at varying of the rotation angle. (b) Average scaling error at varying of the scale factor.

computed taking into account only the images on which all approaches are able to work. Our approach outperforms [6] obtaining a considerable gain both in terms of rotational and scaling accuracy (Tab. 3). Moreover, the performance of our approach significantly improves with the increasing of the extracted feature points (SIFT). On the contrary, the technique in [6] is not able to exploit the additional information coming from the higher number of extracted points.

To further study the contribution of the replicated matchings we performed tests by considering the modified version of [6] in which replicated matchings have been allowed (Tab. 3). Although the modified approach obtains better performance with respect to the original one in terms of rotational accuracy, it is not able to obtain satisfactory results in terms of scaling estimation. The wrong pairs introduced by the replicated matchings cannot be handled by the method. Our approach deals with the problem of wrong pairs combining a filtering based on the SIFT dominant direction (θ) with a robust estimator based on voting.

To better compare the methods, the Regression Error Characteristic Curves (REC) have been employed (Fig. 3). The area over the curve is an estimation of the expected error of a model. The proposed approach obtains the best results.

Additional experiments have been performed to examine the dependence of the average rotational and scaling error with respect to the rotation and scale transformation parameters respectively. Results in Fig. 4(a) show that the rotational estimation error increases with the rotation angle. For the scale transformation, the error has lower values in the proximity of one (no scale change) and increases considering scale factors higher or lower than one (Fig. 4(b)). It should be noted that our approach obtains the best performances in all cases.

4   Conclusions and Future Works

The assessment of the reliability of an image received through the Internet is an important issue in nowadays society. This paper addressed the image alignment task in the context of distributed forensic systems. Specifically, a robust image alignment component which exploits an image signature based on the Bag of Visual Words paradigm has been introduced. The proposed approach has been experimentally tested on a representative dataset of scenes obtaining effective results in terms of estimation accuracy. Future works will concern the extension of the system to allow tampering detection. Moreover, a selection step able to takes into account the spatial distribution of SIFT will be addressed in order to avoid their concentration on high textured regions.

References

1. Farid, H.: Digital doctoring: how to tell the real from the fake. Significance 3(4), 162–166 (2006)
2. Battiato, S., Farinella, G.M., Messina, E., Puglisi, G.: Understanding Geometric Manipulations of Images Through BOVW-Based Hashing. In: IEEE International Workshop on Content Protection & Forensics, held in conjunction with the IEEE International Conference on Multimedia & Expo (2011)
3. Lin, Y.-C., Varodayan, D., Girod, B.: Image authentication based on distributed source coding. In: IEEE International Conference on Image Processing, pp. 3–8 (2007)
4. Roy, S., Sun, Q.: Robust hash for detecting and localizing image tampering. In: IEEE International Conference on Image Processing, pp. 117–120 (2007)

5. Lu, W., Varna, A.L., Wu, M.: Forensic hash for multimedia information. In: IS&T-SPIE Electronic Imaging Symposium - Media Forensics and Security (2010)
6. Lu, W.J., Wu, M.: Multimedia forensic hash based on visual words. In: IEEE International Conference on Image Processing, pp. 989–992 (2010)
7. Szeliski, R.: Image alignment and stitching: A tutorial. Foundations and Trends in Computer Graphics and Computer Vision 2(1), 1–104 (2006)
8. Csurka, G., Dance, C.R., Fan, L., Willamowski, J., Bray, C.: Visual categorization with bags of keypoints. In: ECCV International Workshop on Statistical Learning in Computer Vision (2004)
9. Lowe, D.: Distinctive image features from scale-invariant keypoints. International Journal of Computer Vision 60(2), 91–110 (2004)
10. Lazebnik, S., Schmid, C., Ponce, J.: Beyond bags of features: Spatial pyramid matching for recognizing natural scene categories. In: IEEE Conference on Computer Vision and Pattern Recognition, pp. 2169–2178 (2006)

Focus of Expansion Localization through Inverse C-Velocity

Adrien Bak[1], Samia Bouchafa[1], and Didier Aubert[2]

[1] Institut d'Electronique Fondamentale, Université Paris-Sud, Bat 220
91405, Orsay, France
{adrien.bak,samia.bouchafa}@ief.u-psud.fr
[2] IFSTTAR
58 Bld Lefebvre, 75015 Paris, France
aubert@ifsttar.fr

Abstract. The Focus of Expansion (FoE) sums up all the available information on translational ego-motion for monocular systems. It has also been shown to present interesting features in cognitive research. As such, its localization bears great importance, either for robotic applications, as well as for attention fixation research. It will be shown that the so-called C-Velocity framework can be inversed in order to extract the FoE position from a rough scene structure estimation. This method rely on robust cumulative framework and only exploit the optical flow field relative norm as such, it is robust to angular noise and bias on the absolute optical flow norm.

1 Introduction – State of the Art

The Focus of Expansion holds a particular importance for monocular vision systems. Its position contains all the available information on the translational ego-motion $\boldsymbol{T} = \begin{vmatrix} T_X \\ T_Y \\ T_Z \end{vmatrix}$:

$$\boldsymbol{FoE} = \begin{vmatrix} x_{FoE} = f\frac{T_X}{T_Z} \\ y_{FoE} = f\frac{T_Y}{T_Z} \end{vmatrix} \tag{1}$$

Knowing its position is useful for a wide variety of problems, such as path planification or collision avoidance [2][3]. The importance of the FoE is not limited to computer vision or robotics. Indeed, it has been shown that the FoE holds a strong importance in the study of behaviour, cognition, and more specifically, visual attention [4].

As such, the localization of the FoE has always been an active research topic. Early methods [5][6] relied on the extraction and matching of image primitives. Such a discrete approach was dictated by the computational costs and reliability of early optical flow implementation. As such, those were not very robust due to the fact that they used only partial information.

With the emergence of efficient optical flow algorithms, global methods were introduced. Most methods rely on the fact that the Focus of Expansion is point

G. Maino and G.L. Foresti (Eds.): ICIAP 2011, Part I, LNCS 6978, pp. 484–493, 2011.

of convergence of all optical flow field lines. For example, the authors of [7] use a voting scheme in order to locate it. In a similar manner, the work presented in [8] presents a matched filter that exploits the change of sign of the optical flow component around the FoE to detect it. An interesting approach, based on vector calculus, was unveiled in [9]. However, due to the fact that it exploits the computation of divergence and curl of the optical flow, it is highly sensitive to noise. Finally, methods based purely on geometric concerns used to be prohibitive, due to their computational costs, authors of [10] present a novel way to take geometric constraints into account, but the computation cost is still unfitted to real-time applications.

All these methods make use of both the angular and the radial components of the optical flow, at least implicitly. However, as it will be shown in this study, this is not necessary. Radial component only is sufficient to extract the position of the FoE. Actually, the radial component is only needed up to a scale factor. This makes the proposed method insensitive to noise on the angular component and to bias on the radial component. The former being the prominent one in most real-time capable algorithms, especially on image discontinuities [11], the later being a common pitfall of local optical flow estimation methods.

Contrary to previous works, the method presented here stems from a scene structure estimation method. As such, the method proposed here is the second step toward an integrated monocular system allowing to determine both scene structure and translational ego-motion without relying on feature extraction or additional hardware.

In the following section, the basic hypothesis will be detailed, the initial Direct C-Velocity framework will be shortly described. Then the proposed method, Inverse C-Velocity , will be unveiled. Results obtained on simulated and real data will the be presented. Finally, these results will be discussed and some future developments will be presented.

2   C-Velocity Reminder

2.1   Basis – Notations

A vision sensor, attached to a mobile (*e.g.* a car) is moving in a world consisting of dynamic and static objects.

The sensor is represented by the pinhole model, its focal length is denoted f. Between $t = t_0$ and $t = t_1$ the sensor undergoes an arbitrary translational motion $\vec{T} = \begin{vmatrix} T_X \\ T_Y \\ T_Z \end{vmatrix}$. A static world point M is located (in a sensor-centered frame of reference) in $\begin{vmatrix} X_M \\ Y_M \\ Z_M \end{vmatrix}$ a $t = t_0$. At $t = t_1$, its coordinates are $\begin{vmatrix} X_M - T_X \\ Y_M - T_Y \\ Z_M - T_Z \end{vmatrix}$. Its image at $t = t_0$ being $m = \begin{vmatrix} x \\ y \end{vmatrix}$, the optical flow in m can be expressed as:

$$\mu = \frac{\partial}{\partial t}(f\frac{X_M}{Z_M}) \approx f(\frac{\Delta X_M}{Z_M} - X_M \frac{\Delta Z_M}{Z_M{}^2})$$
$$\nu = \frac{\partial}{\partial t}(f\frac{Y_M}{Z_M}) \approx f(\frac{\Delta Y_M}{Z_M} - Y_M \frac{\Delta Z_M}{Z_M{}^2}) \tag{2}$$

Where $\Delta A = A(t_1) - A(T_0)$.

Under classical assumptions, the optical flow, at a point $m = \begin{vmatrix} x \\ y \end{vmatrix}$ can then be written:

$$\begin{cases} \mu = \frac{xT_Z - fT_X}{Z} \\ \nu = \frac{yT_Z - fT_Z}{Z} \end{cases} \tag{3}$$

Let us now consider a planar surface of equation:

$$nP = d \tag{4}$$

with $n = \begin{vmatrix} n_x \\ n_y \\ n_z \end{vmatrix}$ its normal vector, P the position vector of a particular world point and d the distance between the plane and the origin. The motion field, for this particular plane is given by, when combining 1 and 2:

$$\begin{cases} \mu = \frac{1}{fd}\left(a_1 x^2 + a_2 xy + a_3 fx + a_4 fy + a_5 f^2\right) \\ \nu = \frac{1}{fd}\left(a_1 xy + a_2 y^2 + a_6 fy + a_7 fx + a_8 f^2\right) \end{cases} \tag{5}$$

where:

$$a_1 = T_Z n_x \qquad a_2 = T_Z n_y \qquad a_3 = T_Z n_z - T_X n_x \; a_4 = T_X n_y$$
$$a_5 = T_X n_z \; a_6 = T_Z n_z - T_Y n_y \qquad a_7 = T_Y n_x \qquad a_8 = T_Y n_z$$

2.2 Direct C-Velocity

Direct C-Velocity is a method that aims at detecting particular scene planes[1], representative of an urban environment through an analysis of motion, especially, optical flow:

Plane Type	n	Distance to Origin
Road	$[0,1,0]^T$	d_r
Building	$[1,0,0]^T$	d_b

In the following, a Building plane will be taken as an example, generalization is immediate.

From Equation 5:

$$\begin{cases} \mu = \frac{x}{fd}\left(x - x_{FoE}\right) \\ \nu = \frac{x}{fd}\left(y - y_{FoE}\right) \end{cases} \tag{6}$$

The two equations 6 can be combined by writing: $w = \sqrt{\mu^2 + \nu^2}$

$$w = \frac{T_Z |x|}{fd}\|m - FoE\| \tag{7}$$

---

[1] The original C-Velocity work consider a third plane type 'Obstacle', however, it is not relevant for this study.

At this point, so-called *c-values* can be defined, each being specific of a particular plane hypothesis.

Plane Hypothesis	C-Value	Notation
Building	$\|x\| \, \|\boldsymbol{m} - \boldsymbol{FoE}\|$	$c_{building}$
Road	$\|y\| \, \|\boldsymbol{m} - \boldsymbol{FoE}\|$	c_{road}

These *c-values* depend only on the coordinates of the studied point and on the coordinate of the FoE.

It is noticeable that, in the case of a point that actually belongs to a building plane:

$$w \propto c_{Building} \tag{8}$$

Direct C-Velocity uses this relation in order to establish a generalized hough transform in which, every image point would vote for a set of curves able to enforce its w value. A further processing step (clustering, Hough-Transform) would later on be used in order to extract dominant lines, which represent scene planes in the corresponding C-Velocity voting space. This whole method was based on prior knowledge of the position of the FoE. It is worth noting that, due to numerical constraints ($c - values$ can typically be in the range of hundreds of thousands), the actual relation to be used is:

$$w \propto \kappa^2 \tag{9}$$

with $\kappa = \sqrt{c}$.

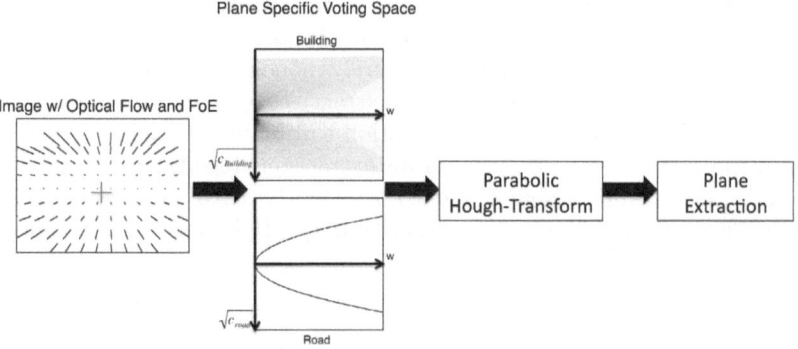

Fig. 1. Direct C-Velocity Workflow

Figure 1 illustrates this. The starting point is a computed optical flow and a known FoE. From these information, voting spaces are populated. In the presented case, only one road plane is present, in the road voting space, it will form a perfect parabola, when in the building voting space, it will generate incoherent noise. Coherent parabolas can then be extracted in order to detect and locate scene planes.

One can refer to [1] for further details.

3   Inverse C-Velocity

Inverse C-Velocity objective is the exact opposite. Instead of using ego-motion prior information to extract scene structure information, scene structure prior information will be used in order to extract translational ego-motion. First, the scene structure information will be briefly discussed, then, the method for localizing the FoE will be exposed.

3.1   Scene Structure Evaluation

This system relies on the identification of sets of coplanar points (at least one) and on the knowledge of their relative normal vectors.The actual plane equations are not needed, neither are the distance from the planes to the sensor. The basic requirement for a useful scene structure evaluation system is that it provides a set of label, their corresponding plane types and a map of these labels in the image space. Not all image points need to be labelled, and not all plane categories need to be represented.

Several systems could fulfil these requirements, just to cite a few: a calibrated LIDAR system [12], a stereo-system [13], or the original C-Velocity Framework [14]. For practical reasons, in this study, a stereo-system will be used. Both V-Disparity [15] and a Generalized Hough Transform, very close to the one presented in [13] will be used in order to detect, respectively, the road plane and building planes.

However, it is important to note that, in this study, stereo-vision was only used as a plane detector, the actual depth of individual pixels was not used. Furthermore, in future developments, a purely monocular version of this system, that uses Direct C-Velocity in order to extract scene structure and Inverse C-Velocity in order to compute the ego-motion will be investigated.

Fig. 2 presents the results of such an extraction. Results, especially on the road plane, might appear sparse. Indeed, the poorly textured nature of the road plane leads to mismatches in the disparity maps. It is important to note that

Fig. 2. Image Extracted from a real sequence. Overlays: Red - the building plane ; Green - the road plane.

Fig. 3. Image extracted from a Sivic-generated sequence. Overlays: Red - the building plane ; Green - the road plane.

it is not relevant, for this study, to strictly identify a plane. What is relevant however, is to ensure that all points sharing the same label (*e.g.* red or green in the above example) belong to the same plane and that this plane validates one of the basic hypothesis (*e.g.* road or building). This drives the setup of the plane detection system toward reliability and not exhaustiveness.

3.2   Focus of Expansion Extraction

The key idea of Inverse C-Velocity is to find a metric that reflects the quality of a given FoE candidate, based on the representation exposed in the Direct C-Velocity formulation. Ideally, such a metric would be convex, allowing the use of traditional optimization techniques. In the following presentation of this evaluation metric, it will be assumed that the observed scene consists of one plane Π of type $T \in \{Building, Road\}$. However, generalization to multiple planes of multiple types is immediate.

For a given FoE candidate, Direct C-Velocity representation of Π can be computed[2]. If the FoE candidate is the actual FoE, the representation of Π will be a perfect parabola. On the other hand, the further the candidate will be from the actual FoE, the less points from Π will accumulate, thus forming a degenerate paraboloid.

It has been exhibited in [14] that the standard deviation of a given plane in the C-Velocity voting space is a function of the distance between the FoE candidate and the actual FoE:

$$\sigma_\Pi = d(\|FoE_{candidate} - FoE_{actual}\| \tag{10}$$

and that this function d is monotonically increasing, Fig. 4 and 5 illustrate this dispersion.

Knowing that the quality of the candidate FoE is directly related to the representation of Π in the C-Velocity voting space, the dispersion of the point cloud representing Π will then be used as an energy function:

$$\epsilon^2 = \sum_{m \in \Pi} (C_{observed} - C_{avg})^2 \tag{11}$$

Where $C_{observed}$ is the observed C-value, and C_{avg} is the average c-value, for a given w. Given the fact that dispersion is convex with respect to the relative position between the candidate and the actual FoE, this metric is also convex.

In this case, the error function is strictly convex with respect to the distance between the candidate and the actual FoE. As such, the problem of localizing the FoE becomes a simple energy minimization problem, for which traditional techniques can be used.

---

[2] In the multi-plane case, a separate voting per plane is used in order to eliminate cross-talk perturbation.

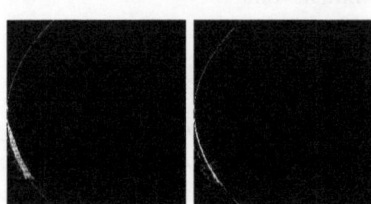

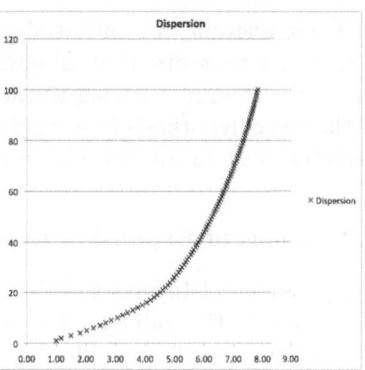

Fig. 4. Building plane voting spaces. Left: FoE hypothesis is located at the center of the image ; Right: FoE hypothesis is the one extracted. Parabolas are the best fit ones, horizontal line only indicates the origin of the c-values.

Fig. 5. Normalized dispersion inside the voting space with respect to the distance between the actual and the candidate FoE (synthetic data)

4   Results

Two sets of experiments were conducted. In the first one, the Sivic simulator [16] was used in order to accurately simulate a monocular system, attached to a car. In a second part, real image sequences were used. Those images are from the french ANR project *Logiciel d'Observation des Vulnérables (LoVE)*.

The experiments were conducted using optical flow provided by the FOLKI [17] method.

The FoE search was conducted using a simple Gradient-Descent approach.

4.1   Simulated Data

The Sivic simulator was used to generate a sequence of 250 640x480 images pairs, presenting a urban environment with moderate traffic. Along this sequence, the trajectory of the vehicle was mostly translational.

Fig. 3 presents one image extracted from this sequence, as well as the planes (overlays) extracted using the procedure described in 3.1.

For the same image, Fig. 6 presents the computed optical flow and the extracted Focus of Expansion, it is important to note that the extracted FoE lies on the line of horizon, which is consistent with the known motion of the camera. More specifically, for this particular image, the error between the estimated Focus of Expansion and the actual FoE was 2.2 pixels.

Over this sequence, the maximum positioning error was 15.6 pixels and the average positioning error was 5.2 pixels. This error can appear large, however, with the simulated optical system (a focal length of $10mm$ and a pixel size of $10\mu m$), an error of one pixel in the localization of the FoE translates to an error of 10^{-3} on the ratio $\frac{\|T\|}{T_Z}$, with: $\boldsymbol{T} = \begin{vmatrix} T_X \\ T_Y \end{vmatrix}$, for an order of magnitude, this represents a lateral translation of $0.6mm$ for a car moving at $50km.h^{-1}$.

The measured error can then be translated to an error on the estimation of the ratio $\frac{\|T\|}{T_Z}$ of 0.8%.

As a comparison, for the same image sequence a standard cumulative approach [7] leads to the following results: a maximum positioning error of 14 pixels and an average positioning error of 10.6 pixels, this can be translated to an error of 1.1% on the estimation of the ratio $\frac{\|T\|}{T_Z}$.

4.2   Real Data

Unfortunately, for the real data evaluation, ground-truth results are not available. As a consequence, quantitative results can not be produced. However, it is possible to qualitatively assess the validity of those results. For instance, Fig. 4 presents a visualization of the building voting space for an arbitrary FoE candidate, and the extracted FoE. This figure particularly illustrates the geometrical sense of the presented metric. The closer the candidate is to the actual FoE, the more the C-Velocity curve will fit the expected parabola.

Fig. 7, Fig.8 and Fig.9 present computed optical flows along with the extracted Focus of Expansion. Fig. 4 illustrates the building voting space for (respectively), the best extracted FoE and the initial candidate, located at the center of the frame. Particularly, it is interesting to note that, in both cases, the extracted FoE is coherent with the motion of the car. Optical Flow field lines can also be used as a visual cue to evaluate the position of the FoE.

Fig. 6. Image extracted from a Sivic sequence, including the computed Optical Flow and the extracted FoE

Fig. 7. Image Extracted from a real sequence, including the computed Optical Flow and the extracted FoE

Fig. 8. Image Extracted from a different sequence, including the computed Optical Flow and the extracted FoE

Fig. 9. Image Extracted from a different sequence, including the computed Optical Flow and the extracted FoE

5   Discussion – Prospective

Through this work, a method for computing the translational ego-motion of a monocular system has been unveiled. This method rely only on relative norm of the optical flow and so, it is insensitive to noisy estimation of its angular component and to bias on its norm. Furthermore, it uses robust cumulative methods which makes it robust toward punctual noise.

As it stems from a method for extraction scene structure information from motion, it will allow the definition of a single, iterative and cumulative framework for scene structure and ego-motion estimation in future developments.

Preliminary results are encouraging, however, some effort should be put into the study of the impact of noise as well as on the development of an absolute evaluation method for real data sets. Furthermore, rotations are currently neglected and should be taken into account.

Finally, the presented system could benefit from a fine analysis of the dispersion in the C-Velocity spaces, due to a shift of the FoE. This will be presented in future works.

Acknowledgements. The authors would like to acknowledge the french cluster for research in new technologies Digiteo for its funding and support to their projects.

References

1. Bouchafa, S., Zavidovique, B.: C-Velocity: A Cumulative Frame to Segment Objects From Ego-Motion. Pattern Recognition and Image Analysis 19, 583–590 (2009)
2. McCarthy, C., Barnes, N., Mahony, R.: A Robust Docking Strategy for a Mobile Robot using Flow Field Divergence. IEEE Transactions on Robotics 24, 832–842 (2008)

3. Ancona, N., Poggio, T.: Optical flow from 1d correlation: Application to a simple time-to-crash detector. In: Proceedings of the International Conference on Computer Vision, pp. 209–214 (1993)
4. Fukuchi, M., Tsuchiya, T., Koch, C.: The focus of expansion in optical flow fields acts as a strong cue for visual attention. Journal of Vision 9(8), article 137 (2009)
5. Prazdny, K.: Motion and Structure from Optical Flow. In: Proceedings of the Sixth International Joint Conference on Artificial Intelligence (1979)
6. Roach, J.W., Aggarwal, J.K.: Determining the Movement of Objects from a Sequence of Images. IEEE Transactions on Pattern Analysis and Machine Intelligence 2, 554–562 (1980)
7. Suhr, J.K., Jung, H.G., Bae, K., Kim, J.: Outlier rejection for cameras on intelligent vehicles. Pattern Recognition Letters 29, 828–840 (2008)
8. Sazbon, D., Rotstein, H., Rivlin, E.: Finding the focus of expansion and estimating range using optical flow images and a matched filter. Machine Vision and Applications 15, 229–236 (2004)
9. Hummel, R., Sundareswaran, V.: Motion-parameter Estiamtion from Global Flow Field Data. IEEE Transactions on Pattern Analysis and Machine Intelligence 15(5), 459–476 (1993)
10. Wu, F.C., Wang, L., Hu, Z.Y.: FOE Estimation: Can Image Measurement Errors Be Totally "Corrected" by the Geometric Method? Pattern Recognition 40(7), 1971–1980 (2007)
11. Baker, S., Scharstein, D., Lewis, J.P., Roth, S., Black, M.J., Szeliski, R.: A Database and Evaluation Methodology for Optical Flow. In: IEEE International Conference on Computer Vision, pp. 1–8 (2007)
12. Rodriguez, S.A., Frémont, F.V., Bonnifait, P.: Extrinsic Calibration Between a Multi-Layer LIDAR and a Camera. In: Proceedings of the IEEE International Conference on Multisensor Fusion and Integration for Intelligent Systems, pp. 214–219 (2008)
13. Iocchi, L., Konolige, K., Bajracharya, M.: Visually Realistic Mapping of a Planar Environment with Stereo. In: Experimental Robotics VII. LNCIS, vol. 271, pp. 521–532 (2001)
14. Bouchafa, S., Patri, A., Zavidovique, B.: Efficient Plane Detection From a Single Moving Camera. In: International Conference on Image Processing, pp. 3493–3496 (2009)
15. Labayrade, R., Aubert, D., Tarel, J.-P.: Real-Time Obstacle Detection on Non-Flat Road Geometry Through V-Disparity Representation. In: Intelligent Vehicles Symposium, pp. 646–651 (2002)
16. Gruyer, D., Royere, C., du Lac, N., Michel, G., Blosseville, J.-M.: SiVIC and RTMaps, interconnected platforms for the conception and the evaluation of driving assistance systems. In: IEEE Conference on Intelligent Transportation Systems (2006)
17. Le Besnerais, G., Champagnat, F.: Dense Optical Flow by Iterative Local Window Registration. In: IEEE International Conference on Image Processing, vol. 1, pp. 137–140 (2005)

Automated Identification of Photoreceptor Cones Using Multi-scale Modelling and Normalized Cross-Correlation

Alan Turpin[1], Philip Morrow[1], Bryan Scotney[1], Roger Anderson[2], and Clive Wolsley[2]

[1] School of Computing and Information Engineering, Faculty of Computing and Engineering
[2] School of Biomedical Sciences, Faculty of Life and Health Sciences,
University of Ulster, Northern Ireland
{turpin-a1,wolsley-c}@email.ulster.ac.uk,
{pj.morrow,bw.scotney,rs.anderson}@ulster.ac.uk

Abstract. Analysis of the retinal photoreceptor mosaic can provide vital information in the assessment of retinal disease. However, visual analysis of photoreceptor cones can be both difficult and time consuming. The use of image processing techniques to automatically count and analyse these photoreceptor cones would be beneficial. This paper proposes the use of multi-scale modelling and normalized cross-correlation to identify retinal cones in image data obtained from a modified commercially available confocal scanning laser ophthalmoscope (CSLO). The paper also illustrates a process of synthetic data generation to create images similar to those obtained from the CSLO. Comparisons between synthetic and manually labelled images and the automated algorithm are also presented.

Keywords: Modelling, Cross Correlation, Multi-Scale, Retinal Cones.

1   Introduction

A number of diseases can affect specific components of the eye, including retinitis pigmentosa, cone/cone-rod dystrophy, glaucoma, age-related macular degeneration and many more. Some of these can be visually detected and early treatment, where available, is applied. However those that affect the retina at the back of the eye, such as cone/cone-rod dystrophy and age-related macular degeneration are significantly harder to detect as it is very difficult to obtain images of the photoreceptor mosaic *in vivo*.

Previously, to view the retinal photoreceptor mosaic a post-mortem donor eye had to be placed under a microscope and cones were then counted. However, with recent advances in technology it is now possible to view the photoreceptor cones through the use of a modified confocal scanning laser ophthalmoscope (CSLO) or adaptive optics (AO) [1] in conjunction with a CSLO. The analysis of such images though can prove time consuming with regard to counting and calculating the density of the retinal cones.

G. Maino and G.L. Foresti (Eds.): ICIAP 2011, Part I, LNCS 6978, pp. 494–503, 2011.
© Springer-Verlag Berlin Heidelberg 2011

In previous work in this area Li and Roorda [2-5] have used AO in conjunction with a CSLO to capture images of the retinal photoreceptor. The images obtained using this technique clearly shows a hexagonal layout of the photoreceptor cones, however the method does not capture images of the rods as they are significantly smaller than the photoreceptor cones. The images that are produced from their approach clearly illustrate that cone brightness can vary across the entire retina. Images that are obtained from a modified CSLO are substantially different from those captured using Li and Roorda's method in both magnification and pixel resolution. Li and Roorda reported a 93-97% accuracy of correct photoreceptor cone identification on AO images.

Algorithms created for automated identification of photoreceptor cones on AO or post-mortem images may not perform as accurately when applied to images captured from a modified CSLO. Wojtas et al [6]achieved an accuracy of 97% on images obtained from a post-mortem donor eye (where the definition of accuracy is the percentage of the total number of cones correctly identified by the algorithm). They applied background removal and local maxima detection to identify the bright points in each image. The use of directional light was applied at 137° which in turn was then used to approximate the distance to the cone centre from the local maxima co-ordinate.

Curcio et al [7] studied the topography of the photoreceptor mosaic in post-mortem donor eyes. They found that the density of the human photoreceptor mosaic reduced when moving away from the fovea. Curcio reported a peak foveal density of 199,000 cones/mm².

Nicholson and Glaeser [8] have illustrated an approach of normalized cross-correlation to detect particles in cryo-electron microscopy images. Although this technique is applied to a different type of image, it may be possible to adapt this approach to retinal photoreceptor images. This process has the potential to be used if a model or series of models were created to mimic the shape and size of the photoreceptor cones.

The remained of this paper is organised as follows. Section 2 provides details of the technologies and methodologies used in this research. Section 3 illustrates the design implementation of our algorithms. Section 4 describes a process of synthetic data generation. Section 5 illustrates experimental results against synthetic and real image data and finally Section 6 provides conclusions and outlines further work to be carried out.

2   Technologies and Methodology

2.1   Technologies Available

The two main technologies that can capture images of the photoreceptor mosaic *in vivo* are adaptive optics (AO) and the use of a modified CSLO. AO was originally designed to correct for atmospheric distortions in telescope images and has subsequently been developed to correct for ocular distortions [9]. Such a system is large and requires a unique setup for each individual subject that is being imaged. This makes the system expensive and a commercial model has yet to be developed. However, the use of a CSLO such as the Heidelberg Retina Tomorgraph (HRT) [10] is widely available and relatively inexpensive.

The standard HRT uses a helium neon diode laser with a wavelength of 670nm. It has a depth resolution of 20-80um and a scanning window of between 10° and 20° field of view (FoV) with a resolution of 256x256 pixels. With these parameters the standard HRT cannot discriminate between the retinal layers and is mostly used to evaluate the optic nerve head and nerve fibre layer. Fig. 1 shows a standard HRT image at 10°x10° FoV. Using the current 10°x10° FoV scan window setting, the pixel size corresponds to around 10um. However, with significant modification of the HRT involving a reduction in the FoV to a 1°x1° scan area, minimizing the scan depth to cover 0.5mm and accurate fine focusing it is possible to capture enface images of the cone photoreceptors. Using these settings the pixel size is reduced to close to 1-2μm and successive scan planes are 0.0156mm apart. Fig. 2 shows an image captured with the modified HRT at 1°x1° FoV. In this image the retinal cones are clearly visible as small, bright circular regions. We can also see retinal blood vessels which appear in the image as the dark region passing from top right to bottom left. The HRT used in this research has been modified with these setting and is located in the Vision Sciences Research labratory at the University of Ulster.

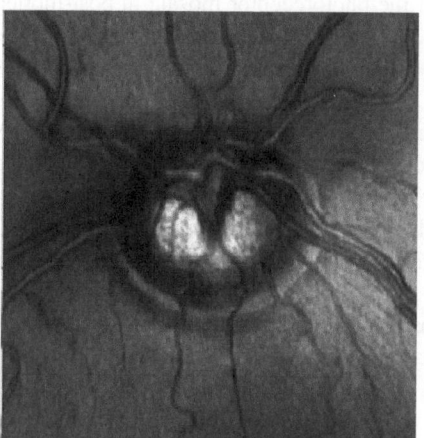 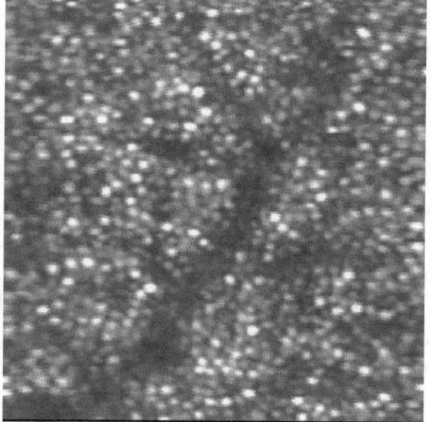

Fig. 1. A standard HRT image captured at 10°x10° FoV

Fig. 2. An example image of retinal cones captured from a modified HRT at 1°x1° FoV

2.2 Methodology

The overall aim of this research is to automatically analyze and detect photoreceptor cones in an image obtained from a modified CSLO. We aim to achieve this by initially modeling the size and shape of the retinal cones using a Gaussian based model. Normalized cross-correlation is then used to apply the Gaussian model to the retinal images. The output of this approach should highlight all the regions of the image that are similar to the shape of the Gaussian. We then apply local maxima detection to identify the highest points of similarity found in the normalized cross-correlation results. To validate our results across a large number of images we have also generated synthetic data through the use of Gaussian modelling and filtering to replicate the density, size and shape of the retinal cones. A small number of real images have also been manually labeled to act as ground truth data for evaluation purposes. Other approaches such as simple thresholding or applying local maxima detection were implemented however these produced poor results due to the low pixel magnification. We therefore decided to use an approach of Gaussian modelling and normalized cross-correlation to identify similarities in the intensity profiles of the images.

3   Implementation

3.1   Cone Modelling

The first step in our process is to generate a series of multi-scale Gaussian models that simulate the average shape, size and height of the retinal cones found in the real image data. After extensive analysis of the images we found that the majority of photoreceptor cones can be classified into one of three different types, small and bright, medium size and brightness or large and dull. Fig. 3 shows an example for each of these groups and Fig. 4 illustrates these cones displayed as a 3D surface.

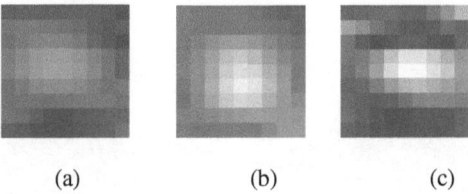

(a)   (b)   (c)

Fig. 3. Three different types of retinal cones, (a) dull and large, (b) medium size and brightness, (c) bright and small

We can generate a family of Gaussian models to approximate these cone shapes and sizes. A 2D Gaussian model can be defined using equation 1.

$$G(x, y) = e^{-\frac{x^2+y^2}{2\sigma^2}} . \tag{1}$$

where x and y are the size of the model and σ controls the spread of the model. The 'sclaing' value that would usually be found in this standard Gaussian formula has been omitted as we wish the values in our models to be always between 0 and 1. The spread determines if the model with be bright with a sharp decrease or dull and flat. The value of σ is specified based on a calculation of the half-height width of the model and is illustrated in equation 2.

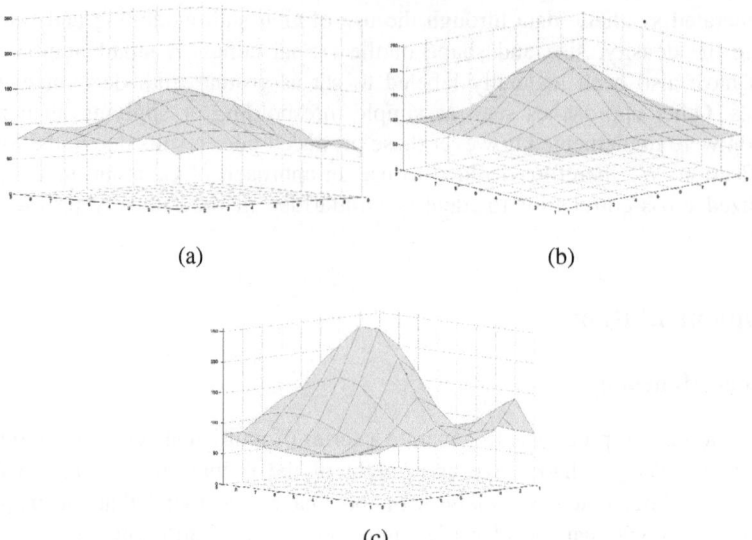

(a)   (b)

(c)

Fig. 4. The example retinal cones from Fig. 3 displayed as 3D surfaces

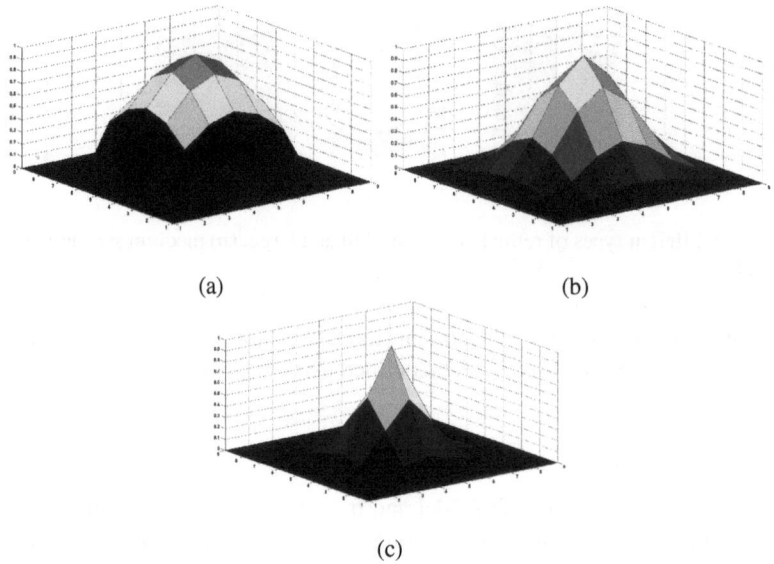

(a)   (b)

(c)

Fig. 5. Gaussian models for each cone type, (a) dull and large, (b) medium size and brightness, (c) bright and small

$$\sigma = \left(-\frac{(m/\gamma)^2}{2\ln(0.5)} \right)^{0.5} .\tag{2}$$

This formula determines the half-height width of the model given a mask size m and a spread γ. After experimenting with a variety of different sizes and spreads a series of Gaussian models were created, and these are illustrated in Fig. 5.

3.2 Normalized Cross-Correlation

Normalized cross-correlation compares a model (illustrated above) to each underlying neighbourhood of an image and returns a value between -1 and 1 at each point in the image. This value illustrates how similar the underlying neighbourhood of the image is when applied with a model. The more similar the underlying neighbourhood is to the model the closer the result will be 1. If the underlying neighbourhood is inverted with respect to the model, the results will be closer to -1. The standard normalized cross-correlation equation is illustrated in equation 3.

$$NCC(u,v) = \frac{\sum_{x,y} \left(f(x,y) - \overline{f}_{u,v} \right)\left(t(x-u, y-v) - \overline{t} \right)}{\left\{ \sum_{x,y} \left(f(x,y) - \overline{f}_{u,v} \right)^2 \sum_{x,y} \left(t(x-u, y-v) - \overline{t} \right)^2 \right\}^{0.5}} .\tag{3}$$

where f is the image, t is the mask, $\overline{t}$ is the mean of the mask and $\overline{f}_{u,v}$ is the mean of the region under the mask. As our approach uses multi-scale models we apply each model in turn. This produces three sets of normalized cross-correlation results from which we identify pixels with the highest similarity in each set of results.

3.3 Centre Identification

To identify the potential photoreceptor cone centres, we analyze the three normalized cross-correlation results. A local maxima detection algorithm is applied to the normalized cross-correlation results. This identifies all the local maximum pixel values in a neighbourhood of a specified size. The size of this neighbourhood is set to the size of the current Gaussian model. As the local maxima algorithm is applied to the three cross-correlation results, we are presented with three lists of co-ordinates. These three lists are then combined by checking the Euclidian distance of each co-ordinate. If two or more co-ordinates are within a Euclidian distance of 2 then the average position of these cones is taken. Also, the co-ordinates that are within the Euclidian distance are also marked as used so that they cannot be counted by any further iterations of the algorithm.

3.4 Post-processing

After the three lists of co-ordinates have been combined there is stil the possibility that some cone positions have been falsely identified and we must then analyse the co-ordinate set to remove those co-ordinates which do not correspond to real cone centres.

To achieve this we firstly average the three normalized cross-correlation results. From this we analyze the co-ordinates in the averaged cross-correlation results, if this co-ordinate is below a defined threshold the co-ordinate is then removed from the list. This threshold has been chosen emperically and is currently set to 0.3.

4 Synthetic Data Generation

The manually labeling of real image data is both difficult and time consuming, therefore a process of synthetic data generation could be very beneficial. This process is applied as an extension of the modelling process. After analysis of the real image data it was found that most images at 1°x1° FoV contain between 600 – 1000 photoreceptor cones varying in size, shape and height. Also there is a degree of noise in the real image data caused during image capture. In our approach to generating synthetic data we randomly place Gaussian models on a 256x256 pixel image ensuring that no two are overlapping. The size and spread of the models are randomly selected at 5, 7 or 9 and 4, 6 or 8 respectively. These random selections are of equal probability and inevitably there are a large number of different cones across the image with varying densities.

The background in real image data also varies. To generate a synthetic background, a 5x5 average filter was passed across a real image a large number of times (typically 100) in order to produce a smoothed image with varying background. The cones are then randomly placed onto this synthetic background. Fig. 6 illustrates 3 images with cone densities of 600, 800 and 1000.

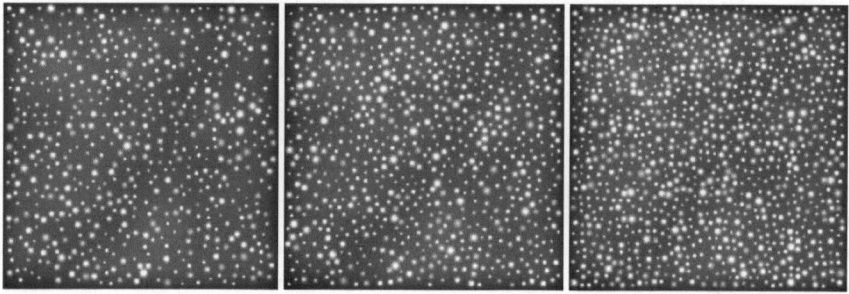

Fig. 6. Three synthetic images with varying densities of 600, 800 and 1000 simulated retinal cones

Using this process of generating ideal images a degree of noise can be applied to simulate real image data. For this we choose two different approaches; the first is to apply a 3x3 Gaussian smoothing algorithm to 70% of the image. When capturing the images, there can be significant eye movement in any direction. To simulate this we implement a process that applies a 3x3 pixel mask to the entire image. This mask will shift the centre pixel either up, down, left, right or not at all. Each pixel in the image has a 20% chance it will be shifted in a direction or not at all. Fig. 7 shows an example of each of these noise filters.

5 Experimental Results

For our experimentation we generated 30 synthetic images, 10 of which are ideal, 10 have a Gaussian smoothing filter and 10 have a random weighted filtering applied. Also three real images have been manually labeled to compare our results against. Fig. 8 illustrates a manually labeled and automatically identified real image. The retinal cones have been identified with a black or white '+' symbol to allow for contrast in the image.

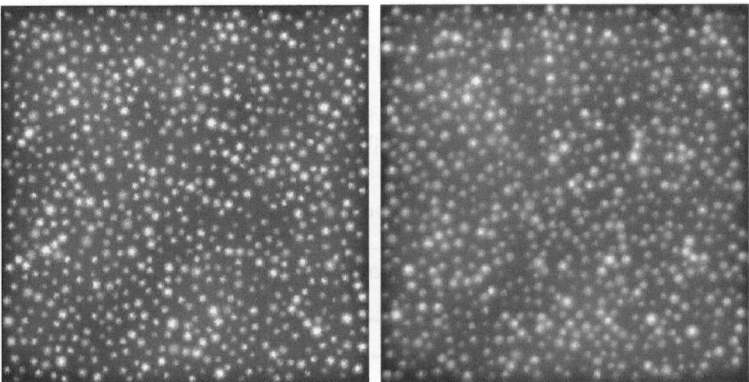

Fig. 7. Two example types of degradation applied to the synthetic images

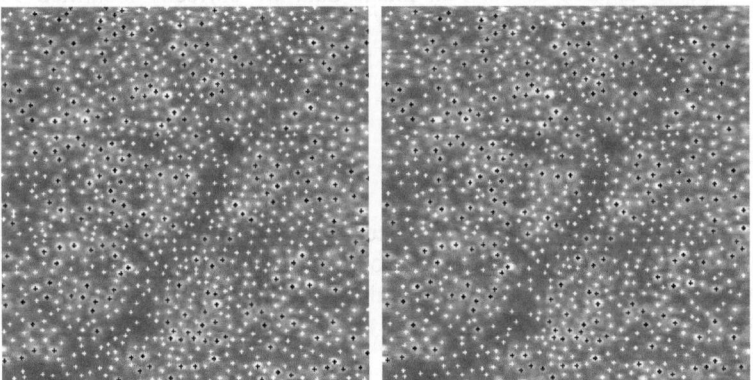

Fig. 8. A manually labelled and an automatically identified real image

Table 1 illustrates the results of our algorithms applied to the synthetic data generated earlier. These results have been separated into three sets, one for each of the synthetic data sets (ideal, Gaussian smoothed and random pixel shifted images).

Table 1. Results of our algorithm against the different types of synthetic data

	Ideal		Gaussian Smoothed		Randomly Shifted	
Manual *Algorithm*	*Cone* *Detected*	*No Cone* *Detected*	*Cone* *Detected*	*No Cone* *Detected*	*Cone* *Detected*	*No Cone* *Detected*
Cone Detected	8425	1	8664	135	8201	882
No Cone Detected	390	-----	422	-----	466	-----

This gives an average accuracy of 96.2% for correct cone identificaton. A relatively low percentage of 3.9% of the total number of photoreceptor cones were missed in our synthetic data. Also 4.8% of the total cones identified were false positives.

Table 2 illustrates the combined results of our algorithms when applied to the three real images (which have been manaually analysed).

Table 2. Results of our algorithm against real image data

Manual *Algorithm*	*Cone* *Detected*	*No Cone* *Detected*
Cone Detected	2364	145
No Cone Detected	172	-----

When this is applied to our real image data we correctly identified 94.2% of photoreceptor cones and missed 5.7%. Of the total cones identified 6.8% were false positives. These results clearly illustrate a good foundation for automatic retinal cone identification.

6 Conclusions

We have illustrated a process of automated photoreceptor cone identification using modelling and normalized cross-correlation. We have also presented a process for synthetic image data generation imitating the images obtained from a modified CSLO such as a HRT. This process has produced an accuracy of 95.2% on synthetic data and 93.2% on real image data.

Further work in this area would include modification of the modelling process in an attempt to increase the number of true positives found. Also the expansion of post-processing algorithms to reduce the number of false positives would be required. The automated identification and masking of retinal blood vessels can also be implemented. With the identification of retinal blood vessels, the ability to automatically align these images on top of a high-resolution Fundus image should be possible. This provides a visual tool to identify how far away from the fovea the image has been captured. Further development of synthetic data may include the random placement of retinal blood vessels and the increased implementation of noise to simulate poor images.

References

1. Miller, D., http://research.opt.indiana.edu/Labs/
   AdaptiveOptics/default.html
2. Li, K.Y., Roorda, A.: Automated Identification of Cone Photoreceptors in Adaptive Optics Retinal Images. J. Opt. Soc. Am. A. 24, 1358–1363 (2007)
3. Roorda, A., Zhang, Y., Duncan, J.L.: High-Resolution In Vivo Imaging of the RPE Mosaic in eyes with Retinal Disease. Invest Ophthalmos. Vis. Sci. 48, 2297–2303 (2007)
4. Roorda, A., Romero-Borja, F., Donnely, W., Queener, H., Hebert, T., Campell, M.: Adaptive Optics Scanning Laser Ophthalmoscopy. Opt. Express. 10, 405–412 (2002)
5. Roorda, A., Metha, A.B., Lennie, P., Williams, D.R.: Packing Arrangement of the Three Cone Classes in Primate Retina. Vision Res. 41, 1291–1306 (2001)
6. Wojtas, D.H., Wu, B., Ahnelt, P.K., Bones, P.J., Millane, R.P.: Automate Analysis of Differential Interference Contrast Microsopy Images of the Foveal Cone Mosaic. J. Opt. Soc. Am. A. 25, 1181–1189 (2008)
7. Curcio, C.A., Sloan, K.R., Kalina, R.E., Hendrickson, A.E.: Human Photoreceptor Topography. JCN 292, 497–523 (1990)
8. Nicholson, W.V., Glaeser, R.M.: Review: Automatic Particle Detection in Electron Microscopy. J. Struct. Biol. 133, 90–101 (2001)
9. Paques, M., Simonutti, M., Roux, M., Bellman, C., Lacombe, F., Grieve, K., Glanc, M., LeMer, Y., Sahel, J.: High-Resolution Imaging of Retinal Cells in the Living Eye 21, S18–S20 (2007)
10. Heidelberg Engineering, http://www.heidelbergengineering.com/

A Finite Element Blob Detector for Robust Features

Dermot Kerr, Sonya Coleman, and Bryan Scotney

[1] School of Computing and Intelligent Systems, University of Ulster, Magee, UK
[2] School of Computing and Information Engineering, University of Ulster, Coleraine, UK
{d.kerr,sa.coleman,bw.scotney}@ulster.ac.uk

Abstract. Traditionally feature extraction is focussed on edge and corner detection, however, more recently points of interest and blob like features have also become prominent in the field of computer vision and are typically used to determine correspondences between two images of the same scene. We present a new approach to a Hessian blob detector, designed within the finite element framework, which is similar to the multi-scale approach applied in the SURF detector. We present performance evaluation that demonstrates the accuracy of our approach in comparison to well known existing algorithms.

Keywords: blob detector, finite element framework.

1 Introduction

Standard corner detectors such as the Harris and Stephens corner detector [4] find corners in an image at one particular scale, but corner points within an image may occur at many natural scales depending on what they represent [7]. To deal with the many natural scales at which features may be present, corner detectors have been developed to work on multiple scales, thereby having the ability to detect all corners. Many of these corner detectors do not only detect actual corner points but also other "interesting points" that may not strictly be recognized as corners [8, 13]. For some particular applications the ability to detect interesting points that are robust to changes within the image is seen as a more desirable characteristic than specifically detection of real corner points. Blobs are interesting features prominent in images. Generally blobs can be thought of as regions in an image that are brighter or darker than the surrounding regions. Blob-type features provide complementary information not obtained from corner detectors [11].

The Laplacian of Gaussian [9] is a popular blob detector, where the image is convolved with a combined Laplacian and Gaussian kernel. However, the main limitation of this detector is that the detector operates at only one particular scale, but features within an image may appear at many different scales. A multi-scale Laplacian of Gaussian detector may be achieved by appropriately adjusting the size of the Gaussian and Laplacian kernels to obtain a set of kernels that are then applied to the image. Thus, a set of features may be detected at multiple scales. However, applying a detector at multiple scales may introduce other issues, as the same feature may be present over a range of scales within the detector's range [10], and by representing the same feature at many scales we increase the difficulty of matching

G. Maino and G.L. Foresti (Eds.): ICIAP 2011, Part I, LNCS 6978, pp. 504–513, 2011.

the detected features. Hence, a scale invariant approach is more appropriate, where the characteristic scale of the feature is identified. This characteristic scale is the scale at which the feature is most strongly represented, and it is not related to the resolution of the image, but rather the underlying structure of the detected feature [10]. By using an operator to measure the response of the same interest point at different scales, the scale at which the peak response is obtained can be identified. The Hessian-Laplace blob detector [11] is based on an approach analogous to this, where second order Gaussian smoothed image derivatives are used to compute the Hessian matrix. This matrix captures the important properties of the image structure. Using a multi-scale approach where kernel sizes are increased, the trace and the determinant of the Hessian matrix are thresholded and blob features detected.

In this paper we present a finite element based Hessian blob detector (FEH) based on techniques borrowed from our own FESID detector [6] and ideas from the SURF [2] and CenSurE detectors [1]. Performance is evaluated with respect to repeatability and feature matching using the evaluation techniques presented in [10], highlighting improvements when compared with other well known interest point detectors and descriptors.

2   Hessian Blob Detector Design

The finite element hessian blob detector for robust features uses second order derivative operators to detect blob-like features that are robust to various transformations.

For the propose of operator design and development, we consider an image to be represented by a rectangular $m{\times}n$ array of samples of a continuous function $u(x,y)$ of image intensity on a domain Ω. The most refined level of finite element discretisation of the image domain Ω is based on considering the image as a rectangular array of pixels. Nodes are placed at the pixel centres, and lines joining these form edges of elements in the domain discretisation. The near-circular operators presented in [12] were based on the use of a "virtual mesh", illustrated in Figure 1, consisting of regular triangular elements, which overlays a regular rectangular pixel array.

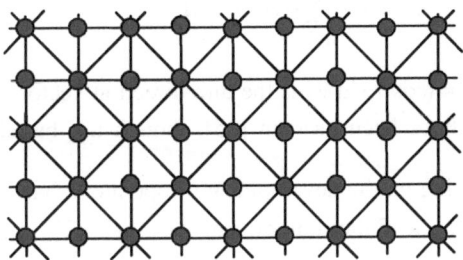

Fig. 1. Virtual mesh of triangular elements

With any node in a mesh, say node i, with co-ordinates (x_i, y_i) we associate a piecewise linear basis function $\phi_i(x, y)$ which has the properties

$$\phi_i(x_j, y_j) = \begin{cases} 1 & \text{if } i = j \\ 0 & \text{if } i \neq j \end{cases} \tag{1}$$

where (x_j, y_j) are the co-ordinates of the nodal point j. $\phi_i(x, y)$ is thus a "tent-shaped" function with support restricted to a small neighbourhood centred on node i consisting of only those elements that have node i as a vertex. For any scale parameter σ, we may define a neighbourhood Ω_i^σ centred on the node i at (x_i, y_i) and consisting of a compact subset of elements. If the set of nodes contained in or on the border of Ω_i^σ is denoted as D_i^σ, then we may approximately represent the image u over the neighbourhood Ω_i^σ by a function

$$U(x, y) = \sum_{j \in D_i^\sigma} U_j \phi_j(x, y) \tag{2}$$

in which the parameters $\{U_j\}$ are mapped from the sampled image intensity values. The approximate image representation is therefore a simple piecewise linear function on each element in the neighbourhood Ω_i^σ and has the sampled intensity value U_j at node j.

To formulate operators involving a weak form of the second order directional derivative in the finite element method, it is required that the image function $u \equiv u(x, y)$ be once differentiable in the sense of belonging to the Hilbert space $H^1(\Omega)$; i.e. the integral $\int_\Omega \left(|\underline{\nabla} u|^2 + u^2 \right) d\Omega$ is finite, where $\underline{\nabla} u$ is the vector $(\partial u / \partial x, \partial u / \partial y)^T$. To obtain a weak form of the second directional derivative, $-\underline{\nabla} \cdot (B \underline{\nabla} u)$, the respective derivative term is multiplied by a test function $v \in H^1$ and the result is integrated on the domain Ω to give

$$Z(u) = -\int_\Omega \underline{\nabla} \cdot (B \underline{\nabla} u) v \, d\Omega . \tag{3}$$

Here $B = \underline{b}\underline{b}^T$ and $\underline{b} = (\cos\theta, \sin\theta)$ is the unit direction vector. Integrated directional derivative operators have been used in [15], though not based on the weak forms introduced above.

Since we are focusing on the development of operators that can explicitly embrace the concept of scale, a finite-dimensional test space $T_\sigma^h \subset H^1$ is employed that explicitly embodies a scale parameter σ. In our design procedure the test space T_σ^h comprises a set of Gaussian basis functions $\psi_i^\sigma(x, y)$, $i = 1, \dots, N$ of the form

$$\psi_i^\sigma(x, y) = \frac{1}{2\pi\sigma^2} e^{-\left(\frac{(x-x_i)^2+(y-y_i)^2}{2\sigma^2}\right)} \tag{4}$$

Each test function $\psi_i^\sigma(x, y)$ is restricted to have support over a neighbourhood Ω_i^σ centred on the node i at (x_i, y_i). The size of the neighbourhood Ω_i^σ to which the support of $\psi_i^\sigma(x, y)$ is restricted is also explicitly related to the scale parameter σ [3]. We construct three operators Dxx_{ij}^σ, Dyy_{ij}^σ, and Dxy_{ij}^σ, representing the second order x- y- and mixed xy- derivatives respectively:

$$Dxx_{ij}^\sigma = \int_{\Omega_i^\sigma} \frac{\partial \phi_j}{\partial x} \frac{\partial \psi_i^\sigma}{\partial x} dxdy, i, j = 1, \dots, N \tag{5}$$

$$Dyy_{ij}^\sigma = \int_{\Omega_i^\sigma} \frac{\partial \phi_j}{\partial y} \frac{\partial \psi_i^\sigma}{\partial y} dxdy, i, j = 1, \dots, N \tag{6}$$

and

$$Dxy_{ij}^\sigma = \int_{\Omega_i^\sigma} \frac{\partial \phi_j}{\partial x} \frac{\partial \psi_i^\sigma}{\partial y} dxdy, i, j = 1, \dots, N \tag{7}$$

The integrals are computed as sums of the element integrals and are computed only over the neighbourhood Ω_i^σ, rather than the entire image domain Ω as ψ_i^σ has support restricted to Ω_i^σ.

3   Blob Detection

For efficient implementation, we have adopted the use of integral images introduced by Viola and Jones [14]; more recently integral images have been a key aspect of the SURF detector and we have previously successfully used integral images with the FESID detector [6] as integral images provide a means of fast computation using small convolution filters.

If an intensity image is represented by an array of $m \times n$ samples of a continuous function $u(x, y)$ of image intensity on a domain Ω, then the integral image value $I_\Sigma(x)$ at a pixel location $\mathbf{x} = (x, y)$ is the sum of all pixel values in the original image I within a rectangular area formed by the origin of the image and location $\mathbf{x}$:

$$I_\Sigma(\mathbf{x}) = \sum_{i=0}^{i \le x} \sum_{j=0}^{j \le y} I(i, j). \tag{8}$$

The time and number of operations required to compute any rectangular area of the integral image is independent of the size of that region, as four memory reads and three additions are required to compute any region,

Using the same multi-scale approach as the SURF detector we select the first filter size of a 9×9 pixel region. In our approach we partition the 9×9 pixel region slightly differently as illustrated in Figure 2.

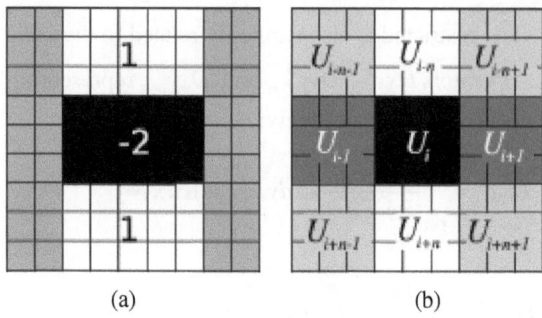

(a)   (b)

Fig. 2. 9×9 filter partitioning for (a) SURF and (b) FEH detector

Our approach differs from the SURF detector in that we need to compute 9-regions for each operator, rather than the 3 or 4 regions that are computed with the SURF detector. The filter partitioning allows the operator values to be simply mapped to the appropriate 3×3 region on the 9×9 filter. The operator values, $Dxx_{ij}^{\sigma}, Dyy_{ij}^{\sigma}$, and Dxy_{ij}^{σ}, (equations 5-7) mapped to the 9×9 regions are then convolved with the sum of the pixel intensities from each of the areas illustrated in Figure 2(b) to form the Hessian matrix

$$H = \begin{bmatrix} Dxx & Dxy \\ Dxy & Dyy \end{bmatrix} \tag{9}$$

The Hessian matrix captures the important properties of the local image structure by describing how the underlying shape varies [11].

The normalised determinant of the Hessian matrix is computed using the formula

$$\det(H) = Dxx\times Dyy + (0.81\times Dxy)^2 \tag{10}$$

where the constant term 0.81 is determined by the size of the filter, i.e., 9×9. This approximated determinant of the Hessian represents the blob response in the image at that particular location.

Similarly, blob responses are computed over further scales by increasing the overall size of the filter, but maintaining the 9 regions. For example, within the first octave filter sizes of 9×9, 15×15, 21×21, and 27×27 are used, and each of these filters has 9 individual regions of size 3×3, 5×5, 7×7, and 9×9 respectively. The blob response is computed over a total of 4 octaves that each contain 4 scale ranges. Blob responses that are not maxima or minima in the immediate neighbourhood of the selected blob are rejected by examining a $3\times3\times3$ neighbourhood (in the x- and y- spatial dimensions, and the scale dimension) around the selected blob. For the remaining blobs that give responses above a specified threshold, these blobs are interpolated in 3D (two spatial dimensions, and scale) to accurately localise the blob [6].

4 Experimental Results

We present results of comparative performance evaluation for our proposed FEH blob detector and other well-known blob detectors, the SURF detector and the Hessian Laplace detector; the detectors used for comparison are limited to those that are most similar to FEH in terms of operation. A full evaluation of various detectors using the same software and images has been carried out in [11], and the reader is referred to this work for full details.

Evaluation of FEH was performed using the set of test images and software provided from the collaborative work between Katholieke Universiteit Leuven, Inria Rhone-Alpes, Visual Geometry Group and the Center for Machine Perception (available for download at [16]). Using the repeatability metric we explicitly compare the geometrical stability of detected points of interest between different images of a scene under different viewing conditions. The test image set consists of real structured and textured images of various scenes, with different geometric and photometric transformations such as viewpoint change, image blur, illumination change, scale and rotation and image compression. For the detectors presented here we describe a circular region with a diameter that is 3× the detected scale of the point of interest, similar to the approach in [10, 11]. The overlap of the circular regions corresponding to an interest point pair in a set of images is measured based on the ratio of intersection and union of the circular regions. Thus, where the error in pixel location is less than 1.5 pixels and the overlap error is below 60%, similar to the evaluation of the SURF detector [2], the points of interest are deemed to correspond. Example images are shown in Figure 3.

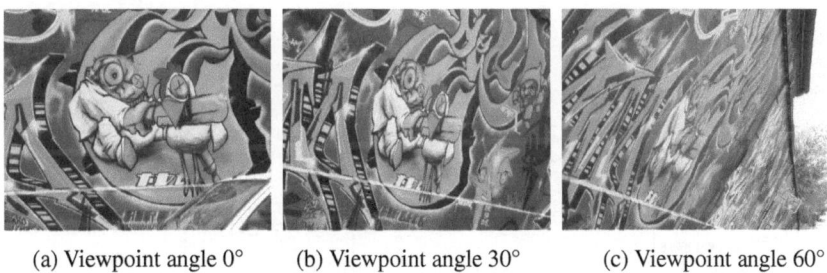

(a) Viewpoint angle 0°   (b) Viewpoint angle 30°   (c) Viewpoint angle 60°

Fig. 3. Viewpoint change - example image sequence [5]

In Figure 4(a) and 4(b) we present comparative evaluation of the detectors using the illumination change scene. The repeatability rate for the FEH detector is consistently better than either SURF or the Hessian-Laplace operators, and generally the number of correspondences is also greater. Figures 4(c) – 4(f) present the evaluation results using the viewpoint change scenes. In Figures 4(c) and 4(d), using the structured viewpoint change, the Hessian-Laplace operator has the best percentage repeatability, and the proposed FEH operator performs similarly to the SURF operator; the FEH has a slightly higher number of correspondences than the Hessian-Laplace operator.

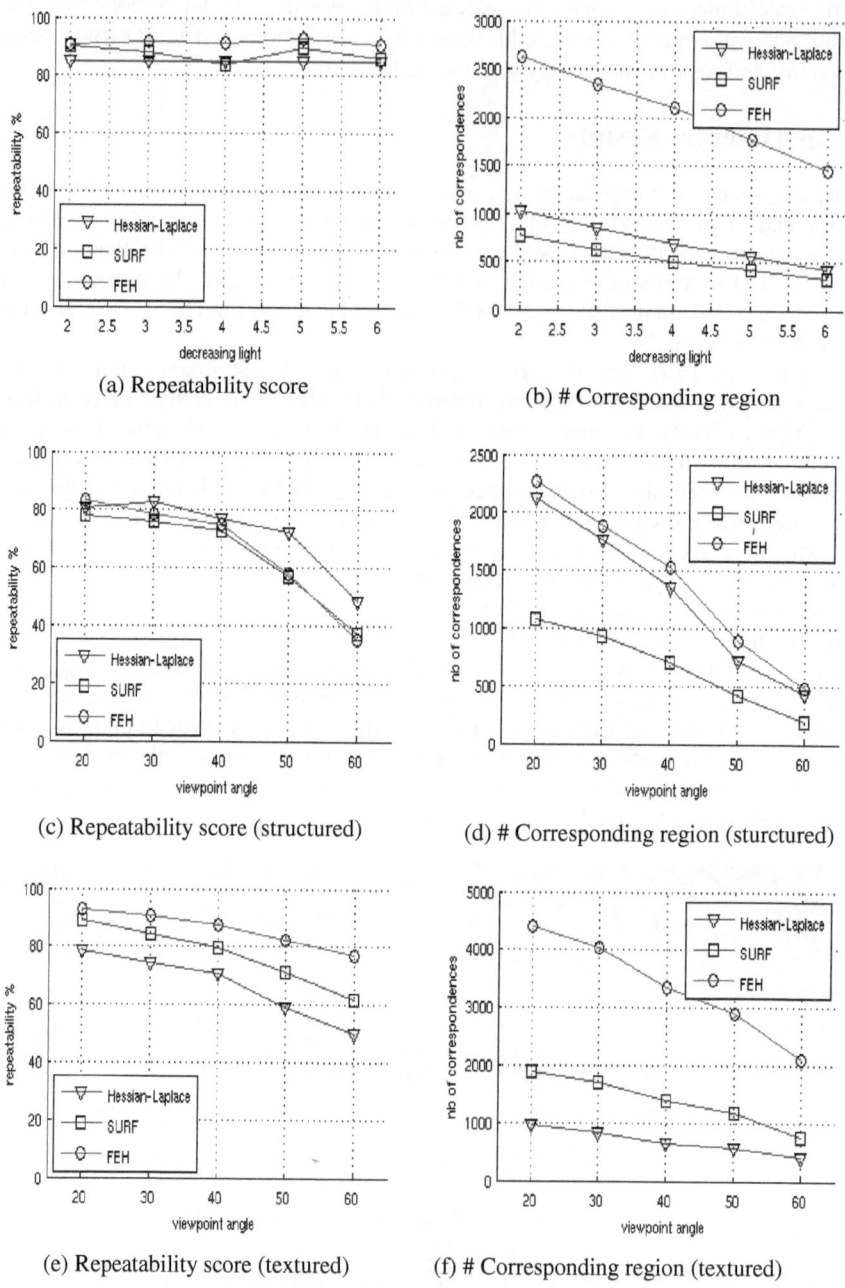

(a) Repeatability score

(b) # Corresponding region

(c) Repeatability score (structured)

(d) # Corresponding region (sturctured)

(e) Repeatability score (textured)

(f) # Corresponding region (textured)

Fig. 4. Repeatability score and number of corresponding regions for image sequences containing illumination change (a)-(b); viewpoint change (c)-(f)

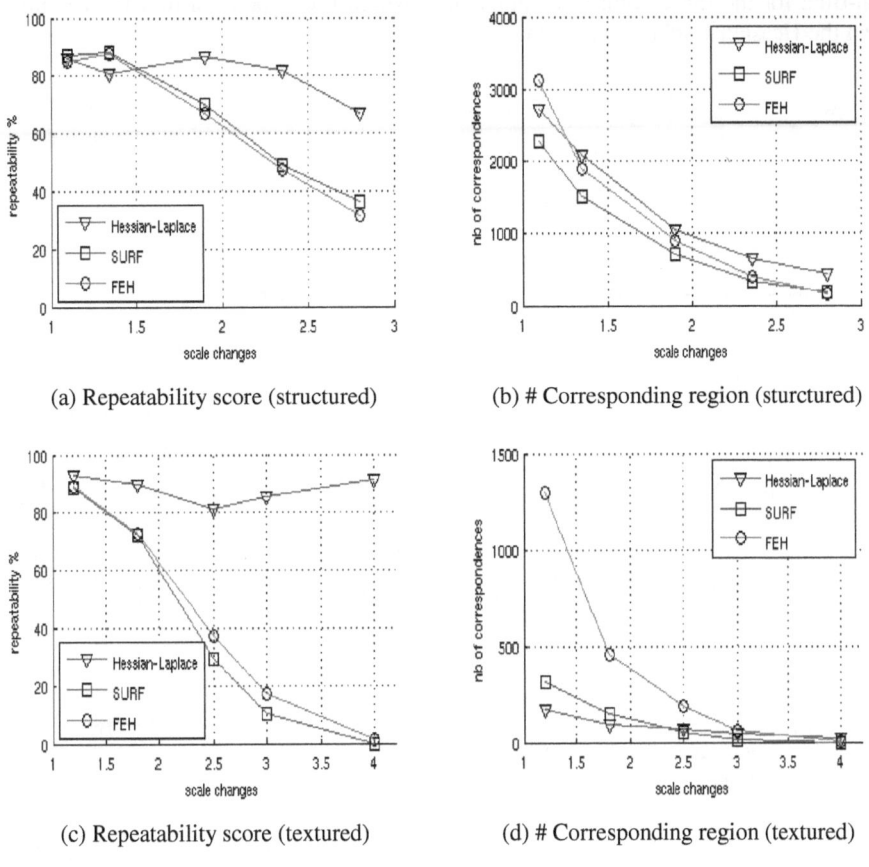

(a) Repeatability score (structured)   (b) # Corresponding region (sturctured)

(c) Repeatability score (textured)   (d) # Corresponding region (textured)

Fig. 5. Repeatability score and number of corresponding regions for image sequences containing scale change

However, in Figure 4(e), using the textured viewpoint change, the proposed FEH operator has the best percentage repeatability, with the Hessian-Laplace operator having the poorest performance; in Figure 4(f), we see that FEH has a significantly larger number of correspondences compared to either the SURF or Hessian-Laplace operators, which possibly leads to the improved repeatability rate.

Figure 5 presents results for the three detectors on image sequences containing scale change. In these images sequences, the Hessian-Laplace operator outperforms the other two operators with respect to repeatability score. However, Figures 5(a) and 5(c) illustrate that the performance of the proposed FEH operator is similar to the SURF operator; the performance of FEH is slightly poorer than SURF in Figure 5(a) and slightly better than SURF in Figure 5(c). Figures 5(b) and 5(d) again illustrate that the number of correspondences is high for the FEH operator. Figure 6 presents results for the three detectors on image sequences containing blur change, and here we see that the FEH and SURF operators perform similarly and better than the Hessian-Laplace operator. In addition to the performance evaluation, we include run-time for the three algorithms in Table 1 where it can be seen that FEH is faster than the Hessian-Laplace approach.

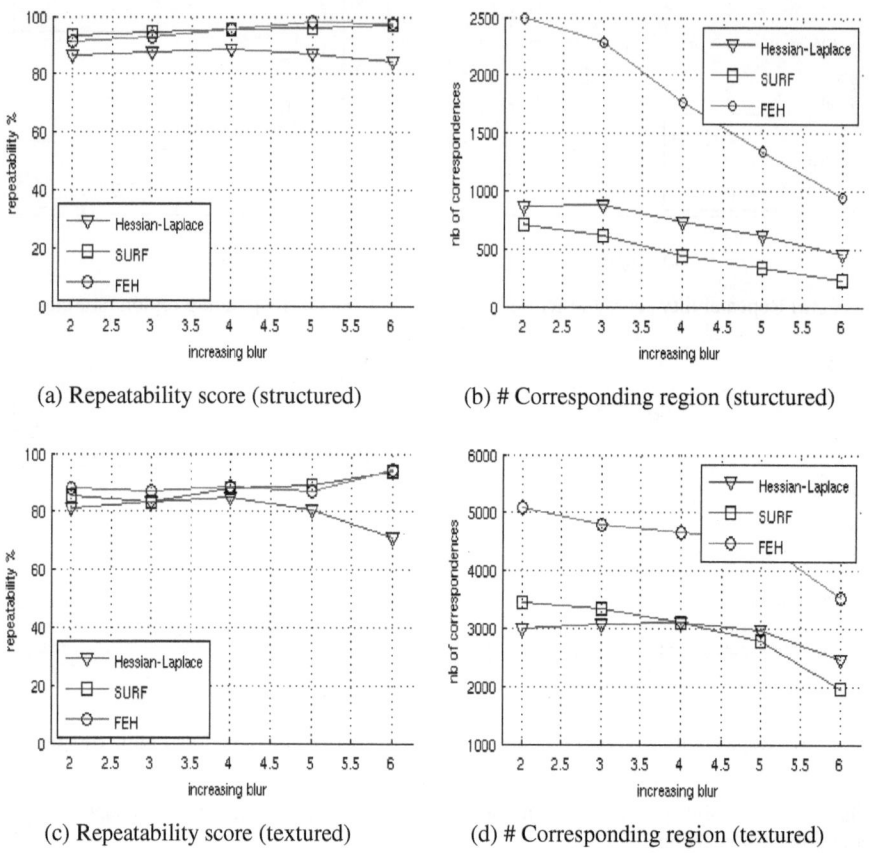

(a) Repeatability score (structured)   (b) # Corresponding region (sturctured)

(c) Repeatability score (textured)   (d) # Corresponding region (textured)

Fig. 6. Repeatability score and number of corresponding regions for image sequences containing blur change

Table 1. Algorithmic run-times

Blob Detector	Run-time(secs)
Hessian-Laplace	0.871
SURF (including SURF-E descriptors	0.636
FEH (including SURF-E descriptors)	0.809

5 Conclusions and Future Work

The results of our comparative performance evaluation indicate that the FEH detector performs better than the SURF detector on most sequences. In some sequences such as the structured viewpoint change the Hessian-Laplace detector performs better than the FEH detector. This is most likely due to the fact that the Hessian-Laplace detector uses large Gaussian derivatives to compute the Hessian matrix rather than the cruder approximations used in the SURF and FEH detectors; however this also makes the Hessian-Laplace operator relatively computationally expensive.

References

[1] Agrawal, M., Konolige, K., Blas, M.R.: CenSurE: Center surround extremas for realtime feature detection and matching. In: Forsyth, D., Torr, P., Zisserman, A. (eds.) ECCV 2008, Part IV. LNCS, vol. 5305, pp. 102–115. Springer, Heidelberg (2008)

[2] Bay, H., Ess, A., Tuytelaars, T., Van Gool, L.: Speeded-Up Robust Features (SURF). CVIU 110, 346–359 (2008)

[3] Davies, E.R.: Design of Optimal Gaussian Operators in Small Neighbourhoods. Image and Vision Computing 5(3), 199–205 (1987)

[4] Harris, C., Stephens, M.: A Combined Corner and Edge Detector. In: Proc. 4th Alvey Vision Conf., pp. 147–151 (1988)

[5] http://www.robots.ox.ac.uk/~vgg/research/affine

[6] Kerr, D., Coleman, S., Scotney, B.: FESID: Finite element scale invariant detector. In: Foggia, P., Sansone, C., Vento, M. (eds.) ICIAP 2009. LNCS, vol. 5716, pp. 72–81. Springer, Heidelberg (2009)

[7] Lindeberg, T.: Scale-Space Theory in Computer Vision. Kluwer, Academic Publisher/Springer, Dordrecht, Netherlands (1994)

[8] Lowe, D.G.: Object recognition from local scale-invariant features. In: Proc. ICCV, pp. 1150–1157 (1999)

[9] Marr, D., Hildreth, E.: Theory of Edge Detection. Proceedings of the Royal Society of London, 215–217 (1980)

[10] Mikolajczyk, K., Schmid, C.: Scale & Affine Invariant Interest Point Detectors. IJCV 60, 63–86 (2004)

[11] Mikolajczyk, K., Tuytelaars, T., Schmid, C., Zisserman, A., Matas, J., Schaffalitzky, F., Kadir, T., Van Gool, L.: A Comparison of Affine Region Detectors. IJCV 65, 43–72 (2005)

[12] Scotney, B.W., Coleman, S.A.: Improving Angular Error via Systematically Designed Near-Circular Gaussian-based Feature Extraction Operators. Pattern Recognition 40(5), 1451–1465

[13] Tuytelaars, T., Mikolajczyk, K.: Local invariant feature detectors: a survey. Foundations and Trends in Computer Graphics and Vision 3(3), 177–280 (2008)

[14] Viola, P., Jones, M.: Rapid object detection using a boosted cascade of simple features. In: CVPR, vol. 1, pp. 511–518 (2001)

[15] Zuniga, O.A., Haralick, R.M.: Integrated Directional Derivative Gradient Operator. IEEE Trans. on Systems, Man, and Cybernetics SMC-17(3), 508–517 (1987)

[16] Mikolajczyk, K.: Affine Covariant Features (July 15, 2007), http://www.robots.ox.ac.uk/~vgg/research/affine/ (March 14, 2011)

Reducing Number of Classifiers in DAGSVM
Based on Class Similarity

Marcin Luckner

Warsaw University of Technology, Faculty of Mathematics and Information Science,
pl. Politechniki 1, 00-661 Warsaw, Poland
mluckner@mini.pw.edu.pl
http://www.mini.pw.edu.pl/~lucknerm/en/

Abstract. Support Vector Machines are excellent binary classifiers. In case of multi–class classification problems individual classifiers can be collected into a directed acyclic graph structure DAGSVM. Such structure implements One-Against-One strategy. In this strategy a split is created for each pair of classes, but, because of hierarchical structure, only a part of them is used in the single classification process.

The number of classifiers may be reduced if their classification tasks will be changed from separation of individual classes into separation of groups of classes. The proposed method is based on the similarity of classes. For near classes the structure of DAG stays immutable. For the distant classes more than one is separated with a single classifier. This solution reduces the classification cost. At the same time the recognition accuracy is not reduced in a significant way. Moreover, a number of SV, which influences on the learning time will not grow rapidly.

Keywords: Classification, Directed Acyclic Graph, Support Vector Machines, One–Against–One.

1 Introduction

Support Vector Machines [11] are excellent binary classifiers, which can also be applied to multi–class problems. One–step solutions [4] are mainly theoretical [6]. Instead, an ensemble of SVMs is used. There are two main strategies for that [6,1].

The first, One–Against–All [3] creates a classifier for each recognized class. The classifier splits a data space between the class and the rest of classes. In the n–classes case the method needs n classifiers. The method can also be implemented in the form of a decision tree [2]. Then, on each level of the tree one class is separated with the rest. This implementation requires only $n-1$ classifiers and the average decision process uses $n-1/2$ classifiers.

The second method is One–Against–One [7]. The method creates a classifier for each pair of classes. In comparison to the One–Against–All strategy the recognition rate grows, the same as the number of created classifiers. In the case of $n$classes the method needs $n(n-1)/2$ classifiers. The method can also

G. Maino and G.L. Foresti (Eds.): ICIAP 2011, Part I, LNCS 6978, pp. 514–523, 2011.

be implemented in the form of a directed acyclic graph [8]. Then only $n - 1$ classifiers have to be used in the classification process. However, this number is constant and independent of a recognizing class. The problem of decrease of the average classification cost is discussed in the section 2.

The aim of this paper is to propose a method for a reduction of classifiers used by the DAGSVM without a significant decrease of the recognition rate. For that the classifiers for a pair of classes will be only created for the similar classes. If the class is separated from the rest of classes then the One–Against–All strategy will be used instead. The creation of the graph with reduced number of classifiers is presented in the section 3.

The graph created in the proposed way will be tested on the handwritten digits recognition task. The results will be compared with the One–Against–All and One–Against–One strategies. The test and its results are presented in the section 4.

2   DAGSVM and Reduction of Classifiers

The DAGSVM has proved it's advantages in many tests [6,2]. The DAGSVM implements the One–Against–One strategy. The SVM classifier is created for each pair of recognized classes. In each step of the classification algorithm one class is rejected. The second one is compared with a next class. The algorithm terminates when the single remaining class is not rejected. In the case of n–classes recognition task, such classification path needs $n - 1$ classifiers. For the n possible results the global number of classifiers, without repetition, is $n(n-1)/2$.

Because a part of classifiers will be the same for different classification paths and the order of comparison in the fixed path is not important, the classifiers can be collected in a directed acyclic graph. Then, a single classification process needs only $n - 1$ classifiers, but this number is constant.

The average number of used classifiers can be reduced when the degenerated tree structure is used. Then the One–Against–All strategy is applied. Despite of reduction of trained classifiers the learning time may grow, because each single classifier solves now a more complex problem [1].

The solution, which lies somewhere between is an Adaptive Directed Acyclic Graph [5]. ADAG uses a tournament tree to reduce the number of classifiers by half on each level of the tree. For that purpose some of the classifiers splits the space between groups of classes instead of single classes.

A different approach is presented in the paper [13], where the gravity of the classifiers is estimated on the base of the probability of positive samples, while the trivial classifiers are ignored.

In this paper the reduced number of classifiers will be achieved by replacing of the splits between classes by the splits between groups. However, the replaced classifiers will be selected on the base of the similarity between classes. Details are given in the next section.

3   Reduction Based on Class Similarity

For the reduction of the number of used classifiers a new structure will be proposed. The structure will also be a directed acyclic graph, therefore some definition about graphs [12] will be formed in the section 3.1. The graph will be based on a similarity. Various definition of the similarity based on a distance are presented in the section 3.2. A structure of the graph is created by grouping of the nearest classes. An algorithm, which builds the graph is presented in the section 3.3. Finally, all nodes of the graph, which are not leaves, have to be connected with the classifiers. In the section 3.4, classification tasks for the individual classifiers are defined.

3.1   Preliminaries

In the presented concept a directed acyclic graph G, which is given by the set of nodes $V(G)$ and the set of edges $E(G)$, is equated with the set of classes. The graph with a single node is an equivalent of a single class C_i, where $i = 1 \ldots n$ is an index of recognized classes. When two classes C_i and C_j, from different nodes, are joined as leaves of the same tree then the tree is equated with the set $\{C_i, C_j\}$. The tree, which groups all equivalents of recognized classes as the leaves is equated with the set $\bigcup_{i=1}^{n} C_i$.

A directed acyclic graphs can be presented as a tree. In such case the number of leaves exceeds n, but some of them will represent the same class. The higher number of leaves increases also the number of nodes on higher levels. Still, those nodes will also be duplicated. For a simplification, in the subsequent deliberation, such tree will be describes as a graph with the leaves $L(G)$ defined as a set of nodes without successors

$$L(G) = \{v_l \in V(G) : \forall_{v \in V(G)} (v_l, v) \notin E(G)\}. \tag{1}$$

In a similar way the root $R(G)$ can be described as a single node without predecessors

$$\exists!_{R(G)} R(G) \in V(G) \wedge \forall_{v \in V(G)} (v, R(G)) \notin E(G). \tag{2}$$

The graph, which collects all recognized classes can be used as an ensemble of classifiers. Each node, which is not a leaf $v \notin L(G)$, is connected with a classifier. The decision of classifier shows the next node in the graph $c(v) = u$. Finally, the decision determines a leaf. When the process starts at the root then the final leaf determines the recognized class $c(v) = C_i$. The decision process can be given as a recursive function

$$C(v) = \begin{cases} C(u), & \text{if } v \notin L(G) \wedge c(v) = u \\ C_i, & \text{if } v \in L(G) \wedge c(v) = C_i \end{cases}. \tag{3}$$

The set of classes C_i , which is equated with the graph G, can be described as

$$C(G) = \{C_i : c(v_l) = i \wedge v_l \in L(G)\}. \tag{4}$$

3.2   Similarity

A similarity between classes is estimated on the base of a distance. The distance between classes $d(C_X, C_Y)$ depends on the distance between elements of those classes $d(x, y)$ and can be defined as the distance between nearest elements

$$d(C_X, C_Y) = \min_{\substack{x \in C_X \\ y \in C_Y}} d(x, y), \tag{5}$$

furthest elements

$$d(C_X, C_Y) = \max_{\substack{x \in C_X \\ y \in C_Y}} d(x, y) \tag{6}$$

or as the average distance between all pair of elements in the two different classes

$$d(C_X, C_Y) = \frac{1}{n_{C_X} n_{C_Y}} \sum_{\substack{x \in C_X \\ y \in C_Y}} d(x, y). \tag{7}$$

All those methods need calculation of the distance between all pair of elements. When the classes have too many members the distance may be approximated as the distance between centroids (the centers of gravity for the classes)

$$d(C_X, C_Y) = d \left(\frac{1}{n_{C_X}} \sum_{x \in C_X} x, \frac{1}{n_{C_Y}} \sum_{y \in C_Y} y \right). \tag{8}$$

In a similar, way a distance between groups of classes can be calculated. If a group is an union of classes $C_X = \bigcup_{i=1}^{k} C_i$ then all members of classes C_i, where $i = 1 \ldots k$, are treated as members of C_X. The distance between such groups can be calculated as (5), (6), (7) or (8). A single class may be compared with a group of classes in the same way.

   The distance between an individual elements of the data space $d(x, y)$ depends on the selected metric. Usually it is the Euclidean metric

$$d(x, y) = \sqrt{\sum_{i=1}^{n} (x_i - y_i)^2}. \tag{9}$$

However, if the recognized elements are described by some specific features it is sometime better to select a different measure.

   Two potential candidates are Manhattan and Chebyshev metrics. the Manhattan distance

$$d(x, y) = \sum_{i=1}^{n} |x_i - y_i| \tag{10}$$

should be calculated when individual features are independent, but their sum may be reated as a rational measure of the similarity. In the Chebyshev distance

$$d(x, y) = \max_{i \in \{1, \ldots, n\}} |x_i - y_i| \tag{11}$$

the similarity will depend on the maximal difference among features.

3.3   Structure

The graph for the n–classes classification task has to group n leaves. The algorithm for creating such graph starts with the set S_G of n graphs with only one node

$$S_G = \{G_i : C(G_i) = \{C_i\} \wedge i = 1 \ldots n\}. \tag{12}$$

In each step a pair (G_i, G_j) of graphs equivalent with the nearest sets of classes is found

$$(G_i, G_j) = \arg\min_{G_i, G_j \in S_G} d(C(G_i), C(G_j)). \tag{13}$$

Both, $C(G_i)$ and $C(G_j)$ are sets of classes and the distance $d(C(G_i), C(G_j))$ can be calculated as (5), (6), (7) or (8).

The graphs G_i and G_j are joined into a single graph G. In the joining process a new node v is added as the root (2) for the new graph. The roots of joined graphs become successors of the new node and the new graph G is given by the set of nodes

$$V(G) = \{v\} \cup V(G_i) \cup V(G_j) \tag{14}$$

and the set of edges

$$E(G) = \{(v, R(G_i)), (v, R(G_j))\} \cup E(G_i) \cup E(G_j). \tag{15}$$

The newly created graph G is added to the set of graphs

$$S_G = S_G \cup \{G\}. \tag{16}$$

The component graphs G_i and G_j are not removed from the set S_G. The graphs are equivalent to classes C_i and C_j respectively. However, on the base of (4) and (14), (15), the graph G is equated with the set $\{C_i, C_j\}$. For that reason the condition (13) has to choose (G_i, G) or (G, G_j) as the nearest graphs. To avoid such situation possibilities of joining have to be limited. Two graphs can be joined if and only if the set of classes represented by one of them (4) is not a subset of the other

$$C(G_i) \nsubseteq C(G_j) \wedge C(G_j) \nsubseteq C(G_i). \tag{17}$$

If the new created graph will not be one of the nearest graphs then graphs G_i and G_j are still the nearest ones. In such situation the algorithm will not stop. For that an additional condition has to be formed. The two graphs can be joined if and only if the union of classes represented by them is not represented by any existed graph

$$\forall_{G \in S_G} C(G) \neq C(G_i) \cup C(G_j). \tag{18}$$

Both conditions (17) and (18) can be used to create a limited set of allowed pairs of graphs

$$S_P = \{(G_i, G_j) : G_i, G_j \in S_G$$
$$\wedge\ C(G_i) \not\subseteq C(G_j) \wedge C(G_j) \not\subseteq C(G_i) \tag{19}$$
$$\wedge\ \forall_{G \in S_G} C(G) \neq C(G_i) \cup C(G_j)\}.$$

Moreover, the common part of classes is ignored when the distance is calculated and the final form of the formula (13) is

$$(G_i, G_j) = \arg\min_{G_i, G_j \in S_P} d(C(G_i) \setminus C(G_i \cap G_j), G_j \setminus C(G_i \cap G_j)), \tag{20}$$

where

$$C(G_i \cap G_j) = C(G_i) \cap C(G_j). \tag{21}$$

In each step of the algorithm, the two allowed graphs G_i, G_j are joined. The algorithm stops when no join can be made

$$\forall_{G_i \in S_G} \exists_{G_j \in S_G} C(G_i) \subseteq C(G_j) \vee C(G_j) \subseteq C(G_i). \tag{22}$$

In such case, the set contains one graph G, which collects all recognized classes

$$C(G) = \{C_i : i = 1 \ldots n\}. \tag{23}$$

In the final graph G some of nodes may be repeated. However, such nodes are leaves or have the same branches and they can be treated as the one. An example of the structure creation process is presented in Fig 1.

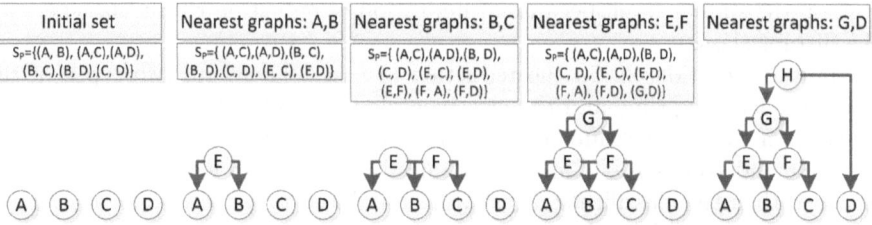

Fig. 1. An example of the creation of the graph structure

3.4   Classifiers

In the DAGSVM each classifier rejects one from the recognized classes. In created structure the number of classifiers may be reduced. For that reason the classifier may reject more than one class.

When a classifier is connected with the node v the function $c(v)$ has to choose one of the successors u, w. Each of them is a root of a graph. When the node u is the root of the graph G_u then the classification function (3) can only reach classes from the set $C(G_u)$. The same with the node w, which represents classes from the set $C(G_w)$. So, the classifier will reject classes

$$Reject(v) = \begin{cases} C(G_u), & \text{if } c(v) = w \\ C(G_w), & \text{if } c(v) = u \end{cases}. \tag{24}$$

The classifier connected with the node v will be trained to separate members of the classes includes in the set $C(G_u)$ from those belonging to the set $C(G_w)$.

However, in the directed acyclic graph, two different nodes can have the same successors. In such case, the function (24) has to be modified to avoid discrimination between members of the same class. The new formula of the rejection function will be

$$Reject(v) = \begin{cases} C(G_u) \setminus C(G_u \cap G_w), & \text{if } c(v) = w \\ C(G_w) \setminus C(G_u \cap G_w), & \text{if } c(v) = u \end{cases}. \tag{25}$$

Now, the classifier connected with the node v will be trained to separate members of classes from the set $C(G_u) \setminus C(G_u \cap G_w)$ with the members of classes from the set $C(G_w) \setminus C(G_u \cap G_w)$ and the problem with discrimination of the same classes will be prevented.

3.5   Discussion

The reduced graph may be compared with the DAGSVM in three aspects. The first one is the recognition accuracy. An increase of the accuracy, when the number of the classifiers is reduced, should not be expected. However, it can stay on the similar level, because of limitation of reduction to the classifiers for the distant classes.

Main expectations lie in the reduction of costs. There are two types of the costs: the learning time and the classification time. The classification time is a function of the average number of classifiers. For the DAGSVM this number is constant and equals $n - 1$ for n–recognized classes. If at least one classifier is removed then the classification cost will be reduced for at least one class. Then the average cost will also be reduced.

The aspect of the learning time is a little more complicated. The number of classifiers, which have to be created is reduced. However, the classification task for a single classifier becomes complex, when the split between groups is examined. The optimization function [11], which calculates coefficients of the discriminant plane, depends on the number of support vectors. Then the number of used SV may be used as an estimator of the learning cost.

The learning process will be faster if the number of the support vectors, which have been used by the rejected classifiers is greater than the number of additional vectors for the new discrimination task. It has to be also noticed that the number

of the support vectors does not influence the classification time, which will be
constant for each SVM classifier.

4   Classification of Digits Using Grouping Trees

The digits recognition task was chosen for testing. Handwritten digits were described with 197 features. The feature set is based on a set of projections, histograms and other fast computed features [9]. The feature vectors create ten
classes. The sizes of classes are unbalanced. Nearly 30 percent of the whole data
set is represented by the class *3*. This follows from a specific character of the
collected data. The digits were extracted from control points on the geodetic
maps [10] dominated by thirties. From 7081 collected digits, 5291 were used as
a learning probe. The rest, 1790, were used for the tests.

The results for the reduced graph was compared to the results of the One–
Against–One strategy. A similar classification error (a percentage of mistaken
cases) was expected. Also, the learning time as well as classification time were
compared. Those two aspect were estimated by a number of support vectors and
the number of binary classifiers respectively.

An additional test was for the One–Against–All strategy. This strategy uses
a minimal number of classifiers and it should give a reference to the reduction
effects. The method was implemented by a tree. The tree, which was created by
adding the nearest class to the existing tree. The tree is initiated by two nearest
classes.

In all tests SVMs with linear kernels, were chosen. The distances between
elements were calculated in the Euclidean metric (9). The tree was built on the
base of the average distance (7) and in case of the reduced DAG, the distance
between centroids (8) was used.

Table 1. Comparison of methods

Classifier	Number of SVM	Average No. of SVM	Number of SV	Average No. of SV	Classification error
DAGSVM	45	9	2176	48	4.25
Reduced graph	26	7	1600	62	4.75
OAA	9	5	1414	157	6.65

The results are presented in the Table 1. For each method, the number of
used classifier is given as well as the average number of classifiers used in the
single classification process. This value allows to estimate the classification time.
For the estimation of the learning time a number of support vector is given.
The average number of SV is calculated for the single classifier. Finally, the
classification error is presented.

The One–Against–One strategy needs 45 classifiers. When the classifiers are grouped in the DAGSVM the average number of used classifiers is 9 for each data distribution. The classification rate, calculated as a percent of positive recognized elements is 95.75 percent.

The One–Against–All strategy requires only 9 classifiers. The average number of used classifiers is about 5 for the balanced case (the number of elements among classes is similar). The method produces a lower classification rate, 93.35 percent of positive recognized elements. However, the method is relatively fast even in the learning process. The effect of increasing of the average number of support vectors is mitigated by the reduce in the number of created classifiers.

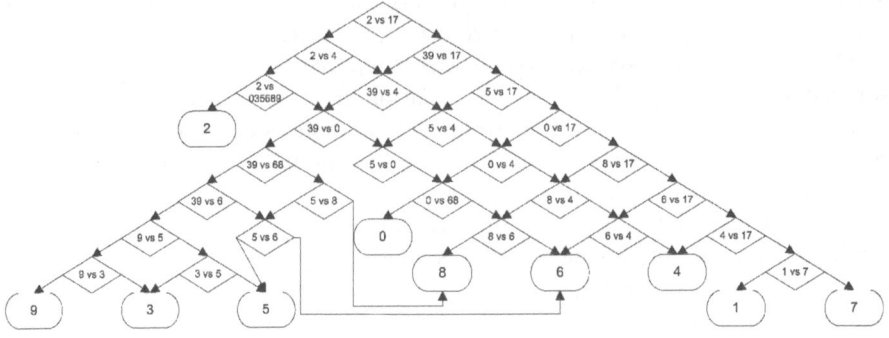

Fig. 2. The reduced DAGSVM for the digits recognition task

The reduced graph (Fig.2) needs 26 classifiers. The average number of used classifiers is 7 in the balanced case. However, some classes, such as *2*, could be reached with 3 classification decisions. The maximal number of used classifiers is 8. The classification process will be definitely faster than for the DAGSVM. At the same time the classification rate is 95.25 percent of positive recognized cases, which is nearly the same as for the One–Against–One strategy.

The last aspect is the learning time. In the reduced graph the global number of support vectors is lower than in the DAGSVM. The average number of SV grows, but many classifiers are eliminated. The increase of SV is not significant, because most of the classifiers split only two classes as in the case of the DSGSVM.

As a conclusion of the test it can be said that the fastest method is One–Against–All (in both the learning time and the classification time). However, the classification error for this method grows in comparison to the One–Against–One strategy. Yet, the One–Against–One strategy is not the best method. The reduced graph has a similar classification rate, but the learning and classification costs are significantly lower.

5 Summary

In this paper a new method, which creates the ensemble of the SVM was presented. The method is based on the DAGSVM. However, it allows to reduce a number of created classifiers. The classifiers, which discriminate distant classes are replaced by the classifiers, which separate groups of classes. The algorithm for creating a new graph structure, was presented.

The algorithm was tested on the handwritten digits recognition task. The recognition rate stays similar to the DAGSVM method, while the reduction of the classifiers was significant. The reduction of the average number of classifiers decreases the classification time. Moreover, the learning time should not be increased despite of increase of the average number of support vectors used by the classifiers.

References

1. Abe, S.: Support Vector Machines for Pattern Classification (Advances in Pattern Recognition). Springer-Verlag New York, Inc., Secaucus (2005)
2. Kumar, M.A., Gopal, M.: A comparison study on multiple binary-class SVM methods for unilabel text categorization. Pattern Recogn. Lett. 32, 9160:311–9160:323 (2010), http://dx.doi.org/10.1007/s11063-010-9160-y
3. Bennett, K.P.: Combining support vector and mathematical programming methods for classification, pp. 307–326. MIT Press, Cambridge (1999)
4. Crammer, K., Singer, Y.: On the learnability and design of output codes for multiclass problems. In: COLT 2000: Proceedings of the Thirteenth Annual Conference on Computational Learning Theory, pp. 35–46. Morgan Kaufmann Publishers Inc., San Francisco (2000)
5. Kijsirikul, B., Ussivakul, N., Meknavin, S.: Adaptive directed acyclic graphs for multiclass classification. In: Ishizuka, M., Sattar, A. (eds.) PRICAI 2002. LNCS (LNAI), vol. 2417, pp. 158–168. Springer, Heidelberg (2002)
6. Kim, H.-C., Pang, S., Je, H.-M., Kim, D., Bang, S.Y.: Constructing support vector machine ensemble. Pattern Recognition 36(12), 2757–2767 (2003)
7. Kressel, U.H.G.: Pairwise classification and support vector machines, pp. 255–268. MIT Press, Cambridge (1999)
8. Platt, J., Cristianini, N., ShaweTaylor, J.: Large margin dags for multiclass classification. In: Solla, S.A., Leen, T.K., Mueller, K.-R. (eds.) Advances in Neural Information Processing Systems 12, pp. 547–553 (2000)
9. Romero, R., Touretzky, D., Thibadeau, R.: Optical chinese character recognition using probabilistic neural networks. Pattern Recognition 8, 1279–1292 (1997)
10. Stapor, K.: Geographic map image interpretation - survey and problems. Machine Graphics & Vision 9(1/2), 497–518 (2000)
11. Vapnik, V.: The Nature of Statistical Learning Theory. Springer, Heidelberg (1995)
12. Wilson, R.J.: Introduction to graph theory. John Wiley & Sons, Inc., New York (1986)
13. Ye, W., Shang-Teng, H.: Reducing the number of sub-classifiers for pairwise multi-category support vector machines. Pattern Recogn. Lett. 28, 2088–2093 (2007)

New Error Measures to Evaluate Features on Three-Dimensional Scenes

Fabio Bellavia and Domenico Tegolo

Department of Mathematics and Computer Science,
University of Palermo, 90123, Palermo, Italy
{fbellavia,domenico.tegolo}@unipa.it

Abstract. In this paper new error measures to evaluate image features in 3D scenes are proposed and reviewed. The proposed error measures are designed to take into account feature shapes, and ground truth data can be easily estimated. As other approaches, they are not error-free and a quantitative evaluation is given according to the number of wrong matches and mismatches in order to assess their validity.

Keywords: Feature detector, feature descriptor, feature matching, feature comparison, overlap error, epipolar geometry.

1 Introduction

Feature-based computer vision applications have been widely used in the last decade [12]. Their spread has increased the focus on feature detectors [8] and feature descriptors [7], as well as sparse matching algorithms [3,13]. Besides, different evaluation strategies to assess their properties have been proposed in [8,7,10,5,4].

The repeatability index introduced in [11] and the matching score [8] are common measures used for comparison. They have been adopted in well-known extensive comparisons for detectors [8], while precision-recall curves have been used for descriptors [7]. Both the error measures described above have been applied to the Oxford dataset [9] which has become a standard de facto. The principal drawback of these approaches is to require a priori knowledge of all the possible correct matches between corresponding points in images. In the case of the planar scenes the Oxford dataset is made of, this can be trivially obtained by computing the planar homography from an exiguous number of hand-taken correspondences [6].

However, the use of features on 3D scenes is the most attractive and interesting topic for which nowadays new features are designed, so a relevant interest has risen in order to understand how they behave and their properties in a fully 3D environment. A strategy to overcome this issue was proposed in [5], where only two further image sequences, which contain fully 3D objects, are added to to extend the Oxford dataset. The trifocal tensor [6] is computed by an intermediate image and ground truth matches are recovered by using a dense matching

G. Maino and G.L. Foresti (Eds.): ICIAP 2011, Part I, LNCS 6978, pp. 524–533, 2011.

strategy [12]. However, no information can be extracted for homogeneous regions or when occlusions are present, which can compromise the evaluation. Moreover, the complexity of the approach increases, which becomes less suitable for adding further image sequences to obtain a better evaluation.

A further evaluation of feature detectors and descriptors on 3D objects is reported in [10]. Differently from the other approaches, triplets of images are used leading to triplets of image correspondences instead of pairs. Correspondences are evaluated by epipolar geometry constrains [6] to build ROC curves used to analyse data and draw out conclusions. Though fundamental matrices [6] used to constrain triplets can be easily obtain by a relative low number of hand-take correspondences, the method is not completely error-free. A relevant number of correct matches can be discarded due to occlusions or by detector failures (about 50-70%), while a few wrong matches can be incorrectly retained, especially when corresponding epipolar lines are near parallel (less than 10%). Moreover this approach does not take into account the feature shapes, but it only considers the distance between the feature centres.

A last method described in [4] uses the pole-polar relationship to build an overlap error measure on segments lying on corresponding feature pairs. Only the fundamental matrix is required to compute the ground truth, but wrong matches can be accepted. According to the authors, this can happen with a low probability they did not experimentally measured.

In this paper new error measures to evaluate image features in 3D scenes are proposed and reviewed, also employing strategies which use image triplets as in [10]. The proposed measures are designed to take into account the feature shapes and, moreover, ground truth data can be easily estimated for 3D scenes. As other approaches, they are not error-free and in order to assess their validity, a quantitative evaluation is given according to the number of wrong matches and mismatches. Moreover, the method described in [4] has been included. The resulting quantitative analysis also provides clues on the possible number of wrong matches retained when these measures are employed to validate matching algorithms. In Sect. 2, the proposed error measures are introduced, while in Sect. 3 the experimental setup to assess their validity and the results are described. Finally, in Sect. 4 conclusions and future works are discussed.

2 The New Error Measures

2.1 Definition

Given a stereo image pair (I_1, I_2) and corresponding points $\mathbf{x}_1 \in I_1$, $\mathbf{x}_2 \in I_2$ in homogeneous coordinates, the fundamental matrix F determines the relation $\mathbf{x}_2^T F \mathbf{x}_1 = 0$. Geometrically the point $\mathbf{x}_1$ is constrained to lie on the epipolar line $l_1 = \mathbf{x}_2^T F$, and in similar way $\mathbf{x}_2$ on $l_2 = \mathbf{x}_1^T F^T$ in the corresponding image. Epipolar lines pass through the epipoles $\mathbf{e}_1 \in I_1$, $\mathbf{e}_2 \in I_2$, which are respectively the right, left null-space of F. The ground truth fundamental matrix F can be extracted by an exiguous number of hand-taken correspondences by using a method described in [6].

Let $\mathcal{R}_1 \in I_1$, $\mathcal{R}_2 \in I_2$ be two feature patches centred in $\mathbf{x}_1$, $\mathbf{x}_2$. Their shapes are elliptical disks, as commonly defined by feature detectors, with minor and major axes respectively α_{min_i}, α_{max_i}, $i \in \{1,2\}$. Let also $d(\cdot,\cdot)$ define the Euclidean distance between points or between a point and a line according to its arguments. The first error measure in the image I_i, $i \in \{1,2\}$, for the a feature pair $(\mathcal{R}_1, \mathcal{R}_2)$ is defined as

$$\xi_i = \min \left(\frac{d(\mathbf{x}_i, \mathbf{l}_i)}{\alpha_{min_i}}, 1 \right) \tag{1}$$

that is, the epipolar distance between the feature centre and its epipolar line computed by using the corresponding point in the other image is normalized by the minor axis of the feature ellipse (see Fig. 1(a)). The error ξ on both the images is

$$\xi = \max_i \xi_i \quad i \in \{1,2\} \tag{2}$$

In similar way the error κ_i in the image I_i is defined as

$$\kappa_i = \min \left(\frac{d(\mathbf{x}_i, \mathbf{l}_i)}{2\,d(\mathbf{x}_i, \mathbf{p}_i)}, 1 \right) \tag{3}$$

where $\mathbf{p}_i$ is the intersection between the feature elliptical boundary and the line perpendicular to the epipolar line $\mathbf{l}_i$ through $\mathbf{x}_i$ (see Fig. 1(b)). Analogously, the error κ on both the image is

$$\kappa = \max_i \kappa_i \quad i \in \{1,2\} \tag{4}$$

Both the error measures ξ and κ take into account the shape of the feature patch, but the former, by considering only the maximum circle inside the feature ellipse, makes a more pessimistic assumption about the correctness of the features as extracted by the feature detector (see Fig. 1(a-b)). The error for both ξ and κ achieves the maximal value of 1 roughly when the reprojected feature ellipse would not touch the considered feature ellipse (see Fig. 1(a-b)). Moreover, the maximal error between the two images is retained instead of their average, as it is the symmetric error, because the former solution is more constraining.

The last error measure extends that introduced in [4] which will be described for clarity. The tangency relation is preserved by perspective projection, which retains incidence relations. Let $\mathbf{l}_{t_1^i}$, $\mathbf{l}_{t_2^i}$ be the epipolar lines in the image I_i, $i \in \{1,2\}$ corresponding respectively to the tangent points $\mathbf{t}_1^{\bar{i}}$, $\mathbf{t}_2^{\bar{i}}$ in $I_{\bar{i}}$ through the epipole $\mathbf{e}_{\bar{i}}$ to the feature ellipse $\mathcal{R}_{\bar{i}}$, where $\bar{i} = 3 - i$ (i.e. I_i and $I_{\bar{i}}$ are the complementary images of the stereo pair). Let $\mathbf{q}_i$ be the line through the tangent points $\mathbf{t}_1^i$, $\mathbf{t}_2^i$. The intersection points $\mathbf{r}_1^i$, $\mathbf{r}_2^i$ of $\mathbf{l}_{t_1^i}$, $\mathbf{l}_{t_2^i}$ with the line $\mathbf{q}_i$ are used to define a linear overlap error ε_{l_i} in the image I_i (see Fig. 1(c)) as follows

$$\varepsilon_{l_i} = 1 - \frac{\max\left(0, \min\left(t_h^i, r_h^i\right) - \max\left(t_l^i, r_l^i\right)\right)}{\max\left(t_h^i, r_h^i\right) - \min\left(t_l^i, r_l^i\right)} \tag{5}$$

where r_h^i and r_l^i are the higher and lower linear coordinates of $\mathbf{r}_1^i$ and $\mathbf{r}_2^i$ on the line $\mathbf{q}_i$ respectively, according to a defined direction, and in similar way t_h^i and

t_l^i for $\mathbf{t}_1^i$ and $\mathbf{t}_2^i$ respectively ($r_h^i \equiv \mathbf{r}_1^i$, $t_h^i \equiv \mathbf{t}_1^i$, $r_l^i \equiv \mathbf{r}_2^i$, $t_h^i \equiv \mathbf{t}_2^i$ in the example of Fig. 1(c)). The final error on both the images is defined as the average error

$$\varepsilon_{l_{avg}} = \frac{\varepsilon_1 + \varepsilon_2}{2} \tag{6}$$

In particular the authors consider a match correct if $\varepsilon_{l_{avg}} < 0.2$ [4]. In the next, for the same motivation described above the follow definition of linear overlap error ε_l will be used instead

$$\varepsilon_l = \max_i \varepsilon_{l_i} \quad i \in \{1, 2\} \tag{7}$$

An extension to the linear overlap error measure ε_l is proposed, by observing that not only the correspondence between epipoles is available, but also fixed correspondences $\left(\mathbf{w}_1^k, \mathbf{w}_2^k\right)$, with $k = \{1, \ldots, m\}$ and $\mathbf{w}_i^k \in I_i$, provided by further hand-taken points used to compute the fundamental matrix. By combining their tangent lines to the feature ellipse it is possible to obtain an inscribed and a circumscribed quadrilaterals, which can be used to approximate the feature ellipse (see Fig. 1(d)). Both the quadrilaterals can be approximately projected through the fundamental matrix, thus an approximate overlap error ε_q between the feature patches can be computed, similar to the standard definition of overlap error ε between surface patches used in [8]

$$\varepsilon\left(\mathcal{R}_1, \mathcal{R}_2\right) = 1 - \frac{\mathcal{R}_1 \cap \mathcal{R}_2}{\mathcal{R}_1 \cup \mathcal{R}_2} \tag{8}$$

In detail, for two pairs of fixed corresponding points $\left(\mathbf{w}_1^k, \mathbf{w}_2^k\right)$ and $\left(\mathbf{w}_1^s, \mathbf{w}_2^s\right)$, $k, s \in \{1, \ldots, m\}$, $k \neq s$, on the image I_i, one can obtain the quadrilateral $\mathcal{Q}_i$ circumscribed to the feature ellipse $\mathcal{R}_i$ by intersecting the tangent lines to $\mathcal{R}_i$ through the fixed points $\mathbf{w}_i^k$, $\mathbf{w}_i^s$ (see Fig. 1(d)). The corresponding tangent points are used instead to get the quadrilateral $\mathcal{Q}_i^\star$ inscribed to the ellipse $\mathcal{R}_i$ (see Fig. 1(d)). With an abuse of notation the area of the ellipse $\mathcal{R}_i$ can be roughly approximated by the average area between the two quadrilaterals $\mathcal{Q}_i$ and $\mathcal{Q}_i^\star$

$$\mathcal{R}_i \approx \frac{\mathcal{Q}_i + \mathcal{Q}_i^\star}{2} \tag{9}$$

Reprojected tangent points $\mathbf{v}_i^z$, $z \in \{1, \ldots, 4\}$ from the feature $\mathcal{R}_{\bar{i}}$ of the complementary image $I_{\bar{i}}$ can be approximated as done for $\mathbf{r}_1^i$, $\mathbf{r}_2^i$ in Equ. 5. The epipole is substituted in turn with the fixed points $\mathbf{w}_i^k$ in the definition of the linear overlap error $\varepsilon_{l_{avg}}$. An approximate reprojection $\mathcal{P}_i^\star$ in I_i of the inscribed quadrilateral $\mathcal{Q}_{\bar{i}}^\star$ is obtained by connecting the points $\mathbf{v}_i^z$ (see Fig. 1(d)). The intersection points of lines through $\mathbf{v}_i^z$ and the corresponding fixed points $\mathbf{w}_i^k$, $\mathbf{w}_i^s$ form instead an approximate reprojection $\mathcal{P}_i$ of the circumscribed quadrilateral $\mathcal{Q}_{\bar{i}}$ (see Fig. 1(d)).

By taking into account the formula to evaluate the approximated area (see Equ. 9), a further approximated overlap error ε_{q_i} in the image I_i is defined as

$$\varepsilon_{q_i} = \frac{\varepsilon\left(\mathcal{Q}_i, \mathcal{P}_i\right) + \varepsilon\left(\mathcal{Q}_i^\star, \mathcal{P}_i^\star\right)}{2} \tag{10}$$

and the corresponding approximate overlap error ε_q on both the image is

$$\varepsilon_q = \max_i \varepsilon_{q_i} \quad i \in \{1,2\} \tag{11}$$

The approximated overlap error ε_q strictly depends on the choice of the two corresponding matches with indexes k, s, and the computation can suffer of numerical instability because the projected tangent points $\mathbf{v}_i^z$ are derived by using epipolar lines which are close together and almost parallel, especially when the epipoles are far away from the image centres. This issue can be alleviated empirically by using only index pairs (k, s) for which the diagonals of the quadrilaterals $\mathcal{Q}_i$, $\mathcal{Q}_i^*$, $\mathcal{P}_i$, $\mathcal{P}_i^*$ form a minimum angle of at least $\pi/3$. Moreover, in order to obtain tangent lines which are not almost parallel, the fixed points $\mathbf{w}_i^k$, $\mathbf{w}_i^s$ are considered if they are three times the semi-major axis of the feature ellipse $\mathcal{R}_i$ far away from the centre $\mathbf{x}_i$. Finally, for each feature pair $(\mathcal{R}_1, \mathcal{R}_2)$ the minimum approximated overlap error ε_q among all the admissible pairs of indexes (k, s) is retained.

2.2   Extension to Image Triplets

All the described error measures ξ, κ, ε_l and ε_q depend on the epipolar constrain, thus they cannot characterize wrong matches when they lie close to the epipolar line of the true corresponding image feature. This however does not happen frequently, as it has been observed in the experimental valuation.

In order to alleviate this issue, three strategies similar to that proposed in [10] have been considered, which make use of triplet of images I_i, $i \in \{1,2,3\}$. Let (I_1, I_2), (I_1, I_3), (I_2, I_3) be the three stereo pairs, and $\gamma \in \{\xi, \kappa, \varepsilon_l, \varepsilon_q\}$ an error measure. Consistent chains of features between the stereo pairs are defined by the chain error γ_{δ_t}, according to different strategies δ_t, $t \in \{1,2,3\}$. This error is associated back to features pair, so that feature chains and pairs are accepted if their respective error is $\gamma_{\delta_t} < 1$.

The first strategy δ_1 acts as follows. If $(\mathcal{R}_{k_1}, \mathcal{R}_{k_2}, \mathcal{R}_{k_3})$ is the k-th triplet among all the possible triplets of feature points , with $\mathcal{R}_{k_i} \in I_i$, let $\gamma(z_i, z_j)$ be the error corresponding to the z-th feature pair $(\mathcal{R}_{z_i}, \mathcal{R}_{z_j})$ belonging to the stereo pair (I_i, I_j), $i, j = \{1,2,3\}$. Clearly, $\gamma(z_i, z_j) = \gamma(z_j, z_i)$ and $\gamma(z_i, z_i) = \gamma(z_j, z_j) = 0$. The maximum among all the errors of the feature pairs inside the triplet, is associated back to the triplet

$$\gamma_{\delta_1}(\mathcal{R}_{k_1}, \mathcal{R}_{k_2}, \mathcal{R}_{k_3}) = \max_{z_i, z_j \in \{k_1, k_2, k_3\}} \gamma(\mathcal{R}_{z_i}, \mathcal{R}_{z_j}) \tag{12}$$

The triplet forms a consistent virtual chain if $\gamma_{\delta_1} < 1$. A pair of features is retained if it belongs to a consistent virtual chain. That is, the relation $\gamma_{\delta_1} < 1$ holds for the pair, defined as

$$\gamma_{\delta_1}(\mathcal{R}_{z_i}, \mathcal{R}_{z_j}) = \min_{z_i, z_j \in \{k_1, k_2, k_3\}} \gamma_{\delta_1}(\mathcal{R}_{k_1}, \mathcal{R}_{k_2}, \mathcal{R}_{k_3}) \tag{13}$$

In similar way two further match selection strategies δ_2, δ_3 are defined, but clues about the correctness of the matches provided by the feature detector and descriptor are used instead to provide more insight. In details, let $\left[\mathcal{R}_{k_i}, \mathcal{R}_{k_j}\right]$ be the k-th match between two features to be evaluated in the stereo pair (I_i, I_j) (e.g. the features have been ranked according to the feature descriptor similarity and a threshold was applied to retain the putative correct ones). Define the error associated to a pair of matches, which form a partial chain, as

$$\gamma_{\delta_2}\left(\left[\mathcal{R}_{k_i}, \mathcal{R}_{k_j}\right], \left[\mathcal{R}_{\overline{k}_j}, \mathcal{R}_{\overline{k}_w}\right]\right) = \max_{z_p, z_q \in \{k_i, k_j, \overline{k}_j, \overline{k}_w\}} \gamma\left(\mathcal{R}_{z_p}, \mathcal{R}_{z_q}\right) \qquad (14)$$

To be noted that virtual matches are considered, e.g. the pair $\left(\mathcal{R}_{k_i}, \mathcal{R}_{\overline{k}_j}\right)$. A partial chain is consistent if $\gamma_{\delta_2} < 1$. A generic pair $\left(\mathcal{R}_{z_p}, \mathcal{R}_{z_q}\right)$ is retained if it belongs to a consistent partial chain, that is $\gamma_{\delta_2} < 1$ for the pair, where

$$\gamma_{\delta_2}\left(\mathcal{R}_{z_p}, \mathcal{R}_{z_q}\right) = \min_{z_p, z_q \in \{k_i, k_j, \overline{k}_j, \overline{k}_w\}} \gamma_{\delta_2}\left(\left[\mathcal{R}_{k_i}, \mathcal{R}_{k_j}\right], \left[\mathcal{R}_{\overline{k}_j}, \mathcal{R}_{\overline{k}_w}\right]\right) \qquad (15)$$

In the last strategy γ_{δ_3} triplets of matches are used instead of pairs to form a full chain of matches, i.e. triplets of the form

$$\left(\left[\mathcal{R}_{k_i}, \mathcal{R}_{k_j}\right], \left[\mathcal{R}_{\overline{k}_j}, \mathcal{R}_{\overline{k}_w}\right], \left[\mathcal{R}_{\underline{k}_w}, \mathcal{R}_{\underline{k}_i}\right]\right) \qquad (16)$$

The relation γ_{δ_3} for both triplets of matches and pair of features is defined analogously.

The proposed strategies δ_i are only used to remove putative wrong matches, they are not used as actual error measures, because they provide very high error values. In particular only matches belonging to consistent chains, i.e $\gamma_{\delta_i} < 1$, are retained and scored according to γ. Though these strategies can increase the quality of the matches classification, it should be noted that when large portions of the images are affected by occlusions, a relevant fraction of good matches can be discarded because consistent chain cannot be formed.

3   Measure Assessment

3.1   Experimental Setup

In order to compare the robustness of proposed error measures $\gamma \in \{\xi, \kappa, \varepsilon_l, \varepsilon_q\}$ a set of 10 sequences, consisting of three images of a 3D scene taken from different points of view, have been used, for a total of 30 stereo image pairs. Different degrees of image transformations are present and their final effect, which depends on different degrees of occlusion, baseline distances and camera orientations, is not quantitatively computable. However the average behaviour of the proposed error measures could be deduced.

The Sampson and the epipolar error measures are not included in this evaluation because they cannot provide an error estimation relative to the feature shape. However, while ξ and κ are normalized forms of the epipolar error, in the case of the Sampson error a straight normalization cannot be deduced.

The ground truth fundamental matrix for each stereo pair was computed using the normalized eight-point algorithms [6] on hand-taken correspondences. Three different users provided each more than 50 homogeneously distributed correspondences for every stereo pair, which have been merged together in order to get a more stable fundamental matrix.

To estimate a ground truth in order to assess the goodness of the error measures γ and of the selection strategies δ_i, $i \in \{1, 2, 3\}$, features extracted by the HarrisZ detector [2] have been ranked using the sGLOH descriptor [1] and matches have been supervised by an user, so that if corresponding features in a match share a minimal region then the match is considered correct. According to this match classification and by taking into account the error measure definitions, every match with $\gamma < 1$ should be classified as correct.

The HarrisZ detector is a corner-based feature detector while the sGLOH descriptor is based on gradient histograms, they both have proven to be robust and stable [2,1]. To be noted that misclassified matches should depend on the inherent structure of stereo pair, not on the detector and the descriptor used in the assessment. In order to validate this statement, the use of further feature detectors and descriptors is planned.

Precision-recall curves have been computed for each stereo pairs. The recall is defined as the fraction of correct matches discarded by $\gamma < 1$, while the precision is the fraction of correct classified matches. In order to draw the curve, matches have been ranked according to increasing error values. A marker underlines an error increase by a step of 0.1 for each curve. To be observed that not only the precision is critical to evaluate features detectors and descriptors as underlined in [10], but also the recall should be taken into consideration.

Moreover, while the denominator in the recall ratio is fixed to the number of ground truth correct matches, the denominator of the precision ratio varies as the error value γ increases. To be noted that the recall monotonically increases as the error threshold increases while this does not hold for the precision which varies in $[0, 1]$. This can give rise to quasi-horizontal oscillations in the precision-recall curves, mostly noticeable for low values of the error threshold, which provide an indicator of the stability of the error measures.

The plots have been computed for the different errors γ, along with selection strategies δ_i. Any combination ε_q, δ_i was omitted, because it requires to compute the error value ε_q for each feature pair combination, not only for matched pairs. This is not feasible due to the huge computational time required.

3.2   Results

Plots are shown in Fig. 2, more detailed results can be found online[1] due to lack of space. A summarizing plot for each error measure on the whole image sequences is not shown, since it would be misleading due to the different and not comparable transformations it should average. According to the plots, all the proposed error measures reach a precision of about 80-95% for a recall around

---

[1] http://www.math.unipa.it/~fbellavia/plots.zip

85-95% on average, from where the curve is stabilized and slowly increases the recall, but also decreases the precision. This point is achieved for error values $\xi, \kappa, \varepsilon_l \approx 0.5$, and $\varepsilon_q \approx 0.9$ respectively, which means that the surface error ε_q better approximates the overlap error ε, while for the other linear errors $\xi, \kappa, \varepsilon_l$ normalization to semi-axes instead of axes is a better choice.

It can also be noted that the plots for the error measures ε_q, ε_l slighty increase faster than ξ, κ. Only for the kermit sequence (see Fig. 2, middle row) a very low precision is achieved, however it happens for a high point view change (about $\pi/2$). Moreover, it seems that ε_l measure degrades when the epipoles are inside the images, as it can be noted in the corridor sequence (see Fig. 2, top row), while this does not occur for the analogous error ε_q.

About the application of the different selection strategies δ_i, the error measures ξ, κ take more benefits with respect to ε_l in terms of faster curve increment, but also in terms of stability. The threshold at 0.5 for the error measures ξ, κ, ε_l is no more requested in order to achieve good precision values, which means that the filtering is effective. As more constrains are added increasing the index i, better precision is achieved. However the decrease of the recall is usually greater and can reach about 50% for $i = 3$ in the experiments. According to these observations, the partial chain strategy δ_2 seems to provide a reasonable compromise.

As further remarks, the two measures ξ, κ are very similar in results, so ξ should be preferred because faster. Also ε_l and ε_q obtain similar results, however while the former is faster, the latter can better handle some epipolar stereo configurations and it seems to provide better values in terms of overlap error approximation. Both ξ, κ are faster than ε_l, ε_q, however the last two error measures provide better results in terms of curve increase.

Though the precision can be relatively low for accurate comparison, qualitative evaluation can be drawn out using the proposed error measures. Moreover it must be remembered that in the experiments a match is correct if a minimum overlap is present, which means that in common test requirements, i.e. an overlap error $\varepsilon \approx 0.5$, the precision is higher with a good recall.

4   Conclusion and Future Works

In this paper new error measures for evaluation of features on 3D scenes have been presented. They make use of the feature shape and ground truth data can be easily estimated. Their robustness have been evaluated according to precision-recall curves, which have shown their effectiveness. Though not error-free, qualitative evaluations can be carried out according to them. Moreover, the analysis of the plots of the proposed error measures can be used as a flexible guideline for their application, which mainly depends on the particular evaluation task.

While the linear overlap approximation ε_l presented in [4] and its proposed extension to surface ε_q provide relatively better approximation of the overlap error ε, the normalized version of the epipolar error κ and ξ are faster, in particular ξ. Furthermore an increase in precision can be obtained by using triplets, however with a lost in recall. The strategy δ_2, which makes use of partial chain of features, seems a good compromise.

Future works will include further experiments using other feature detectors and descriptors to assess the results, as well as the use of finer match ground truth, i.e. instead of consider matches correct if they have a minimal overlap $\varepsilon < 1$, different overlap error could be applied. Moreover, the error measure ξ is very fast and could be applied to retrieve inliers in a RANSAC framework, as it done by MSAC [13], but by taking into account also the feature shape.

Acknowledgments. Thanks to Cesare Valenti, for the useful discussions and for his help in the ground truth estimation of the fundamental matrices.

This work has also been partially supported by a grant "Fondo per il Potenziamento della Ricerca del Dipartimento di Matematica e Applicazioni dellUniversità degli Studi di Palermo".

References

1. Bellavia, F., Tegolo, D., Trucco, E.: Improving SIFT-based descriptors stability to rotations. In: International Conference on Pattern Recognition (2010)
2. Bellavia, F., Tegolo, D., Valenti, C.: Improving Harris corner selection strategy. IET Computer Vision 5(2) (2011)
3. Fischler, M.A., Bolles, R.C.: Random sample consensus: a paradigm for model fitting with applications to image analysis and automated cartography. Communications of the ACM 24(6), 381–395 (1981)
4. Forssén, P., Lowe, D.G.: Shape descriptors for maximally stable extremal regions. In: International Conference on Computer Vision. IEEE Computer Society, Los Alamitos (2007)
5. Fraundorfer, F., Bischof, H.: A novel performance evaluation method of local detectors on non-planar scenes. In: IEEE Conference on Computer Vision and Pattern Recognition, p. 33. IEEE Computer Society, Los Alamitos (2005)
6. Hartley, R., Zisserman, A.: Multiple View Geometry in Computer Vision. Cambridge University Press, Cambridge (2000)
7. Mikolajczyk, K., Schmid, C.: A performance evaluation of local descriptors. IEEE Transactions on Pattern Analysis and Machine Intelligence 27(10), 1615–1630 (2005)
8. Mikolajczyk, K., Tuytelaars, T., Schmid, C., Zisserman, A., Matas, J., Schaffalitzky, F., Kadir, T., Van Gool, L.: A comparison of affine region detectors. International Journal of Computer Vision 65(1-2), 43–72 (2005)
9. Mikolajczyk, K., Tuytelaars, T., et al.: Affine covariant features (2010), http://www.robots.ox.ac.uk/~vgg/research/affine
10. Moreels, P., Perona, P.: Evaluation of features detectors and descriptors based on 3d objects. International Journal of Computer Vision 73(3), 263–284 (2007)
11. Schmid, C., Mohr, R., Bauckhage, C.: Evaluation of interest point detectors. International Journal of Computer Vision 37(2), 151–172 (2000)
12. Szeliski, R.: Computer Vision: Algorithms and Applications. Springer, Heidelberg (2010)
13. Torr, P.H.S., Zisserman, A.: Robust computation and parametrization of multiple view relations. In: International Conference on Computer Vision, p. 727. IEEE Computer Society, Los Alamitos (1998)

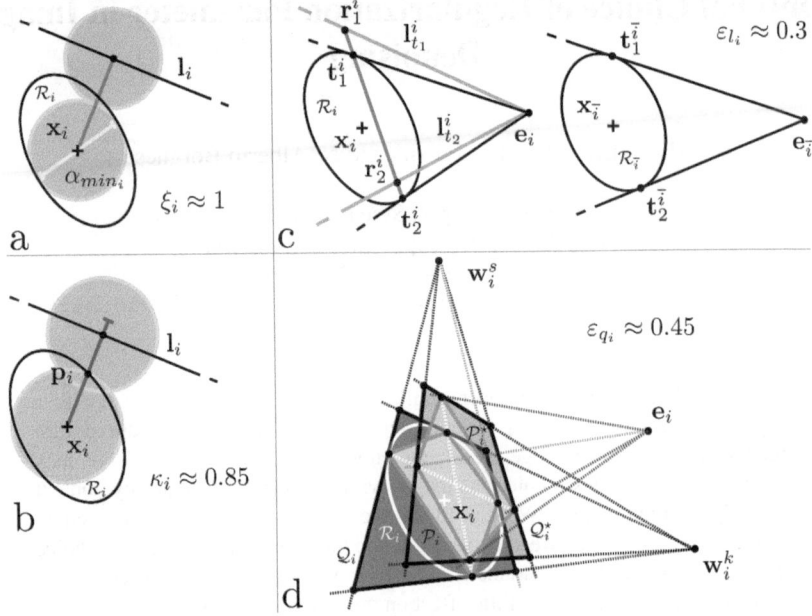

Fig. 1. Error measures ξ (a), κ (b), ε_l (c) and ε_q (d)

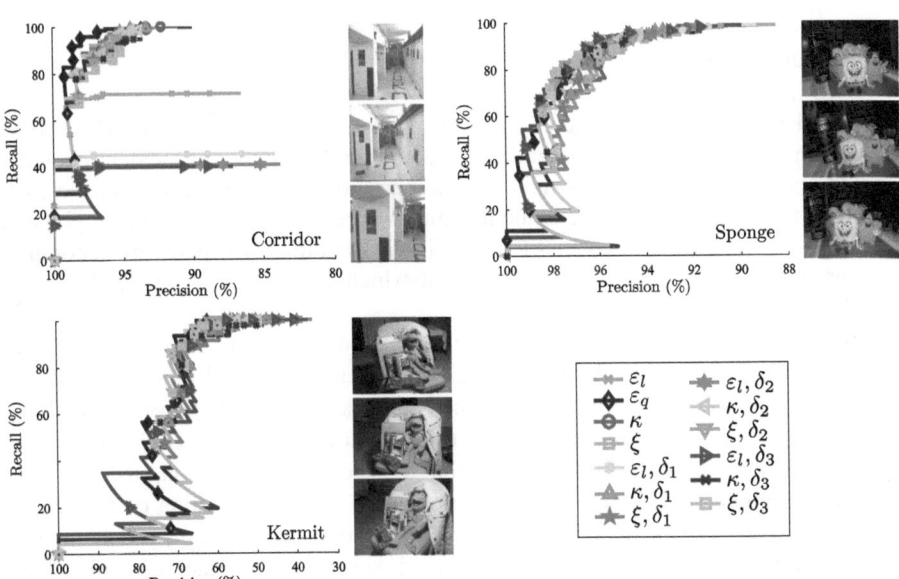

Fig. 2. Precision-recall curves (left) for some image sequences (right). The intermediate image to compute the strategies δ_i is in the corresponding middle row.

Optimal Choice of Regularization Parameter in Image Denoising

Mirko Lucchese, Iuri Frosio, and N. Alberto Borghese

Applied Intelligent System Laboratory
Computer Science Dept., University of Milan
Via Comelico 39/41 – 20135 Milan Italy
{mirko.lucchese,iuri.frosio,alberto.borghese}@unimi.it

Abstract. The Bayesian approach applied to image denoising gives rise to a regularization problem. Total variation regularizers have been introduced with the motivation of being edge preserving. However we show here that this may not always be the best choice in images with low/medium frequency content like digital radiographs. We also draw the attention on the metric used to evaluate the distance between two images and how this can influence the choice of the regularization parameter. Lastly, we show that hyper-surface regularization parameter has little effect on the filtering quality.

Keywords: Denoising, Total Variation Regularization, Bayesian Filtering, Digital Radiography.

1 Introduction

Poisson data-noise models naturally arise in image processing where CCD cameras are often used to measure image luminance counting the number of incident photons. Photon counting process is known to have a measurement error that is modeled by a Poisson distribution [1]. Radiographic imaging, where the number of counted photons is low (e.g. a maximum count of about 10,000 photons per pixel in panoramic radiographies [2]) is one of the domains in which Poisson noise model has been largely adopted.

The characteristics of this kind of noise can be taken into account inside the Bayesian filtering framework, developing an adequate likelihood function which is, apart from a constant term, equivalent to the Kullback–Leibler (KL) divergence [3, 4]. Assuming the a-priori distribution of the solution image of Gibbs type and considering the negative logarithm of the a-posteriori distribution, the estimate problem is equivalent to a regularization problem [5, 6]. The resulting cost function, $J(.)$, is a weighted sum of a negative log-likelihood (data-fit, $J^L(.)$) and a regularization term (associated to the a-priori knowledge on the solution, $J^R(.)$). Tikhonov-like (quadratic) regularization often leads to over-smoothed images and Total Variation (TV) regularizers, proposed by [7] to better preserve edges, are nowadays widely adopted. As the resulting cost-function is non-linear, iterative optimization algorithms have been developed to determine the solution [3, 8]. To get

G. Maino and G.L. Foresti (Eds.): ICIAP 2011, Part I, LNCS 6978, pp. 534–543, 2011.

a differentiable cost function, a parameter, δ, has been introduced into the TV term (known in this case as "hyper-surface regularizer" [11]). The regularization parameter, β, weights the two terms $J^L(.)$ and $J^R(.)$ in $J(.)$ and it strongly influences the characteristics of the filtered image. Some attempts to set its optimal value have been proposed resorting to various forms of the discrepancy principle [9, 10], but the results are not always satisfying.

Aim of this paper is to investigate the adequacy of TV regularization in filtering radiographs and in general images with low photon counts. Results show that when low frequency components are dominant the optimal solution is obtained after a few iterations, while increasing the number of iterations the cost function further decreases but the distance between the true and the filtered image increases. We also investigated how the value of parameters β and δ affects the filtered image and we show that the optimal value of β is influenced by both the signal level and the frequency content of the image; on the other hand, δ has little impact: it slightly increases the regularization effect only when photon count is very low.

2 Denoising Framework

2.1 Definition of the Cost Function

Let $g_{n,j}$ and g_j indicate respectively the gray level of the noisy and noise-free image at pixel j. Aim of any denoising algorithm is to estimate the true image, $\mathbf{g}=\{g_j\}_{j=1..M}$, from the measured, noisy one, $\mathbf{g_n}=\{g_{n,j}\}_{j=1..M}$, where M is the number of image pixels. In the Bayesian framework, the filtered image is obtained maximizing the a-posteriori probability of the image, given by the product of the likelihood of the noisy image given the filtered one, and the a-priori probability of the filtered image. Assuming that each pixel is independent from the others, and without considering the constant terms, the negative log–likelihood function can be written as:

$$J^L_{\mathbf{g_n}}(\mathbf{g}) = \sum_{j=1}^{M} g_j - g_{n,j}\log(g_j) \,, \tag{1}$$

which, apart from constant terms, is equivalent to the KL divergence between $\mathbf{g_n}$ and $\mathbf{g}$. In the TV approach the regularization term is represented by the image TV norm, defined as:

$$J^R(\mathbf{g}) = \|\mathbf{g}\|_{TV} = \sum_{j=1}^{M} \sqrt{(\partial g_j/\partial x)^2 + (\partial g_j/\partial y)^2} \,, \tag{2}$$

where $\partial g_j/\partial x$ and $\partial g_j/\partial y$ are respectively the horizontal and the vertical derivative of the image $\mathbf{g}$ in position j. To get a differentiable cost function even if the image gradient is equal to zero, an additional parameter, δ, is introduced and the regularizer becomes:

$$J^R(\mathbf{g}) = \sum_{j=1}^{M} \sqrt{(\partial g_j/\partial x)^2 + (\partial g_j/\partial y)^2 + \delta^2} \,. \tag{3}$$

For $\delta = 1$ the sum in (3) is equivalent to the area of the surface defined by the image values in each pixel; therefore, this is also called Hyper-surface regularization [9]. Combining eqs. (1) and (3), the following cost function is obtained:

$$J_{g_n}(\mathbf{g}) = J_{g_n}^L(\mathbf{g}) + \beta J^R(\mathbf{g}) = \sum_{j=1}^{N}[g_j - g_{n,j}\log(g_j)] + \beta\sum_{j=1}^{M}\sqrt{(\partial g_j/\partial x)^2 + (\partial g_j/\partial y)^2 + \delta^2} \quad (4)$$

where β is the so called regularization parameter. In the context of digital image processing, the derivatives in eq. (4) are replaced by the discrete differences between each pixel and its neighbors. In this respect, a parallel between eq. (3) and the potential function in a Markov Random Field has been drawn in [11]. In particular, let $N_8(j)$ be the set of indices of the eight first neighbors of jth pixel; eq. (4) becomes:

$$J_{g_n}(\mathbf{g}) = \sum_{j=1}^{N}[g_j - g_{n,j}\log(g_j)] + \beta\sum_{j=1}^{M}\sqrt{\sum_{i\in N_8}[(g_j - g_i)/\rho_{i,j}]^2 + \delta^2} , \quad (5)$$

where $\rho_{i,j}$ is a normalization factor introduced to take into account the different distance between the j-th pixel and its neighbors:

$$\rho_{i,j} = \begin{cases} \sqrt{2} & \text{if } i \text{ and } j \text{ are diagonal neighbors} . \\ 1 & \text{otherwise} \end{cases} \quad (6)$$

2.2 Minimization of the Cost Function

The cost function in eq. (5) is strongly non linear and iterative optimization algorithms are used to minimize it. Since second order methods require the computation and inversion of the Hessian of $J_{gn}(\mathbf{g})$, and since these operations are computationally intensive, minimization is usually performed via first order methods. We use here the recently proposed Scaled Gradient Projection method [4] that has shown fast convergence rate. At each step, the solution is updated as:

$$\mathbf{g}^{k+1} = \mathbf{g}^k + \lambda_k \mathbf{d}^k , \quad (10)$$

where $\mathbf{d}^k$ is a descent direction derived from the Karush – Kuhn – Tucker conditions, and λ_k is determined such that the decrease of the solution is guaranteed to be large along that direction. The procedure is stopped when the normalized variation of the cost function goes below a given threshold τ, that is:

$$\left\|[J_{g_n}(\mathbf{g}^{k+1}) - J_{g_n}(\mathbf{g}^k)]/J_{g_n}(\mathbf{g}^{k+1})\right\| \leq \tau , \quad (11)$$

or when the maximum number of iterations is achieved.

3   Experimental Setup

We have created a set of simulated digital radiographs of 512×512 pixels as follows. First, an absorption coefficients map was created, with coefficients increasing from 0% for the left-most pixels to 100% for the right-most ones. Then, 50 different geometrical figures (circles and rectangles) were randomly positioned inside the image. The radius of the circles and the rectangle sides had length randomly chosen between 1 and 512 pixels. Each time a circle or a rectangle was added to the map, all the absorption coefficients covered by the figure were modified: either they were substituted by their complements with respect to 100%, or they were multiplied by a random value between 0 and 1, or a random value between 0% and 100% was added to them. In the latter case, the resulting absorption coefficients were always clipped to 100%. The choice among the three modalities was random. To control the frequency content, these images were filtered with different moving average filters with size of 41×41, 23×23 and 9×9 pixels to generate respectively low (LF), medium (MF) and high frequency (HF) simulated radiographs. Three sets of images were considered each with a different maximum number of photons reaching the sensor: 10,000, 1,000 and 100 photons. For each image, we considered five realizations of Poisson noise; we explicitly notice that the images with low photon count have noise with a lower standard deviation (and lower signal to noise ratio, SNR) than those at high photon count due to the nature of Poisson noise. Three typical images are shown in Fig. 1.

We filtered the noisy images with the method in Section 2 and we measured the difference between each filtered image and the true one using three different quality indices. The first one is the Root Mean Squared Error (RMSE), which is widely adopted in signal and image processing as it is related to the power of the error:

$$\text{RMSE}(\mathbf{g}_1, \mathbf{g}_2) = \sqrt{\sum_{j=1}^{M} (g_{1,j} - g_{2,j})^2 \Big/ M} \cdot \qquad (12)$$

The second index is the Structural Similarity (SSIM) proposed in [12]. This index compares local patterns of pixel intensities that have been normalized for luminance and contrast and it evaluates the similarity by comparing structural information from processed images. It is defined as:

$$\text{SSIM}(\mathbf{g}_1, \mathbf{g}_2) = \frac{1}{M} \sum_{j=1}^{M} \left[\left(\frac{2\mu_{\mathbf{g}_1}(j) \cdot \mu_{\mathbf{g}_2}(j) + c_1}{\mu_{\mathbf{g}_1}^2(j) + \mu_{\mathbf{g}_2}^2(j) + c_1} \right) \cdot \left(\frac{2\text{cov}_{\mathbf{g}_1,\mathbf{g}_2}(j) + c_2}{\sigma_{\mathbf{g}_1}^2(j) + \sigma_{\mathbf{g}_2}^2(j) + c_2} \right) \right], \qquad (13)$$

where $\mu_{\mathbf{g}}(j)$ is the local mean of $\mathbf{g}$ in the neighborhood of the j-th pixel, $\text{cov}_{g1,g2}(j)$ is the local covariance between $\mathbf{g}_1$ and $\mathbf{g}_2$, $\sigma_{\mathbf{g}}^2(j)$ is the local variance of the image $\mathbf{g}$ and the constants c_1 and c_2 are defined as $c_1 = (k_1 L)^2$ and $c_2 = (k_2 L)^2$ where L is the maximum image gray level (e.g. 255 for 8 bit images) and k_1 and k_2 are user defined constants set equal to 0.01 and 0.03 according to [12].

The last index is the Features Similarity (FSIM) proposed in [13]. This index, based on the phase congruency (PC) model of the Human Visual System (HVS), takes into account the local phase of the Fourier components of the image and the local gradient magnitude. It is defined as:

$$\text{FSIM}(\mathbf{g}_1, \mathbf{g}_2) = \sum_{j=1}^{M} w(j) \left[\left(\frac{2PC_{\mathbf{g}_1}(j) \cdot PC_{\mathbf{g}_2}(j) + T_1}{PC_{\mathbf{g}_1}^2(j) + PC_{\mathbf{g}_2}^2(j) + T_1} \right) \left(\frac{2G_{\mathbf{g}_1}(j) \cdot G_{\mathbf{g}_2}(j) + T_2}{G_{\mathbf{g}_1}^2(j) + G_{\mathbf{g}_2}^2(j) + T_2} \right) \right], \qquad (14)$$

where $PC_{\mathbf{g}}$ is the phase congruency term and $G_{\mathbf{g}}$ is the gradient magnitude in j, computed here by Scharr operator. T_1 and T_2 are user defined constants and, for 8-bit images, they are equal to 0.85 and 160 respectively [13]. In our case T_2 was linearly rescaled proportionally to the maximum gray levels. The weights $w(j)$ are computed to assign a higher weight to the position with high phase congruency [13]:

$$w(j) = PC_m(j) \bigg/ \sum_{j=1}^{M} PC_m(j) \, ,$$ (15)

where $PC_m(j) = \max\{PC_{g1}(j), PC_{g2}(j)\}$.

The difference between the filtered and true image has been evaluated for different values of β (in the range from 0.01 to 0.9) and for $\delta = \{0, \text{eps} = 2.2204 \times 10^{-16}, 0.01, 0.05, 0.1, 0.5, 1\}$. To avoid singular derivatives of $J^R(\mathbf{g})$ for $\delta = 0$, $\partial J^R(\mathbf{g})/\partial g_j$ was assumed equal to zero for all the pixels j whose gradient norm was equal to zero.

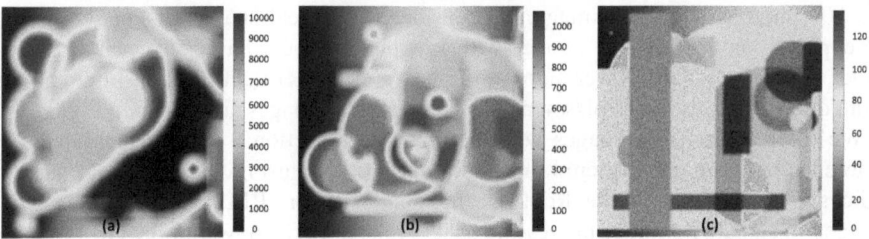

Fig. 1. Panels (a-c) show a LF, MF and HF simulated radiograph whose expected maximum number of photons is equal to, respectively, 10,000, 1,000 and 100

4   Results and Discussion

Figs. 2, 3 and 4 represent the mean value of RMSE, SSIM and FSIM respectively, averaged over five filtered images, as a function of β and δ. The optimal value of β is higher for LF images and lower for MF and HF images (Fig. 5): in practice, when the structures (edges) in the image become sharper, the regularizer should be more edge-preserving. This effect is obtained lowering the value of β in the cost function (5), and therefore increasing the probability of observing high gradients (associated to edges) in the filtered image. The same figures show that the optimal value of β increases when the number of photons decreases. In fact, according to Poisson statistics, the SNR is low when the number of counted photons is low: in this case, the low reliability of the measured data has to be counterbalanced by a high regularization and therefore it calls for high values of β.

Figs. 2-4 also show also that RMSE generally leads to an optimal value of β lower than that suggested by SSIM and FSIM, although the values provided by FSIM and by RMSE are very similar in almost all cases. This reflects different capacities of the indices to quantify the image quality. More in detail, the visual inspection of the images filtered with the optimal β value suggested by different indices reveals that the SSIM generally provides a too regular image, especially when the photon count is low (Fig 6). Overall, although SSIM has been proposed as a principled evaluation metric based on the properties of the HVS, RMSE generally provides a less regularized image with lower noise and more visible edges. In case of low photon count images, FSIM shows a higher capability of identifying the optimal value of β than SSIM (curves in Fig. 4g-i have higher curvature than those in Fig. 3h-i), providing results similar to RMSE. Visual inspection of filtered images with low photons count reveals that images suggested as optimal by RMSE and FSIM definitely appear more similar to the ground truth than the image obtained with SSIM (Fig. 6).

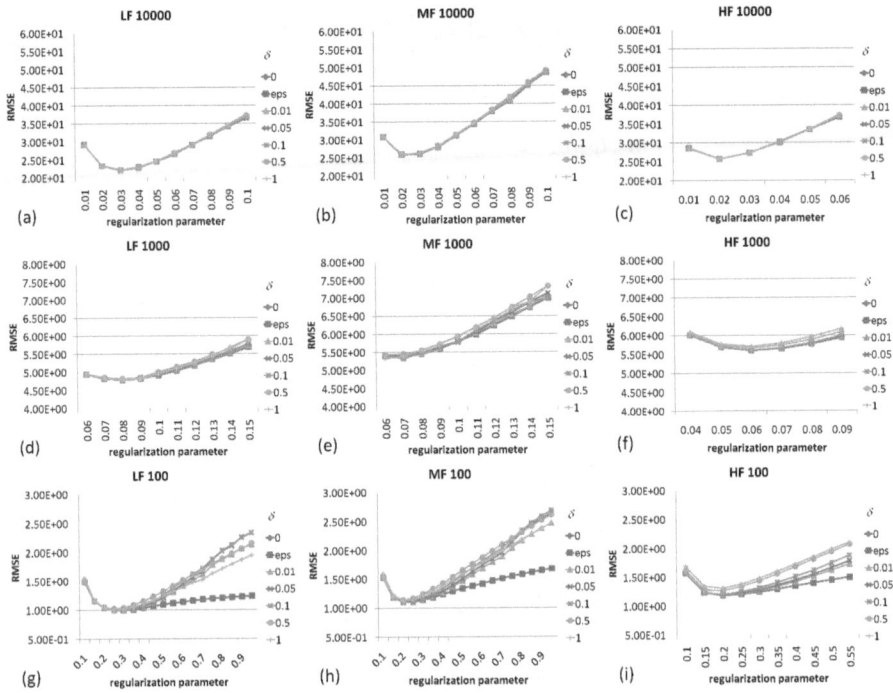

Fig. 2. Mean RMSE as a function of the regularization parameter β. Here, eps = 2.2204×10^{-16}. The title of each graph reports the frequency content of the images and the number of photons.

The similarity indices in figs. 2, 3 and 4 also indicate that the parameter δ has little or even no influence on the quality of the filtered image for medium and high photon counts as far as it assumes a very small value; in fact, the curves reported in these figures show low dependency on the value of δ. For a low number of photons, better results are obtained with small values of δ (0 or eps). This fact is explained considering that, for low counts, the term $[(g_i - g_i)/\rho_{i,j}]^2$ in the TV norm in eq. (5) is low and therefore δ may strongly influence the value of the TV norm.

For LF and MF images, optimization exhibits semi-convergence: the similarity between the filtered image and the ground truth increases in the first iterations, reaches a maximum and then decreases (Fig. 7a-b). This is not the case of HF images, at least up to the 800th iteration (Fig. 7c-d). This fact can be explained considering that TV regularization implicitly assumes that images are composed by flat areas separated by sharp edges. However LF and MF images, that are somehow similar to digital radiographies, are not coherent with such hypothesis. As a result, the filter cuts the image valleys and ridges introducing spurious plateaus in the high photon count areas; the cost function is actually minimized but the image is over-smoothed (Fig. 8b). This phenomenon becomes evident only in the last iterations and it is less evident in the low photon count regions. To analytically explain it, let us consider the derivative of the likelihood term in eq. (1):

$$\partial J_{g_n}^L(\mathbf{g})/\partial g_j = 1 - g_{n,j}/g_j \;. \tag{16}$$

Defining $\Delta g_j = g_j - g_{n,j}$, eq. (16) becomes:

$$\partial J_{g_n}^L(\mathbf{g})/\partial g_j = 1 - g_{n,j}/(g_{n,j} + \Delta g_j) \;. \tag{17}$$

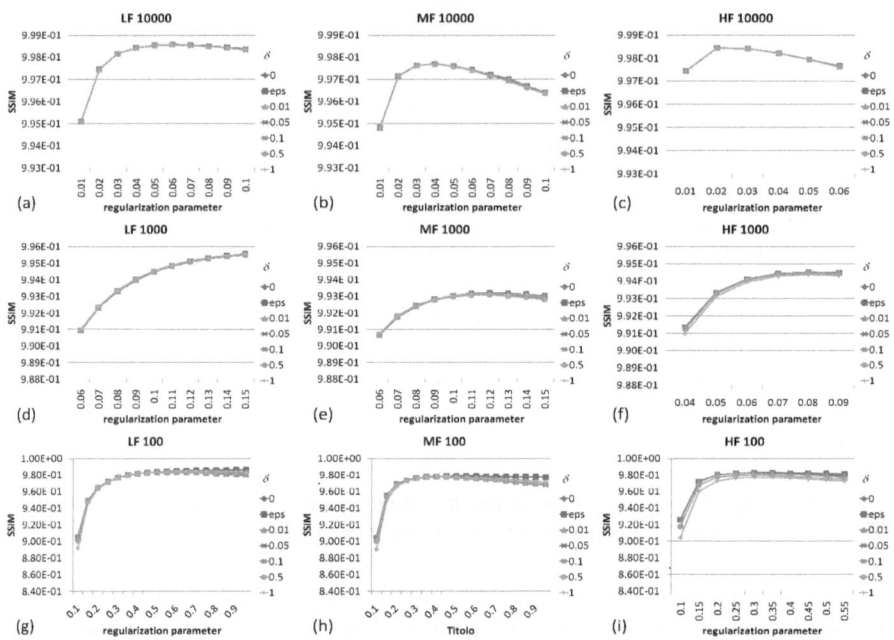

Fig. 3. Mean SSIM index varying the regularization parameter β. Here, eps = 2.2204×10^{-16}. The title of each graph reports the frequency content of the images and the number of photons.

Since $g_{n,j}$ is fixed, eq. (17) approximately expresses the variation of the likelihood as a function of Δg_j. If the noisy gray level, $g_{n,j}$ is high, $\partial J_{g_n}^L(\mathbf{g})/\partial g_j$ is close to zero for large intervals of $|\Delta g_j|$. In other words, the jth element of the likelihood in (5) slightly differs from its minimum even if the filtered image significantly differs from the noisy one. In these regions, the filtering effect depends mainly on the regularization term, which tends to suppress the edges, introducing a plateau where a ridge (or a valley) is present (Fig. 8b). On the other hand, if $g_{n,j}$ is small, a high value of $|\Delta g_j|$ produces a significant increase of $J_{g_n}^L(\mathbf{g})$. In this case, the variation of the cost function is mainly influenced by the likelihood which constrains the solution to be close to the noisy image, thus retaining noise oscillations (Fig. 8d). This further explains the need of a higher value of β for low-photon count images: in this case the SNR is low, the measured values are less reliable, $\left|\partial J_{g_n}^L(\mathbf{g})/\partial g_j\right|$ increases significantly for small $|\Delta g_j|$ and more regularization is necessary to get the denoising effect.

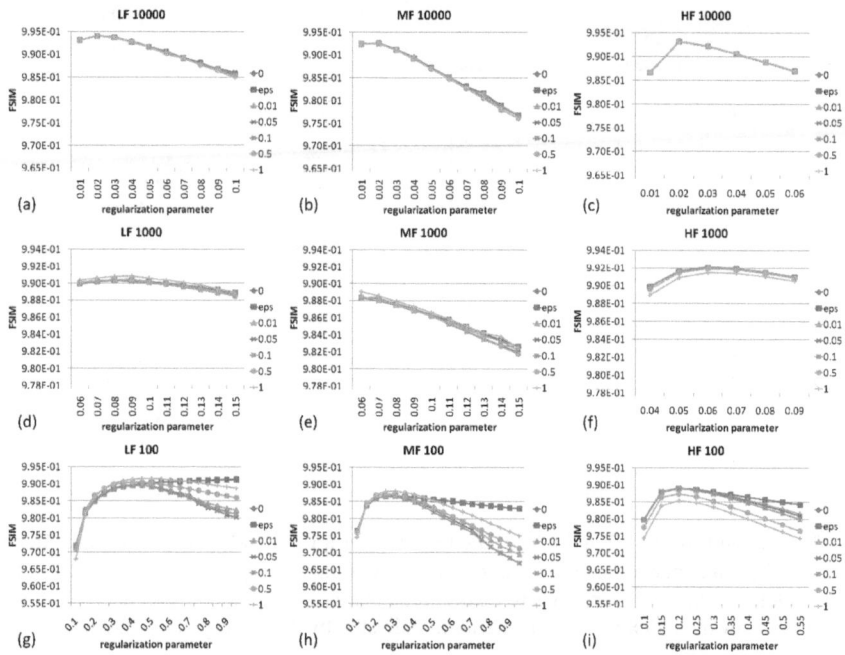

Fig. 4. Mean FSIM index varying the regularization parameter β. Here, eps = 2.2204×10^{-16}. The title of each graph reports the frequency content of the images and the number of photons.

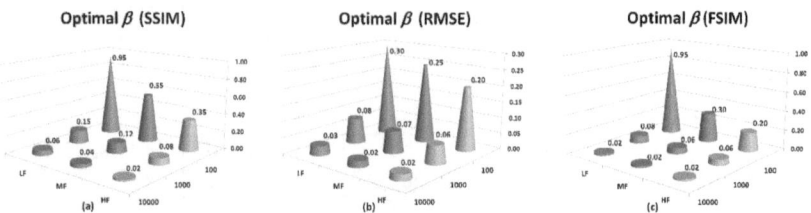

Fig. 5. Optimal values of β associated to the optimal value of SSIM (a), RMSE (b) and FSIM (c) as a function of photon count and image frequency content

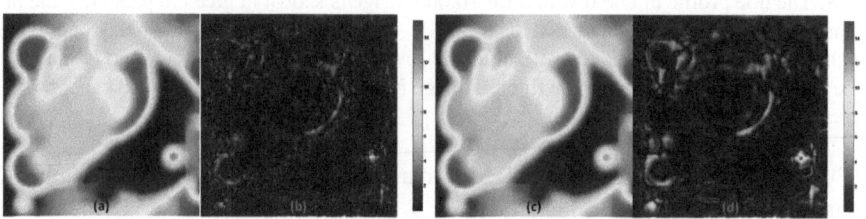

Fig. 6. Filtered LF 100 image with δ = eps and β = 0.3 (a) and β = 0.95 (c), corresponding to the optimal value suggested by RMSE and SSIM respectively. Panels (b) and (d) show the absolute difference between the filtered image and the ground truth.

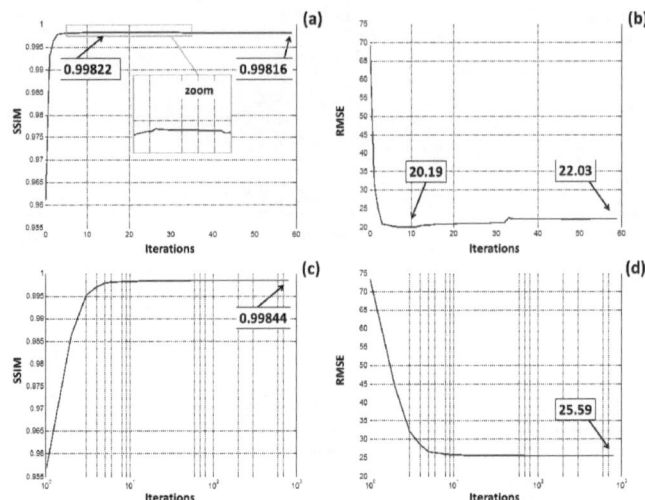

Fig. 7. In (a) and (b), SSIM and RMSE are reported for a LF 10,000 image versus the iteration number. The same indexes are reported for a HF 10,000 image in (c) and (d), for $\tau = 10^{-14}$. Panel (a) includes a zoom of the area where SSIM stops increasing (semi-convergence). Numbers in the panels show the index value at the iteration indicated by the corresponding arrow.

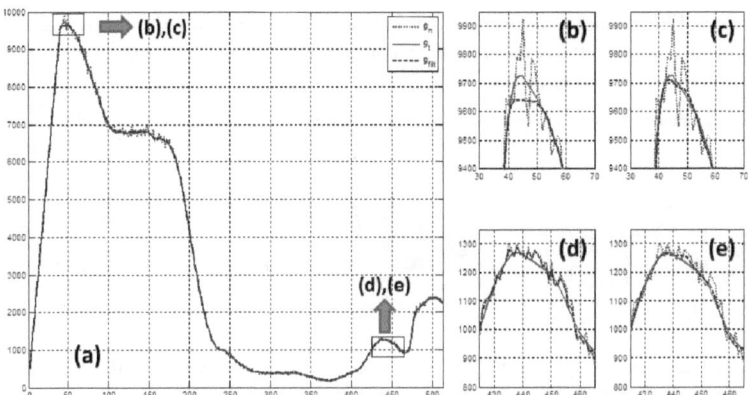

Fig. 8. The true profile of one row of a LF 10,000 image is shown in green in panel (a); the blue line is the filtered ($\beta = 0.03$ and $\tau = 10^{-10}$) profile, the red line represents the noisy one. Panels (b) and (d) show zooms of ridges with high and low photon counts. Panels (c) and (e) show the same areas of (b) and (d), when the filtered image has been obtained with $\beta = 0.64$ and $\tau = 10^{-4}$, corresponding to only four iterations.

Plateau introduction in LF and MF images is prevented by small number of iterations (e.g. $\tau = 10^{-4}$) and high value of β. The ridges in high photon counts areas are preserved by the low number of iterations; the high value of β gives the proper weight to the regularization term in the low photon count areas. This definitely produces an image with effective denoise in the low count regions (Fig. 8e) without cutting ridges and valleys (Fig. 8c) in the high count areas.

5 Conclusion

Image denoising through TV regularization is nowadays widely diffused as a "Holy Graal" of image filtering. However, as shown here, it may not be the best choice, at least for digital radiographs. In fact, the semi-convergence property observed for MF and LF images suggests the need for developing a better model for a-priori term, at least for this kind of images. On the other hand, also the definition of a general, reliable image quality index remains an open issue, as demonstrated by the difference in the optimal value of β obtained when RMSE, SSIM or FSIM measure were used to evaluate the difference between the filtered and the true image.

References

1. Snyder, D.L., Hammoud, A.M., White, R.L.: Image recovery from data acquired with a charge coupled device camera. J. Opt. Soc. Am. A 10, 1014–1023 (1993)
2. Frosio, I., Borghese, N.A.: Statistical Based Impulsive Noise Removal in Digital Radiography. IEEE Trans. on Med. Imag. 28(1), 3–16 (2009)
3. Zanella, R., Boccacci, P., Zanni, L., Bertero, M.: Efficient Gradient Projection methods for edge-preserving removal of Poisson noise. Inverse Problems 25(4), 045010 (2009)
4. Bonettini, S., Zanni, L., Zanella, R.: A scaled gradient projection method for constrained image deblurring. Inverse Problems 25(1), 015002 (2009)
5. Bertero, M., Lanteri, H., Zanni, L.: Iterative image reconstruction: a point of view. In: Proc. IMRT (2008)
6. Tikhonov, A.N., Arsenin, V.Y.: Solutions of Ill-posed Problems. W. H. Winston (1977)
7. Rudin, L., Osher, S., Fatemi, E.: Nonlinear Total Variation based noise removal algorithms. Physica D 60, 259–268 (1992)
8. Carbonetto, P., Schmidt, M., de Freitas, N.: An interior-point stochastic approximation method and an L1-regularized delta rule. In: Proc. NIPS 2008, pp. 112–119 (2008)
9. Bertero, M., Boccacci, P., Talenti, G., Zanella, R., Zanni, L.: A discrepancy principle for Poisson data. Inverse Problems 26(10), 105004 (2010)
10. Lucchese, M., Borghese, N.A.: Denoising of Digital Radiographic Images with Automatic Regularization Based on Total Variation. In: Proc. ICIAP, pp. 711–720 (2009)
11. Geman, S., Geman, D.: Stochastic Relaxation, Gibbs Distributions, and the Bayesian Restoration of Images. IEEE Trans. PAMI 6, 721–741 (1984)
12. Wang, Z., Bovick, A.C., Sheikh, H.R., Simoncelli, E.P.: Image Quality Assessment: From Error Visibility to Structural Similarity. IEEE Tran. Image Proc. 13(4), 600–612 (2004)
13. Zhang, L., Zhang, L., Mou, X., Zhang, D.: FSIM: A Feature Similarity Index for Image Quality Assessment. IEEE Tran. Image Proc. (in press)

Neighborhood Dependent Approximation by Nonlinear Embedding for Face Recognition

Ann Theja Alex, Vijayan K. Asari, and Alex Mathew

Computer Vision and Wide Area Surveillance Laboratory,
Department of Electrical and Computer Engineering,
University of Dayton, Dayton, Ohio
{alexa1,vijayan.asari,mathewa3}@notes.udayton.edu

Abstract. Variations in pose, illumination and expression in faces make face recognition a difficult problem. Several researchers have shown that faces of the same individual, despite all these variations, lie on a complex manifold in a higher dimensional space. Several methods have been proposed to exploit this fact to build better recognition systems, but have not succeeded to a satisfactory extent. We propose a new method to model this higher dimensional manifold with available data, and use a reconstruction technique to approximate unavailable data points. The proposed method is tested on Sheffield (previously UMIST) database, Extended Yale Face database B and AT&T (previously ORL) database of faces. Our method outperforms other manifold based methods such as Nearest Manifold and other methods such as PCA, LDA Modular PCA, Generalized 2D PCA and super-resolution method for face recognition using nonlinear mappings on coherent features.

Keywords: Face Recognition, Manifold Learning, Nonlinear Embedding.

1 Introduction

Face recognition is a challenging problem in computer vision. Variations in face images make this a difficult task. Pose, illumination, expression, occlusion etc are different factors that influence recognition accuracy. Different approaches such as PCA (Principal Component Analysis), LDA (Linear Discriminant Analysis), ICA (Independent Component Analysis), Modular PCA and several manifold based approaches have evolved over time to deal with this issue. None of the methods offers a complete solution. Many a times, the recognition of frontal face achieves a better recognition rate, but as the pose varies, it becomes difficult to effectively identify faces.

The main objective of the current research is to improve the recognition accuracy in varying facial poses, illumination, expression variations and occlusions. Though PCA is a very popular method, it does not offer high accuracy with the variations in pose and illumination. Modular PCA approach is better than PCA with pose and illumination invariance [1]. However, Modular PCA is a

G. Maino and G.L. Foresti (Eds.): ICIAP 2011, Part I, LNCS 6978, pp. 544–553, 2011.

linear approach. This affects testing accuracy and can introduce false alarms. Many studies have proved that faces of a particular person tend to lie on a non-linear manifold in a higher dimensional space [2]. But methods like PCA and LDA use only Euclidean distance and hence fail to see the nonlinear structure [3]. An alternative is to use Locally Linear Embedding (LLE) [4], which allows data to be handled as lying on a linear neighborhood. This is a piecewise linear approach. The rationale behind this method is that every curve can be represented as a connection of linear segments. Generalizing this concept, a higher dimensional manifold can be represented as a combination of hyper-planes. The linear approach of LLE is not enough to yield good results, as the data points have non-linear relationships between them. Although many other methods such as Nearest Manifold, Locally Linear Embedded Eigenspace analysis, Discriminative manifolds, and 2D PCA evolved over time [5-8], recognizing faces under variations is still an open problem.

The proposed algorithm, Neighborhood Dependent Approximation by Nonlinear Embedding (NDANE), aims at addressing this issue and offers a novel method for an effective representation of a manifold. The method is developed from the Hopfield network, LLE and non-linear attractor theories [9].

The Hopfield network is a recurrent network invented by John Hopfield in early 1980s. In Hopfield network, the output of one node is computed as the linear combination of the other inputs. Hopfield suggested that any input can be best approximated as the weighted combination of other related inputs. The Hopfield network supports only linear relationships.

In [4], Locally Linear Embedding is introduced as a dimensionality reduction technique. The approach provides a novel method for nonlinear dimensionality reduction. The method is considered one of the best for nonlinear dimensionality reduction, which is not effectively handled by component analysis methods. In LLE, the inputs are considered as data points in higher dimensional space. All the data points with the same characteristics form a geometric structure called the manifold. As all data points in the manifold belong to a particular class, any data point in the manifold can be redefined as a reconstruction of other data points in the same manifold. In the proposed method, the same observation is used for effective reconstruction of the data points.

In [9], a nonlinear attractor was proposed for learning the relationships in a manifold. This is a variation of the Hopfield network in which a higher dimensional relationship is defined between an input and its neighbors (represented by the other inputs). It is proposed that a higher dimensional polynomial defines the relationship of each image with its neighborhood. This, together with LLE resulted in the development of our method, in which a manifold is represented by a nonlinear approximation. As the manifolds are complex, a nonlinear approach can yield a better representation than a piecewise linear approach.

2   Theoretical Background

The proposed supervised learning algorithm is based on the fact that facial images are data points in a higher dimensional space, whose dimension is determined by the number of pixels in the image. The data points that correspond to the face of a particular person are observed to lie on a single nonlinear manifold. As the images in a manifold belong to the same person, there is a relationship between these images. A relationship can be modeled as proposed by Hopfield and as established by Seow and Asari [9]. In this paper, the same idea is extended so that it can be used to represent closely related points in terms of the neighborhood points. In other words, each image can be represented as the relationship of a few images in the manifold to which it belongs. Since manifolds can have a broad set of images of the same person, we do not consider all the images in the manifold to reconstruct an image. The nearest neighbors to each image are identified using Euclidean distance as the distance measure. Thus we construct small neighborhoods within the same manifold. Each image can be represented as the relationship of the neighborhood images. This relationship is defined as a nonlinear relationship, as it is observed that a nonlinear relationship offers a better representation of the manifold than any piece wise linear approach.

2.1   Architecture

A simple schematic diagram of the proposed algorithm is provided in this section (Fig. 1 and Fig. 2). We have a pipeline approach with two pipelines for handling training and testing phases.

The algorithm is supervised and hence the training input will already be divided into classes. In the training phase, there are two separate stages. The first stage identifies the k nearest neighbors of an input image from the class to which the input image belongs. The nearest neighbors are the neighbors with the minimum Euclidean distances. The set of the nearest neighbors is also called the Proximity set. This step also models a higher dimensional polynomial relationship to represent the data point under consideration, using the nearest neighbors identified. The coefficients of the terms in the polynomial are called the weights. This stage is called the Relationship Modeling stage. The polynomial relationship that is used in this stage is described in more detail in section 2.2. The second stage is the Reconstruction stage, where the actual reconstruction of the data points is done using the neighbors and the weights. As the outcome of this stage, we obtain a reconstructed image. The process is repeated for all images in the training set. A set of reconstructed inputs are generated at the end of this stage.

In the Relationship Modeling stage of the testing phase, for each test image, we find the k nearest neighbors from each class. These neighbors are identified from the reconstructed training set inputs. Thus corresponding to each test input image, there are as many neighborhood sets as there are classes. In the Reconstruction stage, the test images are reconstructed based on the neighbors from all classes. The best reconstruction of the test image comes from the class to which the test image should belong. This concept is used for classification. The testing procedure is described in section 2.3. The geometric foundation of the method is described in section 2.4.

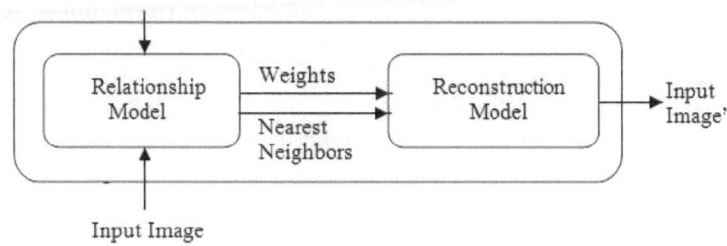

Fig. 1. The schematic represents the stages in training. The inputs are reconstructed as the relationship of the nearest neighbors.

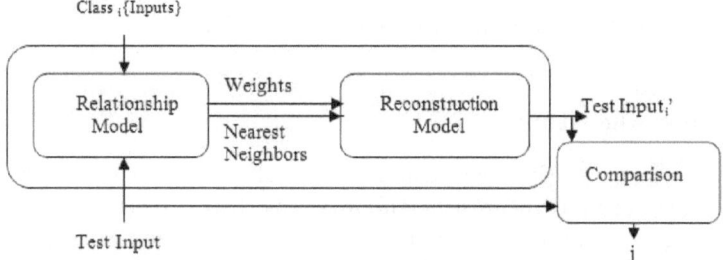

Fig. 2. The schematic represents the stages in testing. The test inputs are reconstructed as the relationship of the nearest neighbors from each class 'i', to get TestInput i'. The best representation is given by the class to which the test image should ideally belong. The best representation is identified and is represented as 'j'.

2.2   Training Algorithm

In the training process, the inputs are first grouped into different classes based on the labels, such that there is one class corresponding to each person. Each class is identified as a manifold and the aim of the training phase is to best approximate a manifold. If the images in a class have a high variance, instead of considering each class as a manifold, the class has to be partitioned into sub manifolds before processing. This partitioning can be repeated several times before the variances with in a sub-manifold are sufficiently low. This can result in a tree like structure of repeated partitioning of the manifolds. However, in this paper, our focus is on the basic framework. Once the classes are identified, the k nearest neighbors of each training inputs in their respective classes are found. This forms a proximity set. Each training input is represented as the weighted sum of elements in the proximity set as given by equation (1).

$$X_i = \sum_{j=1}^{k} \sum_{m=1}^{d} W_{mji} X_j{}^m + L. \tag{1}$$

where k is the number of neighbors and d, the degree of the polynomial.

The degree d can be any positive integer, depending on the nonlinear relationship. If d=2, the resulting polynomial relationship is a second degree equation. We have used d=2 in our experiments. L is a constant term. Since this representation is an approximation, we need to estimate the error in such a representation. The deviation of a representation from the actual data point can be characterized as a squared error given by equation (2).

$$E = (X_i - \sum_{j=1}^{k} \sum_{m=1}^{d} W_{mji} X_j{}^m + L)^2. \tag{2}$$

The weights corresponding to each term are found by minimizing the mean square error given by equation (2). The original training images are replaced with the form given in equation (1). Intuitively, this process incorporates the characteristics of its neighborhood into an image point.

2.3   Testing Algorithm

The first step in the testing algorithm is to find the k nearest neighbors in each class. This gives as many nearest neighbor sets as there are classes. The weights corresponding to each representation are found by minimizing E in equation (2). Each test image is represented as the weighted sum of elements in its proximity set. This gives as many representations as there are classes. Redefining each of the test image as the weighted sum of the neighborhood images gives as many representations as there are classes. The Euclidean distance, D_{class}, of each representation to the test image is computed. The class corresponding to the representation whose Euclidean distance is minimum, is taken as the class to which the test data belongs. This is given by equation (3).

$$j = min[D_{class}]. \tag{3}$$

Here, j is the class which gives the minimum distance. The best reconstruction can only be provided by the set of neighborhood images in the class to which the image can belong. This is because the relationship in equation (1) can model only related images. If the test image belongs to a class, then the images in that particular class are only related to the test image and images from other classes have little or no relationship. So the reconstruction produced by the right class is a better approximation than the ones produced by other classes. Thus the reconstruction that results in minimum Euclidean distance is the one that is produced by the class to which the test image should belong.

2.4 Why NDANE Works?

Fig.3 illustrates two manifolds, M1 and M2, and a data point T. The geometric foundation of the method can be understood by analyzing such a topology. As there are only a few points on the manifold, the manifold is not fully defined. The distance d1 between the data point T and M1 is less than d2, the distance between the data point and the closest data point on M2. In a Euclidean distance based algorithm, the distances d3 (distance between T and the closest data point on M1) and d2 are taken into consideration. This gives an incorrect classification of T as belonging to M2, although the actual distance of T to M1 (d1) is smaller than all other distances. Interpolating the manifold with available data points produces the point P on M1, whose distance to T is less than that to any point on M2. In our method, we complete the manifold by modeling it as a hyper-surface in image space. This completion process allows the calculation of the actual distance of the point to the closest manifold.

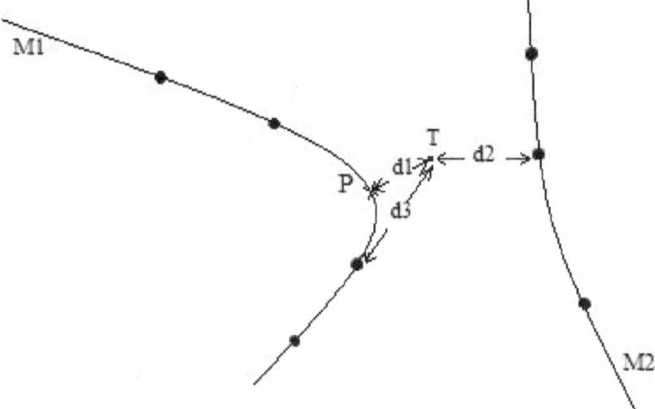

Fig. 3. Embedding a test data point on a manifold

3   Experiments

The proposed algorithm was tested using the pre-cropped images in the Extended Yale Face database B [10][11] , the AT&T (previously ORL) database [12] and Sheffield (previously UMIST) database [13].

3.1 Experiments on Extended Yale Face Database B

The Extended Yale Face Database B has 16128 images of 28 human subjects under 9 poses and 64 illumination conditions [10][11]. Our experiments were conducted on a subset of the database. Images of 5 different people were used for testing and training. 10 images of each person were randomly chosen for training. Another set of 10 images were chosen for testing. The training and testing images do not overlap. In this experiment d = 2 and k=5 in equation (1). This is a five class classification problem. We achieved a 100% recognition in this subset of the Extended Yale Face Database.

3.2   Experiments on AT&T (Previously ORL) Database of Faces

The AT&T database has a collection of 400 images of different people in various illumination conditions, pose and expression variations. The database is divided into 8 subsets. Each subset included 5 classes corresponding to the 5 individuals in the database. Out of the 10 available images per person, 5 were chosen for training and rest 5 for testing. In this experiment d=2 and k=2 in equation (1). The experimental results are given in Fig.4.

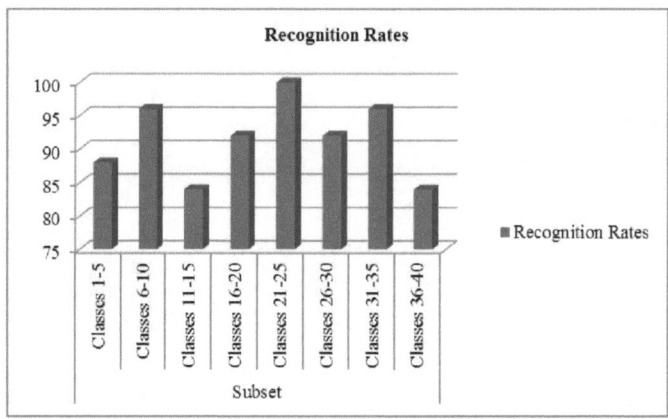

Fig. 4. Recognition rates on each subset of the AT&T database

3.3   Experiments on Sheffield (Previously UMIST) Database

The Sheffield (previously UMIST) database consists of 575 images of 20 individuals. There are images from different races, gender and poses. A sample data set is provided in Fig.5. It shows the reconstructed images which we obtained on reconstructing each image as relationship of its k nearest neighbors. In this experiment, d = 2 and k=5 in equation (1). A sample test set and the corresponding reconstructions are shown in Fig.6. It can be seen that the test image reconstructions are almost the same as the original images.In this database, number of images per person differs, so we have chosen half of the available number for training and the other half for testing. We have trained the system with 290 images and have tested the system with the remaining 285 images.

In all the three databases on which we conducted the experiments, our algorithm yielded a better result than other methods (Table 1 and Fig.7). We have achieved 95.44% recognition for the complete set of UMIST database. These results are better than the best results in [6], [7] and [14].

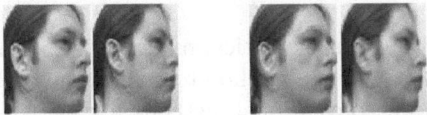

Fig. 5. Sample training set on left and the reconstructed images on right

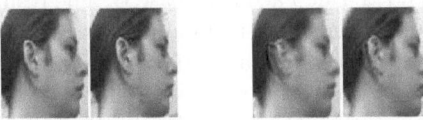

Fig. 6. Sample test set on left and the reconstructed images on right

Table 1. Comparison of Recognition rates (Other methods Vs Proposed method)

Algorithm	%recognition on Sheffield database
PCA [7]	86.87
KPCA [7]	87.64
LDA [7]	90.87
2DPCA [7]	92.9
DCV [7]	91.51
B2DPCA [7]	93.38
K2DPCA [7]	94.77
MLA+NN [6]	94.29
Super Resolution Method [14]	93
Proposed Method	**95.44**

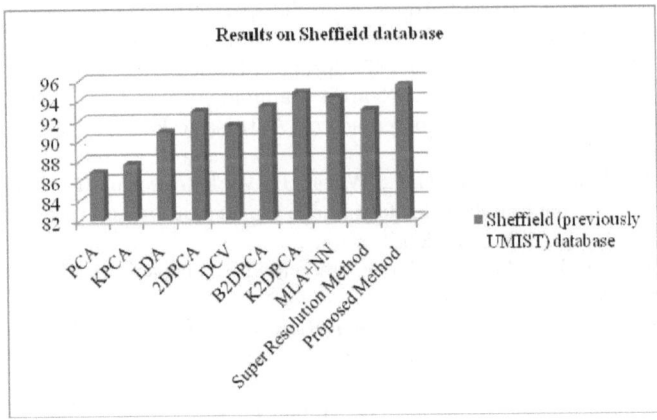

Fig. 7. Recognition rate comparison (Other methods Vs Proposed method)

4   Conclusion

We have proposed a robust mathematical model for representing complex manifolds formed by face images and a method to map test images to these manifolds. Our method outperforms conventional methods like PCA, Modular PCA, LDA and the methods in [6], [7] and [14]. All the experiments in this paper were conducted on the Sheffield (previously UMIST) database, AT&T (previously ORL) database and the Extended Yale Face Database B. The results obtained on these databases substantiate the mathematical foundation of our concept. Further experiments on larger databases such as FRGC database and FERET database are progressing. Research is also in progress to find a method to parameterize the value of k used in equation (1) depending on the data set variance and the data availability.

Acknowledgments. The authors acknowledge the creators of cropped Sheffield (previously UMIST) Face Database, Extended Yale Face Database B and AT&T ORL database for making these databases available for face recognition research.

References

1. Gottumukkal, R., Asari, V.: An improved face recognition technique based on modular PCA approach. Pattern Recognition Letters 25(4), 429–436 (2004)
2. Park, S.W., Savvides, M.: An extension of multifactor analysis for face recognition based on submanifold learning. In: IEEE Conference on Computer Vision and Pattern Recognition (CVPR), pp. 2645–2652 (2010)
3. Wu, Y., Chan, K.L., Wang, L.: Face recognition based on discriminative manifold learning. In: Proceedings of the 17th International Conference on Pattern Recognition, ICPR 2004, vol. 4, pp. 171–174 (2004)
4. Roweis, S., Saul, L.: Nonlinear dimensionality reduction by locally linear embedding. Science 290(5500), 2323–2326 (2000)
5. Yang, Q., Tang, X.: Recent Advances in Subspace Analysis for Face Recognition. In: Li, S.Z., Lai, J.-H., Tan, T., Feng, G.-C., Wang, Y. (eds.) SINOBIOMETRICS 2004. LNCS, vol. 3338, pp. 275–287. Springer, Heidelberg (2004)
6. Zhang, J., Stan, Z.L., Wang, J.: Nearest Manifold Approach for Face Recognition. In: 6th IEEE International Conference on Automatic Face and Gesture Recognition, pp. 223–228 (2004)
7. Kong, H., Wang, L., Teoh, E.K., Wang, J., Venkateswarlu, R.: Generalized 2D principal component analysis for face image representation and recognition. Neural Networks 18(5/6), 585–594 (2005)
8. Fu, Y., Huang, T.S.: Locally Linear Embedded Eigenspace Analysis. In: IFP-TR, UIUC, vol. 2005, pp. 2–5 (2005)
9. Seow, M., Asari, V.: Towards representation of a perceptual color manifold using associative memory for color constancy. Neural Networks, The Official Journal of the International Neural Network Society, European Neural Network Society & Japanese Neural Network Society 22, 91–99 (2009)
10. Lee, K.C., Ho, J., Kriegman, D.: Acquiring Linear Subspaces for Face Recognition under Variable Lighting. IEEE Trans. Pattern Anal. Mach. Intelligence 27(5), 684–698 (2005)

11. Georghiades, A.S., Belhumeur, P.N., Kriegman, D.J.: From Few to Many: Illumination Cone Models for Face Recognition under Variable. IEEE Transactions on Pattern Analysis and Machine Intelligence 23(6), 643–660 (2001)
12. Samaria, F.S., Harter, A.C.: Parameterisation of a stochastic model for human face identification. In: Proceedings of the Second IEEE Workshop on Applications of Computer Vision, vol. 5(7), pp. 138–142 (1994)
13. Graham, D., Allinson, N.M.: Characterizing Virtual Eigensignatures for General Purpose Face Recognition. In: Face Recognition: From Theory to Applications. NATO ASI Series F, Computer and Systems Sciences, vol. 163, pp. 446–456 (1998)
14. Huang, H., He, H.: Super-Resolution Method for Face Recognition Using Nonlinear Mappings on Coherent Features. IEEE Transactions on Neural Networks 22(1), 121–130 (2011)

Ellipse Detection through Decomposition of Circular Arcs and Line Segments

Thanh Phuong Nguyen[1]and Bertrand Kerautret[1,2]

[1] ADAGIo team, LORIA, Nancy University, 54506 Vandoeuvre, France
[2] LAMA (UMR CNRS 5127), University of Savoie, France
{nguyentp,kerautre}@loria.fr

Abstract. In this work we propose an efficient and original method for ellipse detection which relies on a recent contour representation based on arcs and line segments [1]. The first step of such a detection is to locate ellipse candidate with a grouping process exploiting geometric properties of adjacent arcs and lines. Then, for each ellipse candidate we extract a compact and significant representation defined from the segment and arc extremities together with the arc middle points. This representation allows then a fast ellipse detection by using a simple least square technique. Finally some first comparisons with other robust approaches are proposed.

1 Introduction

Shape identification is an important task in image analysis. Ellipse is a basic shape that can appear naturally in images from 3D environment. Therefore, ellipse detection is a key problem in many applications in the field of computer vision or pattern recognition.

In general, we can group existing methods into three main approaches. The first one relies on the Hough transform [2, 3, 4, 5, 6]. These methods transform image into parametric space and then take the peaks in this space as candidate of ellipse. Generally, it requires a parameter space that has five dimensions - contrariwise to two for straight line detection and three for circle detection, so it needs more execution time and memory space than the two last approaches. Some modifications [4, 5, 6] of Hough transform have been proposed to minimize storage space and computation complexity. Daul et al. [4] reduce the problem to two dimensional parametric space. Later, Chia et al. [5] introduced a method based on Hough transform in one dimensional parametric space. Lu et al. [6] proposed an iterative randomized Hough transform (IRHT) for ellipse detection with strong noise.

The second one uses least square fitting technique [7, 8, 9, 10] that minimizes the sum of square error. There are two main types of least square fitting (see [10]) that are characterized by the definition of error distances: algebraic fitting and geometric fitting. Concerning the first type, the error distance is defined by considering the deviation at each point to the expected ellipse described by implicit equation $F(x, a) = 0$ where a is vector of parameters. Contrariwise, for

G. Maino and G.L. Foresti (Eds.): ICIAP 2011, Part I, LNCS 6978, pp. 554–564, 2011.

the second type, the error distance is defined as orthogonal distance from each point to the fitting ellipse.

The third group of approach detects ellipse candidates by using their moment [11, 12, 13, 14, 15].

We propose a new method for ellipse detection based on the decomposition of an edge image into arc and line primitives. The main contribution of this paper is to propose a pre-processing step that allows to speed up the detection of ellipse based on a linear scanning process on the sequence of arc and line primitives. The rest of this paper is organized as follows. The following section recalls a method for the representation of a digital curve by arcs and line segments. Section 3 presents the proposed method for ellipse detection before experimentation.

2   Descriptor Based on Arc and Line Primitives

In this section, we recall a linear method [1] for the decomposition of a digital curve into circular arcs and line segments.

2.1   Tangent Space Representation and Properties of Arc in the Tangent Space

Nguyen and Debled-Rennesson proposed in [16] some properties of arcs in tangent space representation that are inspired from Latecki [17]. Let $C = \{C_i\}_{i=0}^{n}$ be a polygon, l_i - length of segment C_iC_{i+1} and $\alpha_i = \angle(\overrightarrow{C_{i-1}C_i}, \overrightarrow{C_iC_{i+1}})$. If C_{i+1} is on the right of $\overrightarrow{C_{i-1}C_i}$ then $\alpha_i > 0$, otherwise $\alpha_i < 0$ (see illustration of Fig. 1(a)).

Let us consider the transformation that associates a polygon C of $\mathbb{Z}^2$ to a polygon of $\mathbb{R}^2$ constituted by segments $T_{i2}T_{(i+1)1}, T_{(i+1)1}T_{(i+1)2}, 0 \leq i < n$ (see Fig. 1(b)) with:

$T_{02} = (0, 0)$,
$T_{i1} = (T_{(i-1)2}.x + l_{i-1}, T_{(i-1)2}.y)$, i from 1 to n,
$T_{i2} = (T_{i1}.x, T_{i1}.y + \alpha_i)$, i from 1 to $n - 1$.

Nguyen et al. also proposed in [16] some properties of a set of sequential chords of a circle in the tangent space. They are resumed by proposition 1 (see also Fig. 2).

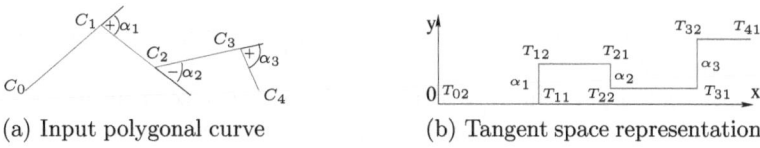

(a) Input polygonal curve   (b) Tangent space representation

Fig. 1. Tangent space representation

Proposition 1. *[16] Let $C = \{C_i\}_{i=0}^n$ be a polygon, $\alpha_i = \angle(\overrightarrow{C_{i-1}C_i}, \overrightarrow{C_iC_{i+1}})$ such that $\alpha_i \leq \alpha_{max} \leq \frac{\pi}{4}$. The length of C_iC_{i+1} is l_i, for $i \in \{1, \ldots, n\}$. We consider the polygon $T(C)$, that corresponds to its representation in the modified tangent space, constituted by the segments $T_{i2}T_{(i+1)1}, T_{(i+1)1}T_{(i+1)2}$ for i from 0 to $n-1$. $MpC = \{M_i\}_{i=0}^{n-1}$ is the midpoint set of $\{T_{i2}T_{(i+1)1}\}_{i=0}^{n-1}$. So, C is a polygon whose vertices are on a real arc only if $MpC = \{M_i\}_{i=0}^{n-1}$ is a set of quasi collinear points.*

From now on, MpC is called the midpoint curve.

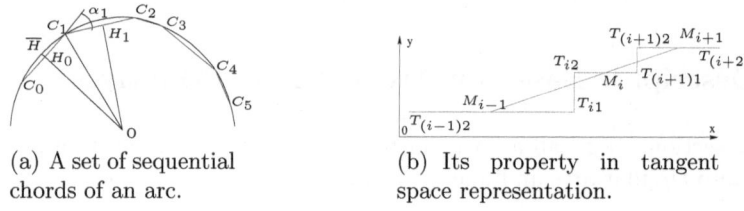

(a) A set of sequential chords of an arc.

(b) Its property in tangent space representation.

Fig. 2. The chords in tangent space

2.2 Arc Line Decomposition

Proposition 1 can be used to decide if a digital curve is a circular circle by detecting straight line segment in the tangent space. Moreover, it is also used for the decomposition of a curve into arcs and line segments. Nguyen introduced the definition below.

Definition 1. *In the curve of midpoints in the tangent space, **an isolated point** is a midpoint satisfying that the differences of ordinate values between it and one of its 2 neighboring midpoints on this curve is higher than the threshold α_{max}. If this condition is satisfied with all 2 neighboring midpoints, it is called **a full isolated point***

Let us consider Fig. 3. In this example, there are all basic configurations among the primitive arc and line: arc-arc, arc-line and line-line. Fig. 4 presents these configurations in detail in the tangent space. Concerning the midpoint curve (MpC) in the tangent space, Nguyen et al. [1] introduced several remarks below.

- An isolated point in MpC corresponds to an extremity among two adjacent primitives in C.
- A full isolated point in MpC corresponds to an line segment in C.
- An isolated point in MpC can be co-linear with a set of co-linear points that corresponds to an arc.

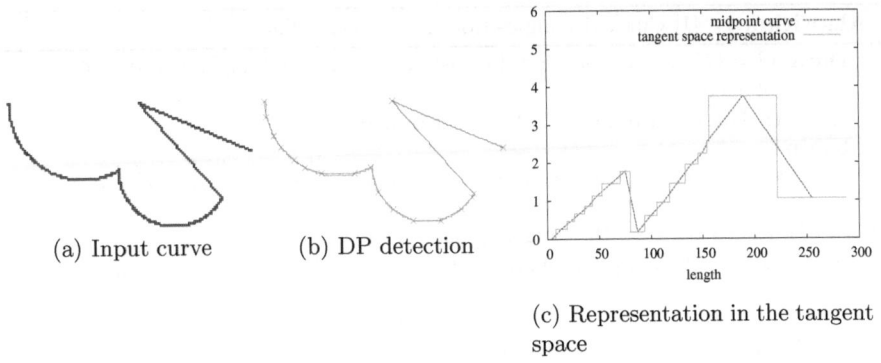

(a) Input curve   (b) DP detection

(c) Representation in the tangent space

Fig. 3. An example of curve

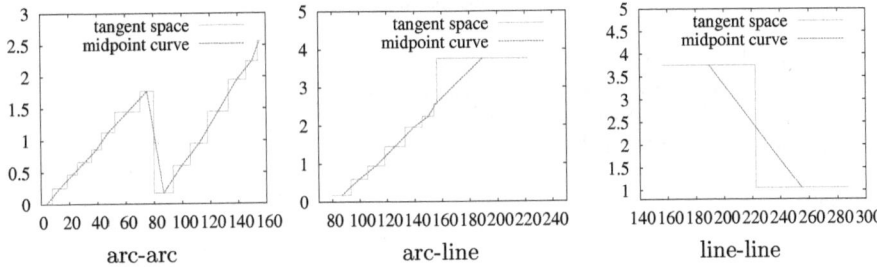

arc-arc   arc-line   line-line

Fig. 4. Configurations on tangent space

In [1], Nguyen et al. proposed an algorithm (see[1] algo. 1) to decompose a curve C into arcs and straight line segments. First, the sequence of dominant points (DpC) of C is computed by using an algorithm presented in [18]. DpC is then transform in the tangent space and the MpC curve is constructed. An incremental process is then used and each point of MpC is tested: if it is not an isolated point (in this case, it corresponds to an arc segment in C), the blurred segment recognition algorithm [19] permits to test if it can be added to the current blurred segment (which corresponds to an arc in C). If it is not possible, a new blurred segment starts with this point.

3   Ellipse Detection

We present hereafter a new method for ellipse extraction from edge map of an image. It is based on three steps:

- Construction of a representation based on arc and line primitive of the edge map of input image.

---

[1] Note this algorithm includes some corrections of algo. 3 of [1].

Algorithm 1. [1] Curve decomposition into arcs and lines [1].

Data: $C = \{C_1, \ldots, C_n\}$-a digital curve, α_{max}- maximal angle, ν-width of
      blurred segments

Result: $ARCs$- set of arcs, $LINEs$- set of lines

begin
    Use [18] to detect the set of dominant points: $DpC = \{D_0, \ldots, D_m\}$;
    $BS = \emptyset$;
    Transform DpC in the tangent space as $T(DpC)$;
    Construct the midpoint curve $MpC = \{M_i\}_{i=0}^{m-1}$ of horizontal segments of
    $T(DpC)$;
    for $i=0$ **to** m-1 **do**
        $\{C_i\}_{b_i}^{e_i}$- part of C wich corresponds to M_i;
        if ($BS \cup M_i$ *is a blurred segment of width* ν *[19]*) *and*
        ($|M_i.y - M_{i-1}.y| < \alpha_{max}$) **then**
           $BS = BS \cup M_i$;

        else
           C'- part of C corresponding to BS;
           Push C' to $ARCs$;
           $BS = \{M_i\}$;
           if ($|M_i.y - M_{i+1}.y| > \alpha_{max}$) **then**
               Push C_{b_i}, C_{e_i} to $LINEs$;
               $BS = \{\emptyset\}$;

end

- Grouping of arcs and lines for detection of ellipse candidate based on geometric properties.
- Fitting of ellipse candidate based on least square fitting.

The first step is done by applying the decomposition of a curve into arcs and lines presented in the above section. We construct the corresponding edge image from input image by using Canny filter. This edge image is considered as a list of digital curves. Thanks to above technique [1], we can obtain a compact representation of this edge image based on arc and line primitives.

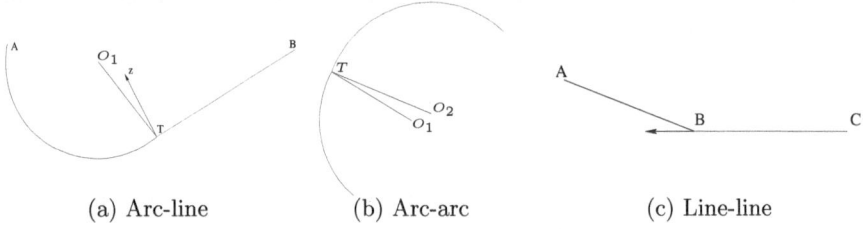

(a) Arc-line   (b) Arc-arc   (c) Line-line

Fig. 5. Arc and line grouping based on geometric property

Grouping of Arcs and Lines Based on Geometric Property. In this step, ellipse candidate is detected by grouping adjacent arcs and lines based on geometric property. Its main idea is to group the adjacent primitives that have a same tangent vector at the common extremity. In practice it is done by verifying the angle between two adjacent primitives.

Let us consider two adjacent primitives. There are three possible configurations: arc-arc, arc-line and line-line. We define the angle among two primitives that depends on its configuration as follows.

- *Arc-line:* Let us see Fig. 5.a. The arc has its center O_1. Two primitives share a common point T. Tz is perpendicular with TB. The angle among two primitives is define as $\angle O_1 Tz$.
- *Arc-arc:* Let us see figure 5.b. Two adjacent arcs whose centers are O_1 and O_2 share common point T. The angle among these arcs is define as $\angle O_1 T O_2$.
- *Line-line:* Let us see figure 5.c. The angle between two line segments AB and BC is $\pi - \angle ABC$.

Thanks to the notion of angle among two adjacent primitives and sequential property of the representation based on arcs and line segments, a linear scanning process is used for grouping adjacent primitives satisfying that the angle among two adjacent primitives does not exceed a fixed threshold. A such group of primitives that contains at least one primitive of arc is called an *ellipse candidate*. To avoid false positive with small detected ellipse, we use two constraints about the arcs in each group of primitives: the maximal radius of arc must be greater than 5 and the total subtending angle of arcs must be greater than $\frac{\pi}{5}$.

Fitting of Ellipse Candidate. For each ellipse candidate constructed from the extracted arcs and lines, we try to fit it by using least square fitting. Contrariwise to existing techniques based on least square fitting, we need a very small set of extracted points for fitting.

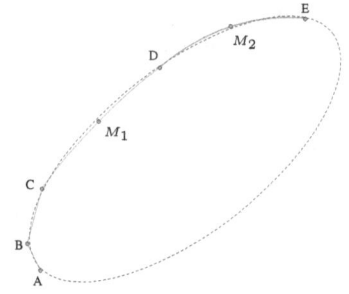

This good property is given by the representation of the curve by arc and line primitives. The set of extracted points for ellipse fitting is constituted from extremities of arcs and lines. Moreover, it contains also midpoints of arcs to reduce the approximated error between the fitting ellipse and the curve reconstructed by arcs and lines. Fig. 6 illustrates this strategy. The ellipse candidate is composed of segments AB, BC and arcs $\overset{\frown}{CD}$, $\overset{\frown}{DE}$. The data set contains A, B, C, D, E the extremities of arc and line primitives, M_1, M_2 the midpoints of arcs $\overset{\frown}{CD}$ and $\overset{\frown}{DE}$.

Fig. 6. Ellipse fitting based on least square fitting

Concerning least square fitting, there are two main categories: algebraic fitting and geometric fitting (see [10]). Let us consider a general conic described by this implicit function: $F(A, X) = A \cdot X = ax^2 + bxy + cy^2 + dx + ey + f = 0$ where

---

Algorithm 2. Ellipse detection

Data: Img- a digital image, α_{max}- maximal angle, ν-width of blurred segments, β_{max} threshold of angle [2]

Result: E - set of ellipses

begin

    Use Canny to detect $edgeImg$ as edge image;

    Consider $edgeImg$ as a list of digital curves;

    Use algorithm 1 to represent $edgeImg$ by $Primitives$ - a sequence of arcs and lines; i=-1; $E = \emptyset$;

    while $i < sizeof(Primitives)$ **do**

        i++; id=i; numArc=0;

        while $determineAngle(Primitives[i], Primitive[i+1]) < \beta_{max}$ **do**

            if $Primitives[i]$ is an arc **then** numArc++; id++;

        if $(numArc > 0)$ & $(id > i)$ & $(maximal\_radius > 5)$ & $(total\_substending\_angle > \frac{\pi}{5})$ **then**

            Collect $\{Primitives[j]\}_{j=i}^{id}$ as an ellipse candidate; $ES = \emptyset$;

            for $j=i$ **to** id **do**

                Add the first point of Primitives[j] to ES;

                if $Primitive[i]$ is an arc **then** Add the middle point of Primitives[j] to ES;

            Add the last point of Primitives[id] to ES;

            Use least square technique to fit ES by a conic curve ζ [20];

            if ζ is an ellipse **then** Add ζ to E;

end

$A = [a, b, c, d, e, f]^t$, $X = [x^2, xy, y^2, x, y]^t$. $F(A, X_i)$ is defined as algebraic (resp. geometric) distance to the conic $F(A, X) = 0$. The least square fitting is used to minimize the sum of squared error distance: $\sum_{i=0}^{m} F(A, X_i)^2$ with none trivial solution $A \neq [0, 0, 0, 0, 0, 0]^t$. Many works have been proposed for minimizing this sum of square error. In our work, the ellipse is fitted from the extracted set of points by using the code of ellipse fitting proposed by Ohad Gal [20].

Proposed Algorithm. Algorithm 2 describes the ellipse detection based on the representation of an edge image by arc and line primitives. The scanning process is applied for the detection of ellipse candidates by grouping adjacent arcs and lines which have their adjacent angles smaller than a threshold (β_{max}). For each ellipse candidate, the algorithm tries to construct an approximated conic by using least square fitting on a small set of points constructed from extremities of arcs and lines and middle points of arcs. If the conic is an ellipse, it is considered as a detected ellipse. In practice, the threshold of angle is set from $\frac{\pi}{8}$ to $\frac{\pi}{5}$.

---

[2] By default, $\alpha_{max} = \frac{\pi}{4}$, $\nu = 1$, $\beta_{max} = \frac{\pi}{5}$ (see algo. 2).

4   Experimentations

We have experimented the proposed method on a 2.26 GHz CPU linux computer, with a 4Go of RAM. The results are illustrated in Fig. 7 and Fig. 8. From the input image (a), the edge image (b) is computed. Then, the detected arcs and lines from algorithm 1 are represented in (c). Afterwards, a scanning process is applied to group arc and line primitives to detect the ellipse candidate (d) by verifying the angle between adjacent primitives. Finally, for each ellipse candidate, the technique of least square fitting is applied to construct the fitting conic on the small set of extracted points (marked points in detected ellipses in figures Fig. 7.e and Fig. 8.e). Fig. 7.f and Fig. 8.f present the results obtained by other methods [21], [22].

Table 1 shows some information about processing of ellipse detection from the previous experiments. Thanks to the representation based on arcs and lines, the number of primitives for processing is reduced from 13699 (resp. 7778) to 308 (resp. 261) for image in Fig. 7 (resp. Fig. 8). In addition, the scanning process on the sequence of primitives is done in linear time due to its sequential property. After applying this process, the number of ellipse candidates is reduced to 27 (resp. 6) for image in Fig. 7 (resp. Fig. 8). Moreover, the average number of extracted points for ellipse fitting of each candidate is only 11.18 and 8.17.

The proposed method has two main advantages that guarantees its fastness:

- *An efficient pre-processing step for the detection of ellipse candidates.* It is based on a fast method to represent the edge image by arcs and line segments and a linear scanning process for the detection of ellipse candidates.
- *A small set of extracted points for ellipse fitting.* Thanks to the representation based on arcs and lines, we don't need all of points corresponding to ellipse candidate for ellipse fitting. We only use extremity points of the primitives (arcs, lines) and midpoints of arcs for this task.

Table 1. Ellipse detection in images in figures 7, 8

Figure	Image size	N° of points in edge image	N° of primitives	N° of candidates	N° of ellipses	N° of points per candidate	Times (ms)
7	584x440	13699	308	27	23	11.18	880
8	508x417	7778	261	6	5	8.17	520

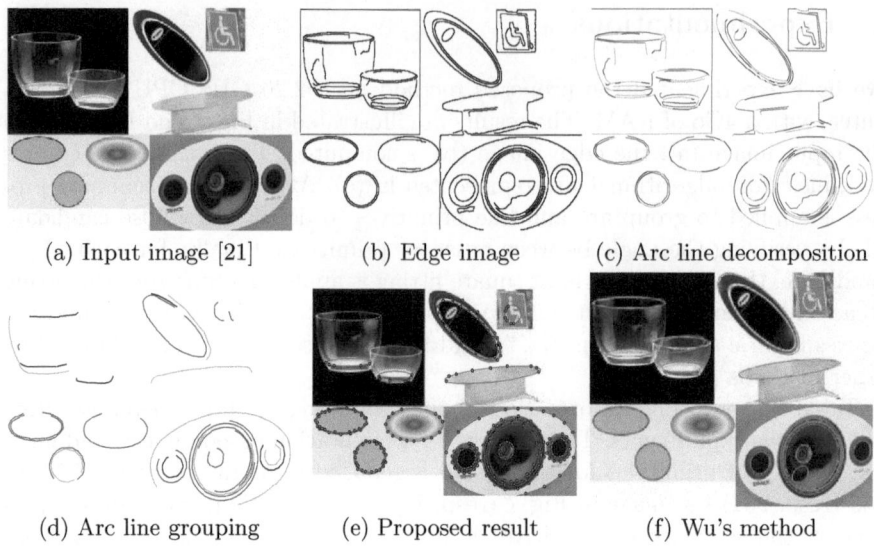

(a) Input image [21]   (b) Edge image   (c) Arc line decomposition

(d) Arc line grouping   (e) Proposed result   (f) Wu's method

Fig. 7. Comparison with Wu's method [21]

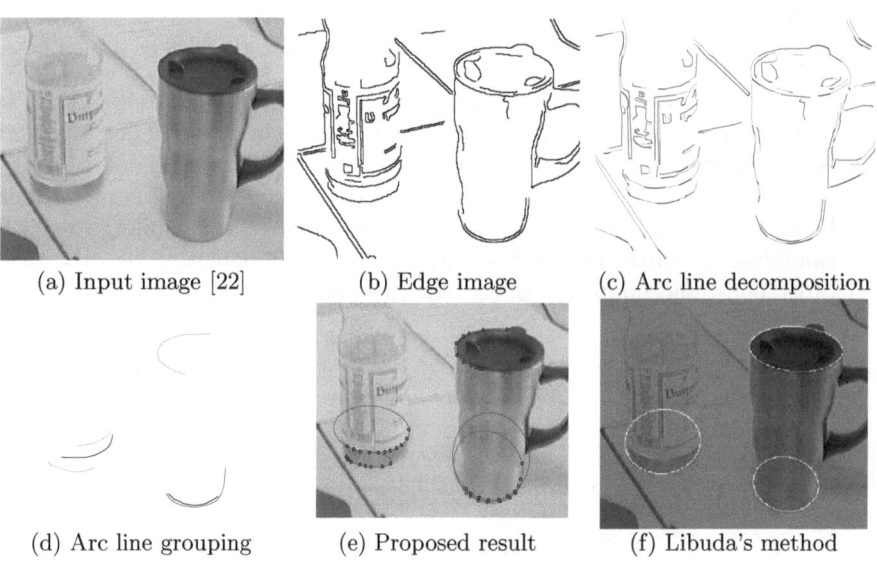

(a) Input image [22]   (b) Edge image   (c) Arc line decomposition

(d) Arc line grouping   (e) Proposed result   (f) Libuda's method

Fig. 8. Comparison with Libuda's method [22]

5   Conclusions

We have presented a promising new method for ellipse detection in images. The scanning process for detection of ellipse candidate is efficient because the representation based on arcs and lines allows us to work with a small number of primitives in relation with the number of points in edge image. In addition, this process is done in linear time. Moreover, for each ellipse candidate, the least square fitting is not directly applied on all the points of the candidate. In future work we plane to perform other comparisons with different approaches and include a recent noise detection defined through the concept of meaningful scale [23].

References

1. Nguyen, T.P., Debled-Rennesson, I.: Decomposition of a curve into arcs and line segments based on dominant point detection. In: Heyden, A., Kahl, F. (eds.) SCIA 2011. LNCS, vol. 6688, pp. 794–805. Springer, Heidelberg (2011)
2. Huang, C.L.: Elliptical feature extraction via an improved hough transform. PRL 10, 93–100 (1989)
3. Aguado, A., Montiel, M., Nixon, M.: On using directional information for parameter space decomposition in ellipse detection. PR 29, 369–381 (1996)
4. Daul, C., Graebling, P., Hirsch, E.: From the Hough Transform to a New Approach for the Detection and Approximation of Elliptical Arcs. CVIU 72, 215–236 (1998)
5. Chia, A., Leung, M., Eng, H., Rahardja, S.: Ellipse detection with hough transform in one dimensional parametric space. In: ICIP, pp. 333–336 (2007)
6. Lu, W., Tan, J.: Detection of incomplete ellipse in images with strong noise by iterative randomized Hough transform (IRHT). PR 41, 1268–1279 (2008)
7. Ahn, S., Rauh, W.: Geometric least squares fitting of circle and ellipse. PRAI 13, 987 (1999)
8. Ciobanu, A., Shahbazkia, H., du Buf, H.: Contour profiling by dynamic ellipse fitting. In: ICPR, vol. 3, pp. 750–753 (2000)
9. Fitzgibbon, A.W., Pilu, M., Fisher, R.B.: Direct least square fitting of ellipses. PAMI 21, 476–480 (1999)
10. Ahn, S., Rauh, W., Warnecke, H.: Least-squares orthogonal distances fitting of circle, sphere, ellipse, hyperbola, and parabola. PR 34, 2283–2303 (2001)
11. Voss, K., Suesse, H.: Invariant fitting of planar objects by primitives. PAMI 19, 80–84 (1997)
12. Lee, R., Lu, P., Tsai, W.: Moment preserving detection of elliptical shapes in gray-scale images. PRL 11, 405–414 (1990)
13. Heikkila, J.: Moment and curvature preserving technique for accurate ellipse boundary detection. In: ICPR, vol. 1, pp. 734–737 (1998)
14. Zunic, J., Sladoje, N.: Efficiency of characterizing ellipses and ellipsoids by discrete moments. PAMI 22, 407–414 (2000)
15. Rosin, P.L.: Measuring shape: ellipticity, rectangularity, and triangularity. Mach. Vis. Appl. 14, 172–184 (2003)
16. Nguyen, T.P., Debled-Rennesson, I.: A linear method for segmentation of digital arcs. In: Computer Analysis of Images and Patterns. LNCS. Springer, Heidelberg (2011)
17. Latecki, L., Lakamper, R.: Shape similarity measure based on correspondence of visual parts. PAMI 22, 1185–1190 (2000)
18. Nguyen, T.P., Debled-Rennesson, I.: A discrete geometry approach for dominant point detection. Pattern Recognition 44, 32–44 (2011)
19. Debled-Rennesson, I., Feschet, F., Rouyer-Degli, J.: Optimal blurred segments decomposition of noisy shapes in linear time. Comp. & Graphics 30, 30–36 (2006)

20. Gal, O.: Matlab code for ellipse fitting (2003),
    http://www.mathworks.com/matlabcentral/fileexchange/3215-fitellipse
21. Wu, J.: Robust real-time ellipse detection by direct least-square-fitting. In: CSSE,
    vol. (1), pp. 923–927 (2008)
22. Libuda, L., Grothues, I., Kraiss, K.: Ellipse detection in digital image data using
    geometric features, pp. 229–239 (2006)
23. Kerautret, B., Lachaud, J.-O.: Multi-scale analysis of discrete contours for unsu-
    pervised noise detection. In: Wiederhold, P., Barneva, R.P. (eds.) IWCIA 2009.
    LNCS, vol. 5852, pp. 187–200. Springer, Heidelberg (2009)

Computing Morse Decompositions for Triangulated Terrains: An Analysis and an Experimental Evaluation

Maria Vitali, Leila De Floriani, and Paola Magillo

University of Genova, Italy
{vitali,deflo,magillo}@disi.unige.it

Abstract. We consider the problem of extracting the morphology of a terrain discretized as a triangle mesh. We discuss first how to transpose Morse theory to the discrete case in order to describe the morphology of triangulated terrains. We review algorithms for computing Morse decompositions, that we have adapted and implemented for triangulated terrains. We compare the the Morse decompositions produced by them, by considering two different metrics.

1 Introduction

Modeling the morphology of a real terrain is a relevant issue in several application domains, including terrain analysis and understanding, knowledge-based reasoning, and hydrological simulation. Morphology consists of feature points (pits, peaks and passes), feature lines (like ridges and ravines), or segmentation of the terrain in regions of influence of minima and maxima or in regions of uniform gradient field. Morphological models of terrains are rooted in Morse theory [26]. Based on Morse theory cellular decompositions of the graph of a scalar field defined on a manifold, called *Morse complexes*, are defined by considering the critical points of the field and the integral lines. The same mathematical and algorithmic tools apply for segmenting 3D shapes based on a scalar field defined over the shape, such as curvature [22,23,24]. Unfortunately, the results from Morse theory apply only to smooth functions, while the most common terrain and shape models are piece-wise linear functions defined over a triangle mesh. Current approaches [4,6] try to simulate Morse complexes in the discrete case, but pose constraints (i.e., flat edges are not allowed), that are usually not satisfied by real terrains or 3D shapes.

The purpose of this work is to review, analyze and compare algorithms for computing Morse complexes on triangulated surfaces built on real data sets. We focus our attention on terrains modeled as piece-wise linear triangulated surfaces with vertices at the sample points, called *Triangulated Irregular Networks (TINs)*. First, we discuss how to overcome the limitations of current discrete approaches by dealing with flat edges effectively. We classify flat edges and then discuss how to eliminate singularities from TINs through a process of edge collapse which may generate new critical points.

G. Maino and G.L. Foresti (Eds.): ICIAP 2011, Part I, LNCS 6978, pp. 565–574, 2011.

In the literature, two major approaches exist for computing Morse decompositions of discrete terrain models: *region-based* algorithms compute the 2-cells of the Morse complexes while *boundary-based* algorithms compute the lines forming their boundaries. We compare results on real datasets, and we consider both noisy data and data after a smoothing process and elimination of the singularities. We compare the segmentations produced by the various algorithms using two metrics.

The remainder of the paper is organized as follows. In Section 2, we present background notions on Morse theory. In Section 3, we present discrete approaches to Morse theory and our new characterization of Morse functions in the discrete, that we call *extended Piecewise Linear (ePL)-Morse* functions. In Section 4, we review algorithms to decompose a terrain into Morse complexes. In Section 5 we discuss the metrics, used in Section 6 to compare the Morse decompositions produced by the different algorithms. Finally, in Section 7, we draw some concluding remarks, and discuss future developments.

2   Background Notions

Let us consider a domain M in the three-dimensional Euclidean space and a smooth real-valued function f defined over M. A point p of M is a *critical point* of f if the gradient of f at p is null. Critical points of f are *minima, saddles* and *maxima*. All other points are called *regular*. A critical point p is *degenerate* if the determinant of the Hessian matrix of f at p is zero. A function f is called a *Morse function* [26], if all its critical points are not degenerate. A Morse function admits only finitely many isolated critical points.

An *integral line* of f is a maximal path which is everywhere tangent to the gradient of f. Integral lines that converge to a minimum [saddle, maximum] p form a 0-cell [1-cell, 2-cell] called the *stable cell* of p. Similarly, integral lines that originate from a minimum [saddles, maximum] p form a 2-cell [1-cell, 0-cell] called the *unstable cell* of p. The stable (unstable) cells decompose M into a Euclidean cell complex, called a *stable (unstable) Morse complex*. If the unstable and the stable cells intersect transversally, function f is said to satisfy the *Smale condition*. If f satisfies the Smale condition, the overlay of stable and unstable Morse complexes is called a *Morse-Smale complex*.

A Morse complex is also related with the concept of watershed transform [29], which provides a decomposition of the domain of a smooth function f into open regions of influence associated with the minima of f, called *catchment basins*. Catchment basins correspond to the cells of the stable Morse complex defined on the same domain. The regions of influence related to the saddles form the *watershed lines*.

3   Discrete Approaches to Morse Theory

In the literature there are two extensions of Morse Theory to a discrete domain. The *discrete Morse theory* proposed by Forman [18] presents an adaptation of

Morse theory that can be applied to any simplicial or cell complex. Forman assumes to have a function value associated with each cell of the complex (not just the vertices). Thus, it is not directly useful in our case. The *piecewise linear Morse theory* by Banchoff [4] extends Morse theory to piecewise-linear functions defined on triangulated surfaces, under the assumption is that every pair of points of the triangulated surface have distinct field values. In order to define the conditions for a vertex v to be critical, the polyhedral surface made by the triangles incident in v is considered, and the number I of intersections between such surface and the plane parallel to the x-y-plane, which is passing through v, is counted. If there are no intersections, v is a maximum or a minimum. If there are two intersections, v is a regular vertex. Otherwise v is a saddle: a *simple saddle* if $I = 4$, or a *multiple saddle* if $I > 4$. Note that I is always even. The *multiplicity* of a saddle is equal $I/2$.

Banchoff's assumption on piece-wise linear Morse functions is quite strong, and thus it is often replaced by the weaker condition that each pair of adjacent vertices (i.e., connected by an edge) have different elevation values [6]. Such condition ensures a decomposition of the domain, which has the same properties as a Morse complex, called a *Piece-wise Linear Morse Complex (PLMC)*. The above condition is still too strong for real data sets. In [15], we have defined a classification of an isolated flat edge e based on the value f_e of function f on e, and on the values f_C and f_D of f at the other vertices of the triangles incident into e. If $f_C < f_e < f_D$ or $f_D < f_e < f_C$, then e is a *regular edge*. If $f_e < \min(f_C, f_D)$ or $f_e > \max(f_C, f_D)$ then e is a *critical edge*. We say that the elevation function defined at the vertices of a TIN is an *extended PL (ePL)-Morse function* if it does not contain critical edges or flat triangles. TINs that present only isolated flat edges can be reduced to TINs for ePL-Morse functions by collapsing such edges.

4   Algorithms for Computing Morse Decompositions

In the literature several approaches have been proposed for providing an approximation of the Morse and Morse-Smale complexes in the discrete.

Boundary-based algorithms (see Section 4.1) extract the Morse-Smale complex by tracing the integral lines, or their approximations, starting from saddle points and converging to minima and maxima. *Region based* algorithms compute the 2-cells of the Morse complexes by growing the neighborhood of minima and maxima (see Section 4.2). *Watershed* algorithms compute the Morse complexes based on the discrete *watershed transform* (see Section 4.3).

4.1   Boundary-Based Algorithms

Boundary-based algorithms [31,3,8,17,27,31] compute the 2-cells of the Morse-Smale complex on a TIN indirectly by computing the 1-cells, i.e., the boundaries of the 2-cells. They first extract the critical points, and then trace the separatrix lines, that are the 1-cells of the Morse-Smale complex.

The basic idea is due to Takahashi et al [31]. According to Banchoff [4], critical vertices are identified by considering the incident triangles in them. Let us consider the radially sorted list of neighbors of a saddle s. Within such sorted list, we consider the maximal sublists formed by vertices with higher (lower) value than s and call them *upper (lower) sequences*. Starting from each saddle p, with multiplicity k, $2k$ path are computed. Given p, we consider the upper sequences and the lower sequences. We select the first point of each path by choosing the highest (lowest) vertex from each upper (lower) sequence. Then, the ascending (descending) paths are computed by choosing, at each step, the highest (lowest) adjacent vertex until a critical point is reached. Saddles play a key role in this algorithm.

Edelsbrunner et al. [17,16] extend the approach by Takahashi et al. [31]. Function f is required to be a piecewise linear Morse function, according to Banchoff's definition, i.e., vertices must have distinct function values. Likewise, ridges and valleys are computed by starting from saddles, but at each step the point with the maximum slope is selected. The Smale condition is simulated a-posteriori by extending the path beyond each saddle, and forcing the path of the critical net not to intersect. The technique is rather complex, it requires edge duplications, and can lead to degenerated 2-cells (see [17] for more details).

4.2   Region-Growing Algorithms

Region growing algorithms [9,12,20] mimic, in the discrete case, the definition of 2-cell for a Morse complex, which is the set of the integral lines that originate or converge to a critical point. We discuss here how the algorithms extract the unstable Morse complex. The computation of the stable one is done in a completely symmetric way.

We review here three region growing methods proposed in our previous work. The algorithm in [12], denoted here as *DIS*, is based on the elevation values at the TIN vertices, while the one in [11], denoted as *GRD*, is based on the gradient associated with the triangles. The algorithms sort the vertices according to their elevation and process them in decreasing elevation order. Let p be p the vertex at the highest elevation among unprocessed vertices. The unstable 2-cell associated with p, that we denote with γ, is initialized with all the triangles incident in p, which have not yet been assigned to any 2-cell. The 2-cell γ is grown by following a predefined criterion until it cannot be extended any more. If the vertex p lies on the boundary of another 2-cell γ', then p is not a local maximum, and thus γ and γ' are merged.

The region growing criterion, according to which a triangle is added to the 2-cell γ, is different in the two algorithms. In the DIS algorithm, the current 2-cell γ is extended by including the triangle $t = ABC$, adjacent to γ along an edge AB, if C is the vertex f t with lowest elevation. Instead, the GRD algorithm incluses t if the gradient of t has the same orientation as the gradient of t', where $t' = ABD$ is the triangle in γ sharing edge AB with t.

In the growing algorithm in [20], denoted here as *STD*, the region growing criterion plays the most important role. Initially, the algorithm labels the highest, middle, and lowest vertex in each triangle t as *source* (S), *through* (T), and *drain* (D), respectively. Then, the minima are found as those vertices labeled D in all their incident triangles. For each minimum m, the 2-cell γ associated with m is initialized with all triangles incident in m. A loop follows, in which, at each iteration, the algorithm selects a set of triangles to be included from an edge e of the boundary of the current 2-cell γ, based on vertex classification.

4.3 Watershed Algorithms

The watershed transform has been first introduced in image processing for gray-scale images. Several definitions, and related algorithms, exist in the discrete case [5,7,21,25,30,32] to compute the watershed transform (see [29] for a survey). Here, we extended and implemented the two major approaches proposed in the literature, based on simulated immersion [32], and on the discretization of the topographic distance [25], to deal with TINs, and in general triangle meshes in 3D space having a scalar value associated with each vertex. We denote the simulated immersion algorithm for TINs as *WVS* and as the one using topographic distance as *WTD*.

5 Metrics for Comparison

In this Section, we describe the metrics we use to compare the decompositions of the same TIN produced by different algorithms. Such metrics provide a number between 0 and 1, where 0 indicates that two decompositions Γ_1 and Γ_2 are completely different, and 1 indicates that they are identical [10].

The first metric, called *Rand index (RI)* [28,10], measures the likelihood that a pair of triangles is either in the same region or in different regions in both decompositions. For each pair of triangles, we mark 1 either if they belong to one region in both decompositions, or if they belong to two different regions in both decomposition. Otherwise, we mark 0. Then, we divide the sum of the values by the number triangle pairs. *RI* metric assesses the similarity among decompositions even in case in which number of regions is widely different and region matching is problematic.

The second metric is a variant of the Hamming distance defined in [19,10]. The general idea is to find a best corresponding 2-cell in decomposition Γ_1 for each 2-cell in decomposition Γ_2, and measure the difference area between the regions. In the original definition, region matching is performed by considering the maximun common area. In our case, we say that two regions (2-cells) are corresponding if they have the same critical point associated. We define as *Hamming distance (HD)* the ratio between the number of the triangles that are assigned to corresponding 2-cells in the two decompositions, and the total number of triangles. The *HD* metric heavily relies on region matching, and is strongly affected even in case of a few mismatches.

6   Experimental Results

We have applied the algorithms presented in Section 4 both on synthetic and real datasets, and compared the Morse decompositions by using the metrics described in Section 5. Real data are affected by sampling errors, and, in most cases, contain flat areas (plains, lakes etc.); thus, real TINs do not correspond to Morse functions.

6.1   Input Data

We have experimented on three groups of real terrain data sets. The datasets of the first group are TINs over subsets of data from the CGIAR-CSI GeoPortal, [1]. Such TINs are composed of vertices distributed on a regular square grid and isosceles right triangles all of the same size. In the following, we refer to them as *regular TINs*. These TINs are available on the Aim@Shape repository [2]. The TINs of second group are extracted from a regular data set but triangulated in a nested triangle structure, called a *hierarchy of diamonds* [34]. The resulting TINs are crack-free triangle meshes extracted from the hierarchy and are composed by isosceles right triangles of different sizes. In the following, we refer to them as *diamonds TINs*. The third group contains irregular triangle meshes, that we call *irregular TINs*.

Table 1. Datasets used in the experiments

	Name	#Vertices	#Triangles	ePL
Regular	MontBlanc	14 400	28 322	reducible
	MontBlanc-ePL	14 394	28 310	yes
	Elba	36 036	70 220	reducible
	Elba-ePL	35 365	69 059	yes
	Monviso	160 000	318 402	reducible
	Monviso-ePL	159 606	317 623	yes
Diamonds	San Bern.	8 022	15 610	reducible
	San Bern-ePL	7 854	15 286	yes
	Marcy0003	214 287	427 242	no
	Marcy003	14 089	27 502	reducible
	Marcy003-ePL	13 722	26 867	yes
	Hawaii0003	120 456	240 840	no
	Hawaii001	15 196	29 599	no
Irregular	Vobbia	6 095	11 838	yes
	Zqn	57 865	114 765	reducible
	Zqn-ePL	57 816	114 695	yes
	Dolomiti	10 004	19 800	reducible
	Dolomiti-ePL	9 983	19 759	yes
	Marcy50	50 004	99 345	no

Fig. 1. Unstable decompositions of Vobbia, produced by TKH (a) and BBD (b)

Fig. 2. Comparison between stable decompositions of Monviso (a) and Monviso ePL (b), produced by WVS and WTD. Red areas represent mismatching

We classify those meshes into (i) ePL-Morse meshes, i.e., meshes that have no critical flat edges, (ii) meshes which can be reduced to ePL-Morse meshes, and (iii) meshes not reducible to ePL-Morse ones. For meshes that are reducible to ePL-Morse meshes, we consider both the original mesh and the mesh obtained by replacing each critical flat component with a component that is not flat, but has the same critical features. We denote those meshes with -*ePL* suffix. Table 1 shows the list of our test TINs. Figure 1 shows unstable decompositions of Vobbia TIN produced by TKH (a) and BBD (b).

6.2   Results

On ePL-meshes, all algorithms find the same number of regions in both the stable and the unstable decompositions. We have compared the algorithms pairwise, according to the Rand-Index (RI) and Hamming-Distance (HD) metrics. RI has, for each pair of algorithms, very high values, almost always in the top 2%. For regular or diamonds ePL-meshes, HD is in the range 80%–99%. In particular, TKH, WVS and STD algorithmd give very similar decompositions (HD is higher than 95%). As soon as the shape of triangles become less regular, decompositions computed by TKH, WVS and STD algorithms start diverging, and HD becomes as low as 50%. GRD, WTD and BBD, that simulate the differentiability by considering in some way the slope of triangles, give always pretty similar results, with HD from 80% to 90%.

Decompositions obtained from non-ePL-meshes can be quite different, even if the number of critical edges is small. Figure 2 shows mismatching between WVS and WTD for a non-ePL-meshe and its ePL version. When we have a large number of critical edges, or wide flat areas, the mesh cannot be reduced to ePL-Morse, and both metrics show very low values. For the Hawaii dataset RI is in the range 57-82%, and HD is in the range 21-43%, mainly due to the fact that the dataset is an island, and the boundary is a minimum composed by a long chain of edges. On Marcy dataset, RI is the range 89-95%, and HD is in the range 20-42%, due to a large flat area representing a river with small islands inside. For full details on the metrics comparing decomposition algorithm pairwise, please refer to [33].

7   Concluding Remarks

We have considered the problem of computing the morphology of triangulated terrains and we have presented and analyzed the various approaches for computing it. In Table 2, we summarize the main properties of the algorithms. GRD and WVS sort vertices and detect maxima and minima during computation, while STD, WTD, TKH and BBD use Banchoff's characterization (see Section 3) to find maxima and minima in a preprocessing step. STD, WVS and TKH are fully discrete and consider only the elevation values at the vertices. Decompositions of regular grid are similar for all fully discrete algorithms. GRD, WTD and BBD, take edge slope into account, but simulate the differentiability in various ways. Decompositions resulting from regular TINs are quite similar. Resulting decompositions on irregular TINs are different, but GRD, WTD and BBD that simulate the differentiability (even if in different ways), give results more similar among them than STD, WVS and TKH algorithms.

GRD and WVS can deal with higher dimensional data. WTD can be extended quite easily, while STD in higher dimension requires handling of a large number of local configurations. BBS and TKH are not extensible to higher dimensions, since they compute separatrix lines, while in higher dimensions we should compute hypersurfaces. See [33] for further details.

Table 2. Main properties of the algorithms

	Extraction of critical points	Simulation of Differentiability	Scale easily to higher dimension
GRD	Sort vertices	Gradient	Yes
STD	Banchoff	Fully Discrete	No
WTD	Banchoff	Top. Distance	No
WVS	Sort vertices	Fully Discrete	Yes
TKH	Sort vertices	Fully Discrete	No
BBD	Sort vertices	Diff. Quotient	No

The size of a morphological representation can be large for common models, which may consist of several millions of triangles. To deal with problem, we have been focusing on a multiresolution representation of the terrain morphology (see [13]), that allows concentrating on the areas of interest and producing approximate representations of the morphology at uniform and variable resolution. In our current work, we are developing a multiresolution model that combines the geometrical and morphological representations, where the morphological representation as an index on the geometrical model [14].

References

1. Reuter, J.A.H., Nelson, A., Guevara, E.: Hole-filled seamless srtm data v4, international centre for tropical agriculture, ciat (2008), http://srtm.csi.cgiar.org
2. AIM@SHAPE. Shape Repository (2006), http://shapes.aimatshape.net
3. Bajaj, C.L., Shikore, D.R.: Topology preserving data simplification with error bounds. Computers and Graphics 22(1), 3–12 (1998)
4. Banchoff, T.: Critical points and curvature for embedded polyhedral surfaces. American Mathematical Monthly 77(5), 475–485 (1970)
5. Beucher, S., Lantuejoul, C.: Use of watersheds in contour detection. In: Int. Workshop on Image Processing: Real-Time Edge and Motion Detection/Estimation (1979)
6. Biasotti, S., De Floriani, L., Falcidieno, B., Frosini, P., Giorgi, D., Landi, C., Papaleo, L., Spagnuolo, M.: Describing shapes by geometrical-topological properties of real functions. ACM Computing Surveys 40(4), 1–87 (2008)
7. Bieniek, A., Moga, A.: A connected component approach to the watershed segmentation. In: Mathematical Morphology and its Application to Image and Signal Processing, pp. 215–222. Kluwer Acad. Publ., Dordrecht (1998)
8. Bremer, P.-T., Edelsbrunner, H., Hamann, B., Pascucci, V.: A topological hierarchy for functions on triangulated surfaces. IEEE Transactions on Visualization and Computer Graphics 10(4), 385–396 (2004)
9. Cazals, F., Chazal, F., Lewiner, T.: Molecular shape analysis based upon the Morse-Smale complex and the connolly function. In: SCG 2003: Proc. 19th annual symposium on Computational geometry, pp. 351–360. ACM Press, New York (2003)
10. Chen, X., Golovinskiy, A., Funkhouser, T.: A benchmark for 3D mesh segmentation. ACM Transactions on Graphics (Proc. SIGGRAPH) 28(3) (2009)
11. Danovaro, E., De Floriani, L., Magillo, P., Mesmoudi, M.M., Puppo, E.: Morphology-driven simplification and multi-resolution modeling of terrains. In: Proc. 11th ACMGIS, pp. 63–70. ACM Press, New York (2003)
12. Danovaro, E., De Floriani, L., Mesmoudi, M.M.: Topological analysis and characterization of discrete scalar fields. In: Asano, T., Klette, R., Ronse, C. (eds.) Geometry, Morphology, and Computational Imaging. LNCS, vol. 2616, pp. 386–402. Springer, Heidelberg (2003)
13. Danovaro, E., De Floriani, L., Vitali, M., Magillo, P.: Multi-scale dual Morse complexes for representing terrain morphology. In: Proc. 15th ACMGIS. ACM Press, New York (2007)
14. Danovaro, E., Floriani, L.D., Magillo, P., Vitali, M.: Multiresolution Morse triangulations. In: Proc. Symposium of Solid and Physical Modeling (2010)

15. De Floriani, L., Magillo, P., Vitali, M.: Modeling and generalization of discrete Morse terrain decompositions. In: Proc. 20th Int. Conf. on Pattern Recognition, ICPR 2010, pp. 999–1002. IEEE Computer Society, Los Alamitos (2010)
16. Edelsbrunner, H., Harer, J., Natarajan, V., Pascucci, V.: Morse-Smale complexes for piecewise linear 3-manifolds. In: Proc. 19th ACM Symposium on Computational Geometry, pp. 361–370 (2003)
17. Edelsbrunner, H., Harer, J., Zomorodian, A.: Hierarchical Morse complexes for piecewise linear 2-manifolds. In: Proc. 17th ACM Symposium on Computational Geometry, pp. 70–79. ACM Press, New York (2001)
18. Forman, R.: Morse theory for cell complexes. Adv. in Math. 134, 90–145 (1998)
19. Huang, Q., Dom, B.: Quantitative methods of evaluating image segmentation. In: ICIP 1995: Proc. Int. Conf. on Image Processing, vol. 3, p. 3053. IEEE Computer Society, Washington, DC, USA (1995)
20. Magillo, P., Danovaro, E., De Floriani, L., Papaleo, L., Vitali, M.: A discrete approach to compute terrain morphology. Computer Vision and Computer Graphics. Theory and Applications 21, 13–26 (2009)
21. Meijster, A., Roerdink, J.: Computation of watersheds based on parallel graph algorithms. In: Mathematical Morphology and its Application to Image Segmentation, pp. 305–312. Kluwer, Dordrecht (1996)
22. Mesmoudi, M., Danovaro, E., De Floriani, L., Port, U.: Surface segmentation through concentrated curvature. In: Proc. ICIAP, pp. 671–676 (2007)
23. Mesmoudi, M., Floriani, L.D., Magillo, P.: Discrete distortion for surface meshes. In: Proc. ICIAP (2009)
24. Mesmoudi, M., Floriani, L.D., Magillo, P.: A geometric approach to curvature estimation on triangulated 3D shapes. In: Int. Conf. on Computer Graphics Theory and Applications (GRAPP), pp. 90–95 (2010)
25. Meyer, F.: Topographic distance and watershed lines. Signal Processing 38, 113–125 (1994)
26. Milnor, J.: Morse Theory. Princeton University Press, Princeton (1963)
27. Pascucci, V.: Topology diagrams of scalar fields in scientific visualization. In: Topological Data Structures for Surfaces, pp. 121–129. John Wiley and Sons Ltd, Chichester (2004)
28. Rand, W.: Objective criteria for the evaluation of clusterings methods. Journal of the American Statistical Association 66(336), 846–850 (1971)
29. Roerdink, J., Meijster, A.: The watershed transform: definitions, algorithms, and parallelization strategies. Fundamenta Informaticae 41, 187–228 (2000)
30. Stoev, S.L., Strasser, W.: Extracting regions of interest applying a local watershed transformation. In: Proc. IEEE Visualization 2000, pp. 21–28. IEEE Computer Society, Los Alamitos (2000)
31. Takahashi, S., Ikeda, T., Kunii, T.L., Ueda, M.: Algorithms for extracting correct critical points and constructing topological graphs from discrete geographic elevation data. Computer Graphics Forum 14(3), 181–192 (1995)
32. Vincent, L., Soile, P.: Watershed in digital spaces: an efficient algorithm based on immersion simulation. IEEE Transactions on Pattern Analysis and Machine Intelligence 13(6), 583–598 (1991)
33. Vitali, M.: Morse Decomposition of Geometric Meshes with Applications. PhD thesis, Computer Science Dept., University of Genova, Italy (2010)
34. Weiss, K., De Floriani, L.: Sparse terrain pyramids. In: Proc. ACM SIGSPATIAL GIS, pp. 115–124. ACM Press, New York (2008)

Spot Detection in Images with Noisy Background

Denis Ferraretti[1], Luca Casarotti[1], Giacomo Gamberoni[2], and Evelina Lamma[1]

[1] ENDIF-Dipartimento di Ingegneria, Università di Ferrara, Ferrara, Italy
{denis.ferraretti,evelina.lamma}@unife.it,
luca.casarotti@student.unife.it
[2] intelliWARE snc, Ferrara, Italy
giacomo@i-ware.it

Abstract. One of the most recurrent problem in digital image processing applications is segmentation. Segmentation is the separation of components in the image: the ability to identify and to separate objects from the background. Depending on the application, this activity can be very difficult and segmentation accuracy is crucial in order to obtain reliable results. In this paper we propose an approach for spot detection in images with noisy background. The overall approach can be divided in three main steps: image segmentation, region labeling and selection. Three segmentation algorithms, based on global or local thresholding technique, are developed and tested in a real-world petroleum geology industrial application. To assess algorithm accuracy we use a simple voting technique: by a visual comparison of the results, three domain experts vote for the best algorithms. Results are encouraging, in terms of accuracy and time reduction, especially for the algorithm based on local thresholding technique.

Keywords: image segmentation, local thresholding, spot detection, petroleum geology application.

1 Introduction

In digital image processing a common task is partitioning an image into multiple areas that collectively cover the entire image. Segmentation subdivides an image into its constituent regions or objects. The level of detail to which the subdivision is carried depends on the problem being solved. That is, segmentation should stop when the objects or regions of interest in an application have been detected[1].

The goal of segmentation is to simplify and/or change the representation of an image into something that is more meaningful and easier to analyse. More precisely, image segmentation is the process of assigning a label to every pixel in an image such that pixels with the same label share certain visual characteristics. Each of the pixels in a region are similar with respect to some characteristic or computed property, such as color, intensity, or texture. Adjacent regions are significantly different with respect to the same characteristics[2].

G. Maino and G.L. Foresti (Eds.): ICIAP 2011, Part I, LNCS 6978, pp. 575–584, 2011.

Segmentation of non trivial images is one of the most difficult tasks in image processing. Segmentation accuracy determines the eventual success or failure of computerized analysis procedures. For this reason, considerable care should be taken to improve the probability of accurate segmentation.

In this paper we propose and experiment three algorithms, based on global and local thresholding techniques, for spot detection in noisy images. Algorithms take as input an image and give as result all the informations (such as position, perimeter and color) about small round areas in contrast with the backgorund, according to some visual criteria. We test the algorithms in a petroleum geology application using digital images provided by a borehole logging tool called FMI[1]. An interpretation of these acquired images (image logs) is usually made, by the petroleum geologist, to locate and quantify potential depth zones containing oil and gas. In this context the segmentation accuracy in spot detection is very important in order to produce reliable evaluation of the studied reservoir.

To assess algorithms accuracy we use a simple voting technique: by the direct observation of the output of chosen images at different depths of the well, three domain experts choose the best algorithm.

The paper is organized as follows: segmentation basic concepts and related work are outlined in Section 3.1. Three developed algorithms for spot detection are introduced in Section 3. A detailed explanation of experimental results over different images is given in Section 4. Finally, section 5 concludes the paper.

2   Segmentation Algorithms

In image analysis, one of the most recurrent problem is the separation of components in the image: the ability to identify and to separate objects from the background. This activity is called image segmentation. Segmentation algorithms are based on one of two basic properties of intensity values: discontinuity and similarity. In the first category, the approach is to partition an image based on abrupt changes in intensity, such as edges (i.e. Canny edge detector[15]). The principal approaches in the second category are based on partitioning an image into regions that are similar according to a set of predefined criteria. Thresholding, region growing (i.e. [17]), and region splitting and merging are examples of methods in this category.

Other proposed recent approaches[3] are segmentation based on the mean shift procedure[4], multiresolution segmentation of low-depth-of-field images[5], a Bayesian-framework-based segmentation involving the Markov chain Monte Carlo technique[6], and an EM-algorithm-based segmentation using a Gaussian mixture model[7]. A sequential segmentation approach that starts with texture features and refines segmentation using color features is explored in[8]. An unsupervised approach for segmentation of images containing homogeneous color/texture regions has been proposed in[9].

---

[1] FMI (Fullbore Formation MicroImager) is the name of the tool used to acquire image logs based on resistivity measures within the borehole.

In our work we focus on segmentation obtained by threshold operations. Let $f(x, y)$ be the function that describes our image. The image consists of a white object on a dark background. The extraction of the object can be achieved by defining a threshold T and then comparing each pixel value with it. If the pixel value exceeds the threshold, the pixel is classified as an *object pixel*, if the value is lower than the threshold, the pixel is classified as a background pixel. The result is typically a binary image, where *object pixels* are represented in white and background pixel are represented in black.

Thresholding is a well known and straightforward technique and can be defined as an operation that involves a test against a T function, which has the following form: $T = T[x, y, p(x, y), f(x, y)]$ where $f(x, y)$ is the function that describes the gray-level intensity for each pixel in the image; $p(x, y)$ describes some local properties for each pixel in the image; (x, y) represents the position of pixels in the image. Depending on T, there are different types of threshold.

Global Threshold. It's the simplest operation: the threshold value T is computed once for the whole image, and the image is thresholded by comparing each pixel value with T, as described above. The result depends on the shape of the image histogram. Many techniques have been proposed for the automatic computation of the threshold value. Otsu's method[10], for example, produces the threshold value that minimizes the intra-classes variance, defined as the weighted sum of the variance of the classes. The class weight correspond to the probability that a pixel belongs to that class.

Local Threshold. A global value for T may not be enough in order to obtain good results in segmentation: the local approach, instead, computes a different threshold value for each pixel in the image, based on local statistical features. A neighbourhood is defined for each pixel: in this neighbourhood some statistical parameters are calculated (i.e.: mean, variance and median), which are used to calculate the threshold value $T(x, y)$. Niblack's algorithm[11] is an example of this type of thresholding.

The simple global threshold method can only be successful if the separation between the two classes (object vs. background) is clear. In real images, this assumption is typically not true. The local threshold method attempts to solve this issue, because the threshold value is not fixed, but calculated for each pixel on the basis of the local image features.

3   Methodology

The overall approach of spot detection in noisy background images can be divided in three steps: segmentation, labeling and selection. Segmentation identifies a set of interesting regions that are eligible to spots. Labeling provides the regions connected components in order to then select only those that are actual objects.

3.1   Image Segmentation

We developed three different segmentation algorithms starting from two main methods. Our algorithms are made by the combinations of modified versions of well known image processing techniques such as image smoothing and thresholding. The first method uses a particular convolution mask and a global thresholding technique. In order to remove noise and unnecessary details, the image is first smoothed with a median filter. The convolution of this image with a circular derivative mask provides a new image where round areas or circular structures, approximately of the same size of the mask, are highlighted. The new image is then thresholded, using two global threshold values: T_{low} and T_{high}. All the (x, y) pixel where $f(x, y) <= T_{low}$ or $f(x, y) >= T_{high}$ are considered *object pixels*, others are background pixels. Using two different threshold is possible to find two types of spots: dark spots in light background and vice versa. Generally we use a percentile value to define two thresholds: T_{low} is the 20th percentile and T_{high} the 80th. In order to remove isolated pixels a opening morphological operator[16] is then applied. This method lead to the implementation of two different algorithm. The difference between these two implementations is in how the convolution manage the image background. In some cases (see Section 4) images can have zones with non-relevant or missing information. Our first algorithm considers these zones as background pixels, conversely in the second algorithm these pixels are considered *null* values (zones with no image).

The second method uses the approach based on local threshold. The first step is the application of a low-pass filter to the image. The purpose of the filter is to reduce the noise in the image. Then, once defined the size of the neighbourhood, intensity mean (μ) and variance (σ) are computed for each pixel. For the calculation of the threshold value, the Niblack's algorithm[11] is applied: $T(x, y) = \mu(x, y) + k\sigma(x, y)$ Mean and variance are calculated in the neighbourhood of each pixel. Here, we are assuming that the image contains white objects on dark background. [2]

In practice, two new images are built, starting from the original: in the first image, the pixel value is replaced with the mean value in the neighbourhood. In the second image, the pixel value is replaced with the variance calculated in the neighbourhood. To apply the Niblack's algorithm to the pixel (x, y) is sufficient to get the pixel value from the original image, and its mean and variance from the new images. The Niblack's algorithm is reinforced with an additional constraint, based on the absolute value of the variance. Variance is related to the image contrast. A small value corresponds to an area fairly uniform in the image. To avoid the detection of false positives, a pixel must belong to a non uniform area: this means that the variance is to assume a high enough value. Hence a threshold value is needed to compare the variance. First the variance image histogram is built, then the threshold is selected as the value corresponding to an arbitrary percentile (for example, the 20th percentile). The pixel for which the variance is

---

[2] The detection of dark objects on light background can be achieved by inverting the original image (doing this causes that dark pixels turn into light pixels and vice versa) and then applying the same algorithm.

lower than this value are automatically classified as background pixel. Niblack's algorithm is applied only to pixels that pass this test.

In order to detect light and dark objects, the method is applied to the original image and to the inverted image. As before, the opening morphological operator is then applied to the binary images, in order to smooth the contours of the regions identified.

3.2   Image Labeling and Region Selection

The second step in the proposed approach is aimed at identifying and labeling the connected components resulting from the segmentation process. Once we obtain a binary image a labeling algorithm is applied to detect all the image regions. The labeling algorithm identifies the connected components in an image and assigns each component a unique label. The algorithm runs an image scan and groups its pixels into regions, based on pixel connectivity. This procedure is often applied to binary images, resulting from segmentation. Once complete, the procedure returns a list of connected regions that were found in the image. Each region should represent an image object.

Finally in the last step, for each identified region a test is applied on the size and shape. In particular, the tested parameters might be: area, roundness and ratio (ratio between maximum height and maximum width). These tests prevent the algorithm from detecting regions which do not correspond to actual objects.

4   Experimental Results

We test our approach in a real-world petroleum geology application: the porosity evaluation of a rock formation in oil and gas reservoir. Image logs are digital images acquired by a special logging tool (FMI tool) within a borehole. They represent resistivity measurements of the rock formation taken by the wellbore surface. Image logs supply fundamental information on the characteristics of the reservoir sections explored and hold important information on the structural, lithological, textural and petrophysical properties of the rocks. Resistivity measurements are converted into gray-level intensity values, and each measurement corresponds to a pixel in the FMI image. This image is the unrolled version of the well surface and it is made by six vertical strips of measurement. There is a strip for each pad of sensors in the FMI tool, see Figure 2(a) for an example.

To estimate the porosity of the rocks from the image, we are interested in the detection of roughly circular areas, in contrast with the background. These spots are called *vugs* or *vacuoles* (see Figure 1 for an example).

Three different algorithms were implemented: the first two (algorithm 1 and 2) are very similar, and use the approach based on convolution. The third (algorithm 3) is an implementation of the local threshold method described in 3.1. All the algorithms are written in JAVA. To determine which method is most suitable for this task, a test was performed on an entire well FMI image. The analysis is carried out through a sliding window technique. From the main image, 300 pixel height windows are extracted, and algorithms are applied directly

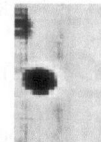

Fig. 1. Dark vug in a light background (on the left) and vice versa (on the right)

to them. Windows are partially overlapping: this is designed to improve the accuracy detection near the edges of the windows. Once completed the analysis on
the entire well, in order to evaluate the results, about ten windows, considered
significant, have been taken: windows, namely, showing the most common situations in which the geologist is interested. For example, a window containing a lot
of small sized vugs was selected, rather than a window with a few large vacuoles.
The chosen windows, and the three results for each of them, were shown to three
geologists: it was asked them, for each window, to vote the algorithm (or the
algorithms) that produced best results. At the end of the procedure, all votes
were collected and a ranking was produced.

In our experiment algorithm 1 and 2 have a 7x7 pixel smoothing filter and
a 9x9 pixel circular derivative convolution mask. Algorithm 3 runs with a 5x5
pixel smoothing filter; the radius of the neighbourhood is 13 pixel and $k = 0.5$
in the Niblack's algorithm. Once each image region is labeled, a test is applied
on the size and shape. In our work the total area of each region must be in the
range 25 - 500 pixel. Roundness is defined as $roundness = \frac{4\pi A}{p^2}$ where A is the
region area and p is the perimeter. All the regions with a *roundness* lower than
0.25 pass the test and can be considered as vugs. The last test is based on the
width-height ratio: for each region the maximum width and height are computed
and only if the ratios *width/height* and *height/width* are greater than 1.8, the
region pass the test. Details on the vote are shown in Table 1.

Table 1. Each geologist votes for the best algorithms (algorithm 1, 2 or 3) for each
well depth. Cells contains geologist choice.

	Geologist A	Geologist B	Geologist C
depth1	1,3	2,3	1
depth2	2,3	2	3
depth3	2,3	2	3
depth4	3	3	3
depth5	2	n.d.	3
depth6	3	3	2
depth7	2,3	2	3
depth8	2	2	3
depth9	3	n.d.	3

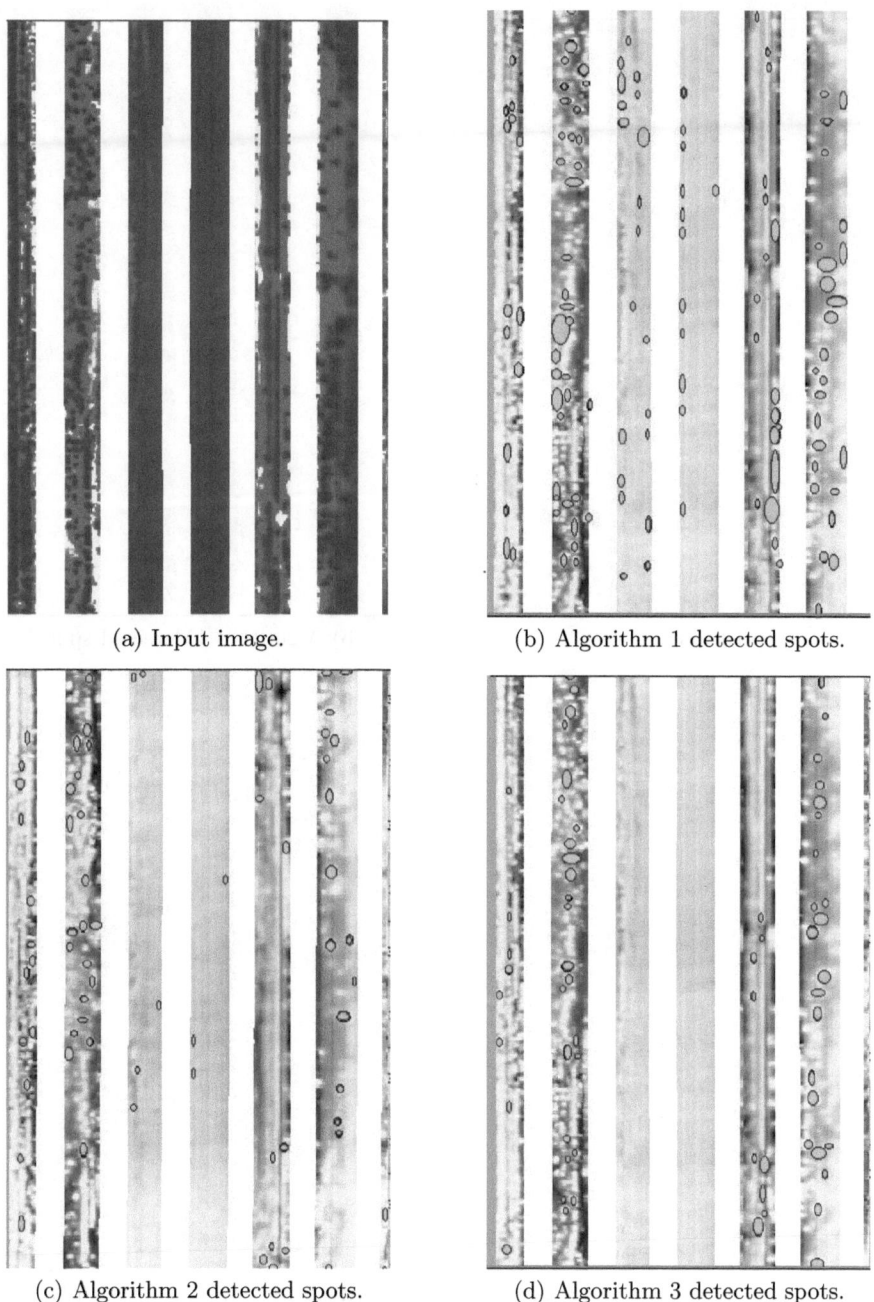

(a) Input image.   (b) Algorithm 1 detected spots.

(c) Algorithm 2 detected spots.   (d) Algorithm 3 detected spots.

Fig. 2. Example of gray-level image input (a) and output (b,c,d) at *depth1*. In output images, detected vugs are round grey area with black thin border.

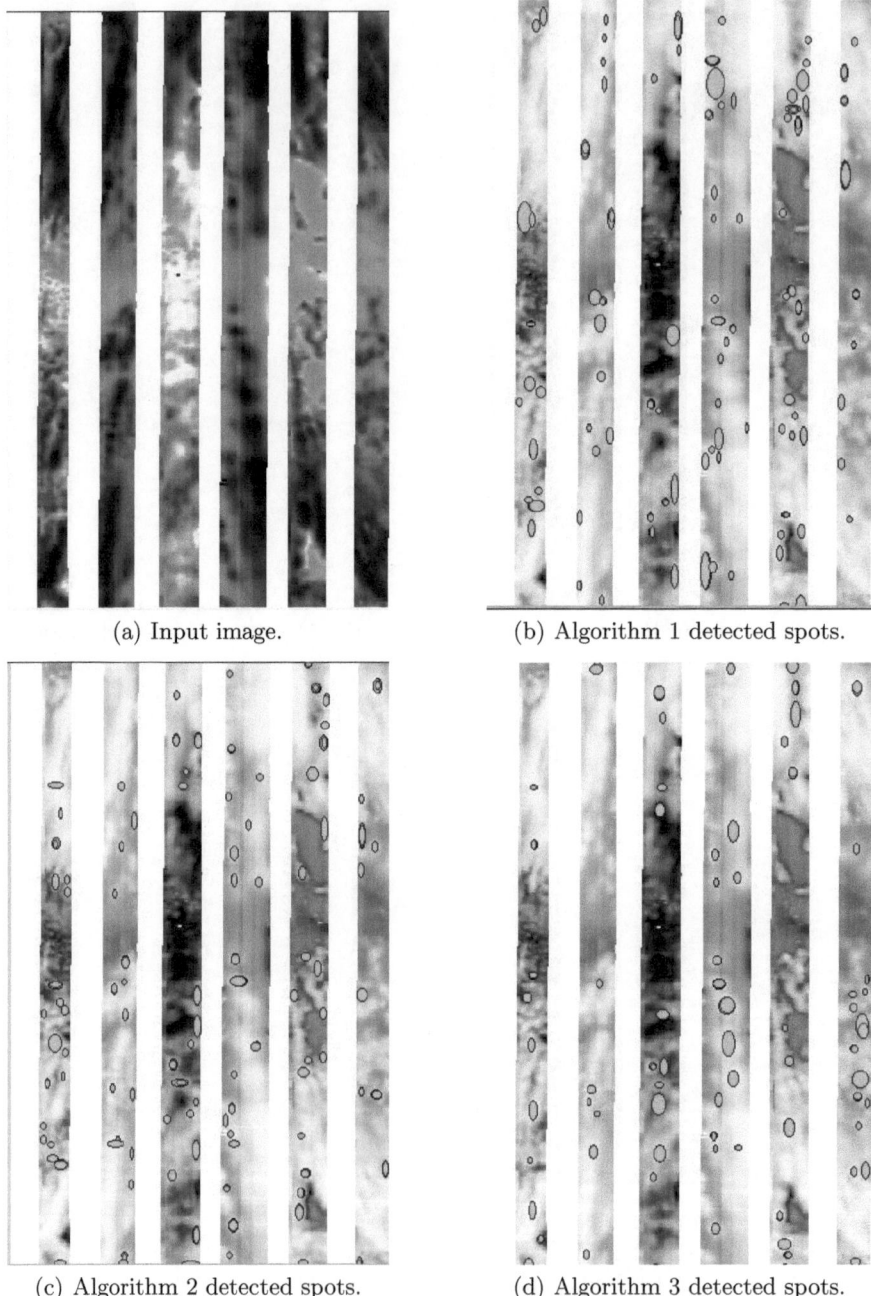

(a) Input image.

(b) Algorithm 1 detected spots.

(c) Algorithm 2 detected spots.

(d) Algorithm 3 detected spots.

Fig. 3. Example of gray-level image input (a) and output (b,c,d) at *depth2*. In output images, detected vugs are round grey area with black thin border.

Final ranking is algorithm 1 has 2 votes, algorithm 2 has 11 votes and algorithm 3 receives 17 votes.

In Figure 2 the input image (*depth1*) shows a lots of small vugs, with a low contrast with respect to the background; two strips in the middle are very dark due to a measurement error [3]. The geologist choice is algorithm 3 with two votes. Although this algorithm detects less vugs than the others, this was preferred because of it provides better results (no false positive) in the dark strips.

Figure 3 shows the image input and output for each algorithm at *depth2*. In this case the input image shows few big vugs and algorithm 2 and 3 give best results, both gaining 2 votes. It is important to note that algorithm 3 shows, in general, a clear output and best accuracy, with a lower number of false positive. Detailed image results can be found in [12]. The algorithm that produced the best overall results was the one based on the local threshold method. The second choice was the algorithm 2. This indicates that, regardless the image shape, the convolution operator gives best results if it considers only actual image zones.

5   Conclusions

An approach for spot detection in images with noisy background has been proposed and tested in a real-world application. It consists of three main steps: image segmentation, image labeling and regions selection. We develop three different algorithms for image segmentation starting from well known techniques. The first and the second are based on convolution with a derivative circular mask and then a global thresholding technique is applied; the third algorithm uses a local threshold. To find actual objects we finally test some parameters (such as size and shape) of the identified regions. The approach was tested detecting vugs (spots in contrast with the noisy background) in a borehole image log. We evaluate our algorithms by a visual comparison of the obtained results, three domain experts then vote for the best algorithms. Results show that the algorithm 3, that uses a local threshold, was preferred by the domain expert. In general it detects less vugs than other algorithms, but it seems to be most suitable in all that cases with a low contrast between spots and background. Vugs detection is very important for the geologist who wants to evaluate the porosity of a rock, in order to quantify potential depth zones containing oil and gas. Our approach helps the geologist reducing the time for detection of vugs in the image logs and improving the detection accuracy. Outcomes from our algorithms can be considered as good starting points for porosity analysis, on which the geologist build his interpretation work.

Acknowledgements. This work has been partially supported by Camera di Commercio, Industria, Artigianato e Agricoltura di Ferrara, under the project "Image Processing and Artificial Vision for Image Classifications in Industrial Applications".

---

[3] This is an unavoidable error and can happens often in these type of image. Due to the complexity and the cost of the image acquisition, it is not possible to repeat the measurement. The final image is made by a single run over the entire well.

References

1. Gonzalez, R.C., Woods, R.E.: Digital Image Processing, 3rd edn., pp. 689–794. Prentice-Hall, Englewood Cliffs (2008)
2. Shapiro, L.G., Stockman, G.C.: Computer Vision, pp. 279–325. Prentice-Hall, Englewood Cliffs (2001)
3. Datta, R., Joshi, D., Li, J., Wang, J.Z.: Image retrieval: Ideas, influences, and trends of the new age. ACM Comput. Surv. 40(2), Article 5 (2008)
4. Comaniciu, D., Meer, P.: Mean Shift: A Robust Approach toward Feature Space Analysis. IEEE Trans. Pattern Analysis Machine Intell. 24(5), 603–619 (2002)
5. Wang, J.Z., Li, J., Gray, R.M., Wiederhold, G.: Unsupervised Multiresolution Segmentation for Images with Low Depth of Field. IEEE Transactions on Pattern Analysis and Machine Intelligence 23(1), 85–90 (2001)
6. Tu, Z., Zhu, S.: Image Segmentation by Data-Driven Markov Chain Monte Carlo. IEEE Trans. on Pattern Analysis and Machine Intelligence 24(5) (May 2002)
7. Carson, C., Belongie, S., Greenspan, H., Malik, J.: Blobworld: Image segmentation using expectation-maximization and its application to image querying. IEEE Trans. Pattern Analysis and Machine Intelligence 24(8), 1026–1038 (2002)
8. Chen, J., Pappas, T., Mojsilovic, A., Rogowitz, B.: Adaptive image segmentation based on color and texture. In: Proceedings of the IEEE International Conference on Image Processing, ICIP (2002)
9. Deng, Y., Majunath, B.: Unsupervised segmentation of color-texture regions in images and video. IEEE Trans. Pattern Anal. Mach. Intell. 23(8), 800–810 (2001)
10. Otsu, N.: A threshold selection method from gray-level histograms. IEEE Trans. Systems, Man, and Cybernetics 9(1), 62–66 (1979)
11. Niblack, W.: An Introduction to Digital Image Processing. Prentice-Hall, Englewood Cliffs (1986)
12. Casarotti, L.: Algoritmi avanzati di analisi delle immagini da pozzi petroliferi (Advanced algorithm for borehole image processing). Master's Thesis, University of Ferrara (2011)
13. Ferraretti, D.: Analisi di immagini da pozzi petroliferi e loro classificazione (Borehole image analysis and classification). Master's Thesis, University of Ferrara (2006)
14. Abramoff, M.D., Magelhaes, P.J., Ram, S.J.: Image Processing with Image. J. Biophotonics International 11(7), 36–42 (2004)
15. Canny, J.: A computational approach to edge detection. IEEE Trans. Pattern Anal. Mach. Intell. 8(6), 679–698 (1986)
16. Serra, J.: Image Analysis and Mathematical Morphology. Academic Press, London (1982)
17. Mancas, M., Gosselin, B., Benoît, M.: Segmentation using a region-growing thresholding. Proceedings of the SPIE 5672, 12–13 (2005)
18. Ferraretti, D., Gamberoni, G., Lamma, E., Di Cuia, R., Turolla, C.: An AI Tool for the Petroleum Industry Based on Image Analysis and Hierarchical Clustering. In: Corchado, E., Yin, H. (eds.) IDEAL 2009. LNCS, vol. 5788, pp. 276–283. Springer, Heidelberg (2009)

Automatic Facial Expression Recognition Using Statistical-Like Moments

Roberto D'Ambrosio, Giulio Iannello, and Paolo Soda

Integrated Research Center, Università Campus Bio-Medico di Roma,
Via Alvaro del Portillo, 00128 Roma, Italy
{r.dambrosio,g.iannello,p.soda}@unicampus.it

Abstract. Research in automatic facial expression recognition has permitted the development of systems discriminating between the six prototypical expressions, i.e. anger, disgust, fear, happiness, sadness and surprise, in frontal video sequences. Achieving high recognition rate often implies high computational costs that are not compatible with real time applications on limited-resource platforms. In order to have high recognition rate as well as computational efficiency, we propose an automatic facial expression recognition system using a set of novel features inspired by statistical moments. Such descriptors, named as statistical-like moments extract high order statistic from texture descriptors such as local binary patterns. The approach has been successfully tested on the second edition of Cohn-Kanade database, showing a computational advantage and achieving a performance recognition rate comparable than methods based on different descriptors.

1 Introduction

Nowadays reliable and intuitive interfaces are one of the major factor influencing acceptance of technological devices. Indeed some researches aim at developing smart human computer interfaces (HCIs) that even could understand user emotion. One way to achieve this goal consists in recognizing facial expressions, one of the most important human communication modality.

The study of facial expressions, started with Darwin and led to a facial expressions encoding by Ekman and Friesen in the 1971. They described elementary facial movements (AUs) roughly corresponding to movement of facial muscles and represented all possible faced expressions as a combination of such AUs. They also proposed six prototypical expressions that are the same among all ethnicity, i.e. anger, disgust, fear, happiness, sadness and surprise.

Facial expression recognition has long been competence of medical doctors and anthropologists but, the necessity of intuitive HCIs stimulate researches on automatic facial expression recognition (AFER) in computer vision and pattern recognition areas.

Researches efforts have been finalized to find the best descriptors and classifiers to recognize facial expressions in images and videos. In this area Gabor

G. Maino and G.L. Foresti (Eds.): ICIAP 2011, Part I, LNCS 6978, pp. 585–594, 2011.

energy filters (GEFs) and local binary patterns (LBPs) provide the best performance. Since the development of real time AFER systems on resource limited devices, i.e. mobile devices, requires that features extraction, selection and classification are characterized by low computation costs, a trade-off between computation costs and recognition rate is an issue still open.

This work presents an AFER system classifying facial expressions into six prototypical expressions. We define a set of features which effectively describes facial expression in frontal view video sequences. They are based on statistical-like moments computed on LBP transformed images. These descriptors reduce the amount of data conveying expressive information and the classification time.

The paper is organized as follow: next section presents an overview of most significant works on AFER. Section 3 discusses system architecture, whereas section 4 present the experimental protocol. I section 5 we discuss the results, and section 6 provides concluding remarks.

2   Background

Reviews of researches carried out until 2009 presented the fundamental approaches for AFER systems development, pointing out research targets and limits [1,2,3]. First systems were inadequate to face real problems due to the exiguous amount of images and videos available to train the classifiers. More recently, the availability of larger databases of facial spontaneous expressions has permitted to develop AFER systems which may be applied to real world situations.

We present now a review of most important AFER systems providing best performance among existing literature. As described below, such systems are based on different methods for both features extraction and selection as well as sample classification.

In [4] the authors present an automatic system which detects faces, normalizes facial images and recognizes 20 facial actions. In the recognition step, features are the outputs of a bank of Gabor energy filters and classifiers are SVM or an Adaboost algorithm.

In [5] different features are tested, i.e. GEFs, box filter (BF), hedge orientation histogram (EOH), BF+EOH and LBPs. In order to test the performance achieved with such descriptors, GentelBoost and SVM are employed as classification methods. Results show that GEFs provides good performance with SVM.

In [6] the authors present a framework for facial expression recognition based on encoded dynamic features. In the first step the system detects faces in each frame, normalize them and computes Haar-like features. Dynamic features units are calculated as a set of Haar-like features located at the same position along a temporal window. The second step consists in a coding phase which analyses each dynamic features distribution generating the corresponding codebook. Such a codebook is used to map a dynamic features into a binary pattern. The last step applies a classifier based on the AdaBoost algorithm.

The system presented in [7] carries out a preprocessing step in which images alignment is manually performed to realign the common regions of the face by the identification of mouth and eyes's coordinates. Then, images are scaled and the first frame of each input sequence is subtracted from the following frames to analyse the changes in facial expression over time. Such an image is named as *delta image*. The system extracts local features from delta images applying Independent component analysis (ICA) over the PCs space. Then, the features classes are separated by fisher linear discriminant analysis. Finally, each local feature is compared with a vector belonging to a codebook generated by a vector quantization algorithm from features of training set samples.

In [8] the authors present a comprehensive study for AFER applications investigating LBP-based features for low resolution facial expression recognition. Furthermore, they propose Boosted-LBP as the most discriminative LBP histograms selected by AdaBoost, reducing the number of features processed in the classification stage.

In [9] the authors use LBP-based features computed using spatio-temporal information, which are named as LBP-TOP. Boosted Haar features are used for automatic coarse face detection and 2D Cascade AdaBoost is applied for localizing eyes in detected faces. The positions of the two eyes, determined in the first frame of each sequence, define the facial area used for further analysis in the whole video sequence. Authors try to achieve real-time performance proposing multi-resolution features, computed from different sized blocks, different neighbouring samplings and different sampling scales of LBPs. AdaBoost is used for features selection, whose training is performed either by one-against-one or all-against-all classes strategies.

In [10] the authors combine the strengths of two-dimensional principal component analysis (2DPCA) and LBP operators for feature extraction. Before LBP computation, authors apply a filter for edge detection aiming at lowering the sensitivity to noise or changes in light conditions of LBP operators, although such operators have proven their robustness to those issue [11]. Finally, the concatenated features are passed as input to a decision direct acyclic graph based multi-class SVM classifier.

In [12] the authors test LBP-based features, histogram of oriented gradient and scale invariant feature transform to characterize facial expressions over 5 yaw rotations angle from frontal to profile views.

In [13] the prototypical expressions are classified at 5 different poses using multi-scale LBP and local gabor binary patterns (LGBP). LGBPs utilize multi-resolution spatial histogram combined with local intensity distributions and spatial information. A SVM is used as classifier. Experiments suggest, that facial expressions recognition is largely consistent across all poses, but the optimal view is dependent on the data and features used.

This review shows that considerable effort has been devoted to find the best combinations of preprocessing steps, descriptors, and classifier to achieve the task of discriminate between the six prototypical expressions. Nevertheless the issue is still open, especially if high recognition performance must be obtained respecting real time and computational constraints.

3   System Architecture

This section first presents an overview of the proposed system, and then focuses on the feature vectors we tested.

The system can be divided into four blocks performing: face and eye detection, image normalization, features extraction, and sample classification (Fig.1).

In the first module each frame of the input video sequence is processed by Haar-like features cascades to find a region of interest containing the face. Then the frame is cropped obtaining the facial image. Next, employing appropriate Haar-cascades, we locate eyes and compute morphological information, such as the distance between the eyes and the degree of head rotation. Since subjects often close their eyes while are showing the facial expressions, an eye tracking algorithm is used to estimate current eyes positions using information on eyes position and speed in previous frames.

In the normalization module the facial image is converted into grey-scale and its histogram is equalized to make the system more robust to light variation. To reduce the computational complexity each facial image is rescaled to 96x96 pixels maintaining a distance between the eyes of 48 pixels. Furthermore to improve system's performance the rescaled facial images are aligned on the basis of eyes position in each frame.

The third module processes facial image with LBP operators and extracts several descriptors, such as histograms and statistical-like moments, used to built the feature vector of the sample (in subsection 3.1).

The fourth module classifies the samples as discussed in section 4.

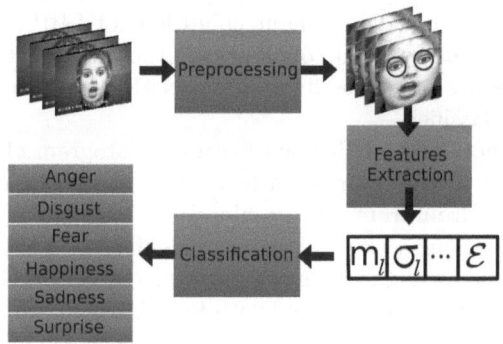

Fig. 1. System architecture

3.1   Feature Vectors

The following two paragraphs describe features commonly used in the literature, whereas the third paragraph presents the novel set of features applied to this work.

I_{LBP} *Histograms.* LBP operators are grey-scale invariant statistical primitives showing good performance in texture classification [11]. Defining g_c the grey value of the central pixel of the neighbourhood and g_p the grey value of the equally spaced pixels on a circle of radius R (with $p = 1, 2, ..., P-1$), LBP operators are defined as:

$$LBP_{P,R} = \sum_{p=0}^{P-1} s(g_p - g_c)2^p \qquad (1)$$

where $s(g_p - g_c)$ is 1 if $g_p - g_c \geq 0$ and 0 otherwise. In our work we use $P = 8$ and $R = 1$. In the following we denoted as I_{LBP} the LBP transformed image computed applying the LBP operators to a grey scale image I.

Features vectors from I_{LBP} are generally obtained collecting several histograms $\mathbf{H}_{\mathbf{LBP}}^{\mathbf{w_i}}$, each one computed on a region w_i of I_{LBP}. Dividing I_{LBP} into n regions, the corresponding feature vector of I_{LBP} is given by:

$$\mathbf{F}_{LBP} = \{\mathbf{H}_{\mathbf{LBP}}^{\mathbf{w_i}}\}_{i=1}^{n} \qquad (2)$$

Since, $\mathbf{H}_{\mathbf{LBP}}^{\mathbf{w_i}}$ is composed of 256 bins [11], $\mathbf{F}_{LBP}$ is composed of $(256 \times n)$ elements.

$I_{LBP^{u2}}$ *Histograms.* A more compact way to describe texture information of an image is to use LBP^{u2} operators. Indeed, information, on about 90%, of the vast majority of all 3x3 patterns obtained when $R = 1$ is captured by a limited number of patterns. $LBPs^{u2}$ work as templates for microstructures such as bright and dark spots, flat areas, and edges. Formally, they are defined by a uniformity measure U corresponding to the number of pattern spatial transitions i.e. bitwise 0/1 changes in binary code. Considering only patterns having U value of at most 2, the following descriptors may be introduced:

$$LBP_{P,R}^{u2} = \begin{cases} \sum_{p=0}^{P-1} s(g_p - g_c) & if \quad U(LBP_{P,R}) \leq 2 \\ P+1 & otherwise \end{cases} \qquad (3)$$

LBP^{u2} can be effectively described by histograms with 59 bins [11]. Hence, the corresponding feature vector is composed of $(59 \times n)$ elements:

$$\mathbf{F}_{LBP^{u2}} = \{\mathbf{H}_{\mathbf{LBP^{u2}}}^{\mathbf{w_i}}\}_{i=1}^{n} \qquad (4)$$

Statistical-like moments. A larger feature vector requires larger time to extract features, to train the classifier and to label test samples. For these reasons, we propose to represent LBP transformed images using very few descriptors. Given I_{LBP} histograms, we may extract statistical descriptors since they are able to catch texture information [14]. In this respect, we consider the mean (m), the variance (σ^2), the skewness (ς), the kurtosis (χ), and the energy (ϵ).

Note that skewness and kurtosis are defined only if $\sigma^2 \neq 0$.

As reported above, features may be extracted not only from the whole I_{LBP}, but also from subregions w_i. In this case, we may find regions with a uniform pixel value, where $\sigma^2 = 0$ and, hence, skewness and kurtosis cannot be computed. In order to overcome this limitation, we introduce in the following a new set of descriptors, named as *statistical-like moments*, inspired by statistical descriptors. To this aim, let us introduce the following notations:

- w is a subregion of I_{LBP};
- n_k is the number of pixels having grey value k in I_{LBP};
- $n_k(w)$ is the number of pixels having grey value k in w;
- S is the number of pixels in I_{LBP};
- $S(w)$ is the number of pixels in w.
- $\mu_n(w)$ is the nth statistical moments of w.

We first introduce the following descriptor named as *local mean* and denoted by $m_l(w)$:

$$m_l(w) \triangleq \sum_{k=0}^{L-1} r_k \frac{n_k(w)}{S} \tag{5}$$

where r denote a discrete variable representing pixel value in the range $[0, L-1]$. Similarly, we define the nth *local moments* as:

$$\mu_{l_n}(w) \triangleq \sum_{k=0}^{L-1} (r_k - m_l(w))^n \frac{n_k(w)}{S} \quad n = 0, 1, \ldots \tag{6}$$

The following relationship between our *local moments* and the conventional statistical moments computed in w holds:

$$\mu_{l_n}(w) = \sum_{i=0}^{n} \binom{n}{i} \mu_{n-i}(w)(m(w) - m_l(w))^{n-i} \phi(w) \quad if \, n > 0 \tag{7}$$

where $\mu_n(w)$ is the nth conventional statistical moment and $m(w)$ is the conventional mean, both in subregion w, and $\phi(w) = S(w)/S$.

Using such quantities we define two further descriptors inspired to skewness and kurtosis and respectively denoted as $\varsigma_l(w)$ and $\chi_l(w)$. They are given by:

$$\varsigma_l(w) = \frac{\mu_{l_3}(w)}{\sigma_l^3(w)}, \qquad \chi_l(w) = \frac{\mu_{l_4}(w)}{\sigma_l^4(w)} \tag{8}$$

where $\sigma_l(w)$ is:

$$\sqrt{\sum_{k=0}^{L-1} (r_k - m_l(w))^2 \frac{n_k(w)}{S}} = \sqrt{\mu_{l_2}(w)} \tag{9}$$

Since it can be easily shown that $\sigma_l(w)$ is different from zero providing that pixels in region w are not all 0, $\varsigma_l(w)$ and $\chi_l(w)$ are always defined.

Using previous descriptors, the features vector is built computing from each subregion the following quantities:

$$\mathbf{f}_l^{w_i} = \{m_l(w_i), \sigma_l(w_i), \varsigma_l(w_i), \chi_l(w_i), \epsilon\} \quad i = 1, \ldots, n \quad \forall w \in I \qquad (10)$$

and collecting them into the final vector which is composed of $(5 \times n)$ elements only:

$$\mathbf{F}_l = \{\mathbf{f}_l^{w_i}\}_{i=1}^n \qquad (11)$$

4 Experimental Protocol

For our experiment we used the second edition of Cohn-Kanade database (CK+) [15] arranging the sequences in two different sets.

One set is composed of 282 video sequences corresponding to 876 images of 95 different subjects containing prototype expressions. According to a general practice reported in the literature, we select the last three frame of each video sequence corresponding to the peak of expression. In the following this set is referred to $\mathbf{D}_p$.

The second set is composed of 4409 images obtained selecting in the 282 video sequences all the frames except the first three ones. In the follow we referred to this set as $\mathbf{D}_s$.

While $\mathbf{D}_p$ permits to test the capability of our system to detect expressions at the maximum of the expressions intensity, $\mathbf{D}_s$ permits to test our system in a scenario near to the real situation where people rarely show expression of maximum intensity.

As already mentioned, features extraction approaches based on both $\mathbf{F}_{LBP}$ and $\mathbf{F}_{LBP^{u2}}$ as well as the proposed statistical-like moments $\mathbf{F}_l$ divide facial image into regions. In the literature, a common approach divides the image into n non-overlapping regions [8,9,10]. Hence, we extract $\mathbf{F}_{LBP}$, $\mathbf{F}_{LBP^{u2}}$ and $\mathbf{F}_l$ dividing facial image into $9, 16, 25, 36, 49$ and 64 non overlapping squared regions. An alternative method may use a shifting window providing partially overlapped subregions. Although this approach cannot be applied to $\mathbf{F}_{LBP}$ and $\mathbf{F}_{LBP^{u2}}$ since their long feature vectors would greatly increase the computational load, the reduced size of our $\mathbf{F}_l$ descriptors enable us to test the shifting window approach. In this case tested windows sizes are 12, 14,16, 20, 24 and 32 pixels. In the rest of paper, the computation of $\mathbf{F}_l$ using a shifting window is denoted as $\mathbf{F}_{l_w}$.

All feature vectors are given to a SVM classifier with an RBF kernel, performing 10-fold cross validation and averaging out the results. We performed a grid search on SVM parameter space, where C ranges from 1 up to 10^4, and γ ranges from 10^{-1} up to 10^{-5}.

Image processing and features extraction stages have been implemented in C++, while classifier training and testing have been performed using Weka-3.0 [16]. Experiments have ran on Hp xw8600 workstation with 8-core $3.00\,GHz$ Intel(R) Xeon(R) CPU and $8.00\,GB$ RAM.

Table 1. Recognition rate (%) measured considering only peak of expression ($\mathbf{D}_p$) and the whole video sequence ($\mathbf{D}_s$). $\mathbf{F}_{LBP}$, $\mathbf{F}_{LBP^{u2}}$, $\mathbf{F}_{stat}$ and $\mathbf{F}_l$ were computed on non overlapping regions. $\mathbf{F}_{l_w}$ was computed on shifting window.

	$\mathbf{D}_p$						$\mathbf{D}_s$					
Features	Side in pixels of squared regions						Side in pixels of squared regions					
	32	24	20	16	14	12	32	24	20	16	14	12
$\mathbf{F}_{LBP}$	98.2	99.1	99.1	99.3	99.1	99.1	95.5	96.2	96.6	96.7	96.6	97.3
$\mathbf{F}_{LBP^{u2}}$	97.9	99.1	99.1	99.4	99.1	99.2	95.4	96.1	96.5	97.0	97.3	97.3
$\mathbf{F}_l$	97.5	97.7	97.1	97.3	97.9	96.0	95.1	96.8	96.8	97.0	96.6	96.5
$\mathbf{F}_{l_w}$	98.7	99.1	99.2	98.9	98.7	94.5	97.7	98.2	98.0	97.5	97.7	96.6

5   Results and Discussion

Table 1 shows the recognition rates achieved using different features on both $\mathbf{D}_p$, and $\mathbf{D}_s$, whereas table 2 shows elapsed time to train and test the classifier on each sample of $\mathbf{D}_p$ (times concerning $\mathbf{D}_s$ convey equivalent information and are not reported for the sake of brevity). As reported above $\mathbf{F}_{LBP}$, $\mathbf{F}_{LBP^{u2}}$ and $\mathbf{F}_l$ are computed from squared, non-overlapping subregions whereas $\mathbf{F}_{l_w}$ are computed using a shifting window. To provide a deeper insight into recognition results, table 3 reports the confusion matrices corresponding to the best and the worst accuracy of $\mathbf{F}_{l_w}$ on $\mathbf{D}_p$. Such best and worst results are achieved using the shifting window with side of 20px and 12px wich are reported in each tabular as the first and second entry, respectively. Other confusion matrices are omitted for brevity.

Turning our attention to results achieved on $\mathbf{D}_p$ (i.e. expression recognition in peak condition) we notice that $\mathbf{F}_l$ descriptors require lower training and test times than LBP-based descriptors (Table 2), whereas their performance is only slightly penalized (Table 1). In particular, focusing on test times, which are the most relevant in real time scenarios, $\mathbf{F}_l$ is 15 and 10 times faster than $\mathbf{F}_{LBP}$, $\mathbf{F}_{LBP^{u2}}$, respectively. Nevertheless, if shifting windows are used, i.e. $\mathbf{F}_{l_w}$, a better trade-off is reached. With respect of LBP-based descriptors, on the one hand, performance are completely comparable and, on the other hand, training and test times are still remarkably low. Indeed, $\mathbf{F}_{l_w}$ is 4 and 3 times faster than $\mathbf{F}_{LBP}$ and $\mathbf{F}_{LBP^{u2}}$, respectively.

Turning our attention on expressions recognition on $\mathbf{D}_s$ (i.e. the whole video sequences) we observe that $\mathbf{F}_l$ performance are comparable with those achieved using $\mathbf{F}_{LBP}$ and $\mathbf{F}_{LBP^{u2}}$. Furthermore, best performance among all tested configurations are achieved by $\mathbf{F}_{l_w}$ descriptors, attaining a recognition rate equal to 99.2%.

This analysis reveals that the introduction of more compact descriptors increases the flexibility in choosing the trade-off between classification performance and computational costs. In particular, our statistical-like moments computed using a shifting window remarkably reduce the computation time and improve classification performance when expressions vary in intensity during a video sequences. This observation suggests that our descriptors may be a good solution for real problems where it is likely to be difficult to isolate peak expression frames.

Table 2. Training and testing time of different descriptors on D_p

Features	Time	Side of squared regions					
		32	24	20	16	14	12
$\mathbf{F}_{LBP}$	train(s)	9.2	12.6	13.6	18.1	18.7	22.6
	test(ms)	11.0	16.9	17.2	23.8	24.4	29.2
$\mathbf{F}_{LBP^{u2}}$	train(s)	4.5	9.1	11.7	12.4	15.1	17.4
	test(ms)	6.1	9.3	14.8	15.3	18.7	21.9
$\mathbf{F}_l$	train(s)	0.2	0.6	1.4	2.4	3.6	3.4
	test(ms)	0.4	1.7	2.1	1.4	1.4	1.9
$\mathbf{F}_{l_w}$	train(s)	2.1	3.6	3.6	5.6	6.8	14.5
	test(ms)	3.5	2.1	3.4	7.3	8.8	12.2

Table 3. Confusion matrix corresponding to the best and the worst accuracy of $\mathbf{F}_{l_w}$ on D_p reported in each tabular as the first and second value, respectively

	Anger	Disgust	Fear	Happiness	Sadness	Surprise
Anger	125-114	0-0	0-0	0-0	0-0	1-12
Disgust	0-1	161-147	1-0	0-0	0-0	0-14
Fear	1-0	0 - 0	65-60	0-2	0-0	0-4
Happiness	0-0	0-0	1-0	197-188	0-0	0-10
Sadness	0-0	2-1	0-0	0-0	76-73	0-4
Surprise	1-0	0-0	0-0	0-0	0-0	245-246

6   Conclusions

In this paper we have presented an automatic facial expression recognition system able to conveying facial expression information using a compact set of descriptors based on statistical properties of LBP transformed images histograms. The small number of elements composing the resulting feature vectors permits us to significantly reduce the classification times making our system suited to real time applications on resource-limited platforms, such as mobile devices. Performance of our descriptors has been compared with that achieved by LBP-based descriptors in their basic configuration. Future works will be directed towards the use of features selection algorithms, such as Adaboost. This will permit us to compare performance of our selected feature vectors with performance achieved by boosted-descriptors known in literature and employed in person-independent system.

Acknowledgements. This work has been (partially) carried out in the framework of the ITINERIS2 project, Codice CUP F87G10000050009, under the financial support of Regione Lazio (Programme "Sviluppo dell'Innovazione Tecnologica nel Territorio Regionale, Art. 182, comma 4, lettera c), L.R. n. 4, 28 Aprile 2006).

References

1. Fasel, B., Luettin, J.: Automatic facial expression analysis: a survey. Pattern Recognition 36(1), 259–275 (2003)
2. Pantic, M., Rothkrantz, L.J.M.: Automatic analysis of facial expression: the state of art. IEEE Trans. on Pattern Analysis and Machine Intelligence (2000)
3. Liu, S.S., Tian, Y.T., Li, D.: New research advances of facial expression recognition. In: Proc. of the Eighth Int. Conf. on Machine Learning and Cybernetics (2009)
4. Bartlett, M.S., Littlewort, G.C., Frank, M.G., Lainscsek, C., Fasel, I.R., Movelland, R.: Automatic recognition of facial actions in spontaneous expressions. Journal of Multimedia (2006)
5. Whitehill, J., Littlewort, G., Fasel, I., Bartlett, M., Movellan, J.: Toward practical smile detection. IEEE Trans. on Pattern Analysis and Machine Intelligence 31, 2106–2111 (2009)
6. Yang, P., Liu, Q., Metaxas, D.N.: Boosting encoded dynamic features for facial expression recognition. Pattern Recognition Letters 30(2), 132–139 (2009), Video-based Object and Event Analysis
7. Uddin, M. Z., Lee, J.J., Kim, T.S.: An enhanced indipendent component-based human facial expression recognition from video. IEEE Trans. on Consumer Electronics (2009)
8. Shan, C., Gong, S., McOwan, P.W.: Facial expression recognition based on local binary patterns: A comprehensive study. Image and Vision Computing 27(6), 803–816 (2009)
9. Zhao, G., Pietikinen, M.: Boosted multi-resolution spatiotemporal descriptors for facial expression recognition. Pattern Recognition Letters 30(12), 1117–1127 (2009), Image/video-based Pattern Analysis and HCI Applications
10. Lin, D.-T., Pan, D.-C.: Integrating a mixed-feature model and multiclass support vector machine for facial expression recognition. Integr. Comput.-Aided Eng. 16, 61–74 (2009)
11. Ojala, T., Pietikäinen, M.: Multiresolution gray-scale and rotation invariant texture classification with local binary patterns. IEEE Trans. on Pattern Analysis and Machine Intelligence 24, 971–987 (2002)
12. Hu, Y., Zeng, Z., Yin, L., Wei, X., Zhou, X., Huang, T.S.: Multi-View Facial Expression Recognition. In: 8th Int. Conf. on Automatic Face and Gesture Recognition (2008)
13. Moore, S., Bowden, R.: Local binary patterns for multi-view facial expression recognition. Computer Vision and Image Understanding 115(4), 541–558 (2011)
14. Gonazalez, C., Woods, E.: Digital Image Processing. Prentice-Hall, Englewood Cliffs (2001)
15. Kanade, T., Cohn, J.F., Tian, Y.: Comprehensive database for facial expression analysis. In: Proc. of the Fourth IEEE Int. Conf. on Automatic Face and Gesture Recognition, pp. 46–53 (2000)
16. Hall, M., Frank, E.: The WEKA data mining software: An update. SIGKDD Explorations 11(1) (2009)

Temporal Analysis of Biometric Template Update Procedures in Uncontrolled Environment

Ajita Rattani, Gian Luca Marcialis, and Fabio Roli

Department of Electrical and Electronic Engineering
University of Cagliari, Italy
{ajita.rattani,marcialis,roli}@diee.unica.it

Abstract. Self-update and co-update algorithms are aimed at gradually adapting biometric templates to the intra-class variations. These update techniques have been claimed to be effective in capturing variations occurring in medium time period but no experimental evaluations have been done in the literature to clearly show this fact. The aim of this paper is the analysis and comparison of these update techniques on the sequence of input batch of samples as available over time, specifically, in the time-span of 1.5 years. Effectiveness of these techniques have been compared in terms of capability to capture significant intra-class variations and the attained performance improvement, over time. Experiments are carried out on DIEE multi-modal dataset, explicitly collected for this aim. This dataset is publicly available by contacting the authors.

Keywords: Biometrics, Face, Fingerprint, Self-update,Co-update.

1 Introduction

A personal biometric verification system consists of two main processes; enrolment and matching. In enrolment, individual's biometric samples are captured, processed and features extracted. These extracted features are labelled with user's ID and is referred to as "template". Matching mode verifies claimed identity by comparing input sample(s) to the enrolled template(s) [1] [2].

The enrolment process typically acquires very few samples, usually a single image, captured under controlled conditions. On the other hand, real time operation in uncontrolled environment encounters large variations in the input data, called "intra-class" variations. These variations can be attributed to factors like human-sensor interaction, illumination conditions, changes in sensor, seasonal variations, occlusions due to user's accessories *etc..* These factors causes "temporary variations" in the biometric data. In addition, biometric traits also undergo gradual ageing process as a result of time lapse [3]. Accordingly, changes in the biometric over time can be termed as "temporal changes" occurring in the medium-long term.

As a consequence of these *temporary* and *temporal* variations, enrolled templates becomes "un-representative". In this paper, we refer to *representative*

G. Maino and G.L. Foresti (Eds.): ICIAP 2011, Part I, LNCS 6978, pp. 595–604, 2011.

templates as *templates with the capability to correctly recognize significant intra-class and inter-class variations (impostor samples)*.

Recently, template update procedures have been introduced aiming to solve the issue of unrepresentative templates by constantly adapting themselves to the intra-class variations of the input data. Most of the existing template update techniques are based on self-update and co-update algorithms [4]. In self-update, biometric system adapts itself on the basis of highly confidently classified input data [5]. Template co-update utilizes the mutual and complementary help of multiple biometric traits to adapt the templates to the variation of the input data [6].

Reference [7] performs experimental comparison of self-update and co-update proving the efficacy of co-update, on utilizing the help of multi-modalities, in capturing intra-class variations available over short-term period (e.g. face expression variations), over the former. However, the experimental analysis and comparison is performed [7] over a limited data set containing only temporary variations.

Till date, no analysis have been done for self and co-update techniques with respect to :

- the dependence of their performances on the representativeness of initial templates,over time;
- their efficiency in capturing intra-class variations as available over time;
- their resulting performance improvement over dataset containing both temporary as well as temporal variations.

To this end, the aim of this paper is to advance the state-of-the-art by a temporal analysis of self and co-update algorithm by using input batches of samples available at different time intervals. In particular, in this paper :

- we propose a conceptual representation of performance of self-update and co-update with respect to representativeness of the initial captured enrolled templates; this representation is supported by the experimental evidence.
- The capability to capture temporal intra-class variations and the obtained performances improvement have been analyzed over time.
- Experiments are conducted on DIEE multimodal dataset,explicitly collected for this aim, containing both temporary as well as temporal variations.

In section II, self-update and co-update techniques are discussed. Experimental evaluations are presented in Section III. Section IV concludes the paper.

2   Self-update and Co-update Algorithms

In this section, we elaborate on self-update and co-update algorithms together with the possible conceptual explanation of their operation.

2.1   Self-updating

In self-updating procedure, matcher is trained on a initial set of enrolled templates, named T. A batch of samples, U, is collected during system's operations over a certain time [5]. Among all the samples in U, only those samples whose match score against the available templates exceeds a given *updating threshold* (i.e., highly confidently classified samples), $thr*$, are used for the adaptation process. This is done to reduce the probability of impostor introduction (false acceptance) into the updated gallery set. This procedure is presented in Algorithm 1.

However, using only highly confident samples may lead to inefficiency in capturing samples representing substantial variations [4]. Thus, the efficiency is dependent on the *representativeness* of the initial enrolled templates. On the other hand, by relaxing the updating threshold, the system may become prone to classification errors (updating using impostor samples) [4] [7].

Algorithm 1. Self updating algorithm

1. Given:
    - $T = \{t_1, ..., t_M\}$ is the template set.
    - $U = \{u_1, ..., u_N\}$
    - $U* = \phi$ is an empty set.
2. Estimate $thr*$ on T
3. For $h = 1, ..., N$
    (a) $s_h = matchscore(u_h, T)$
    (b) If $(s_h > thr*)$, then $U* = U * \cup \{u_h\}$
4. end For
5. $T = T \cup U*$

For the conceptual representation of self-update's operation, let us consider a sample space of a specific user for a given biometric $b0$.

If this user-specific sample space is represented in the form of a directed relational graph G where nodes are the samples and the edges are labelled with the related matching score between any two samples. On removing the edges labelled with the matching score below $thr*$. The sample space may be partitioned into n sub-graphs $G_1^{b0}, ..., G_n^{b0}$. Each of these subgraphs G_i^{b0}, contains only those nodes (samples) connected to other nodes with edges labelled with matching score above $thr*$.

On the availability of input batch U, consisting of random samples forming the part of different subgraphs $G_1^{b0}, ..., G_n^{b0}$ in a user-specific sample space. Let us suppose for simplicity that only single initial enrolled template, t, is available, and belongs to subgraph G_1^{b0}. Considering the self-update behaviour (Algorithm 1), may be only subset $U_{G_1^{b0}} = \{u \in U : u \in G_1^{b0}\}$ of samples belonging to subgraph G_1^{b0} may be inserted into the template gallery. Provided their exists a directed path from template t to sample $u \in U\{G_1\}$. Other samples of $U \notin U\{G_1\}$ will be completely neglected for the adaptation, in this case.

This implies that, self-update will result in limited capture of samples depending on the representativeness of the enrolled templates. In other words, in order to allow self-update to be efficient in capturing significant variations, a careful *a priori* selection of templates must be done by human experts, on the basis of user-specific subgraphs characteristics, with respect to the given user population and the selected threshold $thr*$. Another alternative is to relax the threshold value,$thr*$, which changes the obtained subgraphs partitioning and reduces the number n of subgraphs.

As an example, Figure 1(a), shows the hypothetical diagram with two component subgraphs G_1^{b0} and G_2^{b0} formed from the user-specific sample space, for any modality $b0$. If the enrolled templates are in G_1^{b0}, only samples lying in the subgraph G_1^{b0} can be captured by the self-update process. The samples in G_2^{b0} will be completely neglected for the adaptation process.

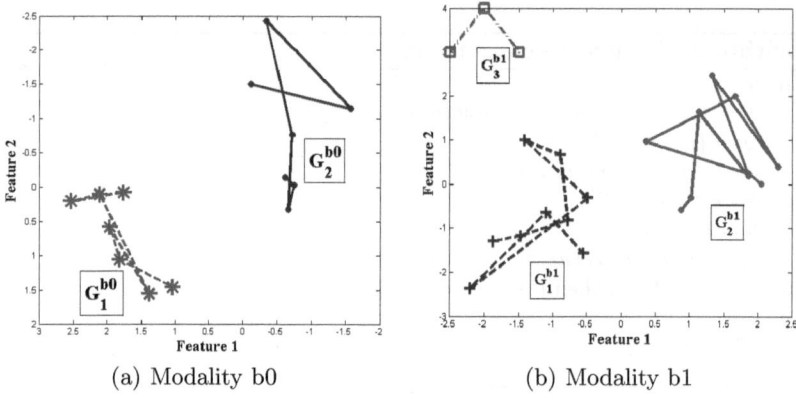

(a) Modality b0   (b) Modality b1

Fig. 1. A hypothetical diagram showing the different component subgraphs G^{b0} and G^{b1}, of a complete directional graph where nodes are the biometric samples hypothetically projected in the feature space, formed from removing the edges labelled with matching score below $thr*$ for the modalities $b0$ and $b1$

2.2   Template Co-updating

In template co-update [6], mutual and complimentary help of two biometrics is utilized for the adaptation of the templates. Specifically, given two modalities, input samples classified as genuine with high confidence by the one modality (for ex, face), together with the corresponding complementary sample from the another modality (for ex, fingerprint), are used for the updating process.

Co-updating procedure is presented in Algorithm 2, where T^{b0} and T^{b1} are the template sets and U^{b0} and U^{b1} are the input batches of samples, for the modalities $b0$ and $b1$ (for example, fingerprint and face). The sample is highly confidently classified if its matching score on comparison with the enrolled templates is above the set threshold $thr^{b0}*$ and $thr^{b1}*$. s_h^{bi} is the confidently classified sample by the modality b_i.

Algorithm 2. Co-updating algorithm

---

1. Given modalities $b1$ and $b2$:
   - $T^{b0} = \{t_1^{b0}, ..., t_M^{b0}\}$ and $T^{b1} = \{t_1^{b1}, ..., t_M^{b1}\}$ as the template sets.
   - $U^{b0} = \{u_1^{b0}, ..., u_N^{b0}\}$ and $U^{b1} = \{u_1^{b1}, ..., u_N^{b1}\}$ as available batches of samples, where sample u_h^{b0} is coupled with u_h^{b1}.
   - $U^{b0}* = \phi$ and $U^{b1}* = \phi$ as empty sets.
2. Estimate $thr^{b0}*$ on T^{b0} and $thr^{b1}*$ on T^{b1}.
3. For $i = 0, 1$
   (a) For $h = 1, ..., N$
      i. $s_h^{bi} = matchscore(u_h^{bi}, T^{bi})$
      ii. If $(s_h^{bi} > thr^{bi}*)$, then $U^{b\bar{i}}* = U^{b\bar{i}} * \cup \{u_h^{b\bar{i}}\}$
4. $T^{b0} = T^{b0} \cup U^{b0}*$
5. $T^{b1} = T^{b1} \cup U^{b1}*$

Following the same convention of directed relational graph and sub-graph partioning of the user-specific sample space, as used for the representation of self-update, the conceptual representation of co-update may be explained as well. Worth noticing that each modality may form different sub-graphs with different number of samples i.e, $G^{b0} = \{G_1^{b0}, ..., G_{n^{b0}}^{b0}\}$ and $G^{b1} = \{G_1^{b1}, ..., G_{n^{b1}}^{b1}\}$ where $n^{b0} \neq n^{b1}$. This is due to the difference in the match score distribution and complementary characteristics of different modalities. Figures 1(a)and 1(b) show the component subgraphs of two independent modalities $b0$ and $b1$.

Irrespective of the induced partitioning (in the form of component subgraphs) of individual biometric modality, template galleries of each modality can be updated with the samples lying in the subgraphs different from the one in which the enrolled templates reside, using the complementary matcher. In other words, biometric $b0$ uses biometric $b1$ to update the template gallery, and vice-versa. This is done thanks to the conditional independence among two biometrics as assumed in co-updating algorithm: each sample of biometric $b0$, u^{b0} may be coupled with any sample, u^{b1}, of $b1$, independently of user-specific sample space partitioning.

3 Experimental Analysis

3.1 Data Set

40 subjects with 50 samples per subjects are used from the DIEE multi-modal face and fingerprint dataset. These 50 samples are acquired in 5 sessions with 10 samples per session (batch) with a gap of minimum three weeks between

two consecutive sessions. These batches are indicated as $B_1, ..., B_5$. The whole collection process span a period of 1.5 years. Each batch B_i consist of fingerprint and face couples for a certain subject. For the co-update process these batches are used as it, however, for self updating process face and fingerprint samples are isolated. For the sake of simplicity, we indicated in the following mono- and multi-modal batches using the same notation.

For face biometrics, variations such as lighting, expression change (i.e., happy, sad, surprised and angry) and eyes half-closed are introduced in every batch. For fingerprint biometrics, rotation, changes in pressure, non-linear deformations and partial fingerprints are introduced as variations [8]. Two consecutive batches, B_i and B_{i+1} are temporally ordered. Figure 2 shows some of the images taken from different sessions of face modality for a randomly chosen subject, where images in different rows represent different sessions.

Fig. 2. Face images from different sessions exhibit ageing as well as other intra-class variations, for a randomnly chosen subject

3.2   Experimental Protocol

- **Training:**
  1. The system is trained with 2 enrolled samples per user from the first session i.e, batch B_1.
  2. Threshold for adaptation is set by estimating genuine distribution according to available templates.
- **Updating:** Remaining user images consisting of eight samples from the batch B_1 and ten samples from the remaining four batches i.e, $B_{2:4}$, are used for adaptation as follows:
  1. For each user, $B1...B4$ are available over time.
  2. Each batch B_i with $i \in 1 : 4$ is used for updating the template set of the respective user by using the self-update and co-update algorithms as mentioned in Section II. These batches B_i corresponds to unlabelled batch U in algorithms 1 and 2.
- **Performance evaluation:**
  1. After updating using batch B_i, i.e., after updating cycle i, batch B_{i+1} is used for testing the system performance. Scores for each test sample are always computed using the max rule [2].
  2. At each updating cycle i, Equal Error Rate of the system, namely, EER_i, is computed as follows: $EER_i = \frac{i-1}{i} * EER_{i-1}^{ave} + \frac{1}{i} * EER_i^*$, as the mean of the EER at previous update cycles i.e., EER_{i-1}^{ave} and the EER (EER_i^*) obtained at the specific update cycle i.
  3. After the evaluation of the system, the same batch B_{i+1} is used for the process of updating.

3.3   Results

The goals of these experiments are to:

- Provide experimental validation of the conceptual representation of the functioning of self-update and co-update as provided in section II. This is done by evaluating the capability of self and co-update techniques in capturing significant intra-class variations and its dependence on the representativeness of the initial enrolled templates over time.
- Evaluation of these techniques over time, following multiple update cycles, that is, using batches $B1, ..., B5$.

3.4   Experiment no. #1

The aim of this experiment is to validate the dependence of self-update on initial templates and to verify the contrary for co-update in capturing large intra-class variations, according to the conceptual representation of Section II.

For sake of space, we present results only for the face biometric, but similar results may be obtained for the fingerprint biometrics as well.

In confirmity with the conceptual representation in section II, structure of data is studied and following steps have been performed, considering all the samples for each user:

- The graph G is formed and partitioned by connecting each sample to only those samples with matching score above the acceptance threshold set at 0.001% FAR of the system. This is a very stringent value in order to avoid presence of impostors in the graph.
- Then, two samples are chosen as templates from the subgraph containing batch B_1 i.e, $G^{face}(B1)$, for face modality.
- Self- and co-update algorithms are applied on batches $B_{1:4}$. Both of these update techniques operate at acceptance threshold set at 0.001% FAR (i.e., thr^*, $thr^{b0}*$ and $thr^{b1}*$ in algorithms 1 and 2) which is same for all the users.
- After four updating cycles using batches $B_{1:4}$, number of captured samples, belonging to subgraph containing enrolled templates and those captured from other different subgraphs are computed and presented in Figures 3(a)-3(b) for each subject. These Figs. (3(a)- 3(b)) clearly show that co-update has the high potential to capture samples with significant variations (that is, belonging to different subgraphs) in contrary to self-update, using exactly the same acceptance threshold as that of self-update (0.001% FAR).
- Finally, using the batch B_5, average mean genuine score over all the users is computed for each updating algorithm and the baseline classifier (without adaptation) for the templates updated till fourth update cycles. In case of baseline classifier, initial enrolled templates are matched against the batch B_5. The relative increase of the genuine mean score for self-update and co-update techniques over the baseline classifier is computed to be 25% and 53%, respectively. This is a direct confirmation of the increase in representativeness of the enrolled templates.

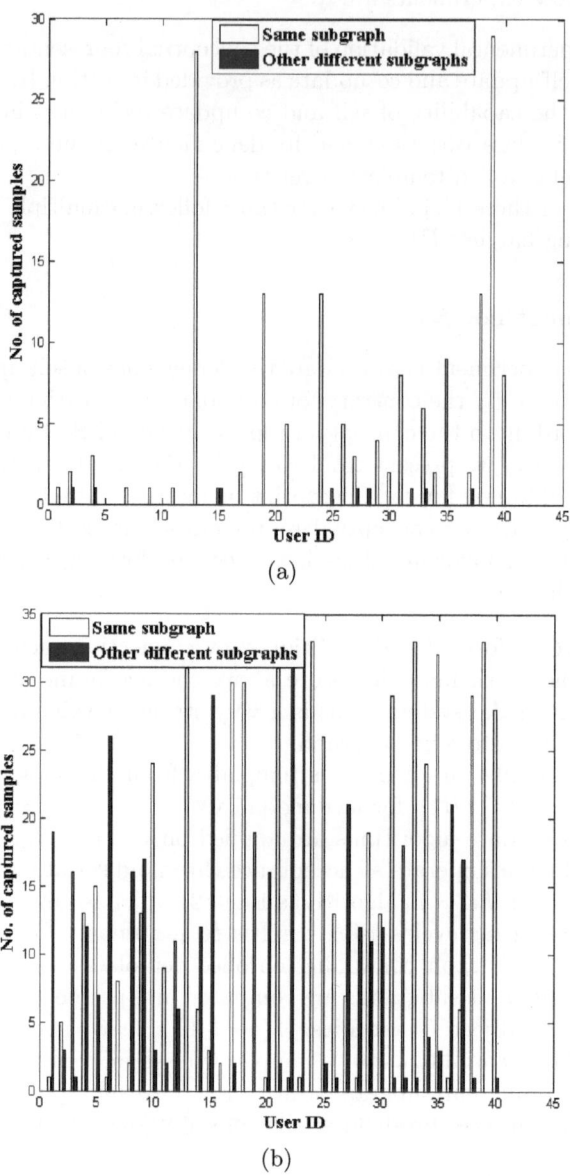

Fig. 3. Number of samples exploited by a) self-update and b)co-update process, from the same sub-graph to which the enrolled templates belong and from the other different sub-graphs. X axis represent different users and y axis represent number of captured samples from the same and different subgraphs.

Table 1. Cumulative percentage of samples in the updated template set till each update cycle i for the self-update and co-update process

Update techniques	(%) B_1	(%) $B_1 \cup B_2$	(%) $B1 \cup ... \cup B_3$	(%) $B1 \cup ... \cup B_4$
Self-update	31	18	17	17
Co-update	77	77	83	86

In addition to above evidences, Table 1, presents the cumulative percentage of samples after the $i - th$ update cycle, that is, the total percentage of data from batches $B_1 \cup ... \cup B_i$ which have been gradually added to the template gallery, again proving the efficacy of co-update.

3.5 Experiment #2

The aim of the experiment is the performance assessment of the self-update and co-update in terms of Equal Error Rate (EER_i) as computed over time (see protocol). For the real time evaluation, five random impostor samples are also inserted in each batch B_i. Figures 4(a) and 4(b) show EER values for baseline, self- and co-update at each update cycle.

It can be seen that, over time, performance of the co-update significantly increases. Differences in performances are due to the intrinsic characteristic of face and fingerprint modalities, as may be deduced from both the figures. This is evident both for the face and fingerprint biometrics. Specifically, after the third update cycle, EER exhibits a decreasing trend. On the other hand, self-update strictly follows the trend of baseline matcher.

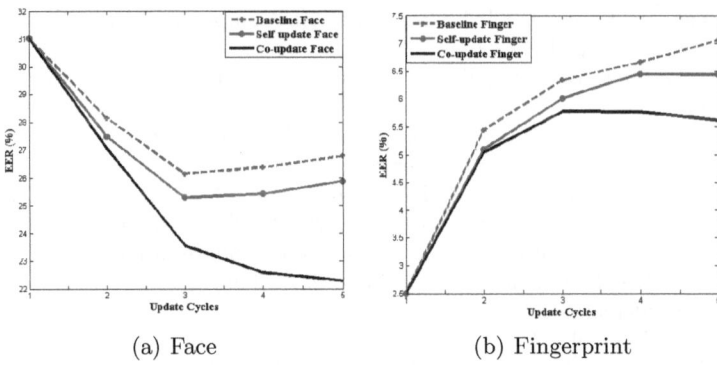

(a) Face   (b) Fingerprint

Fig. 4. EER Curve of the performance evaluation of the co-update in comparison to self-update and baseline classifier over time for a) face and b) fingerprint modalities

4 Conclusions

In this paper, we experimentally evaluated the performance of self- and co-update algorithms on multiple update cycles, available over time.

Existing claims in the literature such as: a) self-update is dependent on the threshold settings and on the representativeness of the initial templates, in capturing significant variations; on the contrary, co-update does not suffer from this limitation and b) self update process results in slow adaptation and limited performance improvement in comparison to co-update over time, were argued but never investigated.

We explicitly did such an analysis, which has been fully confirmed in our opinion, by the proposed conceptual representation supported by experimental evidences on DIEE Multi-modal data set.

Acknowledgments. This work has been partially supported by Regione Autonoma della Sardegna ref. no. CRP2-442 through the Regional Law n.7 for Fundamental and Applied Research, in the context of the funded project Adaptive biometric systems: models, methods and algorithms. Ajita Rattani is partly supported by a grant awarded to Regione Autonoma della Sardegna, PO Sardegna FSE 2007-2013, L.R. 7/2007 "Promotion of the scientific research and technological innovation in Sardinia".

References

1. Jain, A.K., Flynn, P., Ross, A.: Handbook of Biometrics. Springer, Heidelberg (2007)
2. Ross, A., Nandakumar, K., Jain, A.K.: Handbook of Multibiometrics. Springer, Heidelberg (2006)
3. Ling, H., Soatto, S., Ramanathan, N., Jacobs, D.: A study of face recognition as people age. In: Proc. 11th IEEE Int. Conf. on Computer Vision, Rio de Janeiro, Brazil, vol. 2688, pp. 1–8 (2007)
4. Rattani, A., Freni, B., Marcialis, G.L., Roli, F.: Template update methods in adaptive biometric systems: A critical review. In: Tistarelli, M., Nixon, M.S. (eds.) ICB 2009. LNCS, vol. 5558, pp. 847–856. Springer, Heidelberg (2009)
5. Roli, F., Marcialis, G.L.: Semi-supervised PCA-based face recognition using self-training. In: Yeung, D.-Y., Kwok, J.T., Fred, A., Roli, F., de Ridder, D. (eds.) SSPR 2006 and SPR 2006. LNCS, vol. 4109, pp. 560–568. Springer, Heidelberg (2006)
6. Roli, F., Didaci, L., Marcialis, G.L.: Template co-update in multimodal biometric systems. In: Lee, S.-W., Li, S.Z. (eds.) ICB 2007. LNCS, vol. 4642, pp. 1194–1202. Springer, Heidelberg (2007)
7. Rattani, A., Marcialis, G.L., Roli, F.: Capturing arge intra-class variations of biometric data by template coupdate. In: Proc. of IEEE Computer Society Conference on Computer Vision and Pattern Recognition Workshops, Anchorage, Alaska, USA, pp. 1–6 (2008)
8. Diee multimodal database,
   http://prag.diee.unica.it/pra/eng/home

Biologically Motivated Feature Extraction

Sonya Coleman, Bryan Scotney, and Bryan Gardiner

[1] School of Computing and Intelligent Systems, University of Ulster, Magee, UK
[2] School of Computing and Information Engineering, University of Ulster, Coleraine, UK
{sa.coleman,bw.scotney,b.gardiner}@ulster.ac.uk

Abstract. We present a biologically motivated approach to fast feature extraction on hexagonal pixel based images using the concept of eye tremor in combination with the use of the spiral architecture and convolution of non-overlapping Laplacian masks. We generate seven feature maps "a-trous" that can be combined into a single complete feature map, and we demonstrate that this approach is significantly faster than the use of conventional spiral convolution or the use of a neighbourhood address look-up table on hexagonal images.

Keywords: hexagonal images, feature extraction, spiral architecture, eye tremor.

1 Introduction

In order to obtain real-time solutions to problems that require efficient large-scale computation, researchers often seek inspiration from biological systems; for real-time image processing we consider the characteristics of the human visual system. In order for humans to process visual input, the eye captures information that is directed to the retina located on the inner surface of the eye. A small region within the retina, known as the fovea and consisting of a high density of cones, is responsible for sharp vision capture and is comprised of cones that are shaped and placed in a hexagonal arrangement [4, 6, 8]. Additional important characteristics of the human eye are that, within the central fovea, receptive fields of ganglion cells of the same type do not overlap [5], and that the eye can be subjected to three types of eye movement: tremor, drift, and micro-saccades. Furthermore, the human vision system does not process single static images, but instead a series of temporal images that are slightly off-set due to involuntary eye movements. Therefore, the traditional approaches to feature detection using overlapping convolution operators applied to static images do not closely resemble the human visual system.

Recent research has focussed on the use of hexagonal pixel-based images [7, 11] as the hexagonal pixel lattice closely resembles the structure of the human fovea and has many advantages in terms of image capture and analysis. In [10], Sheridan introduced a unique addressing system for hexagonal pixel based images, known as the spiral architecture, that addresses each hexagonal pixel with a single co-ordinate address, rather than the two co-ordinate address scheme typically used with rectangular image structures. Such a one-dimensional addressing scheme potentially provides an

G. Maino and G.L. Foresti (Eds.): ICIAP 2011, Part I, LNCS 6978, pp. 605–615, 2011.

appropriate structure for real-time image processing of hexagonal images. However, with respect to feature extraction via convolution, where typically an operator is applied to a pixel and its neighbours, the process of determining these neighbours in a one-dimensional addressing scheme is not always trivial and can require time consuming special hexagonal and radix-7 addition. In [9] the concept of eye tremor - rhythmic oscillations of the eye - has been exploited for image processing. Instead of applying the operators to every pixel in an overlapping manner typical of standard convolution, Roka et al. used nine overlapping images and applied the masks in a non-overlapping manner; however, they still assume the image to be comprised of rectangular pixels that are addressed in a two-dimensional structure, and they apply standard square image processing masks.

We present a biologically motivated approach to feature detection based on the use of the spiral architecture in conjunction with eye tremor and convolution of non-overlapping Laplacian masks [9]. The Laplacian operator can be considered analogous to the on-off receptive fields found in the retina in which the centre of the receptive field is negative, surrounded by positive values or vice versa. We develop a *cluster operator,* based on the spiral architecture, which can be applied to a one-dimensional spiral image in a fast and efficient way.

2   Spiral Architecture

In the spiral architecture [10] the addressing scheme originates at the centre of the hexagonal image (pixel index 0) and spirals out using one-dimensional indexing. Pixel 0 may be considered as a layer 0 tile. Pixel 0, together with its six immediate neighbours indexed in a clockwise direction (pixels 1, ..., 6) then form a layer 1 super-tile centred at pixel 0.   This layer 1 super-tile may then be combined with its six immediately neighbouring layer 1 super-tiles, the centres of which are indexed as 10, 20, 30, 40, 50 and 60; the remaining pixels in each of these layer 1 super-tiles are then indexed in a clockwise direction in the same fashion as the layer 1 super-tile centred at 0, (e.g., for the layer 1 super-tile centred at 30, the pixel indices are 30, 31, 32, 33, 34, 35 and 36).

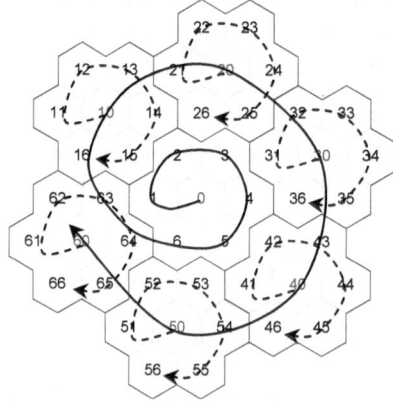

Fig. 1. One-dimensional addressing scheme in the central region of the image, showing one layer 2 super-tile, comprising 7 layer 1 super-tiles

The entire spiral addressing scheme is generated by recursive use of the super-tiles; for example, seven layer 2 super-tiles are combined to form a layer 3 super-tile. Ultimately the entire hexagonal image may be considered to be a layer L super-tile centred at 0 comprising 7^L pixels. Figure 1 shows the spiral addressing scheme for the central portion of an image (up to the layer 2 super-tile).

A major advantage of this addressing scheme is that any location in the image can be represented by a single co-ordinate value. This is advantageous for a number of reasons: it permits full exploitation of the symmetry of the hexagonal lattice; placement of the origin at the centre of the image simplifies geometric transformations such as rotation and translation; and most importantly it allows the spiral image to be stored as a vector [7]. Spatially neighbouring pixels within the 7-pixel layer 1 super-tiles in the image remain neighbouring pixels in the one-dimensional image storage vector. This is a very useful characteristic when performing image processing tasks on the stored image vector, and this contiguity property lies at the heart of our approach to achieve fast and efficient processing for feature extraction.

3   Cluster Operators

We refer to the operator that is applied to a cluster neighbourhood in the spiral architecture as a *cluster operator*. In recent work [2] we have shown how a finite element based approach can be used to create hexagonal operators based on constructing either two independent directional derivative operators aligned in the x- and y- directions, or tri-directional operators aligned along the x-, y- and z- hexagonal axes. Our operators are built using a regular mesh of equilateral triangles with nodes placed at the centres of each hexagonal pixel. With each node s we associate a piecewise linear basis function ϕ_s, with $\phi_s = 1$ at node s and $\phi_s = 0$ at all other nodes $t \neq s$. Each ϕ_s is thus a "tent-shaped" function with support restricted to a small neighbourhood of six triangular elements centred on node s. We represent the image by a function $I = \sum_{q \in Q} I(q)\phi_q$, where Q denotes the set of all nodal addresses; the parameters $\{I(q)\}$ are the image intensity values at the pixel centres.

Feature detection and enhancement operators are often based on first or second order derivative approximations, and we consider a weak form of the second order directional derivative $-\underline{\nabla} \cdot (\mathbf{B}\underline{\nabla}u)$ over small neighbourhoods. To approximate the second directional derivative, $-\underline{\nabla} \cdot (\mathbf{B}\underline{\nabla}u)$, over a λ-neighbourhood $N_\lambda(s)$ centred on the pixel with spiral address s, the respective derivative term is multiplied by a test function $v \in H^1$ and the result is integrated over $N_\lambda(s)$. The neighbourhood size λ corresponds to the layer λ: here we focus only on the neighbourhood $N_1(s)$ containing seven pixels (Figure 1 shows seven layer 1 neighbourhoods with $s = 0, 10, 20, 30, 40, 50$ and 60). Hence at each node s we may obtain a layer λ weak second order directional derivative $D_\lambda(s)$ as

$$D_\lambda(s) = \int_{N_\lambda(s)} (B\underline{\nabla}I) \cdot \underline{\nabla}\psi_s^\lambda \, d\Omega \tag{1}$$

where $\mathbf{B} = \underline{b}\,\underline{b}^{\mathrm{T}}$ and $\underline{b} = (\cos\theta, \sin\theta)$ is the unit direction vector. Our cluster operator is the isotropic form of the second order derivative, namely the Laplacian $-\underline{\nabla}.(\underline{\nabla}u)$; this is equivalent to the general form in which the matrix $\mathbf{B}$ is the identity matrix $\mathbf{I}$.

Although we are not addressing the issue of scale at this stage, it should be noted that each neighbourhood test function ψ_s^λ is restricted to have support over the neighbourhood $N_\lambda(s)$ for any choice of layer $\lambda = 1,2,3,...$ Thus we may write

$$D_\lambda(s) = \sum_{q \in Q} \left(I(q) \int_{N_\lambda(s)} (\underline{\nabla}\phi_q) \underline{\nabla}\psi_s^\lambda \, d\Omega \right) = \sum_{q \in N_\lambda(s)} H_\lambda(q) \times I(q) \tag{2}$$

where H_λ is a hexagonal operator of "size" λ (having the size and shape of a hexagonal λ-neighbourhood). Our general design procedure incorporates a layer-related parameter that enables scale to be addressed in future work. In this paper we have chosen each neighbourhood test function ψ_s^λ to be a Gaussian function centred on node s, parameterised so that 95% of its central cross section falls within $N_\lambda(s)$ and then its tails truncated so that support is restricted to the neighbourhood $N_\lambda(s)$. Figure 2 shows the layer 2 Laplacian cluster operator generated in this way.

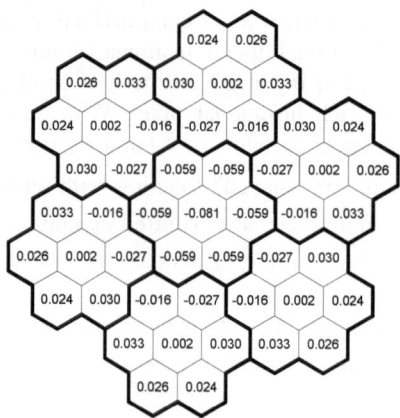

Fig. 2. Layer 2 Laplacian cluster operator

4   Framework for Fast Processing

Our proposed fast image processing framework has three essential elements: simulation of eye tremor by use of a set of slightly off-set images of the same scene; a special definition of sparse spiral convolution; and efficient identification of the spiral architecture pixel addresses within those $\lambda = 2$ neighbourhood clusters on which local operator convolution is required to be performed. We consider each of these three elements in turn below.

4.1 Simulation of Eye Tremor

We consider the hexagonal image I_0 to be the "base" image, and we capture six further images, $I_j, j = 1,...,6$, of the same scene. Each of these additional images is off-set spatially from I_0 by a distance of one pixel in the image plane along one of the three natural hexagonal axis directions. This mechanism simulates the phenomenon of "eye tremor". In each image $I_j, j = 1,...,6$, the pixel with spiral address 0 represents the same spatial location in the scene as the pixel with spiral address j in I_0.

The centre (i.e., the pixel with spiral address zero) of each image $I_j, j = 0,...,6$, is thus located at a pixel within the *layer* $\lambda = 1$ neighbourhood centred at the pixel with spiral address 0 in image I_0, as shown in Figure 3.

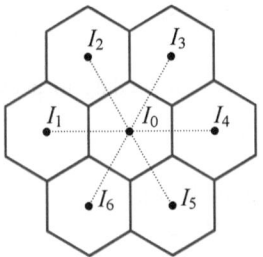

Fig. 3. The 7 image centres in the eye tremor approach

Through use of the spiral architecture for pixel addressing, it is assumed that image I_0 is stored in a one-dimensional vector (with base-7 indexing), as shown in Figure 4. Using the spiral architecture the additional images $I_j, j = 1,...,6$, are stored similarly.

0	1	2	3	4	5	6	10	...	16	20	...	26	...	...	60	...	66	100	...	106	110	...

Fig. 4. One-dimensional storage vector showing address values for image I_0

4.2 Sparse Spiral Convolution

For a given image I_0, convolution of a hexagonal operator H_λ of "size" λ (having the size and shape of a layer λ cluster neighbourhood) across the entire image plane is achieved by convolving the operator sparsely with each of the seven images $I_j, j = 0,...,6$ and then combining the resultant outputs.

In each of the images $I_j, j = 0,...,6$, we apply the operator H_λ only when centred at those pixels with spiral address 0 (*mod* 7). Figure 5 shows a sample of pixels in image I_0 for which the label $j = 0,...,6$ for each pixel indicates in which of the images $I_j, j = 0,...,6$, the pixel address takes the value 0 (*mod* 7). Each pixel in image I_0 may be thus uniquely labeled.

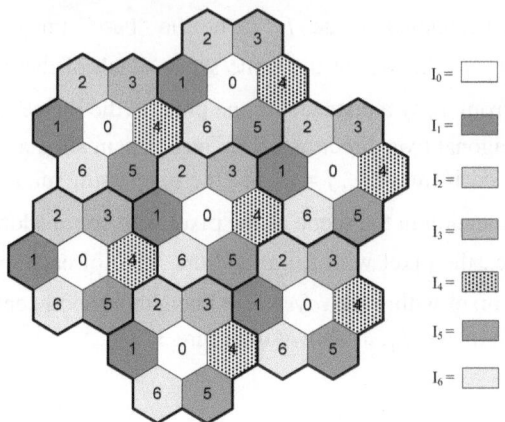

Fig. 5. Pixel positions in image I_0 corresponding to pixels in images $I_j, j = 0,...,6$ with address value 0 (*mod* 7).

We then define "base-7-zero" convolution
$$S_\lambda = H_\lambda \otimes_{7_0} I \tag{3}$$
of the hexagonal operator H_λ with the hexagonal image I by:

$$S_\lambda(s_0) = \sum_{s \in N_\lambda(s_0)} H_\lambda(s) \times I(s) \qquad \forall s_0 \in \{s | s = 0(\mathrm{mod}\,7)\} \tag{4}$$

and $N_\lambda(s)$ denotes the layer λ neighbourhood cluster centred on the pixel with spiral address s in image I. Thus in order to implement "eye tremor" simulation in relation to the base image I_0, we apply "base-7-zero" convolution ($\otimes_{7_0}$) of the hexagonal operator H_λ with each of the hexagonal images $I_j, j = 0,...,6$, thus generating seven output responses:

$$S_\lambda^j = H_\lambda \otimes_{7_0} I_j, j = 0,...,6 \tag{5}$$

that are combined to provide the consolidated response $E_\lambda = H_\lambda \otimes I_0$.

4.3   Layer 2 Cluster Neighbourhood Address Identification

As it is not appropriate to construct a Laplacian operator as small as Layer 1 due to its sensitivity to noise [12], we focus on a layer 2 Laplacian cluster operator. Application of a layer $\lambda = 2$ operator on a neighbourhood $N_2(s_0)$ requires identification of an ordered set of addresses of the centres of the layer 1 cluster neighbourhoods contained in $N_2(s_0)$. As $s_0 = 0 \, (mod \, 7)$ we may determine these 7 centres of layer 1 cluster neighbourhoods as:

$$c_\alpha = s_0 + 10 i_\alpha \, , \quad i_\alpha = 0,...,6 \tag{6}$$

For each layer 1 centre, c_α, the corresponding layer 0 cluster neighbourhood addresses are then simply given by the ordered set

$$\{c_\alpha + j\}_{j=0}^6 \tag{7}$$

From the above it can be seen that the amount of special hexagonal addition required to identify an ordered list of the pixel addresses in a layer 2 cluster neighbourhood $N_2(s_0)$ centred on a pixel with spiral architecture address s_0 ($s_0 = 0, mod \, 7$) is considerably less than would typically be required to identify such an address list for an arbitrary layer 2 cluster neighbourhood $N_2(s)$ with $s \neq 0, mod \, 7$. (For $s \neq 0, mod \, 7$, a full set of 49 special hexagonal additions would be required.) Hence, as demonstrated by the performance evaluation results presented in Section 6, our proposed approach using sparse "base-7-zero" convolution is significantly more efficient than standard spiral convolution.

5 Spiral Implementation

In terms of implementation using the one-dimensional vector structure for the images $I_j, j = 0,...,6$, that is facilitated by the spiral architecture, each output response $S_\lambda^j, j = 0,...,6$ is stored in a one-dimensional vector "*a-trous*" with non-empty values corresponding to the array positions with indices 0 (*mod* 7), as illustrated in Figure 6.

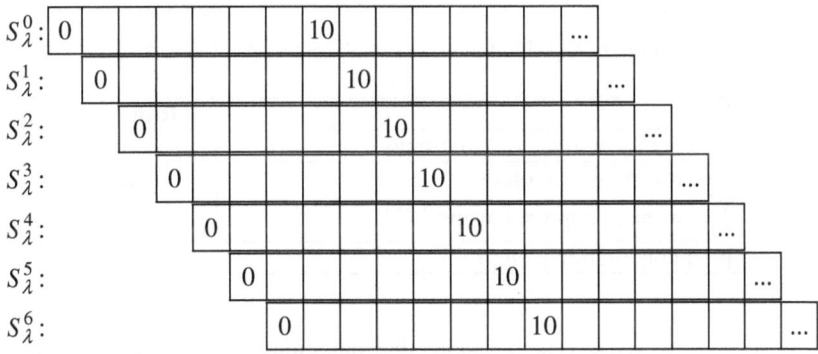

Fig. 6. Assembly of the one-dimensional vectors "*a-trous*" $S_\lambda^j, j = 0,...,6$

The one-dimensional vectors "*a-trous*" $S_\lambda^j, j = 0,...,6$, may then be assembled according to the "shifted" structure illustrated in Figure 6:

$$\forall s_0 \in \left\{s \mid s = 0(\text{mod}\,7)\right\}, \quad E_\lambda(s_0 + k) = S_\lambda^k(s_0) \quad \text{for } k = 0,...,6 \qquad (8)$$

to yield the consolidated output image $E_\lambda(I_0) = H_\lambda \otimes I_0$ as shown in Figure 7.

$S_\lambda^0(0)$	$S_\lambda^1(0)$	$S_\lambda^2(0)$	$S_\lambda^3(0)$	$S_\lambda^4(0)$	$S_\lambda^5(0)$	$S_\lambda^6(0)$	$S_\lambda^0(10)$	$S_\lambda^1(10)$	...	...

Fig. 7. Consolidated output image $E_\lambda(I_0) = H_\lambda \otimes I_0$ resulting from assembly of the vectors "*a-trous*" in Fig 6

6 Performance Evaluation

We present run-times for our proposed biologically motivated approach in comparison with standard convolution of an operator with a spiral image where the pixel neighbour addresses are found in two different ways: (i) via standard spiral convolution using special hexagonal and radix-7 addition; and (ii) neighbours are stored in a look-up table (LUT). The LUT takes 8.001s to generate for the spiral convolution approach and 0.091s for the eye tremor approach in the case of $\lambda = 2$, as the spiral convolution LUT requires all 49 addresses to be stored per record whereas the eye tremor LUT requires only the addresses of the centres of the seven $\lambda = 1$ sub-clusters. Algorithmic run-times were recorded for application of $\lambda=2$ Laplacian cluster operators to the *clock* image shown in Figure 8a. The hexagonal *clock* image was obtained by resampling the original square pixel-based image to a spiral image containing 117649 hexagonal pixels. The edge map and each of the seven "*a-trous*" gradient outputs used to generate it are also shown in Figure 8. The run-times presented in Table 1 are the averages over 100 runs using a workstation with a 2.99Ghz Pentium D processor and 3.50Gb of RAM.

Table 1. Algorithm run-times for 49-point operator ($\lambda = 2$)

Method	Run-time
Biologically motivated "eye tremor" approach	0.106s
Standard spiral convolution	25.087s
"Eye tremor" approach using LUT	0.028s
Spiral convolution using LUT	0.018s

The results in Table 1 demonstrate that our biologically motivated approach is approximately 250 times faster than standard spiral convolution. The implementation of both approaches can be accelerated by use of a LUT; this results in the run-time for the spiral convolution approach being faster than our approach but there is also, of course, the additional overhead of storing a larger LUT, which can be considerable for larger values of λ. More specifically though, the run-time of 0.028s for our approach is a combined time for processing seven "*a-trous*" gradient outputs; however, for video processing, as the spiral addressing in each frame in a sequence can be off-set slightly from its adjacent frames in a cyclic pattern, once the first seven *a-trous* images are processed to generate a complete frame, the addition of each subsequent *a-trous* image will generate a new complete frame. Hence each subsequent frame will be generated in one seventh of the time stated in Table 1, i.e., approx 0.004s. Therefore, our biologically motivated approach will be approximately four times faster than full implementation of the spiral convolution LUT approach when processing a stream of video images.

Additionally, using an adaptation of the Figure of Merit [1], we compare the accuracy of the Layer 2 Laplacian cluster operator (denoted as L2) with two Laplacian operators designed for use on standard rectangular-based images: the 7×7 Marr Hildreth operator (denoted as MH7) and the 7×7 Laplacian near-circular operator [2] (denoted as LNC7), which are both equivalent in size to the Layer 2 operator (L2, MH7 and LNC7 all have 49 mask values). In order to measure accurately the performance of the Laplacian cluster operator, we have modified the

(a)   (b)

(c)

Fig. 8. (a) Original image; (b) Completed spiral edge map; (c) S_0, an example of a corresponding edge map "*a-trous*"

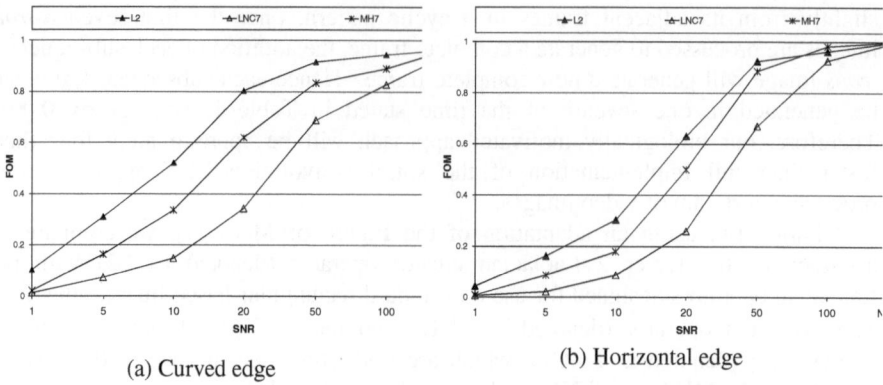

(a) Curved edge

(b) Horizontal edge

Fig. 9. Figure of Merit results comparing the tri-directional spiral operators (L2) with equivalent standard use of square operators (LNC7, MH7) using synthetic images containing (a) a curved edge; (b) a horizontal edge

well-known Figure of Merit technique in [1] to accommodate the use of hexagonal pixel-based images. The Figure of Merit results shown in Figure 9 illustrate that the proposed L2 spiral Laplacian operator has increased edge locational accuracy over the equivalent rectangular operators for all evaluated edge directions.

7 Conclusion

We present a biologically motivated approach to fast feature extraction using the concept of eye tremor. We have presented a design procedure for Laplacian cluster operators for use within our fast framework. The Figure of Merit results show that the Laplacian cluster operator provides better edge detection performance than the other square Laplacian masks of equivalent size. We have demonstrated that the approach of generating feature maps *"a-trous"*, using Layer 2 cluster operators, that can be combined into a single complete feature map is significantly faster than standard convolution or use of a neighbourhood LUT on hexagonal images. Generalisation of the approach to enable application of Laplacian operators at various scales will be the focus of future work.

References

[1] Abdou, I., Pratt, W.: Quantitative design and evaluation of enhancement/thresholding edge detectors. Proceedings of the IEEE 67(5), 753–763 (1979)
[2] Coleman, S.A., Gardiner, B., Scotney, B.W.: Adaptive Tri-Direction Edge Detection Operators based on the Spiral Architecture. In: IEEE ICIP, pp. 141–144 (2010)
[3] Coleman, S.A., Scotney, B.W., Herron, M.G.: A Systematic Design Procedure for Scalable Near-Circular Laplacian of Gaussian Operators. In: ICPR, pp. 700–703 (2004)
[4] Curcio, C.A., et al.: Human Photoreceptor Topography. Journal of Comparative Neurology 292, 497–523 (1990)

[5] Dacey, D.M., Packer, O.S.: Receptive Field Structure of h1 Horizontal Cells in Macaque Monkey Retina. Journal of Vision 2(4), 279–292 (2000)

[6] Hirsch, J., Miller, W.H.: Does Cone Positional Disorder Limit Resolution? Journal Optical Soc. of America A: Optics, Image Sci, and Vision 4, 1481–1492 (1987)

[7] Middleton, L., Sivaswamy, J.: Hexagonal Image Processing; A Practical Approach. Springer, Heidelberg (2005)

[8] Mollon, J.D., Bowmaker, J.K.: The Spatial Arrangement of Cones in the Primate Fovea. Nature 360, 677–679 (1992)

[9] Roka, A., et al.: Edge Detection Model Based on Involuntary Eye Movements of the Eye-Retina System. Acta Polytechnica Hungarica 4(1), 31–46 (2007)

[10] Sheridan, P.: Spiral Architecture for Machine Vision. Ph. D. Thesis, University of Technology, Sydney (1996)

[11] Shima, T., Saito, S., Nakajima, M.: Design and Evaluation of More Accurate Gradient Operators on Hexagonal Lattices. IEEE Trans. PAMI 32(6), 961–973 (2010)

[12] Vernon, D.: Machine Vision. Prentice Hall International (UK) Ltd, Englewood Cliffs (1991)

Entropy-Based Localization of Textured Regions

Liliana Lo Presti and Marco La Cascia

University of Palermo
lopresti@dinfo.unipa.it

Abstract. Appearance description is a relevant field in computer vision that enables object recognition in domains as re-identification, retrieval and classification. Important cues to describe appearance are colors and textures. However, in real cases, texture detection is challenging due to occlusions and to deformations of the clothing while person's pose changes. Moreover, in some cases, the processed images have a low resolution and methods at the state of the art for texture analysis are not appropriate.

In this paper, we deal with the problem of localizing real textures for clothing description purposes, such as stripes and/or complex patterns. Our method uses the entropy of primitive distribution to measure if a texture is present in a region and applies a quad-tree method for texture segmentation.

We performed experiments on a publicly available dataset and compared to a method at the state of the art[16]. Our experiments showed our method has satisfactory performance.

1 Introduction

In many applications, it is required a proper object description to enable recognition and/or classification. When the object is mainly a person, such description is related to the appearance and can be used to solve the people re-identification problem, very common in domain such as surveillance. There are many cues that can be used to perform re-identification; an approach could be to focus on the face and use facial features. However, this approach requires the face is adequately visible, and this is not the general case. Moreover, face descriptors are generally affected by pose and illumination changes so that the re-identification should be performed by using also other cues, i.e colors and textures in clothes.

Indeed, in case of people re-identification, clothing has an important role particularly when it presents some evident texture; e.g. "a person wearing a shirt with white and red stripes". Intuitively, an appearance description aiming to capture such properties would be more discriminative than a simple bag of words [7] color description.

In the following we describe a method to discover "salient" structured areas in an image of a person that can be interpreted as "texture" characterizing person's clothing. Persons' clothing can be described by their colors and characteristics such as stripes, text and, broadly speaking, textures. To describe these kinds of

G. Maino and G.L. Foresti (Eds.): ICIAP 2011, Part I, LNCS 6978, pp. 616–625, 2011.

properties, features robust to illumination changes should be used. For detecting texture we employ edge-based primitives, as they tend to be invariant to illumination. We use an approach inspired in some respects by the method of Kadir and Brady [15] for salient point detection. As we will explain later, we use an entropy based approach for detecting both the scale and the area of the texture in the object instance. In the method of Kadir and Brady, a strong response was obtained for textured areas that would result in false salient points; what was a limitation of their method is an advantage: for detecting interesting regions on a person's clothing that can ease object instance discrimination, the saliency measure is helpful, in the sense that a texture can be interpreted as a salient region; for example, considering a flat region, that is an area where no texture is present, a textured area would be salient inside this region.

The plan of the paper is as follows. In Section 2 we present related works and discuss important applications for the proposed method. In Sections 3 and 4, we present our method and discuss implementation details. In Sections 5 we present experimental results we got on a publicly available dataset and comparison to a method at the state of the art [16]. Finally, in Section 6 we present conclusions and future directions for our work.

2 Related Works

Appearance descriptors have an important role in establishing correspondences in multi-camera system to perform consistent labeling [6]. In [23], each object is represented as a "bag-of-visterms" where the visual words are local features. A model is created for each individual detected in the site. Descriptors consist of 128-dimensional SIFT vectors that are quantized to form visual words using a predefined vocabulary. In [18], appearance is modeled as bag of words in which a latent structure of features is recovered. A latent Dirichlet allocation (LDA) machine is used to describe appearance and discover correspondences between persons' instances.

The person re-identification task is not restricted just to video-surveillance systems but it is a recurrent problem in multimedia database management. A particular case is, for example, photo collection organization. In such application, Content Based Image Retrieval (CBIR) techniques and face features can be integrated in a probabilistic framework to define clusters of photos in order to ease browsing the collection [1]. In some works [5,9,19], once the face is detected, the region under the face is used to compute information about the clothing of the person. In [21], face and clothing information are used for finding persons in many photos. First, a hierarchical clustering method is used for finding the set of persons whose face was detected in the photo sequence. Then, a clothing model is estimated for each person and used to recover mis-detected person's instances.

There are other applications for texture detection and recognition in clothing. Recently, in [24] a new method for clothes matching to help blind or color blind people has been presented. The method handles clothes with multiple colors and complex patterns by using both color and texture information.

In [8], the focus is on learning attributes, which are visual qualities of objects, such as red, striped, or spotted. To minimize the human effort needed to learn an attribute, they train models from web search engines. Once a model is learnt, it is capable of recognizing the attribute and determine its spatial extent in novel images. However, it seems impractical to enumerate all the possible kinds of textures that can arise in images. We believe, instead, the problem should be addressed at a lower visual level.

Texture detection and description are relevant problem, and a huge literature exists on the subject. Some works concern about texture detection for object segmentation [4,25] and classification [20]. A common problem when working with texture is the scale selection [17,13]. A texture can be detected and described by local properties and a proper neighborhood must be chosen. In particular, in [13] an approach based on the entropy of local patches is combined with a measure of the difference between neighboring patches to determine the best textel size for the texture.

In [16], a method to detect and localize instances of repeated elements in a photo is presented. Such elements may represent texels and the method is meant to detect repeated elements such as stripes in person's clothing. In this work, we consider a similar application. We use a multi-scale approach to find the best scale at which a texture can be detected. Texture is localized by using the entropy of some primitive and is segmented by a quad-tree strategy.

3   Problem Definition

Many definitions have been proposed for texture: Haralick [12] defines a texture as an organized area for which it exists a spatial distribution of a "primitive" defining some sort of ordered structure [22]. For Gibson, a texture is a characteristic based on which an area can appear as homogeneous [10].

The main goal in this work is to detect a predominant and well visible texture – for example, text or stripes on clothing – that can be used for describing the person appearance. Methods in literature, as for example [11], are not completely suitable for our application as they are generally conceived for high resolution images; moreover, generally objects are moving and undergoing articulated deformations so that fine details are not generally distinguishable and only rough texture can be discriminated.

Many methods for texture detection consider properties about the pixel intensity [13] or try to recognize specific patterns that can arise. However, the kind of textures we are interested in can be generally detected by exploring color organization with particular attention to their structure repeated over the space. As also proposed in other works [4], this structure can be highlighted, for example, by the edges in the image; therefore, an analysis of the detected edges can put in evidence some properties of the texture itself. Other properties can be explored too, as in the case of Law's energy measures where the basic features are: the average gray level, edges, spots, ripples and waves [2].

To segment a textured area, we need to measure how much a certain pixel belongs to a texture. This problem is strongly connected to the ability to detect the texel and the natural scale of a texture [22]. For natural scale we mean the smallest size of the structure (texel) that can be spatially replicated to reproduce the texture itself [22,2].

In our approach, we use the entropy of a certain primitive to automatically detect the scale to use for segmenting the texture; then we use a split and merge approach for localizing and segmenting the texture.

4   Texture Detection

Detecting a texture requires reference to local properties measured in a suitable neighborhood. The size of this neighborhood is related – but not necessary equal – to the natural scale of the texture. In our approach, the scale of the texture is related to the size of the neighborhood that permits to have the best texture detection. To measure the presence of a texture, it makes sense to look at the local disorder measured for each pixel in an appropriate primitive space with respect to flat and ordered area. The most natural way is to use the entropy of these primitives; intuitively, the entropy will be maximum in textured areas. Other approaches in literature use the entropy for detecting the texture, as for example in [13], where the entropy is combined with other measures to detect textured area based on pixel intensities.

Given the probability distribution P for a set of N primitives, the entropy E is defined as:

$$E = \frac{1}{N} \sum_{i=1}^{N} P_i \cdot \log P_i \tag{1}$$

Algorithm 1 summarizes the general framework we used for detecting pixel candidates to belong to a texture. In general, for each possible scale, we compute pixel per pixel the local entropy by using the statistic of the selected set of primitives in a suitable neighborhood. Points belonging to a texture have a very high entropy that decreases when the area becomes more and more flat. To segment the textured area, we take into account the spatial distribution of the pixels in the image: pixels near and with high entropy are much more probable to belong to a textured region.

4.1   Primitives for Texture Detection

The method described until now is quite general and applicable to different kinds of features as, for example, Gabor filters, Law's energy measures or features based on Local Binary Pattern [20]. In this paper we use as primitive the orientation of the edges, even if any other primitive could be used.

Algorithm 1. Scale and Texture Detection

---

for scale s in range $[1, Max]$ **do**
  for each pixel (x, y) **do**
    compute primitives in Neighborhood $N = \{(x - s, y - s); (x + s, y + s))\}$
    compute entropy $E(x, y)$ of the primitive probability distribution in N
  end for
  apply split and merge algorithm on matrix E to detect textured area
  store area corresponding to maximum average entropy (MAE)
  store MAE
end for
select scale s corresponding to maximum MAE
select the textured area corresponding to s

Algorithm 2. Computing Edge-based Primitives

---

apply Canny detector to compute edges in the image
discard edges in flat area by applying an adaptive threshold on local standard deviations
set orientations to 0 for all pixels in the image
for each pixel on an edge **do**
  compute orientation, assign a value in $[1, 5]$ depending on the estimated direction
end for
for each pixel (x, y) **do**
  compute orientation histogram in Neighborhood $N = \{(x - s, y - s); (x + s, y + s))\}$
end for
return as Primitives the orientation histograms

As we already said, a texture can be detected by analyzing properties of the detected edges. A texture is characterized by the presence of an organization/structure that can be detected considering the orientations of its edges. For each pixel on an edge we compute the orientation and then we quantize it so to consider only 4 predominant orientations: horizontal, vertical and obliques (that is $\pm 45 deg$). Orientations are computed by using the gradient components of the gray level image. Algorithm 2 reports the pseudo-code for the extraction of this kind of primitives. Not all the detected edges belong to a texture but some of them are in flat areas, in correspondence to wrinkles in the clothing or self-occlusions of the person. To improve the edge detection, we filter them to remove edges in areas with homogeneous colors. The remaining edges are then used to compute the statistic of the orientations in a local neighborhood. The size of this neighborhood is related to the scale s the texture is detected to. Given a neighborhood and the orientations of the edges in it, we computed an histogram of 5 bins: 1 for counting pixels in flat areas, and 4 for considering the orientations of the edges within the neighborhood. Based on this histogram, for each pixel we computed the entropy associated to the pixel within a neighborhood of size $2 \cdot s$ (the scale).

4.2   Texture Segmentation

We segment the textures by a split and merge approach to construct a quad-tree. For splitting, we consider how much homogeneous each quad is and we used the local standard deviation of the entropy assigned to each pixel as metric. The minimum quad size during decomposition has been set to 8×8 pixels, while the threshold on the standard deviation has been set to 0.1.

For the merging step, instead, we merge all the neighbor quad regions that have similar entropies. In this case, the threshold has been set to 0.3. The success of the merging step depends strongly on the order with which the quads are merged together. We start the merging phase from the quad presenting the highest entropy.

For selecting the correct scale, we considered the neighborhood size that gives the best texture representation, that is we choose the scale corresponding to the maximum average entropy in the detected area. At the selected scale, we consider the region with the highest entropy, and we apply a threshold τ_E to determine if such region can be classified as texture.

Fig. 1a shows an example with an artificial texture where each quad is 10 pixel large (so that the natural scale should be 20 pixels). Figure 1b shows how the expected entropy in the textured area varies with respect to the scale. As the figure shows, the entropy has a periodic trend depending on the scale of the texture. The selected scale is chosen to be the first peak.

In real case images, however, the trend of the entropy is not so regularly periodic because of the noise in the images, and the deformation to which primitives generally undergo due to different orientations and person's pose. In this case, we consider the scale corresponding to the maximum value of the entropy.

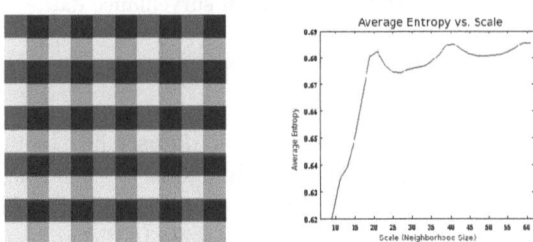

Fig. 1. Artificial texture and average entropy at different scales

4.3   Implementation Details

Fig. 2 shows two examples of texture detection on surveillance data. The goodness of the detection strongly depends on the operator used for the edge detection and the filtering step. In our implementation, we computed edges by considering the maximum response of a Canny detector applied to each color channel. Filtering was performed discarding all the edges in color flat area. This area were determined by looking at the maximum local standard deviation of the channel intensities in a neighborhood of size 5 × 5 pixels. We applied a local threshold to maintain all the edges having standard deviation in the 65% confidence interval. In our implementation, the threshold τ_E on the entropy has been set empirically to 0.4.

Performance of the multi-scale approach has been considerably improved by using integral images when computing the histogram for each orientation. In our experiments, when a texture is detected, the corresponding scale changes according to the distance of the object from the camera; we noted that in sequences of frames, it is possible to track the scale itself. In this case, once a scale is detected, the corresponding value may be used as prior for the next frame to process so that not all the range of possible values for the scale is spanned but only those values in a neighborhood of the prior scale value, speeding up the computation.

We also note that the texture detector fails when there are too strong features in the image not related to a texture, for example when there are too evident wrinkles in clothing or a cluttered background, and the filtering step is not able to classify the area around the edge as homogeneous. Another case of failure arises, of course, in case the edge detector fails.

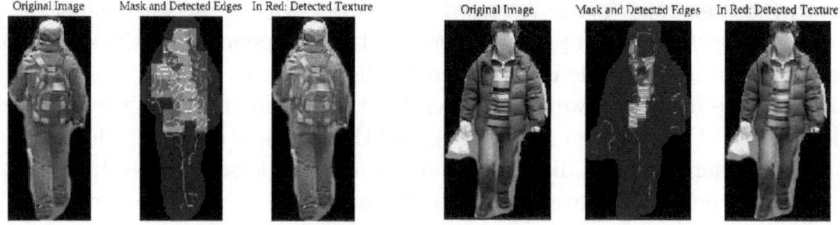

Fig. 2. Textures detected on surveillance data

5   Evaluation

We tested our method on a publicly available dataset [9]. To measure the performance, we randomly choose a subset of images. For each image, we detected the face and considered the region under the face as possible clothing region. We considered a subset of 50 persons. Of these images, 16 did not have any textured region in the clothing, while the remaining 34 had. For each image, we manually detected the region corresponding to the texture in the clothing (see fig. 3).

Fig. 3. Test images and corresponding manually segmented ground-truth

As the detection method works on local properties, we measured the performance by dividing the image in sub-windows (blocks) of 30×30 pixels. For each block, we computed the balanced accuracy defined as:

$$BA = \frac{\text{sensitivity} + \text{specificity}}{2} = \frac{1}{2} \cdot \frac{TP}{TP + FN} + \frac{1}{2} \cdot \frac{TN}{TN + FP} \quad (2)$$

where TP, TN, FP, and FN are true positive, true negative, false positive and false negative respectively. This performance measure treats both classes (the positive and the negative ones) with equal importance [3], and it is particularly appropriated in our case as we want to measure the ability to correctly classify a pixel as belonging to a texture or not.

We considered correctly classified a sub-window for which the balanced accuracy was greater than a threshold τ_a and, for each image, we measured the ratio of the correctly classified blocks over the total number of sub-windows.

For comparison purposes, we tested our method against the one in [16]. This method is conceived to detect and group repeated scene elements representing complex patterns or stripes from an image. First, the method detects interesting elements in the image by analyzing the structure tensor; such candidates are matched against their neighbors estimating the affine transform between them. The elements are grown and iteratively grouped together to form a distinctive unit (for more details, refer to[16]). In Fig. 4, the graph shows the curves of the average classification rate over the whole dataset for different values of the threshold τ_a for our method and for the one in [16]. $\tau_a = 1$ represents the ideal case the texture is always correctly detected and in such a case our method shows a classification rate of about 64% while the other method has a classification rate equals to 51.3%. For a balanced accuracy of about 80% ($\tau_a = 0.8$), our method shows a classification rate of 79.27% while the other method has a classification rate equals to 61.5%. In comparison to [16], our method presents an higher specificity and showed to be more robust when detecting text and complex patterns. The method in [16] suffers from the fact that the scale is not automatically detected and it is unable to detect those regions where the texel size is large. On the contrary, in our method the scale is automatically detected by evaluating the entropy as already explained in section 4.

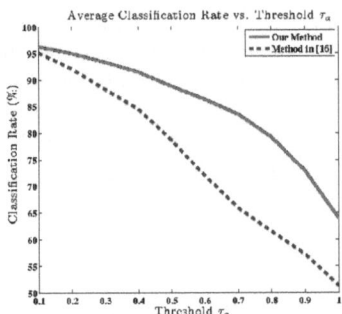

Fig. 4. Comparison between our method and the one in [16]

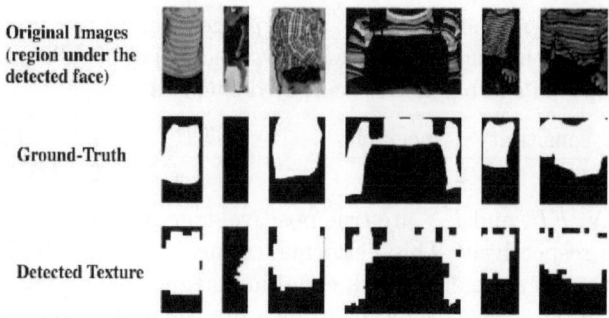

Original Images
(region under the
detected face)

Ground-Truth

Detected Texture

Fig. 5. Results by applying our method on the test images

Fig. 5 shows an example of results we got by applying our method. As the figure shows, the method can be sensitive to the background (see second image in the figure). However, this problem could be limited using proper algorithm to detect persons. In surveillance application, background suppression could be adopted. In photo collections, the method in [9] can be used to approximate the clothing region, then our method can be used to find the textured area.

6   Conclusion and Future Works

In this paper, we proposed a new method to detect textured area in clothes region. Our method computes the predominant orientation in a neighborhood for each pixel and uses entropy to capture the disorder associated to such a primitive distribution. A split and merge approach is then used to segment the textured area. Our method is able to capture 2D structures representing complex (i.e. text) or regular patterns (i.e. stripes) in clothing regions. We deal with the problem of selecting the scale of the texture automatically by adopting a multi-scale approach and using the entropy as measure of the goodness of such selection. In future works, we will study how the entropy distribution can be used also for description purposes and if it is possible to use it to perform classification. We will also study techniques for fusing color and texture information to enhance person re-identification by using probabilistic frameworks.

References

1. Ardizzone, E., La Cascia, M., Vella, F.: A novel approach to personal photo album representation and management. In: Proc. of Multimedia Content Access: Algorithms and systems II. IS&T/SPIE Symp. on EI (2008)
2. Ballard, D.H., Brown, C.: Computer Vision. Prentice-Hall, Englewood Cliffs
3. Brodersen, K.H., Ong, C.S., Stephan, K.E., Buhmann, J.M.: The balanced accuracy and its posterior distribution. In: Proc. of Int. Conf. on Pattern Recognition (ICPR), pp. 3121–3124 (2010)
4. Carson, C., Belongie, S., Greenspan, H., Malik, J.: Blobworld: image segmentation using expectation-maximization and its application to image querying. Trans. on Pattern Analysis and Machine Intelligence 24(8), 1026–1038 (2002)
5. Cooray, S., O'Connor, N.E., Gurrin, C., Jones, G.J.F., O'Hare, N., Smeaton, A.F.: Identifying Person Re-Occurences For Personal Photo Management Applications. In: IET Int. Conf. on Visual Information Engineering (VIE), pp. 144–149 (2006)

6. Doretto, G., Sebastian, T., Tu, P., Rittscher, J.: Appearance-based person reiden-tification in camera networks: problem overview and current approaches. Journal of Ambient Intelligence and Humanized Computing, 1–25 (2010)
7. Fei-Fei, L., Perona, P.: A Bayesian Hierarchical Model for Learning Natural Scene Categories. In: Proc. of Computer Vision and Pattern Recognition, pp. 524–531 (2005)
8. Ferrari, V., Zisserman, A.: Learning visual attributes. Advances in Neural Information Processing Systems (2008)
9. Gallagher, A.C., Chen, T.: Clothing cosegmentation for recognizing people. In: Proc. of. Computer Vision and Pattern Recognition (CVPR), pp. 1–8 (2008)
10. Gibson, J.J.: The Perception of the Visual World. Houghton Mifflin, Boston (1952)
11. Han, B., Yang, C., Duraiswami, R., Davis, L.: Bayesian Filtering and Integral Image for Visual Tracking. In: IEEE Proc. of Int. Workshop on Image Analysis for Multimedia Interactive Services, WIAMIS 2005 (2005)
12. Haralick, R.M.: Statistical and structural approaches to texture. In: Proc. IEEE (1979)
13. Hong, B.-W., Soatto, S., Ni, K., Chan, T.: The scale of a texture and its application to segmentation. In: Proc. of Int. Conf. on Computer Vision and Pattern Recognition (CVPR), pp. 1–8 (2008)
14. Huang, T., Russell, S.: Object Identification in a Bayesian Context. In: Proc. the Int. Joint Conf. on Artificial Intelligence (IJCAI), pp. 1276–1283 (1997)
15. Kadir, T., Brady, M.: Saliency, Scale and Image Description. Int. J. Comput. Vision 45, 83–105 (2001)
16. Leung, T., Malik, J.: Detecting, localizing and grouping repeated scene elements from an image. In: Buxton, B.F., Cipolla, R. (eds.) ECCV 1996. LNCS, vol. 1064, pp. 546–555. Springer, Heidelberg (1996)
17. Lindeberg, T.: Feature Detection with Automatic Scale Selection. Int. Journal of Computer Vision 30(2), 79–116 (1998)
18. Lo Presti, L., Sclaroff, S., La Cascia, M.: Object Matching in Distributed Video Surveillance Systems by LDA-Based Appearance Descriptors. In: Foggia, P., Sansone, C., Vento, M. (eds.) ICIAP 2009. LNCS, vol. 5716, pp. 547–557. Springer, Heidelberg (2009)
19. Lo Presti, L., Morana, M., La Cascia, M.: A Data Association Algorithm for People Re-Identification in Photo Sequences. In: IEEE Proc. of Int. Symposium on Multimedia (ISM), pp. 318–323 (2010)
20. Ojala, T., Pietikainen, M., Harwood, D.: A comparative study of texture measures with classification based on featured distributions. Pattern Recognition 29(1), 51–59 (1996)
21. Sivic, J., Zitnick, C., Szeliski, R.: Finding people in repeated shots of the same scene. Proc. BMVC 3, 909–918 (2006)
22. Schwartz, J.: Studies in Visual Perception, IV - Homogeneous Textures, New York University
23. Teixeira, L.F., Corte-Real, L.: Video object matching across multiple independent views using local descriptors and adaptive learning. Pattern Recogn. Lett. 30(2), 157–167 (2009)
24. Tian, Y., Yuan, S.: Clothes matching for blind and color blind people. In: Proc. of Int. Conference on Computers helping people with special needs, pp. 324–331. Springer, Heidelberg (2010)
25. Voorhees, H., Poggio, T.: Computing texture boundaries from images. Nature 333(6171), 364–367 (1988)

Evaluation of Global Descriptors for Large Scale Image Retrieval

Hai Wang and Shuwu Zhang

Institute of Automation Chinese Academy of Sciences
haiwang@hitic.ia.ac.cn, swzhang@hitic.ia.ac.cn

Abstract. In this paper, we evaluate the effectiveness and efficiency of the global image descriptors and their distance metric functions in the domain of object recognition and near duplicate detection. Recently, the global descriptor GIST has been compared with the bag-of-words local image representation, and has achieved satisfying results. We compare different global descriptors in two famous datasets against mean average precision (MAP) measure. The results show that Fuzzy Color and Texture Histogram (FCTH) is outperforming GIST and several MPEG-7 descriptors by a large margin. We apply different distance metrics to global features so as to see how the similarity measures can affect the retrieval performance. In order to achieve the goal of lower memory cost and shorter retrieval time, we use the Spectral Hashing algorithm to embed the FCTH in the hamming space. Querying an image, from 1.26 million images database, takes 0.16 second on a common notebook computer without losing much searching accuracy.

1   Introduction

There are more and more images in our daily life. It is of great significance to find the one needed among a large number of images. The content based image retrieval (CBIR) may be just a solution to this problem, which is a prosperous researching field. See a recent survey [3] for a deep understanding.

The CBIR retrieval process usually follows a similar pattern. Firstly, an image is represented by features, either a vector of global features like several MPEG-7 image descriptors or a set of local image features like SIFT [7]. After an image is represented by features, a similarity measure is proposed to calculate the similarity between images. Usually, the image representation and the distance measure should be considered simultaneously; Secondly, to tradeoff between effectiveness and efficiency, an indexing scheme has to be proposed to tackle the dilemma of the large scale image database and the requirement of a real-time response time.

Currently, in the field of the near duplicate detection and object recognition, the bag-of-words features based on local image descriptors have gained most of the attention, and have achieved some success, like [11,12,6]. However, the local image features take a long time to extract. When performing the visual key words generation process like the k-means clustering, it will consume a lot of time to deal with large database. At the same time, when the number of visual

G. Maino and G.L. Foresti (Eds.): ICIAP 2011, Part I, LNCS 6978, pp. 626–635, 2011.
© Springer-Verlag Berlin Heidelberg 2011

words is very large, for example, millions or even larger, the new comer image to be retrieved will take lots of time to compare with each visual word in order to get the bag-of-words representation. Although some ingenious methods like hierarchy quantization method Vocabulary Tree [9] have been proposed to reduce the bag-of-words quantization time, the quantization error has also increased. Besides, because each image has a set of local descriptors, ranging from hundreds to thousands of dimensions, the storage space for these features is very huge.

Considering the near duplicate images often share most of the same appearances, only some small parts change significantly. One vector of a global representation may suffice to depict the specific image, which indeed has the merit of easy computing and storage efficient. The global descriptors also have the merits of no need to take a long time and use a large dataset to train the bag-of-words model. In spite of these merits, the global features seem to be forgotten in the domain of object recognition and near duplicate detection.

Recently,the authors [5] evaluate the GIST descriptor [10] in the web-scale image search, which has achieved fairly exciting results. This encourages us to evaluate different global features against two famous datasets with ground truth. The results show that The GIST descriptor is indeed a better choice than several global MPEG-7 descriptors, see [1] for an overview, like Color Layout Descriptor, Edge Histogram Descriptor and Scalable Color Descriptor, but it seems that the FCTH [2] a fuzzy color and texture histogram outperforms GIST by a large margin with fewer dimensions of feature. FCTH feature only needs 72 bytes, while the GIST descriptor needs 960 floating numbers. The FCTH descriptor is also much efficient by using a simple similarity measure compared to the GIST descriptor, which using L_2 similarity measure. Considering this in a context of millions of images to be compared, this little promotion of performance will save a lot of computation resources as well as lots of time, which may make the retrieval to be processed in real time.

In this paper, we compare different global image features using the MAP protocol against two famous datasets with ground truth. We evaluate different similarity measures for two effective global features GIST and FCTH. The results show that the FCTH is outperforming GIST and several MPEG-7 descriptors. We propose to use the L_1 similarity measure for both the GIST and FCTH, considering the better performance and lower computational complexity. At the end, we use the state-of-art Spectral Hashing to represent the FCTH feature in the hamming space. We present the results of the scalability of using Spectral Hashing algorithm in large scale image retrieval context.

The rest of the paper is organized as follows. It starts with the image descriptors and similarity measures in Section 2, and then in Section 3 we give a short introduction to the Spectral Hashing algorithm and use it to derive the hamming features for retrieval. In Section 4 we show the datasets and measure to evaluate the performance of the retrieval results. In Section 5 we list the experiments we are performing and give the evaluation results. Conclusions are presented in Section 6.

2   Descriptors and Similarity Measure

2.1   Image Descriptors

In this section, we give a brief description on the image features and distance functions we are going to evaluate.

FCTH feature, which includes color and texture information in one histogram, is very compact and only needs 72 bytes to characterize it. This feature is derived from the combination of 3 fuzzy systems. To compute this feature, the image is initially segmented into blocks. For each block, a 10-bin histogram is generated from the first fuzzy system. The 10-bin histogram is derived from 10 preselected colors in the HSV color space. This histogram is then expanded to 24-bins using the second fuzzy system by including hue-related information for each color. For each image block, a Haar Wavelet transform is applied to the Y component. After a one-level wavelet transform, each block is decomposed into four frequency bands, and the coefficients of the three high frequency bands HL, LH, and HH are used to compute the texture features. The intuition for using these three high frequency bands is that each of them reflects the texture changing directions. After using the third fuzzy system, the histogram is expanded to 192-bins by integrating the extracted texture information and the 24-bins color information. A quantization is applied to limit the final length of the feature descriptor to 72 bytes per image.

GIST feature is based on a low dimensional representation of the scene, by-passing the segmentation and the processing of individual objects or regions. The authors propose a set of perceptual dimensions (naturalness, openness, roughness, expansion, ruggedness) that represent the dominant spatial structure of a scene. The descriptor is gained as follows: the image is segmented by a 4×4 grids, and the orientation histograms are extracted.

MPEG-7 Color Layout Descriptor(CLD) is designed to represent the spatial color distribution of an image in YCbCr color space. This feature is obtained by applying the discrete cosine transform (DCT) in a 2-D image space. It includes five steps to compute this descriptor: (1) partition image into 8×8 blocks; (2) calculate the dominant color for each of the partitioned blocks; (3) compute the DCT transform; (4) nonlinear quantizate the DCT coefficients; (5) zigzag scan of the DCT coefficients.

MPEG-7 Edge Histogram Descriptor (EHD) is describing spatial distribution of four directional edges and one non-directional edge in the image. An image is divided into non-overlapping 4×4 sub-images. Then, from each sub-image an edge histogram is extracted, each sub-image histogram consists of 5 bins with vertical, horizontal, 45-degree diagonal, 135-degree diagonal, and non-directional edge types. Each image is represented by an edge histogram with a total of 80 ($4\times4\times5$) bins.

MPEG-7 Scalable Color Descriptor(SCD) is a color histogram in HSV color space encoded by Haar Transform. SCD aims at improving storage efficiency and computation complexity. Usually the number of bins can span from 16 to 256.

2.2 Similarity Measure

In terms of the CLD, EHD, SCD and FCTH, we use the excellent image retrieval LIRe [8] framework to extract these features, and for CLD, EHD and SCD, we use the default similarity measure to measure the similarity between images. For the FCTH and GIST features, from the later experiment results, we can clearly see their better performance, so we compare different similarity function including L_1, L_2, Histogram Intersection(HI), Tanimoto (T) [2] and evaluate the retrieval results.

$$L_1(x, y) = \sum_{i=1}^{d} \|x_i - y_i\| \tag{1}$$

$$L_2(x, y) = \sqrt{\sum_{i=1}^{d} (x_i - y_i)^2} \tag{2}$$

$$HI(x, y) = 1 - \frac{\sum_{i=1}^{d} \min(x, y)}{\min(\sum_{i=1}^{d} x_i, \sum_{i=1}^{d} y_i)} \tag{3}$$

$$T(x, y) = \frac{x^T y}{x^T x + y^T y - x^T y} \tag{4}$$

3 Image Indexing Scheme

In this section, we present the image indexing scheme used in this paper to solve the problem of retrieval from a large scale image dataset. We use the state-of-art technique Spectral Hashing [13] to map features into hamming space, and apply the hamming distance to compare image similarities. The computing of hamming distance runs fairly fast in that it only needs bits processing. Furthermore features embedded into the hamming space are very distance preserving, which means that the similar data points in the original feature space will also be mapped nearly in the hamming space. The result will be shown later. Next we give a brief introduction to Spectral Hashing.

3.1 Spectral Hashing

In [13] the authors aim at designing a code which has three properties: (1) is to compute easily for a novel input; (2) is that the code should be compact which only take a small number of bits to represent the feature; (3) maps similar items to similar binary code-words. Considering these properties the authors seek to minimize the average Hamming distance between similar points as follows:

$$\text{Minimize}: \quad \sum_{ij} W_{ij} \|y_i - y_j\|^2 \tag{5}$$

$$y_i \in \{-1, 1\}^k$$
$$Subject\ to: \quad \sum_i y_i = \quad 0$$
$$\frac{1}{n} \sum_i y_i y_i^T = \quad 1$$

Where $\{y_i\}_{i=1}^n$ is the n data-points embedded into hamming space with the length of k, and $W_{n \times n}$ is the distance matrix from the original space. There are three constraints, each of which requires the code should be binary. Every bit has probability 0.5 to equal 1, and the bits should be uncorrelated.

The direct solution to the above optimization is non-trivial since even a single bit binary code is a balanced graph partition problem, which is NP hard. The authors relax the constraints, and the relaxed problem can be efficiently solved by using spectral graph analysis. Further, the authors assume that the data-points are sampled from a multidimensional uniform distribution, which means that the probability distribution p(x) is a separable distribution. After this assumption the out of samples problem can be efficiently solved by a closed form solution not using the Nystrom method which computes linearly by the size of the database for a new point.

The final Spectral Hashing algorithm has two input parameters. One is a list containing n data points, and each one is represented by a d-dimensional vector; the other is the number k, using k binary bits to represent the final embedded hamming feature. The algorithm has three main steps: (1) finding the principal components of the data using PCA; (2) for each coordinate of the final k bits, assume the data distribution are uniform and learn analytical eigenfunction by a sinusoidal function; (3) threshold the analytical eigenfunction to obtain binary codes.

4   Datasets and Evaluation Protocol

4.1   Datasets

We have used two famous evaluation datasets with ground truth, the University of Kentucky dataset and the INRIA Holidays dataset. Apart from the two datasets with ground-truth manual annotations, we also use the large scale IMAGENET dataset as distracting images to evaluate the performance of different image descriptors and the indexing scheme in a large scale dataset.

The University of Kentucky Recognition Benchmark Images [11]. This dataset contains 10200 images altogether, with 4 images in a group to depict either the same object or the same scene from different viewpoints. When searching an image, the first four images should be the images in that group.

INRIA Holidays dataset [6], this dataset mainly contains personal holiday photos. The remaining ones are taken on purpose to test the robustness against various transformations: rotations, viewpoint and illumination changes, blurring, etc. The dataset includes a very large variety of scene types (natural, man-made, water and fire effects, etc.) and images are of high resolution. The dataset contains 500 image groups, each of which represents a distinct scene. The first image of each group is the query image and the correct retrieval results are the other images in the group.

IMAGENET Large Scale Visual Recognition Challenge 2010 [4]. We use a subset of 1256612 images from the datasets training set of the JPEG format. The number of images for each category ranges from 668 to 3047.

4.2 Evaluation Protocol

To evaluate performance we use Average Precision, computed as the area under the precision-recall curve. Precision is the number of retrieved positive images relative to the total number of images retrieved. Recall is the number of retrieved positive images relative to the total number of positives in the database. We compute an Average Precision score for each of the query image, and then average these scores to obtain a Mean Average Precision (MAP) as a single value to evaluate the results. The bigger the number is, the better the performance is.

5 Experiments

5.1 Evaluate Global Features

At first, we evaluate the different global features listed in the Section 2. For the GIST we scale the image to 128×128 pixels, then use the implementation in [10] to extract the 960 dimensions feature vector. For other features, like FCTH, SCD, EHD, and CLD, we use the wonderful package LIRe [8] to extract these features and use the default similarity measures to calculate the similarity between images. The result is shown in the Figure 1. From the figure we can see that in both the Kentucky and the Holidays datasets, the FCTH is much better than the GIST descriptor by a large margin, which is much surprising since the FCTH only use 72 bytes while GIST has to use 960 floating numbers. The GIST descriptor performs almost the same as the Color Layout Descriptor, while both the Edge Histogram Descriptor and Scalable Color Descriptor show unsatisfactory results.

From this graph we can see that all the features from the Kentucky dataset perform better than the Holidays dataset. This is because the images in the former group share most of their appearances, while the latter change a lot in the same group. From the results of Holidays dataset, the best performance is still lower than 0.5, we admit that this is an intrinsic defect of the global features compared to local descriptors. In the Kentucky dataset the result is much encouraging, with the FCTH feature has achieved a MAP score almost close to 0.7. We attribute this to the merit of that FCTH consider both the color and texture feature simultaneously.

5.2 Evaluate Similarity Measures

In this subsection we will evaluate how the different similarity measures can affect the retrieval performance. We choose the FCTH and GIST descriptors to

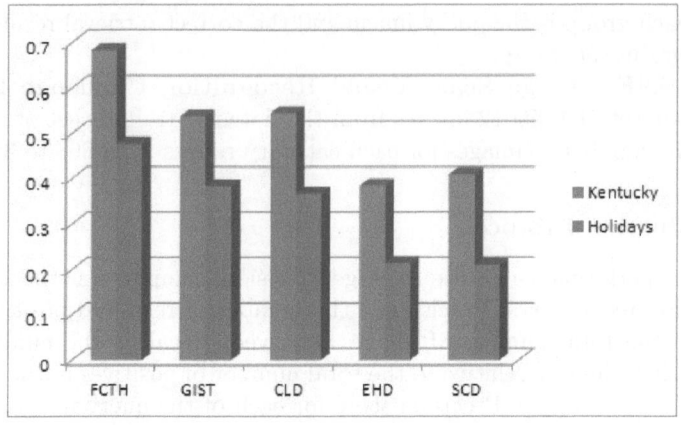

Fig. 1. Evaluate different global descriptors

compare in this round for their better performance in the above experiments. Firstly, we evaluate how the different similarity measures affect the performance of the FCTH feature, and the result is shown in Figure 2, which shows that the Tanimoto, L_1 and L_2 all perform well, achieving almost the same result. In the Kentucky dataset the Tanimoto measure is the best and in the Holidays the L_1 is the best, while in both datasets the Histogram Intersection gives the most unsatisfying results. Then we evaluate how the different similarity measures influence the GIST descriptor, the result of which is also shown in Figure 2. Clearly, the L_1 is the best performer in both datasets, and the L_2 and Tanimoto almost achieve the same score. The Histogram Intersection again performs worst.

The authors in [2] use Tanimoto measure as the similarity measure. Judging from the results it performs well, but it seems that the L_1 measure is much better, not only that they make a draw from the evaluation with the Tanimoto measure, but also it is much computational efficient in the large scale retrieval context, where it requires to compare millions or billions of image features, so a lower complexity will indeed decrease the retrieval time, and promote the user experience. For the GIST feature, no doubt, the L_1 is the best choice, which also contradicts with [10]. The authors use L_2 as similarity measure. From this evaluation the L_1 is indeed better than the L_2 measure because of its performance and its lower complexity.

Now let's compare the best result from the FCTH and the GIST feature similarity measure, the FCTH outperforms GIST in both datasets. So despite of the success of the GIST descriptor in the domain of object recognition and near duplicate detection, it seems much wiser if we can try the FCTH feature to test if they can achieve a better result. Judged from the two famous datasets the FCTH indeed gives a better result.

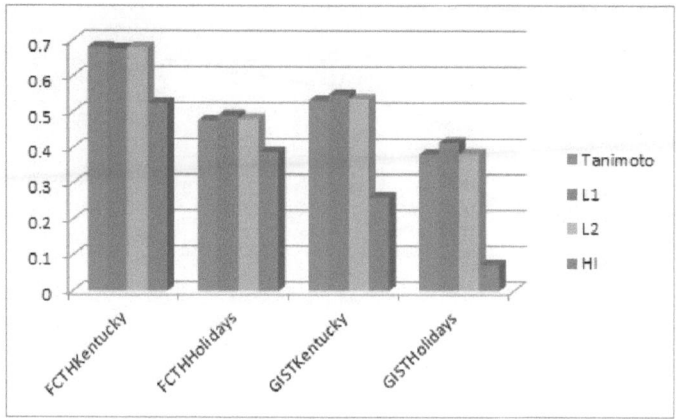

Fig. 2. Evaluate different similarity measures for FCTH and GIST descriptors

5.3   Evaluate Spectral Hashing

From the above experiments we can clearly see that the FCTH descriptor performs better than several MPEG-7 descriptors and even better than the GIST feature. So in this subsection we select the FCTH feature as our final descriptor to evaluate the Spectral Hashing [13] algorithm, to see how the length of embedding bits can affect the results. We use our own implementation of the Spectral Hashing algorithm. The result is shown in Figure 3. Clearly, we can see the trend that the longer hamming bits used as feature, the better performance will achieve in both datasets, which is conform to our intuition. For the different length of the hamming feature from 32 bits to 192 bits, the searching time difference is almost negligible, because this only takes a XOR bits processing, so we use the 192 hamming bits to compare with best performance in the above experiments. First we check the effect on the Kentucky dataset and use the best performance similarity measure. The best MAP score from the four different similarity measures is 0.68, while the 192 bits hamming MAP score is 0.59, decreased by 0.09, but we should also note that the feature is reduced to one third, from the 72 bytes to 24 bytes. In the Holidays dataset the best performance of the L_1 similarity measure is 0.49, decreased to 0.39 with a 192 bits hamming representation. From the above experiments we can conclude that when using the hamming feature derived from the Spectral Hashing, the feature size and retrieval time are reduced significantly, and also can preserve the most of correct results. Later we will mix a large scale of distracting images with each of the two datasets to see how the performance will be.

In this round we will evaluate how a mixture of distracting images will affect the final retrieval performance. We evaluate both datasets. When each of the two datasets is chosen, the IMAGENET dataset with a size of 1256612 images is mixed with the benchmark dataset. We use the 192 bits hamming feature as descriptor. In the Holidays dataset, when mixing with the IMAGENET 1256612

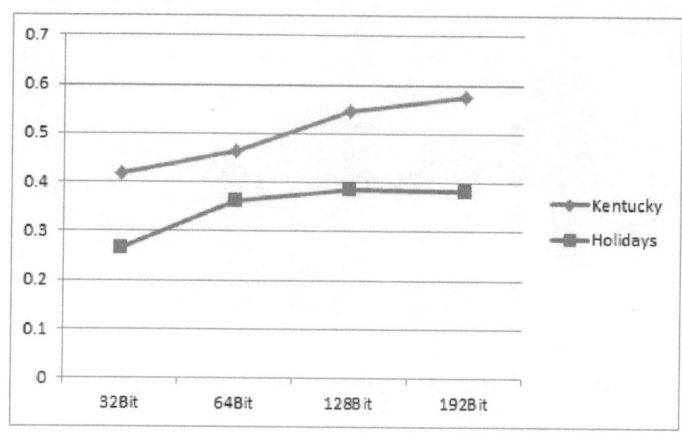

Fig. 3. Evaluate how the length of hamming bits affect results

images, the MAP score is dropping to 0.14 compared to 0.39 without distracting images and also uses the 192 hamming bits as descriptor. In the Kentucky dataset when mixing with the IMAGENET dataset set, the MAP score is from the 0.59 to the 0.39. Although there are some performance dropping, we should also note that we only use the 72 bytes global descriptor to derive the hamming bits features. When using more complicated features, the performance will indeed boost a lot. Also when using these bits features derived from the Spectral Hashing algorithm, the retrieval process can be very efficient, we just exhaustively compare the query image to all the images in the database, and sort the results, without using other indexing methods, the average query response time is 0.16 second from a database of more than 1.26 million images.

6   Conclusions

In this paper, we evaluate the different global features in the domain of object recognition and near duplicate detection against two famous datasets with ground truth. We show the result that FCTH global feature outperforms the state-of-art GIST global feature and several other MPEG-7 global features. This may give the resurgence of the global features when performing some specific image understanding tasks, and may be a complement to the local features to achieve a better result. We also evaluate the different similarity measures to compute the similarity between images, the result of which shows that the L_1 is a better choice for its performance and its low computation complexity for GIST descriptor. To tackle the dilemma of the large scale of image database and the requirement of a real-time response time, we use the Spectral Hashing to embed the feature points to the hamming space, and simply use the hamming distance to efficiently compute similarities between images, which is very efficient because the computation is only the bits processing. This technique is not only efficient but effective with an average query response time of 0.16 second from a database of more than 1.26 million images with a little performance degradation.

References

1. Chang, S.F., Sikora, T., Puri, A.: Overview of the mpeg-7 standard. IEEE Transactions on Circuits and Systems for Video Technology 11(6), 688–695 (2001)
2. Chatzichristofis, S.A., Boutalis, Y.S.: Fcth: Fuzzy color and texture histogram a low level feature for accurate image retrieval. In: 9th International Workshop on Image Analysis for Multimedia Interactive Services, pp. 191–196 (2008)
3. Datta, R., Joshi, D., Li, J., Wang, J.Z.: Image retrieval: Ideas, influences, and trends of the new age. ACM Computing Surveys 40(2) (2008)
4. Deng, J., Dong, W., Socher, R., Li, L.-J., Li, K., Fei-Fei, L.: ImageNet: A Large-Scale Hierarchical Image Database. In: IEEE Conference on Computer Vision and Pattern Recognition (2009)
5. Douze, M., Jegou, H., Sandhawalia, H., Amsaleg, L., Schmid, C.: Evaluation of gist descriptors for web-scale image search. In: ACM International Conference on Image and Video Retrieval, pp. 140–147 (2009)
6. Jegou, H., Douze, M., Schmid, C.: Hamming embedding and weak geometric consistency for large scale image search. In: Forsyth, D., Torr, P., Zisserman, A. (eds.) ECCV 2008, Part I. LNCS, vol. 5302, pp. 304–317. Springer, Heidelberg (2008)
7. Lowe, D.G.: Distinctive image features from scale-invariant keypoints. International Journal of Computer Vision 60(2), 91–110 (2004)
8. Lux, M., Chatzichristofis, S.A.: Lire: Lucene image retrieval - an extensible java cbir library. In: 16th ACM International Conference on Multimedia, pp. 1085–1087 (2008)
9. Nister, D., Stewenius, H.: Scalable recognition with a vocabulary tree. In: IEEE Conference on Computer Vision and Pattern Recognition, pp. 2161–2168 (2006)
10. Oliva, A., Torralba, A.: Modeling the shape of the scene: A holistic representation of the spatial envelope. International Journal of Computer Vision 42(3), 145–175 (2001)
11. Sivic, J., Zisserman, A.: Video google: A text retrieval approach to object matching in videos. In: Ninth IEEE International Conference On Computer Vision, vol. 2, pp. 1470–1477 (2003)
12. Torralba, A., Fergus, R., Weiss, Y.: Small codes and large image databases for recognition. In: 26th IEEE Conference on Computer Vision and Pattern Recognition (2008)
13. Weiss, Y., Torralba, A., Fergus, R.: Spectral hashing. In: Advances in Neural Information Processing Systems (2009)

Improved Content-Based Watermarking Using Scale-Invariant Feature Points

Na Li[1,2,*], Edwin Hancock[2], Xiaoshi Zheng[1], and Lin Han[2]

[1] Shandong Computer Science Center,
Shandong Provincial Key Laboratory of computer Network, Ji'nan, 250014, China
lina@keylab.net
http://www.scsc.cn/
[2] Department of Computer Science, University of York,
Deramore Lane, York, YO10 5GH, UK

Abstract. For most HVS(Human Visual System) perceptual models, the JND(Just Noticeable Difference) values in highly-textured image regions have little difference with those in edge areas. This is not consistent with the characteristics of human vision. In this paper, an improved method is introduced to give a better content-based perceptual mask than traditional ones using the arrangement of scale-invariant feature points. It could decrease the JND values in edge areas of those traditional masks so that they have an obvious difference with values in highly textured areas. Experimental results show the advantages of this improved approach visually, and the enhancement of the invisibility of watermarks.

Keywords: content-based watermarking, scale-invariant feature transform, density-based clustering.

1 Introduction

With the increasing use of the Internet and the effortless copying, tampering and distribution of digital data, copyright protection for multimedia data has become an important issue. Digital watermarking has emerged [1] as a tool for protecting multimedia data from copyright infringement. Efficient techniques for image watermarking must accurately balance two contrasting requirements. On the one hand, the hidden watermark should be imperceivable to the HVS, and on the other hand, a watermark should not be inserted into image regions which are not perceptually important [2]. Furthermore the watermark must be robust against intentional and unintentional attacks. Robustness and imperceptibility of watermarks are at odds with one-another. In order to ensure an optimal trade-off, HVS perceptual models have found widespread applications in watermarking.

* Shandong Province Natural Science Foundation(No.ZR2009GM025,ZR2010FQ018), Shangdong Province Young and Middle-Aged Scientists Research Awards Fund(No. 2008BS01019).

G. Maino and G.L. Foresti (Eds.): ICIAP 2011, Part I, LNCS 6978, pp. 636–649, 2011.
© Springer-Verlag Berlin Heidelberg 2011

A variety of HVS models developed from image compression and quality assessment [3,4,5], have been applied to the design and optimization of digital watermarking since the late 1990's [6,7]. Generally, most of HVS models create a perceptual mask or a JND mask using a multi-channel visual decomposition suggested by psychophysics experiments. A JND mask indicates the maximum amount one can add or subtract at every pixel position of an image without producing any visible difference. This is a so called masking effect. In addition, most watermarking techniques aim at the optimization of the robustness-invisibility trade-off, and are motivated by qualitative perceptual models rather than quantitative visual models.

2   Literature Review

HVS models can be applied in the spatial or frequency domain (DFT, DCT, DWT). In general, watermark embedding can be represented by following equation:

$$y = x + \alpha w \tag{1}$$

where y is the watermarked image, x is the original image. The embedding can be applied in either the time-spatial domain, or the coefficients in the frequency domain. Here α controls the embedding strength. The aim of the HVS model and the perceptual mask is to optimize embedding strength α so that the best trade-off can be obtained between the robustness and invisibility of the watermark.

Delaigle et al. [8] present an additive watermarking technique in the Fourier domain. The core of the embedding process is a masking criterion that guarantees the invisibility of the watermark. This perceptual model is derived from Michelson's contrast C, which is defined as:

$$c = \frac{L_{max} - L_{min}}{L_{max} + L_{min}} \tag{2}$$

where L_{max} and L_{min} are respectively the maximal and minimal luminance value of grating. The masking criterion is depicted as a general expression of the detection threshold contrast.

Another HVS Fourier domain mask is described in [9]. Here Florent et al. introduce a perceptual model by taking into account advanced features of the HVS identified from psychophysics experiments. This HVS model is used to create a perceptual mask and optimize the watermark strength. This is done by combining the perceptual sub-band decomposition of Fourier spectrum with the quantization noise visibility based on the local band limited contrast. Experimental results demonstrate that this method can resist many attacks, including geometrical distortions.

The Watson model [10] is a popular HVS model for DCT domain. It assesses image visual quality by estimating the final perceptual masking threshold used for image compression. According to the mechanisms of the HVS, three factors are considered in the Watson model in order to comprehensively approximate the perceptual quality of an image. The three factors are: a) a frequency sensitivity function, b) a luminance masking function, and c) a contrast masking function. A pooling process is used then to combine all the estimated local perceptual distances together to achieve a global perceptual distance.

Lewis et al. [11] tackle the problem of DWT coefficient quantization for compression and propose to adapt the quantization step of each coefficient according to the local noise sensitivity of the eye. Barni *et al.* [12] make some modifications of the model proposed in [11] in order to better fit the behavior of the HVS to the watermarking problem. A number of factors are taken into account, including luminance, the frequency band, texture and proximity to an edge. Podilchuk *et al.* [13] propose an image adaptive watermarking algorithm for both the DCT and DWT domains. The JND masks applied to the DCT domain are computed from quantization matrices established, while the JND masks used in the DWT domain are computed from visual thresholds given by Watson et al. [14]. For both embedding domains, the watermark robustness has been tested against JPEG compression and cropping.

Huiyan Qi *et al.* [15] design a perceptual mask in the spatial domain using image features such as the brightness, edges, and region activities. In their research, the exact mask values of the cover image can be obtained, guaranteeing the maximum-possible imperceptivity of the watermark. Therefore, the watermark embedding directly substitutes the final mask for watermark strength α. The authors also successfully extend the proposed spatial masking to the DCT domain by searching the extreme value of a quadratic function subject to the bounds on the variable.

In the above-mentioned papers, the computational complexity for the perceptual masks complicates the analysis of the results. Experimental results support the robustness of these approaches. By contrast Voloshynovskiy *et al.* [16] propose and verify a general perceptual mask referred to as the Noise Visibility Function (NVF), which is based on the Maximum Aposteriori Probability (MAP) estimation and Markov random fields. It is simple, practical and has been widely used in many watermarking algorithms, both in the spatial and frequency domains.

In this paper, we describe an improved perceptual mask using an arrangement of scale-invariant feature points. The approach can decrease the JND values in edge regions so that they have given obvious difference in highly textured areas. Therefore, the improved mask is more suitable for human visual characteristics, with low JND values in both edge and flat areas, and high values only in highly textured regions. We choose the perceptual mask in [16] as the prototype. However the proposed improvement also applies to alternative perceptual masks where the JND values in highly textured area have an insignificant difference with the JND values in edge regions. The remainder of this paper is organized as follows. Section 3 introduces the perceptual model in [16]. Section 4 introduces Scale-Invariant Feature Transform (SIFT). Section 5 presents our improved mask and estimation method. Section 6 gives experimental results. The paper is concluded in Section 7.

3   Original HVS Model

The perceptual mask in [16] is based on the computation of NVF that characterizes the local image properties, identifying textured and edge regions. In order to determine the final NVF, the authors consider the watermark as noise and estimate it using a classical MAP image denoising approach. We will not provide the details of this theory here, but rather give the main formula of the method and explain its deficiency.

The authors examine two NVFs, a non-stationary Gaussian model and a stationary generalized Gaussian model, and finally propose a stochastically empirical expression for the optimal NVF. This is widely used in image restoration applications, and calculated using the equation:

$$NVF(i,j) = \frac{1}{1 + \theta \cdot \sigma_x^2(i,j)} \tag{3}$$

where $\sigma_x^2(i,j)$ denotes the grey-scale local variance for neighboring pixels. The parameter θ is used for tuning and plays the role of contrast adjustment in NVF. This version of NVF is the basic prototype for a large quantity of adaptive regularization algorithms. The parameter θ depends on the image variance and is given by:

$$\theta = \frac{D}{\sigma_{max}^2(i,j)} \tag{4}$$

where $D \in [50, 100]$ is an experimentally determined constant, and $\sigma_{max}^2(i,j)$ is the maximum local variance for a given image.

Using NVF, the perceptual mask is:

$$\Lambda = \alpha \cdot (1 - NVF) + \beta \cdot NVF \tag{5}$$

and the watermark embedding equation is:

$$y = x + \alpha \cdot (1 - NVF) \cdot w + \beta \cdot NVF \cdot w \tag{6}$$

where β can be set to 3 for most of real world and computer generated images. The watermark strength parameter α approaches 1 in highly textured areas and approaches 0 in flat region. The third term in equation (6) is added to increase the watermark strength in very flat regions to a level below the visibility threshold. This avoids the problem that the watermark information is (nearly) lost in these areas. The method is illustrated in Fig 1.

As shown in Fig.1, the watermarking rule (Equ.6) embeds the watermark in highly textured areas and areas containing edges stronger than in very flat regions. The deficiency of the method is the relative extraction of the edge and texture information. The JND values in these two regions are very close and the difference between them is not significant for the original HVS model in [16](as illustrated in Fig.1).

From the literature [13,15,16], we note the following two rules consistent with the characteristics of human vision, which are the foundation of the improved method described in this paper:

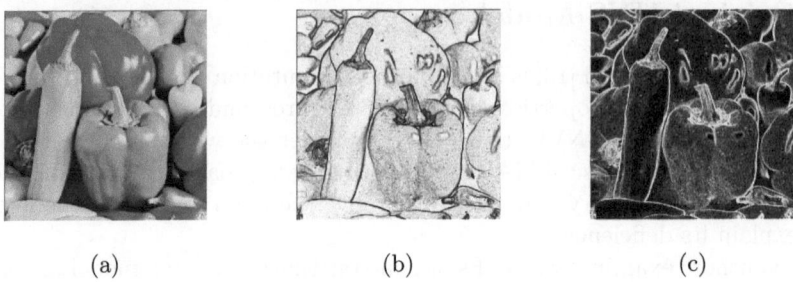

(a)   (b)   (c)

Fig. 1. Effect of HVS model in [16]. (a) original image(size of 512×512); (b) NVF image ; (c) final perceptual mask, here $\alpha = 250$ and $\beta = 3$.

- disturbances are less visible in highly textured regions than in flat areas.
- Edges and contours are more sensitive to noise addition than highly textured regions, and less than but close to the very flat areas.

That is to say, the distortion visibility is low in highly textured areas. However it is high in both edge and very flat region. Therefore, only highly textured areas are strongly suited to watermark embedding and the JND values corresponding to these areas must be high. By contrast, edge areas have the low JND values and very flat areas that contain little (or no) frequency information have the lowest JND values. The perceptual mask in [16] follows just the first of two rules. It is therefore not consistent with the characteristics of human vision.

4   SIFT Theory and Analysis

Affine-invariant features have recently been studied in object recognition and image retrieval applications. These features are highly distinctive and can be matched with high probability under large image distortions. SIFT was proposed by Lowe [17,18] and has been proved to be robust to image rotation, scaling, translation, and to some extend illumination changes, and projective transforms. It has been applied to image forensics [19,20], and digital watermarking [21]. The basic idea of SIFT is to extract features through a staged filtering that identifies stable feature points in the scale-space. In order to extract candidate locations for features, the scale-space is computed using Difference of Gaussian function, where an image is filtered by Gaussian function of different scales and then difference images are calculated. In this scale-space, all local maximum and minimum are retrieved by checking the eight closest neighbors at the same scale and nine neighbors at the scale above and the scale below [22]. Finally the locations and descriptors of feature points are determined, using the scale and orientation changes. Some example images and their corresponding SIFT feature points are shown in Fig.2 (using the program provided by [23]).We can distinguish between edge areas and highly textured regions using the distribution of SIFT points, that is, the small number of points involved in edge areas can be regarded as noise that is supposed to be classified. This idea is the basis of this paper.

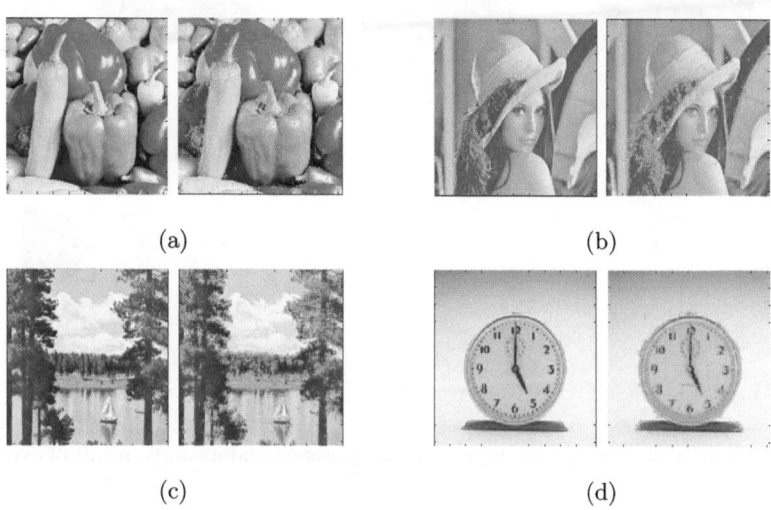

(a)   (b)

(c)   (d)

Fig. 2. Original images and the corresponding SIFT point locations

5   Improved Method

Our improved watermarking method has four steps,and is illustrated as Fig. 3.
Step1. SIFT extractor
   Calculate feature points using the SIFT algorithm and obtain a *binary siftmap* for the arrangement of SIFT key points. We denote the pixel of cover image as $I(i,j), i = 1, 2, \ldots, M, j = 1, 2, \ldots, N$. The location set of feature points that are calculated with SIFT algorithm is denoted as S. The *binary siftmap* is denoted as $\mathbf{I}'$, which is obtained by following formulation:

$$I'(i,j) = \begin{cases} 1 & (i,j) \in S \\ 0 & otherwise \end{cases} \tag{7}$$

Obviously, *binary siftmap* $\mathbf{I}'$ is size of $M \times N$, the same with original cover image $\mathbf{I}$.

Step2. Density-based clustering
   Cluster analysis is a primary method for database mining. The biggest advantage of density-based clustering is that regions indicated clusters may have an

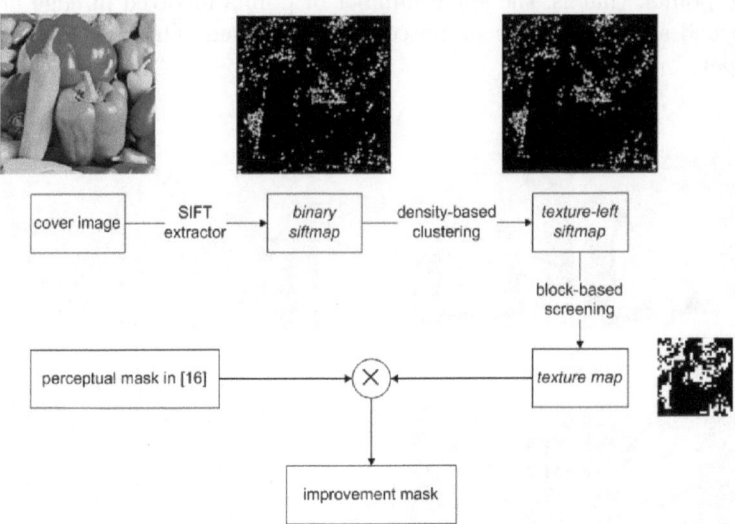

Fig. 3. Flow chart of proposed improvement approach and example result of every step

arbitrary shape and that the points inside a region may be arbitrarily distributed. In density-based clustering, clusters are regarded as regions in the data space in which the objects are dense, and which are separated by regions of low object density (noise). So density-based clustering is very suitable for classification of points on *binary siftmap* in the case of proposed improvement approach.

The first density-based clustering algorithm is proposed by Martin *et al.* in [24], which is called Density-Based Spatial Clustering of Applications with Noise (DBSCAN). It has two important parameters, search radius *Eps* and density search parameter *MinPts*. The details of the algorithm will not be described here and we only give the parameter settings. From adequate experiments we can find, the number of clusters is empirically supposed to be at the range from 6 to 30 in order to denoise to the greatest degree in this step. Moreover, clustering results are just sensitive to *Eps*, and *MinPts* can be constant. Therefore, the parameter *Eps* should be set carefully to make sure of getting a suitable number of clusters. We do experiments for an image database more than 30 and get an empirical range of the optimal search radius, that is, $Eps \in [10, 16, 20, 32, 40]$. The parameter *MinPts* is set to 4.

A certain number of clusters is obtained after implement of DBSCAN. We denote these clusters as $C_1, C_2, \ldots, C_k$ and the number of points in them as $Num_1, Num_2, \ldots, Num_k$. Then we sort the set of Num_i in accordance with descending order, which can be expressed as:

$$sort(\{Num_1, Num_2, \ldots, Num_k\}) = \{Num_1^*, Num_2^*, \ldots, Num_k^*\} \qquad (8)$$

Thus permutation can be made to set of C_i correspondingly and we can get $C_1^*, C_2^*, \ldots, C_k^*$. After all this, we can get *texture-left siftmap*, which is denoted as $\mathbf{I}^*$, through keeping the first t of C_i^* and discarding the points involved in other clusters on *binary siftmap* $\mathbf{I}'$, which t satisfies the following constrain:

$$\sum_{i=1}^{t} Num_i^* \geq p \sum_{i=1}^{k} Num_i \qquad (9)$$

here t is a natural number and $1 < t < k$, p is an given percent.

Step3. Block-based screening

As shown in Fig.4, the *texture-left siftmap* is a point-scattering mask so that it cannot be directly used as a perceptual mask in watermark embedding. We design a simple method to get a final texture map, which is called 'block-based density screening'. First of all, *texture-left siftmap* $\mathbf{I}^*$ (size of $M \times N$) is divided into non-overlapping small blocks. If the size of each block is defined as $m \times n$, the division process can be denoted as this expression:

$$\mathbf{I}^* = \begin{bmatrix} \mathbf{B}_{1,1} & \mathbf{B}_{1,2} & \ldots & \mathbf{B}_{1,\frac{N}{n}} \\ \mathbf{B}_{2,1} & \mathbf{B}_{2,2} & \ldots & \mathbf{B}_{2,\frac{N}{n}} \\ \vdots & \vdots & \ddots & \\ \mathbf{B}_{\frac{M}{m},1} & \mathbf{B}_{\frac{M}{m},2} & \ldots & \mathbf{B}_{\frac{M}{m},\frac{N}{n}} \end{bmatrix} \qquad (10)$$

Then calculate the number of points in every corresponding block, which is expressed as:

$$\begin{bmatrix} Num_{1,1} & Num_{1,2} & \ldots & Num_{1,\frac{N}{n}} \\ Num_{2,1} & Num_{2,2} & \ldots & Num_{2,\frac{N}{n}} \\ \vdots & \vdots & \ddots & \\ Num_{\frac{M}{m},1} & Num_{\frac{M}{m},2} & \ldots & Num_{\frac{M}{m},\frac{N}{n}} \end{bmatrix} \qquad (11)$$

The *texture map*, which is denoted as $\boldsymbol{TM}$, can be obtained using following formulation:

$$TM(i,j) = \begin{cases} 1 & Num_{i,j} > Threshold \\ 0 & Num_{i,j} \leq Threshold \end{cases} \qquad (12)$$

Step4. Improved mask

For our research, the prototype of perceptual mask in [16] is improved using the texture map. Bilinear interpolation and binarization are sequentially applied to texture map to stretch and re-binarize it. Thus, a big texture map can be obtained with the same size of the prototype. In order to avoid the watermark information to be lost when the value of *texture map* $\boldsymbol{TM}$ equals to 0, the final *texture mask* is processed by following rule: the values of pixels which equal to

0 are set to a given small constant t_{min} at the range of $(0, 0.4]$; and the values of pixels which equal to 1 are unchanged. Finally, the perceptual mask (Equ.5) is modified as follows,

$$\tilde{\Lambda} = \alpha \cdot (1 - NVF) * \boldsymbol{TM} + \beta \cdot NVF * \boldsymbol{TM} \tag{13}$$

here the operator $*$ does not represent matrix multiplication but array multiplication.

6   Experiment Results

6.1   Improved Mask

Experiments are performed on a set of images for our improvement approach. The parameter p in the second step of our method is set to 90%, Threshold in the third step is 1, and t_{min} in final step is 0.4. Besides, $\alpha = 250$ and $\beta = 3$ is the same with Fig.1(c). For images in Fig.2, the comparison of perceptual masks is shown in Fig.4.

The advantage of our improved method is directly and clearly illustrated in Fig.4. Our method significantly decreases the JND values in all (or most) edge areas, and makes them have a considerable difference with JND values in highly textured regions. Before implement of proposed method, values of perceptual mask just have two levels, but the entire mask is to be divided into three levels after: the lowest in very flat region, the relatively lower in edge region, and the highest in highly-textured region. This advantage in the test image lake is most evident, particularly in parts of trunks and the small boat. Obviously, our perceptual masks are better for the characteristic of human vision than before.

6.2   Application in Watermark Embedding

We test the improved mask in the direct embedding method with the following equation,

$$y = x + \tilde{\Lambda} \cdot \omega \tag{14}$$

where y denotes watermarked image, x denotes cover image and w denotes watermark. We use a grey-scale image (shown in Fig.5(a)) as watermark and embed it to a group of common test images. The embedding strength parameters α are set to 20, and still $\beta = 3$. The examples of cover image and watermarked image are shown in Fig.5(b) and (c).

The peak signal-to-noise ratio (PSNR) is chosen to measure the visibility of watermark and the quality of cover image. For an 8-bit grey-scale image, the equation of PSNR is as follows:

$$PSNR = 10 \log 10 \frac{255^2}{\frac{1}{M \times N} \sum_{i=1}^{M} \sum_{j=1}^{N} (y(i,j) - x(i,j))^2} \tag{15}$$

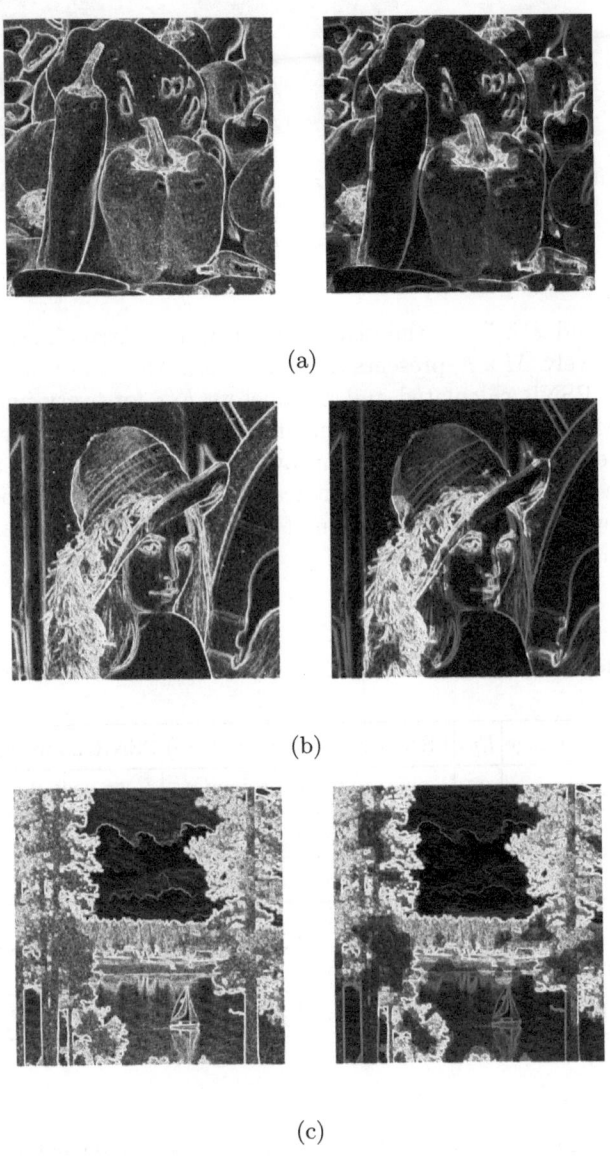

(a)

(b)

(c)

Fig. 4. Comparison of perceptual masks before and after implement of proposed improved method

(a) watermark   (b) cover image   (c) watermarked image

Fig. 5. Watermark embedding

where $y(i,j)$ and $x(i,j)$ are the pixel value in watermarked image and original image respectively. $M \times N$ presents the image size. After watermark embedding, the values of PSNR calculated and parameters *Eps* for every image are listed in Table 1. More test results have been shown in Fig.6 for our image database. From these results we can find, our improved method significantly and generally increases the value of PSNR and enhances the invisibility of embedded watermark. Moreover, the more edge information cover image contains, the more of the PSNR value increases.

Table 1. Comparison of PSNR for mask effect in different methods

cover image	Eps	PSNR(in [16])	PSNR(here)	PSNR(no mask)
peppers	32	32.7540db	35.5923db	25.2939db
lena	24	31.5628db	33.5113db	25.7776db
lake	10	31.2367db	32.6696db	25.7413db
clock	10	33.5889db	35.0734db	26.3042db
cameraman	10	32.6604db	33.5734db	26.3042db
baboon	32	28.4366db	29.1577db	24.7816db

We also do experiments about watermarking attacks and calculate correlation values between original watermark and extracted watermark. The results against noise-adding (Salt & pepper, Gaussian, Speckle) are illustrated in Fig.7.

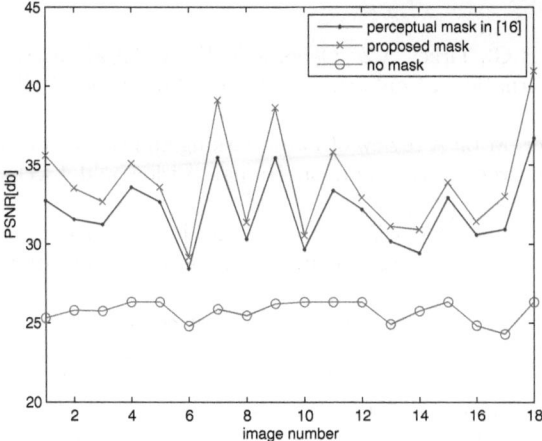

Fig. 6. Comparison of PSNR for image database

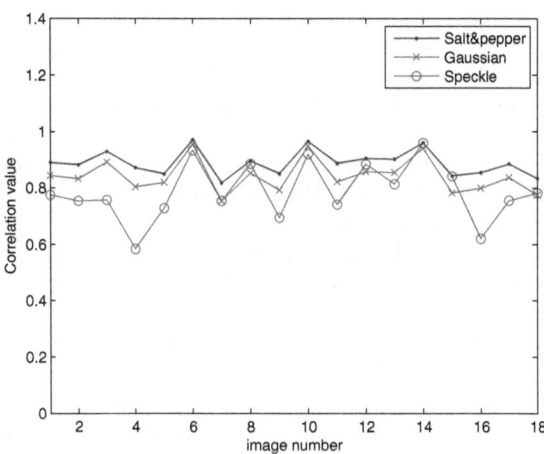

Fig. 7. Detection values of watermarked image after noise-adding attack

7   Conclusion

In this article, we describe an improved method for the perceptual mask using the arrangement of scale-invariant feature key points. It could decrease the JND values in edge areas of those traditional masks so that they give obvious difference with values in highly textured areas. Therefore, the improved mask is more suitable for human visual characteristics. But one of the shortcomings of our method is the imprecision of segment between the highly textured region and the edge areas. This is one of the directions of our future work.

References

1. van Schyndel, R.G., Tirkel, A.Z., Osborne, C.F.: A digital watermark. In: Proceedings of 1st International Conference on Image Processing (ICIP 1994), vol. 2, pp. 86–90 (1994)
2. Cox, I.J., Miller, M.L.: A review of watermarking and the importance of perceptual modelling. In: Proc. SPIE Conf. on Human Vision and Electronic Imaging II, vol. 3016, pp. 92–99 (1997)
3. Nill, N.B.: A Visual Model Weighted Cosine Transform for Image Compression and Quality Assessment. IEEE Transactions on Communications 33(6), 551–557 (1985)
4. Jayant, N., Johnston, J., Safranek, B.: Signal compression based on models of human perception. Proceedings of the IEEE 81(10), 1385–1422 (1993)
5. Eckert, M.P., Bradley, A.P.: Perceptual quality metrics applied to still image compression. Signal Processing 70(3), 177–200 (1998)
6. Kankanhalli, M.S., Rajmohan, Ramakrishnan, K.R.: Content based watermarking of images. In: International Multimedia Conference in: Proceedings of the Sixth ACM International Conference on Multimedia, Bristol, United Kingdom, pp. 61–70 (1998)
7. Delaigle, J.F., De Vleeschouwer, C., Macq, B.: Psychovisual approach to digital picture watermarking. Journal of Electronic Imaging 7(3), 628–640 (1998)
8. Delaigle, J.F., De Vleeschouwer, C., Macq, B.: Watermarking algorithm based on a human visual model. Signal Processing 66(3), 319–335 (1998)
9. Autrusseau, F., Le Callet, P.: A robust image watermarking technique based on quantization noise visibility thresholds. Signal Processing 87(6), 1363–1383 (2007)
10. Andrew, B.: Watson. DCT quantization matrices visually optimized for individual images. In: Proceedings of SPIE: Human vision, Visual Processing and Digital Display IV, vol. 1913, pp. 202–216
11. Lewis, A.S., Knowles, G.: Image compression using the 2-D wavelet transform. IEEE Trans. Image Processing 1, 244–250 (1992)
12. Barni, M., Bartolini, F., Piva, A.: Improved wavelet-based watermarking through pixel-wise masking. IEEE Transaction on Image Processing 10(5), 783–791 (2001)
13. Podilchuk, C.I., Zeng, W.: Image-adaptive watermarking using visual models. IEEE Journal on Selected Areas in Communications 16(4), 525–539 (1998)
14. Watson, A.B., Yang, G.Y., Solomon, J.A., Villasenor, J.D.: Visibility of wavelet quantization noise. IEEE Transactions on Image Processing 6(8), 1164–1175 (1997)
15. Qi, H., Zheng, D., Zhao, J.: Human visual system based adaptive digital image watermarking. Signal Processing 88(1), 174–188 (2008)
16. Voloshynovskiy, S., Herrigel, A., Baumgaertner, N., Pun, T.: A stochastic approach to content adaptive digital image watermarking. In: Pfitzmann, A. (ed.) IH 1999. LNCS, vol. 1768, pp. 212–236. Springer, Heidelberg (2000)
17. David, G.: Lowe. Object Recognition from Local Scale-Invariant Features. In: International Conference on Computer Vision, Corfu, Greece, pp. 1150–1157 (September 1999)
18. Lowe, D.G.: Distinctive Image Features from Scale-Invariant Keypoints. International Journal of Computer Vision 60(2), 91–110 (2004)
19. Amerini, I., Ballan, L., Caldelli, R., Del Bimbo, A., Serra, G.: A SIFT-based forensic method for copy-move attack detection and transformation recovery. IEEE Transactions on Information Forensics and Security (in press, 2011)

20. Pan, X., Lyu, S.: Region Duplication Detection Using Image Feature Matching. IEEE Transactions on Information Forensics and Security 5(4), 857–867 (2010)
21. Nguyen, P.-B., Beghdadi, A., Luong, M.: Robust Watermarking in DoG Scale Space Using a Multi-scale JND Model. In: Muneesawang, P., Wu, F., Kumazawa, I., Roeksabutr, A., Liao, M., Tang, X. (eds.) PCM 2009. LNCS, vol. 5879, pp. 561–573. Springer, Heidelberg (2009)
22. Lee, H.-Y., Lee, C.-h., Lee, H.-K., Nam, J.: Feature-Based Image Watermarking Method Using Scale-Invariant Keypoints. In: Ho, Y.-S., Kim, H.-J. (eds.) PCM 2005. LNCS, vol. 3768, pp. 312–324. Springer, Heidelberg (2005)
23. http://www.cs.ubc.ca/~lowe/keypoints/
24. Ester, M., Kriegel, H.-P., Sander, J., Xu, X.: A Density-Based Algorithm for Discovering Clusters in Large Spatial Databases with Noise. In: Proceedings of 2nd International Conference on Knowledge Discovery and Data Mining (KDD 1996), pp. 226–231 (1996)

Crop Detection through Blocking Artefacts Analysis

A.R. Bruna, G. Messina, and S. Battiato

Image Processing Laboratory
Department of Mathematics and Computer Science
University of Catania
Viale A. Doria 6 - 95125 Catania, Italia
{bruna,gmessina,battiato}@dmi.unict.it
http://iplab.dmi.unict.it

Abstract. In this paper we propose a new method to detect cropped images by analyzing the blocking artefacts produced by a previous block based compression techniques such as JPEG and MPEG family that are the most used compression standards for still images and video sequences. It is useful for image forgery detection, in particular when an image has been cropped. The proposed solution is very fast compared to the previous art and the experimental results show that it is quite reliable also when the compression ratio is low, i.e. the blocking artefact is not visible.

Keywords: crop and paste technique, DCT blocking artefacts analysis, tampered images, image forensic.

1 Introduction

In the last years, the number of forged images has drastically increased due to the spread of image capture devices, especially mobile phones, and the availability of image processing software. Copy and paste is the simplest and most used technique to counterfeit images. It can be used to obtain a completely new image by cropping the interest part or to cover unwanted details. In image forensics it is important to understand (without any doubts) if an image has been modified after the acquisition process.

There are two methodologies to detect tampered images: active protection methods and passive detection methods [1]. The active protection methods make use of a signature inserted in the image [2]. If the signature is no more detectable, the image has been tampered. These techniques are used basically to assess ownership for artworks and/or relative copyrights. Passive detection methods make use of ad-hoc image analysis procedures to detect forgeries. Usually the presence of peculiar artefacts is properly investigated and, in case of anomalies, the image is supposed to be counterfeit. Several algorithms exist in literature as reported in a recent survey [3]. Among others, a lot of methods in the field consider the possibility to exploit the statistical distribution of DCT coefficients in order to reveal the irregularities due to the presence of a superimposed signal over the original one [4, 5, 6, 7]. The usage of

G. Maino and G.L. Foresti (Eds.): ICIAP 2011, Part I, LNCS 6978, pp. 650–659, 2011.

Discrete Cosine Transform (DCT) artefacts analysis have a further advantage due to the fact, that is the most used compression technique; JPEG [8] for still images and MPEG compression family [9] for video sequences make use of block based DCT data compression (usually 8x8 pixel size non-overlapping windows).

While there are a lot of algorithms in literature for the estimation of the quantization (and compression) history [5, 6], there are only a few approaches for cropping detection [7]. In this paper a pipeline is suggested aiming to merge both techniques in order to obtain a more reliable system. In fact, blocking artefacts analysis works well when the image is aligned to the block boundary. But, in case the image has been cropped, this assumption is no more valid and the detection may fail. In the proposed system the quantization detection block is preceded by a cropping detection in order to align the image to the block boundary. It allows to increase the reliability of the detection and to better understand if only some parts of the image are tampered.

In Li et al. [1] the grid extraction is realized by extrapolating the Block Artefact Grid (BAG) embedded in the image where block artefact appears, i.e. the grid corresponding to the blocks boundaries. The authors introduce a measure of blockiness just considering the ratio between the sum of AC components of the first raw (and column) with respect to the DC coefficient in the DCT domain. One of the main drawbacks of the method is related to the fact that it requires to compute 8x8 pixels DCT again (in a dense way) over the image under detection, just to locate properly the correct alignment. Also the detection of tampered regions corresponding to misaligned grid is demanded to a visual inspection of the resulting filtering operation without a clear and objective measure. The same authors propose in [7] an interesting approach based on spatial consideration devoted to locate the BAG just combining together a series of derivative and non-linear filters to isolate blockiness avoiding the influence of textures or strong edges present in the input image. Unfortunately, both the techniques were not properly evaluated just considering a proper dataset (high variability with respect to resolution size) and a sufficient number of misalignments cropping with respect to the 8x8 grid. Also an exhaustive comparison just considering the overall range of JPEG compression factors is lacking.

As well described in [3, 10] the new emerging fields of Digital Forensics require to provide common benchmarks and datasets that are needed for fair comparisons among the numerous proposed techniques and algorithms published in the field.

The rest of the paper is organized as follows. In Section 2 the proposed technique is described; the next Section reports a series of experiments devoted to assess the effectiveness of the method. Finally, some conclusions are given together with a few hints for the future work.

2   Proposed System

The proposed solution can be used as stand-alone algorithm to detect crop operations or it can be inserted in a typical advanced pipeline for complex tampering detection using compression artefacts analysis. The proposed solution aimed to handle images without any further compression. In this case, in fact, the further compression may introduce blocking artefact that may deceive the algorithm. In Figure 1 an example aiming to detect the camera model from an image, after a copy/paste and re-encoding counterfeit process, is shown. In this case, the algorithms described in [5] and [6] can be used for the quantization detection, while the study proposed in [11] can be used for the signature detection block.

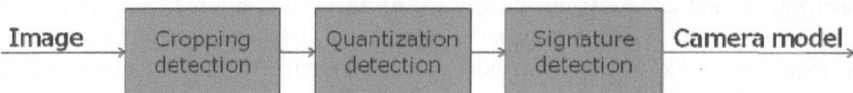

Image → Cropping detection → Quantization detection → Signature detection → Camera model

Fig. 1. Block based schema of the pipeline used to retrieve the camera model from an image

2.1   Algorithm Description

The DCT codec-engines (e.g., JPEG, MPEG, etc.) typically apply a quantization step in the transform domain just considering non-overlapping blocks of the input data. Such quantization is usually achieved by a quantization table useful to differentiate the levels of quantization adapting its behavior to each DCT basis. The JPEG standard allows to use different quantization factors for each of the bi-dimensional DCT coefficients. Usually standard quantization tables are used and a single multiplicative factor is applied to modify the compression ratio [12, 13], obtaining different quality levels. As the tables are included into the image file, they are also customizable as proved by some commercial codec solutions that exploit proprietary tables. Images compressed by DCT codec-engines are affected by annoying blocking artefacts that usually appear like a regular grid superimposed to the signal. In the following, we discuss a simple example based on the Lena image. The picture has been compressed using the *cjpeg* [12] software with a properly managed compression ratio, obtained modifying the quantization tables through the variation of the quality parameter in the range {10, 90} (see Figure 2).

Fig. 2. Blocking artefact example in a JPEG compressed image (quality factor = 40)

It is an annoying artefact visible especially in flat regions. It is also regular, since it depends on the quantization of the DCT coefficients of every 8x8 blocks. Unfortunately, this kind of artefact is not simple to be characterized (i.e., detectable) in the Fourier domain, since image content and the effect of the quantization step of the encoding pipeline mask the regular pattern, as shown in Figure 3.

Fig. 3. Results of the visualization of the Fourier spectrum applied to the compressed Lena image shown in Figure 2

We established to work directly in the spatial domain and, in particular, in the luminance component. The blocking artefact is basically a discrepancy in all the borders between adjacents blocks. It is regularly spaced (8x8 for JPEG and MPEG) and it also affects the image in only two perpendicular directions (horizontal and vertical if the image has not been rotated).

The straightforward way to detect such artefact exploits a derivative filter along the horizontal and vertical direction. The proposed strategy could be easily generalized to consider all possible malicious rotation of the cropped image, just iterating the process at different rotation angles. In Figure 4 the overall schema of the proposed algorithm is depicted.

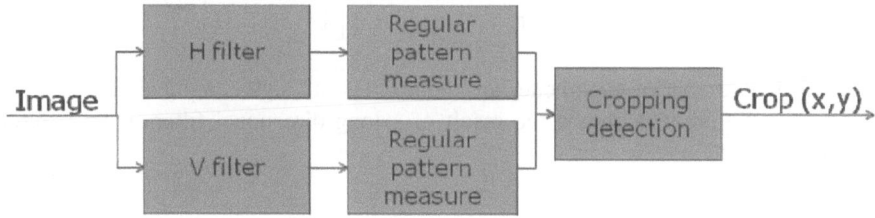

Fig. 4. Block based schema of the cropping detection algorithm

The blocks "*H filter*" and "*V filter*" are derivative filters. Basically they are High Pass Filters (HPF) usually used to isolate image contours. An example is the Sobel filter, with the following basic kernel:

$$H = \begin{bmatrix} -1 & -2 & -1 \\ 0 & 0 & 0 \\ 1 & 2 & 1 \end{bmatrix}; \quad V = \begin{bmatrix} -1 & 0 & 1 \\ -2 & 0 & 2 \\ -1 & 0 & 1 \end{bmatrix};$$

Fig. 5. Sobel filter masks

These filters are able to detect textures, as shown in the image below.

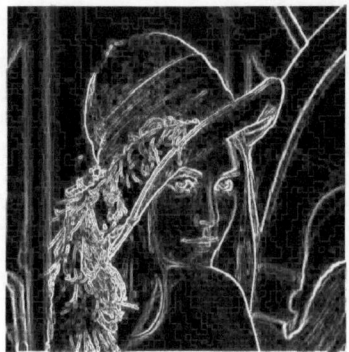

Fig. 6. Effect of the Sobel filter (3x3 kernel size) applied to the Lena image

It is very useful to retrieve textures of the image, but the blocking artefact effect is also masked. In order to detect only regular pattern and discard real edges, a very long taps directional filter has been used It was obtained by properly expanding the following 3x3 filters along the horizontal or vertical direction:

$$H = \begin{bmatrix} 1 & 1 & 1 \\ -1 & -1 & -1 \\ 0 & 0 & 0 \end{bmatrix}; \quad V = \begin{bmatrix} 1 & -1 & 0 \\ 1 & -1 & 0 \\ 1 & -1 & 0 \end{bmatrix};$$

Bigger is the number of taps, better are the results, although computational time also increase. In Figure 7 is shown the result of a long directional filters with different kernel size.

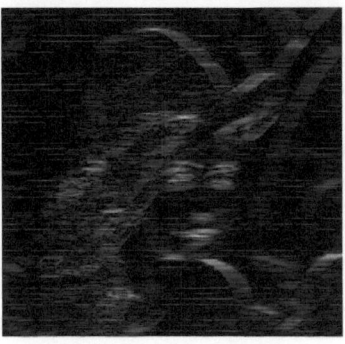 

Fig. 7. 30 taps directional HPF applied to the Lena image. Borders are not considered (thus the vertical size in the left image and the horizontal size in the right images are less than the original size).

We define the *Regular Pattern Measure* (RPM) that computes a measure of the blockiness effect as defined in the following. Let I a *MxN* pixel size image and I^H, I^V the corresponding filtered images obtained by applying a directional HPF as above. For sake of simplicity let suppose to have serialized the image, by simple scan line ordering of the corresponding rows and columns just obtaining the vector $I^{H'}$, $I^{V'}$. The RPM values for both directions is obtained as:

$$RPM_H(i) = \sum_{j=0}^{floor(N/8)} I^{H'}(8 \cdot j + i); \quad i = 1,...,7;$$

$$RPM_V(i) = \sum_{j=0}^{floor(M/8)} I^{V'}(8 \cdot j + i); \quad i = 1,...,7;$$

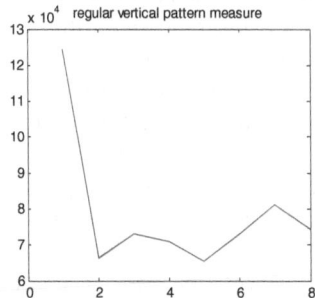 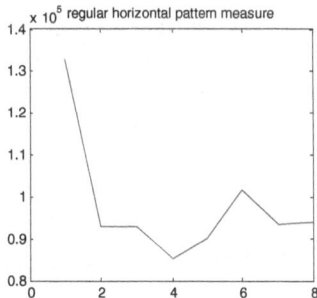

Fig. 8. RPM measure without cropping. The blocking artefact starts with the pixel [1,1].

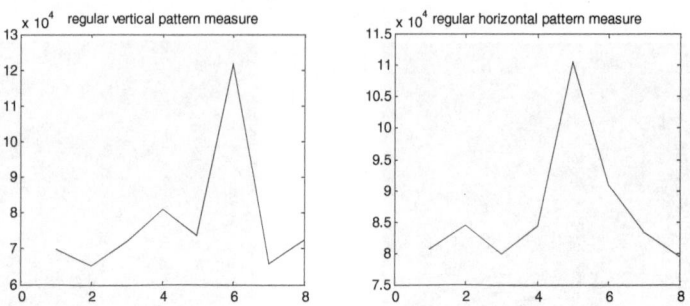

Fig. 9. RPM measure with cropping. The blocking artefact starts with the pixel [6,5].

Experiments have shown that both the RPM_H and RPM_V measures allow discriminating, in a very robust way (e.g., with respect to the main content of the image), the periodicity of the underlying cropping positions. Such values can be extracted by considering a simple order statistic criterion (e.g., the maximum). Figures 8 and 9 show the plot of the two RPM measures, in case of no cropping (e.g., blocking artefact starts with the pixel [1,1]) and in case of malicious cropping at position [6, 5].

3   Experimental Results

To assess the effectiveness of any forgery detection technique, a suitable dataset of examples should be used for evaluation. According to [10] the input dataset contains a number of uncompressed images organized with respect to different resolutions, sizes and camera models. Also the standard dataset from Kodak images and from UCID v.2 have been used. The overall dataset can be downloaded from [14]. It is composed by 114 images with different resolution. Experiments were done varying in an exhaustive way the cropping position and the compression rate.

Moreover the *cjpeg* [12], (i.e., the reference code for the JPEG encoder), has been used to compress the images and the flag *-quality* was used to modify the quality from 10 (high compression ratio) to 90 (low compression ratio) with the step of 10. In particular each image has been cropped in order to test every possible cropping position in the 8x8 block, just to consider the possibility to test the method also in presence of real regular patterns in the image that could influence the results. In Table 1 are reported the overall results, described in terms of correct percentage of the cropping position detection, with respect to the involved compression ratio.

Exhaustive tests have been done for every image, every cropping position and quality factor. Thus the accuracy has been obtained considering 2891 cases.

Experimental results show that performances increase according to the compression rate. It is reasonable, since the blocking artefact increases at higher compression ratio.

Table 1. Results of the proposed method

Quality factor	Accuracy (%)
10	99
20	91
30	80
40	69
50	58
60	46
70	39
80	28
90	16

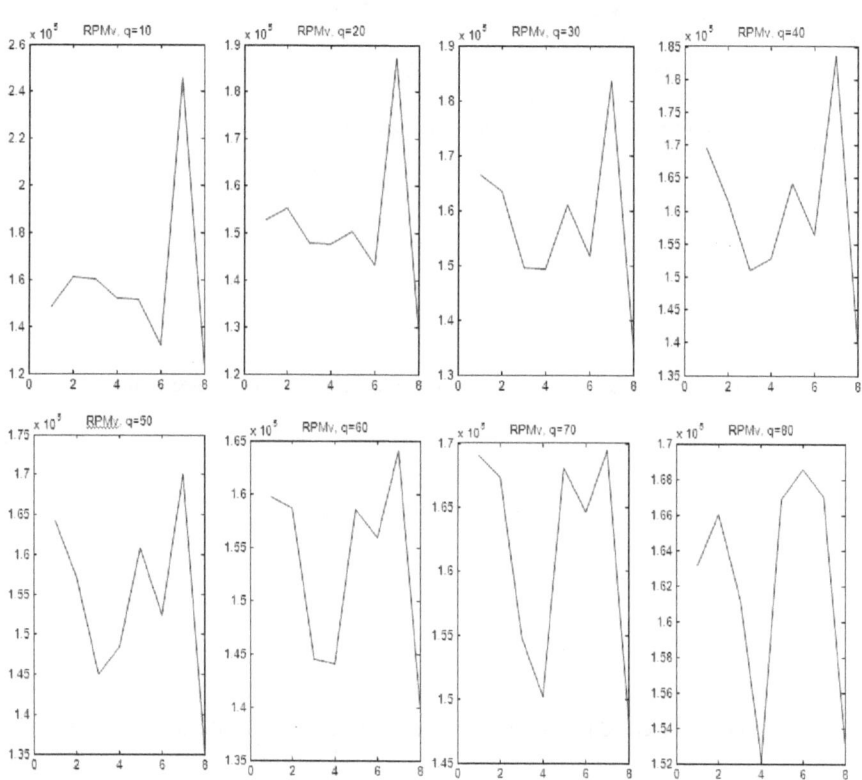

Fig. 10. RPM measure obtained at varying the *quality* factor

In Figure 10 the RPM (only vertical) measure is shown at varying the *quality* factor (real crop position = 7). Reducing the quantization (i.e., increasing the quality factor), the peak is less evident and, in this example, with the *quality=80* the estimation fails, since the effect of a real edge becomes predominant.

The proposed solution was compared to the method described in [1,7]. Unfortunately in these papers the cropping detection is not automatic, but it is supposed a visual inspection at the end of the process. In Figure 11 are shown the results of this method at varying the quality factor from 10 to 90 for the Lena image for a cropping position (3,5). It is evident that the cropping position is detectable up to $q=40$. Above this value it is no more visible. Similar results have been obtained for all the involved dataset.

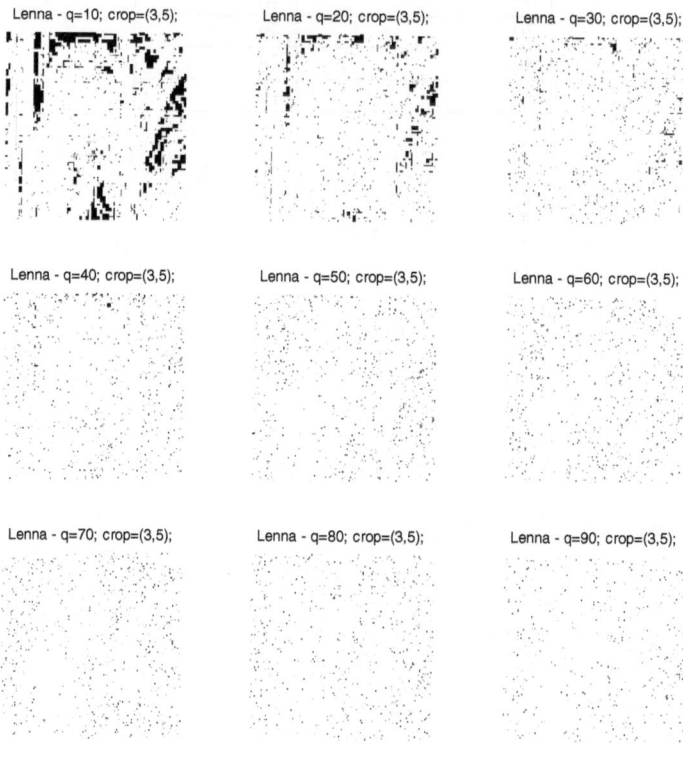

Fig. 11. Li's method [1, 7] applied to Lena image at varying the *quality* factor

4   Conclusions

A new algorithm for cropping detection has been presented. It can be used in forensic applications to detect tampered images affected by cropping pre-compressed images. It can also be used in the pipeline with other blocks to increase the reliability of the results. The method is based on DCT artefacts analysis, in particular on the blocking artefacts that are detected through an adaptive system working in the luminance component. Experimental results show that, according to the blocking artefact behavior, the reliability of the response increase according to the compression ratio.

The main advantages of the proposed technique with respect to the state of the art are the speed (e.g. the [1] requires 50 seconds while the proposed solution requires less that 1 second for each inspection). Moreover it is a fully unsupervised (e.g., not require any visual inspection). Also its reliability is acceptable at lower compression ratio (i.e. when the blocking artefact is almost negligible). Further works will aim to increase the reliability of the system (e.g., by weighting differently the blocks contribution according to a flatness measure) and extending the methodology also to the color component, since these are heavily compressed. Moreover, further research will also devoted to exploit local information, in order to locate discrepancies inside the image (i.e., to discover copy and paste forgery).

References

1. Li, W., Yu, N.I., Yuan, Y.: Doctored JPEG image detection. In: Proceedings of International Conference on Multimedia and Expo, IEEE ICME (2008)
2. Lie, W.N., Lin, G.S., Cheng, S.L.: Dual protection JPEG images based on informed embedding and two-stage watermark extraction techniques. IEEE Transactions on Information Forensics and Security 1, 330–341 (2006)
3. Redi, J.A., Taktak, W., Dugelay, J.-L.: Digital image forensics: a booklet for beginners. Multimedia Tools and Applications 51(1) (January 2011)
4. Battiato, S., Mancuso, M., Bosco, A., Guarnera, M.: Psychovisual and Statistical Optimization of Quantization Tables for DCT Compression Engines. In: IEEE Proceedings of International Conference on Image Analysis and Processing, ICIAP 2001, Palermo, Italy, pp. 602–606 (September 2001)
5. Fan, Z., De Queiroz, R.L.: Identification of bitmap compression history: JPEG detection and quantizer estimation. IEEE Transactions on Image Processing, 230–235 (2003)
6. Huang, F., Huang, J., Shi, Y.Q.: Detecting double JPEG compression with the same quantization matrix. IEEE Transactions on Information Forensics and Security 5(4), 848–856 (2010), art. no. 5560817
7. Li, W., Yuan, Y., Yu, N.: Passive detection of doctored JPEG image via block artefact grid extraction. Signal Processing 89(9), 1821–1829 (2009)
8. http://www.jpeg.org/jpeg/index.html
9. http://www.mpeg.org/MPEG/video/
10. Battiato, S., Messina, G.: Digital Forgery Estimation into DCT Domain - A Critical Analysis. In: Proceedings of ACM Multimedia 2009 – Workshop Multimedia in Forensics, Bejing (China) (October 2009)
11. Farid, H.: Digital Image Ballistics from JPEG Quantization, Technical Report, TR2006-583, Dartmouth College, Computer Science (2006)
12. cjpeg code can be found in, http://www.ijg.org/
13. Bruna, A., Mancuso, M., Capra, A., Curti, S.: Very Fast algorithm for Jpeg Compression Factor Control. In: Proceedings of SPIE Electronic Imaging 2002 - Sensors, Cameras, and Applications for Digital Photography IV, San José, CA, USA (January 2002)
14. DBForgery 1.0, http://iplab.dmi.unict.it/index.php?option=com_content&task=view&id=41&Itemid=118

Structure from Motion and Photometric Stereo for Dense 3D Shape Recovery

Reza Sabzevari, Alessio Del Bue, and Vittorio Murino

Istituto Italiano di Tecnologia
Via Morego 30, 16163 Genova, Italy
{reza.sabzevari,alessio.delbue,vittorio.murino}@iit.it
http://www.iit.it

Abstract. In this paper we present a dense 3D reconstruction pipeline from monocular video sequences using jointly Photometric Stereo (PS) and Structure from Motion (SfM) approaches. The input videos are completely uncalibrated both from the multi-view geometry and photometric stereo aspects. In particular we make use of the 3D metric information computed with SfM from a set of 2D landmarks in order to solve for the bas-relief ambiguity which is intrinsic from dense PS surface estimation. The algorithm is evaluated over the CMU Multi-Pie database which contains the images of 337 subjects viewed under different lighting conditions and showing various facial expressions.

Keywords: Structure from Motion, Photometric Stereo, Dense 3D Reconstruction.

1 Introduction

The 3D inference of the objects shape is of paramount interest in many fields of engineering and life science. However, the inference of the depth from a set of images is one of the most challenging inverse problem and various features has been used in order to extract information of 3D surfaces from images. The field of multi-view geometry [8] uses the information of a set of known 2D correspondences in order to estimate the localization of a set of 3D points lying over the surface. Shape from defocus [6] instead uses the blurring effect of images obtained from varying the distance of the lens with respect to the camera sensor. Shape from texture [4] infers the 3D surface bending given the variation of the texture belonging to an object. Differently, shape from shading [13] and photometric stereo [17] compute dense surfaces by analyzing the variations of a pixel subject to different illumination sources. In our work we will focus exclusively on the multi view geometry and photometric aspects of such problems and present a joint algorithm which obtains reliable 3D reconstruction from an uncalibrated monocular sequence of images.

Regarding the multi-view geometry aspect, the sparse 3D surface estimation from a single video requires the extraction of a set of 2D image points for each frame. These selected points are then uniquely matched for the whole video

G. Maino and G.L. Foresti (Eds.): ICIAP 2011, Part I, LNCS 6978, pp. 660–669, 2011.

sequence thus creating image trajectories. The collection of such 2D trajectories define the motion of the image shape and it is subject to the metric properties of the 3D object and the camera position projecting the 3D points onto the image plane. The localization of the 2D image points is fatally sparse since good features to track and matches are restricted to a few particular regions in the image [14]. However, given a rigid object, very accurate sparse representation of the world can be obtained, even in the presence of interrupted 2D trajectories [12].

Differently, from a photometric perspective, Photometric Stereo (PS) computes dense 3D localization directly from image intensity variations in a single video sequence. Each surface point imaged by a camera reflects a light source with respect to its orientation and surface photometric properties. Thus, if enough views of the same surface point are given at different lighting positions, we may succeed to infer the 3D surface and its photometric properties (i.e. the albedo). However, this approach, in the case of a video sequence, requires the dense matching of each pixel from frame to frame or completely stationary shape with controlled lighting conditions. The latter case is the standard scenario and recent techniques [2] may compute lighting parameter and 3D surface without a prior calibration of the lighting setup. However, a well known problem implicit in the most PS reconstructions is the generalized bas-relief ambiguity. In few words, the same image pixels may correspond to different configurations of 3D surfaces and lighting sources. Choosing the right solution depends generally on a priori information of the 3D shape of the object.

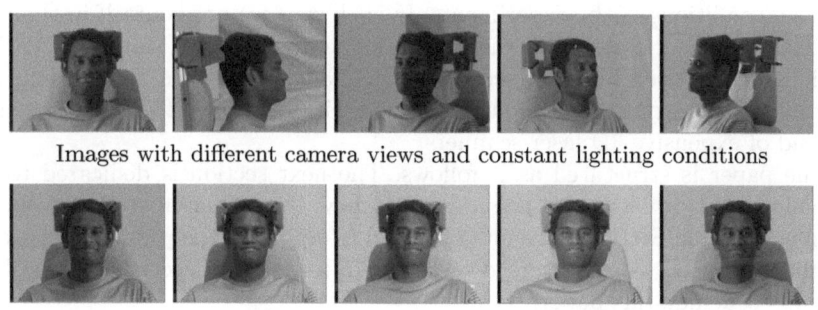

Images with different camera views and constant lighting conditions

Images with same camera view and varying lighting conditions

Fig. 1. An example of the set of images in the Multi-Pie database for subject 42

In this work, we design a 3D dense reconstruction pipeline that joins SfM and PS techniques in order to obtain reliable 3D reconstructions from image sequences. The generalized bas-relief ambiguity is reduced by obtaining a reliable localization of a set of sparse 2D points extracted from the images and used to obtain a metric 3D reconstruction of the shape. In this case, we can define a photo-geometric relation between the 3D sparse model and the dense 3D surface obtained with SfM and PS respectively. By solving for the transformation, we obtain the correct alignment of the two reconstructions up to an overall scale.

There are similar works in the literature that attempt to include SfM constraints into a dense reconstruction problem. Lim et al. [11] tried to recover correct depths from multiple images of a moving object illuminated by time varying lighting. They used multiple views of the object to generate a coarse planar surface based on the recovered 3D points and then they used PS in an iterative process to recover dense surface and align it into the recovered 3D point. Zhang et al. [18] use an iterative algorithm to solve a sub-constrained optical flow formulation. They use SfM to compute the camera motion and initialize the lighting on sparse features. Then, they iteratively recover the shape and lighting in a coarse-to-fine manner using an image pyramid. All such methods have known features and drawbacks. Other works such as [9] uses specific setups with colored lights or [1] active patterns using projectors in order to constraints the photometric ambiguities. Since our proposed method uses solely standard image sequences, we deal with a less constrained case than the two previously mentioned approaches.

The testing framework used to verify the effectiveness of our 3D reconstruction algorithm is the CMU Multi-Pie face database [7]. This database contains more than 750,000 images of 337 subjects recorded in up to four sessions over the span of five months. Subjects were imaged under 15 view points and 19 illumination conditions while displaying a range of facial expressions. Figure 1 shows samples of the database for different views and varying lighting conditions. No information is provided by this database to recover the 3D position of any point on the subjects. Notice that in such scenario, we deliberately choose not to use any a priori information about the calibration for both the cameras and lighting conditions of the experiments. In such way, we pose ourself in the most challenging scenario and with the largest modeling freedom. Many applications could be considered for such scenario, e.g. model-based face recognition, face morphing or creating 3D face databases using inexpensive off-the-shelf facilities instead of expensive 3D laser scanners.

The paper is structured as it follows. The next section is dedicated to the formulation of problem and it is described how the metric upgrade affects the results. Then, in Section 3, results obtained by applying the proposed approach on the MultiPIE database are discussed. And finally, in section 4 some brief remarks conclude this paper.

2   Sparse and Dense 3D Reconstruction

We first formulate the SfM and PS problems in the mathematical context of bilinear matrix factorization. In general, either the 2D image trajectories used by SfM and the image pixel variations in time can be both described by bilinear matrix models. For the case of SfM, the bilinear model contains the 3D shape coordinates and the camera projection matrices. Similarly, the PS case results in two factors that contain the object surface normals with the albedo and the lighting directions.

2.1 Structure from Motion

Structure from Motion algorithms simultaneously reconstruct the 3D position and camera matrices using a set of 2D points extracted from an image sequence. The inter-image relations are linked by the fact that a unique shape is projected into the images by a moving camera. Thus the 2D image trajectories created by this mapping can be used to estimate the 3D position of a shape if a sufficient baseline is given. A set of popular approaches compute simultaneously the 3D structure and camera motion via a factorization approach using solely the collection of such 2D trajectories. In more detail, the 3D structure and the camera projection matrices can be expressed as a bilinear matrix model.

In more detail, by defining the non-homogeneous coordinate of a point j in frame i as the vector $\mathbf{w}_{ij} = (u_{ij} \ v_{ij})^T$, we may write the measurement matrix W that gathers the coordinates of all the points in all the views as:

$$
W = \begin{bmatrix} \mathbf{w}_{11} & \cdots & \mathbf{w}_{1p} \\ \vdots & \ddots & \vdots \\ \mathbf{w}_{g1} & \cdots & \mathbf{w}_{gp} \end{bmatrix} = \begin{bmatrix} W_1 \\ \vdots \\ W_g \end{bmatrix}
\tag{1}
$$

where g is the number of frames and p the number of points. In general, the rank of W is constrained to be rank$\{W\} \leq r$ where $r \ll \min\{2g, p\}$

In the case of a rigid object viewed by an orthographic camera, if we assume the measurements in W are registered to the image centroid, the camera motion matrices R_i and the 3D points $\mathbf{S}_j$ can be expressed as:

$$
R_i = \begin{bmatrix} r_{i1} & r_{i2} & r_{i3} \\ r_{i4} & r_{i5} & r_{i6} \end{bmatrix} \quad \text{and} \quad \mathbf{S}_j = \begin{bmatrix} X_j & Y_j & Z_j \end{bmatrix}^T
\tag{2}
$$

where R_i is a 2×3 matrix that contains the first two rows of a rotation matrix (i.e. $R_i R_i^T = I_{2 \times 2}$) and $\mathbf{S}_j$ is a 3-vector containing the metric coordinates of the 3D point. Thus a 2D point j in a frame i is given by $\mathbf{w}_{ij} = R_i \mathbf{S}_j$. We can collect all the image measurements and their respective bilinear components R_i and $\mathbf{S}_j$ in a global matrix as in Eq. (1). Thus we can formulate the factorization model of the image trajectories as

$$
W = R_{2g \times 3} \ S_{3 \times p}
\tag{3}
$$

where the bilinear components R and S are defined as:

$$
R = \begin{bmatrix} R_1 \\ \vdots \\ R_g \end{bmatrix} \quad \text{and} \quad S_{sfm} = \begin{bmatrix} \mathbf{S}_1 & \cdots & \mathbf{S}_p \end{bmatrix}.
\tag{4}
$$

Expressing the camera projections and 3D points in such matrix form makes evident the rank constraint of W.

Given the rank relation: $\mathrm{rank}(\mathtt{W}) \leq \min\{\mathrm{rank}(\mathtt{R}), \mathrm{rank}(\mathtt{S}_{sfm})\}$, we have that the rank of the measurement is at most equal to three. This constraints is used to obtain a closed form solution for the 3D position of the points and the camera matrices as presented in the seminal paper of Tomasi and Kanade [15]. In the case of different imaging conditions remember that the simplistic assumption of an orthographic camera model has been extended to more complex affine cameras [10] or either projective ones [16].

2.2 Photometric Stereo

The principle at the base of PS is that an object illuminated by a light source will reflect light with respect to the surface orientation, light direction and intrinsic photometric properties of the shape. Thus, we can use a collection of the data representing the lighting variations of the pixels in order to infer the photometric properties of the shape. Notice that in this case we treat the object as being static and the light source moving – the aim here is to find a dense 3D reconstruction (i.e. for each pixel position in the image) of the object shape.

The chosen photometric model is based on a spherical harmonics representation of lighting variations [2] and it allows to frame PS as a factorization problem with normality constraints on one of the bilinear factors. Given a set of images of a Lambertian object with varying illumination, it is possible to extract the dense normals to the surface of the object $\mathbf{n}$, the albedos ρ and the lighting directions $\mathbf{l}$. For a 1st order spherical harmonics approximation, the brightness at image pixel j at frame i can be modeled as:

$$Y_{ij} = \mathbf{l}_i^\top \, \rho_j [1 \ \mathbf{n}_j^\top]^\top = \mathbf{l}_i \mathbf{s}_j$$

where $\mathbf{l}_i \in \mathbb{R}^4$, $\rho_j \in \mathbb{R}$, $\mathbf{z}_j \in \mathbb{R}^3$ with $\mathbf{n}_j^\top \mathbf{n}_j = 1$. A compact matrix form can be obtained for each pixel y_{ij} as:

$$\mathtt{Y} = \begin{bmatrix} y_{11} & \cdots & y_{1t} \\ \vdots & \ddots & \vdots \\ y_{f1} & \cdots & y_{ft} \end{bmatrix} = \begin{bmatrix} \mathbf{l}_1^\top \\ \vdots \\ \mathbf{l}_f^\top \end{bmatrix} \begin{bmatrix} \rho_1 \begin{bmatrix} 1 \\ \mathbf{n}_1 \end{bmatrix} & \cdots & \rho_t \begin{bmatrix} 1 \\ \mathbf{n}_t \end{bmatrix} \end{bmatrix} = \mathtt{L}\,\mathtt{N} \tag{5}$$

where a single image i is represented by the vector $\mathbf{y}_i = \begin{bmatrix} y_{i1} & \cdots & y_{it} \end{bmatrix}^T$. The $f \times 4$ matrix $\mathtt{L}$ contains the collection of the lighting directions while the $4 \times t$ matrix $\mathtt{N}$ the values for the normals and the albedos. Thus the 1st order spherical harmonics model enforces a rank four constraint over the image brightness of the scene. Similarly to the SfM case, it is possible to factorize the pixel values in $\mathtt{Y}$ to obtain a closed form solution that complies with the normal constraints (i.e. $\mathbf{n}_j^\top \mathbf{n}_j = 1$) as presented in [2].

Notice that we solve for the surface normals associated to each pixel. Normals integration is then required to recover the final 3D surface from the surface normals. Thus, after applying the overall PS algorithm we obtain a matrix $\mathtt{S}_{ps}$ of size $3 \times t$ containing the 3D coordinates of the surface. However this final step give a solution which is up to an unknown Generalized Bas Relief (GBR) transformation [3]. Figure 2 shows qualitatively the difference between a correct solution and a metric 3D surface. In order to find an unique solution, we use the SfM 3D metric shape to resolve the GBR ambiguity. How to estimate the correct transformation which respects the shape metric using SfM represents the main novel issue and the core of the work, which will be described in the next session.

Fig. 2. Left image shows the surface before the upgrade. The right image shows the surface after the metric upgrade for subject 42.

2.3   Photometric Stereo Metric Upgrade

The photometric stereo step estimates at each image pixel position the 3D surface S_{ps}. Notice however that a GBR transformation H such as [3]:

$$H = \begin{bmatrix} 1 & 0 & 0 \\ 0 & 1 & 0 \\ u & v & \lambda \end{bmatrix}, \tag{6}$$

can be multiplied to the recovered shape giving $\tilde{S} = HS_{ps}$. The shape $\tilde{S}$ is still a valid solution to the PS problem. Thus, we need to fix the GBR transformation that reflects the correct depth of the surface. If a set $\mathcal{O}$ of metric 3D coordinates in S_{ps} is available, we might be able to estimate the GBR parameters that define the correct metric surface. Such correspondences can be obtained through the mentioned SfM algorithm in Section 2.1. First, we extract a set of 2D points from a multi view image sequence such as the one showed in Figure 3. These points will form the matrix W as in Eq. (1). However notice that not all the points are visible in each view thus the matrix W will have missing entries. This leads to a factorization problem with missing data which can be solved with general purpose optimizers such as the BALM [5]. After this step, we have a set of sparse 3D metric coordinates which can be used to solve for the GBR ambiguity.

A further problem should be definitely solved. First, the SfM 3D shape S_{sfm} is up to an unknown rotation and in general it is not aligned with respect to the 3D surface estimated with PS (i.e. S_{ps}). This can be solved with the assumption that one of the views in the SfM sequence corresponds to the view of the sequence used by the PS algorithm. This is always true for the Multi-Pie database and, more in general, this is a strict assumption of our method. If we have such

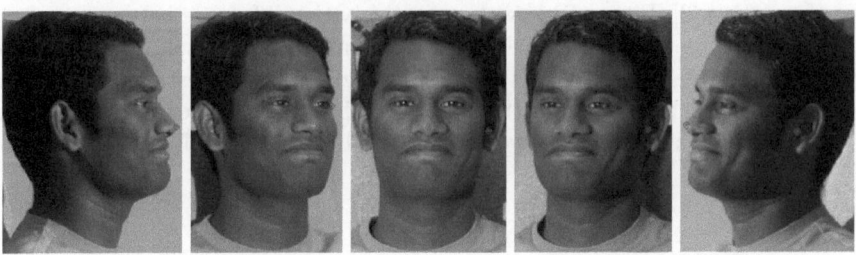

Fig. 3. 2D points on the image (Subject 42, Session 1, Recording 2)

image in common, the correspondence between the image point used by PS and SfM is also given. We call $\bar{S}_{ps}$ as the $3 \times p$ matrix containing the corresponding points between PS and SfM sequences. Thus, we can define the following *photogeometric transformation* A such as:

$$S_{sfm} = H\ R_{rel}\ \bar{S}_{ps} = A\ \bar{S}_{ps} \tag{7}$$

where R_{rel} is a 3×3 rotation matrix that aligns the PS and SfM 3D points. The solution can be found by computing the matrix A with standard Least Squares that simultaneously aligns and solves for the GBR shape ambiguity.

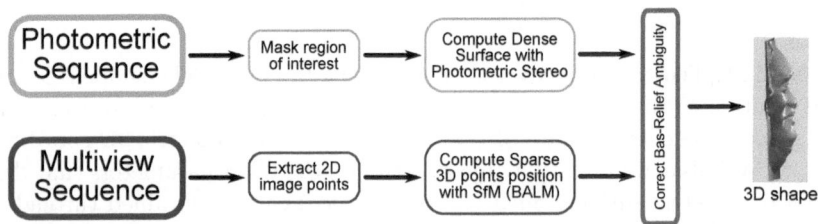

Fig. 4. Scheme of the 3D dense reconstruction algorithm pipeline

In summary, Figure 4 depicts the algorithm pipeline for 3D dense reconstruction. Our approach uses two different sequences: one contains 20 single view images with different illuminations to be used for the PS reconstruction, and the other sequence contains 15 images showing multiple views of the same subject used to extract 2D image points for the BALM algorithm. First, we preprocess the photometric sequence in order to select only the part of the image where the skin is present. A treshholding technique on Hue channel of the sequence with 20 images is used and refined with morphological operators to remove the background and clothing. A dense 3D surface is obtained applying the photometric stereo method on the masked images. On the other hand, in the SfM phase, some corresponding points are marked in 15 images, as it can be seen in Figure 3. As some of the points may be invisible in some views the resulted matrix will have some missing entries. To compute the 3D position of marker points and dealing with such missing data as well, the BALM method is used as the SfM engine, which results in a sparse reconstruction. Finally, the resulted reconstructions, dense and sparse one, are merged to reduce the bas-relief ambiguity effect. At this step, the sparse points are projected on to the image plane and their corresponding points on the surface are extracted. Having these two sets of points resulted by PS and SfM methods, we can solve Eq. (7) for A. As soon as we find the *photo-geometric transformation* A which relates these two point clouds, we can apply it on the dense surface of PS and correct all the points of such surface.

3 Results on Multi-Pie Database

This section shows our results for two sample subjects of the MultiPIE database (42 and 46). Figure 5 presents the dense surface computed with photometric stereo (top left) and the surface with the attached texture (top right). As it is apparent in this figure, the elevation of surface points from the image plane does not comply with the metric condition. Figure 6 shows the dense surface after bas-relief correction.

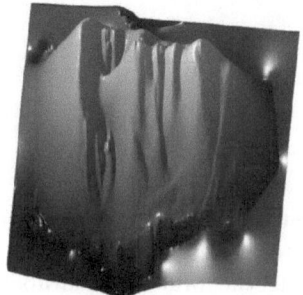 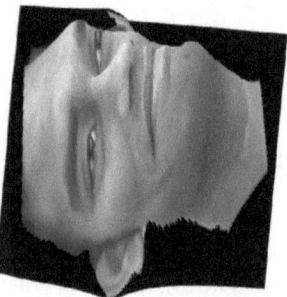

Fig. 5. Dense Surface and texture from Photometric Stereo reconstruction for subject 42

Fig. 6. Surface after bas-relief correction for subject 42

Figures 7 and 8 illustrate the results of proposed approach for another subject in the database. Computed dense surface and the surface with the attached texture are presented in Fig. 7. The dense surface after bas-relief correction is presented in Fig. 8.

 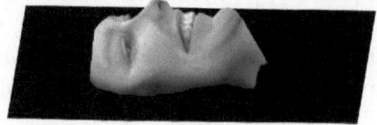

Fig. 7. Dense Surface and texture from Photometric Stereo reconstruction for subject 46

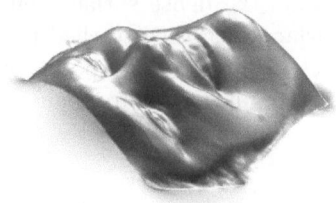

Fig. 8. Surface after bas-relief correction for subject 46

4   Conclusions

In this paper, we have presented a 3D reconstruction pipeline to obtain dense 3D metric surfaces using both Photometric Stereo and Structure from Motion techniques. The method has been tested using the Multi-Pie database in an uncalibrated scenario. The 3D reconstructions are satisfactory, however we plan to use more complex photometric models in order to grasp finer details of the objects that may strongly diverge from the Lambertian surface assumption (e.g. glasses, hairs). Another point for future investigations is to couple more deeply both SfM and PS techniques with the aim to achieve a simultaneous estimation of both photometric and 3D structure components. Such future work will be tested on ground truth data in order to be able to compare our reconstructed surfaces with real ones.

References

1. Aliaga, D.G., Xu, Y.: A self-calibrating method for photogeometric acquisition of 3d objects. IEEE Transactions on Pattern Analysis and Machine Intelligence 32(4), 747–754 (2010)
2. Basri, R., Jacobs, D., Kemelmacher, I.: Photometric stereo with general, unknown lighting. International Journal of Computer Vision 72(3), 239–257 (2007)

3. Belhumeur, P.N., Kriegman, D.J., Yuille, A.L.: The bas-relief ambiguity. International Journal of Computer Vision 35(1), 33–44 (1999)
4. Blostein, D., Ahuja, N.: Shape from texture: Integrating texture-element extraction and surface estimation. IEEE Transactions on Pattern Analysis and Machine Intelligence 11(12), 1233–1251 (2002)
5. Del Bue, A., Xavier, J., Agapito, L., Paladini, M.: Bilinear factorization via augmented lagrange multipliers. In: Daniilidis, K., Maragos, P., Paragios, N. (eds.) ECCV 2010. LNCS, vol. 6314, pp. 283–296. Springer, Heidelberg (2010)
6. Favaro, P., Soatto, S.: A geometric approach to shape from defocus. IEEE Transactions on Pattern Analysis and Machine Intelligence, 406–417 (2005)
7. Gross, R., Matthews, I., Cohn, J., Kanade, T., Baker, S.: Multi-pie. Image and Vision Computing 28(5), 807–813 (2010)
8. Hartley, R., Zisserman, A.: Multiple view geometry, vol. 642. Cambridge University Press, Cambridge (2000)
9. Hernández, C., Vogiatzis, G.: Self-calibrating a real-time monocular 3d facial capture system. In: Proceedings International Symposium on 3D Data Processing, Visualization and Transmission (3DPVT) (2010)
10. Kanatani, K., Sugaya, Y., Ackermann, H.: Uncalibrated factorization using a variable symmetric affine camera. IEICE Transactions on Information and Systems 90(5), 851 (2007)
11. Lim, J., Ho, J., Yang, M., Kriegman, D.: Passive photometric stereo from motion. In: Proceedings of the Tenth IEEE International Conference on Computer Vision, vol. 2, pp. 1635–1642. IEEE Computer Society, Los Alamitos (2005)
12. Marques, M., Costeira, J.: Estimating 3d shape from degenerate sequences with missing data. Computer Vision and Image Understanding 113(2), 261–272 (2009)
13. Prados, E., Faugeras, O.: Shape from shading. In: Handbook of Mathematical Models in Computer Vision, pp. 375–388 (2006)
14. Shi, J., Tomasi, C.: Good features to track. In: Proc. IEEE Conference on Computer Vision and Pattern Recognition, pp. 593–600 (1994)
15. Tomasi, C., Kanade, T.: Shape and motion from image streams under orthography: A factorization approach. International Journal of Computer Vision 9(2) (1992)
16. Triggs, B.: Factorization methods for projective structure and motion. In: Proc. IEEE Conference on Computer Vision and Pattern Recognition, San Francisco, pp. 845–851 (1996)
17. Woodham, R.: Photometric method for determining surface orientation from multiple images. Optical Engineering 19(1), 139–144 (1980)
18. Zhang, L., Curless, B., Hertzmann, A., Seitz, S.M.: Shape and motion under varying illumination: Unifying structure from motion, photometric stereo, and multi-view stereo. In: Proceedings of the Ninth IEEE International Conference on Computer Vision, vol. 2, p. 618. IEEE Computer Society, Los Alamitos (2003)

Genetic Normalized Convolution

Giulia Albanese[1], Marco Cipolla[2], and Cesare Valenti[2]

[1] Dipartimento di Scienze dell'Informazione
Università di Bologna, via Mura Anteo Zamboni 7, 40127, Italy
galbanes@cs.unibo.it
[2] Dipartimento di Matematica e Informatica
Università di Palermo, via Archirafi 34, 90123, Italy
{cvalenti,mcipolla}@math.unipa.it

Abstract. Normalized convolution techniques operate on very few samples of a given digital signal and add missing information, trough spatial interpolation. From a practical viewpoint, they make use of data really available and approximate the assumed values of the missing information. The quality of the final result is generally better than that obtained by traditional filling methods as, for example, bilinear or bicubic interpolations. Usually, the position of the samples is assumed to be random and due to transmission errors of the signal. Vice versa, we want to apply normalized convolution to compress data. In this case, we need to arrange a higher density of samples in proximity of zones which contain details, with respect to less significant, uniform parts of the image. This paper describes an evolutionary approach to evaluate the position of certain samples, in order to reconstruct better images, according to a subjective definition of visual quality. An extensive analysis on real data was carried out to verify the correctness of the proposed methodology.

1 Introduction

Normalized convolution is a signal processing method that allows to reconstruct an image when just a few pixels are available due to the presence, for example, of noise or instrumental error. These pixels, also called certain samples, are assumed to have an uniform random distribution. Adaptive normalized convolution consists in a pipeline of normalized convolutions to improve the overall quality of the resulting reconstruction, though it requires a lot of computation.

Aim of this paper is the robustness evaluation of normalized convolution for data coding and compression purposes. In this case, we want to establish the correct amount of samples, together with their positions, needed to reconstruct images having an high perceived quality. In particular, we present a genetic algorithm able to locate these samples, according to an attentive model based on edges and centers of symmetry that usually correspond to details in the scene. A recent metrics has been applied to measure the subjective quality of the solution, trough the analysis of luminance, contrast and structures.

We experimentally verified on a database of real images that our method outperforms classic normalized convolution. Moreover, it can be used as a pre-processing step to enhance adaptive normalized convolution, too.

G. Maino and G.L. Foresti (Eds.): ICIAP 2011, Part I, LNCS 6978, pp. 670–679, 2011.
© Springer-Verlag Berlin Heidelberg 2011

Section 2 briefly describes both normalized convolution and its adaptive variant. Section 3.1 sketches phase congruency and radial symmetry transforms to detect regions of interest in images. The structural similarity metrics is reported in section 3.4 as part of the fitness function of our genetic algorithm, introduced in the rest of section 3. Experiments and conclusions are reported in sections 4 and 5.

2 Normalized Convolution

Normalized convolution represents an important tool in the digital signal processing field. It was described for the first time by Knutsson and Westin [1] who pointed out the opportunity to provide also a confidence measure of the available samples. Actually, a map should indicate the presence degree of a sample in a given position. In particular, a binary map would indicate just the absence or presence of the signal. The underlying theory is simple and its implementation appears quite fast.

Let S be the positive map that represents certain samples of a digital image I. If we indicate by $\{S \cdot I\}$ the pixelwise product of S and I and by $\{K * I\}$ the usual convolution with a kernel K, then normalized convolution is defined by

$$NC(I, S, K) = \{K * S \cdot I\} / \{K * S\}$$

In other words, to reconstruct the whole image I from its samples specified in S, we just have to weight $\{K * S \cdot I\}$ by the confidence $\{K * S\}$ of the results generated.

The kernel, centered in the origin, is usually defined as a Gaussian-like surface

$$K_{x,y} = \begin{cases} r^{-2} \cos^2 \left(\frac{\pi r}{2 r_{\max}} \right) & \text{if } r < r_{\max} \\ 0 & \text{otherwise} \end{cases}$$

where $r = \sqrt{(x^2 + y^2)}$. To avoid over-smoothing the output image, K should be big enough to contain just some pixels of the input signal. Vice versa, if the distance between the nearest samples in S is greater than the size $r_{\max}$ of K, then the reconstructed image will contain gaps. Without a priori information, $r_{\max}$ is automatically set to the minimal distance among the available samples to reduce artifact effects along edges: at least one pixel lies always within the radius. It is noteworthy that fast implementations can be developed for both convolving images with very few samples and computing the distance among these samples.

A variant of this algorithm is known as adaptive normalized convolution and it modulates both the size and the shape of the kernel K, according to the position of certain samples [2,3]. In this case, implementing an optimized and efficient custom convolution routine can be quite difficult. Indeed, a different filter can be arranged for each pixel of the output image and an estimate of the gradient of the whole image is used to determine this proper kernel. Obviously, this gradient itself is just an approximation since it has to be computed from available samples specified by S. This preprocessing step, known as derivative of normalized convolution or normalized differential convolution, requires a considerable amount of computational time. Actually, we do not to use the adaptive normalized convolution technique because its performances usually does not justify the enhancement of the final output image, which we aim to achieve by better positioning the samples in S. Anyway, adaptive normalized convolution can be applied with the sample positions returned by our method.

3   Genetic Normalized Convolution

Genetic algorithms have been already applied to the inpainting problem (i.e. automatic completing missing areas and spatiotemporal restoration by using image samples) [4], but, to our knowledge, this is the first description of a method that combines normalized convolution and an evolutionary approach. We have not modified this former algorithm which will be used to code and compress data. We implemented a genetic algorithm to locate the best position of certain samples. We desire to hold all details in the image by assigning most information to them. In order to decide if such parts of the image should be considered as a regions of interest, we used models suggested by the Gestalt theory. Anyway, enough samples still have to be devoted to uniform zones, which are usually due to the background. The following section regards basic algorithms helpful to identify important details in images.

3.1   Regions of Interest

Regions of interest detectors are usually applied to selectively process images, to locate peculiar features and to simulate active vision systems. Many definitions of regions of interest do exist and provide complementary information as edges, corners, blobs and symmetries. In particular, we will sketch a fast radial symmetry transform and an effective edge detector. These methods are the basis for a combined detector to manage the positions of certain pixels needed by the normalized convolution process.

A variety of algorithms to locate centers of radial symmetry in digital images have been described in the literature. Some of these methods require custom hardware or do complex filtering within big windows. Often, the size of symmetric zones is unknown, thus multiresolution pyramidal approaches were proposed too [5]. A few methods reduce the computational time by letting the gradient of the image drive the analysis in some way [6]. In particular, we used the fast radial symmetry transform [7] because it returns in real time an output coherent with present attentive human models. The key idea is to use an accumulation map, as for the Hough transform, which highlights the contribution of the gradient vector field. If a point is a center of symmetry, then it receives a degree proportional to the number of gradient vectors heading to it (see figure 1a). Actually, the radius of the scanning window, that is the length of the vectors to consider, should be set by the user, but satisfactory results can be obtained by merging together the outputs corresponding to a set of different radii (figure 2b shows an example with radius from 1 to 11 with an incremental step equal to 2).

In [8] it was presented an algorithm to locate both corners and edges from phase congruency information. This is particularly robust against changes in illumination and contrast because contours have many of their frequency components in the same phase, as in figure 1b. Moreover, these edges present a single response as line features, are less prone to the presence of noise and are largely independent of the local contrast (see figure 2c). These observations are due to the fact that this algorithm is not based on a first-derivative operator (on the contrary of Canny or Sobel filters) which usually exalts also the contribution of negligible details. The whole method is computationally intensive, but fortunately we need to apply both phase congruency and the above symmetry detector just at the beginning of our genetic algorithm to create a regions of interest mask that comprises both edges and centers of radial symmetries.

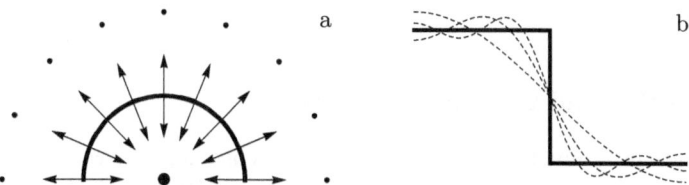

Fig. 1. (a) dots represent the degree of radial symmetry pointed by gradient vectors. (b) Fourier components, as dashed lines, share the same zero crossing of a step edge.

3.2 Population

Given an image I, our algorithm creates a corresponding mask ROI that measures the pixels' 'importance'. This mask is the superimposition, normalized in $[0,1]$, of edges and centers, obtained trough the methods described in section 3.1. To reduce the effect of pointlike noise and to spread the regions of interest, a Gaussian convolution, with radius equal to 19 and standard deviation equal to 6.3, is performed. The pixelwise product of the resulting mask with its thresholded version (on the average value), puts in evidence all details [9,10,11] (figure 2d).

Each genome A of the population encodes co different indices with values in $[1, n^2]$, where the image has exactly $n \times n$ pixels and co samples will be used to reconstruct it. Even if a convenient coding would be the use of whole permutations of n^2 elements, we limited the genome dimension to co, thus reducing drastically the storage complexity of the algorithm. Indeed, we want to remark that usually $co \ll n^2$.

The initial set of ni random individuals have fd genes constrained on details highlighted by the mask ROI. This trick is not necessary and does not follow any evolutionary strategy, but we verified that it reduces the number ng of generations required to reach a stable solution. It must be noted that our algorithm does not constrain the number of samples on details during the evolution of the population. The cardinality ni of the population, the quantity fd of details and further parameters will be considered in section 4.

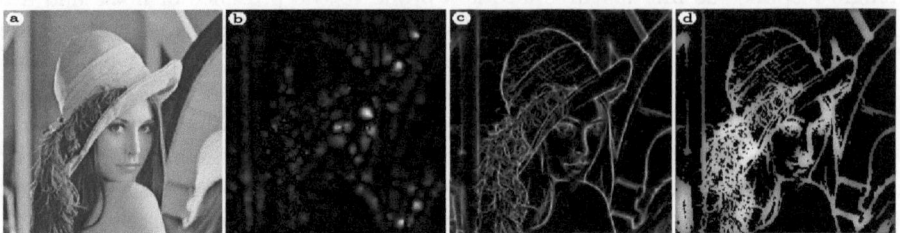

Fig. 2. (a) input image. (b) radial symmetries. (c) phase congruency edges. (d) regions of interest map.

3.3 Selection and Genetic Operators

During the development of our algorithm, we considered a variety of selection methods (e.g. roulette wheel, ranking and random) [12] to pick the chromosomes for crossover. The final version of the system includes only a 4-way tournament, without duplication of already selected individuals, because other strategies returned poor results. An elitist selection is also applied to assure that the worst chromosome will be eliminated and the best one be present as two copies in the next population too. Although this approach reduces the variability of the hereditary characteristics, we ascertained that it increases the result accuracy.

Since our individuals store just part of permutations, we developed ad hoc tools instead of using standard genetic operators for permutations (e.g. CX, OX and PMX). Let us indicate with $A \bigcup B$ and $A \bigcap B$ respectively the union and intersection sets of genes owned by two parents A and B. Two offsprings C and D are obtained by joining the indices in $A \bigcap B$ with half of the indices in $(A \bigcup B) \backslash (A \bigcap B)$, randomly chosen (see figure 3). This operation is not subject to any probability because, due to data structure optimizations, we always force the same percentage pc of chromosomes to undergo crossover.

$$A = (\underline{26}\ 11\ \underline{30}\ \underline{9}\ \underline{15}\ 24\ 14\ 27\ 5\ 34)$$
$$B = (21\ 6\ \underline{30}\ \underline{15}\ \underline{9}\ 29\ 19\ \underline{26}\ 10\ 16)$$
$$P((A \bigcup B) \backslash (A \bigcap B)) = (29\ 16\ 10\ 19\ 27\ 21\ \mid\ 6\ 11\ 34\ 14\ 5\ 24)$$
$$C = (\underline{26}\ 29\ \underline{30}\ \underline{9}\ \underline{15}\ 16\ 10\ 19\ 27\ 21)$$
$$D = (6\ 11\ \underline{30}\ \underline{15}\ \underline{9}\ 34\ 14\ \underline{26}\ 5\ 24)$$

Fig. 3. A crossover example. Common genes in A and B are underlined. P indicates a random permutation of the indices.

In the case of complete permutations, mutation is accomplished usually by swapping a pair of random indices. We cannot apply this method because, while the order of the indices in an individual is irrelevant, we must ensure to replace a gene already coded with another one that is not in the genome yet. This operation corresponds to move just a sample to another position. We introduced the parameter mg to guarantee that an adequate number of genes are modified to significantly change the content of the image. Moving mg samples in a completely random fashion does not improve the overall quality of the reconstructed image, because, on average, the same number of samples remain in the same zones of the image. To slightly change the aspect of the scene (see figure 4), we verified that a better strategy consists in moving mg samples within their own neighborhoods of radius 3. As for crossover, this operation is not subject to any probability because we force always the same percentage mg of genes to change and the same percentage pm of chromosomes to undergo mutation.

$$A = (26\ 11\ 30\ \underline{9}\ 15\ 24\ 14\ \underline{27}\ 5\ 34)$$
$$B = (26\ 11\ 30\ \underline{8}\ 15\ 24\ 14\ \underline{29}\ 5\ 34)$$

Fig. 4. A chromosome A and its mutated version B. Small perturbations, underlined here, usually occur.

3.4   Quality Metrics and Fitness Function

Many metrics were defined to evaluate the similarity between images. For example, mean square error and its variant peak signal-to-noise ratio are widely applied to quantify pixelwise differences between two images, though this approach does not satisfy a real human perception system. That is, it can be proved that very altered versions of the same image still present almost identical PSNR values. It must be noted that the bigger this metrics is, the better the image is perceived. Generally speaking, a medium quality image should have a PSNR value not smaller than 30dB.

A recent subjective metrics, known as structural similarity, was introduced in [13] to compare local patterns of pixel intensities, after luminance and contrast normalization trough a z-score function. This method assumes that the main structures in the scene are independent from luminosity, which should be isolated. The similarity index between two images with ℓ gray levels, means μ and standard deviations σ is given by

$$SSIM(I_1, I_2) = \frac{(2\mu_1\mu_2 - \kappa_\mu)(2\sigma_{12} + \kappa_\sigma)}{(\mu_1^2 + \mu_2^2 + \kappa_\mu)(\sigma_1^2 + \sigma_2^2 + \kappa_\sigma)}$$

where σ_{12} represents the covariance of the images and the constants $\kappa_\mu = \varepsilon_\mu(\ell-1)^2$ and $\kappa_\sigma = \varepsilon_\sigma(\ell-1)^2$, with $0 < \varepsilon_\mu \ll 1$ and $0 < \varepsilon_\sigma \ll 1$, avoid instability when $\mu_1^2 + \mu_2^2$ or $\sigma_1^2 + \sigma_2^2$ is close to 0.

Again, the bigger the value of this metrics is, the better the image is perceived. Assessments derived from SSIM keep count of the whole image and are closer to subjective judgments. Sometimes, PSNR provides very different values in the case of images which can be considered almost identical and, therefore, with similar SSIM values. For the sake of completeness, we compare the reference image and the reconstructed one by both SSIM and PSNR.

Our fitness function f to evaluate a chromosome A is just the inner product, normalized to the number of pixels, between the regions of interest mask (see section 3.1) and the SSIM value of the original image I and its reconstructed version, obtained trough the map of samples coded in A

$$f(A) = \langle ROI(I), \ SSIM(I, NC(I, S(A), K)) \rangle \ / \ n^2$$

Better individuals correspond to higher values of f. Despite its simplicity, this formula provided enough variability to produce good results for all images we considered. The population evolves until a stable solution is found or merely the maximum number of generation is reached.

4   Experimental Setup

We verified the performance of our methodology on a database of gray scale images, freely available on the Internet. We studied 50 images that contain

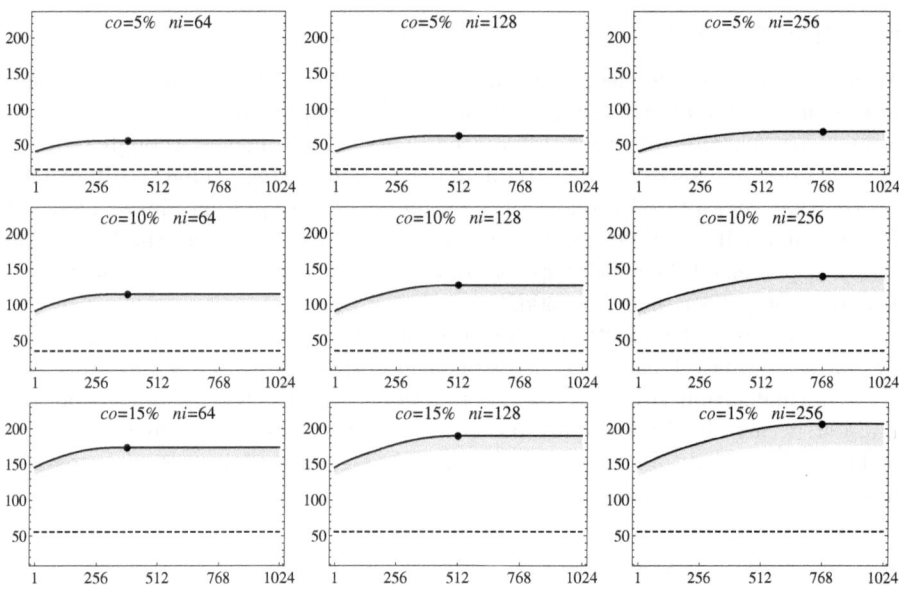

Fig. 5. Average fitness trend versus number of generations, with respect to number ni of individuals and percentage co of confident samples. The values of f lie within the gray band. Average fitness due to random samples is reported as dashed line. Suggested number of generations is indicated by the point.

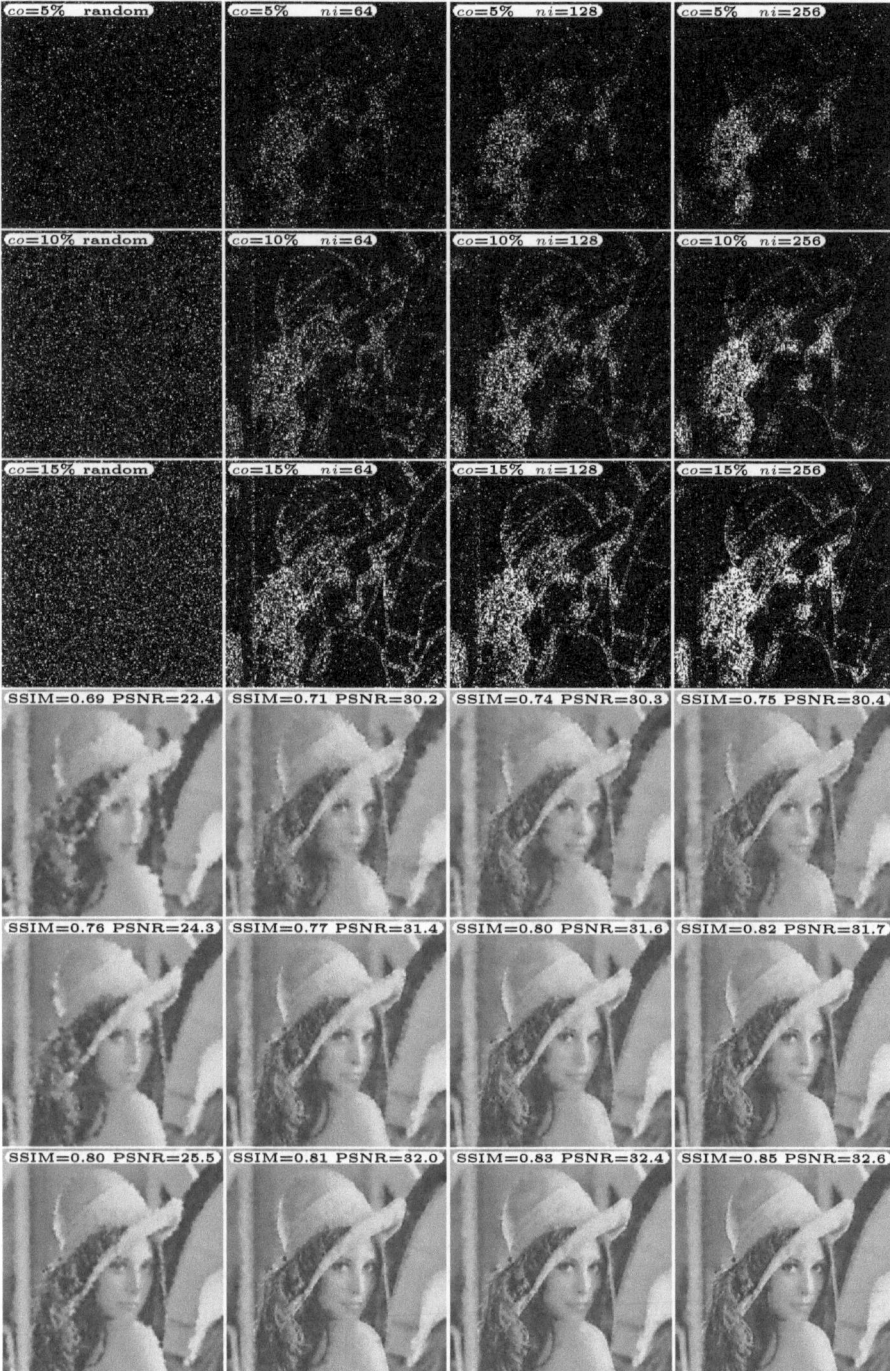

Fig. 6. Comparison among random samples and samples arranged by our algorithm

structured and textured scenes with 256×256 or 512×512 pixels. They were compared with the reconstructed versions by the SSIM quality metrics. Plots in figure 5 show the average fitness during the genetic evolution on the entire database. The gray band represents the minimum and maximum values of f. The average value is close to the best one (i.e. the maximum), thus most chromosomes really converge to the solution found by our method. Further parameters, as the percentage co of certain samples and the number ni of individuals are also reported. The average fitness of an image reconstructed from samples in completely random positions is depicted as a dashed line. Reconstruction examples are shown in figure 6. It is quite evident that our genetic algorithm is able to locate samples that improve the performance of the usual normalized convolution, regardless the values of ni and co.

Various experiments were carried out to fine tune the percentages $mg \in \{0.10, 0.40\}$ of genes to mutate, $fd \in \{0.10, 0.40\}$ of details to maintain, $pc \in \{0.70, 0.95\}$ and $pm \in \{0.05, 0.30\}$ of chromosomes for crossover and mutation, respectively. As a rule of thumb, $mg = 0.25$, $fd = 0.15$, $pc = 0.95$, $pm = 0.05$ allow to reduce the run time and to achieve satisfactory final images. For all these values, we studied the results corresponding to an incremental step equal to 0.05 and a number ni of individuals equal to 64, 128 and 256. In the case of lots of chromosomes, a better solution is expected but many generations are needed to let it stabilize. The following sets of suitable parameters, indicated trough a point in the plots of figure 5, can be used independently of the number co of certain samples (as percentage of the image size): $ni = 64$ and $ng = 384$, $ni = 128$ and $ng = 512$, $ni = 256$ and $ng = 768$.

5   Results and Conclusions

Normalized convolution is an important tool to reconstruct images from few samples and it is particularly useful when part of the information is lost due to noise or when a failure occurs during data transmission. Usually, these samples are assumed to be randomly distributed, in a uniform fashion. This paper considers normalized convolution for data coding and compression. In this case, we are interested in the reconstruction from a given number of certain samples or below an error threshold. A genetic algorithm has been presented to locate the best positions of the samples and an extensive experiment was carried out to verify its robustness on real images. In particular, an attentive model, based on phase congruency and radial symmetry, was used to create a proper initial population, while a structural similarity metrics was adopted to measure the perceived quality of the solution. It is noteworthy that this image can be slightly improved by using the usual adaptive normalized convolution on the samples chosen by our genetic algorithm, in spite of the increased amount of computations.

We developed our system in the interpreted high level MatLab language. This means that a few minutes are still necessary to elaborate a 256×256 pixels image, though a C/MPI compiled version of the program should be able to reach a feasible solution very quickly. We are already working on an improved, parallel version of our genetic algorithm to get benefit from new multicore CPUs, available on present computers. We experimentally verified that the time complexity is roughly linear with respect to both the number of individuals and the size of the image to reconstruct. On the other hand, the number of certain samples does not cause any overload, due to the reduced quantity of genes coded by the chromosomes.

References

1. Knutsson, H., Westin, C.: Normalised and differential convolution. In: IEEE Proc. CVPR on Computer Society Conf., vol. 15-17, pp. 515–523 (1993)
2. Pham, T.Q., van Vliet, L.J.: Normalized averaging using adaptive applicability functions with applications in image reconstruction from sparsely and randomly sampled data. In: Bigun, J., Gustavsson, T. (eds.) SCIA 2003. LNCS, vol. 2749, pp. 485–492. Springer, Heidelberg (2003)
3. Pham, T.Q., van Vliet, L.J., Schutte, K.: Robust Fusion of Irregularly Sampled Data using Adaptive Normalized Convolution. Journal Applied Signal Processing (2006)
4. Kim, E.Y., Jung, K.: Object Detection and Removal Using Genetic Algorithms. In: Zhang, C., W. Guesgen, H., Yeap, W.-K. (eds.) PRICAI 2004. LNCS (LNAI), vol. 3157, pp. 411–421. Springer, Heidelberg (2004)
5. Di Gesù, V., Valenti, C.: Detection of regions of interest via the Pyramid Discrete Symmetry Transform. In: Proceedings of International Workshop on Theoretical Foundations of Computer Vision (1997)
6. Reisfeld, D., Wolfson, H., Yeshurun, Y.: Context Free Attentional Operators: The Generalized Symmetry Transform. Int'l J. Computer Vision 14, 119–130 (1995)
7. Loy, G., Zelinsky, A.: Fast Radial Symmetry for Detecting Points of Interest. IEEE Trans. on Pattern Analysis and Machine intelligence 25(8) (2003)
8. Kovesi, P.: Phase Congruency Detects Corners and Edges. In: The Australian Pattern Recognition Society Conference: DICTA Sydney 2003, pp. 309–318 (2003)
9. Lowe, D.G.: Object recognition from local scale-invariant features. In: Proceedings of the 7th International Conference on Computer Vision, pp. 1150–1157 (1999)
10. Lindeberg, T.: Feature detection with automatic scale selection. International Journal of Computer Vision 30(2), 79–116 (1998)
11. Mikolajczyk, K., Schmid, C.: Scale and affine invariant interest point detectors. International Journal on Computer Vision 60(1), 63–86 (2004)
12. Michalewicz, Z.: Genetic Algorithms + Data Structures = Evolution Programs. Springer, Heidelberg (1996)
13. Wang, Z., Bovik, A.C., Sheikn, H.R., Simoncelli, E.P.: Image Quality Assessment: From Error Visibility to Structural Similarity. IEEE Transactions on Image Processing 13(4), 600–612 (2004)

Combining Probabilistic Shape-from-Shading and Statistical Facial Shape Models

Touqeer Ahmad[1,2,*], Richard C. Wilson[1], William A.P. Smith[1], and Tom S.F. Haines[3]

[1] Department of Computer Science, The University of York, UK
[2] School of Science and Engineering, LUMS, Pakistan
[3] Queen Mary University of London, UK
{touqeer,wilson,wsmith}@cs.york.ac.uk, tom.haines@eecs.qmul.ac.uk

Abstract. Shape-from-shading is an interesting approach to the problem of finding the shape of a face because it only requires one image and no subject participation. However, SfS is not accurate enough to produce good shape models. Previously, SfS has been combined with shape models to produce realistic face reconstructions. In this work, we aim to improve the quality of such models by exploiting a probabilistic SfS model based on Fisher-Bingham 8-parameter distributions (FB_8). The benefits are two-fold; firstly we can correctly weight the contributions of the data and model where the surface normals are uncertain, and secondly we can locate areas of shadow and facial hair using inconsistencies between the data and model. We sample the FB_8 distributions using a Gibbs sampling algorithm. These are then modelled as Gaussian distributions on the surface tangent plane defined by the model. The shape model provides a second Gaussian distribution describing the likely configurations of the model; these distributions are combined on the tangent plane of the directional sphere to give the most probable surface normal directions for all pixels. The Fisher criterion is used to locate inconsistencies between the two distributions and smoothing is used to deal with outliers originating in the shadowed and specular regions. A surface height model is then used to recover surface heights from surface normals. The combined approach shows improved results over the case when only surface normals from shape-from-shading are used.

1 Introduction

Shape-from-shading(SfS) is a method of finding the shape of a surface from a single image, and has been used previous for finding 3D face models. Faces are quite suitable for this approach because they have a fairly constant albedo over most of the surface (excluding eyes, facial hair and so on) and the shape varies smoothly. Capturing a face in this way only requires one image and no subject participation and so it is non-invasive and can be done at a distance. However,

---

* Touqeer Ahmad is thankful to EURECA scholarship scheme for fully funding his M.Sc. at University of York.

G. Maino and G.L. Foresti (Eds.): ICIAP 2011, Part I, LNCS 6978, pp. 680–690, 2011.
© Springer-Verlag Berlin Heidelberg 2011

even state-of-the-art SfS methods are not accurate enough to produce correct face shapes on their own.

In order to improve the quality of the face shape, a model is required. These may vary from specific constraints on the shape to complete models of the shape of faces; for example Prados and Faugeras[16] enforce convexity through the location of singular points. Zhao and Chellappa[17] use the symmetry of faces as a constraint. Samaras and Metaxas[18] use a deformable model of face-shape. Blanz et al.[6] exploit a full 3D morphable head model built from laser range scans which also includes texture. By minimising the difference between the rendered model and the measured face image, they can estimate face-shape, pose and lighting at the same time. Smith and Hancock[15] combined SfS with a statistical model of face shape. This face model was derived from 3D face range scans and provided sufficient constraints on the shape to produce realistic models of faces. This method exploited the SfS method of Worthington and Hancock[3] However, Haines and Wilson [1] presented a more reliable SfS algorithm which use directional statistics; specifically Bingham, Bingham-Mardia, von-Mises-Fisher and Fisher-Bingham 8-parameter distributions (FB_8) [10] to model different entities involved in any SfS algorithm Cone constraint [3], boundary and gradient information and resulting surface normals [1]. In our work we make use of advantages of both of these models and combine the statistical face shape models of Smith and Hancock with probabilistic Shape-from-Shading method of Haines and Wilson in a probabilistic framework to give a more reliable facial Shape-from-Shading method.

These two models produce results on different spaces and using different probability distributions. The SfS algorithm of Haines and Wilson outputs the surface normal directions on the unit sphere as Fisher-Bingham distributions (FB_8) for every pixel whereas statistical model of Smith and Hancock gives a Normal distribution on the tangent plane for each pixel. The two models can only be combined when they are in the same space and are represented using tractable distributions.

2   Background

This research work aims to improve the quality of SfS face-shape models by combining a probabilistic Shape-from-Shading method with a statistical shape models of human faces. The goal is to exploit the distributions provided by the two methods to locate a maximum-likelihood shape which accounts for the uncertainties in both models. To commence, we briefly describe the formulation of the two models, and the exponential map, which form the key elements of our framework.

2.1   Shape-from-Shading

Haines and Wilson presented a probabilistic Shape-from-Shading algorithm based on Markov random fields and belief propagation [1]. Directional statistics, specifically FB_8 distributions[9] were used for the probabilistic representation of surface orientation. In particular, the method produces a FB$_8$ distribution describing the surface normal at each point on the image

$$p_{\mathrm{FB}}(\hat{\mathbf{x}}; \mathbf{A}) \propto \exp(\mathbf{b}^T \hat{\mathbf{x}} + \hat{\mathbf{x}}^T \mathbf{A} \hat{\mathbf{x}}) \tag{1}$$

where $\hat{\mathbf{x}}$ is the surface normal direction, $\mathbf{b}$ is the Fisher parameter and $\mathbf{A}$ is Bingham matrix and is symmetric. In this method, the irradiance and smoothness constraints are defined in terms of directional distributions. The global distribution of normals is described using a Markov random field model. Haines and Wilson used the hierarchical belief propagation method of Felzenszwalb and Huttenlocher[11] to find the marginals of the directional distributions; once the belief propagation has converged the Shape-from-Shading algorithm gives surface orientation for each pixel using an FB$_8$ distribution. Haines and Wilson's approach generally performed well as compared to Lee & Kuo[2] and Worthington & Hancock[3] and has the advantage of characterizing the uncertainty in normal direction at any point.

2.2   The Exponential Map

Directional data can be naturally modelled as points on the surface of a unit sphere. The direction is given by the unit vector from the origin to the point on the surface. This is a non-linear manifold which highlights some of the difficulties in computing statistics with surface normals.

The exponential map is a map from points on the manifold to points on a tangent space of the manifold. As the tangent space is flat (i.e. Euclidean), we can calculate quantities on the tangent space in a straightforward way. The map has an origin, which defines the point at which we construct the tangent space of the manifold. Formally, the definition of these properties as follows: Let T_M be the tangent space at some point M on the manifold, P be a point on the manifold and X a point on the tangent space. We have

$$X = \mathrm{Log}_M P \tag{2}$$

$$P = \mathrm{Exp}_M X \tag{3}$$

The Log and Exp notation defines a log-map from the manifold to the tangent space and an exp-map from the tangent space to the manifold. This is a formal notation and does not imply the normal log and exp functions - although they do co-incide for some types of data, they are not the same for the spherical space. M is the origin of the map and is mapped onto the origin of the tangent space.

For the spherical manifold, which is used to represent directional data, the log-map corresponds to the azimuthal equidistant projection, and the exp-map to its inverse. We define a point P on the sphere as a unit vector $\mathbf{p}$. Similarly, the point M is represented by the unit vector $\mathbf{m}$ which is the origin of the map. The maps are then

$$\mathbf{x} = \frac{\theta}{\sin \theta}(\mathbf{p} - \mathbf{m} \cos \theta) \tag{4}$$

$$\mathbf{p} = \mathbf{m} \cos \theta + \frac{\sin \theta}{\theta} \mathbf{x} \tag{5}$$

where θ is the angle between the vectors. The vector $\mathbf{x}$ is the image of P and lies in the tangent space of the sphere at point $\mathbf{m}$, and the image of M is at the origin of the tangent space. $\mathbf{x}$ is intrinsically 2-dimensional as it lies in the tangent plane, and we can characterise it by two variables if we define an orthogonal coordinate system for the tangent plane using the unit vectors $\{\mathbf{u}_1, \mathbf{u}_2\}$; the two parameters are then $\alpha = \mathbf{x}^T\mathbf{u}_1$ and $\beta = \mathbf{x}^T\mathbf{u}_2$.

2.3   Facial Shape Model

Smith[4] constructed a statistical model of needle maps using range images from 3DFS database[5] and Max Plank database[6] based on Principal Geodesic Analysis[7]. Each range image in the training data is converted to a field of surface normals and then PGA is applied to the data. Smith calculates the *spherical median* $\mu(x, y)$ at each pixel location. At each point in the image, a tangent plane to the sphere is defined by the direction $\mu(x, y)$ and the surface normals projected onto this tangent plane using the log-map. The covariance matrix calculated using PGA then describes the distribution of surface normals on the union of all of these tangent planes. As a result, the statistical model consisted of the set of spherical medians $\mu(x, y)$ and a covariance matrix on the union of tangent planes $\mathbf{L}$.

Smith also constructed a statistical surface height model which gives surface height from surface normals implicitly without following any explicit surface integration method. Like the statistical surface normal model Smith's height model extracts the principal modes of variation from a joint model of surface normal and height. Least-squares can then be used to locate the best height model for a particular configuration of surface normals.

2.4   Outline of the Method

Our goal is to combine these two probabilistic descriptions of the set of surface normals, one from SfS and one from the shape model. The FB_8 distribution is difficult to manipulate; for example finding maximum-likelihood estimates for the parameters is a non-trivial problem. Instead, we convert the SfS distributions to distributions on the tangent planes of the shape model using sampling. We commence with the the FB_8 distributions delivered by SfS and the Smith and Hancock statistical face model. The first step is to sample the FB_8 distributions. From these samples we compute the spherical median at each point. We then project the samples onto the tangent plane of the directional sphere and compute the Normal distribution on the tangent plane. We then combine this with the Smith and Hancock model on the tangent plane to provide a combined distribution of surface normals. We identify and eliminate SfS outliers using the Fisher criterion. Finally we smooth and integrate the resulting surface normals to obtain a shape estimate.

In section 3, we describe a sampling method for the FB_8 distribution. In section 4 we explain how to model the samples on the tangent plane and combine them with the shape model. Section 5 describes how to identify outliers where the SfS is ineffective, and how to construct a final face shape. Finally, we present some results in Section 6.

3   Sampling of FB$_8$ Distributions

The FB$_8$ distribution[9] is a multivariate normal distribution that is constrained to lie on the surface of a unit sphere. The Fisher-Bingham distribution is used to model the directional data on spheres and sometimes for shape analysis[1]. If $\mathbf{x} = (x_0, x_1, x_2)$ is a random variable from this distribution then according to the unit norm constraint $\|x\|^2 = 1$. Hence, (x_0^2, x_1^2, x_2^2) lies on a simplex.

Our goal is to generate a set of samples from the distribution using the method of Kume and Walker[9] which we briefly describe here. The key idea of Kume and Walker is to transform $\mathbf{x}$ to (ω, s) where $s_i = x_i^2$ and $\omega_i = \frac{x_i}{\|x_i\|}$, so ω_i can either be 1 or -1. They then study the marginal and conditional distributions of ω and s.

The FB$_8$ distribution is

$$p_{FB}(\hat{\mathbf{x}}; \mathbf{A}) = N(\mathbf{b}, \mathbf{A}) \exp(\mathbf{b}^T \hat{\mathbf{x}} + \hat{\mathbf{x}}^T \mathbf{A} \hat{\mathbf{x}}) \qquad (6)$$

where $N(\mathbf{b}, \mathbf{A})$ is the normalizing constant, $\mathbf{x} \in R^3$ and $\mathbf{x}^T \mathbf{x} = 1$. We can diagonalize the Bingham matrix $\mathbf{A}$ by a suitable orthogonal transform so we may assume $\mathbf{A}$ is diagonal without any loss of generality i.e. $\mathbf{A} = \text{diag}(\lambda_0, \lambda_1, \lambda_2)$. Using the parameters $a_i = \lambda_i - \lambda_0$ for $i = 1, 2$ and $s = 1 - s_0 = s_1 + s_2$. We can write the density as

$$P_{FB}(\omega, s) \propto \exp\left[\sum_{i=0}^{2} (a_i s_i + b_i \omega_i \sqrt{s_i})\right]$$
$$\times \exp\left[b_0 \omega_0 \sqrt{1-s}\right] \qquad (7)$$
$$\times \prod_{i=0}^{2} \frac{1}{\sqrt{s_i}} \frac{1}{\sqrt{1-s}} 1(s \leq 1)$$

where, $1(s \leq 1)$ is the *indicator* variable and c is a constant. Kume and Walker then introduced three latent variables (u, v, w) and construct a joint density of

$$f(\omega, s, u, v, w) \propto 1\left[u < \exp\left(\sum_{i=1}^{2} (a_i s_i + b_i \omega_i \sqrt{s_i})\right)\right]$$
$$\times 1\left[v < \exp\left(b_0 \omega_0 \sqrt{1-s}\right)\right] \qquad (8)$$
$$\times 1\left[w < \frac{1}{\sqrt{1-s}}\right] \times \prod_{i=1}^{2} \frac{1}{\sqrt{s_i}} 1(s \leq 1)$$

Finally, Gibbs sampling may be used to draw samples using the conditional distributions of all of these variables. The values of s and ω conform to the original Fisher-Bingham distribution and can be converted back into samples of

$\hat{\mathbf{x}}$. In the interests of space we omit the details of the sampling process which can be found in [9].

We use this slice sampling process to sample a set of surface normals from the FB_8 distribution produced by Haines and Wilson shape-from-shading. Typically we sample of the order of 100 normals, which we use in the next section to compute a distribution on the tangent plane. We analyze the number of samples required in the experimental section.

4 Combining the Normals

The shape model and the SfS process both provide a distribution for the surface normals at each point in the image. We now explain how we combine these distributions. Quantities from the shape model will be denoted with the subscript 1 and those from the SfS with subscript 2. These calculations are applied at each pixel.

4.1 Acquiring the Mean and Covariance Matrices

Finding the appropriate mean and covariance from Smith's model is straightforward. The mean surface normal is already defined as μ_1. We use this direction to define the tangent plane on which we combine the models, because the normals from SfS could potentially be outliers.

The covariance matrix $\mathbf{L}$ from Smith's model describes the covariance on the union of tangent planes at all points. To extract the covariance matrix at a single point, we need only pull out the appropriate components from $\mathbf{L}$ to find the 2-by-2 covariance. We represent this by a precision matrix $\mathbf{P}_1$ for convenience:

$$\mathbf{C}_1 = \begin{bmatrix} \mathbf{L}_{xy\alpha,xy\alpha} & \mathbf{L}_{xy\alpha,xy\beta} \\ \mathbf{L}_{xy\alpha,xy\beta} & \mathbf{L}_{xy\beta,xy\beta} \end{bmatrix} \tag{9}$$

$$\mathbf{P}_1 = \mathbf{C}_1^{-1} \tag{10}$$

where $xy\alpha$ refers to the α component of the tangent-plane normals at pixel (x, y).

For Haines and Wilson SfS, we have k samples from the FB_8 distribution for each pixel $\{\mathbf{n}_1, \mathbf{n}_2, \ldots, \mathbf{n}_k\}$ as described in section 3. We begin by computing the the extrinsic mean of the surface normal samples using, $\hat{\mathbf{p}}_0 = \frac{1}{K}\sum_{k=1}^{K}\mathbf{n_k}$. We then use this extrinsic mean to calculate the spherical median using the iterative process below:

$$\hat{\mathbf{p}}_{j+1} = \mathrm{Exp}_{\hat{\mathbf{p}}_j}\left(\frac{1}{K}\sum_{i=1}^{K}\mathrm{Log}_{\hat{\mathbf{p}}_j}(\mathbf{n_i})\right) \tag{11}$$

We use $5-10$ iterations to calculate the spherical median. Once the spherical median is found we use this surface normal $\mu_2 = \hat{\mathbf{p}}_{final}$ as the base point to convert the k samples $[\mathbf{n}_1, \mathbf{n}_2, \ldots, \mathbf{n}_k]^T$ from last iteration to the vectors $[\mathbf{v}_1, \mathbf{v}_2, \ldots, \mathbf{v}_k]^T$ on the tangent plane using the log-map:$\mathbf{v}_k = \mathrm{Log}_{\mu_2}(\mathbf{n}_k)$. From these samples, we compute a tangent-plane covariance $\mathbf{C}_2$ and corresponding precision $\mathbf{P}_2 = \mathbf{C}_2^{-1}$

4.2 Combining Gaussian Distributions

We now have two Gaussian distributions on two different tangent planes corresponding to the SfS model and the shape model. We combine these on a single tangent plane defined by the shape model surface normal as the SfS normal may in some cases be an outlier. The mean of the shape model distribution is of course at the origin on the tangent plane, but we must project the SfS mean onto this tangent plane using the log-map i.e. $\mathbf{v}_{\mu 2} = \mathrm{Log}_{\mu_1}\mu_2$. We may then combine the distributions using the normal rules for combining Gaussians:

$$\mathbf{P}^* = \mathbf{P}_1 + \mathbf{P}_2 \tag{12}$$

$$\mathbf{v}^* = (\mathbf{P}^*)^{-1}(\mathbf{P}_2\mathbf{v}_{\mu 2}) \tag{13}$$

The normal on the tangent plane is then converted back on the unit-sphere with exponential mapping i.e. $\mu^* = \mathrm{Exp}_{\mu_1}\mathbf{v}^*$. $\mathbf{v}^*$ is then our maximum-likelihood estimate of the surface normal.

4.3 Dealing with Outliers

Outliers arise in the regions where surface normals resulting from Smith's model do not comply with the surface normals generated through the sampling of FB_8 distribution. This is due to the fact that we are so far ignoring the effects of albedo i.e. the irradiance equation assumes unit albedo which does not hold in the eyes, areas of facial hair or shadow regions. To detect these outliers we use Fisher criterion. The Fisher criterion is ideal for our algorithm due to the fact that we have separate class distributions to represent surface normals from shape model and FB sampling.

The Fisher criterion is a linear pattern classifier that evaluates between-class variance relative to the within-class variance [13]. The idea of Fisher criterion lies in finding such a vector $\mathbf{d}$ that the patterns belonging to opposite classes would be optimally separated after projecting them onto $\mathbf{d}$ [14]. The Fisher criterion for two classes can be expressed as

$$F(\mathbf{d}) = \frac{\mathbf{d}^T\mathbf{S}\mathbf{d}}{\mathbf{d}^T\Sigma\mathbf{d}} \tag{14}$$

where, $\mathbf{S}$ is between class scatter matrix given by $\mathbf{S} = \Delta\Delta^T$ and $\Delta = \mu_1 - \mu_2$. μ_1, μ_2 are the mean vectors and Σ_1, Σ_2 are the covariance matrices of classes 1 and 2. $\Sigma = p_1\Sigma_1 + p_2\Sigma_2$ where p_1 and p_2 are the a priori probabilities of the two classes which are taken as 0.5 while giving equal weightage to both classes. The optimal value of Fisher discriminant $\mathbf{d}_{opt}$ maximises the value of F. Here we are interested in the optimal value of $F(\mathbf{d}_{opt})$ as it gives a good measure of the separation of the two classes. We use the $F(\mathbf{d}_{opt})$ as a threshold to deal with the outliers; the outliers exist at the pixels where the two surface normals from FB sampling and Smith's model do not register with each other. The distributions will therefore be well separated and give a large value of the Fisher criterion. As we already have the necessary class means and covariances to hand, it is a straightforward task to compute the criterion. We decide a threshold value for the criterion and then replace that particular pixel's surface normal μ^* with Smith's surface normal μ_1 at that specific location if the value of $F(\mathbf{d}_{opt})$ is higher than decided threshold.

4.4 Surface Reconstruction

After the surface normals have been combined using product normal distribution we perform smoothing on these normals. Smoothing is required as it finds the normal's best fit to the statistical model. The normals are first converted on the tangent plane using the Smith's statistical normals as the base points using Log map i.e.

$$u(x,y) = \text{Log}_{\mu_1(x,y)}\left(\mu^*(x,y)\right) \qquad (15)$$

All the image points $[u(1,1), u(1,2),\ldots]$ are then stacked together to give a $N \times 2$ vector $\mathbf{U}_c$ of combined normals on the tangent plane. The best fit vector on the tangent plane is then found using[15],

$$\mathbf{U}_r = \mathbf{P}\mathbf{P}^T\mathbf{U}_c \qquad (16)$$

where $\mathbf{P}$ is the matrix of principal directions from Smith's statistical model. $\mathbf{U}_c$ is the vector comprised of combined surface normals on the tangent plane and $\mathbf{U}_r$ is the vector of recovered smoothed surface normals. The surface normals on the tangent plane are then brought back to the unit sphere through exponential mapping using Smith's statistical needle map as base points. We will show in experiments that smoothing helps in reducing the RMS error.

Finally we use Smith's height model to recover surface height from surface normals, as described in section 2.3.

5 Experimental Results

We have tested our algorithm on the same data set that has been used to construct Smith's statistical models [4]. Iterative Closest Point method [12] was used to compute the RMS distance between ground truth normals and recovered normals and between ground truth surface height and recovered surface height.

5.1 Effects of Number of Samples

The only time consuming component in our technique is the slice sampling of the FB_8 distributions. For each pixel first 200 samples are used for burn in period then extrinsic mean is calculated with next 100 samples. We use $5-10$ iterations to calculate intrinsic mean; each iteration in turn may use 25, 100, 250 or 500 samples. Increasing the number of samples increases the global consistency of sampled FB_8 surface normals resulting from SfS at the expense of time. Figure 1 shows results for one of the test images that has been used as input to the SfS algorithm; the outputted FB_8 distributions from SfS were sampled using Kume & Walker [9]. In figure 1 row 1 we have shown the resulting sampled illuminated surface normals when 25, 100, 250 and 500 samples were used per iteration for computing spherical median. Row 2 of figure 1 shows the combined normals when SfS sampled normals are combined with statistical model of needle maps. No fisher criterion or smoothing yet have been performed.

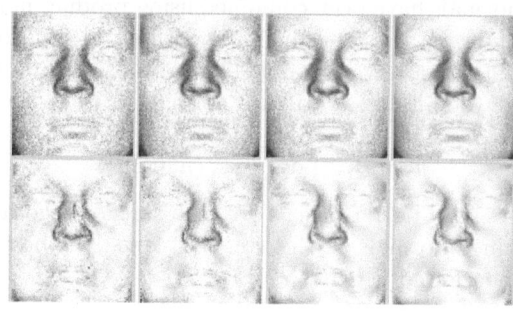

Fig. 1. Illuminated FB Sampled (Row1) and Combined Normals (Row2)

5.2   Surface Normals and Surface Height Errors

Figure 2 on the left shows the ground truth normals (Row1) for 5 subjects from our test data, the Shape-from-Shading FB sampled normals (Row2) and resultant Combined normals (Row3). In right half of figure 2 we have shown the illuminated version of these normals when they are illuminated with a fronto-parallel unit light source [0,0,1]. For subjects shown in figure 2 ; 500 samples per iteration have been used for computing the spherical median.

We have computed error distances for 200 synthetic images for surface normals and surface heights using Iterative Closest Point algorithm. Figure 3 on the left shows the Root Mean Squared error computed between Groundtruth surface normals and SfS FB sampled normals ($E_{FBSampled}$ shown in red); and between Groundtruth normals and resultant Combined normals ($E_{Combined}$ shown in green). In the right halft the RMS distance profile computed for surface heights is shown. The number of subjects have been adjusted according to the ascending $E_{FBSampled}$ error. Only 25 samples per iteration have been used for spherical median computations for the purpose of computing these errors.

From the error profiles it is apparent that the Root Mean Squared error has reduced significantly when combined surface normals have been used instead of using the surface normals computed through the SfS model alone. So the surface heights constructed from these normals have shown improvements as well as apparent from figure 3.

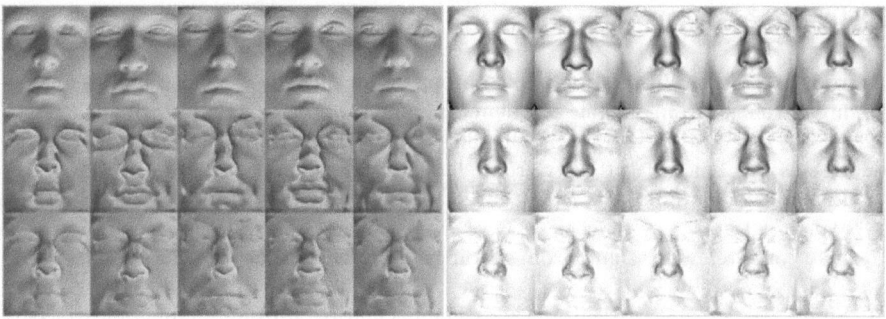

Fig. 2. Groundtruth Surface Normals (Row1), SfS FB Sampled Normals (Row2) and Combined Normals (Row3)

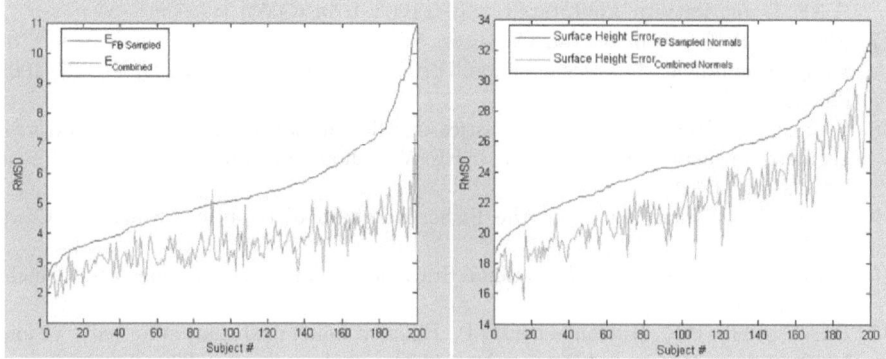

Fig. 3. Profiles of Surface Normal (Left) and Surface Hieght (Right) Errors

6   Conclusions

We have presented a probabilistic framework in which Fisher-Bingham 8-parameter distributions arising from a probabilistic Shape-from-Shading method are sampled using a slice sampling algorithm; surface normals are calculated from these samples using the machinery of spherical median and log/exp mapping. These normals are then combined with normals resulting from a statistical facial shape model using individual Gaussian distributions on the tangent plane. Fisher criterion and smoothing are used to deal with outliers. From our experiments we have shown that the error distances reduce when combined normals are used; even when less samples from FB sampling were used to compute spherical medians. In our future work we will try to fit Fisher-Bingham distributions to the statistical normals instead of fitting Gaussian distribution to the sampled SfS normals on the tangent plane and then combine the resulting Fisher-Bingham distributions. Since, Fisher-Bingham distribution models directional data more precisely than Gaussian distribution; we hope this will give more reliable combined normals.

References

1. Haines, T.S.F., Wilson, R.C.: Belief propagation with directional statistics for solving the shape-from-shading problem. In: Forsyth, D., Torr, P., Zisserman, A. (eds.) ECCV 2008, Part III. LNCS, vol. 5304, pp. 780–791. Springer, Heidelberg (2008)
2. Lee, K.M., Kuo, C.J.: Shape from shading with perspective projection. CVGIP: Image Understanding 59(2), 202–212 (1994)
3. Worthington, P.L., Hancock, E.R.: New constraints on data-closeness and needle map consistency for shape-from-shading. IEEE Trans. Pattern Anal. Intell. 21(12), 1250–1267 (1999)
4. Smith, W.A.P.: Statistical Methods For Facial Shape-from-Shading and Recognition. PhD thesis, University of York (2007)
5. USF HumanID 3D Face Database, Courtesy of Sundeep. Sarkar, University of South Florida, Tampa, FL
6. Blanz, V., Vetter, T.: Face Recognition based on fitting a 3D morphabale model. IEEE Trans. Pattern Anal. Intell. 25(9), 1063–1074 (2003)
7. Fletcher, P.T., Joshi, S., Lu, C., Pizer, S.M.: Principal geodesic analysis for the study of nonlinear statistics of shape. IEEE Trans. Med. Imaging. 23(8), 995–1005 (2004)
8. Pennec, X.: Probabilities and statistics on Riemannian manifolds: basic tools for geometric measurements. In: Proc. IEEE Workshop on Nonlinear Signal and Image Processing (1999)
9. Kume, A., Walker, S.G.: On the Fisher-Bingham distribution. Stat. and Comput. 19, 167–172 (2009)
10. Mardia, K.V., Jupp, P.E.: Directional Statistics. John Wiley and Sons Ltd., Chichester (2000)
11. Felzenszwalb, P.F., Huttenlocher, D.P.: Efficient belief propagation for early vision. In: Computer Vision and Pattern Recognition, vol. 1, pp. 261–268 (2004)
12. Phillips, J.M., Liu, R., Tomasi, C.: Outlier Robust ICP for Minimizing Fractional RMSD. In: 6th International Conference on 3-D Digital Imaging and Modeling, pp. 427–434 (2007)
13. Luebke, K., Weihs, C.: Improving Feature Extraction by Replacing the Fisher Criterion by an Upper Error Bound. Pattern Recognition 38(2005), 2220–2223 (2005)
14. Maciej, S., Witold, M.: Versatile Pattern Recognition System Based on Fisher Criterion. In: Proceedings of the KOSYR, pp. 343–348 (2003)
15. William, A.P.: Smith and Edwin R. Hancock. Facial Shape-from-shading and Recognition Using Principal Geodesic Analysis and Robust Statistics. International Journal of Computer Vision 76(1), 71–91 (2008)
16. Prados, E., Faugeras, O.: Unifying approaches and removing unrealistic assumptions in shape from shading: Mathematics can help. In: Pajdla, T., Matas, J(G.) (eds.) ECCV 2004. LNCS, vol. 3024, pp. 141–154. Springer, Heidelberg (2004)
17. Zhao, W.Y., Chellappa, R.: Symmetric shape-from-shading using self-ratio image. International Journal of Computer Vision 45, 55–75 (2001)
18. Samaras, D., Metaxas, D.: Illumination constraints in deformable models for shape and light direction estimation. IEEE Transactions on Pattern Analysis and Machine Intelligence 25(2), 247–264 (2003)

Visual Saliency by Keypoints Distribution Analysis

Edoardo Ardizzone, Alessandro Bruno, and Giuseppe Mazzola

Dipartimento di Ingegneria Chimica, Gestionale, Informatica e Meccanica.
Università degli Studi di Palermo, viale delle Scienze ed. 6, 90128, Palermo, Italy
ardizzon@unipa.it,
{bruno,mazzola}@dinfo.unipa.it

Abstract. In this paper we introduce a new method for Visual Saliency detection. The goal of our method is to emphasize regions that show rare visual aspects in comparison with those showing frequent ones. We propose a bottom up approach that performs a new technique based on low level image features (texture) analysis. More precisely, we use SIFT Density Maps (SDM), to study the distribution of keypoints into the image with different scales of observation, and its relationship with real fixation points. The hypothesis is that the image regions that show a larger distance from the mode (most frequent value) of the keypoints distribution over all the image are the same that better capture our visual attention. Results have been compared to two other low-level approaches and a supervised method.

Keywords: saliency, visual attention, texture, SIFT.

1 Introduction

One of the most challenging issues in Computer Vision field is the detection of salient regions in an image. Psychovisual experiments [1] suggest that, in absence of any external guidance, attention is directed to visually salient locations in the image. Visual Saliency or Saliency mainly deal with identifying fixation points that a human viewer would focus on at the first glance. Visual saliency usually refers to a property of a "point" in an image (scene), which makes it likely to be fixated. Most models for visual saliency detection are inspired by human visual system and tend to reproduce the dynamic modifications of cortical connectivity for scene perception. In scientific literature Saliency approaches can be subdivided in three main groups: Bottom-up, Top-down, Hybrid.

In Bottom-up approaches (stimulus driven) human attention is considered a cognitive process that selects most unusual aspects of an environment while ignoring more common aspects. In [2] the method is based on parallel extraction of various feature maps using center-surround differences. In [3] multiscale image features are combined into a single topographical saliency map. A dynamical neural network then selects attended locations in order of decreasing saliency. Harel et al. in [4] proposed graph based activation maps.

In Top-down approaches [5,6] the visual attention process is considered task dependent, and the observer's goal in scene analysis is the reason why a point is fixed

G. Maino and G.L. Foresti (Eds.): ICIAP 2011, Part I, LNCS 6978, pp. 691–699, 2011.

rather than others. Object and face detection are examples of high level tasks that guide the human visual system in top-down view.

Generally Hybrid systems for saliency use the combination of the two levels, bottom-up and top-down. In hybrid approaches [7,8] Top-down layer usually cleans the noisy map extracted from Bottom-up layer. In [7] top-down component is face detection. Chen et al. [8] used a combination of face and text detection and they found the optimal solutions through branch and bound technique.

A common problem for many of these models is that they often don't match real fixation maps of a scene. A newer kind of approach was proposed by Judd et al. [9] who built a database [10] of eye tracking data from 15 viewers. Low, middle and high-level features of this data have been used to learn a model of saliency. In our work we aimed to further study this problem. We decided to investigate about the relationship between real fixation points and computer generated distinctive points. Our method performs a new measure of visual saliency based on image low level features, particularly through the distribution of keypoints extracted by SIFT algorithm, as descriptor of texture variations into the image. In this work we are not interested in color information. Our method is totally unsupervised and it belongs to bottom-up saliency methods. We measured method effectiveness comparing resulting maps with real fixation maps of the reference database [10] and with two of the most important bottom–up approaches [3][4] and a hybrid method[9].

2   Proposed Saliency Measure

Our method propose a new measure of Visual Saliency, focusing on low level image features such as texture. What's the matter for which we use texture information for detecting visual saliency? The answer is that texture gives us important information about image "behavior". The base for extracting salient regions, according to our method, is to emphasize texture rare event. We decide to study the spatial distribution of keypoints inside an image to describe texture variations all over the image. The levels of roughness of both fine and coarse regions can be very different (in a fine region we will find a larger number of keypoints than in coarse regions), so we use keypoints density, to find various texture events and to identify the most salient regions. In this work we use SIFT algorithm to extract keypoints from an image. Then we introduce the concept of SIFT density maps (SDM) which are used to compute the final saliency map.

2.1   SIFT Feature

SIFT (Scale Invariant Feature Transform) descriptors [11] are generated by finding interesting local keypoints, in a greyscale image, by locating the maxima end the minima of Difference-of-Gaussian in the scale-space pyramid. SIFT algorithm takes different levels (octaves) of Gaussian blur on the input image, and computes the difference between the neighboring octaves. Information about orientation vector is then computed for each keypoint, and for each scale. Briefly, a SIFT descriptor is a 128-dimensional vector, which is computed by combining the orientation histograms of locations closely surrounding the keypoint in scale-space. The most important advantage of SIFT descriptors is that they are invariant to scale and rotation, and relatively robust to perspective changes. SIFT can be very useful for many computer vision application: image registration, mosaicing, object recognition and tracking, etc. Their main drawback is the relatively high dimensionality which make them less suitable for nearest neighbor lookups against a training dataset.

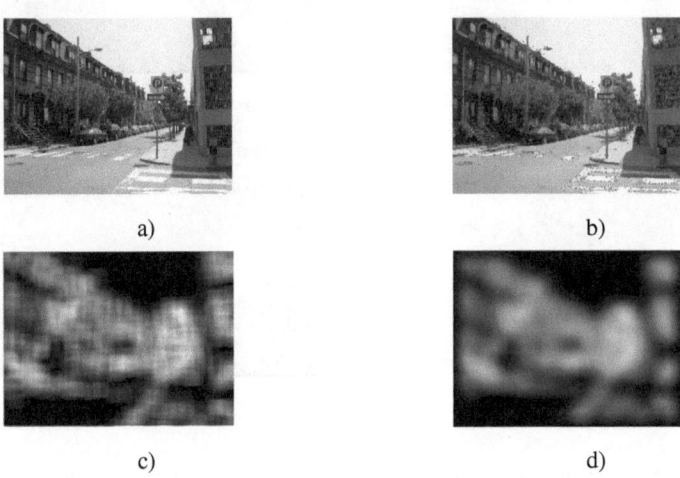

a)   b)

c)   d)

Fig. 1. Original Image (a), SIFT keypoints (b), SIFT Density Map (k=64) (c), final Saliency Map (d)

2.2 SIFT Density Maps

A SIFT Density Map (SDM) is a representation of the density of keypoints in an image, and can give essential information about the regularity of its texture. A SIFT Density Map SDM(k) is built by counting the number of keypoints into a sliding window of size k x k, which represent our scale of observation. Each point in the SDM(k) indicates the number of keyponts into a squared area of size k x k, centered in corresponding point of the image. It is evident that density values are strictly related to the value of k, and are limited by the window size. In fact smaller windows should be sensible to texture variations at a finer level, while larger windows will emphasize coarser deviations. In section 3 we will discuss the sensibility of the results with k.

In real scenes, the simultaneous presence of many elements (the sky, the urban habitations, the urban green spaces) will show many kinds of texture. From a SIFT distribution point of view, the homogeneous surface of the sky has almost null values, the urban green spaces has mean density while urban habitations have high concentration of keypoints. (fig. 1)

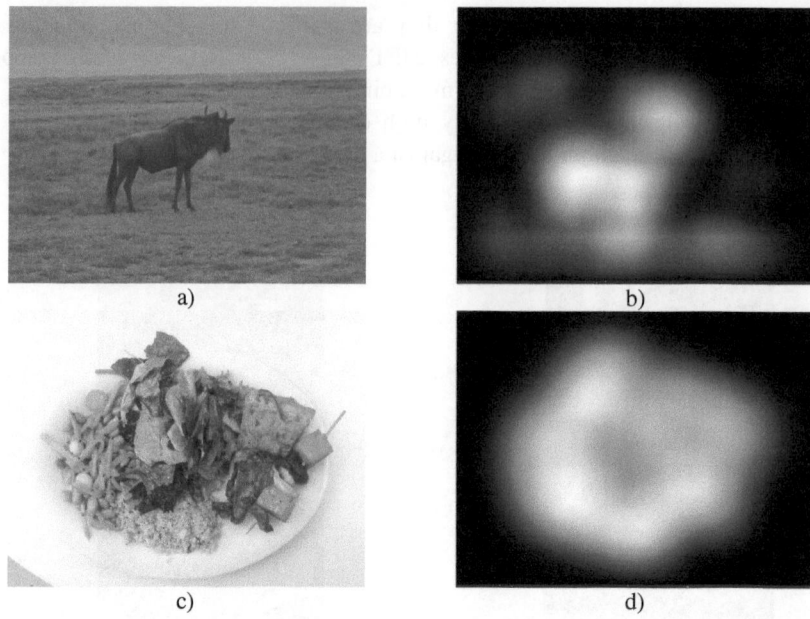

Fig. 2. Two image examples: a homogeneous subject in a textured scene (a) and the corresponding Saliency Map; a textured object in a homogeneous background (c) and the corresponding Saliency Map (d)

2.3 Saliency Map

Our saliency map SM, for a given k, is built as the absolute difference between the SDM values and the most frequent value MV of the map:

$$SM(k) = |SDM(k) - MV(SDM(k))| \tag{1}$$

which is further normalized with respect to the maximum value to restrict SM values to [0,1].

The most salient areas into the image are those related to the SDM values with the maximum deviation from the most frequent value, typically the most rare texture events in the image. This measure emphasizes both the case in which a textured object is the salient region, as it is surrounded by homogeneous areas (the most frequent value near to 0), and the case in which a homogeneous area is surrounded by textured parts (a higher most frequent value). (fig. 2)

In addition, for a smoother representation of the saliency map, we apply to the SM an average filter which has a window size that is a half of that used to build the map (k).

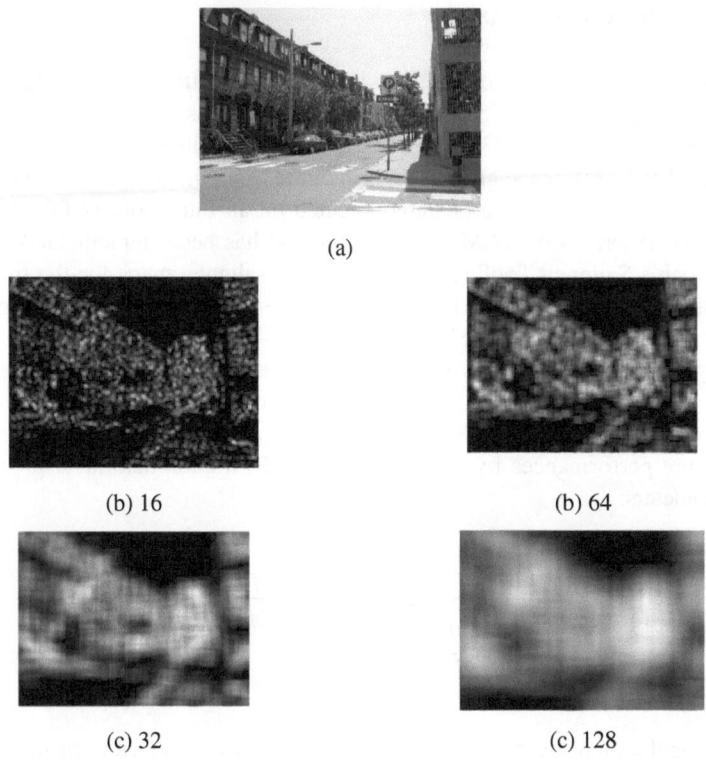

(a)

(b) 16

(b) 64

(c) 32

(c) 128

Fig. 3. Original Image (a), SIFT Density Maps with different values of k (16,32,64,128)

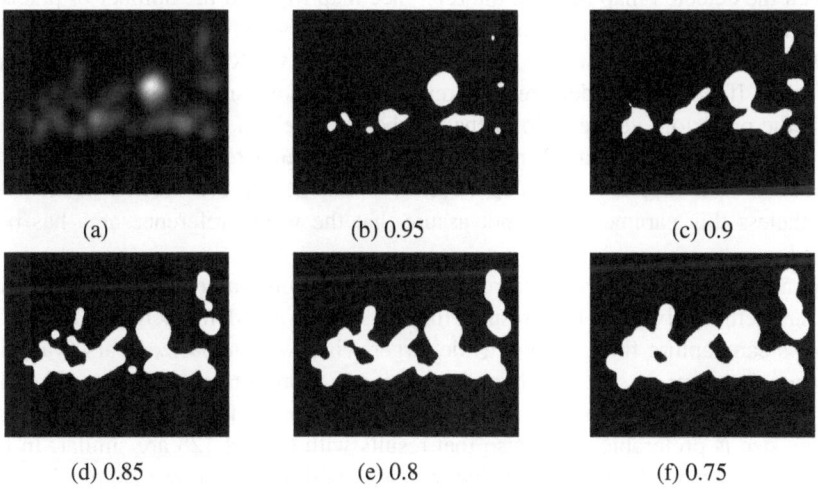

(a)

(b) 0.95

(c) 0.9

(d) 0.85

(e) 0.8

(f) 0.75

Fig. 4. Fixation Map (a) of the image in fig. 3.a, Binary maps with different thresholds (0.95, 0.9, 0.85, 0.8, 0.75)

3   Experimental Results

In this section we compare our results with those of Itti-Koch[3], Harel's Graph Based Visual Saliency (GBVS) [4] and Judd [9] methods. Tests were made on [10] dataset which consists of 1003 images and the corresponding maps of fixation points, which are taken as reference groundtruth (in our tests all the images have been resized down by a factor of two). Tests were executed on an Intel Core i7 PC (4 CPU, 1.6 GHz per processor, 4 GB RAM), and our method has been implemented in Matlab. We use Koch's Saliency Toolbox[12] to compute saliency maps for the methods [3] and [4], and the maps given in [10] for the Judd's method. Tests were repeated for different values of window size (16, 32, 64, 128 - fig.3), with the aim to study the sensibility of the results to this parameter. To compare our results with the other methods, and to the groundtruth, we discard from the saliency maps the less N% salient pixels (with N= 95, 90, 85, 80, 75 - fig.4) to create a set of binary maps. We then measure performances by using as metric (SP) a combination of precision and recall parameters:

$$R = \frac{n(M_D \cap M_R)}{n(M_R)}; P = \frac{n(M_D \cap M_R)}{n(M_D)}$$

$$SP = R \cdot P$$

(2)

where M_D is the binary version of the detected saliency map (with our method or the others), while M_R is the binary version of the reference fixation map.

R is the recall, i.e. the ratio between the number of pixels in the intersection between the detected map M_D and the reference map M_R, and the number of pixels in M_R. When it tends to 1, M_D covers the whole M_R, but we have no information about pixels outside M_R (a map made of only salient pixels gives R=1 if compared with any other map). If it tends to 0 detected and reference map have smaller intersection.

P is the precision, i.e the ratio of the number of pixels in the intersection between M_D and M_R, and the number of pixels in M_D. When P tends to 0, the whole M_D has no intersection with M_R. If it tends to 1, fewer pixels of M_D are labeled outside M_R. Nevertheless this parameter will not assure that the whole reference area has been covered.

Fig. 5 shows average precision results versus different values of thresholds. Note that our method gives its best results for k=128. As noted in section 2.2, smaller windows can capture finer details, while larger windows emphasize coarse variation of texture. In terms of saliency, human attention is more attracted by areas in which there are large texture variations, rather than by small deviation. Then a larger window size is preferable. Note also that results with 64 and 128 are similar. In fact we observed that while the recall value increases with the window size, precision, in case of very large window, does not increase as well.

In the comparison with the other methods, we must first underline some fundamental issues:

- Judd method is supervised, and uses 9/10 of the whole dataset for training and 1/10 for testing. Judd results are averaged only on the 100 testing images. It uses both color and texture information.
- Itti-Koch and GBVS method are unsupervised method and use both color and texture information.
- Our method is unsupervised and use only texture information to build the saliency map.

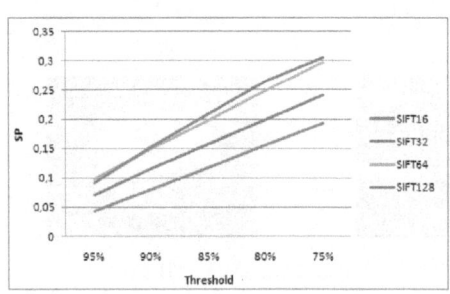 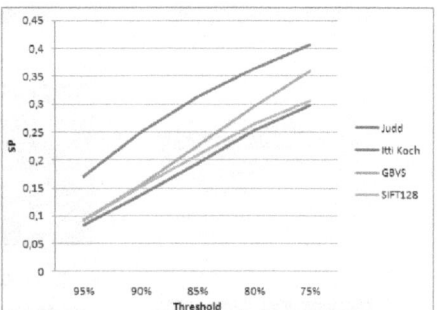

Fig. 5. Average Saliency Precision (SP) vs. Threshold for our method with different window sizes (left), and for our method (k=128) compared to the other methods.

Our saliency map gives better results than Itti-Koch, for all the threshold values, even if we use only texture information. Results are similar to GBVS for higher threshold values (0.95 and 0.9), which give information about the most salient pixels, while our precision does not increase as well for lower values of threshold (0.85, 0.8). As expected Judd method achieves best results, as it is a supervised method, while all the other methods are unsupervised. Furthermore Judd tests refers only within a small selected subset of images (100 testing images), while other methods have been tested within the whole dataset. Judd results are reported only as asymptotic values to be compared with. Fig. 6 shows some examples of saliency maps with all the discussed methods. Regarding temporal efficiency, our method takes less than 10s to build a saliency map, and it is comparable with Itti-Koch and GBVS method for medium images (300 x 600). Most of the time (70% ca) is spent to extract keypoints, but it depends on image complexity, i.e. the number of keypoints extracted.

Fig. 6. Some visual results. Original images (a,g,m), fixation maps (b,h,n), Judd maps (c,i,o), Itti-Koch maps (d,j,p), GBVS maps (e,k,q), our method (f,l,r) window size 128.

4 Conclusions

Visual saliency has been investigated for many years but it is still an open problem, especially if the aim is to investigate the relationship between synthetic maps and points, in a real scene, that attract a viewer attention.

The purpose of this paper was to study how computer generated keypoints are related to real fixation points. No color information has been used to build our saliency maps, as keypoints are typically related only to image texture property.

Even if we use only texture information, experimental results show that our method is very competitive with respect of two of the most cited low-level approaches. Judd's method achieves better results as it is a supervised method which has been trained with the fixation maps within the selected dataset.

In our future works we want to study new color based saliency techniques to be integrated with our proposed approach, to improve experimental results.

References

1. Constantinidis, C., Steinmetz, M.A.: Posterior parietal cortex automatically encodes the location of salient stimuli. The Journal of Neuroscience 25(1), 233–238 (2005)
2. Koch, C., Ullman, S.: Shifts in selective visual attention: towards the underlying neural circuitry. Human Neurobiology 4, 219–227 (1985)
3. Itti, L., Koch, C., Niebur, E.: A model of saliency-based visual attention for rapid scene analysis. IEEE Transactions on Pattern Analysis and Machine Intelligence 20(11), 1254–1259 (1998)
4. Harel, J., Koch, C., Perona, P.: Graph-based visual saliency. In: Advances in Neural Information Processing Systems 19, pp. 545–552. MIT Press, Cambridge (2007)
5. Luo, J.: Subject content-based intelligent cropping of digital photos. In: IEEE International Conference on Multimedia and Expo (2007)
6. Sundstedt, V., Chalmers, A., Cater, K., Debattista, K.: Topdown visual attention for efficient rendering of task related scenes. In: In Vision, Modeling and Visualization, pp. 209–216 (2004)
7. Itti, L., Koch, C.: Computational modeling of visual attention. Nature Reviews Neuroscience 2(3) (2001)
8. Chen, L.-Q., Xie, X., Fan, X., Ma, W.-Y., Zhang, H.-J., Zhou, H.-Q.: A visual attention model for adapting images on small displays. ACM Multimedia Systems Journal 9(4) (2003)
9. Judd, Y., Ehinger, K., Durand, F., Torralba, A.: Learning to predict where humans look. In: IEEE 12th International Conference on Computer Vision, pp. 2106–2133 (2009)
10. http://people.csail.mit.edu/tjudd/WherePeopleLook/index.html
11. Lowe, D.G.: Distinctive Image Features from Scale-Invariant Keypoints. International Journal of Computer Vision 60(2), 91–110 (2004)
12. http://www.saliencytoolbox.net

From the Physical Restoration for Preserving to the Virtual Restoration for Enhancing

Elena Nencini and Giuseppe Maino

Faculty of Preservation of the Cultural Heritage, University of Bologna,
Ravenna site, 5, via Mariani, Ravenna, Italy
e.nencini@gmail.com, giuseppe.maino@unibo.it
ENEA: Italian National agency for new technologies, Energy and sustainable economic
development, 4, via Martiri di Montesole, Bologna, Italy
giuseppe.maino@enea.it

Abstract. Digital image processing techniques are increasingly applied to the
study of cultural heritage, namely to the analysis of paintings and
archaeological artefacts, in order to identify particular features or patterns and
to improve the readability of the artistic work. Digital or 'virtual' restoration
provides a useful framework where comparisons can be made and hypotheses
of reconstruction proposed without action or damage for the original object,
according to the adopted general rules for practical restoration.

Keywords: Geometrical distortion, image enhancement, virtual restoration,
mosaics, cultural heritage.

1 Introduction

The cultural heritage field has undergone profound changes in recent decades, linked
to an increasingly demanding public, but also to the transformation of culture in a real
'good' that can generate wealth and employment. The computer expert that can
manage virtual restoration programs as now joined the more traditional professional
figure of the restorer, in a still experimental way.

These are applications that, in this specific field, provide a number of interesting
proposals: the virtual restoration does not act on the art work, but simulates a visual
and aesthetic improvement of the work, so enhancing it. It also gives the possibility to
choose a series of solutions, before technical operations.

The case study object of this paper is an experimental project of virtual restoration
carried out on a mosaic from the church of San Severo in Classe, near Ravenna. The
construction of San Severo dates from the late sixth century, was consecrated in 582
and was pulled down and abandoned in the early '20s of the XIX century. The floor of
the church was formed by a rich mosaic, which was only partially found.

At the end of the 1966 excavation campaign, a mosaic carpet - 4.50x2.75 m – was
discovered at the center of the main nave. This mosaic showed a grid of rows of
tangent pised square containing figures of birds made of glass paste, of great elegance
and extremely naturalistic; around this main schema runs a shaded three strand
guillocheon on a black ground (see fig.1).

G. Maino and G.L. Foresti (Eds.): ICIAP 2011, Part I, LNCS 6978, pp. 700–709, 2011.

The mosaic was found at a height of 0.40 m beneath the floor of the church, almost in the middle of the main nave, but unlike all the other mosaics, it was not aligned and oriented north to south direction. These data have suggest that it might be connected to an *ecclesia domestica*, an oratory that might have been part of the Roman *domus*, whose remains are beneath the church. A place of great devotional importance, perhaps, the place itself where the saint had exercised (carried out) his apostolate.

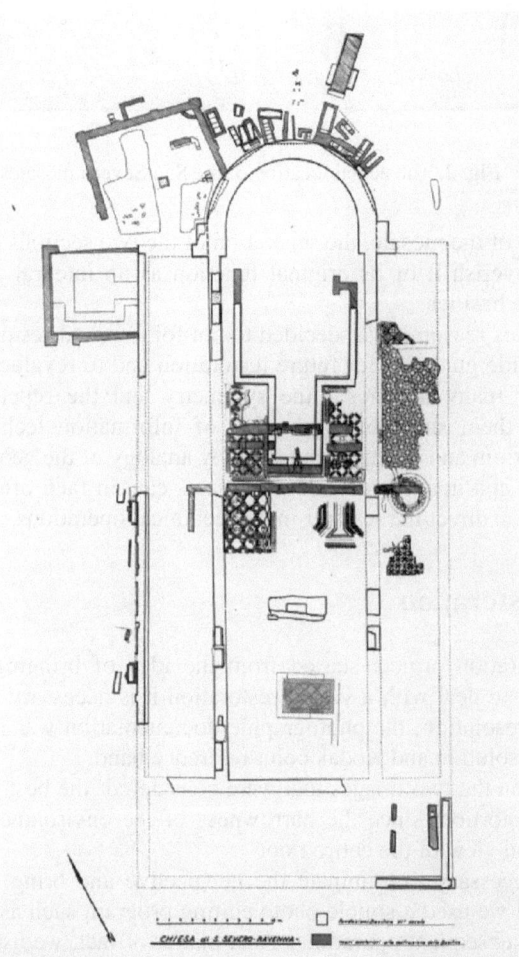

Fig. 1. Plan of the San Severo church. The original location of the mosaic considered in this work is represented by the rectangle in the lower part of the building, near the entrance.

Today the floor, lifting and relaiding on cement mortar in the late 60s of last century, is at the Museo Nazionale in Ravenna, divided into two sections, placed far apart on a wall in a narrow and dark lobby, as shown in fig. 2.

Fig. 2. The actual location of the San Severo mosaics

The verticality of the mosaic, the separation of the two sections decontextualize the mosaic and impoverish it of its original function as an integral part of the internal architecture of the basilica.

Precisely for this reason it was decided to opt for a virtual restoration project, with the intent to provide guidance for future restoration and to revalue this art piece. The characteristics of many mosaics - the symmetry and the repetition of geometric patterns - make them suitable for the use of information technology both for a simulated integratiom and for the reconstruct by analogy of the geometric pattern. The use of computers and image processing programs can, in fact, offer a preview of the restoration, that can direct the restorer in the techmical operations.

2   Virtual Restoration

Our virtual restoration project started from the idea of bringing together the two sections. In order to deal with a virtual restoration it is necessary to have the images scanned at high resolution; the photographic documentation was made with a Nikon D90 at 300 dpi resolution and Kodak color reference band.

We have chosen the two images that were considered the best for color reliability and for minor distortion, since the narrowness of the environment did not allow a clear global frontal view of the entire floor.

It was then necessary to eliminate the perspective and bring all images in 1:10 scale. To this end we used a simple photo editing program such as GIMP, which was also used for all subsequent operations. As a matter of fact, we used very simple and easily provided software to perform this work since our purpose was to give an example of virtual processing useful for conservators and restorers as well as for scholars (archaeologists, art historians, etc.) that can be carried out without specific expertise and computer skills.

After the images were acquired in digital format, we were able to proceed with the actual reconstruction of the floor, taking into account the geometric motifs that make up the mosaic assembly and that made the recomposition the most reliable. It was not possible to use excavation maps as they had no metric references at all.

Once we had a basic image we went on to reconstruct by analogy the geometric motif of the floor, which covered both the outer frame with its shaded three strand guillocheon, as well as the internal allocation of the carpet, formed by a grid of partly overlapping circles, in alternating colors.

Fig. 3. Front image of the two mosaic fragments; the parts inside rectangles refer to fragments mistakenly placed in the museum relocation

Once the virtual reconstruction was completed, some fragments of the mosaic did not fit perfectly, this probably because of injuries and of deformations caused by the tearing up and repositioning the mosaic on mortar support, therefore small corrections were needed, with minimal and calibrated geometric modifications (fig. 4).

In addition to the reconstruction of geometric patterns, shown in fig. 5, a proposal was made concerning the integration of the figurative lacunae that are not reinstated, consisting of three tables.

The lacunae have been reintegrated, respectively:

- with a neutral color;
- with a deliberately discordant color;
- with a camouflage integration.

These different choices are shown in figs. 6-8 and make an evaluation possible of the obtained results from an aesthetic and conservative points of view.

The virtual restoration offers a number of possibilities in the field of mosaic restoration, although to this date it is not yet so widely adopted. In the case under consideration - the Gallinelle mosaics - it allowed, at low cost and with great ease, to create a graphic reconstruction of the mosaic floors, so that when a conservative restoration will be carried out, the restorers will have more significant information and different solutions for the integrations. In addition, the handling of mosaic fragments - often very bulky and made heavy by concrete support – certainly does not not make easy delicate operations such as recompositions.

But the virtual restoration not only offers to restorers the possibility of recomposing a floor: it also provides the opportunity of simulating on the digital image with a photo editing program, the type of integration and the color so that scholars, restorers and conservators may evaluate the final appearance of the work and the different aesthetic choices.

Fig. 4. Front image of the two mosaic fragments where the smaller pieces have been placed in the correct positions

3 Concluding Remarks

The virtual restoration is also an essential tool for the revaluation of cultural heritage. The new kind of consumer must in fact find the most suitable means to understand what she/he sees. A mosaic floor with large lacunae may in fact resemble a ruin rather than remind the magnificent mansions that it had been part of. Therefore a mosaic floor, decontextualized, hung on a wall and with large lacunae deprives the visitor of vital information.

The virtual restoration then, at very low cost, may intervene by proposing the reconstruction of the entire floor, thus playing an important didactic role and giving back to the mosaic its readability. In the case of the Gallinelle's mosaic, split in two fragments, attached on a wall in a dark, narrow room, the visitor can not understand its importance: she/he, certainly, can not imagine that same floor within the church where its position probably indicated an important ancient worship area.

Fig. 5. The mosaic with virtual reconstruction of the geometric motifs

In this case the presence of the digital reconstruction of the floor with camouflage mimetic integration could to revalue the importance of this mosaic floor with its delicate birds, made more precious by the glass tesserae, and give back, albeit partially, its integrity and dignity as a work of art, while waiting for the conservative restoration.

Fig. 6. The 'restored' mosaic with neutral color integration

Fig. 7. The 'restored' mosaic with discordant color integration

Fig. 8. The virtually restored mosaic with camouflage integration

The case subject of this work highlights the importance of a preliminary study, as a preparation of a restoration before performing surgery. But the virtual restoration not only offers the possibility to the restorer of reconstructing a mosaic: It also offers the ability to simulate on suitable digital images, by means of a photo editing program, the type of integration and the relevant color. Therefore, scholars, conservators and Superintendents can evaluate the final appearance of the work and the different aesthetic choices.

In the case of fig. 7, for example, we deliberately chose a red-violet great impact in order to emphasize the integration. In the words of Cesare Brandi – the first director of Italian National Institute of Restoration - the color of integration must be reduced to the background level and should not compose directly with the color distribution of the surface of the work (as seen in fig. 8).

Last but not least, the virtual restoration is an essential tool for promotion of cultural heritage. In the face of increasing and diverse audiences with different levels of cultural education, museums and archaeological sites seek more and more to create accurate and comprehensive educational courses. The new user should in fact find the most appropriate ways to understand what sees.

A mosaic floor with large gaps may well resemble a ruin rather than refer to the magnificent mansions for which it had been made. The mosaic floors were, in fact, in the Roman domus as well as later in churches, closely related to architecture and function of rooms: The 'triclinium, for example, i.e. the floor space that would house the beds had no drawing, while, at the center, was placed an emblem or a representation, positioned so as to be watched by the landlord and the guests when they were eating. In every room, then, the drawings of the floor were turned towards the entrance, to be admired by those who entered. Then the virtual restoration, with a very low cost, may intervene proposing the reconstruction of the entire floor, thus playing an important educational function and returning to the mosaic readability.

References

1. Russ, J.C.: The Image Processing Handbook. CRC Press, Boca Raton (1999)
2. Brandi, C.: L'Istituto Centrale del Restauro in Roma e la Ricostituzione degli affreschi. Phoebus I(3/4), 165–172 (1947); idem, Il Mantegna ricostruito. In: L'Immagine 3, 179–180 (1963); idem, Teoria del Restauro, Milan (1984)
3. Maioli, M.G.: La basilica di San Severo a Classe, scavo e architettura. In: Santi, Banchieri, Re. Ravenna e Classe nel VI secolo. San Severo il tempio ritrovato, Milan (2006)
4. Racagni, P.: Del distacco dei mosaici e della loro conservazione. In: La basilica ritrovata. I restauri dei mosaici antichi di San Severo a Classe, Ravenna, Città di Castello (2010)

Author Index

Abbate, Maurizio   II-209
Achard, Catherine   I-68
Adya, Ashok K.   I-423
Ahmad, Touqeer   I-680
Ahmad Fauzi, Mohammad Faizal   I-247
Akhtar, Zahid   I-159
Albanese, Giulia   I-670
Alex, Ann Theja   I-544
Al-Hamadi, Ayoub   I-227
Anderson, Roger   I-494
Antúnez, Esther   I-327
Ardizzone, Edoardo   I-237, I-691
Arezoomand, Mehdi   II-1
Asari, Vijayan K.   I-544
Aubert, Didier   I-484
Avola, Danilo   II-414
Avraham, Tamar   I-38

Bak, Adrien   I-484
Baldassarri, Paola   II-79
Baltieri, Davide   I-197
Bandera, Antonio   I-327
Bartocci, Daniele   II-99
Battiato, Sebastiano   I-473, I-650
Bellavia, Fabio   I-524
Bellon, Olga   I-374
Ben-Abdallah, Hanène   II-454
Ben-Ari, Rami   I-19
Benvegna, Francesco   II-434
Bertozzi, Massimo   II-424
Bevilacqua, Alessandro   II-404
Bispo, Aline   II-333
Bloomberg, Dan   I-149
Boccalini, Gionata   II-424
Boccignone, Giuseppe   I-187
Bombini, Luca   I-217, II-374
Borghesani, Daniele   I-443
Borghese, N. Alberto   I-534
Borzeshi, Ehsan Zare   II-19
Bouchafa, Samia   I-484
Bräuer-Burchardt, Christian   II-265,
   II-363
Broggi, Alberto   I-217, II-374, II-424
Brun, Anders   I-403

Bruna, Arcangelo R.   I-337, I-650
Bruno, Alessandro   I-691
Bushra, Jalil   II-11
Buzzoni, Michele   I-217

Calnegru, Florina-Cristina   I-9
Camastra, Francesco   I-365
Campadelli, Paola   I-433
Canazza, Sergio   II-219
Cancela, Brais   II-50
Canetta, Elisabetta   I-423
Capodiferro, Licia   II-199
Carrión, Pilar   II-303
Casarotti, Luca   I-575
Casiraghi, Elena   I-433
Castellani, Umberto   I-413
Chibani, Youcef   II-248
Cho, Siu-Yeung   II-29
Cinque, Luigi   I-58, II-158, II-414
Cipolla, Marco   I-670
Cohen, Shimon   I-19
Coleman, Sonya   I-504, I-605
Colombo, Carlo   II-323
Comanducci, Dario   II-323
Conte, Dajana   I-178
Cossu, Rossella   I-58
Costantini, Luca   II-199
Cristani, Marco   II-140
Cucchiara, Rita   I-197, I-443

Dacal-Nieto, Angel   II-303
D'Alessando, Antonino   II-434
D'Ambrosio, Roberto   I-585
Debattisti, Stefano   II-374
Deboeverie, Francis   II-109
De Floriani, Leila   I-565
Del Bimbo, Alberto   II-199
Del Bue, Alessio   I-660
D'Elia, Ciro   II-209
De Marsico, Maria   II-313
De Stefano, Claudio   I-393
Di Girolamo, Marco   II-414
Doggaz, Narjes   II-229
Dondi, Piercarlo   II-89, II-158

Dragoni, Aldo Franco   II-79
Durou, Jean-Denis   I-286

El Merabet, Youssef   II-394
Eric, Fauvet   II-11

Faez, Karim   II-343
Farinella, Giovanni Maria   I-473
Ferjani, Imene   II-229
Fernández, Alba   II-50
Fernández-Delgado, Manuel   II-303
Ferone, Alessio   I-29
Ferraretti, Denis   I-575
Ferraro, Mario   I-187
Ferretti, Marco   II-99
Flancquart, Amaury   I-68
Fleming, Roland W.   I-128
Foggia, Pasquale   I-178
Fontanella, Francesco   I-393
Foresti, Gian Luca   II-178
Formella, Arno   II-303
Frejlichowski, Dariusz   I-356, II-285,
    II-294
Frosio, Iuri   I-534
Frucci, Maria   II-168
Fumera, Giorgio   I-98, I-159, II-130,
    II-140
Furuse, Tatsuhiko   I-276
Fusco, Roberta   I-48

Gallea, Roberto   I-237
Gamberoni, Giacomo   I-575
Gardiner, Bryan   I-605
Gattal, Abdeldjalil   II-248
Gherardi, Alessandro   II-404
Giacinto, Giorgio   I-139
G. Penedo, Manuel   II-50
Grana, Costantino   I-443
Guha, Prithwijit   II-69
Guida, Claudio   II-323
Gumhold, Stefan   I-128
Gurvich, Ilya   I-38

Haines, Tom S.F.   I-680
Han, Lin   I-636
Hancock, Edwin R.   I-1, I-267, I-636
Hast, Anders   II-275
He, Xiangjian   II-444
Heinze, Matthias   II-265

Hiura, Shinsaku   I-276
Hoang, Hiep Van   II-150

Iannello, Giulio   I-585
Impedovo, Donato   II-241

Kerautret, Bertrand   I-554
Kerr, Dermot   I-504
Khoudour, Louahdi   I-68
Krylov, Andrey S.   II-40, II-384
Kühmstedt, Peter   II-265, II-363
Kumar, Sanjeev   II-178

La Cascia, Marco   I-237, I-616
Lacroix, Vinciane   I-318
Laganiere, Robert   II-1
Lamberti, Luigi   I-365
Lamma, Evelina   I-575
Landi, Marco   II-464
Layher, Georg   I-227
Le, Duy-Dinh   I-108, II-150
Lecca, Michela   I-296
Leung, M.K.H.   II-29
Li, Na   I-636
Liebau, Hendrik   I-227
Lindenbaum, Michael   I-38
Lo Bosco, Giosuè   II-434
Lombardi, Gabriele   I-433
Lombardi, Luca   II-89, II-158
Lo Presti, Liliana   I-616
Lucchese, Mirko   I-534
Luckner, Marcin   I-514
Luszczkiewicz-Piatek, Maria   I-347
Luzio, Dario   II-434

Maciol, Ryszard   I-306
Magillo, Paola   I-565
Mahdi, Walid   II-454
Mahony, Robert   I-78
Maino, Giuseppe   I-700, II-464, II-475,
    II-486
Maleika, Wojciech   II-285
Maniaci, Marilena   I-393
Marana, Aparecido Nilceu   I-169
Marasco, Emanuela   II-255
Marchetti, Andrea   II-275
Marcialis, Gian Luca   I-159, I-595
Marfil, Rebeca   I-327
Mariano, Paola   II-209
Marrocco, Claudio   I-118, I-384
Martel, Luc   II-1

Martinel, Niki   II-189
Mathew, Alex   I-544
Maugeri, Valentina   I-337
Mazzei, Luca   II-424
Mazzola, Giuseppe   I-691
Mecca, Roberto   I-286
Medici, Paolo   I-217
Menghi, Roberta   II-475
Messelodi, Stefano   I-207, I-296
Messina, Enrico   I-473
Messina, G.   I-650
Meurie, Cyril   II-394
Michaelis, Bernd   I-227
Micheloni, Christian   II-178, II-189
Molinara, Mario   I-384
Montangero, Manuela   I-443
Monti, Mariapaola   II-486
Moran, Bill   II-119
Morana, Marco   I-237
Morrow, Philip   I-494
Mufti, Faisal   I-78
Mukerjee, Amitabha   II-69
Munkelt, Christoph   II-265
Murino, Vittorio   I-413, I-660, II-140

Nabney, Ian T.   I-306
Nappi, Michele   II-313
Nasonov, Andrey   II-40
Nencini, Elena   I-700
Neumann, Heiko   I-227
Ngo, Thanh Duc   I-108
Nguyen, Quang Hong   II-150
Nguyen, Thanh Phuong   I-554
Niese, Robert   I-227
Notni, Gunther   II-265, II-363
Nyström, Ingela   I-403

Olivier, Laligant   II-11
Ortega, Marcos   II-50
Osaku, Daniel   I-169

Pal, Sankar Kumar   I-29
Pałczyński, Michał   II-285
Panah, Amir   II-343
Panebarco, Marianna   II-475
Papa, João Paulo   I-169
Pelaes, Evaldo   II-333
Percannella, Gennaro   I-178, II-353
Petrillo, Antonella   I-48
Petrosino, Alfredo   I-29

Philips, Wilfried   II-109
Piccardi, Massimo   II-19
Piciarelli, Claudio   II-189
Pillai, Ignazio   I-98, II-130
Pinello, Luca   II-434
Piras, Luca   I-139
Pirlo, Giuseppe   II-241
Poncin, Guillaume   I-149
Porta, Marco   II-89
Puglisi, Giovanni   I-473

Qin, Zengchang   II-60

Rahman, Shah Atiqur   II-29
Rani, Asha   II-178
Rattani, Ajita   I-595
Ren, Peng   I-1
Ricci, Elisa   I-207
Riccio, Daniel   II-313
Rodà, Antonio   II-219
Rodin, A.S.   II-384
Roli, Fabio   I-98, I-159, I-595, II-130, II-140
Rosa, Marco   I-433
Rozza, Alessandro   I-433
Ruichek, Yassine   II-394

Saboune, Jamal   II-1
Sabzevari, Reza   I-660
Saman, Gule   I-267
Sanniti di Baja, Gabriella   II-168
Sansone, Carlo   I-48, II-255
Sansone, Mario   I-48
Santos Junior, Jurandir   I-374
Sato, Kosuke   I-276
Satoh, Shin'ichi   I-108, II-150
Satta, Riccardo   II-130, II-140
Sbihi, Abderrahmane   II-394
Schüffler, Peter   I-413
Scotney, Bryan   I-494, I-504, I-605
Scotto di Freca, Alessandra   I-393
Sebe, Nicu   I-207
See, John   I-247
Seidenari, Lorenzo   II-199
Semashko, Alexander S.   II-384
Sergeev, Nikolai   I-463
Serra, Giuseppe   II-199
Silva, Luciano   I-374
Simeone, Paolo   I-118
Sintorn, Ida-Maria   I-403

Siravenha, Ana Carolina   II-333
Smith, William A.P.   I-680
Smolka, Bogdan   I-347
Snidaro, Lauro   II-219
Soda, Paolo   I-585, II-353
Sousa, Danilo   II-333
Spasojevic, Nemanja   I-149
Spehr, Marcel   I-128
Stanco, Filippo   I-337
Svensson, Lennart   I-403
Svoboda, David   I-453

Tanasi, Davide   I-337
Tegolo, Domenico   I-524, II-434
Tortora, Genny   II-313
Tortorella, Francesco   I-118, I-384
Touahni, Rajaa   II-394
Truong Cong, Dung Nghi   I-68
Tschechne, Stephan   I-463
Tufano, Francesco   I-178
Turpin, Alan   I-494

Ulaş, Aydın   I-413

Valenti, Cesare   I-670
Vallesi, Germano   II-79
Vazquez-Fernandez, Esteban   II-303

Veelaert, Peter   II-109
Venkatesh, K.S.   II-69
Vento, Mario   I-178, II-353
Vezzani, Roberto   I-197
Visentini, Ingrid   II-219
Vitali, Maria   I-565
Vrubel, Alexandre   I-374

Wan, Tao   II-60
Wanat, Robert   II-294
Wang, Hai   I-626
Wang, Han   I-257
Wilson, Richard C.   I-1, I-680
Wolsley, Clive   I-494

Xu, Richard   II-19

Ying, Ying   I-257
Yuan, Yuan   I-306

Zdunek, Rafal   I-88
Zen, Gloria   I-207
Zhang, Shuwu   I-626
Zheng, Lihong   II-444
Zheng, Xiaoshi   I-636
Zlitni, Tarek   II-454
Zulkifley, Mohd Asyraf   II-119